Solutions Manual

to accompany
Petrucci and Harwood's
General Chemistry

Sixth Edition

Student Solutions Manual
to accompany
Petrucci and Harwood's

GENERAL
CHEMISTRY

SIXTH EDITION

Robert K. Wismer
Millersville University
Millersville, PA

MACMILLAN PUBLISHING COMPANY
New York

MAXWELL MACMILLAN CANADA, INC.
Toronto

MAXWELL MACMILLAN INTERNATIONAL
New York   Oxford   Singapore   Sydney

Macmillan Publishing Company
113 Sylvan Avenue, Englewood Cliffs, NJ 07632

ISBN   0-02-428851-9

Printing: 3 4 5 6 7 8   Year: 4 5 6 7 8 9 0 1 2

To the Student

One of the main benefits that you will obtain from your study of chemistry is the ability to explain a large number of phenomena by learning and applying a relatively small number of fundamental principles. But it is very difficult to learn these principles and how to apply them by merely reading about chemistry. Instead you must practice using them, both in the laboratory and by solving problems.

The purpose of this supplement is to help you in mastering the use of many principles through problem solving. In it there are solutions to all of the Review Questions and the Exercises in the companion textbook: Petrucci and Harwood, General Chemistry, Principles and Modern Application, 6th ed., Macmillan, New York, 1993. Below are given a few suggestions to help you use this book so that you obtain the maximum benefit from it.

You should attempt the Review Questions and the Exercises only after you have carefully studied the text in the chapter. If you try to short circuit this process, by attempting to solve the problems before you have studied the text, you will find yourself constantly returning to the chapter, paging through it for ideas that you need to approach the problem. And this search will be inefficient, because you are not familiar with the chapter. Worse, you may simply look back for an example that is similar to the problem you are attempting to work. This latter technique does not help you learn how to solve problems; it simply develops your ability to copy someone else's solution.

You should attempt to solve the Review Questions and the Exercises with a pencil and paper (and probably an electronic calculator as well). For each chapter, you should attempt to solve all of the Review Questions and a representative sampling of the Exercises, so that you know you have mastered the principles presented in that chapter. Only after you have made a determined effort on each problem should you turn to the solutions herein. If you simply look at the statement of the problem, think about it briefly, and then look at its solution, you will have fooled yourself into believing that you would have come up with that solution on your own. You will never develop the ability to solve problems. And you will miss something else: arriving at your own method of solving the problem. There are many ways of solving problems and, in some of the solutions herein, some of these alternate solutions are suggested.

Be aware that you are *not* trying to "get the right answer." The correct answers to these problems are known. Rather, you are trying to learn how to solve problems, so that when you are confronted by new problems (either on an examination or in another course—or your job—that requires chemistry) you will know how to approach them. Sometimes even the wrong answer is arrived at through a process that is correctly based on chemical principles with perhaps one small error, such as misplacing a decimal point.

There is, of course, a lot more to chemistry than solving mathematical problems. And many problems of a quantitative nature are presented in the textbook. These problems involve correctly defining terms; contrasting two similiar but distinct concepts; explaining chemical phenomena; predicting the products of chemical reactions; and representing chemical entities—atoms, ions, and molecules—through names, formula, and sketches; and so on. Remember to work on this aspect of your chemical education as well.

But what if you make a mistake? First, be aware that if your answer agrees with the one given herein in only the last digit, you have probably not made a mistake at all; the discrepancy is simply due to rounding. The procedure followed in solving these problems is to round the result at each step in the calculation where an intermediate answer has been written down. Where an extra significant figure has been retained until the end of the calculation, that digit is written as a subscript (such as 1.5_2, rather than the misleading 1.52). Second, you answer may differ simply by a power of ten or by a simple factor (such as 2 or 3). Many times these discrepancies are the result of not pushing the correct buttons on the calculator, or due to some small error in logic. Finally, you may have a serious error. If you wrote down your technique for solving the problem in detail, you can compare it with the solution given here and see where you went wrong. If your technique differs substantially from that of the solution, you may be able to get some ideas of what you did wrong from reading the solution carefully. Or your technique may be radically different from the one presented here and still be correct, except for a small flaw. At this point you should ask your instructor for assistance.

The important point is to work on problem solving yourself. You are the one who is trying to learn chemistry, not your instructor and not the chemistry major down the hall who seems to know all the answers. If you make the mistakes, and you fix them, you will learn chemistry much better than if someone else shows you the right answer. And, of course, you will have the satisfaction of having done it yourself.

Millersville, Pennsylvania   Robert K. Wismer

This book is dedicated to the three most active learners in my family: Michael, Mary, and Karen (who is so proud that "I can do it myself," with reading and writing as her most recent achievement).

Contents

ACKNOWLEDGEMENTS

Any book is the product of many individuals and this one is also. Many improvements have resulted from their efforts; the errors that remain are mine. I owe special thanks to Professor Ralph H. Petrucci for his frequent comments throughout the writing of this edition, and for checking many of the solutions.

The students of Introductory Chemistry I and Physical Chemistry I at Millersville University during the fall semester of 1992 have shown great patience with a professor preoccupied with writing. This is also true of my colleagues on the faculty of the Chemistry Department. I appreciate their support and encouragement.

Paul Corey, chemistry editor at Macmillan Publishing Company, supervised this book. Elisabeth Belfer of college production at Macmillan has made the entire writing process as painless as possible with her advice and encouragement throughout the task of producing this and previous versions. Any beauty in the printed pages herein is largely the result of her guidance.

This book was written with Microsoft Word on a Macintosh Plus computer, using SuperPaint and Thunderscan for most pictures. Dr. Jan M. Shepherd assisted in producing many chemical structures with ChemDraw. My thanks to Dave Measel and Russ Baker at Microsoft for their technical advice as I was learning.

But I still owe my largest debt of gratitude to my wife Debbie and our children Michael, Mary, and Karen. They have understood a husband and father who always is thinking of and working on "a book," and have constantly loved, supported, and encouraged him.

Robert K. Wismer

Solutions Manual

to accompany
Petrucci and Harwood's
General Chemistry

Sixth Edition

1 MATTER—ITS PROPERTIES AND MEASUREMENT

1. **a.** A **m³** is a cubic meter. As a regular solid, it is a cube one meter on a side, very crudely in volume equal to one cubic yard.

b. **% by mass** is read "percent by mass." It is the mass in grams of one portion of a sample present in precisely 100 grams of that sample.

c. **°C** is the temperature of substance, expressed on a scale (the Celsius scale) where the freezing point of water has a value of "zero" and the boiling point of water has a value of "one hundred."

d. **Density** is the concentration of the mass of a material. It is calculated as the mass of the material (in grams) divided by its volume (in mL or cm³).

e. An **element** is a substance that cannot be altered or decomposed chemically. Each element has a definite name and a specific position on the periodic table.

2. **a.** The seven **SI base units** are those from which all other units are derived. Among them are the meter for length, the kilogram for mass, the kelvin for temperature, the second for time, and the mole for amount of substance.

b. **Significant figures** are those digits in a number that are the result of experimental measurement, or are derived from such a measurement.

c. A **natural law** is a summary of experimental results or of observations.

d. The **conversion factor method of problem solving** is a technique that converts the units of the information given into those requested for the answer, while simultaneously obtaining the correct numerical result.

3. **a.** The **mass** of an object is a measure of the amount of material in that object. Its **weight**, on the other hand, is the force that the object exerts due to gravitational attraction.

b. An **extensive property** is one that depends on the amount of material present; an **intensive** property does not so depend.

c. A **substance** is a pure form of matter; it is either an element or a compound. A **mixture** is a blend of two or more substances, in no particular proportion.

d. **Precision** refers to the reproducibility of an experimental measurement; **accuracy** describes the agreement between the measurement and an accepted value of the same property.

e. A **hypothesis** is a tentative explantation of a natural law. A theory is a hypothesis that has survived the test of repeated experiments.

4. **a.** $4.15 \text{ kg} \times \dfrac{1000 \text{ g}}{1 \text{ kg}} = 4.15 \times 10^3 \text{ g}$ **b.** $375 \text{ g} \times \dfrac{1 \text{ kg}}{1000 \text{ g}} = 0.375 \text{ kg}$

c. $1485 \text{ mm} \times \dfrac{1 \text{ cm}}{10 \text{ mm}} = 148.5 \text{ cm}$ **d.** $0.135 \text{ cm} \times \dfrac{10 \text{ mm}}{1 \text{ cm}} = 1.35 \text{ mm}$

5. **a.** $0.0787 \text{ L} \times \dfrac{1000 \text{ mL}}{1 \text{ L}} = 78.7 \text{ mL}$ **b.** $47.2 \text{ mL} \times \dfrac{1 \text{ L}}{1000 \text{ mL}} = 0.0472 \text{ L}$

c. $2721 \text{ cm}^3 \times \dfrac{1 \text{ L}}{1000 \text{ cm}^3} = 2.721 \text{ L}$ **d.** $2.65 \text{ dm}^3 \times \dfrac{1 \text{ L}}{1 \text{ dm}^3} \times \dfrac{1000 \text{ mL}}{1 \text{ L}} = 2.65 \times 10^3 \text{ mL}$

6. **a.** $42.5 \text{ in} \times \dfrac{2.54 \text{ cm}}{1 \text{ in.}} \times \dfrac{1 \text{ m}}{100 \text{ cm}} = 1.08 \text{ m}$

b. $108 \text{ ft} \times \dfrac{12 \text{ in.}}{1 \text{ ft}} \times \dfrac{2.54 \text{ cm}}{1 \text{ in.}} \times \dfrac{1 \text{ m}}{100 \text{ cm}} = 32.9 \text{ m}$

c. $0.76 \text{ lb} \times \dfrac{453.6 \text{ g}}{1 \text{ lb}} = 3.4 \times 10^2 \text{ g}$ **d.** $812 \text{ lb} \times \dfrac{0.4536 \text{ kg}}{1 \text{ lb}} = 368 \text{ kg}$

e. $1.77 \text{ gal} \times \dfrac{4 \text{ qt}}{1 \text{ gal}} \times \dfrac{0.9464 \text{ L}}{1 \text{ qt}} \times \dfrac{1000 \text{ mL}}{1 \text{ L}} = 6.70 \times 10^3 \text{ mL}$

f. $6.10 \text{ qt} \times \dfrac{0.9464 \text{ L}}{1 \text{ qt}} = 5.77 \text{ L}$

7. **(a)** $1.00 \text{ km}^2 \times \left(\dfrac{1000 \text{ m}}{1 \text{ km}}\right)^2 = 1.00 \times 10^6 \text{ m}^2$ **(b)** $1.00 \text{ m}^2 \times \left(\dfrac{100 \text{ cm}}{1 \text{ m}}\right)^2 = 1.00 \times 10^4 \text{ cm}^2$

(c) $1.00 \text{ mi}^2 \times \left(\dfrac{5280 \text{ ft}}{1 \text{ mi}} \times \dfrac{12 \text{ in.}}{1 \text{ ft}} \times \dfrac{2.54 \text{ cm}}{1 \text{ in.}} \times \dfrac{1 \text{ m}}{100 \text{ cm}}\right)^2 = 2.59 \times 10^6 \text{ m}^2$

8. The boiling point of water can serve as our reference. 204°F is below the 212°F boiling point of water, while 102°C is above the 100°C boiling point of water. Thus, 102°C is the higher temperature.

9. We use the appropriate equation from Example 1-1 to convert each temperature.
 a. $(165°\text{F} - 32) \, ^5/_9 = 73.9°\text{C}$ **b.** $(48.6°\text{F} - 32.0) \, ^5/_9 = 9.22°\text{C}$
 c. $(12.8°\text{C} \times \, ^9/_5) + 32.0 = 55.0°\text{F}$ **d.** $(-21.2°\text{C} \times \, ^9/_5) + 32.0 = -6.2°\text{F}$
 e. $(-31.2°\text{F} - 32.0) \, ^5/_9 = -35.1°\text{C}$ **f.** $(-12.6°\text{F} - 32.0) \, ^5/_9 = -24.7°\text{C}$

10. 100.0 mL of benzene, with a density less than 1 g/mL must have a mass less than 100.0 g (actually it is 87 g). But 90.0 mL of carbon disulfide, with a density of 1.26 g/mL, should have a mass somewhat in excess of 100.0 g (actually it is 113 g). Thus 90.0 mL of carbon disulfide is the most massive.

11. $\text{Density} = \dfrac{\text{mass}}{\text{volume}} = \dfrac{3153 \text{ g}}{2.50 \text{ L}} \times \dfrac{1 \text{ L}}{1000 \text{ mL}} = 1.26 \text{ g/mL}$

12. We need to convert the mass in kg to a mass in g in order to determine the density.
 $\text{Density} = \dfrac{1.60 \text{ kg}}{515. \text{ mL}} \times \dfrac{1000 \text{ g}}{1 \text{ kg}} = 3.11 \text{ g/mL}$

13. **(a)** $\text{mass} = 717 \text{ mL} \times \dfrac{1.11 \text{ g}}{1 \text{ mL}} = 796 \text{ g ethylene glycol}$

 (b) $\text{mass} = 12.0 \text{ L} \times \dfrac{1000 \text{ mL}}{1 \text{ L}} \times \dfrac{1.11 \text{ g}}{1 \text{ mL}} \times \dfrac{1 \text{ kg}}{1000 \text{ g}} = 13.3 \text{ kg}$

 (c) $\text{volume} = 50.0 \text{ g} \times \dfrac{1 \text{ mL}}{1.11 \text{ g}} = 45.0 \text{ mL ethylene gylcol}$

 (d) $\text{volume} = 33.4 \text{ kg} \times \dfrac{1000 \text{ g}}{1 \text{ kg}} \times \dfrac{1 \text{ mL}}{1.11 \text{ g}} \times \dfrac{1 \text{ L}}{1000 \text{ mL}} = 30.1 \text{ L ethylene glycol}$

14. $\text{Ethanol mass} = 7.50 \text{ L antifreeze} \times \dfrac{1000 \text{ mL}}{1 \text{L}} \times \dfrac{0.937 \text{ g antifreeze}}{1 \text{ mL antifreeze}} \times \dfrac{39.0 \text{ g ethanol}}{100.0 \text{ g antifreeze}} \times \dfrac{1 \text{ kg}}{1000 \text{ g}}$
 $= 2.74 \text{ kg ethanol}$

15. $\text{Acetic acid mass} = 1.00 \text{ lb vinegar} \times \dfrac{453.6 \text{ g}}{1 \text{ lb}} \times \dfrac{5.1 \text{ g acetic acid}}{100.0 \text{ g vinegar}} = 23 \text{ g acetic acid}$

16. $\text{Solution mass} = 1.00 \text{ kg sucrose} \times \dfrac{1000 \text{ g}}{1 \text{ kg}} \times \dfrac{100.00 \text{ g solution}}{8.60 \text{ g sucrose}} = 1.16 \times 10^4 \text{ g solution}$

17. $\text{Fertilizer mass} = 1.00 \text{ lb nitrogen} \times \dfrac{0.4536 \text{ kg N}}{1.00 \text{ lb N}} \times \dfrac{100 \text{ kg fertilizer}}{21 \text{ kg N}} = 2.2 \text{ kg fertilizer}$

18. **a.** $9,200. = 9.200 \times 10^3$ **b.** $1760. = 1.760 \times 10^3$ **c.** $0.240 = 2.40 \times 10^{-1}$
 d. $0.063 = 6.3 \times 10^{-2}$ **e.** $1267.3 = 1.2673 \times 10^3$ **f.** $315,622 = 3.15622 \times 10^5$

19. **a.** $4.19 \times 10^{-3} = 0.00419$ **b.** $6.17 \times 10^{-2} = 0.0617$ **c.** $19.9 \times 10^{-5} = 0.000199$
 d. $371 \times 10^{-3} = 0.371$

20. **a.** 625 has three significant figures; non-zero digits are significant.
 b. 45.3 has three signifcant figures; non-zero digits are significant.
 c. 0.033 has two significant digits; leading zeros are not significant.
 d. 820.03 has five significant digits; zeros surrounded by non-zero digits are significant.

    e. 0.04050 has four significant digits; trailing zeros to the right of the decimal are significant.
    f. 0.007 has one significant figure; leading zeros are not significant.
    g. 700. has three significant figures; the decimal point indicates that the preceeding zeros are significant.
    h. 67,000,000 may have two to eight significant figures; there is not way to determine which, if any, of the zeros are significant.

21. Each of the following is expressed with four significant figures.

    a. $7218.7 \approx 7219$ **b.** $802.07 \approx 802.1$ **c.** $186{,}000 = 1.860 \times 10^5$

    d. $21{,}700 \approx 2.170 \times 10^4$ **e.** $7.705 \times 10^3 \approx 7705$ **f.** $8.0827 \times 10^{-4} \approx 8.083 \times 10^{-4}$

22. **a.** $0.065 \times 0.0700 = 4.6 \times 10^{-3}$ **b.** $0.0120 \times 48.15 \times 0.0087 = 5.0 \times 10^{-3}$

    c. $0.109 + 0.09 - 0.055 = 1.4 \times 10^{-1}$ **d.** $12.12 + 0.065 - 0.955 = 1.123 \times 10^1$

23. **a.** $\dfrac{4400 \times 13.7}{0.0090} = 6.7 \times 10^6$ **b.** $\dfrac{350 \times 0.10 \times 678{,}000}{0.0017 \times 6.3} = 2.2 \times 10^9$

    c. $\dfrac{12.2 + 0.086 - 0.0034}{162.2} = 7.57 \times 10^{-2}$ **d.** $\dfrac{0.0882}{35.88 - 0.066 + 0.0085} = 2.46 \times 10^{-3}$

24. The calculated volume of the block is converted to its mass with the density of iron.

    Mass $= 12.34$ cm $\times 5.68$ cm $\times 2.44$ cm $\times 7.86 \dfrac{\text{g}}{\text{cm}^3} = 1.34 \times 10^3$ g iron

25. The calculated volume of the cylinder is converted to its mass with the density of steel.

    Mass $= \pi r^2 h \times d = 3.14159 (3.55 \text{ cm})^2 \; 12.05$ cm $\times 7.75 \dfrac{\text{g}}{\text{cm}^3} = 3.70 \times 10^3$ g steel

EXERCISES

Scientific Method

26. No. The greater the number of experiments that conform to the predictions of the law, the more confidence we have in the law. However, there is no point at which the law is ever verified with certainty.

27. One theory is preferred to another if it can predict a wider range of phenomena and if it has fewer assumptions.

28. A given set of conditions, a cause, is expected to produce a certain result, an effect. Although these cause-and-effect relationships may be difficult to establish at times ("God is subtle"), they nevertheless do exist ("he is not malicious").

29. All of the pennies minted before 1982 weigh more than 3.00 g, while all of those minted after 1982 weigh less than 2.60 g. One might infer that the composition of a penny changed in 1982. In fact, pennies minted prior to 1982 are composed of almost pure copper (about 96% pure). Those minted after 1982 are composed of zinc with a thin copper cladding. Some pennies of each type were minted in 1982.

30. The scientific method requires that *all* observations, the results of *all* experiments, be consistent with the predictions of a theory ("the rule"). Even one exception is sufficient reason to challenge a theory and to search for a modification of the theory to explain that exception.

Properties and Classification of Matter

31. An object displaying a physical property retains its basic chemical identity. Display of a chemical property is accompanied by a change in composition.
    (a) Physical: The iron nail is not changed in any significant way when it is attracted to a magnet. Its basic chemical identity is unchanged.
    (b) Chemical: The liquid lighter fluid is converted into a gas (carbon dioxide) and water vapor, along with the evolution of considerable energy.
    (c) Chemical: The green patina is the result of the combination of water, oxygen, and carbon dioxide with the copper in the bronze to produce basic copper carbonate.

(d) Physical: Neither the block of wood nor the water has changed its identity.

32. (a) Substance: Table sugar is the chemical compound, sucrose, in a high state of purity.
   (b) Heterogeneous mixture: We can easily see the chunks of chicken in the soup.
   (c) Homogeneous mixture: Premium gasoline consists of a large number of different hydrocarbon liquids and some special additives.
   (d) Heterogeneous mixture: We easily see the pieces of spices in the dressing. Even if the dressing appears smooth and uniform, as French dressing, distinct droplets of oil can be detected with a microscope.
   (e) Homogeneous mixture: Tap water contains dissolved gases and minerals.
   (f) Heterogeneous mixture: Pieces of garlic can be distinguished from those of salt by careful examination.
   (g) Heterogeneous mixture: The pieces of cocoa can be seen through a microscope.
   (h) Substance: Ice is simply solid water.

33. (a) Air is a simple mixture of gases—oxygen, nitrogen, and traces of others. The individual gases can be separated by the *physical* process of liquefaction followed by that of distillation.
   (b) To obtain the element chlorine from the compound sodium chloride requires a *chemical* change.
   (c) To obtain the element sulfur from the compound sulfuric acid requires a *chemical* change.
   (d) Seawater is a homogeneous mixture containing a large number of substances (principally sodium chloride) dissolved in water. Any process that separates the water from the dissolved substances is purely *physical*.

34. (a) If a magnet is drawn through the mixture, the iron filings will be attracted to the magnet and the wood will be left behind.
   (b) When the sugar-sand mixture is mixed with water, the sugar will dissolve but the sand will not. The water then can be evaporated from the solution to produce pure sugar.
   (c) Water and gasoline do not mix with each other. Hence, simply draw off the gasoline, which floats on top, perhaps with an eyedropper.
   (d) The gold flakes will settle to the bottom if the mixture is left undisturbed. The water then can be decanted, that is, carefully poured off.

35. (a) Extensive: The mass of air depends on the size of the balloon.
   (b) Intensive: The temperature of the ice cube is fixed, regardless of the size of the ice cube. However, it does not have to be at 0 °C, unless liquid water is in contact with the ice.
   (c) Extensive: The time required to bring the water to a boil depends on the amount of water present.
   (d) Intensive: The color given off by the lamp is the same, regardless of the size of the lamp.

Exponential Arithmetic (see Appendix A)

36. Think first of how each number would be written in decimal form. Then convert the number to its exponential form.
   (a) 186 thousand = 186,000 = 1.86×10^5 mi/s
   (b) 5 to 6 quadrillion = 5,000,000,000,000,000 to 6,000,000,000,000,000
$$= 5 \times 10^{15} \text{ to } 6 \times 10^{15} \text{ tons}$$
   (c) 173 thousand trillion = 173,000,000,000,000,000 = 1.73×10^{17} W
   (d) ten millionths = $10 \times 0.000\ 001 = 1 \times 10^{-5}$ m = 10 μm

37. (a) $\dfrac{(2.2 \times 10^3) + 4.7 \times 10^2}{4.6 \times 10^{-2}} = \dfrac{2.7 \times 10^3}{4.6 \times 10^{-2}} = 5.9 \times 10^4$

   (b) $\dfrac{3.15 \times 10^4 \times (2.6 \times 10^{-3})^2}{0.060 + (2.2 \times 10^{-2})} = \dfrac{0.21}{0.082} = 2.6$

Significant figures

38. (a) An exact number—12.
   (b) The capacity of an irregularly shaped tank can only be determined by experiment and thus is subject to error.

(c) The distance between any pair of planetary bodies can only be determined through certain measured quantities. These measurements are subject to error.

(d) An exact number. Although the number of days can vary from one month to another (say from January to February), the month of January always has 31 days.

(e) The area is determined by calculations based on measurements. These measurements are subject to error.

39. (a) $45.6 \times 10^3 \times 1.25 \times 10^5 = 5.70 \times 10^9$ (b) $\dfrac{7.34 \times 10^2 \times 3.18 \times 10^{-4}}{(3.1 \times 10^{-3})^2} = 2.4 \times 10^4$

(c) $35.24 + 36.3 - 1.08 = 70.5$

(d) $(1.561 \times 10^3) - (1.80 \times 10^2) + (2.02 \times 10^4 \times 3.18 \times 10^{-2}) = 2.023 \times 10^3$

40. (a) $3.15 \times 10^5 \times 6.17 \times 10^{-3} \times 8.2 \times 10^{-4} \times 9.15 \times 10^{-3} = 1.5 \times 10^{-2}$

(b) $\dfrac{1711 \times (0.0033 \times 10^4) \times 1.25 \times 10^{-3}}{(6.15 \times 10^{-4})^3} = 3.0 \times 10^{11}$

41. (a) The average speed is obtained by dividing the distance travelled (in miles) by the elapsed time (in hours). First, we need to obtain the elapsed time, in hours.

$9 \text{ days} \times \dfrac{24 \text{ h}}{1 \text{ d}} = 216.000 \text{ h}$ $3 \text{ min} \times \dfrac{1 \text{ h}}{60 \text{ min}} = 0.050 \text{ h}$ $44 \text{ s} \times \dfrac{1 \text{ h}}{3600 \text{ s}} = 0.012 \text{ h}$

Total time $= 216.000 + 0.050 \text{ h} + 0.012 \text{ h} = 216.062 \text{ h}$

average speed $= \dfrac{25{,}012 \text{ mi}}{216.062 \text{ h}} = 115.76 \text{ mi/h}$

(b) First compute the mass of fuel remaining

mass $= 14 \text{ gal} \times \dfrac{4 \text{ qt}}{1 \text{ gal}} \times \dfrac{0.9464 \text{ L}}{1 \text{ qt}} \times \dfrac{1000 \text{ mL}}{1 \text{ L}} \times \dfrac{0.70 \text{ g}}{1 \text{ mL}} \times \dfrac{1 \text{ lb}}{453.6 \text{ g}} = 82 \text{ lb}$

Then determine the mass of fuel used, and finally, the fuel consumption.

mass of fuel used $= 9000 \text{ lb} - 82 \text{ lb} = 9000 \text{ lb}$ fuel consumption $= \dfrac{25{,}012 \text{ mi}}{9000 \text{ lb}} = 2.8 \text{ mi/lb}$

42. If the proved reserve truly was an estimate, rather than an actual measurement, it would have been difficult to estimate it to the nearest trillion cubic feet. A statement such as 2,911,000 trillion cubic feet (or 2.911×10^{15} ft^3) would have more accurately reflected the precision with which the proved reserve was known.

43. **a.** 0.1 m in 76.5 m is equivalent to **b.** 0.1 mL in 250.0 mL is equivalent to

$\dfrac{0.1}{76.5} \times 100\% = 0.1 \%$ precision. $\dfrac{0.1}{250.0} \times 100\% = 0.04\%$ precision.

c. 0.01 g in 182.15 g is equivalent to **d.** 0.1 s in 3726.1 s is equivalent to

$\dfrac{0.01}{182.15} \times 100\% = 0.005\%$ precision. $\dfrac{0.1}{3726.1} \times 100\% = 0.003\%$ precision.

Units of Measurement

44. Express both masses in the same units for comparison. $1956 \text{ μg} \times \dfrac{1 \text{ g}}{10^6 \text{ μg}} \times \dfrac{10^3 \text{ mg}}{1 \text{ g}} = 1.956 \text{ mg}$, which is larger than 0.00205 mg.

45. height $= 16 \text{ hands} \times \dfrac{4 \text{ in.}}{1 \text{ hand}} \times \dfrac{2.54 \text{ cm}}{1 \text{ in.}} \times \dfrac{1 \text{ m}}{100 \text{ cm}} = 1.6 \text{ m}$

46. We use the rate of 100 yd/9.3 s as a conversion factor.

$100.0 \text{ m} \times \dfrac{100 \text{ cm}}{1 \text{ m}} \times \dfrac{1 \text{ in.}}{2.54 \text{ cm}} \times \dfrac{1 \text{ yd}}{36 \text{ in.}} \times \dfrac{9.3 \text{ s}}{100 \text{ yd}} = 10.2 \text{ s}$

47. We begin with 1.00 link and convert that length to inches.

length (in.) $= 1.00 \text{ link} \times \dfrac{1 \text{ chain}}{100 \text{ link}} \times \dfrac{1 \text{ furlong}}{10 \text{ chain}} \times \dfrac{1 \text{ mi}}{8 \text{ furlong}} \times \dfrac{5280 \text{ ft}}{1 \text{ mi}} \times \dfrac{12 \text{ in.}}{1 \text{ ft}} = 7.92 \text{ in.}$

48. **a.** mass (mg) $= 2 \text{ tablets} \times \dfrac{5.0 \text{ gr}}{1 \text{ tablet}} \times \dfrac{1.0 \text{ g}}{15 \text{ gr}} \times \dfrac{1000 \text{ mg}}{1 \text{ g}} = 6.7 \times 10^2 \text{ mg} = 0.67 \text{ g}$

b. dosage rate $= \dfrac{6.7 \times 10^2 \text{ mg}}{155 \text{ lb}} \times \dfrac{1 \text{ lb}}{453.6 \text{ g}} \times \dfrac{1000 \text{ g}}{1 \text{ kg}} = 9.5$ mg aspirin/kg body weight

c. time $= 1.0 \text{ lb} \times \dfrac{453.6 \text{ g}}{1 \text{ lb}} \times \dfrac{2 \text{ tablets}}{0.67 \text{ g}} \times \dfrac{1 \text{ day}}{2 \text{ tablets}} = 6.8 \times 10^2$ days

49. 1 hectare $= 1 \text{ hm}^2 \times \left(\dfrac{100 \text{ m}}{1 \text{ hm}} \times \dfrac{100 \text{ cm}}{1 \text{ m}} \times \dfrac{1 \text{ in.}}{2.54 \text{ cm}} \times \dfrac{1 \text{ ft}}{12 \text{ in.}} \times \dfrac{1 \text{ mi}}{5280 \text{ ft}}\right)^2 \times \dfrac{640 \text{ acres}}{1 \text{ mi}^2}$

$= 2.47$ acres

50. (a) $\dfrac{65 \text{ mi}}{h} \times \dfrac{5280 \text{ ft}}{1 \text{ mi}} \times \dfrac{12 \text{ in.}}{1 \text{ ft}} \times \dfrac{2.54 \text{ cm}}{1 \text{ in.}} \times \dfrac{1 \text{ m}}{100 \text{ cm}} \times \dfrac{1 \text{ km}}{1000 \text{ m}} = 1.0 \times 10^2$ km/h

(b) $\dfrac{33 \text{ ft}}{s} \times \dfrac{12 \text{ in.}}{1 \text{ ft}} \times \dfrac{2.54 \text{ cm}}{1 \text{ in.}} \times \dfrac{1 \text{ m}}{100 \text{ cm}} \times \dfrac{1 \text{ km}}{1000 \text{ m}} \times \dfrac{3600 \text{ s}}{1 \text{ h}} = 36$ km/h

(c) $\dfrac{0.235 \text{ lb}}{\text{in.}^3} \times \left(\dfrac{1 \text{ in.}}{2.54 \text{ cm}}\right)^3 \times \dfrac{453.6 \text{ g}}{1 \text{ lb}} = 6.50$ g/cm³

(d) $\dfrac{17.5 \text{ lb}}{\text{in.}^2} \times \left(\dfrac{1 \text{ in.}}{2.54 \text{ cm}} \times \dfrac{100 \text{ cm}}{1 \text{ m}}\right)^2 \times \dfrac{453.6 \text{ g}}{1 \text{ lb}} \times \dfrac{1 \text{ kg}}{1000 \text{ g}} = \dfrac{1.23 \times 10^4 \text{ kg}}{\text{m}^2}$

51. speed $= \text{Mach } 1.27 \times \dfrac{1130 \text{ ft/s}}{\text{Mach 1}} \times \dfrac{3600 \text{ s}}{1 \text{ h}} \times \dfrac{12 \text{ in.}}{1 \text{ ft}} \times \dfrac{2.54 \text{ cm}}{1 \text{ in.}} \times \dfrac{1 \text{ m}}{100 \text{ cm}} \times \dfrac{1 \text{ km}}{1000 \text{ m}} = 1.57 \times 10^3$ km/h

Temperature Scales

52. high: $\quad °C = \tfrac{5}{9}(°F - 32) = \tfrac{5}{9}(118°F - 32) = 47.8°C$

low: $\quad °C = \tfrac{5}{9}(°F - 32) = \tfrac{5}{9}(17°F - 32) = -8.3°C$

53. Determine the Celsius temperature that corresponds to the highest Fahrenheit temperature, 240°F.

$°C = \tfrac{5}{9}(°F - 32) = \tfrac{5}{9}(240°F - 32) = 116°C$ $\qquad$ Since 116°C is above the range of the thermometer, this thermometer cannot be used in this candy making assignment.

54. $°F = (\tfrac{9}{5})°C + 32 = (\tfrac{9}{5} \times (-273.15)) + 32 = -459.67°F$

Density

55. The mass of glycerol is the difference in masses between empty and fillaed masses.

Density $= \dfrac{653.4 \text{ lb} - 75.0 \text{ lb}}{55.0 \text{ gal}} \times \dfrac{453.6 \text{ g}}{1 \text{ lb}} \times \dfrac{1 \text{ gal}}{3.785 \text{ L}} \times \dfrac{1 \text{ L}}{1000 \text{ mL}} = 1.26$ g/mL

56. Density is a conversion factor. $\quad$ volume $= (283.2 \text{ g filled} - 121.3 \text{ g empty}) \times \dfrac{1 \text{ mL}}{1.59 \text{ g}} = 102$ mL

57. Determine the density of a specific substance, such as water, 1.000 g/cm³, in each of the other units.

density (kg/m³) $= 1.000 \text{ g/cm}^3 \times \dfrac{1 \text{ kg}}{1000 \text{ g}} \times \left(\dfrac{100 \text{ cm}}{1 \text{ m}}\right)^3 = 1.000 \times 10^3$ kg/m³

density (lb/ft³) $= 1.000 \text{ g/cm}^3 \times \dfrac{1 \text{ lb}}{453.6 \text{ g}} \times \left(\dfrac{2.54 \text{ cm}}{1 \text{ in.}} \times \dfrac{12 \text{ in.}}{1 \text{ ft}}\right)^3 = 62.4$ lb/ft³

density (oz/gal) $= 1.000 \text{ g/cm}^3 \times \dfrac{1 \text{ lb}}{453.6 \text{ g}} \times \dfrac{16 \text{ oz}}{1 \text{ lb}} \times \dfrac{1000 \text{ cm}^3}{1 \text{ L}} \times \dfrac{0.9464 \text{ L}}{1 \text{ qt}} \times \dfrac{4 \text{ qt}}{1 \text{ gal}}$

$= 133.5$ oz/gal

Density has the greatest magnitude when expressed in kg/m³.

58. Determine the mass of each item.

(a) mass of iron $= (71.8 \text{ cm} \times 1.8 \text{ cm} \times 1.2 \text{ cm}) \times 7.86 \text{ g/cm}^3 = 1.2 \times 10^3$ g iron

(b) mass of aluminum $= (11.35 \text{ m} \times 4.85 \text{ m} \times 0.002 \text{ cm}) \times \left(\dfrac{100 \text{ cm}}{1 \text{ m}}\right)^2 \times 2.70 \text{ g/cm}^3$

$= 3.0 \times 10^3$ g aluminum

(c) mass of water $= 2.716 \text{ L} \times \dfrac{1000 \text{ cm}^3}{1 \text{ L}} \times 0.998 \text{ g/cm}^3 = 2.711 \times 10^3$ g water

In order of increasing mass, the items are: $\quad$ iron bar < water < aluminum foil

59. First determine the volume of the aluminum foil. Then determine its area. And finally determine its thickness.

$$\text{volume} = 1.863 \text{ g} \times \frac{1 \text{ cm}^3}{2.70 \text{ g}} = 0.690 \text{ cm}^3 \qquad \text{area} = \left(8.0 \text{ in.} \times \frac{2.54 \text{ cm}}{\text{in.}}\right)^2 = 4.1 \times 10^2 \text{ cm}^2$$

$$\text{thickness} = \frac{\text{volume}}{\text{area}} = \frac{0.690 \text{ cm}^3}{4.1 \times 10^2 \text{ cm}^2} \times \frac{10 \text{ mm}}{1 \text{ cm}} = 1.7 \times 10^{-2} \text{ mm}$$

60. $$\text{milk volume} = 2.50 \text{ kg} \times \frac{1000 \text{ g}}{1 \text{ kg}} \times \frac{1 \text{ mL}}{1.03 \text{ g}} \times \frac{1 \text{ L}}{1000 \text{ mL}} \times \frac{1 \text{ gal}}{3.785 \text{ L}} \times \frac{4 \text{ qt}}{1 \text{ gal}} = 2.56 \text{ qt}$$

Since 1 pint = 0.500 qt, this volume is 2 quarts, 1 pint, and 0.06 qt.

$$\text{no. fl. oz.} = 0.06 \text{ qt} \times \frac{32 \text{ fl. oz.}}{1 \text{ qt}} = 2 \text{ fl. oz.} \qquad \text{Total volume} = 2 \text{ quarts, 1 pint, 2 fl. oz.}$$

61. The vertical piece of steel has a volume $= 10.22 \text{ cm} \times 1.35 \text{ cm} \times 2.75 \text{ cm} = 37.9 \text{ cm}^3$

The horizontal piece of steel has a volume $= 8.10 \text{ cm} \times 1.35 \text{ cm} \times 2.75 \text{ cm} = 30.1 \text{ cm}^3$

Thus, the total volume $= 37.9 \text{ cm}^3 + 30.1 \text{ cm}^3 = 68.0 \text{ cm}^3$

mass $= 68.0 \text{ cm}^3 \times 7.78 \text{ g/cm}^3 = 529 \text{ g of steel}$

62. Total volume of 100 pieces of shot $= 8.8 \text{ mL} - 8.4 \text{ mL} = 0.4 \text{ mL}$

$$\frac{\text{mass}}{\text{shot}} = \frac{0.4 \text{ mL}}{100 \text{ shot}} \times \frac{8.92 \text{ g}}{\text{cm}^3} = 0.04 \text{ g/shot}$$

63. The percent of students with each grade is obtained by dividing the number of students with that grade by the total number of students.

$$\%A = \frac{8 \text{ A's}}{66 \text{ students}} \times 100\% = 12 \% \text{ A} \qquad \%B = \frac{16 \text{ B's}}{66 \text{ students}} \times 100\% = 24 \% \text{ B}$$

$$\%C = \frac{30 \text{ C's}}{66 \text{ students}} \times 100\% = 45 \% \text{ C} \qquad \%D = \frac{9 \text{ D's}}{66 \text{ students}} \times 100\% = 14 \% \text{ D}$$

$$\%F = \frac{3 \text{ F's}}{66 \text{ students}} \times 100\% = 5 \% \text{ F}$$

64. Use the percent composition as a conversion factor.

$$\text{mass of sucrose} = 3.05 \text{ L} \times \frac{1000 \text{ mL}}{1 \text{ L}} \times \frac{1.059 \text{ g soln}}{1 \text{ mL}} \times \frac{15.0 \text{ g sucrose}}{100 \text{ g soln}} = 484 \text{ g sucrose}$$

65. Again, percent composition is used as a conversion factor. We are careful to label both numerator and denominator.

$$\text{soln volume, L} = 1.00 \text{ kg sodium hydroxide} \times \frac{1000 \text{ g}}{1 \text{ kg}} \times \frac{100.0 \text{ g soln}}{12.0 \text{ g sodium hydroxide}}$$

$$\times \frac{1 \text{ mL}}{1.131 \text{ g soln}} = 7.37 \times 10^3 \text{ mL soln} \times \frac{1 \text{ L}}{1000 \text{ mL}} = 7.37 \text{ L soln}$$

66. The information in the first two sentences provides the density of the wine, in kg/L.

$$\text{mass of ethyl alcohol} = 400. \text{ mL} \times \frac{1 \text{ L}}{1000 \text{ mL}} \times \frac{4.55 \text{ kg} - 1.70 \text{ kg}}{3.00 \text{ L}} \times \frac{1000 \text{ g}}{1 \text{ kg}}$$

$$\times \frac{11.0 \text{ g ethyl alcohol}}{100 \text{ g wine}} \times \frac{1 \text{ lb}}{453.6 \text{ g}} \times \frac{16 \text{ oz}}{1 \text{ lb}} = 1.47 \text{ oz ethyl alcohol}$$

2 ATOMS AND THE ATOMIC THEORY

REVIEW QUESTIONS

1. **a.** Z is the symbol for atomic number. This equals the number of protons in the nucleus, and also the positive charge of the nucleus.
   b. A β particle refers to an electron ejected by the nucleus, and is one of the three forms of natural radioactivity.
   c. An isotope is one of at least two forms of an atom of an element which have the same number of protons in the nucleus, but different numbers of electrons.
   d. ^{16}O is the symbol for the isotope of oxygen that has 16 nucleons in its nucleus: 8 protons (characteristic of the element oxygen) and 8 neutrons.
   e. Molar mass is the mass of a quantity of an element (or a compound) that contains Avogadro's number (6.022×10^{23}) of atoms (or formula units).

2. **a.** The law of conservation of mass states that there is no gain or loss of mass during a chemical reaction.
   b. The atom as described by Rutherford consists of a very small (approximately 10^{-13} cm diameter), positively charged, and massive (more than 99.5% of the mass) nucleus; surrounded by a quite large (approximately 10^{-8} cm diameter), tenuous (less than 0.5% of the mass), and negatively charged cloud of electrons.
   c. The atomic mass that appears in the periodic table for each element is a weighted average, with contributions from each isotope of the element, each weighted by the relative abundance of that element.
   d. Radioactivity refers to the spontaneous emission from the nucleus of an atom of energy (γ radiation) or particles (α or β particles).

3. **a.** Although cathode rays and beta rays both refer to electrons, cathode rays are produced by electrical current, while beta rays are emitted directly from atomic nuclei as the result of radioactivity.
   b. Protons and neutrons are both particles in the nucleus of the atom, and both have a mass of approximately 1 u. However, protons are positively charged, while neutrons have no electric charge.
   c. The nuclear charge of an atom is a positive charge equal to the number of protons in the nucleus. The ionic charge equals the nuclear charge minus the number of electrons; as a consequence, the ionic charge may be negative.
   d. Avogadro's constant is equal to the number of particles of any type that are present in a mole.

4. By the law of conservation of mass, all of the magnesium initially present and all of the oxygen that reacted are present in the product. Thus, the mass of oxygen that has reacted is obtained by difference.
$$\text{mass of oxygen} = 0.423 \text{ g magnesium oxide} - 0.255 \text{ g magnesium}$$
$$= 0.168 \text{ g oxygen}$$

5. Again we use the law of conservation of mass. The mass of the starting materials equals the mass of substances present after the reaction is complete.
$$\text{mass of sodium} + \text{mass of chlorine} = \text{mass of sodium chloride} + \text{mass of unreacted chlorine}$$
$$0.750 \text{ g sodium} + 2.050 \text{ g chlorine} = \text{mass of sodium chloride} + 0.893 \text{ g unreacted chlorine}$$
$$\text{mass of sodium chloride} = (0.750 \text{ g} + 2.050 \text{ g}) - 0.893 \text{ g} = 1.907 \text{ g sodium chloride}$$

6. If the two elements combine in the ratio 1:1, there will be one atom of sodium present for each atom of chlorine. To determine the mass percent chlorine, we simply convert these atomic quantities to masses (in u) and convert the resulting ratio to a percent.

$$\text{percent Na} = \frac{1 \text{ Na atom} \times \frac{22.99 \text{ u Na}}{1 \text{ na atom}}}{\left(1 \text{ atom Na} \times \frac{22.99 \text{ u Na}}{1 \text{ Na atom}}\right) + \left(1 \text{ Cl atom} \times \frac{35.45 \text{ u Cl}}{1 \text{ Cl atom}}\right)} \times 100\% = 39.34\% \text{ Na}$$

7. The observations cited do not necessarily violate the law of conservation of mass. The oxide formed when iron rusts is a solid and remains with the solid iron, increasing the mass of the solid by the mass of the oxygen that combined. The oxide formed when a match burns is a gas and will not remain with the solid product (the ash); the mass of the ash thus is less than that of the reactants. We would have to collect all reactants and all products and weigh them to determine if the law of conservation of mass is obeyed or violated.

8. (a) The mass of oxygen present in 0.166 g magnesium oxide is the remainder when the 0.100 g magnesium is deducted, or 0.066 g oxygen. Hence, there is 0.066 g oxygen/0.166 g magnesium oxide.

   (b) From the numbers we have already obtained, we see that there is 0.066 g oxygen/0.100 g magnesium, or 0.66 g oxygen/1 g magnesium.

   (c) % Mg, by mass $= \dfrac{0.100 \text{ g Mg}}{0.166 \text{ g magnesium oxide}} \times 100\% = 60.2 \text{ \% Mg}$

9. (a) We can determine whether carbon dioxide has a fixed composition by determining the % C in each sample. (In the calculations below, the abbreviation "cmpd" is short for "compound.")

   $\%C = \dfrac{1.48 \text{ g C}}{5.42 \text{ g cmpd}} \times 100\% = 27.3\% \text{ C}$ $\%C = \dfrac{2.06 \text{ g C}}{7.55 \text{ g cmpd}} \times 100\% = 27.3 \text{ \% C}$

   $\%C = \dfrac{3.17 \text{ g C}}{11.62 \text{ g cmpd}} \times 100\% = 27.3\% \text{ C}$

   Since all three samples have the same percent of carbon, these data do establish that carbon dioxide has a fixed composition.

   (b) Carbon dioxide contains only carbon and oxygen. The percent of oxygen in carbon dioxide is obtained by difference.   % O = 100.0% – 27.3% C = 72.7% O

10. By knowing that all of the 5.00 g of magnesium reacts, producing only magnesium bromide and leaving excess bromine unreacted, we are unable at this point to calculate the mass of magnesium bromide produced. In order to perform this calculation, we need to know how many moles of bromine are combined with each mole of magnesium in the compound.

11.

Name	Symbol	Number protons	Number electrons	Number neutrons	Mass number
sodium	^{23}Na	11	11	12	23
silicon	^{28}Si	14	14[a]	14	28
rubidium	^{85}Rb	37	37[a]	48	85
potassium	^{40}K	19	19	21	40
arsenic[a]	^{75}As	33[a]	33	42	75
neon	^{20}Ne^{2+}	10	8	10	20
bromine[b]	^{80}Br	35	35	45	80
lead[b]	^{208}Pb	82	82	126	208

[a]This result assumes that a neutral atom is involved.

[b]The information given is not enough to characterize a specific nuclide; several possibilities exist.

The minimum information needed is the atomic number (or some way to obtain it: the name or the symbol of the element involved), the number of electrons (or some way to obtain it, such as the charge on the species); and the mass number (or the number of neutrons).

12. (a) Since all of these species are neutral atoms, the number of electrons are the atomic numbers, the subscript numbers. The symbols must be arranged in order of increasing value of these subscripts.

   $^{40}_{18}\text{Ar} < ^{39}_{19}\text{K} < ^{58}_{27}\text{Co} < ^{59}_{29}\text{Cu} < ^{120}_{48}\text{Cd} < ^{112}_{50}\text{Sn} < ^{122}_{52}\text{Te}$

(b)   The number of neutrons is given by the difference between the mass number and the atomic number, $A - Z$. Thus it is the difference between superscript and subscript, given in parentheses after each element in the following list.

$^{39}_{19}K(20) < {}^{40}_{18}Ar(22) < {}^{59}_{29}Cu(30) < {}^{58}_{27}Co(31) < {}^{112}_{50}Sn(62) < {}^{122}_{52}Te(70) < {}^{120}_{48}Cd(72)$

(c)   Here the nuclides are arranged by increasing mass number, given by the superscripts.

$^{39}_{19}K < {}^{40}_{18}Ar < {}^{58}_{27}Co < {}^{59}_{29}Cu < {}^{112}_{50}Sn < {}^{120}_{48}Cd < {}^{122}_{52}Te$

13. (a)   cobalt-60 $^{60}_{27}Co$ (b)   phosphorus-32 $^{32}_{15}P$

  (c)   iodine-131 $^{131}_{53}I$ (d)   sulfur-35 $^{35}_{16}S$

14. The nucleus of $^{133}_{55}Cs$ contains 55 protons and (133 – 55 =) 78 neutrons. Thus, the percent of nucleons that are neutrons is given by   % neutrons $= \dfrac{78 \text{ neutrons}}{133 \text{ nucleons}} \times 100 = 59\%$ neutrons

15. The weighted-average atomic mass of the element indium is just slightly less than 115 u. The mass of the first isotope is approximately 113 u. Hence, the mass of the second isotope must be approximately 115 u; that isotope must be ^{115}In.

16. If we let x represent the number of protons, then $x + 5$ is the number of neutrons. The mass number is the sum of the number of protons and the number of neutrons: $63 = x + (x + 5) = 2x + 5$. We solve this expression for x, and obtain $x = 29$. This is the number of protons of the nuclide and equals the atomic number. Reference to the periodic table indicates that 29 is the atomic number of the element copper.

17. Each of the listed isotopic masses is divided by the isotopic mass of ^{12}C, 12.00000 u.

(a) $^{27}Al \div {}^{12}C = 26.98153 \text{ u} \div 12.00000 \text{ u} = 2.248461$

(b) $^{40}Ca \div {}^{12}C = 39.96259 \text{ u} \div 12.00000 \text{ u} = 3.330216$

(c) $^{197}Au \div {}^{12}C = 196.9666 \text{ u} \div 12.00000 \text{ u} = 16.41388$

18. We need to work through the mass ratios in sequence to determine the mass of ^{81}Br.

mass of ^{19}F = mass of $^{12}C \times 1.5832 = 12.00000 \text{ u} \times 1.5832 = 18.998$ u

mass of ^{35}Cl = mass of $^{19}F \times 1.8406 = 18.998 \text{ u} \times 1.8406 = 34.968$ u

mass of ^{81}Br = mass of $^{35}Cl \times 2.3140 = 34.968 \text{ u} \times 2.3140 = 80.916$ u

19. Each of the isotopic masses is multiplied by its fractional abundance. The resulting products are summed to obtain the average atomic mass.

  contribution from $^{238}U = 238.05 \text{ u} \times 0.9927 = 236.3$ u

  contribution from $^{235}U = 235.04 \text{ u} \times 0.0072 = \quad 1.7$ u

  contribution from $^{234}U = 234.04 \text{ u} \times 0.00006 = \quad 0.01$ u

  average atomic mass of uranium = 236.3 u + 1.7 u + 0.01 u = 238.0 u

Of course, this calculation can be combined in one step.

  $(238.05 \text{ u} \times 0.9927) + (235.04 \text{ u} \times 0.0072) + (234.04 \text{ u} \times 0.00006) = 238.0$ u

20. One mole of any element contains 6.022×10^{23} atoms, the Avogadro constant.

(a)   34.5 mol Mg $\times \dfrac{6.022 \times 10^{23} \text{ Mg atoms}}{1 \text{ mol Mg}} = 2.08 \times 10^{25}$ Mg atoms

(b)   0.0123 mol He $\times \dfrac{6.022 \times 10^{23} \text{ He atoms}}{1 \text{ mol He}} = 7.41 \times 10^{21}$ He atoms

(c) 6.1×10^{-12} mol Np $\times \dfrac{6.022 \times 10^{23} \text{ Np atoms}}{1 \text{ mol Np}} = 3.7 \times 10^{12}$ Np atoms

21. In these problems we use the Avogadro constant and the fact that one mole of atoms of an element has a weight in grams equal to its atomic weight.

(a)   no. moles Zn = 9.32×10^{25} Zn atoms $\times \dfrac{1 \text{ mol Zn}}{6.022 \times 10^{23} \text{ Zn atoms}} = 155$ mol Zn

(b)   mass of Ar, g = 3.27 mol Ar $\times \dfrac{39.95 \text{ g Ar}}{1 \text{ mol Ar}} = 131$ g Ar

(c) mass of Ag, mg = 3.07×10^{20} Ag atoms $\times \dfrac{1 \text{ mol Ag}}{6.022 \times 10^{23} \text{ atoms}} \times \dfrac{107.87 \text{ g Ag}}{1 \text{ mol Ag}} \times \dfrac{1000 \text{ mg}}{1 \text{ g}}$

$\qquad\qquad\qquad$ = 55.0 mg Ag

(d) no. Fe atoms = $46.5 \text{ cm}^3 \text{ Fe} \times \dfrac{7.86 \text{ g}}{1 \text{ cm}^3} \times \dfrac{1 \text{ mol Fe}}{55.85 \text{ g Fe}} \times \dfrac{6.022 \times 10^{23} \text{ atoms}}{1 \text{ mol Fe}}$

$\qquad\qquad\qquad$ = 3.94×10^{24} Fe atoms

22. Since the molar mass of sulfur is 32.1 g/mol, 50.0 g S is considerably more than 1 mole, while 6.02×10^{23} Fe atoms is almost precisely a mole, and 55.0 g Fe (55.8 g/mol Fe) is slightly less than a mole. Finally, 5.0 cm³ Fe has a mass of less than 40 g, and contains considerably less than a mole of atoms. Thus, 50.0 g S contains the greatest number of atoms.

23. We first determine the number of Pb atoms of all types in 1.57 g of Pb, and then use the percent abundance to determine the number of ^{204}Pb atoms present.

$\qquad$ no. ^{204}Pb atoms = $1.57 \text{ g Pb} \times \dfrac{1 \text{ mol Pb}}{207.2 \text{ g Pb}} \times \dfrac{6.022 \times 10^{23} \text{ atoms}}{1 \text{ mol Pb}} \times \dfrac{14 \ ^{204}\text{Pb atoms}}{1000 \text{ Pb atoms}}$

$\qquad\qquad$ = $6.39 \times 10^{19} \ ^{204}$Pb atoms

24. The remaining 8.0% of the alloy must be Cd.

$\qquad$ mass of alloy = 1.50×10^{23} Cd atoms $\times \dfrac{1 \text{ mol Cd}}{6.022 \times 10^{23} \text{ Cd atoms}} \times \dfrac{112.4 \text{ g Cd}}{1 \text{ mol Cd}} \times \dfrac{100.0 \text{ g alloy}}{8.0 \text{ g Cd}}$

$\qquad\qquad$ = 3.5×10^2 g alloy

EXERCISES

Law of Conservation of Mass

25. The magnesium that is burned in air combines with some of the oxygen in the air and this oxygen (which, of course, was not weighed when the magnesium metal was weighed) adds its mass to the mass of the magnesium, making the magnesium oxide product weigh more than did the original magnesium. When this same reaction is carried out in a photoflash bulb, the oxygen (in fact, some excess oxygen) that will combine with the magnesium is already present in the bulb before the reaction. Consequently, the product contains no unweighed oxygen.

26. We compute the mass of the reactants and compare that with the mass of the products.

$\qquad$ reactant mass = mass of calcium carbonate + mass of hydrochloric acid solution

$\qquad\qquad$ = 10.00 g calcium carbonate + 100.0 mL soln $\times \dfrac{1.148 \text{ g}}{1 \text{ mL soln}}$

$\qquad\qquad$ = 10.00 g calcium carbonate + 114.8 g solution

$\qquad\qquad$ = 124.8 g reactants

$\qquad$ product mass = mass of solution + mass of carbon dioxide

$\qquad\qquad$ = 120.40 g soln + 2.22 L gas $\times \dfrac{1.9769 \text{ g}}{1 \text{ L gas}}$

$\qquad\qquad$ = 120.40 g soln + 4.39 g carbon dioxide

$\qquad\qquad$ = 124.79 g products

We see that the mass of the products agrees with the mass of the reactants within experimental error. The law of conservation of mass was obeyed in this case.

27. The product mass differs from that of the reactants by (5.62 – 2.50 =) 3.12 grains. In order to determine the percent gain in mass, we need to convert the reactant mass entirely to grains.

$\qquad$ 13 onces $\times \dfrac{8 \text{ gros}}{1 \text{ once}}$ = 104 onces $\qquad$ (104 + 2) gros $\times \dfrac{72 \text{ grains}}{1 \text{ gros}}$ = 7632 grains

$\qquad$ % mass increase = $\dfrac{3.12 \text{ grains increase}}{(7632 + 2.50) \text{ grains original}} \times 100\%$ = 0.0409% mass increase

The sensitivity of Lavoisier's balance can be as little as 0.01 grain, which seems to be the limit of the readability of the balance, or it can be as large as 3.12 grains, which assumes that all of the error in the experiment is due to the (in)sensitivity of the balance. Let us convert 0.01 grains to a mass in grams.

$$\text{minimum error} = 0.01 \text{ gr} \times \frac{1 \text{ gros}}{72 \text{ gr}} \times \frac{1 \text{ once}}{8 \text{ gros}} \times \frac{1 \text{ livre}}{16 \text{ once}} \times \frac{30.59 \text{ g}}{1 \text{ livre}} = 3 \times 10^{-5} \text{ g} = 0.03 \text{ mg}$$

$$\text{maximum error} = 3.12 \text{ gr} \times \frac{3 \times 10^{-5} \text{ g}}{0.01 \text{ g}} = 9 \times 10^{-3} \text{ g} = 9 \text{ mg}$$

The maximum error is close to that of a somewhat common modern laboratory balance, which typically has a sensitivity of 1 mg. The minimum error is approximated by a good quality analytical balance.

Law of Constant Composition

28. In the first experiment, 2.18 g of sodium produces 5.54 g of sodium chloride. In the second experiment, 2.10 g of chlorine produces 3.46 g of sodium chloride. The amount of sodium contained in this second sample of sodium chloride is given by

mass of sodium = 3.46 g sodium chloride – 2.10 g chlorine = 1.36 g sodium

We now have sufficient information to determine the % Na in each of the samples of sodium chloride.

$$\%\text{Na} = \frac{2.18 \text{ g Na}}{5.54 \text{ g cmpd}} \times 100\% = 39.4\% \text{ Na} \qquad \%\text{Na} = \frac{1.36 \text{ g Na}}{3.46 \text{ g cmpd}} \times 100\% = 39.3\% \text{ Na}$$

Thus, the two samples of sodium chloride have the same composition. Recognize that, according to the interpretation of numbers based on significant figures, each percent has an uncertainty of ±0.1%.

29. If the two samples of water have the same %H, the law of constant composition is demonstrated. Notice that, in the second experiment, the mass of the compound is equal to the sum of the masses of the elements produced from it.

$$\% \text{ H} = \frac{3.06 \text{ g}}{27.35 \text{ g H}_2\text{O}} \times 100\% = 11.2\% \text{ H} \qquad \%\text{H} = \frac{1.45 \text{ g H}}{(1.45 + 11.51)\text{g H}_2\text{O}} \times 100\% = 11.2\% \text{ H}$$

The results are consistent with the law of constant composition.

30. The mass of sulfur (0.312 g) needed to produce 0.623 g sulfur dioxide provides the information for the conversion factor.

$$\text{sulfur mass} = 0.842 \text{ g sulfur dioxide} \times \frac{0.312 \text{ g sulfur}}{0.623 \text{ g sulfur dioxide}} = 0.422 \text{ g sulfur}$$

31. (a) From the first experiment we see that 1.16 g of compound is produced per gram of Hg. These masses enable us to determine the mass of compound produced from 1.50 g Hg.

$$\text{mass of cmpd} = 1.50 \text{ g Hg} \times \frac{1.16 \text{ g cmpd}}{1.00 \text{ g Hg}} = 1.74 \text{ g cmpd}$$

(b) Since the compound weighs 0.24 g more than the mass of mercury (1.50 g) that was used, 0.24 g of sulfur must have reacted. Thus, the unreacted sulfur has a mass of 0.76 g (=1.00 g initially present – 0.24 g reacted).

Fundamental Particles

32. A fundamental particle would be expected to be found in all samples of matter. This is demonstrated with electrons by the fact that cathode rays have the same properties no matter how they are prepared. These properties are independent of the material that was used to construct the cathode ray tube, of the gas that filled the tube when it was constructed (and has since been pumped out), and of the method used to generate electricity.

33. The detection and characterization of electrons was based on the fact that they are charged particles. For instance, in Thompson's experiment, a beam of electrons is made to curve by the force between a magnetic field and a beam of charged particles. A neutron, however, does not have a charge and thus cannot be detected or characterized by the methods used for electrons.

34. The particles produced in a cathode ray tube are electrons and cation. Electrons are fundamental particles. Among cations, only the hydrogen ion, H^+, is a fundamental particle. The reason for the production of cations in a cathode ray tube rather than anions is a result of the relatively high velocity of the electron beam. These high velocity electrons are much more likely to knock an electron out of an atom than to add to the atom and form an anion.

Fundamental Charges and Mass-to-Charge Ratios

35. We can calculate the charge on each drop, express each in terms of 10^{-19} C, and finally express each in terms of $e = 1.60 \times 10^{-19}$ C.

drops 1 & 10:	1.28×10^{-18} C	$= 12.8 \times 10^{-19}$ C	$= 8\,e$
drops 2 & 3:	$1.28 \times 10^{-18} \div 2 = 0.640 \times 10^{-18}$ C	$= 6.40 \times 10^{-19}$ C	$= 4\,e$
drop 4:	$1.28 \times 10^{-18} \div 4 = 0.320 \times 10^{-18}$ C	$= 3.20 \times 10^{-19}$ C	$= 2\,e$
drop 5:	$1.28 \times 10^{-18} \times 4 = 5.12 \times 10^{-18}$ C	$= 51.2 \times 10^{-19}$ C	$= 32\,e$
drops 6 & 7:	$1.28 \times 10^{-18} \times 3 = 3.84 \times 10^{-18}$ C	$= 38.4 \times 10^{-19}$ C	$= 24\,e$
drops 8 & 9:	$1.28 \times 10^{-18} \times 2 = 2.56 \times 10^{-18}$ C	$= 25.6 \times 10^{-19}$ C	$= 16\,e$

We see that these values are consistent with the charge that Millikan found for that of the electron. However, Millikan could not have inferred the correct charge from these data, since they are all a multiple of $2\,e$. With only these data, Millikan would have probably arrived at a value of 3.20×10^{-19} C for the charge on the electron.

36. (a) Determine the ratio of the mass of a hydrogen atom to that of an electron. We use the mass of a proton plus that of an electron for the mass of a hydrogen atom.

$$\frac{\text{mass of proton} + \text{mass of electron}}{\text{mass of electron}} = \frac{1.0073\ u + 0.00055\ u}{0.00055\ u} = 1.8 \times 10^3$$

or
$$\frac{\text{mass of electron}}{\text{mass of proton} + \text{mass of electron}} = \frac{1}{1.8 \times 10^3} = 5.6 \times 10^{-4}$$

(b) The only two mass-to-charge ratios that we can determine from the data of Table 2-1 are those for the proton, a hydrogen ion, H^+ (given first below); and that for the electron (given second).

For the proton: $\dfrac{\text{mass}}{\text{charge}} = \dfrac{1.673 \times 10^{-24}\ g}{1.602 \times 10^{-19}\ C} = 1.044 \times 10^{-5}$ g/C

For the electron: $\dfrac{\text{mass}}{\text{charge}} = \dfrac{9.109 \times 10^{-28}\ g}{1.602 \times 10^{-19}\ C} = 5.686 \times 10^{-9}$ g/C

The hydrogen ion is the lightest positive ion available. We see that the mass-to-charge ratio for a positive particle is considerably larger than that for an electron.

37. We do not have the precise isotopic masses of the two ions, and thus the values of the mass-to-charge ratios are only approximate. As a consequence, we have used a three-significant figure mass for a nucleon, rather than the more precisely known masses of proton and neutron. (Recall that the term "nucleon" refers to a nuclear particle—either a proton or a neutron.)

$^{127}I^-$ $\quad \dfrac{m}{e} = \dfrac{127\ \text{nucleons}}{1\ \text{electron}} \times \dfrac{1\ e}{1.602 \times 10^{-19}\ C} \times \dfrac{1.67 \times 10^{-24}\ g}{1\ \text{nucleon}} = 1.32 \times 10^{-3}$ g/C

$^{32}S^{2-}$ $\quad \dfrac{m}{e} = \dfrac{32\ \text{nucleons}}{2\ \text{electrons}} \times \dfrac{1\ e}{1.602 \times 10^{-19}\ C} \times \dfrac{1.67 \times 10^{-24}\ g}{1\ \text{nucleon}} = 1.67 \times 10^{-4}$ g/C

Atomic Number, Mass Number, and Isotopes

38. In the symbol $^A_Z X$, the term X is the symbol for the element. Z represents the atomic number, the number of protons in the nucleus. (Of course, since each element is uniquely associated with a given number of protons, Z only needs to be included for emphasis.) Finally, A is the mass number, the number of protons plus the number of neutrons; this is the number of nucleons.

39. (a) An atom of $^{138}_{56}$Ba contains 56 protons, and thus also 56 electrons, since an atom is neutral and there is no difference between the number of protons and electrons. Since there are 138 nucleous in the nucleus, the number of neutrons is 82 (= 138 nucleons – 56 protons).

(b) The ratio of the two masses is determined as follows. $\quad \dfrac{^{138}\text{Ba}}{^{12}\text{C}} = \dfrac{137.9050\ u}{12.00000\ u} = 11.49208$

(c) Example 2-4 gives the ratio of the mass of ^{16}O to that of ^{12}C as 1.33291. That is, $\frac{^{16}O}{^{12}C}$ = 1.33291.

Thus, we find the ratio of ^{138}Ba to ^{16}O as follows (or use the ^{16}O mass of 15.9949 from Example 2-4).

$$\frac{^{138}Ba}{^{16}O} = \frac{^{138}Ba}{^{12}C} \times \frac{^{12}C}{^{16}O} = \frac{11.49208}{1.33291} = 8.62180 \qquad \left(\frac{^{138}Ba}{^{16}O} = \frac{137.9050 \text{ u}}{15.9949 \text{ u}} = 8.62181\right)$$

Note that the second ratio we used above is the inverse of the ratio from Example 2-4.

40. The mass of ^{16}O is 15.9948 u. isotopic mass = 15.9948 u × 6.68374 (times ^{16}O) = 106.905 u

ratio = $\dfrac{\text{mass of isotope}}{\text{mass of }^{12}C} = \dfrac{106.905 \text{ u}}{12.0000 \text{ u}}$ = 8.90875

41. $^{35}_{17}Cl$ indicates a chlorine atom that contains 17 protons and 35 nucleons. The symbol ^{35}Cl indicates a chlorine atom that contains 37 nucleons. But we know that all atoms of the element chlorine contain 17 protons. Thus, including the atomic number in the symbol for an element is done either for emphasis or convenience. On the other hand, the symbol $_{17}Cl$ does not indicate the total number of nucleons and hence does not convey sufficient information to distinguish among the isotopes of chlorine.

42. (a) Species with equal numbers of electrons and neutrons will also have a mass number that is not greatly larger than the atomic number. The following species are suitable (with numbers of electrons and neutrons in parentheses). $^{24}_{12}Mg^{2+}$ (10 e, 12 n), $^{47}_{24}Cr$ (24 e, 23 n), $^{59}_{27}Co^{2+}$ (25 e, 32 n), and $^{35}_{17}Cl^-$ (18 e, 18 n). Of these four nuclides, only $^{35}_{17}Cl^-$ has as many electrons as neutrons.

(b) A species in which protons have more than 50% of the mass must have a mass number smaller than twice the atomic number. Of these species, only in $^{47}_{24}Cr$ is more than 50% of the mass contributed by the protons.

(c) To answer this question, we list the number of protons, neutrons, and electrons of each species.

species:	$^{24}_{12}Mg^{2+}$	$^{47}_{24}Cr$	$^{59}_{27}Co^{2+}$	$^{35}_{17}Cl^-$	$^{124}_{50}Sn^{2+}$	$^{226}_{90}Th$	$^{90}_{38}Sr$
no. protons	12	24	27	17	50	90	38
no. neutrons	12	23	32	18	74	136	52
no. electrons	10	24	25	18	48	90	38

Note that $^{124}_{50}Sn^{2+}$ has a number of neutrons (74) equal to its number of protons (50) plus one-half its number of electrons (48÷2 = 24)

Atomic Mass Units, Atomic Masses

43. We use the mass of a proton, expressed in both sets of units, to determine the mass in grams of 1.000 u.

$$1.000 \text{ u} \times \frac{1.673 \times 10^{-24} \text{ g}}{1.0073 \text{ u}} = 1.661 \times 10^{-24} \text{ g}$$

Notice that the inverse of this relationship, that is 6.021×10^{23} u/g, is numerically equal to Avogadro's constant. Alternatively, we can solve the problem based on ^{12}C atoms, as follows.

$$\frac{12.00000 \text{ g } ^{12}C}{1 \text{ mol } ^{12}C} \times \frac{1 \text{ mol } ^{12}C}{6.02214 \times 10^{23} \, ^{12}C \text{ atoms}} \times \frac{1 \, ^{12}C \text{ atom}}{12.00000 \text{ u}} = \frac{1.66054 \times 10^{-24} \text{ g}}{1 \text{ u}}$$

44. It is exceedingly unlikely that another nuclide would have an exact integral mass. The mass of carbon-12 is *defined* as precisely 12.00000 u. Each nuclidic mass is close to integral, but none that we have encountered in this chapter are precisely integral. The reason is that each nuclide is composed of protons, neutrons, and electrons, none of which have integral masses, and there is a small quantity of the mass of each nucleon (nuclear particle) lost in the binding energy holding the nuclide together. It would be highly unlikely that all of these effects would add up to a precisely integral mass.

45. There are no copper atoms that have a mass of 63.546 u. The masses of individual atoms are close to integers and this mass (63.546 u) is about midway between two integers. It is an average atomic mass, the result of averaging two (or more) isotopic masses, each weighted by its natural abundance.

46. To determine the average atomic mass, we use the following expression (where Σ indicates the sum of the individual products). average atomic mass = Σ (isotopic mass × fractional natural abundance) Each of the three percents given is converted to a fractional abundance by dividing it by 100.

Mg atomic mass = (23.985042 u × 0.7899) + (24.985837 u × 0.1000) + (25.982593 u × 0.1101)
= 18.95 u + 2.499 u + 2.861 u = 24.31 u

47. We use the expression for determining the weighted-average atomic mass.
107.8682 u = (106.905092 u × 0.5184) + (^{109}Ag × 0.4816) = 55.42 u + 0.4816 ^{109}Ag

107.8682 u − 55.42 u = 0.4816 ^{109}Ag = 52.45 u $\qquad$ ^{109}Ag = $\dfrac{52.45 \text{ u}}{0.4816}$ = 108.9 u

48. The percent abundances of the two isotopes must add to 100.00%, since there are only two naturally occurring isotopes of bromine. Thus, we can determine the percent natural abundance of the second isotope by difference. $\qquad$ % second isotope = 100.00% − 50.69% = 49.31%
From the periodic table, we see that the weighted-average atomic mass of bromine is 79.904 u. We use this value in the expression for determining the weighted-average atomic mass, along with the isotopic mass of ^{79}Br and the fractional abundances of the two isotopes (the percent abundances divided by 100).
79.904 u = (0.5069 × 78.918336 u) + (0.4931 × other isotope) = 40.00 u + (0.4931 × other isotope)
other isotope = $\dfrac{79.904 \text{ u} - 40.00 \text{ u}}{0.4931}$ = 80.92 u = mass of ^{81}Br, the other isotope

49. Since the three precent abundances total 100%, the percent abundance of ^{40}K is found by difference.
% ^{40}K = 100.0000% − 93.2581% − 6.7302% = 0.0117%
Then the expression for the weighted-average atomic mass is used, with the percent abundances converted to fractional abundances by dividing by 100. The average atomic mass of potassium is 39.0983 u.
39.0983 u = (0.932581 × 38.963707 u) + (0.000117 × 39.963999 u) + (0.067302 × ^{41}K)
= 36.3368 u + 0.00468 u + (0.067302 × ^{41}K)
mass of ^{41}K = $\dfrac{39.0983 \text{ u} - (36.3368 \text{ u} + 0.00468 \text{ u})}{0.067302}$ = 40.962 u

Mass spectrometry

50. Use the expression for determining average atomic weight; the sum of products of nuclidic mass times fractional abundances (from Figure 2-14).
^{198}Hg: $\qquad$ 197.9668 × 0.1002 = 19.84
^{200}Hg: $\qquad$ 199.9683 × 0.2313 = 46.25
^{202}Hg: $\qquad$ 201.9706 × 0.2980 = 60.19
^{196}Hg: $\qquad$ 195.9658 × 0.00146 = 0.286
^{199}Hg: $\qquad$ 198.9683 × 0.1684 = 33.51
^{201}Hg: $\qquad$ 200.9703 × 0.1322 = 26.57
^{204}Hg: $\qquad$ 203.9735 × 0.0685 = 14.0
Atomic weight = 0.286 + 19.84 + 33.51 + 46.25 + 26.57 + 60.19 + 14.0 = 200.6
This is in good agreement with the value of the atomic weight given on the periodic table.

51. As closely as we can estimate, there are 21 atoms of ^{70}Ge, 27 atoms of ^{72}Ge, 8 atoms of ^{73}Ge, 31 atoms of ^{74}Ge, and 8 atoms of ^{76}Ge. This totals 95 atoms. Then the average atomic weight is obtained as follows.
$$\dfrac{(21 \text{ atoms} \times 70) + (27 \text{ atoms} \times 72) + (8 \text{ atoms} \times 73) + (31 \text{ atoms} \times 74) + (8 \text{ atoms} \times 76)}{195 \text{ atoms total}}$$
$\dfrac{1470 + 1944 + 584 + 2294 + 608}{95}$ = 72.$_6$ = average atomic weight of germanium
The result is only approximately correct both because of the difficulty of determining the relative abundances from the graphical presentation with any more accuracy than two significant digits, and because the isotopic masses are given to only two significant figures. Thus, only a two-significant-figure result can be obtained.

52. **(a)** There are six possible types of molecule: ^{1}H^{35}Cl, ^{2}H^{35}Cl, ^{3}H^{35}Cl, ^{1}H^{37}Cl, ^{2}H^{37}Cl, and $\qquad$ ^{3}H^{37}Cl
(b) The mass numbers of the six different possible types of molecules are obtained by summing the mass numbers of the two atoms in each molecule:
^{1}H^{35}Cl has A = 36 $\qquad$ ^{2}H^{35}Cl has A = 37 $\qquad$ ^{3}H^{35}Cl has A = 38
^{1}H^{37}Cl has A = 38 $\qquad$ ^{2}H^{37}Cl has A = 39 $\qquad$ ^{3}H^{37}Cl has A = 40
(c) The most abundant molecule contains the most abundant of each element's isotope. It is ^{1}H^{35}Cl. The second most abundant molecule is ^{1}H^{37}Cl
(d) The relative abundance of each type of molecule is determined by multiplying together the fractional abundances of the two isotopes that are present in that type. Because ^{3}H is so scarce, it will be difficult to detect. Relative abundances of the molecules are as follows.

$^1H^{35}Cl$: 75.52%
$^2H^{35}Cl$: 0.011%
$^3H^{35}Cl$: < 0.001%
$^1H^{37}Cl$: 24.47%
$^2H^{37}Cl$: 0.004%
$^3H^{37}Cl$: < 0.001%

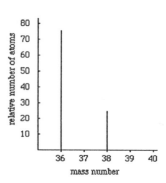

The Avogadro Constant and the Mole

53. One method of establishing this relationship was presented in the solution to Exercise 43. Another method follows, based on more fundamental constants.

$$1.000 \text{ g C} \times \frac{1 \text{ mol C}}{12.0000 \text{ g C}} \times \frac{6.022 \times 10^{23} \text{ C atoms}}{1 \text{ mol C}} \times \frac{12.0000 \text{ u}}{1 \text{ C atom}} = 6.022 \times 10^{23} \text{ u}$$

54. Each of these calculations employs the average atomic mass as a conversion factor.

(a) amount of Na = $135.0 \text{ g Na} \times \dfrac{1 \text{ mol Na}}{22.990 \text{ g Na}} = 5.872 \text{ mol Na}$

(b) number of S atoms = $245.0 \text{ kg S} \times \dfrac{1000 \text{ g}}{1 \text{ kg}} \times \dfrac{1 \text{ mol S}}{32.066 \text{ g}} \times \dfrac{6.022 \times 10^{23} \text{ S atoms}}{1 \text{ mol S}}$

$= 4.601 \times 10^{27} \text{ S atoms}$

(c) Cu mass = $1.0 \times 10^{12} \text{ Cu atoms} \times \dfrac{1 \text{ mol Cu}}{6.022 \times 10^{23} \text{ Cu atoms}} \times \dfrac{63.546 \text{ g Cu}}{1 \text{ mol Cu}}$

$= 1.1 \times 10^{-10} \text{ g Cu} = 0.11 \text{ ng Cu}$

55. We determine the mass of Ag in the piece of jewelry and then the number of Ag atoms.

no. Ag atoms = $38.7 \text{ g sterling} \times \dfrac{92.5 \text{ g Ag}}{100.0 \text{ g sterling}} \times \dfrac{1 \text{ mol Ag}}{107.9 \text{ g Ag}} \times \dfrac{6.022 \times 10^{23} \text{ atoms}}{1 \text{ mol Ag}}$

$= 2.00 \times 10^{23} \text{ Ag atoms}$

56. no. Mg-24 atoms = $1 \text{ mol Mg} \times \dfrac{6.0221 \times 10^{23} \text{ atoms}}{1 \text{ mol Mg}} \times \dfrac{78.99 \text{ Mg-24 atoms}}{100 \text{ Mg atoms}}$

$= 4.757 \times 10^{23} \text{ Mg-24 atoms}$

The number of Mg-24 atoms comes from the following.
% Mg-24 = 100.00% Mg − 10.00% Mg-25 − 11.01% Mg-26 = 78.99% Mg-24

57. We use the average atomic mass of lead, 207.2 g/mol.

(a) $\dfrac{30 \text{ μg Pb}}{1 \text{ dL}} \times \dfrac{1 \text{ dL}}{0.1 \text{ L}} \times \dfrac{1 \text{ g Pb}}{10^6 \text{ μg Pb}} \times \dfrac{1 \text{ mol Pb}}{207.2 \text{ g}} = 1.4 \times 10^{-6} \text{ mol Pb/L}$

(b) $\dfrac{1.4 \times 10^{-6} \text{ mol Pb}}{\text{L}} \times \dfrac{1 \text{ L}}{1000 \text{ cm}^3} \times \dfrac{6.022 \times 10^{23} \text{ atoms}}{1 \text{ mol}} = 8.4 \times 10^{14} \text{ Pb atoms/cm}^3$

58. The concentration of Pb in air provides the principal conversion factor. Other conversion factors are needed to convert to and from its units, beginning with the 0.500-L volume, and ending with the number of atoms.

no. Pb atoms = $0.500 \text{ L} \times \dfrac{1 \text{ m}^3}{1000 \text{ L}} \times \dfrac{3.01 \text{ μg Pb}}{1 \text{ m}^3} \times \dfrac{1 \text{ g Pb}}{10^6 \text{ μg Pb}} \times \dfrac{1 \text{ mol Pb}}{207.2 \text{ g Pb}}$

$\times \dfrac{6.022 \times 10^{23} \text{ Pb atoms}}{1 \text{ mol Pb}} = 4.37 \times 10^{12} \text{ Pb atoms}$

59. The pivotal conversion factor is the concentration of gold in seawater. Other conversion factors transform the starting mass into tons of seawater, and the mass of gold into number of gold atoms.

no. Au atoms = $300. \text{ g seawater} \times \dfrac{1 \text{ lb}}{454 \text{ g}} \times \dfrac{1 \text{ ton}}{2000 \text{ lb}} \times \dfrac{0.15 \text{ mg Au}}{1 \text{ ton seawater}} \times \dfrac{1 \text{ g Au}}{1000 \text{ mg}} \times \dfrac{1 \text{ mol Au}}{197 \text{ g}}$

$$= 2.5 \times 10^{-10} \text{ mol Au} \times \frac{6.02 \times 10^{23} \text{ atoms}}{1 \text{ mol Au}} = 1.5 \times 10^{14} \text{ Au atoms}$$

60. The numbers total 21 (= 10 + 6 + 5). Thus, in one mole of the alloy there is $^{10}/_{21}$ mol Bi, $^{6}/_{21}$ mol Pb, and $^{5}/_{21}$ mol Sn. The mass of this mole of material is figured in a similar fashion to computing an average atomic mass from isotopic masses.

$$\text{mass of alloy} = \left(\frac{10}{21} \text{ mol Bi} \times \frac{209.0 \text{ g}}{1 \text{ mol Bi}}\right) + \left(\frac{6}{21} \text{ mol Pb} \times \frac{207.2 \text{ g}}{1 \text{ mol Pb}}\right) + \left(\frac{5}{21} \text{ mol Sn} \times \frac{118.7 \text{ g}}{1 \text{ mol Sn}}\right)$$

$$= 99.52 \text{ g Bi} + 59.20 \text{ g Pb} + 28.26 \text{ g Sn} = 186.98 \text{ g alloy}$$

3 CHEMICAL COMPOUNDS

REVIEW QUESTIONS

1. (a) The formula unit of a compound is a group of atoms that has atoms of the same type and number as they are present in the formula of that compound. For example, if "Na_2" appears in the formula of the compound, then there will be two sodium atoms in the formula unit of that compound.

   (b) S_8 is a molecule of elemental sulfur. In the same manner as several other elements (including H_2, F_2, Cl_2, Br_2, I_2, N_2, O_2, and P_4) elemental sulfur exists as molecules rather than as isolated atoms.

   (c) An ionic compound is one that is composed of (positively charged) cations and (negatively charged) anions. Most binary ionic compounds are composed of a metal (which becomes the cation) and a nonmetal (which becomes the anion).

   (d) An oxoacid is an acid that contains the element oxygen, in addition to some other element and the element hydrogen. H_2SO_4, HNO_3, and $HClO_4$ all are oxoacids, in which the "other element" is S, N, and Cl, respectively.

   (e) A hydrate is a compound that contains water rather loosely bound. Usually this water of hydration can be driven off by mild heating of the compound.

2. (a) A molecule of an element refers to the smallest independent particle of that element. Usually this is an atom, but in some cases (notably H_2, F_2, Cl_2, Br_2, I_2, N_2, O_2, S_8, and P_4) it is a grouping of two or more atoms.

   (b) The structural formula of a compound not only indicates which atoms are present in the formula unit, but also how they are joined together, the bonding in the molecule, for instance.

   (c) The oxidation state of an element in a compound is an indication of how many electrons each atom of that element has lost (positive oxidation state) or gained (negative). Since oxidation state is determined by a set of rules, rather than by experiment, its connection to the number of electrons actually transferred is rather tenuous. Still, the concept is useful in naming compounds and balancing some types of chemical equations.

   (d) The determination of the carbon-hydrogen-oxygen content of a compound by combustion analysis involves realizing that all of the carbon has formed carbon dioxide, all of the hydrogen has formed water, and the amount of oxygen present in the original compound must be determined by difference.

3. (a) A chemical symbol is the symbol for one element, while a chemical formula is the "word" made up by several symbols and indicates the composition of a compound.

   (b) An empirical formula indicates the simplest grouping of atoms that has the same ratio of elements as are present in the compound. The molecular formula indicates the actual number of atoms of each type present in the molecule. The molecular formula is an integral multiple of the empirical formula.

   (c) A metal is an element that readily forms a cation; a nonmetal readily forms an anion. The nonmetals are found in the upper right region of the periodic table; the metals in the rest of the table.

   (d) The systematic name for a compound is based on the elements present in it and gives an indication of its composition. The trivial or common name is simply a label for the substance.

   (e) A binary acid consists of hydrogen and one other element. A ternary acid consists of hydrogen, the other element, and the element oxygen: three elements in all.

4. (a) The atomic mass of oxygen is the mass of one (average) atom, 15.994 u.

   (b) The molecular mass of oxygen is the mass of one (average) molecule of O_2, 31.9988 u.

(c) The molar mass of molecular oxygen is the mass of one mole of oxygen molecules, 31.9988 g. That of atomic oxygen is the mass of one mole of oxygen atoms, 15.9994 g.

5. (a) A TNT molecule, $C_7H_5(NO_2)_3$, contains 7 C atoms, 5 H atoms, 3 N atoms, and 3×2 = 6 O atoms, for a total of (7 + 5 + 3 + 6 =) 21 atoms

(b) Each molecule of NH_3 contains 3 H atoms and 1 N atom, 4 atoms total

$$\text{number of atoms} = 0.0505 \text{ mol } NH_3 \times \frac{6.022 \times 10^{23} \text{ } NH_3 \text{ molecules}}{1 \text{ mol } NH_3} \times \frac{4 \text{ atoms}}{1 \text{ } NH_3 \text{ molecule}}$$

$$= 1.22 \times 10^{23} \text{ atoms}$$

(c) number of F atoms = $2.35 \text{ mol } C_2HBrClF_3 \times \frac{3 \text{ mol F}}{1 \text{ mol } C_2HBrClF_3} \times \frac{6.022 \times 10^{23} \text{ F atom}}{1 \text{mol F atoms}}$

$$= 4.25 \times 10^{24} \text{ F atoms}$$

6. (a) To convert amount in moles to mass, we need the mole weight of SO_2.

$$\text{mol wt. } SO_2 = \left(1 \text{ mol S} \times \frac{32.07 \text{ g S}}{1 \text{ mol S}}\right) + \left(2 \text{ mol } O_2 \times \frac{16.00 \text{ g O}}{1 \text{ mol O}}\right) = 64.07 \text{ g/mol } SO_2$$

$$\text{mass } SO_2 = 3.65 \text{ mol } SO_2 \times \frac{64.07 \text{ g } SO_2}{1 \text{ mol}} = 234 \text{ g } SO_2$$

(b) mass of O_2 = $2.50 \times 10^{24} \text{ } O_2 \text{ molecules} \times \frac{1 \text{ mol } O_2}{6.022 \times 10^{23} \text{ molecules}} \times \frac{32.00 \text{ g } O_2}{1 \text{ mol } O_2} = 133 \text{ g } O_2$

(c) mol wt. $CuSO_4 \cdot 5H_2O$ = 63.5 g Cu + 32.1 g S + (9 × 16.0 g O) + (10 × 1.01 g H)

$$= 249.7 \text{ g/mol } CuSO_4 \cdot 5H_2O$$

$$\text{mass of } CuSO_4 \cdot 5H_2O = 22.5 \text{ mol} \times \frac{249.7 \text{ g } CuSO_4 \cdot 5H_2O}{1 \text{ mol}} = 5.62 \times 10^3 \text{ g } CuSO_4 \cdot 5H_2O$$

(d) molar mass C_2H_5OH = (2 × 12.01 g C) + (6 × 1.01 g H) + 16.0 g O = 46.08 g/mol C_2H_5OH

$$\text{mass of } C_2H_5OH = 5.00 \times 10^{25} \text{ molecules} \times \frac{1 \text{ mol}}{6.022 \times 10^{23} \text{ molec.}} \times \frac{46.08 \text{ g}}{1 \text{ mol } C_2H_5OH}$$

$$= 3.83 \times 10^3 \text{ g } C_2H_5OH$$

7. (a) amount of Br_2 = $3.52 \times 10^{22} \text{ } Br_2 \text{ molecules} \times \frac{1 \text{ mole } Br_2}{6.022 \times 10^{23} \text{ } Br_2 \text{ molecules}} = 0.0585 \text{ mol } Br_2$

(b) amount of Br_2 = $4.86 \times 10^{24} \text{ Br atoms} \times \frac{1 \text{ } Br_2 \text{ molecule}}{2 \text{ Br atoms}} \times \frac{1 \text{ mole } Br_2}{6.022 \times 10^{23} \text{ } Br_2 \text{ molecules}}$

$$= 4.04 \text{ mol } Br_2$$

(c) amount of Br_2 = $188 \text{ g } Br_2 \times \frac{1 \text{ mol } Br_2}{159.8 \text{ g } Br_2} = 1.18 \text{ mol } Br_2$

(d) amount of Br_2 = $12.6 \text{ mL } Br_2 \times \frac{3.10 \text{ g } Br_2}{1 \text{ mL } Br_2} \times \frac{1 \text{ mol } Br_2}{159.8 \text{ g } Br_2} = 0.244 \text{ mol } Br_2$

8. (a) molec.ms. $C_5H_{11}NO_2S$ = (5 × 12.0 u C) + (11 × 1.01 u H) + 14.0 u N + (2 × 16.0 u O) + 32.1 u S

$$= 149.2 \text{ u/}C_5H_{11}NO_2S \text{ molecule}$$

(b) Since there are 11 H atoms in each $C_5H_{11}NO_2S$ molecule, there are 11 moles of H atoms in each mole of $C_5H_{11}NO_2S$ molecules

(c) mass C = $1 \text{ mol } C_5H_{11}NO_2S \times \frac{5 \text{ mol C}}{1 \text{ mol } C_5H_{11}NO_2S} \times \frac{12.011 \text{ g C}}{1 \text{ mol C}} = 60.055 \text{ g C}$

(d) no. C atoms = $3.18 \text{ mol } C_5H_{11}NO_2S \times \frac{5 \text{ mol C}}{1 \text{ mol } C_5H_{11}NO_2S} \times \frac{6.022 \times 10^{23} \text{ atoms}}{1 \text{ mol C}}$

$$= 9.57 \times 10^{24} \text{ C atoms}$$

9. The information obtained in the course of calculating the molar mass is used to determine the mass percent of H in octane.

$$\text{molar mass } C_8H_{18} = \left(\frac{8 \text{ mol C}}{1 \text{ mol } C_8H_{18}} \times \frac{12.011 \text{ g C}}{1 \text{ mol C}}\right) + \left(\frac{18 \text{ mol H}}{1 \text{ mol } C_8H_{18}} \times \frac{1.00794 \text{ g H}}{1 \text{ mol H}}\right)$$

$$= \frac{96.088 \text{ g C}}{1 \text{ mol } C_8H_{18}} + \frac{18.14292 \text{ g H}}{1 \text{ mol } C_8H_{18}} = \frac{114.231 \text{ g}}{1 \text{ mol } C_8H_{18}}$$

$$\%H = \frac{18.14292 \text{ g H/mol octane}}{114.231 \text{ g C}_8\text{H}_{18}\text{/mol octane}} \times 100\% = 15.8827\% \text{ H} = 15.88\% \text{ H (to two decimal places)}$$

10. We first determine the mass of O in each mol of $Cu_2(OH)_2CO_3$ and the molar mass of the compound.

$$\text{mass O/mol Cu}_2(\text{OH})_2\text{CO}_3 = \frac{5 \text{ mol O}}{1 \text{ mol Cu}_2(\text{OH})_2\text{CO}_3} \times \frac{16.00 \text{ g O}}{1 \text{ mol O}} = 80.00 \text{ g O/mol Cu}_2(\text{OH})_2\text{CO}_3$$

$$\text{mol wt Cu}_2(\text{OH})_2\text{CO}_3 = (2 \times 63.55 \text{ g Cu}) + (5 \times 16.00 \text{ g O}) + (2 \times 1.01 \text{ g H}) + 12.01 \text{ g C}$$
$$= 221.13 \text{ g/mol Cu}_2(\text{OH})_2\text{CO}_3$$

$$\% \text{ O} = \frac{80.00 \text{g O/mol malachite}}{221.13 \text{ g Cu}_2(\text{OH})_2\text{CO}_3\text{/mol malachite}} \times 100\% = 36.18\% \text{ O}$$

11. Determine the molar mass of $ZnSO_4 \cdot 7H_2O$, and then the mass of water per mole of $ZnSO_4 \cdot 7H_2O$.

$$\text{mol wt. ZnSO}_4 \cdot 7\text{H}_2\text{O} = 65.39 \text{ g Zn} + 32.07 \text{ g S} + (11 \times 16.00 \text{ g O}) + (14 \times 1.01 \text{ g H})$$
$$= 287.60 \text{ g/mol ZnSO}_4 \cdot 7\text{H}_2\text{O}$$

$$\text{mass H}_2\text{O} = \frac{7 \text{ mol H}_2\text{O}}{1 \text{ mol ZnSO}_4 \cdot 7\text{H}_2\text{O}} \times \frac{18.02 \text{ g H}_2\text{O}}{1 \text{ mol H}_2\text{O}} = 126.1 \text{ g H}_2\text{O/mol ZnSO}_4 \cdot 7\text{H}_2\text{O}$$

$$\frac{126.1 \text{ g H}_2\text{O/mol ZnSO}_4 \cdot 7\text{H}_2\text{O}}{287.60 \text{ g/mol ZnSO}_4 \cdot 7\text{H}_2\text{O}} \times 100\% = 43.85\% \text{ H}_2\text{O}$$

12. In each case, we first determine the molar mass of the compound, and then the mass of the indicated element in one mole of the compound. Finally, we determine the percent by mass of the indicated element to four significant figures..

(a) $\text{mol wt. Pb(C}_2\text{H}_5)_4 = 207.2 \text{ g Pb} + (8 \times 12.01 \text{ g C}) + (20 \times 1.008 \text{ g H}) = 323.4 \text{ g/mol Pb(C}_2\text{H}_5)_4$

$$\text{mass Pb/mol Pb(C}_2\text{H}_5)_4 = \frac{1 \text{ mol Pb}}{1 \text{ mol Pb(C}_2\text{H}_5)_4} \times \frac{207.2 \text{ g Pb}}{1 \text{ mol Pb}} = 207.2 \text{ g Pb/mol Pb(C}_2\text{H}_5)_4$$

$$\% \text{ Pb} = \frac{207.2 \text{ g Pb}}{323.4 \text{ g Pb(C}_2\text{H}_5)_4} \times 100\% = 64.07\% \text{ Pb}$$

(b) $\text{mol wt. Fe}_4[\text{Fe(CN)}_6]_3 = (7 \times 55.85 \text{ g Fe}) + (18 \times 12.01 \text{ g C}) + (18 \times 14.01 \text{ g N})$
$$= 859.3 \text{ g/mol Fe}_4[\text{Fe(CN)}_6]_3$$

$$\text{mass Fe/mol Fe}_4[\text{Fe(CN)}_6]_3 = \frac{7 \text{ mol Fe}}{1 \text{ mol Fe}_4[\text{Fe(CN)}_6]_3} \times \frac{55.85 \text{ g}}{1 \text{ mol Fe}} = 391.0 \text{ g Fe/mol Fe}_4[\text{Fe(CN)}_6]_3$$

$$\% \text{ Fe} = \frac{391.0 \text{ g Fe}}{859.3 \text{ g Fe}_4[\text{Fe(CN)}_6]_3} \times 100\% = 45.50\% \text{ Fe}$$

13. $MgSO_4$ contains a smaller %S than does SO_3 because, while 1 mole of each compound contains 1 mole of S, each mole of SO_3 contains 3 mol O, while each mole of $MgSO_4$ contains 4 mol of O and 1 mol Mg—more extraneous material. In similar fashion, SO_3 contains a smaller %S than does SO_2, since SO_2 contains only two mol O for each mole S, while SO_3 contains 3 mol O for each mol S. Finally, Li_2S contains a greater %S than does SO_2, since each mole of S accompanies 32.0 g extraneous material (2 mol O) in SO_2, while 1 mol S is accompanied by only 15.9 g extraneous material (2 mol Li) in Li_2S. Therefore Li_2S contains the greatest %S of the compounds listed.

14. Express each percent as a mass in grams, convert each mass to moles of that element, the determine the simplest ratio of numbers of moles by dividing these mole numbers by the smallest.

$$\text{no. mol Co} = 71.06 \text{ g Co} \times \frac{1 \text{ mol Co}}{58.93 \text{ g Co}} = 1.206 \text{ mol Co}$$

$$\text{no. mol O} = 28.94 \text{ g O} \times \frac{1 \text{ mol O}}{16.00 \text{ g O}} = 1.808 \text{ mol O}$$

Divide each number by 1.206 to obtain 1.000 mol Co and 1.499 mol O. Empirical formula is Co_2O_3.

15. Determine the percent oxygen by difference first. $\% \text{ O} = 100.00\% - 59.96\% \text{ C} - 13.42\% \text{ H} = 26.62\% \text{ O}$

$$\text{no. mol O} = 26.62 \text{ g} \times \frac{1 \text{ mol O}}{16.00 \text{ g O}} = 1.664 \text{ mol O} \qquad \div 1.664 \longrightarrow 1.000 \text{ mol O}$$

$$\text{no mol C} = 59.96 \text{ g C} \times \frac{1 \text{ mol C}}{12.01 \text{ g C}} = 4.993 \text{ mol C} \qquad \div 1.664 \longrightarrow 3.000 \text{ mol C}$$

$$\text{no. mol H} = 13.42 \text{ g H} \times \frac{1 \text{ mol H}}{1.008 \text{ g H}} = 13.31 \text{ mol H} \qquad \div 1.664 \longrightarrow 7.999 \text{ mol H}$$

empirical formula is C_3H_8O

16. We base our calculation on 100.0 g MSG.

$$13.6 \text{ g Na} \times \frac{1 \text{ mol Na}}{22.99 \text{ g Na}} = 0.592 \text{ mol Na} \quad \div 0.592 \longrightarrow 1.00 \text{ mol Na}$$

$$35.5 \text{ g C} \times \frac{1 \text{ mol Cl}}{12.01 \text{ g C}} = 2.96 \text{ mol C} \quad \div 0.592 \longrightarrow 5.00 \text{ mol C}$$

$$4.8 \text{ g H} \times \frac{1 \text{ mol H}}{1.01 \text{ g H}} = 4.8 \text{ mol H} \quad \div 0.592 \longrightarrow 8.1 \text{ mol H}$$

$$8.3 \text{ g N} \times \frac{1 \text{ mol N}}{14.01 \text{ g N}} = 0.59 \text{ mol N} \quad \div 0.592 \longrightarrow 1.0 \text{ mol N}$$

$$37.8 \text{ g O} \times \frac{1 \text{ mol O}}{16.00 \text{ g O}} = 2.36 \text{ mol O} \quad \div 0.592 \longrightarrow 3.99 \text{ mol O}$$

Empirical formula: $NaC_5H_8NO_4$

17. First determine the empirical formula. Begin by determining the percent oxygen by difference.

% O = 100% − 57.83% C − 3.64% H = 38.53% O

$$\text{no. mol O} = 38.53 \text{ g} \times \frac{1 \text{ mol O}}{16.00 \text{ g O}} = 2.408 \text{ mol O} \quad \div 2.408 \longrightarrow 1.000 \text{ mol O}$$

$$\text{no mol C} = 57.83 \text{ g C} \times \frac{1 \text{ mol C}}{12.01 \text{ g C}} = 4.819 \text{ mol C} \quad \div 2.408 \longrightarrow 2.001 \text{ mol C}$$

$$\text{no. mol H} = 3.64 \text{ g H} \times \frac{1 \text{ mol H}}{1.008 \text{ g H}} = 3.61 \text{ mol H} \quad \div 2.408 \longrightarrow 1.50 \text{ mol H}$$

empirical formula is $C_4H_3O_2$. The weight of an empirical formula unit is

emp. mol wt. = (4 × 12.0 u C) + (3 × 1.0 u H) + (2 × 16.0 u O) = 83.0 u

This empirical molecular weight is one-half of the measured molecular weight. Thus, the molecular formula of terephthalic acid is twice the empirical formula.   Molecular formula: $C_8H_6O_4$

18. (a) First determine the masses of carbon and hydrogen in the original sample.

$$\text{mass C} = 3.425 \text{ g CO}_2 \times \frac{1 \text{ mol CO}_2}{44.010 \text{ g}} \times \frac{1 \text{ mol C}}{1 \text{ mol CO}_2} = 0.07782 \text{ mol C} \times \frac{12.011 \text{ g C}}{1 \text{ mol C}} = 0.9347 \text{ g C}$$

$$\text{mass H} = 0.971 \text{ g H}_2O \times \frac{1 \text{ mol H}_2O}{18.02 \text{ g H}_2O} \times \frac{2 \text{ mol H}}{1 \text{ mol H}_2O} = 0.108 \text{ mol H} \times \frac{1.008 \text{ g H}}{1 \text{ mol H}} = 0.109 \text{ g H}$$

Then the percents of the two elements in the compound are computed.

$$\% \text{ C} = \frac{0.9347 \text{ g C}}{1.235 \text{ g cmpd}} \times 100\% = 75.68\% \text{ C} \qquad \% \text{ H} = \frac{0.109 \text{ g H}}{1.235 \text{ g cmpd}} \times 100\% = 8.82\% \text{ H}$$

The % O is determined by difference. % O = 100% − 75.68% C − 8.82% H = 15.50% O

(b) In part (a), we determined the number of moles of C and H in the original sample of the compound. We can determine the mass of oxygen in that sample by difference, and then the number of moles of oxygen in that sample. We divide each of these numbers of moles by the smallest number to determine the empirical formula.

mass O = 1.235 g cmpd − 0.9347 g C − 0.109 g H = 0.191 g O

$$\text{mol O} = 0.191 \text{ g O} \times \frac{1 \text{ mol O}}{16.00 \text{ g O}} = 0.0119 \text{ mol O} \quad \div 0.0119 \longrightarrow 1.00 \text{ mol O}$$

$$0.108 \text{ mol H} \quad \div 0.0119 \longrightarrow 9.08 \text{ mol H}$$

$$0.07782 \text{ mol C} \quad \div 0.0119 \longrightarrow 6.54 \text{ mol C}$$

Thus, the empirical formula of ibuprofen is $C_{13}H_{18}O_2$.

19. The element chromium has an atomic weight of 52.0 u. Thus, there can only be one chromium atom per formula unit of the compound. (Two atoms of chromium weigh 104 u, more than the formula weight of the compound.) The remaining three of the four atoms in the formula unit must be oxygen. Thus, the oxide is CrO_3, chromium(VI) oxide.

20. (a) Sn^{2+} tin(II)   (b) Co^{3+} cobalt(III)   (c) Mg^{2+} magnesium
    (d) Cr^{2+} chromium(II)   (e) IO_3^- iodate   (f) ClO_2^- chlorite
    (g) Au^{3+} gold(III)   (h) HSO_4^- bisulfate or hydrogen sulfate
    (i) HCO_3^- hydrogen carbonate   (j) OH^- hydroxide

21. (a) LiI   lithium iodide   (b) $CaCl_2$ calcium chloride
    (c) ICl_3 iodine trichloride   (d) N_2O_3 dinitrogen trioxide
    (e) PCl_5 phosphorus pentachloride

22. **(a)** NaCN   sodium cyanide   **(b)** HClO   hypochlorous acid
      (c) NH_4NO_3 ammonium nitrate   **(d)** $KBrO_3$ potassium bromate

23. The desired oxidation state is given first, followed by the method of assigning that oxidation state.
   (a) Al = 0   The oxidation state (O. S.) of an uncombined element is 0.
   (b) S = –2 in Li_2S The O. S. of Li in its compounds is +1. Remember that oxidation states in a
                             compound must sum to zero.
   (c) N = +4 in NO_2 The O. S. of O in its compounds is –2.
   (d) N = +3 in HNO_2 The O. S. of H in its compounds is +1; that of O is –2.
   (e) V = +3 in V^{3+} For monatomic ions, O. S. = ionic charge.
   (f) P = +5 in HPO_4^{2-} The O. S. of H in its compounds is +1; that of O is –2. Remember that O.S.
                             in a polyatomic ion must sum to the charge on that ion.

24. **(a)** MgO   magnesium oxide   **(b)** BaF_2 barium fluoride
      (c) $Hg(NO_3)_2$ mercury(II) nitrate   **(d)** $Fe_2(SO_4)_3$ iron(III) sulfate
      (e) $Sr(ClO_4)_2$ strontium perchlorate   **(f)** $KHCO_3$ potassium hydrogen carbonate
      (g) NCl_3 nitrogen trichloride   **(h)** BrF_5 bromine pentafluoride

25. **(a)** HBr   hydrobromic acid   **(b)** $HClO_2$ chlorous acid
      (c) HIO_3 iodic acid   **(d)** H_2SO_3 sulfurous acid
      (e) H_3PO_4 phosphoric acid   **(f)** H_2Se hydroselenic acid
      (g) $HClO_3$ chloric acid   **(h)** HNO_2 nitrous acid

EXERCISES

The Avogadro Constant and the Mole

26. 9×10^{-4} μmol of C_2H_6S (62.0 g/mol, 0.84 g/cm³) need to be present in each m³ to be detected.

no. molec. = $1500 \text{ m}^3 \times \dfrac{9 \times 10^{-4} \text{ μmol}}{1 \text{ m}^3} \times \dfrac{1 \text{ mol}}{10^6 \text{ μmol}} \times \dfrac{6.022 \times 10^{23} \text{ molec.}}{1 \text{ mol}}$

$= 8 \times 10^{17} \ C_2H_6S$ molecules

27. The greatest number of N atoms is contained in the compound with the greatest number of moles of N.

The molar mass of N_2O is (2 mol N × 14.0 g N) + (1 mol O × 16.0 g O) = 44.0 g/mol N_2O. Thus, 50.0 g N_2O is slightly more than 1 mole of N_2O, and contains slightly more than 2 moles of N.

Each mole of N_2 contains 2 moles of N.

The molar mass of NH_3 is 17.0 g. Thus, there is 1 mole of NH_3 present, which contains 1 mole of N.

The molar mass of pyridine is (5 mol C × 12.0 g C) + (5 mol H × 1.01 g H) + 14.0 g N = 79.1 g/mol. Since each mole of pyridine contains 1 mole of N, we need slightly more than 2 moles of pyridine to have more N than in the N_2O. But that would be a mass of about 158 g pyridine, and 150 mL has a mass of less than 150 g.

Thus, the greatest mass of N is present in 50.0 g N_2O.

28. **(a)** P mass = 3.56×10^{-2} mol $P_4 \times \dfrac{4 \text{ mol P}}{1 \text{ mol } P_4} \times \dfrac{30.97 \text{ g P}}{1 \text{ mol P}} = 4.41$ g P

(b) First we need the molar mass of $C_{18}H_{36}O_2$, stearic acid:

molar mass = (18 mol C × 12.01 g C) + (36 mol H × 1.01 g H) + (2 mol O × 16.00 g O)
        = 284.5 g/mol

Stearic acid mass = 6.62×10^{22} molecules $\times \dfrac{1 \text{ mole}}{6.022 \times 10^{23} \text{ molecules}} \times \dfrac{284.5 \text{ g}}{1 \text{ mole } C_{18}H_{36}O_2}$

= 31.3 g stearic acid.

(c) molar mass = (6 mol C × 12.01 g C) + (14 mol H × 1.01 g H) + (2 mol N × 14.00 g N)
        + (2 mol O × 16.00 g O) = 146.2 g/mol

lysine mass = 1.50 mol N $\times \dfrac{1 \text{ mol } C_6H_{14}N_2O_2}{2 \text{ mol N}} \times \dfrac{146.2 \text{ g lysine}}{1 \text{ mol lysine}} = 109.7$ g lysine

(d) H_3PO_4 molar mass = (3 mol H × 1.01 g C) + (1 mol P × 30.97 g P) + (4 mol O × 16.00 g O)

$$= 98.00 \text{ g/mol}$$

$$\text{H}_3\text{PO}_4 \text{ mass} = 2.33 \text{ mol SO}_3 \times \frac{3 \text{ mol O}}{1 \text{ mol SO}_3} \times \frac{1 \text{ mol H}_3\text{PO}_4}{4 \text{ mol O}} \times \frac{98.00 \text{ g H}_3\text{PO}_4}{1 \text{ mol H}_3\text{PO}_4} = 171 \text{ g H}_3\text{PO}_4$$

29. (a) $\text{mol S}_8 = 2.10 \times 10^{-3} \text{ cm}^3 \times \frac{2.07 \text{ g}}{1 \text{ cm}^3} \times \frac{1 \text{ mol S}}{32.07 \text{ g}} \times \frac{1 \text{ mol S}_8}{8 \text{ mol S}} = 1.69 \times 10^{-5} \text{ mol S}_8$

(b) $\text{no. S atoms} = 1.69 \times 10^{-5} \text{ mol S}_8 \times \frac{8 \text{ mol S}}{1 \text{ mol S}_8} \times \frac{6.022 \times 10^{23} \text{ atoms}}{1 \text{ mol S}} = 8.14 \times 10^{19} \text{ S atoms}$

30. $\text{no. Fe atoms} = 6 \text{ L blood} \times \frac{1000 \text{ mL}}{1 \text{ L}} \times \frac{15.5 \text{ g hem}}{100 \text{ mL blood}} \times \frac{1 \text{ mol hem}}{64,500 \text{ g}} \times \frac{4 \text{ mol Fe}}{1 \text{ mol hem}}$

$$\times \frac{6.022 \times 10^{23} \text{ atoms}}{1 \text{ mol Fe}} = 3 \times 10^{22} \text{ Fe atoms}$$

Chemical Formulas

31. For glucose (blood sugar), $\text{C}_6\text{H}_{12}\text{O}_6$,

(a) FALSE   The percentages by mass of C and O are *different* than in CO. For one thing, CO contains no hydrogen.

(b) TRUE   In dihydroxyacetone, $(\text{CH}_2\text{OH})_2\text{CO}$ or $\text{C}_3\text{H}_6\text{O}_3$, the ratio of $\text{C} : \text{H} : \text{O} = 3 : 6 : 3 = 1 : 2 : 1$. In glucose, this ratio is $\text{C} : \text{H} : \text{O} = 6 : 12 : 6 = 1 : 2 : 1$. Thus, the ratios are the *same*.

(c) FALSE   The proportions, by number of atoms, of C and O are the same in glucose. Since, however, C and O have different mole weights, their proportions by mass must be *different*.

(d) FALSE   Each mole of glucose contains $(12 \times 1.01 =) 12.1$ g H. But each mole also contains 72.0 g C and 96.0 g O. Thus, the highest percentage, by mass, is that of O. The highest percentage, by number of atoms, is that of H.

32. (a) C_2H_6 is a molecular formula, since the stoichiometric subscripts could both be divided by two and retain the same proportion of each element in the compound.

(b) Cl_2O could be an empirical or a molecular formula. The molecules could consist of two Cl atoms and an O atom; then it would be a molecular formula. However, if the molecules consisted of more atoms, such as Cl_4O_2 or Cl_6O_3, then the formula Cl_2O would be an empirical formula.

(c) CH_4O could be an empirical or a molecular formula. The reasoning is similar to that of part (b).

(d) N_2O_4 is a molecular formula. The reasoning is the same as that of part (a).

33. (a) The formula of the halothane molecule is CHBrClCF_3. Each formula unit (molecule) contains two C atoms, three F atoms, and one atom each of H, Cl, and Br, for a total of 8 atoms per formula unit.

(b) There are three F atoms for every two C atoms, for a ratio of $\frac{3}{2}$.

(c) The ratio of Br to F by mass is determined as follows.
$$\frac{\text{mass Br}}{\text{mass F}} = \frac{1 \text{ mol Br}}{3 \text{ mol F}} \times \frac{79.904 \text{ g Br}}{1 \text{ mol Br}} \times \frac{1 \text{ mol F}}{18.9984 \text{ g F}} = 1.4019 \text{ g Br/g F}$$

(d) $\text{mass} = 1.00 \text{ g F} \times \frac{1 \text{ mol F}}{19.00 \text{ g F}} \times \frac{1 \text{ mol C}_2\text{HBrClF}_3}{3 \text{ mol F}} \times \frac{197.4 \text{ g C}_2\text{HBrClF}_3}{1 \text{ mol C}_2\text{HBrClF}_3} = 3.46 \text{ g C}_2\text{HBrClF}_3$

34. (a) A formula unit of $\text{Ge}[\text{S}(\text{CH}_2)_4\text{CH}_3]_4$ contains:

1 Ge atom   4 S atoms $4 (4 + 1) = 20$ C atoms $4[4(2) + 3] = 44$ H atoms

For a total of $1 + 4 + 20 + 44 = 69$ atoms per formula unit

(b) $\frac{\text{no. C atoms}}{\text{no. H atoms}} = \frac{20 \text{ C atoms}}{44 \text{ H atoms}} = \frac{5 \text{ C atoms}}{11 \text{ H atoms}} = 0.454 \text{ C atom/H atom}$

(c) $\frac{\text{mass Ge}}{\text{mass Se}} = \frac{1 \text{ mol Ge} \times \dfrac{72.6 \text{ g Ge}}{1 \text{ mol Ge}}}{4 \text{ mol S} \times \dfrac{32.07 \text{ g S}}{1 \text{ mol S}}} = \frac{72.6 \text{ g Ge}}{128.3 \text{ g S}} = 0.566 \text{ g Ge/g S}$

(d) mol wt $= 72.6 \text{ g Ge} + (4 \times 32.1 \text{ g S}) + (20 \times 12.01 \text{ g C}) + (44 \times 1.01 \text{ g H}) = 485.4 \text{ g/mol}$

mass of S $= 1 \text{ mol Ge}[\text{S}(\text{CH}_2)_4\text{CH}_3]_4 \times \frac{4 \text{ mol S}}{1 \text{ mol Ge}[\text{S}(\text{CH}_2)_4\text{CH}_3]_4} \times \frac{32.07 \text{ g S}}{1 \text{ mol S}} = 128.3 \text{ g S}$

(e) mass of $\text{Ge}[\text{S}(\text{CH}_2)_4\text{CH}_3]_4 = 1.00 \text{ g Ge} \times \frac{1 \text{ mol Ge}}{72.61 \text{ g Ge}} \times \frac{1 \text{ mol Ge}[\text{S}(\text{CH}_2)_4\text{CH}_3]_4}{1 \text{ mol Ge}}$

$$\times \frac{485.4 \text{ g}}{1 \text{ mol Ge}[\text{S}(\text{CH}_2)_4\text{CH}_3]_4} = 6.69 \text{ g Ge}[\text{S}(\text{CH}_2)_4\text{CH}_3]_4$$

(f) no. C atoms = $33.10 \text{ g cmpd} \times \dfrac{1 \text{ mol cmpd}}{485.4 \text{ g cmpd}} \times \dfrac{20 \text{ mol C}}{1 \text{ mol cmpd}} \times \dfrac{6.022 \times 10^{23} \text{ C atoms}}{1 \text{ mol C}}$

$= 8.213 \times 10^{23}$ C atoms

Percent Composition of Compounds

35. The information obtained in the course of calculating the molar mass is used to determine the mass percent of each element in stearic acid, $C_{18}H_{36}O_2$, abbreviated as SA below.

molar mass C_8H_{18}

$$= \left(\dfrac{18 \text{ mol C}}{1 \text{ mol SA}} \times \dfrac{12.011 \text{ g C}}{1 \text{ mol C}}\right) + \left(\dfrac{36 \text{ mol H}}{1 \text{ mol SA}} \times \dfrac{1.00794 \text{ g H}}{1 \text{ mol H}}\right) + \left(\dfrac{2 \text{ mol O}}{1 \text{ mol SA}} \times \dfrac{15.9994 \text{ g O}}{1 \text{ mol O}}\right)$$

$$= \dfrac{216.20 \text{ g C}}{1 \text{ mol SA}} + \dfrac{36.2858 \text{ g H}}{1 \text{ mol SA}} + \dfrac{31.9988 \text{ g O}}{1 \text{ mol SA}} = \dfrac{284.48 \text{ g}}{1 \text{ mol SA}}$$

$\%C = \dfrac{216.20 \text{ g C/mol SA}}{284.48 \text{ g SA/mol SA}} \times 100\% = 75.998\%$ C; $\%H = \dfrac{36.2858 \text{ g H/mol SA}}{284.48 \text{ g SA/mol SA}} \times 100\% = 12.755\%$ H

$\%O = \dfrac{31.9988 \text{ g O/mol SA}}{284.48 \text{ g SA/mol SA}} \times 100\% = 11.248\%$ O

36. The method of solution of Exercise 35 is abbreviated in the following problem solution.

molar mass = $(20 \text{ mol C} \times 12.011 \text{ g C}) + (24 \text{ mol H} \times 1.00794 \text{ g H}) + (2 \text{ mol N} \times 14.0067 \text{ g N})$

$+ (2 \text{ mol O} \times 15.9994 \text{ g O})$

$= 240.22 \text{ g C} + 24.1906 \text{ g H} + 28.0134 \text{ g N} + 31.9988 \text{ g O} = 324.42 \text{ g/mol}$

$\%C = \dfrac{240.22}{324.42} \times 100\% = 74.046 \%C$ $\%H = \dfrac{24.1906}{324.42} \times 100\% = 7.4566 \%H$

$\%N = \dfrac{28.0134}{324.42} \times 100\% = 8.6349 \%N$ $\%O = \dfrac{31.9988}{324.42} \times 100\% = 9.8634 \%O$

37. molar mass of axinite = $(1 \text{ mol H} \times 1.00794 \text{ g H}) + (3 \text{ mol Ca} \times 40.078 \text{ g Ca}) + (2 \text{ mol Al} \times 26.9815 \text{ g Al})$

$+ (1 \text{ mol B} \times 10.811 \text{ g B}) + (4 \text{ mol Si} \times 28.0855 \text{ g Si}) + (16 \text{ mol O} \times 15.9994 \text{ g O})$

$= 1.00794 \text{ g H} + 120.23 \text{ g Ca} + 53.9630 \text{ g Al} + 10.811 \text{ g B} + 112.342 \text{ g Si} + 255.990 \text{ g O}$

$= 554.34 \text{ g/mol}$

$\%B = \dfrac{10.811 \text{ g B/mol axinite}}{554.34 \text{ g/mol axinite}} \times 100\% = 1.9502\%$ B

38. (a) $\%Zr = \dfrac{1 \text{ mol Zr}}{1 \text{ mol ZrSiO}_4} \times \dfrac{1 \text{ mol ZrSiO}_4}{183.31 \text{ g ZrSO}_4} \times \dfrac{91.224 \text{ g Zr}}{1 \text{ mol Zr}} \times 100\% = 49.766\%$ Zr

(b) $\%Be = \dfrac{3 \text{ mol Be}}{1 \text{ mol Be}_3\text{Al}_2\text{Si}_6\text{O}_{18}} \times \dfrac{1 \text{ mol Be}_3\text{Al}_2\text{Si}_6\text{O}_{18}}{537.502 \text{ g Be}_3\text{Al}_2\text{Si}_6\text{O}_{18}} \times \dfrac{9.01218 \text{ g Be}}{1 \text{ mol Be}} \times 100\% = 5.03004\%$ Be

(c) $\%Fe = \dfrac{3 \text{ mol Fe}}{1 \text{ mol Fe}_3\text{Al}_2\text{Si}_6\text{O}_{18}} \times \dfrac{1 \text{ mol Fe}_3\text{Al}_2\text{Si}_6\text{O}_{18}}{678.006 \text{ g Fe}_3\text{Al}_2\text{Si}_6\text{O}_{18}} \times \dfrac{55.847 \text{ g Fe}}{1 \text{ mol Fe}} \times 100\% = 24.711\%$ Fe

(d) $\%S = \dfrac{1 \text{ mol S}}{1 \text{ mol Na}_4\text{SSi}_3\text{Al}_3\text{O}_{12}} \times \dfrac{1 \text{ mol Na}_4\text{SSi}_3\text{Al}_3\text{O}_{12}}{481.219 \text{ g Na}_4\text{SSi}_3\text{Al}_3\text{O}_{12}} \times \dfrac{32.066 \text{ g S}}{1 \text{ mol S}} \times 100\% = 6.6353\%$ S

39. We need to express all three on the same percent basis. We use % Cl by mass.

$14\% \text{ NaOCl} \times \dfrac{1 \text{ mol NaOCl}}{74.5 \text{ g NaOCl}} \times \dfrac{1 \text{ mol Cl}}{1 \text{ mol NaOCl}} \times \dfrac{35.5 \text{ g Cl}}{1 \text{ mol Cl}} = 6.7\%$ Cl

$10\% \text{ OCl} \times \dfrac{1 \text{ mol OCl}}{51.5 \text{ g OCl}} \times \dfrac{1 \text{ mol Cl}}{1 \text{ mol OCl}} \times \dfrac{35.5 \text{ g Cl}}{1 \text{ mol Cl}} = 6.9\%$ Cl

Thus, the material that is 7% available chlorine is the best buy, assuming that 7% can be interpreted as 7.0%, and also assuming that all three solutions have about the same density.

40. The oxide with the largest %Cr will have the largest number of moles of Cr per mol of oxygen.

CrO $\dfrac{1 \text{ mol Cr}}{1 \text{ mol O}} = 1 \text{ mol Cr/mol O}$ Cr_2O_3 $\dfrac{2 \text{ mol Cr}}{3 \text{ mol O}} = 0.667 \text{ mol Cr/mol O}$

CrO_2 $\dfrac{1 \text{ mol Cr}}{2 \text{ mol O}} = 0.500 \text{ mol Cr/mol O}$ CrO_3 $\dfrac{1 \text{ mol Cr}}{3 \text{ mol O}} = 0.333 \text{ mol Cr/mol O}$

Arranged in order of increasing %Cr: $CrO_3 < CrO_2 < Cr_2O_3 < CrO$

41. In each case a P atom is associated with four O atoms. The species with the largest percent phosphorus will have the smallest mass (per P atom) of atoms other than P and O.

In H_3PO_4, the mass of those other atoms is (3 H atoms × 1 u/H atom) = 3 u

In Na_3PO_4, their mass is (3 Na atoms × 23 u/Na atom) = 69 u

In $(NH_4)_2HPO_4$, their mass is (2 N atoms × 14 u/N atom) + (9 H atoms × 1 u/H atom) = 37 u

In $Ca(H_2PO_4)_2$, their mass is ($^1/_2$ Ca atom × 40 u/Ca atom) + (2 H atoms × 1 u/H atom) = 22 u

Thus, arranged in order of increasing %P: $Na_3PO_4 < (NH_4)_2HPO_4 < Ca(H_2PO_4)_2 < H_3PO_4$

Chemical Formulas from Percent Composition

42. Convert each of the percentages into the mass in 100.00 g, and then to the amount in moles of that element.

$$93.71 \text{ g C} \times \frac{1 \text{ mol C}}{12.01 \text{ g C}} = 7.803 \text{ mol C} \qquad \div 6.23 \longrightarrow 1.25 \text{ mol C}$$

$$6.29 \text{ g H} \times \frac{1 \text{ mol H}}{1.01 \text{ g H}} = 6.23 \text{ mol H} \qquad \div 6.23 \longrightarrow 1.00 \text{ mol H}$$

The empirical formula is C_5H_4, which has an empirical molar mass of [(5 × 12.0 g C) + (4 × 1.0 g H)] = 64.0 g/mol. Since this empirical molar mass is one-half of the 128 g/mol correct molar mass, the molecular formula must be twice the empirical formula. Molecular formula: $C_{10}H_8$

43. The percent of selenium in each oxide is found by difference.

First oxide: % Se = 100.0% − 28.8% O = 71.2% Se

$$28.8\% \text{ O} \times \frac{1 \text{ mol O}}{16.0 \text{ g O}} = 1.80 \text{ mol O} \qquad \div 0.901 \longrightarrow 2.00 \text{ mol O}$$

$$71.2\% \text{ Se} \times \frac{1 \text{ mol Se}}{79.0 \text{ g Se}} = 0.901 \text{ mol Se} \qquad \div 0.901 \longrightarrow 1.00 \text{ mol Se}$$

The empirical formula is SeO_2. An appropriate name is selenium dioxide.

Second oxide: % Se = 100.0% − 37.8% O = 62.2% Se

$$37.8\% \text{ O} \times \frac{1 \text{ mol O}}{16.0 \text{ g O}} = 2.36 \text{ mol O} \qquad \div 0.787 \longrightarrow 3.00 \text{ mol O}$$

$$62.2\% \text{ Se} \times \frac{1 \text{ mol Se}}{79.0 \text{ g Se}} = 0.787 \text{ mol Se} \qquad \div 0.787 \longrightarrow 1.00 \text{ mol Se}$$

The empirical formula is SeO_3. An appropriate name is selenium trioxide.

44. (a) $\quad 74.01 \text{ g C} \times \dfrac{1 \text{ mol C}}{12.01 \text{ g C}} = 6.162 \text{ mol C} \qquad \div 1.298 \longrightarrow 4.747 \text{ mol C}$

$\quad 5.23 \text{ g H} \times \dfrac{1 \text{ mol H}}{1.01 \text{ g H}} = 5.18 \text{ mol H} \qquad \div 1.298 \longrightarrow 3.99 \text{ mol H}$

$\quad 20.76 \text{ g O} \times \dfrac{1 \text{ mol O}}{16.00 \text{ g O}} = 1.298 \text{ mol O} \qquad \div 1.298 \longrightarrow 1.00 \text{ mol O}$

Now multiply each of the mole numbers by 4 to obtain an empirical formula of $C_{19}H_{16}O_4$.

(b) $\quad 30.20 \text{ g C} \times \dfrac{1 \text{ mol C}}{12.01 \text{ g C}} = 2.515 \text{ mol C} \qquad \div 0.6288 \longrightarrow 4.000 \text{ mol C}$

$\quad 5.07 \text{ g H} \times \dfrac{1 \text{ mol H}}{1.01 \text{ g H}} = 5.02 \text{ mol H} \qquad \div 0.6288 \longrightarrow 7.98 \text{ mol H}$

$\quad 44.58 \text{ g Cl} \times \dfrac{1 \text{ mol Cl}}{35.45 \text{ g Cl}} = 1.258 \text{ mol Cl} \qquad \div 0.6288 \longrightarrow 2.001 \text{ mol Cl}$

$\quad 20.16 \text{ g S} \times \dfrac{1 \text{ mol S}}{32.06 \text{ g H}} = 0.6288 \text{ mol S} \qquad \div 0.6288 \longrightarrow 1.000 \text{ mol S}$

The empirical formula is $C_4H_8Cl_2S$.

(c) $\quad 95.21 \text{ g C} \times \dfrac{1 \text{ mol C}}{12.01 \text{ g C}} = 7.928 \text{ mol C} \qquad \div 4.74 \longrightarrow 1.67 \text{ mol C}$

$\quad 4.79 \text{ g H} \times \dfrac{1 \text{ mol H}}{1.01 \text{ g H}} = 4.74 \text{ mol H} \qquad \div 4.74 \longrightarrow 1.00 \text{ mol H}$

Multiply each of the mole numbers by 3 to obtain an empirical formula of C_5H_3

45. Each percent is numerically equal to the mass of that element present in 100.00 g of the compound. These masses then are converted to amounts of the elements, in moles.

$$\text{amount C} = 38.37 \text{ g C} \times \frac{1 \text{ mol C}}{12.01 \text{ g C}} = 3.195 \text{ mol C} \quad \div 0.491 \longrightarrow 6.51 \text{ mol C}$$

$$\text{amount H} = 1.49 \text{ g H} \times \frac{1 \text{ mol H}}{1.01 \text{ g H}} = 1.48 \text{ mol H} \quad \div 0.491 \longrightarrow 3.01 \text{ mol H}$$

$$\text{amount Cl} = 52.28 \text{ g Cl} \times \frac{1 \text{ mol Cl}}{35.45 \text{ g Cl}} = 1.475 \text{ mol Cl} \quad \div 0.491 \longrightarrow 3.004 \text{ mol Cl}$$

$$\text{amount O} + 7.86 \text{ g O} \times \frac{1 \text{ mol O}}{16.0 \text{ g O}} = 0.491 \text{ mol O} \quad \div 0.491 \longrightarrow 1.00 \text{ mol O}$$

Multiply each of these numbers of moles by 2 to obtain the empirical formula: $C_{13}H_6Cl_6O_2$.

46. In each 100 g of the compound there are 65 g of F and 35 g of X. The number of moles of X is given by

$$\text{no. mol X} = 65 \text{ g F} \times \frac{1 \text{ mol F}}{19.0 \text{ g}} \times \frac{1 \text{ mol X}}{3 \text{ mol F}} = 1.14 \text{ mol X}$$

Thus, the molar mass of $X = \frac{35 \text{ g X}}{1.14 \text{ g X}} = 31 \text{ g/mol X}$

47. The atomic weight of element X has the units of grams per mole. We can determine the amount, in moles of Cl and convert that to the amount of X, equivalent to 25.0 g of X.

$$\text{atomic weight} = \frac{25.0 \text{ g X}}{75.0 \text{ g Cl}} \times \frac{35.453 \text{ g Cl}}{1 \text{ mol Cl}} \times \frac{4 \text{ mol Cl}}{1 \text{ mol X}} = \frac{47.3 \text{ g X}}{1 \text{ mol X}}$$

This atomic weight is close to that of the element titanium, which therefore is identified as element X.

48. A freon derived from CH_4 contains 1 mole of carbon per mole of the freon, and at most four moles of fluorine. Let us consider a number of possibilities, and determine % F in each one, looking for 31.43% F.

CH_3F 34.0 g/mol $\% F = \dfrac{19.0 \text{ g F}}{34.0 \text{ g cmpd}} \times 100\% = 55.9\% \text{ F}$ The mole weight must be greater.

CH_2ClF 68.5 g/mol $\% F = \dfrac{19.0 \text{ g F}}{68.5 \text{ g cmpd}} \times 100\% = 27.7\% \text{ F}$ More F is needed.

$CHClF_2$ 86.5 g/mol $\% F = \dfrac{38.0 \text{ g F}}{86.5 \text{ g cmpd}} \times 100\% = 43.9\% \text{ F}$ The mole weight must be greater.

CCl_2F_2 121.0 g/mol $\% F = \dfrac{38.0 \text{ g F}}{121.0 \text{ g cmpd}} \times 100\% = 31.4\% \text{ F}$ This is the compound in question.

Law of Multiple Proportions

49. The modern atomic weights are 12.0 for carbon and 1.0 for hydrogen. Thus, we establish the following ratio with the formulas C_2H_4 (in the numerator) and CH_4 (in the denominator).

$$\frac{(2 \times 12.0 \text{ g C})/(4 \times 1.0 \text{ g H})}{12.0 \text{ g C}/(4 \times 1.0 \text{ g H})} = \frac{24.0 \text{ g C}/4.0 \text{ g H}}{12.0 \text{ g C}/4.0 \text{ g H}} = \frac{2}{1}$$

This is the same ratio that Dalton obtained, as follows, with CH in the numerator, and CH_2 in the denominator.

$$\frac{12.0 \text{ g C}/(1 \times 1.0 \text{ g H})}{12.0 \text{ g C}/(2 \times 1.0 \text{ g H})} = \frac{12.0 \text{ g C}/1.0 \text{ g H}}{12.0 \text{ g C}/2.0 \text{ g H}} = \frac{12.0 \text{ g C}/1.0 \text{ g H}}{6.00 \text{ g H}/1.0 \text{ g H}} = \frac{2}{1}$$

50. To verify the law of multiple proportions, we determine the mass of nitrogen that combines with 1.00 g of oxygen in each compound.   In NO: $\dfrac{14.0 \text{ g N}}{16.0 \text{ g O}} = 0.875 \text{ g N/g O}$

In NO_2: $\dfrac{14.0 \text{ g N}}{2 \times 16.0 \text{ g O}} = 0.4375 \text{ g N/g O}$ In N_2O: $\dfrac{2 \times 14.0 \text{ g N}}{16.0 \text{ g O}} = 1.75 \text{ g N/g O}$

We demonstrate that these three ratios are related as small whole numbers by dividing each of them by the smallest, 0.4375.

0.875 : 0.4375 : 1.75   Divide by 0.4375 to obtain   2 : 1 : 4   a series of small whole numbers.

51. We determine the ration of the mass of O to the mass of H in each compound. In hydrogen peroxide, the percent of oxygen is determined by difference: %O = 100.00% − 5.93% H = 94.07% O

For hydrogen peroxide: $\dfrac{94.07 \text{ g O}}{5.93 \text{ g H}} = 15.9 \text{ g O/g H}$

For water, H_2O: $\dfrac{1\ mol\ O \times \dfrac{15.9994\ g\ O}{1\ mol\ O}}{2\ mol\ H \times \dfrac{1.00794\ g\ H}{1\ mol\ H}} = 7.93668\ g\ O/g\ H$

Then compare these two combining weights: $\dfrac{15.9\ g\ O/g\ H}{7.93668\ g\ O/g\ H} = 2.00$ which is a small whole number.

52. If we assume samples of 100.0 g of each compound, we can determine the mass of mercury that combines with 1.00 g of oxygen.

$\dfrac{mass\ Hg}{g\ O} = \dfrac{96.2\ g\ Hg}{3.8\ g\ O} = \dfrac{25._3\ g\ Hg}{1\ g\ O}$ $\qquad$ $\dfrac{mass\ Hg}{g\ O} = \dfrac{92.6\ g\ Hg}{7.4\ g\ O} = \dfrac{12._5\ g\ Hg}{1\ g\ O}$

Since the two combining masses are related approximately as 2 to 1, the data conform to the law of multiple proportions. The compounds could be Hg_2O and HgO, since the first has twice as much mercury for a specific mass of oxygen as does the second.

Combustion Analysis

53. (a) Determine the mass of carbon and of hydrogen present in the sample. A hydrocarbon only contains the two elements hydrogen and carbon.

$0.6260\ g\ CO_2 \times \dfrac{1\ mol\ CO_2}{44.010\ g} \times \dfrac{1\ mol\ C}{1\ mol\ CO_2} = 0.01422\ mol\ C \times \dfrac{12.011\ g\ C}{1\ mol\ C} = 0.1708\ g\ C$

$0.1602\ g\ H_2O \times \dfrac{1\ mol\ H_2O}{18.015\ g} \times \dfrac{2\ mol\ H}{1\ mol\ H_2O} = 0.01779\ mol\ H \times \dfrac{1.0079\ g\ H}{1\ mol\ H} = 0.01793\ g\ H$

The % C and % H then are found.

$\%\ C = \dfrac{0.1708\ g\ C}{0.1888\ g\ cmpd} \times 100\% = 90.47\%\ C$ $\qquad$ $\%\ H = \dfrac{0.01793\ g\ H}{0.1888\ g\ cmpd} \times 100\% = 9.497\%\ H$

(b) Use the moles of C and H from part (a), and divide both by the smallest.

$0.01422\ mol\ C \div 0.01422 = 1.000\ mol\ C$ $\qquad$ $0.01779\ mol\ H \div 0.01422 = 1.251\ mol\ H$

The empirical formula is obtained by multiplying these mole numbers by 4. It is C_4H_5.

(c) The empirical formula C_4H_5 has an empirical molar mass of [(4 × 12.0 g C) + (5 × 1.0 g H) =] 53.0 g/mol. This value is one-half of the actual molar mass. Hence, the molecular formula is twice the empirical formula. Molecular formula: C_8H_{10}.

54. Determine the mass of carbon and of hydrogen present in the sample.

$1.1518\ g\ CO_2 \times \dfrac{1\ mol\ CO_2}{44.01\ g} \times \dfrac{1\ mol\ C}{1\ mol\ CO_2} = 0.02617\ mol\ C \times \dfrac{12.01\ g\ C}{1\ mol\ C} = 0.3143\ g\ C$

$0.2694\ g\ H_2O \times \dfrac{1\ mol\ H_2O}{18.02\ g} \times \dfrac{2\ mol\ H}{1\ mol\ H_2O} = 0.02990\ mol\ H \times \dfrac{1.008\ g\ H}{1\ mol\ H} = 0.03014\ g\ H$

Then determine the mass of oxygen by difference, and its amount in moles.

mass O = 0.4039 g cmpd – 0.3143 g C – 0.03014 g H = 0.0595 g O

$0.0595\ g\ O \times \dfrac{1\ mol\ O}{16.0\ g\ O} = 0.00372\ mol\ O \div 0.00372 \longrightarrow 1.00\ mol\ O$

$0.02617\ mol\ C \div 0.00372 \longrightarrow 7.03\ mol\ C$ $\qquad$ This gives an empirical

$0.02990\ mol\ H \div 0.00372 \longrightarrow 8.04\ mol\ H$ $\qquad$ formula of C_7H_8O.

55. First determine the mass of carbon and hydrogen present in the sample.

$0.458\ g\ CO_2 \times \dfrac{1\ mol\ CO_2}{44.0\ g} \times \dfrac{1\ mol\ C}{1\ mol\ CO_2} = 0.0104\ mol\ C \times \dfrac{12.0\ g\ C}{1\ mol\ C} = 0.125\ g\ C$

$0.374\ g\ H_2O \times \dfrac{1\ mol\ H_2O}{18.0\ g} \times \dfrac{2\ mol\ H}{1\ mol\ H_2O} = 0.0416\ mol\ H \times \dfrac{1.01\ g\ H}{1\ mol\ H} = 0.0420\ g\ H$

Then the mass of N that this sample would have produced is determined. (Note that this is also the mass of N_2 produced in the reaction.) $\qquad$ $0.244\ g\ N \times \dfrac{0.312\ g\ 1st\ sample}{0.525\ g\ 2nd\ sample} = 0.145\ g\ N$

We also can determine this mass of N by difference. 0.312 g cmpd – 0.125 g C – 0.0420 g H = 0.145 g N

Then we can calculate the relative number of moles of each element.

$0.145\ g\ N \times \dfrac{1\ mol\ N}{14.0\ g\ N} = 0.0104\ mol\ N \div 0.0104 \longrightarrow 1.00\ mol\ N$

$0.0104\ mol\ C \div 0.0104 \longrightarrow 1.00\ mol\ C$ $\qquad$ Thus, the empirical

$0.0416\ mol\ H \div 0.0104 \longrightarrow 4.00\ mol\ H$ $\qquad$ formula is CH_4N.

56. **(a)** The compound that will produce the greatest mass of H_2O per 1.00 mole of the compound is the one that has the greatest number of moles of H in one mole of the compound. 1 mol CH_4 contains 4 mol H, 1 mol C_2H_5OH and 1 mol C_6H_5OH each contain 6 mol H. 1 mol $C_{10}H_8$ contains 8 mol H. Thus, $C_{10}H_8$ will produce the largest mass of water per mole of the compound.

(b) The compound that produces the largest mass of water per gram of the compound will have the largest amount of hydrogen per gram of the compound. Thus, we need to compare the ratios of amount of hydrogen per mole to the mole weight for each compound.

Notice that C_2H_5OH has the same amount of H per mole as does C_6H_5OH, but C_6H_5OH has a higher mole weight. Thus, C_2H_5OH will produce more H_2O per gram than will C_6H_5OH. Notice also that CH_4 has 4 H's per C, while $C_{10}H_8$ has 8 H's per 10 C's or 0.8 H per C. Thus CH_4 will produce more H_2O than will $C_{10}H_8$.

Thus, we are left with comparing CH_4 to C_2H_5OH. The O in the second compound has about the same mass (16 u) as does C (12 u). Thus, in CH_4 there are 4 H's per C, while in C_2H_5OH there are about 2 H's per C. CH_4 will produce more water per gram, on combustion, the most of all four compounds.

57. One way to solve this problem is to determine both the mole weight of BHA, $C_{11}H_{16}O_2$, and the mass of CO_2 produced by the combustion of one mole.

mol wt. $C_{11}H_{16}O_2$ = (11 × 12.011 g C) + (16 × 1.008 g H) + (2 × 15.999 g O) = 180.25 g/mol $C_{11}H_{16}O_2$.

The combustion reaction follows. $\quad C_{11}H_{16}O_2 + 14\ O_2 \longrightarrow 11\ CO_2 + 8\ H_2O$

mass CO_2 = 1 mol $C_{11}H_{16}O_2 \times \dfrac{11\ \text{mol CO}_2}{1\ \text{mol C}_{11}\text{H}_{16}\text{O}_2} \times \dfrac{44.010\ \text{g CO}_2}{1\ \text{mol CO}_2}$ = 484.11 g CO_2

Then we determine the mass of $C_{11}H_{16}O_2$ needed to produce 0.500 g CO_2.

mass $C_{11}H_{16}O_2$ = 0.5000 g $CO_2 \times \dfrac{1\ \text{mol C}_{11}\text{H}_{16}\text{O}_2}{484.11\ \text{g CO}_2} \times \dfrac{180.25\ \text{g C}_{11}\text{H}_{16}\text{O}_2}{1\ \text{mol C}_{11}\text{H}_{16}\text{O}_2}$ = 0.1862 g $C_{11}H_{16}O_2$.

Oxidation States

58. The oxidation state (O. S.) is given first, followed by the explanation for its assignment.
    (a) C = –4 in CH_4 H has an oxidation state of +1 in its compounds (Remember that the sum of the oxidation states in a compound equals 0.)
    (b) S = +4 in SF_4 F has O. S. = –1 in its compounds.
    (c) O = –1 in Na_2O_2 Na has O. S. = +1 in its compounds.
    (d) C = 0 in $C_2H_3O_2^-$ H has O. S. = +1 in its compounds; that of O = –2. (Remember that the sum of the oxidation states in a polyatomic ion equals the charge on that ion.)
    (e) Fe = +6 in FeO_4^{2-} O has O. S. = –2 in its compounds.
    (f) S = +2.5 in $S_4O_6^{2-}$ O has O. S. = –2 in its compounds.

59. First we determine the oxidation state of sulfur in each species, remembering that the oxidation state of O is –2 in its compounds.
         S = +4 in $S_2O_3^{2-}$ S = +2 in $S_2O_3^{2-}$ S = +7 in $S_2O_8^2$
         S = +6 in HSO_4^- S = –2 in HS^- S = +2.5 in $S_4O_6^{2-}$

Thus, ranked in order of increasing oxidation state of sulfur, the anions are
$$HS^- < S_2O_3^{2-} < S_4O_6^{2-} < S_2O_3^{2-} < HSO_4^- < S_2O_8^{2-}$$

60. Remember that oxygen has an oxidation state of –2 in its compounds.
     N = +1 in N_2O N = +2 in NO   N = +3 in N_2O_3 N = +4 in NO_2 N = +5 in N_2O_5

Nomenclature

61.

(a)	BaS	barium sulfide	**(b)**	ZnO	zinc oxide
(c)	K_2CrO_4	potassium chromate	**(d)**	Cs_2SO_4	cesium sulfate
(e)	Cr_2O_3	chromium(III) oxide	**(f)**	$FeSO_4$	iron(II) sulfate
(g)	$Mg(HCO_3)_2$	magnesium hydrogen carbonate	**(h)**	$(NH_4)_2HPO_4$	ammonium hydrogen phosphate
(i)	$Ca(HSO_3)_2$	calcium hydrogen sulfite	**(j)**	$Cu(OH)_2$	copper(II) hydroxide
(k)	HNO_3	nitric acid	**(l)**	$KClO_4$	potassium perchlorate
(m)	KIO	potassium hypoiodite	**(n)**	LiCN	lithium cyanide
(o)	$HBrO_3$	bromic acid	**(p)**	H_3PO_3	phosphorus acid

62. (a) ICl   iodine monochloride   (b) ClF_3 chlorine trifluoride
    (c) SF_4 sulfur tetrafluoride   (d) BrF_5 bromine pentafluoride
    (e) CS_2 carbon disulfide   (f) $SiCl_4$ silicon tetrafluoride
    (g) N_4S_4 tetranitrogen tetrasulfide   (h) SF_6 sulfur hexafluoride

63. (a) $Al_2(SO_4)_3$ aluminum sulfate   (b) $(NH_4)_2CrO_4$ ammonium chromate
    (c) SiF_4 silicon tetrafluoride   (d) $Pb(C_2H_3O_2)_2$ lead(II) acetate
    (e) Fe_2O_3 iron(III) oxide   (f) C_3S_2 tricarbon disulfide
    (g) $Co(NO_3)_2$ cobalt(II) nitrate   (h) $Sr(NO_2)_2$ strontium nitrite
    (i) SnO_2 tin(IV) oxide   (j) $Ca(H_2PO_4)_2$ calcium dihydrogen phosphate
    (k) HBr   hydrobromic acid   (l) HIO_3 iodic acid
    (m) $AlPO_4$ aluminum phosphate   (n) PCl_2F_3 phosphorus dichloride trifluoride
    (o) S_4N_2 tetrasulfur dinitride

64. (a) $Fe_2(SO_4)_3$ iron(III) sulfate   has Fe in the oxidation state +3
    (b) HNO_2 nitrous acid   has N in the oxidation state +3
    (c) Cl_2O_7 dichlorine heptoxide   has Cl in the oxidation state +7

Hydrates

65. The hydrates with the greatest % H_2O will be the one with the highest ratio of the molar amount of water (in a mole of the compound) to the mass of one mole of the anhydrous compound.

$$\frac{5\ H_2O}{CuSO_4} = \frac{5\ mol\ H_2O}{159.6\ g} = 0.03133 \qquad \frac{6\ H_2O}{MgCl_2} = \frac{6\ mol\ H_2O}{95.3\ g} = 0.0630$$

$$\frac{18\ H_2O}{Cr_2(SO_4)_3} = \frac{18\ mol\ H_2O}{392.3\ g} = 0.04588 \qquad \frac{2\ H_2O}{LiC_2H_3O_2} = \frac{2\ mol\ H_2O}{65.9\ g} = 0.0303$$

The hydrate with the greatest % H_2O therefore is $MgCl_2 \cdot 6\ H_2O$

66. A mole of this hydrate will contain about the same mass of H_2O and of Na_2SO_3.

mol wt. $Na_2SO_3 = (2 \times 23.0\ g\ Na) + 32.1\ g\ S + (3 \times 16.0\ g\ O) = 126.1\ g/mol$

no. mol $H_2O = 126.1\ g \times \dfrac{1\ mol\ H_2O}{18.0\ g\ H_2O} = 7.01\ mol\ H_2O$

Thus, the formula of the hydrate is $Na_2SO_3 \cdot 7H_2O$.

67. mol wt. $CuSO_4 = 63.5\ g\ Cu + 32.1\ g\ S + (4 \times 16.0\ g\ O) = 159.6\ g\ CuSO_4/mol$

The mass of this solid needed to combine with 8.5 mL of water is depends on the absorption of 5 moles of water by one mole of the solid, based on the formula $CuSO_4 \cdot 5H_2O$.

mass $CuSO_4 = 8.5\ mL\ H_2O \times \dfrac{1.00\ g\ H_2O}{1\ mL\ H_2O} \times \dfrac{1\ mol\ H_2O}{18.0\ g\ H_2O} \times \dfrac{1\ mol\ CuSO_4}{5\ mol\ H_2O} \times \dfrac{159.6\ g\ CuSO_4}{1\ mol\ CuSO_4}$

$= 15\ g\ CuSO_4$ required to remove all the water

68. The method of solution is similar to determining the empirical formula of a compound. The mass of water in the original sample is found by difference.

mass H_2O = mass of hydrate − mass of anhydrate = 8.129 g − 3.967 g = 4.162 g H_2O

mol $MgSO_4 = 3.967\ g\ MgSO_4 \times \dfrac{1\ mol\ MgSO_4}{120.4\ g} = 0.03295\ mol \qquad \div 0.03295 \longrightarrow 1.000\ mol\ MgSO_4$

mol $H_2O = 4.162\ g\ H_2O \times \dfrac{1\ mol\ H_2O}{18.02\ g\ H_2O} = 0.2310\ mol \qquad \div 0.03295 \longrightarrow 7.011\ mol\ H_2O$

Thus, the formula of the hydrate is $MgSO_4 \cdot 7H_2O$.

69. The increase in mass of the solid is related to each mole of the solid absorbing ten moles of water.

increase in mass $= 3.50\ g\ Na_2SO_4 \times \dfrac{1\ mol\ Na_2SO_4}{142.0\ g\ Na_2SO_4} \times \dfrac{10\ mol\ H_2O\ added}{1\ mol\ Na_2SO_4} \times \dfrac{18.015\ g\ H_2O}{1\ mol\ H_2O}$

$= 4.44\ g\ H_2O\ added$

4 CHEMICAL REACTIONS AND THE CHEMICAL EQUATION

REVIEW QUESTIONS

1. (a) The symbol "$\xrightarrow{\Delta}$" indicates that a reaction mixture is heated to produce the reaction.
   (b) The symbol "(aq)" indicates that the species preceeding this symbol is dissolved in aqueous solution, that is, a solution with water as the solvent.
   (c) The stoichiometric coefficient is the number that appears in a chemical equation immediately before the chemical formula of a species.
   (d) The net equation is the chemical equation that remains after species that appear on both sides of an equation are "cancelled." The term also is used to describe an equation that describes the overall result of a process consisting of several reactions.

2. (a) One "balances a chemical equation" by inserting stoichiometric coefficients into the unbalanced expression, so that the resulting equation has the same number and type of atoms on each side.
   (b) When one "prepares a solution by dilution" one begins with a more concentrated solution (a homogeneous mixture with a larger concentration of solute) and adds solvent, thus producing a less concentrated (or more dilute) solution.
   (c) One "determines the limiting reagent in a reaction" by discovering which reactant will produce the smallest quantity of product. That reactant will limit the quantity of product that can be formed from the other reactants.

3. (a) A chemical formula is a short-hand representation of a chemical species: atom, ion, or molecule. A chemical equation is a written representation of a chemical reaction; it involves two or more species. Whereas a chemical formula is rather analogous to a "word", chemical equations parallel "sentences."
   (b) A combustion reaction described the burning of a compound in oxygen; oxides are the usual products. A synthesis reaction is one in which two or more substances combine to form a third.
   (c) The solute is the substance that is dispersed in a solution. The solvent is the substance that does the dispersing. Usually, a solution is of the same physical state (solid liquid, or gas) as the solvent, and the solvent is the component present in larger amount.
   (d) The actual yield of a chemical reaction is the quantity of product that actually was formed. The percent yield relates the actual yield to the quantity of product that was calculated to be produced, assuming that all reagents produced only one set of products and the reaction continued until one reagent was exhausted.

4. (a) $Na_2SO_4(s) + 2\,C(s) \longrightarrow Na_2S(s) + 2\,CO_2(g)$ (b) $4\,HCl(g) + O_2(g) \longrightarrow 2\,H_2O(l) + 2\,Cl_2(g)$
   (c) $PCl_3(l) + 3\,H_2O(l) \longrightarrow H_3PO_3(aq) + 3\,HCl(aq)$
   (d) $3\,PbO(s) + 2\,NH_3(g) \longrightarrow 3\,Pb(s) + N_2(g) + 3\,H_2O(l)$
   (e) $Mg_3N_2(s) + 6\,H_2O(l) \longrightarrow 3\,Mg(OH)_2(s) + 2\,NH_3(g)$

5. (a) $2\,Mg(s) + O_2(g) \longrightarrow 2\,MgO(s)$ (b) $S(s) + O_2(g) \longrightarrow SO_2(g)$
   (c) $CH_4(g) + 2\,O_2(g) \longrightarrow CO_2(g) + 2\,H_2O(l)$ (d) $Ag_2SO_4(aq) + BaI_2(aq) \longrightarrow BaSO_4(s) + 2\,AgI(s)$

6. (a) $C_5H_{12}(l) + 8\,O_2(g) \longrightarrow 5\,CO_2(g) + 6\,H_2O(l)$
   (b) $2\,C_2H_6O_2(l) + 5\,O_2(g) \longrightarrow 4\,CO_2(g) + 6\,H_2O(l)$ (c) $HI(aq) + NaOH(aq) \longrightarrow NaI(aq) + H_2O(l)$
   (d) $2\,KI(aq) + Pb(NO_3)_2(aq) \longrightarrow PbI_2(s) + 2\,KNO_3(aq)$

7. Since KCl is potassium chloride, the stated product, while KClO is potassium hypochlorite, the last reaction is incorrect. The other two reactions are incorrect because $O(g)$ is not normally produced in chemical reactions; $O_2(g)$ is a more stable species. The correct equation is $2\,KClO_3(s) \longrightarrow 2\,KCl(s) + 3\,O_2(g)$

8. For the reaction $2\,H_2S + SO_2 \longrightarrow 3\,S + 2\,H_2O$
  (1) FALSE   3 moles of S is produced per *two* mol H_2S.
  (2) FALSE   3 *moles* S is produced for every *mole* of SO_2 consumed.
  (3) TRUE   1 mol H_2O *is* produced per mol H_2S consumed.
  (4) TRUE   Two-thirds of the S produced *does* come from the H_2S.
  (5) FALSE   There are *five* moles of products and *three* moles of reactants.

9. The conversion factor is obtained from the balanced chemical equation.

$$\text{no. mol } FeCl_3 = 6.12 \text{ mol } Cl_2 \times \frac{2 \text{ mol } FeCl_3}{3 \text{ mol } Cl_2} = 4.08 \text{ mol } FeCl_3$$

10. The pivotal conversion factor, from the balanced equation, enables one to related the amounts of O_2 and $KClO_3$.

$$\text{mass } O_2 = 53.5 \text{ g } KClO_3 \times \frac{1 \text{ mol } KClO_3}{122.5 \text{ g } KClO_3} \times \frac{3 \text{ mol } O_2}{2 \text{ mol } KClO_3} \times \frac{32.00 \text{ g } O_2}{1 \text{ mol } O_2} = 20.9 \text{ g } O_2$$

11. Each calculation uses the stoichiometric coefficients from the balanced chemical equation and the mole weight of the reactant.

$$\text{mass } Cl_2 = 0.426 \text{ mol } PCl_3 \times \frac{6 \text{ mol } Cl_2}{4 \text{ mol } PCl_3} \times \frac{70.91 \text{ g } Cl_2}{1 \text{ mol } Cl_2} = 45.3 \text{ g } Cl_2$$

$$\text{mass } P_4 = 0.426 \text{ mol } PCl_3 \times \frac{1 \text{ mol } P_4}{4 \text{ mol } PCl_3} \times \frac{123.9 \text{ g } P_4}{1 \text{ mol } P_4} = 13.2 \text{ g } P_4$$

12. (a) $\text{no. mol } H_2 = 127 \text{ g } CaH_2 \times \dfrac{1 \text{ mol } CaH_2}{42.10 \text{ g } CaH_2} \times \dfrac{2 \text{ mol } H_2}{1 \text{ mol } CaH_2} = 6.03 \text{ mol } H_2(g)$

  (b) $\text{mass } H_2O = 56.2 \text{ g } CaH_2 \times \dfrac{1 \text{ mol } CaH_2}{42.10 \text{ g } CaH_2} \times \dfrac{2 \text{ mol } H_2O}{1 \text{ mol } CaH_2} \times \dfrac{18.02 \text{ g } H_2O}{1 \text{ mol } H_2O} = 48.1 \text{ g } H_2O$

  (c) $\text{mass } CaH_2 = 8.12 \times 10^{24} \text{ molec.} \times \dfrac{1 \text{ mol } H_2O}{6.022 \times 10^{23}} \times \dfrac{1 \text{ mol } CaH_2}{2 \text{ mol } H_2O} \times \dfrac{42.10 \text{ g } CaH_2}{1 \text{ mol } CaH_2} = 284 \text{ g } CaH_2$

13. (a) C_2H_5OH molarity $= \dfrac{3.28 \text{ mol } C_2H_5OH}{7.16 \text{ L}} = 0.458 \text{ M}$

  (b) CH_3OH molarity $= \dfrac{8.67 \times 10^{-3} \text{ mol } CH_3OH}{50.00 \text{ mL}} \times \dfrac{1000 \text{ mL}}{1 \text{ L}} = 0.173 \text{ M}$

  (c) $(CH_3)_2CO$ molarity $= \dfrac{12.3 \text{ g } (CH_3)_2CO}{225 \text{ mL}} \times \dfrac{1 \text{ mol } (CH_3)_2CO}{58.08 \text{ g } (CH_3)_2CO} \times \dfrac{1000 \text{ mL}}{1 \text{ L}} = 0.941 \text{ M}$

  (d) molarity $= \dfrac{14.5 \text{ mL } C_3H_5(OH)_3}{425 \text{ mL soln}} \times \dfrac{1.26 \text{ g}}{1 \text{ cm}^3} \times \dfrac{1 \text{ mol } C_3H_5(OH)_3}{92.0 \text{ g } C_3H_5(OH)_3} \times \dfrac{1000 \text{ mL}}{1 \text{ L}} = 0.467 \text{ M}$

14. (a) $\text{no. mol } KCl = 1.75 \times 10^3 \text{ L} \times \dfrac{0.115 \text{ mol } KCl}{1 \text{ L soln}} = 201 \text{ mol } KCl$

  (b) $\text{mass } Na_2CO_3 = 535 \text{ mL} \times \dfrac{1 \text{ L}}{1000 \text{ mL}} \times \dfrac{0.412 \text{ mol } Na_2CO_3}{1 \text{ L}} \times \dfrac{106.0 \text{ g } Na_2CO_3}{1 \text{ mol } Na_2CO_3} = 23.4 \text{ g } Na_2CO_3$

  (c) $\text{no. mg } NaOH = 1.00 \text{ mL} \times \dfrac{0.183 \text{ mmol } NaOH}{1 \text{ mL}} \times \dfrac{40.0 \text{ mg } NaOH}{1 \text{ mmol } NaOH} = 7.32 \text{ mg } NaOH$

15. A 1.00 M $NaNO_3$ solution contains 1 mol $NaNO_3$ per liter of solution. The molar mass of $NaNO_3$ is 85.0 g. Thus, a 1.00 M $NaNO_3$ solution contains 85.0 g $NaNO_3$ per liter of solution.
  "1.00 L … containing 100 g" is incorrect; 1.00 L should contain 85.0 g.
  "500 mL … containing 85 g" also is incorrect; 85 g should be contained in 1000 mL.
  "1.00 mL … containing 8.5 mg" in incorrect; there should be 85 mg.
  5.00 L of 1.00 M $NaNO_3$ contains five times the mass of solute as does 1.00 L of this solution, 425 g. The last description is correct.

16. volume dilute KOH soln $= 250.0 \text{ mL conc'd soln} \times \dfrac{0.448 \text{ mmol } KOH}{1 \text{ mL conc'd soln}} \times \dfrac{1 \text{ mL dilute soln}}{0.750 \text{ mmol } KOH}$

$$= 149 \text{ mL dilute KOH soln}$$

17. The same amount in moles of $MgSO_4$ is present in both solutions. That amount is given in thenumerator of the following expression.

$$MgSO_4 \text{ molarity} = \frac{135 \text{ mL} \times \dfrac{0.224 \text{ mmol } MgSO_4}{1 \text{ mL}}}{105 \text{ mL}} = 0.288 \text{ M}$$

18. The balanced chemical equation provides a conversion factor between the two compounds.

$$\text{mass } CuCO_3 = 325 \text{ mL} \times \frac{1 \text{ L}}{1000 \text{ mL}} \times \frac{0.305 \text{ mol } Cu(NO_3)_2}{1 \text{ L}} \times \frac{1 \text{ mol } CuCO_3}{1 \text{ mol } Cu(NO_3)_2} \times \frac{123.6 \text{ g } CuCO_3}{1 \text{ mol } CuCO_3}$$
$$= 12.3 \text{ g } CuCO_3$$

19. After determining the amount of $CaCO_3$ (100.09 g/mol), we find the volume of HCl.

$$\text{volume HCl(aq)} = 1.35 \text{ g } CaCO_3 \times \frac{1 \text{ mol } CaCO_3}{100.09 \text{ g}} \times \frac{2 \text{ mol HCl}}{1 \text{ mol } CaCO_3} \times \frac{1 \text{ L soln}}{2.50 \text{ mol HCl}} \times \frac{1000 \text{ mL}}{1 \text{ L}}$$
$$= 10.8 \text{ mL HCl(aq) solution}$$

20. In this situation, since 3 mol H_2O are required per mol $CaCN_2$, 1.00 mol H_2O is the limiting reagent. Thus, the amount in moles of NH_3 can be computed as follows.

$$\text{amount } NH_3 = 1.0 \text{ mol } H_2O \times \frac{2 \text{ mol } NH_3}{3 \text{ mol } H_2O} = 0.67 \text{ mol } NH_3$$

21. Determine the number of moles of NO produced by each reagent. The one producing the smaller amount of NO is the limiting reagent.

$$\text{no. mol NO} = 0.496 \text{ mol Cu} \times \frac{2 \text{ mol NO}}{3 \text{ mol Cu}} = 0.331 \text{ mol NO}$$

$$\text{no. mol NO} = 128 \text{ mL } HNO_3\text{(aq)} \times \frac{1 \text{ L}}{1000 \text{ mL}} \times \frac{6.0 \text{ mol } HNO_3}{1 \text{ L}} \times \frac{2 \text{ mol NO}}{8 \text{ mol } HNO_3} = 0.192 \text{ mol NO}$$

Since HNO_3(aq) is the limiting reagent, it will be completely consumed, leaving some Cu unreacted.

22. (a) Since the stoichiometry indicates that 1 mol CCl_2F_2 is produced per mol CCl_4, the use of 2.00 mol CCl_4 will produce 2.00 mol CCl_2F_2. This is the theoretical yield of the reaction.

(b) The actual yield of the reaction is the amount actually produced, 1.70 mol CCl_2F_2.

(c) $\% \text{ yield} = \dfrac{1.70 \text{ mol } CCl_2F_2 \text{ obtained}}{2.00 \text{ mol } CCl_2F_2 \text{ calculated}} \times 100\% = 85.0\% \text{ yield}$

23. (a) $\text{mass } C_6H_{10} = 100.0 \text{ g } C_6H_{11}OH \times \dfrac{1 \text{ mol } C_6H_{11}OH}{100.16 \text{ g } C_6H_{11}OH} \times \dfrac{1 \text{ mol } C_6H_{10}}{1 \text{ mol } C_6H_{11}OH} \times \dfrac{82.146 \text{ g } C_6H_{10}}{1 \text{ mol } C_6H_{10}}$

$= 82.01 \text{ g } C_6H_{10} = \text{theoretical yield}$

(b) $\text{percent yield} = \dfrac{69.0 \text{ g } C_6H_{10} \text{ produced}}{82.01 \text{ g } C_6H_{10} \text{ calculated}} \times 100\% = 84.1\% \text{ yield}$

(c) $\text{mass } C_6H_{11}OH = 100.0 \text{ g } C_6H_{10} \text{ produced} \times \dfrac{1.000 \text{ g calculated}}{0.841 \text{ g produced}} \times \dfrac{1 \text{ mol } C_6H_{10}}{82.15 \text{ g } C_6H_{10}}$

$\times \dfrac{1 \text{ mol } C_6H_{11}OH}{1 \text{ mol } C_6H_{10}} \times \dfrac{100.2 \text{ g } C_6H_{11}OH}{1 \text{ mol } C_6H_{11}OH} = 145 \text{ g } C_6H_{11}OH \text{ needed}$

24. $\text{mass } CaCO_3 = 0.818 \text{ g } CO_2 \times \dfrac{1 \text{ mol } CO_2}{44.01 \text{ g } CO_2} \times \dfrac{1 \text{ mol } CaCO_3}{1 \text{ mol } CO_2} \times \dfrac{100.1 \text{ g } CaCO_3}{1 \text{ mol } CaCO_3} = 1.86 \text{ g } CaCO_3$

$\% \ CaCO_3 = \dfrac{1.86 \text{ g } CaCO_3}{2.58 \text{ g sample}} \times 100\% = 72.1\% \ CaCO_3$

EXERCISES

Writing and Balancing Chemical Equations

25. (a) $SiCl_4(l) + 2 H_2O(l) \longrightarrow SiO_2(s) + 4 HCl(g)$

(b) $CaC_2(s) + 2 H_2O(l) \longrightarrow Ca(OH)_2(s) + C_2H_2(g)$

(c) $2 Na_2HPO_4(s) \longrightarrow Na_4P_2O_7(s) + H_2O(l)$

(d) $NCl_3(g) + 3 H_2O(l) \longrightarrow NH_3(g) + 3 HOCl(aq)$

26. (a) $6 P_2H_4(l) \longrightarrow 8 PH_3(g) + P_4(s)$ **(b)** $3 NO_2(g) + H_2O(l) \longrightarrow 2 HNO_3(aq) + NO(g)$

(c) $6 S_2Cl_2 + 16 NH_3 \longrightarrow N_4S_4 + 12 NH_4Cl + S_8$

(d) $SO_2Cl_2 + 8 HI \longrightarrow H_2S + 2 H_2O + 2 HCl + 4 I_2$

27. (a) $2 C_6H_6 + 15 O_2(g) \longrightarrow 12 CO_2(g) + 6 H_2O(l)$

   (b) $2 C_3H_7OH + 9 O_2(g) \longrightarrow 6 CO_2(g) + 8 H_2O(l)$

   (c) $2 C_6H_5COOH(s) + 15 O_2(g) \longrightarrow 14 CO_2(g) + 6 H_2O(l)$

   (d) $C_6H_5COSH(s) + 9 O_2(g) \longrightarrow 7 CO_2(g) + 3 H_2O(l) + SO_2(g)$

28. (a) $Hg(NO_3)_2(s) \xrightarrow{\text{heat}} Hg(l) + 2 NO_2(g) + O_2(g)$

   (b) $Na_2CO_3(aq) + 2 HCl(aq) \longrightarrow 2 NaCl(aq) + H_2O(l) + CO_2(g)$

   (c) $3 NO_2(g) + H_2O(l) \longrightarrow 2 HNO_3(aq) + NO(g)$

   (d) $CaCO_3(s) + H_2O(l) + CO_2(aq) \longrightarrow Ca(HCO_3)_2(aq)$

Stoichiometry of Chemical Reactions

29. (a) no. mol O_2 = 26.3 g $KClO_3 \times \dfrac{1\text{ mol }KClO_3}{122.6\text{ g }KClO_3} \times \dfrac{3\text{ mol }O_2}{2\text{ mol }KClO_3} = 0.322$ mol O_2

   (b) no. molec. O_2 = 0.322 mol $O_2 \times \dfrac{6.022 \times 10^{23}\text{ molec.}}{1\text{ mol }O_2} = 1.94 \times 10^{23}$ O_2 molecules

   (c) mass KCl = 26.3 g $KClO_3 \times \dfrac{1\text{ mol }KClO_3}{122.6\text{ g }KClO_3} \times \dfrac{2\text{ mol }KCl}{2\text{ mol }KClO_3} \times \dfrac{74.55\text{ g }KCl}{1\text{ mol }KCl} = 16.0$ g KCl

30. (a) amount H_2 = 64.4 g $Fe \times \dfrac{1\text{ mol }Fe}{55.85\text{ g }Fe} \times \dfrac{4\text{ mol }H_2}{3\text{ mol }Fe} = 1.54$ mol H_2

   (b) mass H_2O = 76.3 g $Fe \times \dfrac{1\text{ mol }Fe}{55.85\text{ g }Fe} \times \dfrac{4\text{ mol }H_2O}{3\text{ mol }Fe} \times \dfrac{18.02\text{ g }H_2O}{1\text{ mol }H_2O} = 32.8$ g H_2O

   (c) mass Fe_3O_4 = 9.02 mol $H_2 \times \dfrac{1\text{ mol }Fe_3O_4}{4\text{ mol }H_2} \times \dfrac{231.55\text{ g }Fe_3O_4}{1\text{ mol }Fe_3O_4} = 522$ g Fe_3O_4

31. Balance the given equation, then solve the problem. $2 Ag_2CO_3(s) \longrightarrow 4 Ag(s) + 2 CO_2(g) + O_2(g)$

   mass Ag_2CO_3 = 56.2 g $Ag \times \dfrac{1\text{ mol }Ag}{107.87\text{ g }Ag} \times \dfrac{2\text{ mol }Ag_2CO_3}{4\text{ mol }Ag} \times \dfrac{275.75\text{ g }Ag_2CO_3}{1\text{ mol }Ag_2CO_3} = 71.8$ g Ag_2CO_3

32. mass SO_2 = 107 g mixt. $\times \dfrac{0.694\text{ g }CaSO_3}{1.000\text{ g mixt.}} \times \dfrac{1\text{ mol }CaSO_3}{120.2\text{ g }CaSO_3} \times \dfrac{1\text{ mol }SO_2}{1\text{ mol }CaSO_3} \times \dfrac{64.07\text{ g }SO_2}{1\text{ mol }SO_2}$

      = 39.6 g SO_2

33. mass Fe_2O_3 = 486 kg $Fe \times \dfrac{1\text{ kmol }Fe}{55.85\text{ kg }Fe} \times \dfrac{1\text{ kmol }Fe_2O_3}{2\text{ kmol }Fe} \times \dfrac{159.7\text{ kg }Fe_2O_3}{1\text{ kmol }Fe_2O_3} = 695$ kg Fe_2O_3

   % Fe_2O_3 in ore = $\dfrac{695\text{ kg }Fe_2O_3}{812\text{ kg ore}} \times 100\% = 85.6\%$ Fe_2O_3

34. The following reaction occurs: $2 Ag_2O(s) \xrightarrow{\text{heat}} 4 Ag(s) + O_2(g)$

   mass Ag_2O = 0.187 g $O_2 \times \dfrac{1\text{ mol }O_2}{32.0\text{ g }O_2} \times \dfrac{2\text{ mol }Ag_2O}{1\text{ mol }O_2} \times \dfrac{231.8\text{ g }Ag_2O}{1\text{ mol }Ag_2O} = 2.71$ g Ag_2O

   % Ag_2O = $\dfrac{2.71\text{ g }Ag_2O}{3.13\text{ g sample}} \times 100\% = 86.6\%$ Ag_2O

35. $2 Al(s) + 6 HCl(aq) \longrightarrow 2 AlCl_3(aq) + 3 H_2(g)$ First determine the mass of Al in the piece of foil.

   mass Al = $(4.04 \text{ in.} \times 2.17 \text{ in.} \times 0.0235 \text{ in.}) \times \left(\dfrac{2.54\text{ cm}}{1\text{ in.}}\right)^3 \times \dfrac{2.70\text{ g}}{1\text{ cm}^3} = 9.12$ g Al

   mass H_2 = 9.12 g $Al \times \dfrac{1\text{ mol }Al}{26.98\text{ g }Al} \times \dfrac{3\text{ mol }H_2}{2\text{ mol }Al} \times \dfrac{2.016\text{ g }H_2}{1\text{ mol }H_2} = 1.02_2$ g H_2

36. We can compare yields based on stoichiometry. In each reaction, one mol $O_2(g)$ is produced for every two moles of reactant. Thus, the reaction that gives the highest yield of $O_2(g)$ per gram of reactant is the one in which the reactant has the *lowest* mole weight. In parts (1), (2), and (4) a mole of the reactant contains two moles of N, but in parts (1) and (4) there are atoms—other than O—in addition to those of N, making the mole weights greater. Finally, (3) has a greater mole weight than (2) because Ag has a greater atomic weight than N. Thus, the reaction that produces the maximum quantity of $O_2(g)$ per gram of reactant is (2).

$$2 N_2O \xrightarrow{\Delta} 2 N_2(g) + O_2(g).$$

37. First write the balanced chemical equation for each reaction.

$2 \text{ Na(s)} + 2 \text{ HCl(aq)} \longrightarrow 2 \text{ NaCl(aq)} + \text{H}_2(\text{g})$ $\qquad$ $\text{Mg(s)} + 2 \text{ HCl(aq)} \longrightarrow \text{MgCl}_2(\text{aq}) + \text{H}_2(\text{g})$

$2 \text{ Al(s)} + 6 \text{ HCl(aq)} \longrightarrow 2 \text{ AlCl}_3(\text{aq}) + 3 \text{ H}_2(\text{g})$ $\qquad$ $\text{Zn(s)} + 2 \text{ HCl(aq)} \longrightarrow \text{ZnCl}_2(\text{aq}) + \text{H}_2(\text{g})$

Three of the reactions—those of Na, Mg, and Zn—produce 1 mole of $\text{H}_2(\text{g})$. The one of these three that produces the most hydrogen per gram of metal is the one for which the metal's atomic weight is the smallest, remembering to compare twice the atomic weight for Na. The atomic weights are: 2×23 u for Na, 24.3 u for Mg, and 65.4 u for Zn. Thus, among these three Mg produces the most H_2 per gram of metal, specifically 1 mol H_2 per 24.3 g Mg. In the case of Al, 3 moles of H_2 are produced by 2 moles of the metal, or 54 g Al. This reduces as follows: 3 mol H_2/54 g Al = 1 mol H_2/18 g Al.

Thus, Al produces the largest amount of H_2 per gram of metal.

38. $2 \text{ KMnO}_4 + 10 \text{ KI} + 8 \text{ H}_2\text{SO}_4 \longrightarrow 6 \text{ K}_2\text{SO}_4 + 2 \text{ MnSO}_4 + 5 \text{ I}_2 + 8 \text{ H}_2\text{O}$

$$\text{soln volume} = 9.13 \text{ g KI} \times \frac{1 \text{ mol KI}}{166.0 \text{ g KI}} \times \frac{2 \text{ mol KMnO}_4}{10 \text{ mol KI}} \times \frac{158.0 \text{ g KMnO}_4}{1 \text{ mol KMnO}_4} \times \frac{1 \text{ L soln}}{12.6 \text{ g KMnO}_4}$$

$$\times \frac{1000 \text{ mL}}{1 \text{ L}} = 138 \text{ mL soln}$$

39. There are two ways to solve this problem. The first is to determine the masses of $\text{CO}_2(\text{g})$ and $\text{H}_2\text{O(l)}$ that are produced in the combustion of 0.875 g $\text{C}_6\text{H}_{14}\text{O}_4$. $\qquad$ $2 \text{ C}_6\text{H}_{14}\text{O}_4 + 15 \text{ O}_2(\text{g}) \longrightarrow 12 \text{ CO}_2(\text{g}) + 14 \text{ H}_2\text{O}$. These masses turn out to be 1.54 g $\text{CO}_2(\text{g})$ and 0.735 g $\text{H}_2\text{O(l)}$. The second way is to determine the mass of $\text{O}_2(\text{g})$ needed to react completely with 0.875 g $\text{C}_6\text{H}_{14}\text{O}_4$, and then make use of the fact that the total mass of the products equals the total mass of the reactants.

$$\text{mass O}_2 = 0.875 \text{ g C}_6\text{H}_{14}\text{O}_4 \times \frac{1 \text{ mol C}_6\text{H}_{14}\text{O}_4}{150.0 \text{ g C}_6\text{H}_{14}\text{O}_4} \times \frac{15 \text{ mol O}_2}{2 \text{ mol C}_6\text{H}_{14}\text{O}_4} \times \frac{32.00 \text{ g O}_2}{1 \text{ mol O}_2} = 1.40 \text{ g O}_2$$

mass products = mass reactants = 0.875 + 1.40 = 2.28 g

40. Determine the amount of HCl needed to react with each component of the mixture.

$\text{Mg(OH)}_2 + 2 \text{ HCl} \longrightarrow \text{MgCl}_2 + 2 \text{ H}_2\text{O}$ $\qquad$ $\text{MgCO}_3 + 2 \text{ HCl} \longrightarrow \text{MgCl}_2 + \text{H}_2\text{O} + \text{CO}_2$

$$\text{no. mol HCl} = 387 \text{ g mixt.} \times \frac{32.8 \text{ g MgCO}_3}{100.0 \text{ g mixt.}} \times \frac{1 \text{ mol MgCO}_3}{84.3 \text{ g MgCO}_3} \times \frac{2 \text{ mol HCl}}{1 \text{ mol MgCO}_3} = 3.01 \text{ mol HCl}$$

$$\text{no. mol HCl} = 387 \text{ g mixt.} \times \frac{67.2 \text{ g Mg(OH)}_2}{100.0 \text{ g mixt.}} \times \frac{1 \text{ mol Mg(OH)}_2}{58.3 \text{ g Mg(OH)}_2} \times \frac{2 \text{ mol HCl}}{1 \text{ mol MgCO}_3} = 8.92 \text{ mol HCl}$$

$$\text{mass HCl} = (3.01 + 8.92) \text{ mol HCl} \times \frac{36.45 \text{ g HCl}}{1 \text{ mol HCl}} = 435 \text{ g HCl}$$

41. Determine the amount of CO_2 produced from each reactant.

$\text{C}_3\text{H}_8(\text{g}) + 5 \text{ O}_2(\text{g}) \longrightarrow 3 \text{ CO}_2(\text{g}) + 4 \text{ H}_2\text{O(l)}$ $\qquad$ $2 \text{ C}_4\text{H}_{10}(\text{g}) + 13 \text{ O}_2(\text{g}) \longrightarrow 8 \text{ CO}_2(\text{g}) + 10 \text{ H}_2\text{O(l)}$

$$\text{no. mol CO}_2 = 406 \text{ g mixt.} \times \frac{70.7 \text{ g C}_3\text{H}_8}{100.0 \text{ g mixt.}} \times \frac{1 \text{ mol C}_3\text{H}_8}{44.10 \text{ g C}_3\text{H}_8} \times \frac{3 \text{ mol CO}_2}{1 \text{ mol C}_3\text{H}_8} = 19.6 \text{ mol CO}_2$$

$$\text{no. mol CO}_2 = 406 \text{ g mixt.} \times \frac{29.3 \text{ g C}_4\text{H}_{10}}{100.0 \text{ g mixt}} \times \frac{1 \text{ mol C}_3\text{H}_8}{58.12 \text{ g C}_4\text{H}_{10}} \times \frac{8 \text{ mol CO}_2}{2 \text{ mol C}_4\text{H}_{10}} = 8.20 \text{ mol CO}_2$$

total no. mol $\text{CO}_2 = (19.6 + 8.20)$ mol $\text{CO}_2 = 27.7 \text{ CO}_2$

42. The molar ratios given by the stoichiometric coefficients in the balanced chemical equations are used in the solution.

$$\text{amount Cl}_2 = 1.00 \times 10^3 \text{ g CCl}_2\text{F}_2 \times \frac{1 \text{ mol CCl}_2\text{F}_2}{120.91 \text{ g CCl}_2\text{F}_2} \times \frac{1 \text{ mol CCl}_4}{1 \text{ mol CCl}_2\text{F}_2} \times \frac{4 \text{ mol Cl}_2}{1 \text{ mol CCl}_4} = 33.1 \text{ mol Cl}_2$$

43. $\text{mass C}_3\text{H}_8 = 0.506 \text{ g BaCO}_3 \times \dfrac{1 \text{ mol BaCO}_3}{197.3 \text{ g BaCO}_3} \times \dfrac{1 \text{ mol CO}_2}{1 \text{ mol BaCO}_3} \times \dfrac{1 \text{ mol C}_3\text{H}_8}{3 \text{ mol CO}_2} \times \dfrac{44.10 \text{ g C}_3\text{H}_8}{1 \text{ mol C}_3\text{H}_8}$

$= 0.0377 \text{ g C}_3\text{H}_8$

Molarity

44. **(a)** $\text{C}_{12}\text{H}_{22}\text{O}_{11} \text{ concn} = \dfrac{100.0 \text{ g C}_{12}\text{H}_{22}\text{O}_{11}}{500.0 \text{ mL soln}} \times \dfrac{1000 \text{ mL}}{1 \text{ L}} \times \dfrac{1 \text{ mol C}_{12}\text{H}_{22}\text{O}_{11}}{342.3 \text{ g C}_{12}\text{H}_{22}\text{O}_{11}} = 0.5843 \text{ M}$

(b) $\text{CO(NH}_2)_2 \text{ concn} = \dfrac{87.8 \text{ mg solid}}{5.00 \text{ mL soln}} \times \dfrac{98.7 \text{ mg CO(NH}_2)_2}{100.0 \text{ mg solid}} \times \dfrac{1 \text{ mmol CO(NH}_2)_2}{60.06 \text{ mg CO(NH}_2)_2} = 0.289 \text{ M}$

(c) CH_3OH concn $= \dfrac{250.0 \text{ mL } CH_3OH}{12.0 \text{ L soln}} \times \dfrac{0.792 \text{ g}}{1 \text{ mL}} \times \dfrac{1 \text{ mol } CH_3OH}{32.04 \text{ g } CH_3OH} = 0.515 \text{ M}$

(d) $(C_2H_5)_2O$ concn $= \dfrac{8.8 \text{ mg } (C_2H_5)_2O}{3.00 \text{ gal soln}} \times \dfrac{1 \text{ gal}}{3.78 \text{ L}} \times \dfrac{1 \text{ g}}{1000 \text{ mg}} \times \dfrac{1 \text{ mol } (C_2H_5)_2O}{74.12 \text{ g } (C_2H_5)_2O} = 1.0 \times 10^{-5} \text{ M}$

45. (a) volume CH_3OH = 2.50 L soln $\times \dfrac{0.515 \text{ mol}}{1 \text{ L}} \times \dfrac{32.04 \text{ g } CH_3OH}{1 \text{ mol } CH_3OH} \times \dfrac{1 \text{ mL}}{0.792 \text{ g}} = 52.1 \text{ mL } CH_3OH$

(b) volume C_2H_5OH = 55.0 gal soln $\times \dfrac{3.78 \text{ L}}{1 \text{ gal}} \times \dfrac{1.65 \text{ mol } C_2H_5OH}{1 \text{ L}} \times \dfrac{46.07 \text{ g } C_2H_5OH}{1 \text{ mol } C_2H_5OH} \times \dfrac{1 \text{ mL}}{0.789 \text{ g}}$

$\times \dfrac{1 \text{ L}}{1000 \text{ cm}^3} \times \dfrac{1 \text{ gal}}{3.78 \text{ L}} = 5.30 \text{ gal } C_2H_5OH$

46. The molarity of the final solution is the total amount of the solute divided by the total volume of the soln.

total volume = 225 mL + 565 mL = 790. mL soln

total no. mmol $= \left(225 \text{ mL} \times \dfrac{1.50 \text{ mol}}{1 \text{ L}}\right) + \left(565 \text{ mL} \times \dfrac{1.25 \text{ mol}}{1 \text{ L}}\right) = 338 \text{ mmol} + 706 \text{ mmol} = 532 \text{ mmol}$

$[C_{12}H_{22}O_{11}] = \dfrac{1044 \text{ mmol } C_{12}H_{22}O_{11}}{790. \text{ mL}} = 1.32 \text{ M}$

47. vol conc = 15.0 L $\times \dfrac{0.346 \text{ mol HCl}}{1 \text{ L}} \times \dfrac{36.45 \text{ g HCl}}{1 \text{ mol HCl}} \times \dfrac{100 \text{ g soln}}{36.0 \text{ g HCl}} \times \dfrac{1 \text{ mL soln}}{1.18 \text{ g}} = 445 \text{ mL conc HCl}$

48. We calculate the $[C_2H_5OH]$ in the white wine and compare it with 1.71 M C_2H_5OH, the concentration of the solution described in Example 4-8.

$[C_2H_5OH] = \dfrac{11.0 \text{ g } C_2H_5OH}{100.0 \text{ g soln}} \times \dfrac{0.95 \text{ g soln}}{1 \text{ cm}^3} \times \dfrac{1000 \text{ cm}^3}{1 \text{ L}} \times \dfrac{1 \text{ mol } C_2H_5OH}{46.1 \text{ g } C_2H_5OH} = 2.3 \text{ M } C_2H_5OH$

The white wine has a greater ethyl alcohol content.

49. Let us compute how many mL of dilute ($_d$) solution we obtain from each mL of concentrated ($_c$) solution.

$V_c \times C_c = V_d \times C_d$ becomes   1.00 mL × 0.250 M = x mL × 0.0125 M   and x = 20

Thus, the ratio of the volume of the volumetric flask to that of the pipet would be 20:1. We could use a 100.0-mL flask and a 5.00-mL pipet, a 1000.0-mL flask and a 50.0-mL pipet, or a 500.0-mL flask and a 25.00-mL pipet

50. (a) The mass of solute needed is the difference between the mass of solute that will be present in the more concentrated solution and the mass that is present in the dilute solution.

mass KCl, conc = 20.0 L soln $\times \dfrac{0.500 \text{ mol KCl}}{1 \text{ L soln}} \times \dfrac{74.551 \text{ g KCl}}{1 \text{ mol KCl}} = 746 \text{ g KCl}$

mass KCl, dil = 20.0 L soln $\times \dfrac{0.496 \text{ mol KCl}}{1 \text{ L soln}} \times \dfrac{74.551 \text{ g KCl}}{1 \text{ mol KCl}} = 740. \text{ g KCl}$

6 g KCl must be added.

(b) Calculate the volume the solution must be for 740. g KCl to give a solution concentration of 0.500 M.

solution volume = 740. g KCl $\times \dfrac{1 \text{ mol KCl}}{74.551 \text{ g KCl}} \times \dfrac{1 \text{ L soln}}{0.500 \text{ mol KCl}} = 19.9 \text{ L soln}$

0.1 L of solvent must be evaporated.

Alternatively: final volume = 20.0 L $\times \dfrac{0.496 \text{ M}}{0.500 \text{ M}} = 19.8 \text{ L}$ 0.2 L of solvent to be evaporated.

It certainly would be easier to add 6 g KCl than to wait for 0.1 L (100 mL) of solvent to evaporate. Adding the solute also is more likely to give accurate results for the reason that it will be somewhat difficult to measure a volume change of 0.1 L in a 20-L solution, while it is quite easy to measure a mass of 6 g.

51. Compute the amount of $AgNO_3$ in the solution on hand and the amount of $AgNO_3$ in the desired solution. the difference is the amount of $AgNO_3$ that must be added; simply convert this amount to a mass.

amount $AgNO_3$ present = 50.00 mL $\times \dfrac{0.0500 \text{ mmol } AgNO_3}{1 \text{ mL soln}} = 2.50 \text{ mmol } AgNO_3$

amount $AgNO_3$ desired = 100.0 mL $\times \dfrac{0.0750 \text{ mmol } AgNO_3}{1 \text{ mL soln}} = 7.50 \text{ mmol } AgNO_3$

mass $AgNO_3$ = (7.50 − 2.50) mmol $AgNO_3 \times \dfrac{1 \text{ mol } AgNO_3}{1000 \text{ mmol } AgNO_3} \times \dfrac{169.9 \text{ g Ag NO}_3}{1 \text{ mol } AgNO_3} = 0.849 \text{ g } AgNO_3$

Chemical Reactions in Solutions

52. (a) mass $Ca(OH)_2 = 485$ mL $\times \dfrac{1 \text{ L}}{1000 \text{ mL}} \times \dfrac{0.886 \text{ mol HCl}}{1 \text{ L soln}} \times \dfrac{1 \text{ mol Ca(OH)}_2}{2 \text{ mol HCl}} \times \dfrac{74.1 \text{ g Ca(OH)}_2}{1 \text{ mol Ca(OH)}_2}$

$= 15.9$ g $Ca(OH)_2$

(b) mass $Ca(OH)_2 = 465$ L $\times \dfrac{1.15 \text{ kg}}{1 \text{ L}} \times \dfrac{30.12 \text{ kg HCl}}{100.00 \text{ kg soln}} \times \dfrac{1 \text{ kmol HCl}}{36.45 \text{ kg HCl}} \times \dfrac{1 \text{ kmol Ca(OH)}_2}{2 \text{ kmol HCl}}$

$\times \dfrac{74.10 \text{ kg Ca(OH)}_2}{1 \text{ kmol Ca(OH)}_2} = 164$ kg $Ca(OH)_2$

53. (a) We know that the Al forms the $AlCl_3$.

no. mol $AlCl_3 = 1.87$ g Al $\times \dfrac{1 \text{ mol Al}}{26.98 \text{ g Al}} \times \dfrac{1 \text{ mol AlCl}_3}{1 \text{ mol Al}} = 0.0693$ mol $AlCl_3$

(b) $[AlCl_3] = \dfrac{0.0693 \text{ mol AlCl}_3}{23.8 \text{ mL}} \times \dfrac{1000 \text{ mL}}{1 \text{ L}} = 2.91$ M $AlCl_3$

54. The balanced chemical equation indicates that 4 mol $NaNO_2$ are formed from 2 mol Na_2CO_3.

$[NaNO_2] = \dfrac{175 \text{ g Na}_2CO_3}{1.15 \text{ L soln}} \times \dfrac{1 \text{ mol Na}_2CO_3}{106.0 \text{ g Na}_2CO_3} \times \dfrac{4 \text{ mol NaNO}_2}{2 \text{ mol Na}_2CO_3} = 2.87$ M $NaNO_2$

55. The molarity unit can be interpreted as millimoles of solute per milliliter of solution.

volume K_2CrO_4(aq) $= 415$ mL $\times \dfrac{0.186 \text{ mmol AgNO}_3}{1 \text{ mL soln}} \times \dfrac{1 \text{ mmol K}_2CrO_4}{2 \text{ mmol AgNO}_3} \times \dfrac{1 \text{ mL K}_2CrO_4(aq)}{0.650 \text{ mmol K}_2CrO_4}$

$= 59.4$ mL K_2CrO_4(aq)

56. mass Na $= 125$ mL soln $\times \dfrac{1 \text{ L}}{1000 \text{ mL}} \times \dfrac{0.250 \text{ mol NaOH}}{1 \text{ L soln}} \times \dfrac{2 \text{ mol Na}}{2 \text{ mol NaOH}} \times \dfrac{22.99 \text{ g Na}}{1 \text{ mol Na}}$

$= 0.718$ g Na

57. We determine the amount of HCl present initially, and the amount desired.

amount HCl present $= 250.0$ mL $\times \dfrac{1.019 \text{ mmol HCl}}{1 \text{ mL soln}} = 254.8$ mmol HCl

amount HCl desired $= 250.0$ mL $\times \dfrac{1.000 \text{ mmol HCl}}{1 \text{ mL soln}} = 250.0$ mmol HCl

mass Mg $= (254.8 - 250.0)$ mmol HCl $\times \dfrac{1 \text{ mmol Mg}}{2 \text{ mmol HCl}} \times \dfrac{24.3 \text{ mg Mg}}{1 \text{ mmol Mg}} = 58.$ mg Mg

58. The mass of oxalic acid enables us to determine the amount of NaOH in the solution.

$[NaOH] = \dfrac{0.2048 \text{ g H}_2C_2O_4}{24.87 \text{ mL soln}} \times \dfrac{1000 \text{ mL}}{1 \text{ L soln}} \times \dfrac{1 \text{ mol H}_2C_2O_4}{90.04 \text{ g H}_2C_2O_4} \times \dfrac{2 \text{ mol NaOH}}{1 \text{ mol H}_2C_2O_4} = 0.1829$ M

59. The amount of HCl that reacted with the $CaCO_3$ is the original amount of HCl less the amount that reacted with the $Ba(OH)_2$(aq).

original amount HCl $= 25.00$ mL $\times \dfrac{1.250 \text{ mmol HCl}}{1 \text{ mL soln}} = 31.25$ mmol HCl

remaining amount HCl $= 45.76$ mL $\times \dfrac{0.01221 \text{ mmol Ba(OH)}_2}{1 \text{ mL soln}} \times \dfrac{2 \text{ mmol HCl}}{1 \text{ mmol Ba(OH)}_2} = 1.117$ mmol HCl

The mass of $CaCO_3$ is determined from the amount of HCl that reacted.

mass $CaCO_3 = (31.25 - 1.117)$ mmol HCl $\times \dfrac{1 \text{ mmol CaCO}_3}{2 \text{ mmol HCl}} \times \dfrac{1 \text{ mol CaCO}_3}{1000 \text{ mmol CaCO}_3} \times \dfrac{100.09 \text{ g}}{1 \text{ mol CaCO}_3}$

$= 1.508$ g $CaCO_3$

Determining the Limiting Reagent

60. There are equal number of moles of each reactant present, but more O_2 is needed than NH_3. Thus, O_2(g) is the limiting reagent, and all of the O_2(g) is consumed. The mass of product produced from 4.0 mol O_2(g) is then calculated.

mass NO(g) $= 1.00$ mol $O_2 \times \dfrac{4 \text{ mol NO(g)}}{5 \text{ mol O}_2(g)} \times \dfrac{30.01 \text{ g NO(g)}}{1 \text{ mol NO(g)}} = 24.0$ g NO(g)

61. (a) It is easy to see that 1.00 mol NaOH is the limiting reagent in this case, since twice the amount of NaOH is required as of CS_2.

$$1.00 \text{ mol NaOH} \times \frac{2 \text{ mol Na}_2\text{CS}_3}{6 \text{ mol NaOH}} = 0.333 \text{ mol Na}_2\text{CS}_3$$

$$1.00 \text{ mol NaOH} \times \frac{1 \text{ mol Na}_2\text{CO}_3}{6 \text{ mol NaOH}} = 0.167 \text{ mol Na}_2\text{CO}_3$$

$$1.00 \text{ mol NaOH} \times \frac{3 \text{ mol H}_2\text{O}}{6 \text{ mol NaOH}} = 0.500 \text{ mol H}_2\text{O}$$

(b) Determine the amount of Na_2CS_3 produced from each of the reactants.

$$\text{amount Na}_2\text{CS}_3 = 88.0 \text{ mL CS}_2 \times \frac{1.26 \text{ g}}{1 \text{ mL}} \times \frac{1 \text{ mol CS}_2}{76.14 \text{ g CS}_2} \times \frac{2 \text{ mol Na}_2\text{CS}_3}{3 \text{ mol CS}_2} = 0.971 \text{ mol Na}_2\text{CS}_3$$

$$\text{amount Na}_2\text{CS}_3 = 3.12 \text{ mol NaOH} \times \frac{2 \text{ mol Na}_2\text{CS}_3}{6 \text{ mol NaOH}} = 1.04 \text{ mol Na}_2\text{CS}_3$$

Thus, the mass produced is $0.971 \text{ mol Na}_2\text{CS}_3 \times \frac{154.3 \text{ g Na}_2\text{CS}_3}{1 \text{ mol Na}_2\text{CS}_3} = 150. \text{ g Na}_2\text{CS}_3$

62. Determine the mass of H_2 produced from each of the reactants. The smaller mass is that produced by the limiting reagent and is the mass actually formed.

$$\text{mass H}_2 = 1.84 \text{ g Al} \times \frac{1 \text{ mol Al}}{26.98 \text{ g Al}} \times \frac{3 \text{ mol H}_2}{2 \text{ mol Al}} \times \frac{2.016 \text{ g H}_2}{1 \text{ mol H}_2} = 0.206 \text{ g H}_2$$

$$\text{mass H}_2 = 75.0 \text{ mL} \times \frac{1 \text{ L}}{1000 \text{ mL}} \times \frac{2.95 \text{ mol HCl}}{1 \text{ L}} \times \frac{3 \text{ mol HCl}}{6 \text{ mol HCl}} \times \frac{2.016 \text{ g H}_2}{1 \text{ mol H}_2} = 0.223 \text{ g H}_2$$

Thus, 0.206 g H_2 is produced.

63. Since the two reactants combined in an equimolar basis, the one present with the fewer number of moles is the limiting reagent and determine the mass of the products.

$$\text{no. mol ZnSO}_4 = 275 \text{ mL} \times \frac{1 \text{ L}}{1000 \text{ mL}} \times \frac{0.350 \text{ mol ZnSO}_4}{1 \text{ L soln}} = 0.0962 \text{ mol ZnSO}_4$$

$$\text{no. mol BaS} = 325 \text{ mL} \times \frac{1 \text{ L}}{1000 \text{ mL}} \times \frac{0.280 \text{ mol BaS}}{1 \text{ L soln}} = 0.0910 \text{ mol BaS}$$

Thus, BaS is the limiting reagent and 0.0910 mol of each of the products will be produced.

$$\text{mass products} = \left(0.0910 \text{ mol BaSO}_4 \times \frac{233.4 \text{ g BaSO}_4}{1 \text{ mol BaSO}_4}\right) + \left(0.0910 \text{ mol ZnS} \times \frac{97.46 \text{ g ZnS}}{1 \text{ mol ZnS}}\right)$$

$$= 30.1 \text{ g product mixture}$$

64. (a) $2 H_2(g) + O_2(g) \longrightarrow 2 H_2O(l)$

(b) $\text{no. mol H}_2\text{O} = 4.800 \text{ g H}_2 \times \frac{1 \text{ mol H}_2}{2.016 \text{ g H}_2} \times \frac{1 \text{ mol H}_2\text{O}}{1 \text{ mol H}_2} = 2.381 \text{ mol H}_2\text{O}$

$$\text{no. mol H}_2\text{O} = 36.40 \text{ g O}_2 \times \frac{1 \text{ mol O}_2}{32.00 \text{ g O}_2} \times \frac{2 \text{ mol H}_2\text{O}}{1 \text{ mol O}_2} = 2.275 \text{ mol H}_2\text{O}$$

Thus, the mass of H_2O produced is $\quad \text{mass H}_2\text{O} = 2.275 \text{ mol H}_2\text{O} \times \frac{18.02 \text{ g H}_2\text{O}}{1 \text{ mol H}_2\text{O}} = 41.00 \text{ g H}_2\text{O}$

All of the 36.40 g of O_2 is used up; only H_2 remains. Since the mass of the products must equal the mass of the reactants, the unreacted mass of H_2 is computed as follows.

mass unreacted H_2 = mass of reactants – mass H_2O produced

$$= (4.800 \text{ g H}_2 + 36.40 \text{ g O}_2) - 41.00 \text{ g H}_2\text{O} = 0.20 \text{ g H}_2$$

(c) We used the principle of equality of masses of products and reactants to solve part (b). We can demonstrate the validity of this principle by computing the mass of H_2 unreacted in another way.

$$\text{mass H}_2 \text{ unreacted} = (2.381 - 2.275) \text{ mol H}_2\text{O} \times \frac{1 \text{ mol H}_2}{1 \text{ mol H}_2\text{O}} \times \frac{2.016 \text{ g H}_2}{1 \text{ mol H}_2} = 0.214 \text{ g H}_2$$

The agreement of this result with the one from part (b) demonstrates the validity of the principle of equality of the masses of products and reactants.

65. $Ca(OH)_2(s) + 2 NH_4Cl(s) \longrightarrow CaCl_2(aq) + 2 H_2O + 2 NH_3(g)$

Compute the amount of NH_3 formed from each reagent in this limiting reagent problem.

$$\text{amount NH}_3 = 25.0 \text{ g NH}_4\text{Cl} \times \frac{1 \text{ mol NH}_4\text{Cl}}{53.49 \text{ g NH}_4\text{Cl}} \times \frac{2 \text{ mol NH}_3}{2 \text{ mol NH}_4\text{Cl}} = 0.467 \text{ mol NH}_3$$

$$\text{amount NH}_3 = 25.0 \text{ g Ca(OH)}_2 \times \frac{1 \text{ mol Ca(OH)}_2}{74.09 \text{ g Ca(OH)}_2} \times \frac{2 \text{ mol NH}_3}{1 \text{ mol Ca(OH)}_2} = 0.675 \text{ mol NH}_3$$

0.467 mol NH_3 is produced. $\quad \text{mass NH}_3 = 0.467 \text{ mol NH}_3 \times \frac{17.03 \text{ g NH}_3}{1 \text{ mol NH}_3} = 7.95 \text{ g NH}_3$

Theoretical, Actual, and Percent Yields

66. (a) We first need to solve the limiting reagent problem involved here.

$$\text{no. mol } C_4H_9Br = 13.0 \text{ g } C_4H_9OH \times \frac{1 \text{ mol } C_4H_9OH}{74.12 \text{ g } C_4H_9OH} \times \frac{1 \text{ mol } C_4H_9Br}{1 \text{ mol } C_4H_9OH} = 0.175 \text{ mol } C_4H_9Br$$

$$\text{no. mol } C_4H_9Br = 21.6 \text{ g } NaBr \times \frac{1 \text{ mol } NaBr}{102.9 \text{ g } NaBr} \times \frac{1 \text{ mol } C_4H_9Br}{1 \text{ mol } NaBr} = 0.210 \text{ mol } C_4H_9Br$$

$$\text{no. mol } C_4H_9Br = 33.8 \text{ g } H_2SO_4 \times \frac{1 \text{ mol } H_2SO_4}{98.1 \text{ g } H_2SO_4} \times \frac{1 \text{ mol } C_4H_9Br}{1 \text{ mol } H_2SO_4} = 0.345 \text{ mol } C_4H_9Br$$

Thus, mass $C_4H_9Br = 0.175$ mol $C_4H_9Br \times \dfrac{136.9 \text{ g } C_4H_9Br}{1 \text{ mol } C_4H_9Br} = 24.0$ g $C_4H_9Br =$ theoretical yield

(b) The actual yield is the mass obtained, 16.8 g C_4H_9Br.

(c) Then $\quad$ % yield $= \dfrac{16.8 \text{ g } C_4H_9Br \text{ produced}}{24.0 \text{ g } C_4H_9Br \text{ expected}} \times 100\% = 70.0\%$ yield

67. Again, we solve the limiting reagent problem first.

$$\text{amount } (C_6H_5N)_2 = 0.10 \text{ L } C_6H_5NO_2 \times \frac{1000 \text{ mL}}{1 \text{ L}} \times \frac{1.20 \text{ g}}{1 \text{ mL}} \times \frac{1 \text{ mol } C_6H_5NO_2}{123.1 \text{ g } C_6H_5NO_2} \times \frac{1 \text{ mol}(C_6H_5N)_2}{2 \text{ mol } C_6H_5NO_2}$$

$$= 0.49 \text{ mol } (C_6H_5N)_2$$

$$\text{amount } (C_6H_5N)_2 = 0.30 \text{ L } C_6H_{14}O_4 \times \frac{1000 \text{ mL}}{1 \text{ L}} \times \frac{1.12 \text{ g}}{1 \text{ mL}} \times \frac{1 \text{ mol } C_6H_{14}O_4}{150.0 \text{ g } C_6H_{14}O_4} \times \frac{1 \text{ mol } (C_6H_5N)_2}{4 \text{ mol } C_6H_{14}O_4}$$

$$= 0.56 \text{ mol } (C_6H_5N)_2$$

$$\text{mass } (C_6H_5N)_2 = 0.49 \text{ mol } (C_6H_5N)_2 \times \frac{182.2 \text{ g } (C_6H_5N)_2}{1 \text{ mol } (C_6H_5N)_2} = 89 \text{ g } (C_6H_5N)_2$$

$$\text{percent yield} = \frac{55 \text{ g } (C_6H_5N)_2 \text{ produced}}{89 \text{ g } (C_6H_5N)_2 \text{ expected}} \times 100\% = 62\% \text{ yield}$$

68. Balanced equation: $\quad 3 C_2H_4O_2 + PCl_3 \longrightarrow 3 C_2H_3OCl + H_3PO_3$

$$\text{mass acid} = 55 \text{ g } C_2H_3OCl \times \frac{100.0 \text{ g calculated}}{71 \text{ g produced}} \times \frac{1 \text{ mol } C_2H_3OCl}{78.5 \text{ g } C_2H_3OCl} \times \frac{3 \text{ mol } C_2H_4O_2}{3 \text{ mol } C_2H_3OCl}$$

$$\times \frac{60.1 \text{ g pure } C_2H_4O_2}{1 \text{ mol } C_2H_4O_2} \times \frac{100 \text{ g commercial}}{97 \text{ g pure } C_2H_4O_2} = 61 \text{ g commercial } C_2H_4O_2$$

69. $\text{mass } CH_2Cl_2 = 112 \text{ g } CH_4 \times \dfrac{1 \text{ mol } CH_4}{16.04 \text{ g } CH_4} \times \dfrac{1 \text{ mol } CH_3Cl}{1 \text{ mol } CH_4} \times \dfrac{0.92 \text{ mol } CH_3Cl \text{ produced}}{1.00 \text{ mol } CH_3Cl \text{ expected}}$

$$\times \frac{1 \text{ mol } CH_2Cl_2}{1 \text{ mol } CH_3Cl} \times \frac{85.0 \text{ g } CH_2Cl_2}{1 \text{ mol } CH_2Cl_2} \times \frac{0.92 \text{ g } CH_2Cl_2 \text{ produced}}{1.00 \text{ g } CH_2Cl_2 \text{ calculated}} = 5.0 \times 10^2 \text{ g } CH_2Cl_2$$

70. A less-than-100% yield of desired product in synthesis reactions is almost always the case. This is because of side reactions that yield products other than the desired one (by-products) and because of the loss of material in various steps of the synthesis process. A main criterion for choosing a synthesis reaction is how economically it can be run. In the analysis of a compound, on the other hand, it is essential that all of the material present be detected. Therefore, a 100% yield is required; none of the material present in the sample can be lost during the analysis. Therefore analysis reactions are carefully chosen to meet this criterion; they need not be economical to run.

71. The theoretical yield is 2.07 g Ag_2CrO_4. If the mass actually obtained is less than this, it is likely that some of the pure material was lost in the reaction, perhaps stuck to the walls of the flask in which the reaction occurred, or suspended in the solution. But the maximum mass of Ag_2CrO_4 that can be produced is 2.07 g. If the precipitate weighs more than 2.07 g, the extra mass must be impurities.

5 INTRODUCTION TO REACTIONS IN AQUEOUS SOLUTIONS

REVIEW QUESTIONS

1. **(a)** The symbol "$\rightleftharpoons$" means that a chemical reaction reaches a point of balance or equilibrium somewhere short of the complete consumption of (the limiting) reactant.

   (b) The square brackets, [ ], surrounding the formula of a species, are the symbol for the molarity of that species in solution.

   (c) A "spectator" ion is one that is present in a solution in which a reactant takes place but is not indicated in the net ionic equation for that reaction because this ion is unaffected by the reaction.

   (d) A weak acid is a species that produces hydrogen ion in aqueous solution (an acid), but that does not dissociate completely into its ions (a weak electrolyte).

2. **(a)** In the half-reaction method of balancing redox equations, each species that is oxidized or reduced is the basis of a balanced half-equation. These half-equations then are combined to produce the balanced net ionic equation for the redox reaction.

   (b) A disproportionation reaction is one in which the same species is both oxidized and reduced.

   (c) Titration is the procedure of adding a measured amount of one material to a measured amount of another, in such a way that chemically equivalent amounts of substances are present at the end of the titration. The concentration of substance in one of the two materials is known; the technique permits the determination of the other concentration.

   (d) Standardization of a solution refers to the determination of its concentration by titration.

3. **(a)** A strong electrolyte is a substance that dissociates completely into its ions when it is dissolved in aqueous solution. A strong acid is a strong electrolyte that produces hydrogen ions and anions when it dissociates.

   (b) An oxidizing agent is a species that causes another species to be oxidized, that is, to donate electrons and thereby have the oxidation state of one of its elements increased. A reducing agent is a pecies that causes another species to be reduced: accept electrons and have the oxidation state of one of its elements decreased.

   (c) A precipitation reaction is one in which an insoluble substance is formed when solutions of two soluble substances are mixed. A neutralization reaction is the reaction of an acid with a base, the normal products are water and a salt that has the same cation as the base and the same anion as the acid.

   (d) A half-equation is a equation that expresses just the oxidation or just the reduction aspect of a redox reaction. A net equation is either a "net ionic equation," that is, an equation with the spectator ions removed or it is the net result of several reactions that, together, make up a process.

4. **(a)** The best electrical conductor is the solution of the strong electrolyte: 0.10 M NaCl. In each liter of this solution, there is 0.10 mol Na^+ ions and 0.10 mol Cl^- ions.

   (b) The poorest electrical conductor is the solution of the nonelectrolyte: 0.10 M C_2H_5OH. In this solution, the concentration of ions is almost nonexistent.

5. **(a)** Na_2SO_4 is the *salt* of sodium hydroxide (NaOH) and sulfuric acid (H_2SO_4).

   (b) KOH is a *strong base*, one of the common ones listed in Table 5-1.

   (c) $CaCl_2$ is the *salt* of calcium hydroxide [$Ca(OH)_2$] and hydrochloric acid (HCl).

   (d) H_2SO_3 is an *acid*, since its formula begins with hydrogen. It is not listed in Table 5-1, so it must be a *weak acid*.

   (e) HI is a *strong acid*, listed in Table 5-1.

(f) HNO_2 is an acid since its formula begins with hydrogen, but it is not listed in Table 5-1. It is a *weak acid*.

(g) NH_3 is the *weak base*, ammonia.

(h) NH_4I is the *salt* of ammonia (NH_3) and hydroiodic acid (HI).

(i) $Ca(OH)_2$ is a *strong base*, one of the common ones listed in Table 5-1.

6.   For all these solutes but one—$Al_2(SO_4)_3$—there is one sulfate ion per formula unit. Consequently, the concentration of the compound and the solufate ion concentration in that compound's aqueous solution will be the same. This makes 0.15 M $CuSO_4$ the solution with the highest $[SO_4^{2-}]$ among these four. But there are three sulfate ions per formula unit of $Al_2(SO_4)_3$. Thus $[SO_4^{2-}]$ in the $Al_2(SO_4)_3$ solution is three times the concentration of the solute, or $[SO_4^{2-}] = 3 \times 0.060$ M = 0.180 M, making this solution the one with the highest $[SO_4^{2-}]$.

7.   (a) $[K^+] = \dfrac{0.187 \text{ mol KNO}_3}{1 \text{ L soln}} \times \dfrac{1 \text{ mol K}^+}{1 \text{ mol KNO}_3} = 0.187$ M K^+

(b) $[NO_3^-] = \dfrac{0.026 \text{ mol Ca(NO}_3)_2}{1 \text{ L soln}} \times \dfrac{2 \text{ mol NO}_3^-}{1 \text{ mol Ca(NO}_3)_2} = 0.052$ M NO_3^-

(c) $[Al^{3+}] = \dfrac{0.112 \text{ mol Al}_2(SO_4)_3}{1 \text{ L soln}} \times \dfrac{2 \text{ mol Al}^{3+}}{1 \text{ mol Al}_2(SO_4)_3} = 0.224$ M Al^{3+}

(d) $[Na^+] = \dfrac{0.283 \text{ mol Na}_3PO_4}{1 \text{ L soln}} \times \dfrac{3 \text{ mol Na}^+}{1 \text{ mol Na}_3PO_4} = 0.849$ M Na^+

8.   The amount of chloride ion in each solution in millimoles is computed.

Cl^- amount = 200.0 mL $\times \dfrac{0.25 \text{ mmol NaCl}}{1 \text{ mL soln}} \times \dfrac{1 \text{ mmol Cl}^-}{1 \text{ mmol NaCl}} = 50.$ mmol Cl^-

Cl^- amount = 500.0 mL $\times \dfrac{0.055 \text{ mmol MgCl}_2}{1 \text{ mL soln}} \times \dfrac{2 \text{ mmol Cl}^-}{1 \text{ mmol MgCl}_2} = 55$ mmol Cl^-

Cl^- amount = 1.00 L $\times \dfrac{1000 \text{ mL}}{1 \text{ L}} \times \dfrac{0.058 \text{ mmol HCl}}{1 \text{ mL}} \times \dfrac{1 \text{ mmol Cl}^-}{1 \text{ mmol HCl}} = 58$ mmol Cl^-

The 1.00 L of 0.058 M HCl contains the largest amount of Cl^-.

9. $[OH^-] = \dfrac{0.164 \text{ g Ba(OH)}_2 \cdot 8H_2O}{325 \text{ mL soln}} \times \dfrac{1000 \text{ mL}}{1 \text{ L}} \times \dfrac{1 \text{ mol Ba(OH)}_2 \cdot 8H_2O}{315.5 \text{ Ba(OH)}_2 \cdot 8H_2O} \times \dfrac{2 \text{ mol OH}^-}{1 \text{ mol Ba(OH)}_2 \cdot 8 \text{ H}_2O}$

$= 3.20 \times 10^{-3}$ M OH^-

10. $[K^+] = \dfrac{0.118 \text{ mol KCl}}{1 \text{ L soln}} \times \dfrac{1 \text{ mol K}^+}{1 \text{ mol KCl}} = 0.118$ M

$[Mg^{2+}] = \dfrac{0.186 \text{ mol MgCl}_2}{1 \text{ L soln}} \times \dfrac{1 \text{ mol Mg}^{2+}}{1 \text{ mol MgCl}_2} = 0.186$ M

Now determine the amount of Cl^- in 1.00 L of the solution.

no. mol Cl^- = $\left(\dfrac{0.118 \text{ mol KCl}}{1 \text{ L soln}} \times \dfrac{1 \text{ mol Cl}^-}{1 \text{ mol KCl}}\right) + \left(\dfrac{0.186 \text{ mol MgCl}_2}{1 \text{ L soln}} \times \dfrac{2 \text{ mol Cl}^-}{1 \text{ mol MgCl}_2}\right)$

$= 0.118$ mol Cl^- + 0.372 mol Cl^- = 0.490 mol Cl^-

$[Cl^-] = \dfrac{0.490 \text{ mol Cl}^-}{1 \text{ L soln}} = 0.490$ M Cl^-

11.   Determine the amount of I^- in the solution as it now exists, and the amount of I^- in the solution of the desired concentration. The different in these two amounts is the amount of I^- that must be added. Convert this amount to a mass of MgI_2 in grams.

I^- amount in final solution = 250.0 mL $\times \dfrac{1 \text{ L}}{1000 \text{ mL}} \times \dfrac{0.1000 \text{ mol I}^-}{1 \text{ L soln}} = 0.02500$ mol I^-

I^- amount in KI solution = 250.0 mL $\times \dfrac{1 \text{ L}}{1000 \text{ mL}} \times \dfrac{0.0952 \text{ mol KI}}{1 \text{ L soln}} \times \dfrac{1 \text{ mol I}^-}{1 \text{ mol KI}} = 0.0238$ mol I^-

mass MgI_2 to be added = $(0.02500 - 0.0249_5)$ mol $I^- \times \dfrac{1 \text{ mol MgI}_2}{2 \text{ mol I}^-} \times \dfrac{278.11 \text{ g MgI}_2}{1 \text{ mol MgI}_2} \times \dfrac{1000 \text{ mg}}{1 \text{ g}}$

$= 1.7 \times 10^2$ mg MgI_2

12.   Since all nitrates and all alkali metal compounds are soluble in water, both $Ba(NO_3)_2$ and K_3PO_4 are soluble.

Most halides are soluble in water; $ZnCl_2$ and NaI are soluble in water.

Although most sulfates are soluble in water, $PbSO_4(s)$ is not soluble in water.

Only a few hydroxides are soluble in water; $Fe(OH)_3(s)$ is not soluble in water.

13. HCl(aq) reacts with active metals and some anions to produce a gas.

Mg is an active metal: $Mg(s) + 2 HCl(aq) \longrightarrow MgCl_2(aq) + H_2(g)$

HSO_3^- produces a gas with and acid: $KHSO_3(s) + HCl(aq) \longrightarrow KCl(aq) + H_2O + SO_2(g)$

14. In each case, each available cation is paired with the available anions, one at a time, to determine if a compound is produced that is insoluble, based on the solubility rules of Chapter 5. Then a net ionic equation is written to summarize this information.

(a) $Pb^{2+}(aq) + 2 Br^-(aq) \longrightarrow PbBr_2(s)$ **(b)** No reaction occurs.

(c) $Fe^{3+}(aq) + 3 OH^-(aq) \longrightarrow Fe(OH)_3(s)$ **(d)** $Ca^{2+}(aq) + CO_3^{2-}(aq) \longrightarrow CaCO_3(s)$

(e) $Ba^{2+}(aq) + SO_4^{2-}(aq) \longrightarrow BaSO_4(s)$ **(f)** No reaction occurs.

15. The type of reaction is given first, followed by the net ionic equation.

(a) Neutralization: $OH^-(aq) + HC_2H_3O_2(aq) \longrightarrow H_2O(l) + C_2H_3O_2^-(aq)$

(b) No reaction occurs.

(c) Gas evolution: $S^{2-}(aq) + 2 H^+(aq) \longrightarrow H_2S(g)$

(d) Gas evolution: $HCO_3^-(aq) + H^+(aq) \longrightarrow$ "$H_2CO_3(aq)$" $\longrightarrow H_2O(l) + CO_2(g)$

(e) Redox: $2 Al(s) + 6 H^+(aq) \longrightarrow 2 Al^{3+}(aq) + 3 H_2(g)$

(f) No reaction occurs, based on the information in Table 5-3.

16. The problem is most easily solved with amounts in millimoles.

$$\text{volume NaOH(aq)} = 10.00 \text{ mL HCl(aq)} \times \frac{0.114 \text{ mmol HCl}}{1 \text{ mL HCl(aq)}} \times \frac{1 \text{ mmol H}^+}{1 \text{mmol HCl}} \times \frac{1 \text{ mmol OH}^-}{1 \text{ mmol H}^+}$$

$$\times \frac{1 \text{ mmol NaOH}}{1 \text{ mmol OH}^-} \times \frac{1 \text{ mL NaOH(aq)}}{0.0567 \text{ mmol NaOH}} = 20.1 \text{ mL NaOH(aq) soln}$$

17. $[NaOH] = \dfrac{10.00 \text{ mL acid} \times \dfrac{0.06852 \text{ mmol H}_2SO_4}{1 \text{ mL acid}} \times \dfrac{2 \text{ mmol NaOH}}{1 \text{ mmol H}_2SO_4}}{22.19 \text{ mL base}} = 0.06176 \text{ M}$

18. The net ionic equation for the reaction of KOH, a strong base, with HCl, a strong acid, is:

$$OH^-(aq) + H^+(aq) \longrightarrow H_2O$$

Thus, the reactant that produces the smaller amount of ions is the limiting reagent. More to the point, the difference between the larger amount of ions and the smaller amount, determines whether the reaction is acidic or basic. If the difference is zero, the solution is neutral.

$$\text{amount OH}^- = 24.67 \text{ mL KOH(aq)} \times \frac{0.1278 \text{ mmol KOH}}{1 \text{ mL KOH(aq)}} \times \frac{1 \text{ mmol OH}^-}{1 \text{ mmol KOH}} = 3.153 \text{ mmol OH}^-$$

$$\text{amount H}^+ = 25.13 \text{ mL HCl(aq)} \times \frac{0.1205 \text{ mmol HCl}}{1 \text{ mL HCl(aq)}} \times \frac{1 \text{ mmol H}^+}{1 \text{ mmol HCl}} = 3.028 \text{ mmol H}^+$$

excess ion = 3.153 mmol OH$^-$ – 3.028 mmol H$^+$ = 0.125 mmol OH$^-$ The solution is basic or alkaline.

19. $\text{mass MgO(s)} = 125 \text{ mL acid} \times \dfrac{1 \text{ L}}{1000 \text{ mL}} \times \dfrac{2.12 \text{ mol HNO}_3}{1 \text{ L acid}} \times \dfrac{1 \text{ mol H}^+}{1 \text{ mol HNO}_3} \times \dfrac{1 \text{ mol MgO}}{2 \text{ mol H}^+}$

$$\times \frac{40.3 \text{ g MgO}}{1 \text{ mol MgO}} = 5.34 \text{ g MgO}$$

20. (a) The O.S. of H is +1, that of O is –2, and that of Zn is +2 on each side of this equation. This is not a redox equation.

(b) The O.S. of Cl_2 is 0 on the left- and –1 on the right-hand side of this equation. The O.S. of Br is –1 of the left- and 0 on the right-hand side of this equation. This is the equation of a redox reaction.

(c) The O.S. of Cu is 0 on the left- and +2 on the right-hand side of this equation. The O.S. of N is +5 on the left- and +2 on the right-hand side of this equation. This is the equation of a redox reaction.

(d) The O.S. of O is –2, that of Ag is +1, and that of Cr is +6 on each side of this equation. This is not a redox equation.

21. (a) The O.S. of O is –2 on both sides of this equation. The O.S. of H is 0 on the left- and +1 on the right hand side of this equation; H is oxidized and thus NO must be an oxidizing agent. The O.S. of N is +2 on the left- and –3 on the right-hand side of this equation; N is reduced and thus H_2 must be a reducing agent.

(b) The O.S. of O is –2 and that of H is +1 on both sides of this equation. The O.S. of Cu is 0 on the left- and +2 on the right-hand side of this equation; Cu is oxidized and thus NO_3^- must be an oxidizing agent. The O.S. of N is +5 on the left- and +2 on the right-hand side of this equation; N is reduced and thus Cu must be a reducing agent.

(c) The O.S. of O is –2 and that of H is +1 on both sides of this equation. The O.S. of Cl is 0 on the left-hand side of this equation; on ther right-hand side, the O.S. of Cl is –1 in Cl^- and it is +5 in ClO_3^-. Cl is both oxidized and reduced and Cl_2 serves as both an oxidizing agent and as a reducing agent.

22. (a) Reduction: $S_2O_8^{2-}(aq) + 2\ e^- \longrightarrow 2\ SO_4^{2-}(aq)$

(b) Reduction: $2\ NO_3^-(aq) + 10\ H^+(aq) + 8\ e^- \longrightarrow N_2O(g) + 5\ H_2O(l)$

(c) Oxidation: $Br^-(aq) + 3\ H_2O(l) \longrightarrow BrO_3^-(aq) + 6\ H^+(aq) + 6\ e^-$

(d) Reduction: $NO_3^-(aq) + 6\ H_2O + 8\ e^- \longrightarrow NH_3(aq) + 9\ OH^-$

23. (a) Oxidation $Zn(s) \longrightarrow Zn^{2+}(aq) + 2\ e^-$ $\times 4$

Reduction: $2\ NO_3^-(aq) + 10\ H^+(aq) + 8\ e^- \longrightarrow N_2O(g) + 5\ H_2O$ $\times 1$

Net: $4\ Zn(s) + 2\ NO_3^-(aq) + 10\ H^+(aq) \longrightarrow 4\ Zn^{2+}(aq) + N_2O(g) + 5\ H_2O$

(b) Oxidation: $Zn(s) \longrightarrow Zn^{2+}(aq) + 2\ e^-$ $\times 4$

Reduction: $NO_3^-(aq) + 10\ H^+(aq) + 8e^- \longrightarrow NH_4^+(aq) + 3\ H_2O(l)$ $\times 1$

Net: $4\ Zn(s) + NO_3^-(aq) + 10\ H^+(aq) \longrightarrow 4\ Zn^{2+}(aq) + NH_4^+(aq) + 3\ H_2O(l)$

(c) Oxidation: $Fe^{2+}(aq) \longrightarrow Fe^{3+}(aq) + e^-$ $\times 6$

Reduction $Cr_2O_7^{2-}(aq) + 14\ H^+(aq) + 6\ e^- \longrightarrow 2\ Cr^{3+}(aq) + 7\ H_2O$ $\times 1$

Net: $Cr_2O_7^{2-}(aq) + 14\ H^+(aq) + 6\ Fe^{2+}(aq) \longrightarrow 6\ Fe^{3+}(aq) + 2\ Cr^{3+}(aq) + 7\ H_2O$

(d) Oxidation: $H_2O_2(aq) \longrightarrow O_2(g) + 2\ H^+(aq) + 2\ e^-$ $\times 5$

Reduction: $MnO_4^-(aq) + 8\ H^+(aq) + 5\ e^- \longrightarrow Mn^{2+}(aq) + 4\ H_2O$ $\times 2$

Net: $2\ MnO_4^-(aq) + 6\ H^+(aq) + 5\ H_2O_2(aq) \longrightarrow 2\ Mn^{2+}(aq) + 8\ H_2O + 5\ O_2(g)$

24. (a) Oxidation: $MnO_2(s) + 4\ OH^-(aq) \longrightarrow MnO_4^-(aq) + 2\ H_2O + 3\ e^-$ $\times 2$

Reduction: $ClO_3^-(aq) + 3\ H_2O + 6\ e^- \longrightarrow Cl^-(aq) + 6\ OH^-(aq)$ $\times 1$

Net: $2\ MnO_2(s) + ClO_3^-(aq) + 2\ OH^-(aq) \longrightarrow 2\ MnO_4^-(aq) + Cl^-(aq) + H_2O$

(b) Oxidation: $Fe(OH)_3(s) + 5\ OH^-(aq) \longrightarrow FeO_4^{2-}(aq) + 4\ H_2O + 3\ e^-$ $\times 2$

Reduction: $OCl^-(aq) + H_2O + 2\ e^- \longrightarrow Cl^-(aq) + 2\ OH^-(aq)$ $\times 3$

Net: $2\ Fe(OH)_3(s) + 3\ OCl^-(aq) + 4\ OH^-(aq) \longrightarrow 2\ FeO_4^{2-}(aq) + 3\ Cl^-(aq) + 5\ H_2O$

(c) Oxidation: $ClO_2 + 2\ OH^-(aq) \longrightarrow ClO_3^-(aq) + H_2O(l) + e^-$ $\times 5$

Reduction: $ClO_2 + 2\ H_2O(l) + 5\ e^- \longrightarrow Cl^- + 4\ OH^-(aq)$ $\times 1$

Net: $6\ ClO_2 + 6\ OH^-(aq) \longrightarrow 5\ ClO_3^-(aq) + Cl^-(aq) + 3\ H_2O$

25. Oxidation: $C_2O_4^{2-}(aq) \longrightarrow 2\ CO_2(g) + 2\ e^-$ $\times 5$

Reduction: $MnO_4^-(aq) + 8\ H^+(aq) + 5\ e^- \longrightarrow Mn^{2+}(aq) + 4\ H_2O$ $\times 2$

Net: $2\ MnO_4^-(aq) + 16\ H^+(aq) + 5\ C_2O_4^{2-}(aq) \longrightarrow 2\ Mn^{2+}(aq) + 8\ H_2O + 10\ CO_2(g)$

$$[MnO_4^-] = \frac{0.3502\ \text{g}\ Na_2C_2O_4 \times \dfrac{1\ \text{mol}\ Na_2C_2O_4}{134.00\ \text{g}\ Na_2C_2O_4} \times \dfrac{1\ \text{mol}\ C_2O_4^{2-}}{1\ \text{mol}\ Na_2C_2O_4} \times \dfrac{2\ \text{mol}\ MnO_4^-}{5\ \text{mol}\ C_2O_4^{2-}}}{24.16\ \text{mL soln} \times \dfrac{1\ \text{L soln}}{1000\ \text{mL soln}}}$$

$= 0.04237\ M$ Thus, $0.04237\ M = KMnO_4$ molarity.

EXERCISES

Strong Electrolytes, Weak Electrolytes, and Nonelectrolytes

26. **(a)** Since its formula begins with hydrogen, $HC_7H_5O_2$ is an acid. It is not listed in Table 5-1, so it is a weak acid. A weak acid is a *weak electrolyte*.

(b) Cs_2SO_4 is an ionic compound, that is, a salt. A salt is a *strong electrolyte*.

(c) $CaCl_2$ also is a salt, a *strong electrolyte*.

(d) $(CH_3)_2CO$ is a covalent compound whose formula does not begin with H. Thus it is neither an acid nor a salt. It also is not built around nitrogen, making it not a weak base. This is a *nonelectrolyte*.

(e) $H_2C_3H_2O_4$ is a *weak electrolyte*. Same reasoning as part (a).

27. $NH_3(aq)$ is a weak base and $HC_2H_3O_2(aq)$ is a weak acid. Their reaction produces a solution of ammonium acetate, $NH_4C_2H_3O_2(aq)$, a salt and a strong electrolyte. $NH_3(aq) + HC_2H_3O_2(aq) \longrightarrow NH_4C_2H_3O_2(aq)$

Ion Concentrations

28. **(a)** $[Ca^{2+}] = \dfrac{400.0 \text{ mg } Ca^{2+}}{1 \text{ L}} \times \dfrac{1 \text{ g}}{1000 \text{ mg}} \times \dfrac{1 \text{ mol } Ca^{2+}}{40.078 \text{ g } Ca^{2+}} = 0.009981 \text{ M } Ca^{2+}$

(b) $[K^+] = \dfrac{380.5 \text{ mg } K^+}{1 \text{ L}} \times \dfrac{1 \text{ g}}{1000 \text{ mg}} \times \dfrac{1 \text{ mol } K^+}{39.098 \text{ g } K^+} = 0.009732 \text{ M } K^+$

(c) $[Zn^{2+}] = \dfrac{0.0104 \text{ mg } Zn^{2+}}{1 \text{ L}} \times \dfrac{1 \text{ g}}{1000 \text{ mg}} \times \dfrac{1 \text{ mol } Zn^{2+}}{65.39 \text{ g } Zn^{2+}} = 1.59 \times 10^{-7} \text{ M } Zn^{2+}$

29. Let us determine the concentration of each solution in mg Na^+/mL.

(a) mg Na^+/mL $= \dfrac{0.124 \text{ mmol } Na_2SO_4}{1 \text{ mL}} \times \dfrac{2 \text{ mmol } Na^+}{1 \text{ mmol } Na_2SO_4} \times \dfrac{23.0 \text{ mg } Na^+}{1 \text{ mmol } Na^+} = 5.70 \text{ mg } Na^+/mL$

(b) mg Na^+/mL $= \dfrac{1.50 \text{ g } NaCl}{100 \text{ mL}} \times \dfrac{1 \text{ mol } NaCl}{58.5 \text{ g } NaCl} \times \dfrac{1 \text{ mol } Na^+}{1 \text{ mol } NaCl} \times \dfrac{23.0 \text{ g } Na^+}{1 \text{ mol } Na^+} \times \dfrac{1000 \text{ mg}}{1 \text{ g}}$
$= 5.90 \text{ mg } Na^+/mL$

(c) The solution with 13.4 mg Na^+/mL has the highest concentration of Na^+.

30. amount $NO_3^- = 0.305 \text{ L} \times \dfrac{0.277 \text{ mol } KNO_3}{1 \text{ L soln}} \times \dfrac{1 \text{ mol } NO_3^-}{1 \text{ mol } KNO_3}$
$+ 0.476 \text{ L} \times \dfrac{0.387 \text{ mol } Mg(NO_3)_2}{1 \text{ L soln}} \times \dfrac{2 \text{ mol } NO_3^-}{1 \text{ mol } Mg(NO_3)_2}$
$= 0.0845 \text{ mol } NO_3^- + 0.368 \text{ mol } NO_3^-$
$[NO_3^-] = \dfrac{(0.0845 + 0.368) \text{ mol } NO_3^-}{0.305 \text{ L} + 0.476 \text{ L} + 0.825 \text{ L}} = 0.282 \text{ M}$

31. Determine $[OH^-]$ in the saturated solution, remembering that $Ba(OH)_2$ is a strong base.
$[OH^-] = \dfrac{39 \text{ g } Ba(OH)_2 \cdot 8H_2O}{1 \text{ L soln}} \times \dfrac{1 \text{ mol } Ba(OH)_2 \cdot 8H_2O}{315.4 \text{ g } Ba(OH)_2 \cdot 8H_2O} \times \dfrac{2 \text{ mol } OH^-}{1 \text{ mol } Ba(OH)_2 \cdot 8H_2O} = 0.25 \text{ M}$
Thus, a solution of $Ba(OH)_2$ with a $[OH^-]$ equal to 0.100 M can be produced by diluting a saturated solution.

32. NH_3 is a weak base and would have a very low $[H^+]$; the answer is not (d). HCl and H_2SO_4 are both strong acids and would have higher $[H^+]$ than the weak acid $HC_2H_3O_2$; the answer is not (a). H_2SO_4 has two ionizable protons per mole while HCl has but one. Thus, H_2SO_4 would have the highest $[H^+]$ in a 0.010 M aqueous solution.

33. We calculate $[OH^-]$ in the initial solution: $[OH^-] = \dfrac{0.0250 \text{ mol } Ba(OH)_2}{1 \text{ L soln}} \times \dfrac{2 \text{ mol } OH^-}{1 \text{ L soln}} = 0.0500 \text{ M}$

Now we can determine the ratio of the dilute (volumetric flask) to the concentrated (pipet) solutions.
$V_c \times C_c = V_d \times C_d = V_c \times 0.0500 \text{ M} = V_d \times 0.0100 \text{ M}$ $\quad \dfrac{V_d}{V_c} = \dfrac{0.0500 \text{ M}}{0.0100 \text{ M}} = 5.00$

If we pipet one sample of 0.0250 M $Ba(OH)_2$ with a 50.00-mL pipet into a 250.0-mL flask, and fill this flask, with mixing to the mark with distilled water, the resulting solution will be 0.0100 M OH^-.

34. CaF_2 molarity $= \dfrac{0.9 \text{ mg F}}{1 \text{ L soln}} \times \dfrac{1 \text{ g}}{1000 \text{ mg}} \times \dfrac{1 \text{ mol F}}{18.998 \text{ g F}} \times \dfrac{1 \text{ mol } CaF_2}{2 \text{ mol F}} = 2 \times 10^{-5} \text{ M}$

Writing Ionic Equations

35.

	Mixture	Result (net ionic equation)
(a)	$NaI(aq) + ZnSO_4(aq) \rightarrow$	No reaction occurs.
(b)	$CuSO_4(aq) + Na_2CO_3(aq) \rightarrow$	$Cu^{2+}(aq) + CO_3{}^{2-}(aq) \longrightarrow CuCO_3(s)$
(c)	$Zn(OH)_2(s) + HCl(aq) \rightarrow$	$Zn(OH)_2(s) + 2 H^+(aq) \longrightarrow Zn^{2+}(aq) + 2 H_2O(l)$
(d)	$AgNO_3(aq) + CuCl_2(aq) \rightarrow$	$Ag^+(aq) + Cl^-(aq) \longrightarrow AgCl(s)$
(e)	$BaS(aq) + CuSO_4(aq) \rightarrow$	$Ba^{2+}(aq) + S^{2-}(aq) + Cu^{2+}(aq) + SO_4{}^{2-}(aq) \rightarrow BaSO_4(s) + CuS(s)$
(f)	$Na_2CO_3(s) + HCl(aq) \rightarrow$	$CO_3{}^{2-}(s) + 2 H^+(aq) \longrightarrow CO_2(g) + H_2O(l)$
(g)	$NH_3(aq) + HNO_3(aq) \rightarrow$	$NH_3(aq) + H^+(aq) \longrightarrow NH_4{}^+(aq)$

36. (a) $2 Na(s) + 2 H_2O(l) \longrightarrow 2 Na^+(aq) + 2 OH^-(aq) + H_2(g)$

    (b) $Fe^{3+}(aq) + 3 OH^-(aq) \longrightarrow Fe(OH)_3(aq)$

    (c) $Fe(OH)_3(s) + 3 H^+(aq) \longrightarrow Fe^{3+}(aq) + 3 H_2O(l)$

37. (a) $NaHCO_3(s) + H^+(aq) \longrightarrow Na^+(aq) + H_2O(l) + CO_2(g)$

    (b) $CaCO_3(s) + 2 H^+(aq) \longrightarrow Ca^{2+}(aq) + H_2O(l) + CO_2(g)$

    (c) $Mg(OH)_2(s) + 2 H^+(aq) \longrightarrow Mg^{2+}(aq) + 2 H_2O(l)$

    (d) $Mg(OH)_2(s) + 2 H^+(aq) \longrightarrow Mg^{2+}(aq) + 2 H_2O(l)$

        $Al(OH)_3(s) + 3 H^+(aq) \longrightarrow Al^{3+}(aq) + 3 H_2O(l)$

    (e) $NaAl(OH)_2CO_3(s) + 4 H^+(aq) \longrightarrow Al^{3+}(aq) + Na^+(aq) + 3 H_2O(l) + CO_2(g)$

38. As a salt: $NaHSO_4(aq) \longrightarrow Na^+(aq) + HSO_4{}^-(aq)$

    As an acid: $HSO_4{}^-(aq) + OH^-(aq) \longrightarrow H_2O(l) + SO_4{}^{2-}(aq)$

39. Dissolve $Na_2CrO_4(s)$ and $ZnSO_4(s)$ each in H_2O to produce solutions of, respectively, Na^+ and Zn^{2+}. Dissolve each of $BaCO_3(s)$ and $Al(OH)_3(s)$ in $HCl(aq)$ to produce solutions of, respectively, $Ba^{2+}(aq)$ and $Al^{3+}(aq)$.

40. (a) Add $K_2SO_4(aq)$; $BaSO_4(s)$ will form. $NaCl(s)$ will dissolve.

        $BaCl_2(s) + K_2SO_4(aq) \longrightarrow BaSO_4(s) + 2 KCl(aq)$

    (b) Add $H_2O(l)$; $Na_2CO_3(s)$ will dissolve, but $MgCO_3(s)$ will not dissolve.

        $Na_2CO_3(s) \xrightarrow{\text{water}} 2 Na^+(aq) + CO_3{}^{2-}(aq)$

    (c) Add $KCl(aq)$; $AgCl(s)$ will form, while $KNO_3(s)$ will dissolve.

        $AgNO_3(s) + KCl(aq) \longrightarrow AgCl(s) + KNO_3(aq)$

    (d) Add H_2O. $Cu(NO_3)_2(s)$ will dissolve, while $PbSO_4(s)$ will not.

        $Cu(NO_3)_2(s) \xrightarrow{\text{water}} Cu^{2+}(aq) + 2 NO_3{}^-(aq)$

41.

	Mixture	Net ionic equation
(a)	$Sr(NO_3)_2(aq) + K_2SO_4(aq) \longrightarrow$	$Sr^{2+}(aq) + SO_4{}^{2-}(aq) \longrightarrow SrSO_4(s)$
(b)	$Mg(NO_3)_2(aq) + NaOH(aq) \longrightarrow$	$Mg^{2+}(aq) + 2 OH^-(aq) \longrightarrow Mg(OH)_2(s)$
(c)	$BaCl_2(aq) + K_2SO_4(aq) \longrightarrow$	$Ba^{2+}(aq) + SO_4{}^{2-}(aq) \longrightarrow BaSO_4(s)$ $KCl(aq)$ solution left.
(d)	$KCl(aq, \text{ from (c)}) + Ag_2SO_4(s) \longrightarrow$	$Ag_2SO_4(aq) + 2 Cl^-(aq) \longrightarrow 2 AgCl(s) + SO_4{}^{2-}(aq)$

        $BaCl_2(s) + Ag_2SO_4(s)$ will simply give a mixed precipitate: $AgCl(s) + BaSO_4(s)$

Oxidation–Reduction (Redox) Equations

42. (a) In this reaction, iron is reduced from $Fe^{3+}(aq)$ to $Fe^{2+}(aq)$ *and* manganese is reduced from a +7 O.S. in $MnO_4{}^-(aq)$ to a +2 O.S. in $Mn^{2+}(aq)$. Thus, there are two reductions and no oxidation, an impossibility.

(b) In this reaction, chlorine is oxidized from an O.S. of 0 in $Cl_2(aq)$ to an O.S. of +1 in $ClO^-(aq)$ *and* oxygen is oxidized from an O.S. of –1 in $H_2O_2(aq)$ to an O.S. of 0 in $O_2(g)$. There thus are two oxidations and no reduction, also an impossibility.

43. (a) Oxidation: $2\ I^-(aq) \longrightarrow I_2(s) + 2\ e^-$ $\times 5$

Reduction: $MnO_4^-(aq) + 8\ H^+(aq) + 5\ e^- \longrightarrow Mn^{2+}(aq) + 4\ H_2O(l)$ $\times 2$

Net: $10\ I^-(aq) + 2\ MnO_4^-(aq) + 16\ H^+(aq) \longrightarrow 5\ I_2(s) + 2\ Mn^{2+}(aq) + 8\ H_2O(l)$

(b) Oxidation: $I^-(aq) + 3\ H_2O(l) \longrightarrow IO_3^-(aq) + 6\ H^+(aq) + 6\ e^-$ $\times 1$

Reduction: $Cl_2(g) + 2\ e^- \longrightarrow 2\ Cl^-(aq)$ $\times 3$

Net: $I^-(aq) + 3\ Cl_2(aq) + 3\ H_2O(l) \longrightarrow IO_3^-(aq) + 6\ H^+(aq) + 6\ Cl^-(aq)$

(c) Oxidation: $N_2H_4(l) \longrightarrow N_2(g) + 4\ H^+(aq) + 4\ e^-$ $\times 3$

Reduction: $BrO_3^-(aq) + 6\ H^+(aq) + 6\ e^- \longrightarrow Br^-(aq) + 3\ H_2O(l)$ $\times 2$

Net: $3\ N_2H_4(l) + 2\ BrO_3^-(aq) \longrightarrow 3\ N_2(g) + 2\ Br^-(aq) + 6\ H_2O(l)$

(d) Oxidation: $Fe^{2+}(aq) \longrightarrow Fe^{3+}(aq) + e^-$ $\times 1$

Reduction: $VO_4^{3-}(aq) + 6\ H^+(aq) + e^- \longrightarrow VO^{2+}(aq) + 3\ H_2O(l)$ $\times 1$

Net: $Fe^{2+}(aq) + VO_4^{3-}(aq) + 6\ H^+(aq) \longrightarrow Fe^{3+}(aq) + VO^{2+}(aq) + 3\ H_2O(l)$

(e) Oxidation: $UO^{2+}(aq) + H_2O(l) \longrightarrow UO_2^{2+}(aq) + 2\ H^+(aq) + 2\ e^-$ $\times 3$

Reduction: $NO_3^-(aq) + 4\ H^+(aq) + 3\ e^- \longrightarrow NO(g) + 2\ H_2O(l)$ $\times 2$

Net: $3\ UO^{2+}(aq) + 2\ NO_3^-(aq) + 2\ H^+(aq) \longrightarrow 3\ UO_2^{2+}(aq) + 2\ NO(g) + 9\ H_2O(l)$

44. (a) Oxidation: $CN^-(aq) + 2\ OH^-(aq) \longrightarrow CNO^-(aq) + H_2O + 2\ e^-$ $\times 3$

Reduction: $MnO_4^-(aq) + 2\ H_2O(l) + 3\ e^- \longrightarrow MnO_2(s) + 4\ OH^-(aq)$ $\times 2$

Net: $3\ CN^-(aq) + 2\ MnO_4^-(aq) + H_2O(l) \longrightarrow 3\ CNO^-(aq) + 2\ MnO_2(s) + 2\ OH^-(aq)$

(b) Oxidation: $N_2H_4(l) + 4\ OH^-(aq) \longrightarrow N_2(g) + 4\ H_2O + 4\ e^-$ $\times 1$

Reduction: $[Fe(CN)_6]^{3-}(aq) + e^- \longrightarrow [Fe(CN)_6]^{4-}(aq)$ $\times 4$

Net: $4\ [Fe(CN)_6]^{3-}(aq) + N_2H_4(l) + 4\ OH^-(aq) \longrightarrow 4\ [Fe(CN)_6]^{4-}(aq) + N_2(g) + 4\ H_2O(l)$

(c) Oxidation: $Fe(OH)_2(s) + OH^-(aq) \longrightarrow Fe(OH)_3(s) + e^-$ $\times 4$

Reduction: $O_2(g) + 2\ H_2O(l) + 4\ e^- \longrightarrow 4\ OH^-(aq)$ $\times 1$

Net: $4\ Fe(OH)_2(s) + O_2(g) + 2\ H_2O(l) \longrightarrow 4\ Fe(OH)_3(s)$

(d) Oxidation: $C_2H_5OH(aq) + 5\ OH^-(aq) \longrightarrow C_2H_3O_2^-(aq) + 4\ H_2O(l) + 4\ e^-$ $\times 3$

Reduction: $MnO_4^-(aq) + 2\ H_2O(l) + 3\ e^- \longrightarrow MnO_2(s) + 4\ OH^-(aq)$ $\times 4$

Net: $3\ C_2H_5OH(aq) + 4\ MnO_4^-(aq) \longrightarrow 3\ C_2H_3O_2^-(aq) + 4\ MnO_2(s) + OH^-(aq) + 4\ H_2O(l)$

45. (a) Oxidation: $Cl_2(g) + 2\ e^- \longrightarrow 2\ Cl^-(aq)$ $\times 5$

Reduction: $Cl_2(g) + 12\ OH^-(aq) \longrightarrow 2\ ClO_3^-(aq) + 6\ H_2O + 10\ e^-$ $\times 1$

Net: $6\ Cl_2(g) + 12\ OH^-(aq) \longrightarrow 10\ Cl^-(aq) + 2\ ClO_3^-(aq) + 6\ H_2O$

Or: $3\ Cl_2(g) + 6\ OH^-(aq) \longrightarrow 5\ Cl^-(aq) + ClO_3^-(aq) + 3\ H_2O$

(b) Oxidation: $S_2O_4^{2-}(aq) + 2\ H_2O \longrightarrow 2\ HSO_3^-(aq) + 2\ H^+(aq) + 2\ e^-$

Reduction: $S_2O_4^{2-}(aq) + 2\ H^+(aq) + 2\ e^- \longrightarrow S_2O_3^{2-}(aq) + H_2O$

Net: $2\ S_2O_4^{2-}(aq) + H_2O \longrightarrow 2\ HSO_3^-(aq) + S_2O_3^{2-}(aq)$

(c) Oxidation: $MnO_4^{2-}(aq) \longrightarrow MnO_4^-(aq) + e^-$ $\times 2$

Reduction: $MnO_4^{2-}(aq) + 2\ H_2O + 2\ e^- \longrightarrow MnO_2(s) + 4\ OH^-(aq)$

Net: $3\ MnO_4^{2-}(aq) + 2\ H_2O \longrightarrow 2\ MnO_4^-(aq) + MnO_2(s) + 4\ OH^-(aq)$

(d) Oxidation: $P_4(s) + 8\ OH^-(aq) \longrightarrow 4\ H_2PO_2^-(aq) + 4\ e^-$ $\times 3$

Reduction: $P_4(s) + 12 H_2O + 12 e^- \longrightarrow 4 PH_3(g) + 12 OH^-(aq)$

Net: $\quad 4 P_4(s) + 12 OH^-(aq) + 12 H_2O \longrightarrow 12 H_2PO_2^-(aq) + 4 PH_3(g)$

46. **(a)** Oxidation: $P_4(s) + 16 H_2O \longrightarrow 4 H_2PO_4^-(aq) + 24 H^+(aq) + 20 e^- \qquad \times 3$

Reduction: $NO_3^-(aq) + 4 H^+(aq) + 3 e^- \longrightarrow NO(g) + 2 H_2O \qquad \times 20$

Net: $\quad 3 P_4(s) + 20 NO_3^-(aq) + 8 H_2O + 8 H^+(aq) \longrightarrow 12 H_2PO_4^-(aq) + 20 NO(g)$

(b) Oxidation: $S_2O_3^{2-}(aq) + 5 H_2O(l) \longrightarrow 2 SO_4^{2-}(aq) + 10 H^+(aq) + 8 e^- \quad \times 5$

Reduction: $MnO_4^-(aq) + 8 H^+(aq) + 5 e^- \longrightarrow Mn^{2+}(aq) + 4 H_2O \qquad \times 8$

Net: $\quad 5 S_2O_3^{2-}(aq) + 8 MnO_4^-(aq) + 14 H^+(aq) \longrightarrow 10 SO_4^{2-}(aq) + 8 Mn^{2+}(aq) + 7 H_2O(l)$

(c) Oxidation: $2 HS^-(aq) + 3 H_2O(l) \longrightarrow S_2O_3^{2-}(aq) + 8 H^+(aq) + 8 e^- \qquad \times 1$

Reduction: $2 HSO_3^-(aq) + 4 H^+(aq) + 4 e^- \longrightarrow S_2O_3^{2-}(aq) + 3 H_2O \qquad \times 2$

Net: $\quad 2 HS^-(aq) + 4 HSO_3^-(aq) \longrightarrow 3 S_2O_3^{2-}(aq) + 3 H_2O(l)$

(d) Oxidation: $2 NH_3OH^+(aq) \longrightarrow N_2O(g) + H_2O(l) + 6 H^+(aq) + 4 e^- \qquad \times 1$

Reduction: $Fe^{3+}(aq) + e^- \longrightarrow Fe^{2+}(aq) \qquad\qquad\qquad\qquad\qquad \times 4$

Net: $\quad 4 Fe^{3+}(aq) + 2 NH_3OH^+(aq) \longrightarrow 4 Fe^{2+}(aq) + N_2O(g) + H_2O(l) + 6 H^+(aq)$

(e) Oxidation: $4 OH^-(aq) \longrightarrow O_2(g) + 2 H_2O(l) + 4 e^- \qquad \times 3$

Reduction: $O_2^-(aq) + 2 H_2O(l) + 3 e^- \longrightarrow 4 OH^-(aq) \qquad \times 4$

Net: $\quad 4 O_2^-(aq) + 2 H_2O(l) \longrightarrow 3 O_2(g) + 4 OH^-(aq)$

47. **(a)** Oxidation: $S_2O_3^{2-}(aq) + 5 H_2O \longrightarrow 2 SO_4^{2-}(aq) + 10 H^+(aq) + 8 e^-$

Reduction: $Cl_2(g) + 2 e^- \longrightarrow 2 Cl^-(aq) \qquad\qquad\qquad\qquad \times 4$

Net: $\quad S_2O_3^{2-}(aq) + 5 H_2O(l) + 4 Cl_2(g) \longrightarrow 2 SO_4^{2-}(aq) + 8 Cl^-(aq) + 10 H^+(aq)$

(b) Oxidation: $NO_2^-(aq) + H_2O \longrightarrow NO_3^-(aq) + 2 H^+(aq) + 2 e^- \qquad \times 5$

Reduction: $MnO_4^-(aq) + 8 H^+(aq) + 5 e^- \longrightarrow Mn^{2+}(aq) + 4 H_2O \qquad \times 2$

Net: $\quad 2 MnO_4^-(aq) + 6 H^+(aq) + 5 NO_2^-(aq) \longrightarrow 2 Mn^{2+}(aq) + 5 NO_3^-(aq) + 3 H_2O(l)$

(c) Oxidation: $S_8(s) + 24 OH^-(aq) \longrightarrow 4 S_2O_3^{2-}(aq) + 12 H_2O(l) + 16 e^-$

Reduction: $S_8(s) + 16 e^- \longrightarrow 8 S^{2-}(aq)$

Net: $\quad 2 S_8(s) + 24 OH^-(aq) \longrightarrow 8 S^{2-}(aq) + 4 S_2O_3^{2-}(aq) + 12 H_2O(l)$

(d) Oxidation: $Fe_2S_3(s) + 6 OH^-(aq) \longrightarrow 2 Fe(OH)_3(s) + 3 S(s) + 6 e^- \qquad \times 2$

Reduction: $O_2(g) + 2 H_2O(l) + 4 e^- \longrightarrow 4 OH^-(aq) \qquad\qquad\qquad \times 3$

Net: $\quad 2 Fe_2S_3(s) + 3 O_2(g) + 6 H_2O(l) \longrightarrow 4 Fe(OH)_3(s) + 6 S(s)$

(e) Oxidation: $\quad As_2S_3(s) + 40 OH^-(aq) \longrightarrow 2 AsO_4^{3-}(aq) + 3 SO_4^{2-}(aq) + 20 H_2O + 28 e^-$

Reduction: $\quad H_2O_2(aq) + 2 e^- \longrightarrow 2 OH^-(aq) \qquad\qquad\qquad \times 14$

Net: $\quad As_2S_3(s) + 12 OH^-(aq) + 14 H_2O_2(aq) \longrightarrow 2 AsO_4^{3-}(aq) + 3 SO_4^{2-}(aq) + 20 H_2O$

48. **(a)** Oxidation: $NO_2^-(aq) + H_2O(l) \longrightarrow NO_3^-(aq) + 2 H^+(aq) + 2 e^- \qquad \times 5$

Reduction: $MnO_4^-(aq) + 8 H^+(aq) + 5 e^- \longrightarrow Mn^{2+}(aq) + 4 H_2O(l) \qquad \times 2$

Net: $\quad 5 NO_2^-(aq) + 2 MnO_4^-(aq) + 6 H^+(aq) \longrightarrow 5 NO_3^-(aq) + 2 Mn^{2+}(aq) + 3 H_2O(l)$

(b) Oxidation: $H_2S(aq) \longrightarrow S(s) + 2 H^+(aq) + 2 e^- \qquad\qquad \times 2$

Reduction: $SO_2(aq) + 4 H^+(aq) + 4 e^- \longrightarrow S(s) + 2 H_2O(l) \quad \times 1$

Net: $\quad 2 H_2S(aq) + SO_2(aq) \longrightarrow 3 S(s) + 2 H_2O(l)$

(c) $2 Na(s) + 2 HI(aq) \longrightarrow 2 NaI(aq) + H_2(g)$

49. Each of the reactions is treated as if it takes place in acidic aqueous solution, for the purpose of balancing its redox equation.

(**a**) Oxidation: $H_2(g) \longrightarrow 2\ H^+(aq) + 2\ e^-$ $\qquad\qquad\qquad\qquad$ $\times\ 5$

$\qquad$ Reduction: $2\ NO(g) + 10\ H^+(aq) + 10\ e^- \longrightarrow 2\ NH_3(g) + 2\ H_2O$

$\qquad$ Net: $\qquad 5\ H_2(g) + 2\ NO(g) \longrightarrow 2\ NH_3(g) + 2\ H_2O(g)$

(**b**) Oxidation: $2\ H_2O \longrightarrow O_2(g) + 4\ H^+(aq) + 4\ e^-$

$\qquad$ Reduction: $NO_3^- + 2\ H^+(aq) + e^- \longrightarrow NO_2(g) + H_2O$ $\qquad\qquad$ $\times\ 4$

$\qquad$ Net: $\qquad 4\ NO_3^- + 4\ H^+(aq) \longrightarrow 4\ NO_2(g) + O_2(g) + 2\ H_2O$

$\qquad$ Add $2\ O^{2-}$: $4\ NO_3^- + \cancel{2\ H_2O} \longrightarrow 2\ O^{2-} + 4\ NO_2(g) + O_2(g) + \cancel{2\ H_2O}$

(**c**) Oxidation: $2\ NH_4^+ \longrightarrow N_2(g) + 8\ H^+(aq) + 6\ e^-$

$\qquad$ Reduction: $Cr_2O_7^{2-} + 8\ H^+(aq) + 6\ e^- \longrightarrow Cr_2O_3(s) + 4\ H_2O$

$\qquad$ Net: $\qquad (NH_4)_2Cr_2O_7(s) \longrightarrow N_2(g) + Cr_2O_3(s) + 4\ H_2O$

Oxidizing and Reducing Agents

50. The oxidizing agents experience a decrease in the oxidation state of one of their elements, while the reducing agents experience an increase in the oxidation state of one of their elements.

(**a**) $SO_3^{2-}(aq)$ is the reducing agent; the O.S. of S = +4 in SO_3^{2-} and = +6 in SO_4^{2-}.
$\qquad$ $MnO_4^-(aq)$ is the oxidizing agent; the O.S. of Mn = +7 in MnO_4^- and = +2 in Mn^{2+}.

(**b**) $H_2(g)$ is the reducing agent; the O.S. of H = 0 in $H_2(g)$ and = +1 in $H_2O(g)$.
$\qquad$ $NO_2(g)$ is the oxidizing agent; the O.S. of N = +4 in $NO_2(g)$ and = −3 in NH_3

(**c**) $[Fe(CN)_6]^{4-}(aq)$ is the reducing agent; the O.S. of Fe = +2 in $[Fe(CN)_6]^{4-}$ and = +3 in $[Fe(CN)_6]^{3-}$
$\qquad$ $H_2O_2(aq)$ is the oxidizing agent; the O.S. of O = −1 in H_2O_2 and = −2 in H_2O.

51. (**a**) $2\ S_2O_3^{2-}(aq) + I_2(s) \longrightarrow S_4O_6^{2-}(aq) + 2\ I^-(aq)$

$\qquad$ (**b**) $S_2O_3^{2-}(aq) + 4\ Cl_2(g) + 5\ H_2O(l) \longrightarrow 2\ HSO_4^-(aq) + 8\ Cl^-(aq) + 8\ H^+(aq)$

$\qquad$ (**c**) $S_2O_3^{2-}(aq) + 4\ OCl^-(aq) + 2\ OH^-(aq) \longrightarrow 2\ SO_4^{2-}(aq) + 4\ Cl^-(aq) + H_2O(l)$

Neutralization and Acid-Base Titrations

52. $NaOH(aq) + HCl(aq) \longrightarrow NaCl(aq) + H_2O$ is the titration reaction.

$$[NaOH] = \frac{0.02627\ L \times \dfrac{0.1107\ mol\ HCl}{1\ L\ soln} \times \dfrac{1\ mol\ NaOH}{1\ mol\ HCl}}{0.02500\ L\ sample} = 0.1163\ M$$

53. (**a**) $[NH_3] = \dfrac{31.08\ mL\ acid \times \dfrac{0.9928\ mmol\ HCl}{1\ mL\ acid} \times \dfrac{1\ mmol\ H^+}{1\ mmol\ HCl} \times \dfrac{1\ mmol\ NH_3}{1\ mmol\ H^+}}{5.00\ mL\ sample}$

$\qquad\qquad = 6.17\ M\ NH_3$

(**b**) % $NH_3 = \dfrac{6.17\ mmol\ NH_3}{1.00\ mL\ soln} \times \dfrac{1\ mL\ soln}{0.96\ g} \times \dfrac{17.0\ mg\ NH_3}{1\ mmol\ NH_3} \times \dfrac{1\ g}{1000\ mg} \times 100\% = 11\%\ NH_3$

54. Titration reaction: $\qquad Ba(OH)_2(aq) + 2\ HNO_3(aq) \longrightarrow Ba(NO_3)_2(aq) + 2\ H_2O(l)$

$\qquad$ volume base = $50.00\ mL\ acid \times \dfrac{0.0526\ mol\ HNO_3}{1\ ml\ acid} \times \dfrac{1\ mmol\ Ba(OH)_2}{2\ mmol\ HNO_3} \times \dfrac{1\ mL\ base}{0.0884\ mmol\ Ba(OH)_2}$

$\qquad\qquad = 14.9\ mL\ base$

55. The mass of acetylsalicylic acid can be converted to the amount of NaOH, in mmols, that will react with it.

$\qquad$ $[NaOH] = \dfrac{0.32\ g\ HC_9H_7O_4}{23\ mL\ NaOH(aq)} \times \dfrac{1\ mol\ HC_9H_7O_4}{180.1\ g\ HC_9H_7O_4} \times \dfrac{1\ mol\ NaOH}{1\ mol\ HC_9H_7O_4} \times \dfrac{1000\ mmol\ NaOH}{1\ mol\ NaOH}$

$\qquad\qquad = 0.077\ M\ NaOH$

56. (**a**) vol conc = $20.0\ L \times \dfrac{0.25\ mol\ HCl}{1\ L\ soln} \times \dfrac{36.5\ g\ HCl}{1\ mol\ HCl} \times \dfrac{100\ g\ conc}{38\ g\ HCl} \times \dfrac{1\ mL}{1.19\ g\ conc}$

$\qquad\qquad = 4.0 \times 10^2\ mL\ conc.\ acid$

(**b**) The titration reaction is $HCl(aq) + NaOH(aq) \longrightarrow NaCl(aq) + H_2O(l)$

$$[HCl] = \frac{29.87 \text{ mL base} \times \dfrac{0.2029 \text{ mmol NaOH}}{1 \text{ mL base}} \times \dfrac{1 \text{ mmol HCl}}{1 \text{ mmol NaOH}}}{25.00 \text{ mL acid}} = 0.2424 \text{ M HCl}$$

(c) First of all the volume of the dilute solution (20 L) is known to a precision of at best two significant figures. Secondly, HCl is somewhat volatile (we can smell its odor above the solution) and some may be lost during the process of producing the solution.

57. The equation for the reaction is $HNO_3(aq) + KOH(aq) \longrightarrow KNO_3(aq) + H_2O(l)$ This equation shows that equal numbers of moles are needed for a complete reaction. We compute the amount of each reagent.

no. mmol HNO_3 = 25.00 mL acid $\times \dfrac{0.144 \text{ mmol } HNO_3}{1 \text{ mL acid}}$ = 3.60 mmol HNO_3

no. mmol KOH = 10.00 mL acid $\times \dfrac{0.408 \text{ mmol KOH}}{1 \text{ mL base}}$ = 4.08 mmol KOH

There is more base present than acid. Thus, the resulting solution is alkaline (basic).

58. volume base = 5.00 mL vinegar $\times \dfrac{1.01 \text{ g vinegar}}{1 \text{ mL}} \times \dfrac{4.0 \text{ g } HC_2H_3O_2}{100.0 \text{ g vinegar}} \times \dfrac{1 \text{ mol } HC_2H_3O_2}{60.0 \text{ g } HC_2H_3O_2}$

$\times \dfrac{1 \text{ mol NaOH}}{1 \text{ mol } HC_2H_3O_2} \times \dfrac{1 \text{ L base}}{0.1000 \text{ mol NaOH}} \times \dfrac{1000 \text{ mL}}{1 \text{ L}}$ = 34 mL base

59. The titration reaction is $2 NaOH(aq) + H_2SO_4(aq) \longrightarrow Na_2SO_4(aq) + 2 H_2O$
It is most convenient to consider molarity as millimoles per milliliter when solving this problem.

$$[H_2SO_4] = \frac{46.40 \text{ mL base} \times \dfrac{0.875 \text{ mmol NaOH}}{1 \text{ mL base}} \times \dfrac{1 \text{ mmol } H_2SO_4}{2 \text{ mmol NaOH}}}{5.00 \text{ mL battery acid}} = 4.06 \text{ M}$$

The battery acid is *not* sufficiently concentrated.

Stoichiometry of Oxidation-Reduction Equations

60. $[KMnO_4] = \dfrac{0.1156 \text{ g } As_2O_3 \times \dfrac{1 \text{ mol } As_2O_3}{197.84 \text{ g } As_2O_3} \times \dfrac{4 \text{ mol } MnO_4^-}{5 \text{ mol } As_2O_3} \times \dfrac{1 \text{ mol } KMnO_4}{1 \text{ mol } MnO_4^-}}{27.08 \text{ mL} \times \dfrac{1 \text{ L}}{1000 \text{ mL}}}$

= 0.01726 M $KMnO_4$

61. The balanced equation: $5 SO_3^{2-}(aq) + 2 MnO_4^-(aq) + 6 H^+(aq) \longrightarrow 5 SO_4^{2-}(aq) + 2 Mn^{2+}(aq) + 3 H_2O(l)$

$[SO_3^{2-}] = \dfrac{34.08 \text{ mL} \times \dfrac{0.01964 \text{ mmol } KMnO_4}{1 \text{ mL soln}} \times \dfrac{1 \text{ mmol } MnO_4^-}{1 \text{ mmol } KMnO_4} \times \dfrac{5 \text{ mmol } SO_3^{2-}}{2 \text{ } MnO_4^-}}{25.00 \text{ mL } SO_3^{2-} \text{ soln}}$

= 0.06693 M SO_3^{2-}

62. First determine the mass of Fe, then the percentage of iron in the ore.

mass Fe = 29.43 mL $\times \dfrac{1 \text{ L}}{1000 \text{ mL}} \times \dfrac{0.04212 \text{ mol } Cr_2O_7^{2-}}{1 \text{ L soln}} \times \dfrac{6 \text{ mol } Fe^{2+}}{1 \text{ mol } Cr_2O_7^{2-}} \times \dfrac{55.85 \text{ g Fe}}{1 \text{ mol } Fe^{2+}}$

= 0.4154 g Fe

% Fe = $\dfrac{0.4154 \text{ g Fe}}{0.8765 \text{ g ore}} \times 100\%$ = 47.39 g Fe

63. First balance the equation.

Oxidation: $Mn^{2+}(aq) + 4 OH^-(aq) \longrightarrow MnO_2(s) + 2 H_2O(l) + 2 e^-$ $\times 3$
Reduction: $MnO_4^-(aq) + 2 H_2O(aq) + 3 e^- \longrightarrow MnO_2(s) + 4 OH^-(aq)$ $\times 2$

Net: $3 Mn^{2+}(aq) + 2 MnO_4^-(aq) + 4 OH^-(aq) \longrightarrow 5 MnO_2(s) + 2 H_2O(l)$

$[Mn^{2+}] = \dfrac{34.77 \text{ mL titrant} \times \dfrac{0.05876 \text{ mmol } MnO_4^-}{1 \text{ mL titrant}} \times \dfrac{3 \text{ mmol } Mn^{2+}}{2 \text{ mmol } MnO_4^-}}{25.00 \text{ mL soln}}$ = 0.1226 M Mn^{2+}

64. Oxidation: $C_2O_4^{2-}(aq) \longrightarrow 2 CO_2(g) + 2 e^-$ $\times 5$
Reduction: $MnO_4^-(aq) + 8 H^+(aq) + 5 e^- \longrightarrow Mn^{2+}(aq) + 4 H_2O(l)$ $\times 2$

Net: $5 C_2O_4^{2-}(aq) + 2 MnO_4^-(aq) + 16 H^+(aq) \longrightarrow 10 CO_2(g) + 2 Mn^{2+}(aq) + 8 H_2O(l)$

$$\text{mass } Na_2C_2O_4 = 1.00 \text{ L satd soln} \times \frac{1000 \text{ mL}}{1 \text{ L}} \times \frac{25.8 \text{ mL}}{50.0 \text{ mL satd soln}} \times \frac{0.02140 \text{ mol KMnO}_4}{1000 \text{ mL}}$$

$$\times \frac{1 \text{ mol MnO}_4^-}{1 \text{ mol KMnO}_4} \times \frac{5 \text{ mol C}_2O_4^{2-}}{2 \text{ mol MnO}_4^-} \times \frac{1 \text{ mol Na}_2C_2O_4}{1 \text{ mol C}_2O_4^{2-}} \times \frac{134.0 \text{ g Na}_2C_2O_4}{1 \text{ mol Na}_2C_2O_4}$$

$$= 3.70 \text{ g Na}_2C_2O_4$$

65. Balanced equation: $\quad 3 \text{ S}_2O_4^{2-}(aq) + 2 \text{ CrO}_4^{2-}(aq) + 4 \text{ H}_2O(l) \longrightarrow 6 \text{ SO}_3^{2-}(aq) + 2 \text{ Cr(OH)}_3(s) + 2 \text{ H}^+(aq)$

(a) $\quad \text{mass Cr(OH)}_3 = 100. \text{ L soln} \times \dfrac{0.0108 \text{ mol CrO}_4^{2-}}{1 \text{ L soln}} \times \dfrac{2 \text{ mol Cr(OH)}_3}{2 \text{ mol CrO}_4^{2-}} \times \dfrac{103.0 \text{ g Cr(OH)}_3}{1 \text{ mol Cr(OH)}_3}$

$\qquad = 111 \text{ g Cr(OH)}_3$

(b) $\quad \text{mass Na}_2S_2O_4 = 100. \text{ L soln} \times \dfrac{0.0108 \text{ mol CrO}_4^{2-}}{1 \text{ L soln}} \times \dfrac{3 \text{ mol S}_2O_4^{2-}}{2 \text{ mol CrO}_4^{2-}} \times \dfrac{1 \text{ mol Na}_2S_2O_4}{1 \text{ mol S}_2O_4^{2-}}$

$\qquad \times \dfrac{174.1 \text{ g Na}_2S_2O_4}{1 \text{ mol Na}_2S_2O_4} = 282 \text{ g Na}_2S_2O_4$

6 GASES

REVIEW QUESTIONS

1. (a) "atm" is the abbreviation for "atmosphere," a unit of pressure equal to 760 mmHg, 101,325 Pa, or 14.7 lb/in.^{2}.
    (b) "STP" is the abbreviation for "standard temperature and pressure:" 25°C and 1 atm pressure.
    (c) R is the symbol for the ideal gas constant. It has a value of 0.08206 L atm mol^{-1} K^{-1}.
    (d) Partial pressure is the pressure that one of the gases in a mixture of gases would exert if it were present in the container by itself under the same conditions.
    (e) "u_{rms}" is the abbreviation for the root mean square speed of a number of moving objects, molecules in our considerations. It is the square root of the average of the squares of the speeds. It also is the median speed: half the molecules are travelling faster than this speed, and half are travelling slower.

2. (a) The absolute zero of temperature, –273.15°C, is the lowest temperature possible. All molecular motion ceases at this temperature.
    (b) A gas is collected over water by bubbling the gas into an updie-down container that is filled with water. The gas rises to the top of the countainer, displacing the water, and being trapped in the container.
    (c) Effusion of a gas refers to the very slow leakage of the gas out of a small hole in a container and into a vacuum.
    (d) The law of combining volumes states that gases, at the same temperature and pressures, react in volumes that are related as small whole numbers. These small whole numbers turn out to be the stoichioimetric coefficients.

3. (a) A barometer is a device used to measure absolute pressures; it measures the pressure between some gas and a vacuum. A manometer measures the difference in pressures between two gases.
    (b) Celsius and Kelvin temperatures both have the same size degree, large enough that 100 degrees span the temperature range from the boiling point to the freezing point of water. The zero point of Celsius temperature is the freezing point of water, that of Kelvin is –273.15°C.
    (c) The ideal gas equation relates the properties of pressure, volume, temperature, and amount for one gas. The general gas equation relates the initial and final values of these four properties, or the properties for two gases.
    (d) An ideal gas is one that obeys the ideal gas law. On a molecular level, the molecules of such a gas are dimensionless points that exert no forces of attraction or repulsion. The molecules of a real gas occupy some space (not much compared to the volume of a gas container) and exert weak attractions on each other. At room temperature and pressure the differences in the properties of real and ideal gases are almost insignificant.

4. (a) $P = 748 \text{ mmHg} \times \dfrac{1 \text{ atm}}{760 \text{ mmHg}} = 0.984 \text{ atm}$ (b) $P = 63.1 \text{ cm Hg} \times \dfrac{1 \text{ atm}}{76 \text{ cmHg}} = 0.830 \text{ atm}$
    (c) $P = 1044 \text{ torr} \times \dfrac{1 \text{ atm}}{760 \text{ torr}} = 1.374 \text{ atm}$ (d) $P = 52 \text{ psi} \times \dfrac{1 \text{ atm}}{14.7 \text{ psi}} = 3.5 \text{ atm}$

5. (a) $h = 1.38 \text{ atm} \times \dfrac{760 \text{ mmHg}}{1 \text{ atm}} = 1.05 \times 10^3 \text{ mmHg}$ (b) $h = 934 \text{ torr} = 934 \text{ mmHg}$
    (c) $h = 126 \text{ ft H}_2\text{O} \times \dfrac{12 \text{ in.}}{1 \text{ ft}} \times \dfrac{2.54 \text{ cm}}{1 \text{ in.}} \times \dfrac{10 \text{ mmH}_2\text{O}}{1 \text{ cm H}_2\text{O}} \times \dfrac{1 \text{ mmHg}}{13.6 \text{ mmH}_2\text{O}} \times \dfrac{1 \text{ mHg}}{1000 \text{ mmHg}} = 2.82 \text{ mHg}$
          $= 9.26 \text{ ft Hg}$

6. The atmospheric pressure is less than the pressure of the gas. The difference in pressures is
    $\Delta P = 18$ mmHg $- 6$ mmHg $= 12$ mmHg
    $P_{gas} = P_{atm} + \Delta P = 744$ mmHg $+ 12$ mmHg $= 756$ mmHg

7. (a) $V = 28.3\ \text{L} \times \dfrac{753\ \text{mmHg}}{335\ \text{mmHg}} = 63.6\ \text{L}$ (b) $V = 28.3\ \text{L} \times \dfrac{753\ \text{mmHg}}{2.07\ \text{atm} \times \dfrac{760\ \text{mmHg}}{1\ \text{atm}}} = 13.5\ \text{L}$

8. Apply Charles's law: $V = k\,T$. $T_i = 28 + 273 = 301$ K
    (a) $T = 273 + 77 = 350$ K $V = 733\ \text{mL} \times \dfrac{350\ \text{K}}{301\ \text{K}} = 852\ \text{mL}$

    (b) $T = 273 - 10 = 263$ K $V = 733\ \text{mL} \times \dfrac{263\ \text{K}}{301\ \text{K}} = 6.40 \times 10^2\ \text{mL}$

9. Apply Charles's law. $T_f = (21 + 273)\ \text{K} \times \dfrac{153\ \text{mL}}{68.2\ \text{mL}} = 6.60 \times 10^2\ \text{K} = 387°\text{C}$

10. $V = n \times V_m = \left(56.8\ \text{g C}_2\text{H}_2 \times \dfrac{1\ \text{mol C}_2\text{H}_2}{26.04\ \text{g}}\right) \times \dfrac{22.414\ \text{L at STP}}{1\ \text{mol}} = 48.9\ \text{L C}_2\text{H}_2$

11. $V = n \times V_m = \left(100.0\ \text{g Cl}_2 \times \dfrac{1\ \text{mol Cl}_2}{70.905\ \text{g}}\right) \times \dfrac{22.414\ \text{L at STP}}{1\ \text{mol}} = 31.61\ \text{L Cl}_2$

12. The gas with the greatest density at STP is the one with the highest mole weight: $\text{Cl}_2 = 70.9$ g/mol, $\text{SO}_3 = 80.1$ g/mol, $\text{N}_2\text{O} = 44.0$ g/mol; and $\text{PF}_3 = 88.0$ g/mol. Thus, PF_3 would have the highest STP density of the four gases listed.

13. Assume that the $\text{CO}_2(g)$ behaves ideally and use the ideal gas law: $PV = nRT$
    $V = \dfrac{nRT}{P} = \dfrac{\left(77.6\ \text{g} \times \dfrac{1\ \text{mol CO}_2}{44.01\ \text{g}}\right) 0.08206\ \dfrac{\text{L atm}}{\text{mol K}}\ (33 + 273.2)\text{K}}{728\ \text{mmHg} \times \dfrac{1\ \text{atm}}{760\ \text{mmHg}}} = 46.2\ \text{L}$

14. $P = \dfrac{nRT}{V} = \dfrac{\left(315\ \text{g} \times \dfrac{1\ \text{mol SO}_2}{64.07\ \text{g}}\right) 0.08206\ \dfrac{\text{L atm}}{\text{mol K}}\ (26 + 273.2)\text{K}}{36.7\ \text{L}} = 3.29\ \text{atm}$

15. Use the ideal gas law to determine the amount in moles of the given quantity of gas.
    $n = \dfrac{PV}{RT} = \dfrac{\left(743\ \text{mmHg} \times \dfrac{1\ \text{atm}}{760\ \text{mmHg}}\right)\left(355\ \text{mL} \times \dfrac{1\ \text{L}}{1000\ \text{mL}}\right)}{0.08206\ \dfrac{\text{L atm}}{\text{mol K}}\ (273.2 + 98.7)\text{K}} = 0.0114\ \text{mol gas}$

    $\mathfrak{M} = \dfrac{0.341\ \text{g}}{0.0114\ \text{mol}} = 29.9\ \text{g/mol}$

16. Density, d (g/L) = molar mass, $\mathfrak{M}$ (g/mol) ÷ molar volume, V/n (L/mol)
    $V/n = \dfrac{RT}{P} = \dfrac{0.08206\ \dfrac{\text{L atm}}{\text{mol K}} \times (26.8 + 273.2)\text{K}}{764\ \text{mmHg} \times \dfrac{1\ \text{atm}}{760\ \text{mmHg}}} = 24.5\ \text{L/mol}$ $d = \dfrac{44.01\ \text{g/mol}}{24.5\ \text{L/mol}} = 1.80\ \text{g/L}$

17. Each mole of gas occupies 22.414 L at STP.
    H_2 STP volume $= 1.00000\ \text{mol HCl} \times \dfrac{3\ \text{mol H}_2(g)}{6\ \text{mol HCl(aq)}} \times \dfrac{22.414\ \text{L H}_2(g)\ \text{at STP}}{1\ \text{mol H}_2} = 11.207\ \text{L H}_2(g)$

18. Determine first the amount of $\text{CO}_2(g)$ that can be removed. Then use the ideal gas law.
    no. mol $\text{CO}_2 = 1.00\ \text{kg LiOH} \times \dfrac{1000\ \text{g}}{1\ \text{kg}} \times \dfrac{1\ \text{mol LiOH}}{23.95\ \text{g LiOH}} \times \dfrac{1\ \text{mol CO}_2}{2\ \text{mol LiOH}} = 20.9\ \text{mol CO}_2$

    $V = \dfrac{nRT}{P} = \dfrac{20.9\ \text{mol} \times 0.08206\ \dfrac{\text{L atm}}{\text{mol K}} \times (24.8 + 273.2)\text{K}}{746\ \text{mmHg} \times \dfrac{1\ \text{atm}}{760\ \text{mmHg}}} = 521\ \text{L CO}_2(g)$

19. Determine the total amount of gas; then use the ideal gas law, assuming that the gases behave ideally.

$$\text{no. moles gas} = \left(14.8 \text{ g Ne} \times \frac{1 \text{ mol Ne}}{20.18 \text{ g Ne}}\right) + \left(37.6 \text{ g Ar} \times \frac{1 \text{ mol Ar}}{39.95 \text{ g Ar}}\right)$$

$$= 0.733 \text{ mol Ne} + 0.941 \text{ mol Ar} = 1.674 \text{ mol gas}$$

$$V = \frac{nRT}{P} = \frac{1.674 \text{ mol} \times 0.08206 \frac{\text{L atm}}{\text{mol K}} \times (31.2 + 273.2)\text{K}}{14.5 \text{ atm}} = 2.88 \text{ L gas}$$

20. 2.24 L $H_2(g)$ at STP is 0.100 mol $H_2(g)$. After 0.10 mol He is added, the container holds 0.20 mol gas.

$$V = \frac{nRT}{P} = \frac{0.20 \text{ mol} \times 0.0821 \frac{\text{L atm}}{\text{mol K}} (273 + 100)\text{K}}{1.00 \text{ atm}} = 6.1 \text{ L gas}$$

21. (a) The total pressure is the sum of the partial pressure of $O_2(g)$ and the vapor pressure of water.

$P_{total} = P_{O2} + P_{H2O} = 751 \text{ mmHg} = P_{O2} + 21 \text{ mmHg}$ $P_{O2} = (751 - 21) \text{ mmHg} = 730. \text{ mmHg}$

(b) The volume percent is equal to the pressure percent.

$$\%O_2(g) \text{ by volume} = \frac{V_{O2}}{V_{total}} \times 100\% = \frac{P_{O2}}{P_{total}} \times 100\% = \frac{730. \text{ mmHg of } O_2}{751 \text{ mmHg total}} \times 100\% = 97.2\%$$

22. $P_{oxygen} = P_{total} - P_{water} = 738 \text{ mmHg} - 19.8 \text{ mmHg} = 718 \text{ mmHg}$

Determine the mass of $O_2(g)$ collected by multiplying the amount of O_2 collected, in moles, by the molar mass of $O_2(g)$.

$$\text{Mass } O_2 = \frac{PV}{RT} \mathcal{M} = \frac{\left(718 \text{ mmHg} \times \frac{1 \text{ atm}}{760 \text{ mmHg}}\right)\left(91.2 \text{ mL} \times \frac{1 \text{ L}}{1000 \text{ mL}}\right)}{0.08206 \frac{\text{L atm}}{\text{mol K}} (22 + 273)\text{K}} \times \frac{32.00 \text{ g } O_2}{1 \text{ mol } O_2}$$

$$= 0.114 \text{ g } O_2$$

23. (1) is the true statement; average molecular kinetic energy depends only on absolute temperature, which is the same for these two gases. **(2)** is incorrect, since average molecular speed depends also on mole weight, which is different for these two gases. **(3)** also is incorrect, since volumes at the same temperature and pressure depend on the amount in moles, which is different for these two gases. And **(4)** is incorrect; the effusion rates at the same temperature and pressure depends inversely on the square root of the mole weights, which differ for these two gases.

24. $\dfrac{\text{rate Cl}_2}{\text{rate N}_2O} = \dfrac{\text{mol Cl}_2/\text{time Cl}_2}{\text{mol N}_2O/\text{time N}_2O} = \dfrac{\text{time N}_2O}{\text{time Cl}_2} = \dfrac{\text{time N}_2O}{28.6 \text{ s}} = \sqrt{\dfrac{44.0 \text{ g/mol N}_2O}{71.0 \text{ g/mol Cl}_2}} = 0.787$

time $N_2O = 0.787 \times 28.6 \text{ s} = 22.5 \text{ s}$

25. The best choice for ideal behavior is **(3)**, 200°C and 0.50 atm. Gases are most nearly ideal at high temperatures where the molecules are moving rapidly, and low pressures where the molecules are relatively far apart.

EXERCISES

Pressure and Its Measurement

26. (a) $P = 1127 \text{ mmHg} \times \dfrac{1 \text{ atm}}{760 \text{ mmHg}} = 1.483 \text{ atm}$ **(c)** $P = 231 \text{ kPa} \times \dfrac{1 \text{ atm}}{101.325 \text{ kPa}} = 2.28 \text{ atm}$

(b) $P = \dfrac{6.78 \text{ kg}}{1 \text{ cm}^2} \times \dfrac{2.205 \text{ lb}}{1 \text{ kg}} \times \left(\dfrac{2.54 \text{ cm}}{1 \text{ in}}\right)^2 \times \dfrac{1 \text{ atm}}{14.7 \text{ lb/in}^2} = 6.56 \text{ atm}$

(d) $P = 912 \text{ mb} \times \dfrac{1 \text{ atm}}{1013.25 \text{ mb}} = 0.900 \text{ atm}$ **(e)** $P = 2.35 \times 10^5 \text{ N/m}^2 \times \dfrac{1 \text{ atm}}{101,325 \text{ N/m}^2} = 2.32 \text{ atm}$

27. (a) We use: $h_{gly}d_{gly} = h_{CCl4}d_{CCl4}$ $h_{gly} = 2.82 \text{ m CCl}_4 \times \dfrac{1.59 \text{ g/cm}^3 \text{ CCl}_4}{1.26 \text{ g/cm}^3 \text{ glycerol}} = 3.56 \text{ m glycerol}$

(b) We use: $h_{bnz}d_{bnz} = h_{Hg}d_{Hg}$ $h_{bnz} = 2.09 \times 10^4 \text{ N/m}^2 \times \dfrac{1 \text{ atm}}{101,325 \text{ N/m}^2} \times \dfrac{0.760 \text{ m Hg}}{1 \text{ atm}}$

$$\times \dfrac{13.6 \text{ g/cm}^3 \text{ Hg}}{0.879 \text{ g/cm}^3 \text{ benzene}} = 2.42 \text{ m benzene}$$

(c) A 15.0-ft column will have a mass of 12.5 lb. (density = mass ÷ volume)

$$\text{density} = \frac{12.5 \text{ lb}}{1.00 \text{ in.} \times 1.00 \text{ in.} \times \left(15.0 \text{ ft} \times \frac{12 \text{ in.}}{1 \text{ ft}}\right)} = 0.0694 \text{ lb/in.}^3 \times \frac{454 \text{ g/lb}}{\left(\frac{2.54 \text{ cm}}{1 \text{ in}}\right)^3} = 1.92 \text{ g/cm}^3$$

28. The standard atmosphere is defined as the pressure exerted by a certain height (760 mm) of mercury. The density of mercury (13.5951 g/cm³) needs to be included in this definition and, because the density of mercury, like that of other liquids, varies somewhat with temperature, the pressure will vary if the density varies. The acceleration due to gravity (9.80665 m/s²) needs to be included because pressure is a *force* per unit area, not just a mass per unit area, and this force depends on the acceleration applied to the given mass: force = mass × acceleration.

29. The mercury level difference equals the difference in pressure between the container and the atmosphere. This mercury level difference is $\Delta P = 283 \text{ mmHg} - 38 \text{ mmHg} = 245 \text{ mmHg}$. Since the mercury level in the arm open to the atmosphere is higher than in the arm connected to the container, the pressure in the container is higher than atmospheric pressure. $P = \Delta P + P_{\text{atm}} = 245 \text{ mmHg} + 753.5 \text{ mmHg} = 999 \text{ mmHg}$

30. The gas inside the container is at a lower pressure than barometric pressure.

$$\Delta P = P_{\text{bar.}} - P_{\text{gas}} = 3.8 \text{ cm H}_2\text{O} \times \frac{10 \text{ mmH}_2\text{O}}{1 \text{ cm H}_2\text{O}} \times \frac{1 \text{ mmHg}}{13.6 \text{ mm H}_2\text{O}} = 2.8 \text{ mmHg}$$

$$P_{\text{gas}} = 753.5 \text{ mmHg} - 2.8 \text{ mmHg} = 750.7 \text{ mmHg}$$

The Simple Gas Laws

31. $V_f = V_i \times \dfrac{P_i}{P_f} = 909 \text{ mL} \times \dfrac{734 \text{ mmHg} \times \dfrac{1 \text{ atm}}{760 \text{ mmHg}}}{2.66 \text{ atm}} = 330. \text{ mL}$

32. $P_i = P_f \times \dfrac{V_f}{V_i} = 708 \text{ mmHg} \times \dfrac{1 \text{ atm}}{760 \text{ mmHg}} \times \dfrac{28.2 \text{ L} + 2195 \text{ L}}{28.2 \text{ L}} = 73.4 \text{ atm} \ (= 5.58 \times 10^4 \text{ mmHg})$

33. $T_f = T_i \times \dfrac{P_f}{P_i} = (24.1 + 273.2)\text{K} \times \dfrac{100 \text{ kPa} \times \dfrac{1 \text{ atm}}{101.325 \text{ kPa}}}{822 \text{ mmHg} \times \dfrac{1 \text{ atm}}{760 \text{ mmHg}}} = 271 \text{ K} = -2°\text{C}$

34. (a) $\text{mass} = 27.6 \text{ mL} \times \dfrac{1 \text{ L}}{1000 \text{ mL}} \times \dfrac{1 \text{ mol}}{22.414 \text{ L STP}} = 0.00123 \text{ mol PH}_3 \times \dfrac{34.0 \text{ g PH}_3}{1 \text{ mol PH}_3} \times \dfrac{1000 \text{ mg}}{1 \text{ g}}$

$$= 41.8 \text{ mg PH}_3$$

(b) $\text{molec. PH}_3 = 0.00123 \text{ mol PH}_3 \times \dfrac{6.022 \times 10^{23} \text{ molecules}}{1 \text{ mol PH}_3} = 7.41 \times 10^{20} \text{ molecules}$

35. We can obtain the ratio of the number of moles both before and after heating, using $V_f = V_i$, and $P_f = P_i$. This will be the same as the ratio of the masses before and after, because the mole weight of the gas is the same. $\quad \dfrac{n_f}{n_i} = \dfrac{P_f V_f / R T_f}{P_i V_i / R T_i} = \dfrac{T_i}{T_f} = \dfrac{(22 + 273)\text{K}}{(202 + 273)\text{K}} = 0.621 = \dfrac{m_f}{m_i} = \dfrac{m_f}{12.5 \text{ g}}$

$m_f = 0.621 \times 12.5 \text{ g} = 7.8 \text{ g}$ $\qquad$ Thus, mass that escapes = 12.5 g − 7.8 g = 4.7 g of gas

36. At the higher elevation of the mountains, the atmospheric pressure is lower than at the beach. However, the bag is leak proof; no gas escapes. Thus, the gas inside the bag expands in the lower pressure until the bag is filled to nearly bursting. (It would have been difficult to have predicted this result. The temperature in the mountains is also lower than at the beach. The lower temperature would *decrease* the pressure of the gas.)

37. We let P represent barometric pressure, and solve the Boyle's law expression below for P.

$P \times 58.0 \text{ cm}^3 = (P + 125 \text{ mmHg}) \times 49.6 \text{ mL}$ $\qquad\qquad$ $58.0\, P = 49.6\, P + 6.20 \times 10^3$

$P = \dfrac{6.20 \times 10^3}{58.0 - 49.6} = 7.4 \times 10^2 \text{ mmHg}$

38. $V_f = 4.12 \text{ L} + 35.1 \text{ L} = 39.1 \text{ L}$ $\quad$ Then apply Boyle's law.

$$P_f = P_i \times \frac{V_i}{V_f} = 4.71 \text{ atm} \times \frac{4.12 \text{ L}}{39.2 \text{ L}} = 0.495 \text{ atm}$$

Ideal Gas Equation

39. $P = \dfrac{nRT}{V} = \dfrac{\left(46.7 \text{ g O}_2 \times \dfrac{1 \text{ mol O}_2}{32.00 \text{ g O}_2}\right) \times \dfrac{0.08206 \text{ L atm}}{\text{mol K}} \times (35 + 273.2) \text{ K}}{14.7 \text{ L}} = 2.51 \text{ atm}$

40. $T = \dfrac{PV}{nR} = \dfrac{3.17 \text{ atm} \times 65.3 \text{ L}}{2.55 \text{ mol} \times 0.08206 \dfrac{\text{L atm}}{\text{mol K}}} = 989 \text{ K}$ $t(^\circ\text{C}) = 989 - 273 = 716^\circ\text{C}$

41. $\text{mass} = n \times \mathfrak{M} = \dfrac{PV}{RT} \mathfrak{M} = \dfrac{9.56 \text{ atm} \times 21.4 \text{ L} \times 83.80 \text{ g/mol}}{0.08206 \dfrac{\text{L atm}}{\text{mol K}} (30.5 + 273.2) \text{K}} = 688 \text{ g Kr}$

42. Since the number of moles of gas does not change, $\dfrac{P_i \times V_i}{T_i} = n R = \dfrac{P_f \times V_f}{T_f}$ is obtained from the ideal gas equation. This expression can be rearranged as follows.

$$V_f = \frac{V_i \times P_i \times T_f}{P_f \times T_i} = \frac{3.87 \text{ L} \times 743 \text{ mmHg} \times (273.2 + 24.4)\text{K}}{755 \text{ mmHg} \times (273.2 + 26.8)\text{K}} = 3.78 \text{ L}$$

43. First determine the mass of O_2 in the cylinder under the final conditions.

$$\text{mass O}_2 = n \times \mathfrak{M} = \frac{PV}{RT} \mathfrak{M} = \frac{1.24 \text{ atm} \times 34.0 \text{ L} \times 32.0 \text{ g/mol}}{0.08206 \dfrac{\text{L atm}}{\text{mol K}} (21 + 273)\text{K}} = 55.9 \text{ g O}_2$$

mass of O_2 to be released = 212 g − 55.9 g = 156 g O_2

44. We first compute the pressure at 35°C as a result of the additional gas. Of course, the gas pressure increases proportionally to the increase in the mass of gas, since the mass is proportional to the number of moles of gas. $P = \dfrac{12.3 \text{ g}}{10.0 \text{ g}} \times 762 \text{ mmHg} = 937 \text{ mmHg}$

Now we compute the pressure resulting from increasing the temperature.

$$P = \frac{(51 + 273) \text{ K}}{(35 + 273) \text{ K}} \times 937 \text{ mmHg} = 986 \text{ mmHg}$$

45. We determine the amount of N_2 under the initial and the final conditions.

$$P = 751 \text{ mmHg} \times \frac{1 \text{ atm}}{760 \text{ mmHg}} = 0.988 \text{ atm} \qquad T_1 = 21.5 + 273.2 = 294.7 \text{ K}$$

$$n_1 = \frac{PV}{RT} = \frac{0.988 \text{ atm} \times 1.98 \text{ L}}{0.08206 \text{ L atm mol}^{-1} \text{ K}^{-1} \times 294.7 \text{ K}} = 0.0809 \text{ mol N}_2$$

$$T_2 = 99.8 + 273.2 = 373.0 \text{ K} \qquad n_2 = \frac{PV}{RT} = \frac{0.988 \text{ atm} \times 1.98 \text{ L}}{0.08206 \dfrac{\text{L atm}}{\text{mol K}} \times 373.0 \text{ K}} = 0.0639 \text{ mol N}_2$$

mass of released N_2 = (0.0809 − 0.0639) mol $N_2 \times \dfrac{28.0 \text{ g N}_2}{1 \text{ mol N}_2} = 0.48 \text{ g N}_2$ released

Determining Molar Mass

46. We first determine the empirical formula of propylene.

no. mol C = $85.63 \text{ g C} \times \dfrac{1 \text{ mol C}}{12.01 \text{ g C}} = 7.130 \text{ mol C}$ $\div 7.130 \longrightarrow 1.000 \text{ mol C}$

no. mol H = $14.37 \text{ g H} \times \dfrac{1 \text{ mol H}}{1.008 \text{ g H}} = 14.26 \text{ mol H}$ $\div 7.130 \longrightarrow 2.000 \text{ mol H}$

Thus, the empirical formula is CH_2 and the empirical mole weight is 14.0 g/mol. The molecular weight of propylene is 42.08 g/mol, three times the empirical mole weight. Thus the molecular formula is C_3H_6, three times the empirical formula.

47. $\mathfrak{M} = \dfrac{mRT}{PV} = \dfrac{0.190 \text{ g} \times 0.08206 \dfrac{\text{L atm}}{\text{mol K}} \times (26 + 273)\text{K}}{\left(743 \text{ mmHg} \times \dfrac{1 \text{ atm}}{760 \text{ mmHg}}\right)\left(113 \text{ mL} \times \dfrac{1 \text{ L}}{1000 \text{ mL}}\right)} = 42.2 \text{ g/mol}$

The formula contains 3 atoms of carbon. (4 atoms of carbon gives a mole weight of 48—too high—and 2 C atoms gives a mole weight of 24—to low to be made up by adding H's.) To produce a mole weight of 42 with 3 carbons requires the inclusion of 6 atoms of H in the formula of the compound: C_3H_6.

48. We first determine the molar mass of the gas.

$T = 25.0 + 273.2 = 298.2 \text{ K}$ $\qquad\qquad$ $P = 745 \text{ mmHg} \times \dfrac{1 \text{ atm}}{760 \text{ mmHg}} = 0.980 \text{ atm}$

$\mathfrak{M} = \dfrac{mRT}{PV} = \dfrac{0.312 \text{ g} \times 0.08206 \text{ L atm mol}^{-1} \text{ K}^{-1} \times 298.2 \text{ K}}{0.980 \text{ atm} \times \left(185 \text{ mL} \times \dfrac{1 \text{ L}}{1000 \text{ mL}}\right)} = 42.1 \text{ g/mol}$

Then we determine the empirical formula of the gas, basing our calculations on a 100.0-g sample.

amount C $= 85.6 \text{ g C} \times \dfrac{1 \text{ mol C}}{12.01 \text{ g C}} = 7.13 \text{ mol C}$ $\qquad \div 7.13 \longrightarrow 1.00 \text{ mol C}$

amount H $= 14.4 \text{ g H} \times \dfrac{1 \text{ mol H}}{1.008 \text{ g H}} = 14.3 \text{ mol H}$ $\qquad \div 7.13 \longrightarrow 2.01 \text{ mol H}$

The empirical formula thus is CH_2, which has an empirical molar mass of 14.0 g/mol. Since this is one-third of the experimentally determined molar mass, the molecular formula is three times as great, or C_3H_6.

49. $\mathfrak{M} = \dfrac{mRT}{PV} = \dfrac{2.650 \text{ g} \times 0.0821 \dfrac{\text{L atm}}{\text{mol K}} \times (24.3 + 273.2)\text{K}}{\left(742.3 \text{ mmHg} \times \dfrac{1 \text{ atm}}{760 \text{ mmHg}}\right)\left(428 \text{ mL} \times \dfrac{1 \text{ L}}{1000 \text{ mL}}\right)} = 155 \text{ g/mol}$

no. mol C $= 15.5 \text{ g C} \times \dfrac{1 \text{ mol C}}{12.0 \text{ g C}} = 1.29 \text{ mol C}$ $\qquad \div 0.648 \longrightarrow 1.99 \text{ mol C}$

no. mol Cl $= 23.0 \text{ g Cl} \times \dfrac{1 \text{ mol Cl}}{35.5 \text{ g Cl}} = 0.648 \text{ mol Cl}$ $\qquad \div 0.648 \longrightarrow 1.00 \text{ mol Cl}$

no. mol F $= 61.5 \text{ g F} \times \dfrac{1 \text{ mol F}}{19.0 \text{ g F}} = 3.24 \text{ mol F}$ $\qquad \div 0.648 \longrightarrow 5.00 \text{ mol F}$

The empirical formula is C_2ClF_5 for which the empirical molar mass is 154.5 g/mol. Thus, C_2ClF_5 is also the molecular formula.

50. First, we obtain the volume of the container, then the mass of acetylene, and finally acetylene's molar mass.

volume $= (264.2931 \text{ g} - 56.1035 \text{ g}) \dfrac{1 \text{ mL}}{1.576 \text{ g}} = 132.1 \text{ mL}$

mass of acetylene $= 56.2445 \text{ g} - 56.1035 \text{ g} = 0.1410 \text{ g acetylene}$

$\mathfrak{M} = \dfrac{mRT}{PV} = \dfrac{0.1410 \text{ g} \times 0.082057 \dfrac{\text{L atm}}{\text{mol K}} \times (20.02 + 273.15)\text{K}}{\left(749.3 \text{ mmHg} \times \dfrac{1 \text{ atm}}{760 \text{ mmHg}}\right)\left(132.1 \text{ mL} \times \dfrac{1 \text{ L}}{1000 \text{ mL}}\right)} = 26.04 \text{ g/mol}$

Gas Densities

51. $d = \dfrac{\mathfrak{M}P}{RT}$ becomes $\quad P = \dfrac{dRT}{\mathfrak{M}} = \dfrac{1.45 \text{ g/L} \times 0.08206 \dfrac{\text{L atm}}{\text{mol K}} \times (25 + 273)\text{K}}{28.0 \text{ g/mol}} = 1.27 \text{ atm}$

52. $d = \dfrac{\mathfrak{M}P}{RT}$ becomes $\quad \mathfrak{M} = \dfrac{dRT}{P} = \dfrac{2.64 \text{ g/L} \times 0.0821 \dfrac{\text{L atm}}{\text{mol K}} \times (310 + 273)\text{K}}{775 \text{ mmHg} \times \dfrac{1 \text{ atm}}{760 \text{ mmHg}}} = 124 \text{ g/mol}$

Since the atomic mass of phosphorus is 31.0, the formula of phosphorus molecules in the vapor is P_4. (4 atoms/molecule $\times$ 31.0 = 124)

53. We first determine the molar mass of the gas, then its empirical formula. These two pieces of information are combined to obtain the molecular formula of the gas.

$$\mathcal{M} = \frac{dRT}{P} = \frac{2.35 \text{ g/L} \times 0.08206 \frac{\text{L atm}}{\text{mol K}} \times 298 \text{ K}}{752 \text{ mmHg} \times \frac{1 \text{ atm}}{760 \text{ mmHg}}} = 58.1 \text{ g/mol}$$

no. mol C = $82.7 \text{ g C} \times \frac{1 \text{ mol C}}{12.0 \text{ g C}} = 6.89 \text{ mol C} \div 6.89 \longrightarrow 1.00 \text{ mol C}$

no. mol H = $17.3 \text{ g H} \times \frac{1 \text{ mol H}}{1.01 \text{ g H}} = 17.1 \text{ mol H} \div 6.89 \longrightarrow 2.48 \text{ mol H}$

Multiply both of these mole numbers by 2 to obtain the empirical formula, C_2H_5, which has an empirical molar mass of 29.0 g/mol. Since the molar mass (calculated as 58.1 g/mol above) is twice this empirical molar mas, twice the empirical formula is the molecular formula: C_4H_{10}.

54. (a) $d = \frac{\mathcal{M}P}{RT} = \frac{28.96 \text{ g/mol} \times 1.00 \text{ atm}}{0.0821 \frac{\text{L atm}}{\text{mol K}} \times (273 + 25)\text{K}} = 1.18 \text{ g/L air}$

(b) $d = \frac{\mathcal{M}P}{RT} = \frac{44.01 \text{ g/mol N}_2\text{O} \times 1.00 \text{ atm}}{0.08206 \frac{\text{L atm}}{\text{mol K}} \times (273 + 25)\text{K}} = 1.80 \text{ g/L N}_2\text{O}$

Since this density is greater than air's, the balloon will not rise in air when filled with N_2O at 25°C.

(c) $d = \frac{\mathcal{M}P}{RT}$ becomes $T = \frac{\mathcal{M}P}{Rd} = \frac{44.01 \text{ g/mol} \times 1.00 \text{ atm}}{0.08206 \frac{\text{L atm}}{\text{mol K}} \times 1.18 \text{ g/L}} = 455 \text{ K} = 182°C$

Gases in Chemical Reactions

55. Balanced equation: $C_3H_8(g) + 5 O_2(g) \longrightarrow 3 CO_2(g) + 4 H_2O(l)$

Use the law of combining volumes. O_2 volume = $45.8 \text{ L C}_3\text{H}_8 \times \frac{5 \text{ L O}_2}{1 \text{ L C}_3\text{H}_8} = 229 \text{ L O}_2$

56. Determine the moles of $SO_2(g)$ produced and then use the ideal gas equation.

$3.5 \times 10^6 \text{ lb coal} \times \frac{2.12 \text{ lb S}}{100.00 \text{ lb coal}} \times \frac{454 \text{ g S}}{1 \text{ lb S}} \times \frac{1 \text{ mol S}}{32.1 \text{ g S}} \times \frac{1 \text{ mol SO}_2}{1 \text{ mol S}} = 1.0 \times 10^6 \text{ mol SO}_2$

$V = \frac{nRT}{P} = \frac{1.0 \times 10^6 \text{ mol SO}_2 \times 0.0821 \frac{\text{L atm}}{\text{mol K}} \times 298 \text{ K}}{738 \text{ mmHg} \times \frac{1 \text{ atm}}{760 \text{ mmHg}}} = 2.6 \times 10^7 \text{ L SO}_2$

57. We first use the law of combining volumes, and then the general gas law.

$3 CO(g) + 7 H_2(g) \longrightarrow C_3H_8(g) + 3 H_2O(l)$ $30.0 \text{ L CO(g)} \times \frac{7 \text{ L H}_2(g)}{3 \text{ L CO(g)}} = 70.0 \text{ L H}_2(g)$

volume $H_2(g) = 70.0 \text{ L H}_2(g) \times \frac{760 \text{ mmHg}}{745 \text{ mmHg}} \times \frac{(22 + 273)\text{K}}{273 \text{ K}} = 77.2 \text{ L H}_2(g)$

58. Determine the amount of O_2, and then the mass of $KClO_3$ that produced this amount of O_2.

no. mol $O_2 = \dfrac{\left(727 \text{ mmHg} \times \frac{1 \text{ atm}}{760 \text{ mmHg}}\right)\left(89.8 \text{ mL} \times \frac{1 \text{ L}}{1000 \text{ mL}}\right)}{0.08206 \frac{\text{L atm}}{\text{mol K}} \times (21.8 + 273.2)\text{K}} = 0.00355 \text{ mol O}_2$

mass $KClO_3 = 0.00355 \text{ mol O}_2 \times \frac{2 \text{ mol KClO}_3}{3 \text{ mol O}_2} \times \frac{122.6 \text{ g KClO}_3}{1 \text{ mol KClO}_3} = 0.290 \text{ g KClO}_3$

% $KClO_3 = \frac{0.290 \text{ g KClO}_3}{2.92 \text{ g sample}} \times 100\% = 9.93\% \text{ KClO}_3$

59. Determine the amount of O_2 liberated, and then its volume. $2 H_2O_2(aq) \longrightarrow 2 H_2O + O_2(g)$

no. mol $O_2 = 1.00 \text{ L soln} \times \frac{1000 \text{ mL}}{1 \text{ L}} \times \frac{1.11 \text{ g}}{1 \text{ mL}} \times \frac{0.300 \text{ g H}_2\text{O}_2}{1 \text{ g soln}} \times \frac{1 \text{ mol H}_2\text{O}_2}{34.0 \text{ g H}_2\text{O}_2} \times \frac{1 \text{ mol O}_2}{2 \text{ mol H}_2\text{O}_2}$

$= 4.90 \text{ mol O}_2$

$$V = \frac{4.90 \text{ mol O}_2 \times 0.08206 \frac{\text{L atm}}{\text{mol K}} \times (26 + 273)\text{K}}{746 \text{ mmHg} \times \frac{1 \text{ atm}}{760 \text{ mmHg}}} = 122 \text{ L O}_2$$

60. **(a)** volume $NH_3 = 313 \text{ L H}_2 \times \frac{2 \text{ L NH}_3}{3 \text{ L H}_2} = 209 \text{ L NH}_3$

(b) no. mol $NH_3 = \frac{525 \text{ atm} \times 313 \text{ L}}{0.08206 \frac{\text{L atm}}{\text{mol K}} \times (515 + 273)\text{K}} = 2.54 \times 10^3 \text{ mol H}_2 \times \frac{2 \text{ mol NH}_3}{3 \text{ mol H}_2}$

$$= 1.69 \times 10^3 \text{ mol NH}_3$$

$$V = \frac{1.69 \times 10^3 \text{ mol NH}_3 \times 0.08206 \frac{\text{L atm}}{\text{mol K}} \times 298\text{K}}{727 \text{ mmHg} \times \frac{1 \text{ atm}}{760 \text{ mmHg}}} = 4.32 \times 10^4 \text{ L NH}_3$$

61. **(a)** This is a limiting reagent problem. Thus, determine the amount of SO_2 produced by each reactant gas.

$$\text{no. mol SO}_2 = \frac{\left(735 \text{ mmHg} \times \frac{1 \text{ atm}}{760 \text{ mmHg}}\right) 1.50 \text{ L}}{0.08206 \frac{\text{L atm}}{\text{mol K}} (273.2 + 23.0)\text{K}} \times \frac{2 \text{ mol SO}_2}{2 \text{ mol H}_2\text{S}} = 0.0597 \text{ mol SO}_2$$

$$\text{no. mol SO}_2 = \frac{\left(751 \text{ mmHg} \times \frac{1 \text{ atm}}{760 \text{ mmHg}}\right) 4.45 \text{ L}}{0.08206 \frac{\text{L atm}}{\text{mol K}} (273.2 + 26.1)\text{K}} \times \frac{2 \text{ mol SO}_2}{3 \text{ mol O}_2} = 0.119 \text{ mol SO}_2$$

Thus, H_2S is the limiting reagent, and 0.0597 mol SO_2 is produced. The amount of O_2 originally present can be computed from the amount of SO_2 that it would have produced, 0.119 mol SO_2.

no. mol O_2 present initially $= 0.119 \text{ mol SO}_2 \times \frac{3 \text{ mol O}_2}{2 \text{ mol SO}_2} = 0.179 \text{ mol O}_2$ present initially

The amount of O_2 consumed can be computed from the amount of product SO_2 that actually was produced, 0.0597 mol SO_2.

no. mol O_2 consumed $= 0.0597 \text{ mol SO}_2 \times \frac{3 \text{ mol O}_2}{2 \text{ mol SO}_2} = 0.0896 \text{ mol O}_2$ consumed

The amount of O_2 remaining is the difference between these two computed amounts.

no. mol O_2 remaining $= 0.179 \text{ mol O}_2$ present initially $- 0.0896 \text{ mol O}_2$ consumed $= 0.089 \text{ mol O}_2$ left

(b) The total amount of gas present after reaction is 0.0597 mol SO_2 + 0.0597 mol H_2O + 0.089 mol O_2 = 0.208 mol gas

$$V = \frac{0.208 \text{ mol} \times 0.08206 \frac{\text{L atm}}{\text{mol K}} \times (120.0 + 273.2)\text{K}}{748 \text{ mmHg} \times \frac{1 \text{ atm}}{760 \text{ mmHg}}} = 6.82 \text{ L gas}$$

Mixtures of Gases

62. The two pressures are related as are the number of moles of $N_2(g)$ and the total number of moles.

amount $N_2 = \frac{PV}{RT} = \frac{31.8 \text{ atm} \times 48.5 \text{ L}}{0.08206 \frac{\text{L atm}}{\text{mol K}} \times (23 + 273)\text{K}} = 63.5 \text{ mol N}_2$

total amount of gas $= 63.5 \text{ mol N}_2 \times \frac{75.0 \text{ atm}}{31.8 \text{ atm}} = 150. \text{ mol gas}$

mass Ne $= (150. \text{ mol total} - 63.5 \text{ mol N}_2) \times \frac{20.18 \text{ g Ne}}{1 \text{ mol Ne}} = 1.7 \times 10^3 \text{ g Ne}$

63. Solve a Boyle's law problem for each gas and add the resulting partial pressures.

$P_{H_2} = 756 \text{ mm Hg} \times \frac{1.37 \text{ L}}{4.35 \text{ L}} = 238 \text{ mmHg}$ $P_{He} = 722 \text{ mmHg} \times \frac{2.98 \text{ L}}{4.35 \text{ L}} = 495 \text{ mmHg}$

$P_{total} = P_{H_2} + P_{He} = 238 \text{ mmHg} + 495 \text{ mmHg} = 733 \text{ mmHg}$

64. The volume of the original mixture equals the volume of the 10.0 g of added H_2 ($= V_2$)

$$\left(\left(4.0 \text{ g } H_2 \times \frac{1 \text{ mol } H_2}{2.0 \text{ g } H_2}\right) + n_{He}\right)\frac{RT}{P} = V_{mixt.} = V_2 = \left(10.0 \text{ g } H_2 \times \frac{1 \text{ mol } H_2}{2.0 \text{ g } H_2}\right)\frac{RT}{P}$$

$$2.0 + n_{He} = 5.0 \qquad n_{He} = 5.0 - 2.0 = 3.0 \text{ mol He} \qquad \text{mass He} = 3.0 \text{ mol} \times \frac{4.0 \text{ g He}}{1 \text{ mol He}} = 12 \text{ g He}$$

65. (a) Density should be related to average molar mass. We expect the average molar mass of air to be between the molar masses of its two principal constituents, N_2 (28.0 g/mol) and O_2 (32.0 g/mol). The average molar mass of normal air is approximately 28.9 g/mol. Expired air would be made more dense by the presence of more CO_2 (44.0 g/mol) and less dense by the presence of more H_2O (18.0 g/mol). The change might be minimal. In fact, it is, as the following calculation shows.

$$\mathcal{M}_{exp.air} = \left(0.742 \text{ mol } N_2 \times \frac{28.013 \text{ g } N_2}{1 \text{ mol } N_2}\right) + \left(0.152 \text{ mol } O_2 \times \frac{31.999 \text{ g } O_2}{1 \text{ mol } O_2}\right)$$

$$+ \left(0.038 \text{ mol } CO_2 \times \frac{44.01 \text{ g } CO_2}{1 \text{ mol } CO_2}\right) + \left(0.059 \text{ mol } H_2O \times \frac{18.02 \text{ g } H_2O}{1 \text{ mol } H_2O}\right)$$

$$+ \left(0.009 \text{ mol } Ar \times \frac{39.9 \text{ g } Ar}{1 \text{ mol } Ar}\right) = 28.7 \text{ g/mol of expired air}$$

(b) We see from equation (6.17) that the mole ratio, the volume ratio, and the ratio of partial pressures all are equal. Thus, to determine the ratio of partial pressure, we can use the ratios of volumes. The $\%CO_2$ in ordinary air is 0.03%, while from the data of this problem, the $\%CO_2$ in expired air is 3.8%.

$$\frac{P\{CO_2 \text{ expired air}\}}{P\{CO_2 \text{ ordinary air}\}} = \frac{3.8\% \text{ } CO_2}{0.03\% \text{ } CO_2} = 1 \times 10^2 \text{ } CO_2 \text{ (expired air to ordinary air)}$$

66. The volume percents are also equal to the mole percents and to the partial pressure percents. We use the mole percents, converted to mole fractions, to compute a molar mass of producer gas.

$$\text{molar mass} = 0.080 \text{ mol } CO_2 \times \frac{44.01 \text{ g } CO_2}{1 \text{ mol } CO_2} + 0.232 \text{ mol } CO \times \frac{28.01 \text{ g } CO}{1 \text{ mol } CO}$$

$$+ 0.177 \text{ mol } H_2 \times \frac{2.016 \text{ g } H_2}{1 \text{ mol } H_2} + 0.011 \text{ mol } CH_4 \times \frac{16.04 \text{ g } CH_4}{1 \text{ mol } CH_4} + 0.500 \text{ mol } N_2 \times \frac{28.01 \text{ g } N_2}{1 \text{ mol } N_2}$$

$$= 3.5 \text{ g } CO_2 + 6.50 \text{ g } CO + 0.357 \text{ g } H_2 + 0.18 \text{ g } CH_4 + 14.0 \text{ g } N_2$$

$$= 24.6 \text{ g/mole of producer gas}$$

(a) $P = 752 \text{ mmHg} \times \frac{1 \text{ atm}}{760 \text{ mmHg}} = 0.989 \text{ atm} \qquad T = 25 + 273 = 298 \text{ K}$

$$\text{density} = \frac{m}{V} = \frac{\mathcal{M}P}{RT} = \frac{24.6 \text{ g mol}^{-1} \times 0.989 \text{ atm}}{0.08206 \text{ L atm mol}^{-1} \text{ K}^{-1} \times 298 \text{ K}} = 0.995 \text{ g/L}$$

(b) The partial pressure of CO equals its mole fraction times the total pressure.

$$P_{CO} = 0.232 \times 1.00 \text{ atm} = 0.232 \text{ atm}$$

Collecting Gases over Liquids

67. The pressure of the liberated $H_2(g)$ is 738 mmHg – 25.2 mmHg = 713 mmHg

$$V = \frac{nRT}{P} = \frac{\left(1.76 \text{ g Al} \times \frac{1 \text{ mol Al}}{26.98 \text{ g}} \times \frac{3 \text{ mol } H_2}{2 \text{ mol Al}}\right) 0.08206 \frac{\text{L atm}}{\text{mol K}} (273 + 26)\text{K}}{713 \text{ mmHg} \times \frac{1 \text{ atm}}{760 \text{ mmHg}}} = 2.56 \text{ L } H_2(g)$$

This is the total volume of both gases, each with a different partial pressure.

68. If the gas were measured at the same volume before and after passing it through H_2O, its pressure would increase by 23.8 mmHg from 751 mmHg to 775 mmHg (= 751 + 23.8 mmHg). Since it is measured at the same pressure, we can determine its new volume with Boyle's law. What we are doing is to assume that the gas is collected at the same volume but at 775 mmHg. We then reduce the pressure to 751 mmHg by expanding the volume of the gas. $\text{volume} = 367 \text{ mL} \times \frac{775 \text{ mmHg}}{751 \text{ mmHg}} = 379 \text{ mL}$

69. We first determine the pressure of the gas collected. This would be its "dry gas" pressure and, when added to 23.8 mmHg, gives the barometric pressure.

$$P = \frac{nRT}{V} = \frac{\left(1.58 \text{ g} \times \frac{1 \text{ mol } O_2}{32.0 \text{ g } O_2}\right) 0.08206 \frac{\text{L atm}}{\text{mol K}} \times 298 \text{ K}}{1.28 \text{ L}} \times \frac{760 \text{ mmHg}}{1 \text{ atm}} = 717 \text{ mmHg}$$

barometric pressure = 717 mm Hg + 23.8 mmHg = 741 mmHg

70. We first determine the "dry gas" pressure of hexane. This pressure, subtracted from the barometric pressure of 738.6 mmHg, gives the vapor pressure of hexane at 25°C.

$$P = \frac{nRT}{V} = \frac{\left(1.072 \text{ g} \times \dfrac{1 \text{ mol He}}{4.003 \text{ g He}}\right) 0.08206 \dfrac{\text{L atm}}{\text{mol K}} \times 298.2 \text{ K}}{8.446 \text{ L}} \times \frac{760 \text{ mmHg}}{1 \text{ atm}} = 589.7 \text{ mmHg}$$

vapor pressure = 738.6 − 589.7 = 148.9 mmHg

Kinetic molecular theory

71. $u_{rms} = \sqrt{\dfrac{3RT}{\mathcal{M}}} = \sqrt{\dfrac{3 \times 8.314 \dfrac{\text{J}}{\text{mol K}} \times 298 \text{ K}}{\dfrac{70.91 \times 10^{-3} \text{ kg Cl}_2}{1 \text{ mol Cl}_2}}} = 324 \text{ m/s}$

72. $u_{rms} = \sqrt{\dfrac{3RT}{\mathcal{M}}} = 1.84 \times 10^3 \text{ m/s}$ 　　　 Solve this equation for temperature with u_{rms} doubled.

$$T = \frac{\mathcal{M} \, u_{rms}^2}{3 \, R} = \frac{2.016 \times 10^{-3} \text{ kg/mol} (2 \times 1.84 \times 10^3 \text{ m/s})^2}{3 \times 8.314 \dfrac{\text{J}}{\text{mol K}}} = 1.09 \times 10^3 \text{ K}$$

73. (a) $\mathcal{M} = \dfrac{3RT}{u_{rms}^2} = \dfrac{3 \times 8.314 \dfrac{\text{J}}{\text{mol K}} \times 298 \text{ K}}{\left(2180 \dfrac{\text{mi}}{\text{hr}} \times \dfrac{1 \text{ hr}}{3600 \text{ sec}} \times \dfrac{5280 \text{ ft}}{1 \text{ mi}} \times \dfrac{12 \text{ in.}}{1 \text{ ft}} \times \dfrac{1 \text{ m}}{39.37 \text{ in.}}\right)^2} = 0.0078 \text{ kg/mol}$

　　　 = 7.8 g/mol　The molecular weigh of the gas is 7.8.

(b) A noble gas with molecules having u_{rms} at 25°C greater than that of a rifle bullet will have a mole weight less than 7.8 g/mol. He is the only possibility. A noble gas with a slower u_{rms} will have a molecular weight greater than 7.8 g/mol; any one of the other noble gases will have a slower u_{rms}.

74. We equate the two expressions for root mean square speed, cancel the common factors, and solve for the temperature of Ne. Note that the units of molar masses do not have to be in kg/mol in this calculation; they simply must be expressed in the same units. 　　$\sqrt{\dfrac{3RT}{\mathcal{M}}} = \sqrt{\dfrac{3R \times 300 \text{ K}}{4.003}} = \sqrt{\dfrac{3R \times T_{Ne}}{20.18}}$

$\dfrac{300 \text{ K}}{4.003} = \dfrac{T_{Ne}}{20.18}$ 　　　 $T_{Ne} = 300 \text{ K} \times \dfrac{20.18}{4.003} = 1.51 \times 10^3 \text{ K}$

75. u_m, the modal speed, is the speed that occurs most often, 50 mi/h

Average speed $= \dfrac{40 + 42 + 45 + 48 + 50 + 50 + 55 + 57 + 58 + 60}{10} = 50.5 \text{ mi/h} = \overline{u}$

Root mean square speed $= \sqrt{\dfrac{40^2 + 42^2 + 45^2 + 48^2 + 50^2 + 50^2 + 55^2 + 57^2 + 58^2 + 60^2}{10}}$

$= \sqrt{\dfrac{1600 + 1764 + 2025 + 2304 + 2500 + 2500 + 3025 + 3249 + 3364 + 3600}{10}}$

$= \sqrt{\dfrac{25931}{10}} = 50.9 \text{ mi/h}$

Diffusion and Effusion of Gases

76. (a) $\dfrac{\text{rate (H}_2)}{\text{rate (O}_2)} = \sqrt{\dfrac{\mathcal{M}(\text{O}_2)}{\mathcal{M}(\text{H}_2)}} = \sqrt{\dfrac{32.00}{2.02}} = 3.98$ 　　　 **(b)** $\dfrac{\text{rate (H}_2)}{\text{rate (D}_2)} = \sqrt{\dfrac{\mathcal{M}(\text{D}_2)}{\mathcal{M}(\text{H}_2)}} = \sqrt{\dfrac{4.0}{2.02}} = 1.41$

(c) $\dfrac{\text{rate}(^{14}\text{CO}_2)}{\text{rate}(^{12}\text{CO}_2)} = \sqrt{\dfrac{\mathcal{M}(^{12}\text{CO}_2)}{\mathcal{M}(^{14}\text{CO}_2)}} = \sqrt{\dfrac{44.0}{46.0}} = 0.978$

(d) $\dfrac{\text{rate}(^{235}\text{UF}_6)}{\text{rate}(^{238}\text{UF}_6)} = \sqrt{\dfrac{\mathcal{M}(^{238}\text{UF}_6)}{\mathcal{M}(^{235}\text{UF}_6)}} = \sqrt{\dfrac{352}{349}} = 1.004$

77. $\dfrac{\text{rate (NO}_2)}{\text{rate (N}_2\text{O)}} = \sqrt{\dfrac{\mathfrak{M} \ (\text{N}_2\text{O})}{\mathfrak{M} \ (\text{NO}_2)}} = \sqrt{\dfrac{44.02}{46.01}} = 0.9781 = \dfrac{x \ \text{mol NO}_2/t}{0.00312 \ \text{mol N}_2\text{O}/t}$

mol $\text{NO}_2 = 0.00312 \ \text{mol} \times 0.9781 = 0.00305 \ \text{mol NO}_2$

78. $\dfrac{\text{rate (N}_2)}{\text{rate (unk)}} = \dfrac{\text{mol (N}_2)/38 \ \text{s}}{\text{mol (unk)}/55 \ \text{s}} = \dfrac{55 \ \text{s}}{38 \ \text{s}} = 1.45 = \sqrt{\dfrac{\mathfrak{M} \ (\text{unk})}{\mathfrak{M} \ (\text{N}_2)}}$

$\mathfrak{M} \ (\text{unk}) = (1.45)^2 \ \mathfrak{M} \ (\text{N}_2) = (1.45)^2 \ (28.01 \ \text{g/mol}) = 59 \ \text{g/mol}$

79. The distances that the two molecules travel are related as are their rates of effusion.

$\dfrac{\text{distance NH}_3}{\text{distance HCl}} = \sqrt{\dfrac{\mathfrak{M} \ (\text{HCl})}{\mathfrak{M} \ (\text{NH}_3)}} = \sqrt{\dfrac{36.5}{17.0}} = 1.46 = \dfrac{x}{1.00 - x}$, if we let distance $\text{NH}_3 = x$.

$1.46 - 1.46 \ x = x \qquad\qquad 2.46 \ x = 1.46 \qquad\qquad x = \dfrac{1.46}{2.46} = 0.593$ meter

The $\text{NH}_4\text{Cl(s)}$ cloud will form about 600 cm from the NH_3 and 400 cm from the HCl.

80. For 1.000 mol $\text{Cl}_2\text{(g)}$, $n^2a = 6.49 \ \text{L}^2$ atm and $nb = 0.0562$ L. $P_{\text{vdW}} = \dfrac{nRT}{V - n b} - \dfrac{n^2a}{V^2}$

At 0°C, $P_{\text{vdW}} = 9.9$ atm and $P_{\text{ideal}} = 11.2$ atm, off by 1.3 atm or +13%

(a) At 100°C $P_{\text{ideal}} = \dfrac{nRT}{V} = \dfrac{1.00 \ \text{mol} \times \dfrac{0.08206 \ \text{L atm}}{\text{mol K}} \times 373 \ \text{K}}{2.00 \ \text{L}} = 15.3$ atm

$P_{\text{vdW}} = \dfrac{1.00 \ \text{mol} \times \dfrac{0.08206 \ \text{L atm}}{\text{mol K}} \times T}{(2.00 - 0.0562) \ \text{L}} - \dfrac{6.49 \ \text{L}^2 \ \text{atm}}{(2.00 \ \text{L})^2} = 0.0422_2 \ T \ \text{atm} - 1.62_3 \ \text{atm}$

$= 0.0422_2 \times 373 \ \text{K} - 1.62_3 = 14.1_3$ atm $\qquad P_{\text{ideal}}$ is off by 1.2 atm or +8.2%

(b) At 200°C $P_{\text{ideal}} = \dfrac{nRT}{V} = \dfrac{1.00 \ \text{mol} \times \dfrac{0.08206 \ \text{L atm}}{\text{mol K}} \times 473 \ \text{K}}{2.00 \ \text{L}} = 19.4$ atm

$P_{\text{vdW}} = 0.0422_2 \times 473 \ \text{K} - 1.62_3 = 18.3_5$ atm $\qquad P_{\text{ideal}}$ is off by 1.0 atm or +6.7%

(c) At 400°C $P_{\text{ideal}} = \dfrac{nRT}{V} = \dfrac{1.00 \ \text{mol} \times \dfrac{0.08206 \ \text{L atm}}{\text{mol K}} \times 673 \ \text{K}}{2.00 \ \text{L}} = 27.6$ atm

$P_{\text{vdW}} = 0.0422_2 \times 673 \ \text{K} - 1.62_3 = 26.7_9$ atm $\qquad P_{\text{ideal}}$ is off by 0.8 atm or +3._0%

7 THERMOCHEMISTRY

1. (a) ΔH, the enthalpy cahnge for a process, is the quantity of heat absorbed when the process occurs at constant pressure.

   (b) $P\Delta V$, the change in volume multiplied by a constant pressure, is the expression for the pressure-volume work in a process that occurs at constant pressure.

   (c) ΔH_f°, the standard enthalpy of formation, is the heat absorbed at constant pressure when 1 mole of product is formed from the stable form of the elements, with reactants and products in their standard states.

   (d) Standard state is defined as a pressure of exactly 1 atm and a pure substance, or an aqueous solute at a 1 M concentration.

   (e) A fossil fuel is a material that can be burned for heat and that was produced by living material that lived eons ago.

2. (a) The law of conservation of energy states that energy is neither created nor destroyed during a process.

   (b) Bomb calorimetry is the technique of running a chemical reaction in a constant-volume container and measuring the heat absorbed by the process.

   (c) A function of state is a measureable property that depends only on the initial and final conditions of a process and not on its path.

   (d) An enthalpy diagram represents the enthalpy values of reactants and products by their vertical positions, and the progress of the reaction horizontally.

   (e) Hess's law states that if several reactions can be combined to form a net reaction, the enthalpy changes of those reactions combine in the same way to produce the enthalpy change of the net reaction.

3. (a) The system is that part of the universe that we are considering, in which we are interested. The surroundings are the rest of the universe, particularly the rest that influences the system.

   (b) Heat is energy in transport that is associated with either a change in temperature or a phase change (such as solid to liquid) of a material. Work is organized energy that has the ability to move a force through a distance.

   (c) The specific heat of a substance is the quantity of heat needed to raise the temperature of one *gram* of that substance by 1.00°C. The (molar) heat capacity of a substance the the quantity of heat needed to raise the temperature (of one *mole*) of that substance by 1.00°C.

   (d) An endothermic reaction is one that absorbs heat from the surroundings. An exothermic reaction is one that evolves heat to the surroundings.

4. (a) $q = 12.5 \text{ L} \times \dfrac{1000 \text{ cm}^3}{1 \text{ L}} \times \dfrac{1.00 \text{ g}}{1 \text{ cm}^3} \times \dfrac{1.00 \text{ cal}}{1 \text{ g °C}} \times (33.7°\text{C} - 22.0°\text{C}) \times \dfrac{1 \text{ kcal}}{1000 \text{ cal}} = +146 \text{ kcal}$

   (b) $q = 6.15 \text{ kg} \times \dfrac{1000 \text{ g}}{1 \text{ kg}} \times \dfrac{0.473 \text{ J}}{\text{g °C}} \times (-42.0 °\text{C}) \times \dfrac{1 \text{ kJ}}{1000 \text{ J}} = -122 \text{ kJ}$

5. $\text{heat} = \text{mass} \times \text{sp ht} \times \Delta t \qquad \text{becomes} \qquad \Delta t = \dfrac{\text{heat}}{\text{mass} \times \text{sp ht}}$

   (a) $\Delta t = \dfrac{+719 \text{ J}}{14.8 \text{ g} \times 4.18 \text{ J g}^{-1} \text{ °C}^{-1}} = +12.7°\text{C} \qquad t_f = t_i + \Delta t = 21.7°\text{C} + 11.6°\text{C} = 33.3°\text{C}$

(b) $\Delta t = \dfrac{-105 \text{ kcal} \times \dfrac{1000 \text{ cal}}{1 \text{ kcal}}}{\left(6.52 \text{ kg} \times \dfrac{1000 \text{ g}}{1 \text{ kg}}\right) 0.173 \dfrac{\text{cal}}{\text{g °C}}} = -93.1\text{°C}$ $t_f = t_i + \Delta t = 67.3\text{°C} - 93.1\text{°C} = -25.8\text{°C}$

6. (a) $\text{sp.ht} = \dfrac{\text{heat}}{\text{mass} \times \Delta t} = \dfrac{192 \text{ J}}{20.0 \text{ g} \times (30.8 - 25.2)} = 1.7 \text{ J g}^{-1} \text{ °C}^{-1}$

(b) $\Delta t = \dfrac{\text{heat}}{\text{mass} \times \text{sp.ht}} = \dfrac{-3.50 \text{ kcal} \times \dfrac{1000 \text{ cal}}{1 \text{ kcal}}}{\left(1.50 \text{ kg} \times \dfrac{1000 \text{ cal}}{1 \text{ kcal}}\right) 1.00 \dfrac{\text{cal}}{\text{g °C}}} = -2.33\text{°C}$

$t_f = t_i + \Delta t = 25.2\text{°C} - 2.33\text{°C} = 22.9\text{°C}$

7. The heat capacities of the two substances are added and then multiplied by the temperature change.

$\Delta H = \left(135 \text{ g Cu} \times \dfrac{0.393 \text{ J}}{\text{g °C}} + 235 \text{ g H}_2\text{O} \times \dfrac{4.18 \text{ J}}{\text{g °C}}\right)(67.6\text{°C} - 22.7\text{°C}) \times \dfrac{1 \text{ kJ}}{1000 \text{ J}} = +46.5 \text{ kJ}$

8. Heat is transferred from the iron to the water.

$q_{\text{water}} = 825 \text{ g} \times \dfrac{4.18 \text{ J}}{\text{g °C}} \times (39.6 - 23.3)\text{°C} = 5.62 \times 10^4 \text{ J} = -q_{\text{iron}}$

$q_{\text{iron}} = -5.62 \times 10^4 \text{ J} = 1.35 \text{ kg} \times \dfrac{1000 \text{ g}}{1 \text{ kg}} \times \text{sp ht} \times (39.6 - 132.0)\text{°C} = -1.25 \times 10^5 \text{ sp ht}$

$\text{sp ht} = -\dfrac{-5.62 \times 10^4}{1.25 \times 10^5} = \dfrac{0.450 \text{ J}}{\text{g °C}}$

9. If the two volumes were the same, a straight average of the temperatures would be the final temperature. But that is not true, so option (2) [40°C] is incorrect. In fact, there is more of the cooler water (100.0 mL) than of the warmer water (75.0 mL). Thus, the final temperature will not be warmer than the straight average; option (1) [45°C] is incorrect. The cooler side of the straight average should be somewhat favored, but not grossly so; option (4) [23°C] is incorrect. The correct final temperature is likely to be 37°C, option (3).

10. (a) $\Delta E = q + w = 67 \text{ J heat} - 67 \text{ J work} = 0 \text{ J}$ (b) $\Delta E = q + w = 356 \text{ J heat} - 592 \text{ J work} = -236 \text{ J}$
(c) $\Delta E = q + w = -38 \text{ J heat} + 171 \text{ J work} = +133 \text{ J}$ (d) $\Delta E = q + w = 0 \text{ J heat} - 416 \text{ J work} = -416 \text{ J}$

11. (a) $q = \dfrac{-5.36 \text{ kJ}}{0.107 \text{ g C}_2\text{H}_2} \times \dfrac{26.04 \text{ g C}_2\text{H}_2}{1 \text{ mol C}_2\text{H}_2} = -1.30 \times 10^3 \text{ kJ/mol C}_2\text{H}_2$

(b) $q = \dfrac{-2.601 \text{ kcal}}{1.030 \text{ g CO(NH}_2)_2} \times \dfrac{4.184 \text{ kJ}}{1 \text{ kcal}} \times \dfrac{60.056 \text{ g CO(NH}_2)_2}{1 \text{ mol CO(NH}_2)_2} = -634.5 \text{ kJ/mol CO(NH}_2)_2$

(c) $q = \dfrac{-26.0 \text{ kJ}}{1.05 \text{ mL (CH}_3)_2\text{CO}} \times \dfrac{1 \text{ mL}}{0.791 \text{ g}} \times \dfrac{58.08 \text{ g (CH}_3)_2\text{CO}}{1 \text{ mol (CH}_3)_2\text{CO}} = -1.82 \times 10^3 \text{ kJ/mol (CH}_3)_2\text{CO}$

12. $\text{heat capacity} = \dfrac{\text{heat absorbed}}{\Delta t} = \dfrac{4708 \text{ cal}}{3.44\text{°C}} \times \dfrac{4.184 \text{ J}}{1 \text{ cal}} \times \dfrac{1 \text{ kJ}}{1000 \text{ J}} = 5.73 \text{ kJ/°C}$

13. In each case, heat absorbed by calorimeter $(= -q_{\text{comb}} \times \text{no. mol}) = \text{heat cap} \times \Delta t$ or $\Delta t = \dfrac{q_{\text{comb}} \times \text{no. mol}}{\text{heat cap}}$

(a) $\Delta t = \dfrac{\left(890.7 \dfrac{\text{kcal}}{\text{mol}} \times 4.184 \dfrac{\text{kJ}}{\text{kcal}}\right)\left(0.5187 \text{ g} \times \dfrac{1 \text{ mol C}_6\text{H}_{12}\text{O}}{100.2 \text{ g C}_6\text{H}_{12}\text{O}}\right)}{4.881 \text{ kJ/°C}} = 3.952\text{°C}$

$t_f = t_i + \Delta t = 24.62\text{°C} + 3.952\text{°C} = 28.57\text{°C}$

(b) $\Delta t = \dfrac{2246 \dfrac{\text{kJ}}{\text{mol}} \left(1.75 \text{ mL} \times \dfrac{0.901 \text{ g}}{1 \text{ mL}} \times \dfrac{1 \text{ mol C}_4\text{H}_8\text{O}_2}{88.11 \text{ g C}_4\text{H}_8\text{O}_2}\right)}{4.881 \text{ kJ/°C}} = 8.23\text{°C}$

$t_f = 24.62\text{°C} + 8.23\text{°C} = 32.85\text{°C}$

14. (a) $\dfrac{\text{heat}}{\text{mass}} = \dfrac{\text{heat cap.} \times \Delta t}{\text{mass}} = \dfrac{4.947 \text{ kJ/°C} \times (35.17 - 24.88)\text{°C}}{1.036 \text{ g}} = 49.14 \text{ kJ/g isobutane}$

$\Delta H = \text{heat given off/g} \times \mathfrak{M} \text{ (g/mol)} = \dfrac{-49.14 \text{ kJ}}{1 \text{ g (CH}_3)_3\text{CH}} \times \dfrac{58.12 \text{ g (CH}_3)_3\text{CH}}{1 \text{ mol}}$

$= -2856 \text{ kJ/mol (CH}_3)_3\text{CH}$

   (b) $(CH_3)_3CH(g) + \frac{13}{2} O_2(g) \longrightarrow 4 CO_2(g) + 5 H_2O(l)$ $\Delta H = -2856$ kJ

15. (a) Since the temperature of the mixture decreases, the reaction (the system) must have absorbed heat from the reaction mixture (the surroundings). Consequently, the reaction must be endothermic.

   (b) We assume that the specific heat of the solution is 4.18 J g⁻¹ °C⁻¹. The enthalpy change in kJ/mol NH_4NO_3 is obtained by the heat absorbed per gram NH_4NO_3.

$$\Delta H = -\frac{(0.50 + 35.0)\text{g} \; \frac{4.18 \text{ J}}{\text{g °C}} \; (21.6 - 22.7)°C}{0.50 \text{ g } NH_4NO_3} \times \frac{1 \text{ kJ}}{1000 \text{ J}} \times \frac{80.0 \text{ g } NH_4NO_3}{1 \text{ mol } NH_4NO_3} = +26 \text{ kJ/mol}$$

16. As indicated by the positive sign for the enthalpy change, this is an endothermic reaction; the temperature of the system should decrease.

$$q_{rxn} = 0.102 \text{ mol KI} \times \frac{21.3 \text{ kJ}}{1 \text{ mol KI}} \times \frac{1000 \text{ J}}{1 \text{ kJ}} = 2.17 \times 10^3 \text{ J} = -q_{calorim}$$

Now we assume that the density of water is 1.00 g/mL, the specific heat of the solution in the calorimeter is 4.18 J g⁻¹ °C⁻¹, and no heat is absorbed by the calorimeter.

$$q_{calorim} = -7.50 \times 10^3 \text{ J} = \left(\left(475 \text{ mL} \times \frac{1.00 \text{ g}}{1 \text{ mL}}\right) + \left(0.102 \text{ mol KI} \times \frac{166.00 \text{ g}}{1 \text{ mol KI}}\right)\right) \times \frac{4.18 \text{ J}}{\text{g °C}} \times \Delta t$$

$$= 2.06 \times 10^3 \, \Delta t \qquad\qquad \Delta t = \frac{-2.17 \times 10^3}{2.06 \times 10^3} = -1.05°C$$

$t_{final} = t_{initial} + \Delta t = 23.8°C - 1.05°C = 22.7°C$

17. (a) $N_2(g) + 2 H_2(g) \longrightarrow N_2H_4(l)$ $\Delta H° = +50.63$ kJ

   (b) $C(\text{graphite}) + \frac{1}{2} O_2(g) + Cl_2(g) \longrightarrow COCl_2$ $\Delta H° = -209$ kJ

   (c) $C_3H_8O_3 + \frac{7}{2} O_2(g) \longrightarrow 3 CO_2(g) + 4 H_2O(l)$ $\Delta H° = -1.66 \times 10^3$ kJ

18. (a) heat evolved $= 1.100 \text{ g } C_3H_8 \times \dfrac{1 \text{ mol } C_3H_8}{44.097 \text{ g } C_3H_8} \times \dfrac{2220. \text{ kJ}}{1 \text{ mol } C_3H_8} = 55.38$ kJ

   (b) heat evolved $= 35.0 \text{ L}_{STP} \text{ } C_3H_8 \times \dfrac{1 \text{ mol } C_3H_8}{22.414 \text{ L}_{STP} \text{ } C_3H_8} \times \dfrac{2220. \text{ kJ}}{1 \text{ mol } C_3H_8} = 3.47 \times 10^3$ kJ

   (c) Use the ideal gas equation to determine the amount of propane in moles and multiply this amount by 2220. kJ heat produced per mole.

$$\text{heat evolved} = \frac{\left(749 \text{ mmHg} \times \frac{1 \text{ atm}}{760 \text{ mmHg}}\right) \times 15.5 \text{ L}}{\frac{0.08206 \text{ L atm}}{\text{mol K}} \times (273.2 + 22.8) \text{ K}} \times \frac{2220. \text{ kJ}}{1 \text{ mol } C_3H_8} = 1.40 \times 10^3 \text{ kJ}$$

19. The formation reaction for $NH_3(g)$ is $\frac{1}{2} N_2(g) + \frac{3}{2} H_2(g) \longrightarrow NH_3(g)$. The given reaction is two-thirds the reverse of the formation. The sign of the enthalpy is changed and it is multiplied by two-thirds. Thus, the enthalpy of the given reaction is $-(-46.11 \text{ kJ}) \times \frac{2}{3} = +30.74$ kJ.

20. $-(1)$ $CO(g) \longrightarrow C(\text{graphite}) + \frac{1}{2} O_2(g)$ $\Delta H = +110.54$ kJ

     $+(2)$ $C(\text{graphite}) + O_2(g) \longrightarrow CO_2(g)$ $\Delta H = -393.51$ kJ

             $CO(g) + \frac{1}{2} O_2(g) \longrightarrow CO_2(g)$ $\Delta H = -282.97$ kJ

21. $-(3)$ $3 CO_2(g) + 5 O_2(g) \longrightarrow C_3H_8(g) + 5 O_2(g)$ $\Delta H = +2220. $ kJ

     $+(2)$ $C_3H_4(g) + 4 O_2(g) \longrightarrow 3 CO_2(g) + 2 H_2O(l)$ $\Delta H = -1941$ kJ

     $2(1)$ $2 H_2(g) + O_2(g) \longrightarrow 2 H_2O(l)$ $\Delta H = -\;571.6$ kJ

             $C_3H_4(g) + 2 H_2(g) \longrightarrow C_3H_8(g)$ $\Delta H = -293$ kJ

22. The second reaction is the only one in which NO(g) appears; it must be run twice to produce 2 NO(g).

        $2 NH_3(g) + \frac{5}{2} O_2(g) \longrightarrow 2 NO(g) + 3 H_2O(l)$ $2 \Delta H_2$

The first reaction is the only one that can eliminate 2 NH₃(g); it must be run twice to eliminate 2 NH₃(g)

        $N_2(g) + 3 H_2(g) \longrightarrow 2 NH_3(g)$ $2 \Delta H_1$

We triple and reverse the third reaction to eliminate 3 H₂(g).

$$3 H_2O(l) \longrightarrow 3 H_2(g) + \tfrac{3}{2} O_2(g) \qquad\qquad -3\,\Delta H_3$$

Result: $\quad N_2(g) + O_2(g) \longrightarrow 2 NO(g) \qquad \Delta H_{rxn} = 2\,\Delta H_1 + 2\,\Delta H_2 - 3\,\Delta H_3$

23. (a) $\Delta H^\circ = \Delta H_f^\circ[C_2H_6(g)] + \Delta H_f^\circ[CH_4(g)] - \Delta H_f^\circ[C_3H_8(g)] - \Delta H_f^\circ[H_2(g)]$

$\qquad = -84.68 - 74.81 - (-103.8) - 0.00 = -55.7$ kJ

(b) $\Delta H^\circ = 2\,\Delta H_f^\circ[SO_2(g)] + 2\,\Delta H_f^\circ[H_2O(l)] - 2\,\Delta H_f^\circ[H_2S(g)] - 3\,\Delta H_f^\circ[O_2(g)]$

$\qquad = 2\,(-296.8) + 2\,(-285.8) - 2\,(-20.63) - 3\,(0.00) = -1123.9$ kJ

24. $\Delta H^\circ = \Delta H_f^\circ[H_2O(l)] + \Delta H_f^\circ[NH_3(g)] - \Delta H_f^\circ[NH_4^+(aq)] - \Delta H_f^\circ[OH^-(aq)]$

$\qquad = -285.8 + (-46.11) - (-132.5) - (-230.0) = +30.6$ kJ

25. $ZnO(s) + SO_2(g) \longrightarrow ZnS(s) + \tfrac{3}{2} O_2(g) \qquad \Delta H^\circ = -(-880.\ kJ)/2 = +440.\ kJ$

$440.\ kJ = \Delta H_f^\circ[ZnS(s)] + \tfrac{3}{2}\Delta H_f^\circ[O_2(g)] - \Delta H_f^\circ[ZnO(s)] - \Delta H_f^\circ[SO_2(g)]$

$\qquad = \Delta H_f^\circ[ZnS(s)] + \tfrac{3}{2}(0.00) - (-348.3) - (-296.8)$

$\Delta H_f^\circ[ZnS(s)] = 440. - 348.3 - 296.8 = -205$ kJ/mol

EXERCISES

Work

26. $P = 735\ \text{mmHg} \times \dfrac{1\ \text{atm}}{760\ \text{mmHg}} = 0.967\ \text{atm}$ 　　(a) $w = P\Delta V = 0.967\ \text{atm} \times 5.0\ \text{L} = 4.8\ \text{L atm}$

(b) $w = 4.8\ \text{L atm} \times \dfrac{8.314\ \text{J}}{0.08206\ \text{L atm}} = 4.9 \times 10^2$ J

(c) $w = 4.8\ \text{L atm} \times \dfrac{1.9872\ \text{cal}}{0.08206\ \text{L atm}} = 1.2 \times 10^2$ cal

Heat Capacity (Specific Heat)

27. heat gained by the water = heat lost by the metal; each heat = mass × sp.ht. × Δt

(a) $50.0\ \text{g} \times 4.184\ \dfrac{J}{g\ ^\circ C}\ (39.0 - 22.0)^\circ C = 3.56 \times 10^3\ J = -150.0\ \text{g} \times \text{sp.ht.} \times (39.0 - 100.0)^\circ C$

$\quad$ sp.ht. $= \dfrac{3.56 \times 10^3\ J}{150.0\ \text{g} \times 61.0^\circ C} = 0.389\ J\ g^{-1}\ C^{\circ -1} \quad$ for Zn

(b) $50.0\ \text{g} \times 4.184\ \dfrac{J}{g\ ^\circ C}\ (28.9 - 22.0)^\circ C = 1.4 \times 10^3\ J = -150.0\ \text{g} \times \text{sp.ht.} \times (28.9 - 100.0)^\circ C$

$\quad$ sp.ht. $= \dfrac{1.4 \times 10^3\ J}{150.0\ \text{g} \times 71.1^\circ C} = 0.13\ J\ g^{-1}\ C^{\circ -1} \quad$ for Pt

(c) $50.0\ \text{g} \times 4.184\ \dfrac{J}{g\ ^\circ C}\ (52.8 - 22.0)^\circ C = 6.44 \times 10^3\ J = -150.0\ \text{g} \times \text{sp.ht.} \times (52.8 - 100.0)^\circ C$

$\quad$ sp.ht. $= \dfrac{6.44 \times 10^3\ J}{150.0\ \text{g} \times 47.2^\circ C} = 0.910\ J\ g^{-1}\ C^{\circ -1} \quad$ for Al

28. $q_{water} = 525\ \text{g} \times 4.18\ \dfrac{J}{g\ ^\circ C}\ (92 - 24)^\circ C = 1.4_9 \times 10^5\ J = -q_{iron}$

$q_{iron} = -1.4_9 \times 10^5\ J = 625\ \text{g} \times 0.45\ \dfrac{J}{g\ ^\circ C}\ (92 - t_i) = 2.5_9 \times 10^4\ J - 2.8_1 \times 10^2\ t_i$

$t_i = \dfrac{-1.4_9 \times 10^5 - 2.5_9 \times 10^4}{-2.8_1 \times 10^2} = 6.2 \times 10^2\ ^\circ C$

The number of significant figures in the final answer is limited by the two significant figures in the value of the specific heat of iron. Only two significant figures throughout the calculation yields $t_i = 6.4 \times 10^2\ ^\circ C$

29. heat gained by steel = −heat lost by steel = heat gained by water

$-m \times 0.50\ \dfrac{J}{g\ ^\circ C}\ (40.3 - 152)^\circ C = 56\ m = 125\ \text{mL} \times \dfrac{1.00\ g}{1\ \text{mL}} \times 4.18\ \dfrac{J}{g\ ^\circ C}\ (40.3 - 24.8)^\circ C = 8.10 \times 10^3\ J$

$m = \dfrac{8.10 \times 10^3}{56} = 1.4 \times 10^2$ g stainless steel.

The accuracy of this calculation is limited by the obedience of the experiment to the assumption that no heat leaks out of the system. When we deal with temperatures far above (or far below) room temperature, this assumption becomes less and less accurate.

On the other hand, the precision of the calculation is limited to two significant figures by the specific heat of the steel. If the two specific heats were known to an unlimited number of significant figures, then the temperature difference would determine the final precision of the calculation. It is unlikely that we could readily measure temperature more precisely than $\pm 0.01°C$, without expensive equipment. The mass of steel in this case would be known to four significant figures, to ± 0.1 g.

30. heat gained by copper = –heat lost by copper = heat gained by glycerol

$$-74.8 \text{ g} \times \frac{0.393 \text{ J}}{\text{g °C}} \times (31.1°C - 143.2°C) = 165 \text{ mL} \times \frac{1.26 \text{ g}}{1 \text{ mL}} \times \text{sp ht} \times (31.1°C - 24.8°C)$$

$$3.30 \times 10^3 = 1.3 \times 10^3 \text{ sp ht} \qquad \text{sp ht} = \frac{3.30 \times 10^3}{1.3 \times 10^3} = 2.5 \text{ J g}^{-1} °C^{-1}$$

$$\text{molar heat capacity} = 2.5 \text{ J g}^{-1} °C^{-1} \times \frac{92.1 \text{ g}}{1 \text{ mol } C_3H_8O_3} = 2.3 \times 10^2 \text{ J mol}^{-1} °C^{-1}$$

31. The additional water simply acts as a heat transfer medium. The essential relationship is

heat gained by iron = –heat lost by iron = heat gained by water (of unknown mass)

$$-\left(1.05 \text{ kg} \times \frac{1000 \text{ g}}{1 \text{ kg}}\right) 0.452 \frac{\text{J}}{\text{g °C}} (25.6 - 60.5)°C = x \text{ g } H_2O \times 4.18 \frac{\text{J}}{\text{g °C}} (25.6 - 15.5)°C$$

$$1.66 \times 10^4 \text{ J} = 42.2 \, x \qquad x = \frac{1.66 \times 10^4}{42.2} = 393 \text{ g } H_2O \times \frac{1 \text{ mL } H_2O}{1.00 \text{ g } H_2O} = 393 \text{ mL } H_2O$$

32. heat gained by Mg = –heat lost by Mg = heat gained by water

$$-\left(1.00 \text{ kg Mg} \times \frac{1000 \text{ g}}{1 \text{ kg}}\right) 1.04 \frac{\text{J}}{\text{g °C}} (t_f - 40.0°C)$$

$$= \left(1.00 \text{ L} \times \frac{1000 \text{ cm}^3}{1 \text{ L}} \times \frac{1.00 \text{ g}}{1 \text{ cm}^3}\right) 4.184 \frac{\text{J}}{\text{g °C}} (t_f - 20.0°C)$$

$$- 1.04 \times 10^3 \, t_f + 4.16 \times 10^4 = 4.184 \times 10^3 \, t_f - 8.37 \times 10^4$$

$$4.16 \times 10^4 + 8.37 \times 10^4 = (4.184 \times 10^3 + 1.04 \times 10^3) \, t_f = 12.53 \times 10^4 = 5.22 \times 10^3 \, t_f$$

$$t_f = \frac{12.53 \times 10^4}{5.22 \times 10^3} = 24.0°C$$

33. heat gained by the water = heat gained by the brass = –heat lost by the brass

$$138 \text{ g} \times 4.18 \frac{\text{J}}{\text{g °C}} \times (t_f - 23.7°C) = -\left(14.5 \text{ cm}^3 \times \frac{8.40 \text{ g}}{1 \text{ cm}^3}\right) 0.385 \frac{\text{J}}{\text{g °C}} (t_f - 152°C)$$

$$577 \, t_f - 1.37 \times 10^4 = -46.9 \, t_f + 7.13 \times 10^3 \qquad t_f = \frac{1.37 \times 10^4 + 7.13 \times 10^3}{577 + 46.9} = 33.4°C$$

34. (a) Si $\quad$ 28.1 g/mol $\times$ 0.17 cal g^{-1} °C^{-1} = 4.8 cal mol^{-1} °C^{-1}

$\quad\quad$ P $\quad$ 31.0 g/mol $\times$ 0.19 cal g^{-1} °C^{-1} = 5.9 cal mol^{-1} °C^{-1}

$\quad\quad$ Pb $\quad$ 207 g/mol $\times$ 0.031 cal g^{-1} °C^{-1} = 6.4 cal mol^{-1} °C^{-1}

We see that the law of Dulong and Petit is most valid for the metallic element lead. In fact, the Law is generally held to be valid for solid metallic elements.

(b) sp ht $= \dfrac{107 \text{ cal}}{75.0 \text{ g} \times 15°C} = \dfrac{0.095 \text{ cal}}{\text{g °C}} \qquad$ atomic weight $= \dfrac{6.4 \text{ cal mol}^{-1} °C^{-1}}{0.095 \text{ cal g}^{-1} °C^{-1}} = 67$ g/mol

The metal could be Zn or Ga

(c) sp ht Sn $= \dfrac{6.4 \text{ cal mol}^{-1} °C^{-1}}{119 \text{ g/mol}} = 0.054$ cal g^{-1} °C^{-1}

–heat lost by Sn = heat gained by Sn = heat gained by water

$$-315 \text{ g} \times \frac{0.054 \text{ cal}}{\text{g °C}} \times (t - 65.0°C) = 100.0 \text{ cm}^3 \times \frac{1.00 \text{ g}}{1 \text{ cm}^3} \times \frac{1.00 \text{ cal}}{\text{g °C}} \times (t - 25.0°C)$$

$$-17t + 1.1_1 \times 10^3 = 100.t - 2.50 \times 10^3 \qquad t = \frac{2.50 \times 10^3 + 1.1_1 \times 10^3}{100. + 17} = 31°C$$

Heats of reaction

35. $\text{heat} = 375 \text{ kg} \times \dfrac{1000 \text{ g}}{1 \text{ kg}} \times \dfrac{1 \text{ mol Ca(OH)}_2}{74.10 \text{ g Ca(OH)}_2} \times \dfrac{65.2 \text{ kJ}}{1 \text{ mol Ca(OH)}_2} = 3.30 \times 10^5 \text{ kJ of heat evolved.}$

36. The heat of combustion of 1.00 L (STP) of synthesis gas produces 11.13 kJ of heat. The volume of synthesis gas needed to heat 30.0 gal of water is given by first determining the quantity of heat needed to raise the temperature of the water.

$\text{heat water} = \left(30.0 \text{ gal} \times \dfrac{3.785 \text{ L}}{1 \text{ gal}} \times \dfrac{1000 \text{ mL}}{1 \text{ L}} \times \dfrac{1.00 \text{ g}}{1 \text{ mL}}\right) 4.18 \dfrac{\text{J}}{\text{g }^\circ\text{C}} (61.6 - 22.3)^\circ\text{C}$

$= 1.87 \times 10^7 \text{ J} \times \dfrac{1 \text{ kJ}}{1000 \text{ J}} = 1.87 \times 10^4 \text{ kJ}$

$\text{gas volume} = 1.87 \times 10^4 \text{ kJ} \times \dfrac{1 \text{ L (STP)}}{11.13 \text{ kJ of heat}} = 1.68 \times 10^3 \text{ L at STP}$

37. **(a)** $\text{mass} = 1.00 \times 10^6 \text{ kJ} \times \dfrac{1 \text{ mol CH}_4}{890.3 \text{ kJ}} \times \dfrac{16.04 \text{ g CH}_4}{1 \text{ mol CH}_4} \times \dfrac{1 \text{ kg}}{1000 \text{ g}} = 18.0 \text{ kg CH}_4$

(b)   First determine the moles of CH_4 present, with the ideal gas law.

$\text{no. mol CH}_4 = \dfrac{\left(748 \text{ mmHg} \times \dfrac{1 \text{ atm}}{760 \text{ mmHg}}\right) 1.03 \times 10^3 \text{ L}}{0.08206 \dfrac{\text{L atm}}{\text{mol K}} \times (21.8 + 273.2) \text{ K}} = 41.9 \text{ mol CH}_4$

$\text{heat} = 41.9 \text{ mol CH}_4 \times \dfrac{-890.3 \text{ kJ}}{1 \text{ mol CH}_4} = -3.73 \times 10^4 \text{ kJ of heat}$

(c) $\text{volume H}_2\text{O} = \dfrac{3.73 \times 10^4 \text{ kJ} \times \dfrac{1000 \text{ J}}{1 \text{ kJ}}}{4.18 \dfrac{\text{J}}{\text{g }^\circ\text{C}} (60.8 - 22.7)^\circ\text{C}} \times \dfrac{1 \text{ mL H}_2\text{O}}{1 \text{ g}} = 2.34 \times 10^5 \text{ mL} = 234 \text{ L H}_2\text{O}$

38. $\text{heat} = 1.00 \text{ gal} \times \dfrac{3.785 \text{ L}}{1 \text{ gal}} \times \dfrac{1000 \text{ mL}}{1 \text{ L}} \times \dfrac{0.703 \text{ g}}{1 \text{ mL}} \times \dfrac{1 \text{ mol C}_8\text{H}_{18}}{114.2 \text{ g C}_8\text{H}_{18}} \times \dfrac{-5.48 \times 10^3 \text{ kJ}}{1 \text{ mol C}_8\text{H}_{18}}$

$= -1.28 \times 10^5 \text{ kJ}$

39. Since the mole weight of H_2 (2.0 g/mol) is $\frac{1}{16}$ of the mole weight of O_2 (32.0 g/mol) and only twice as many moles of H_2 are needed as O_2, we see that $O_2(g)$ is the limiting reagent in this reaction.

$\dfrac{105}{2} \text{ g O}_2 \times \dfrac{1 \text{ mol O}_2}{32.0 \text{ g O}_2} \times \dfrac{241.8 \text{ kJ heat}}{0.500 \text{ mol O}_2} = 793 \text{ kJ heat}$

40. The amounts of the two reactants are the same as their stoichiometric coefficients in the balanced equation. Thus 852 kJ of heat is given off by the reaction. We can use this quantity of heat, along with the specific heat of the mixture to determine the temperature change that will occur if all of the heat is retained in the reaction mixture.

$\text{heat} = 852 \text{ kJ} \times \dfrac{1000 \text{ J}}{1 \text{ kJ}} = 852 \times 10^3 \text{ J} = \text{mass} \times \text{sp.ht.} \times \Delta t$

$= \left(\left(1 \text{ mol Al}_2\text{O}_3 \times \dfrac{102 \text{ g Al}_2\text{O}_3}{1 \text{ mol Al}_2\text{O}_3}\right) + \left(2 \text{ mol Fe} \times \dfrac{55.8 \text{ g Fe}}{1 \text{ mol Fe}}\right)\right) 0.8 \dfrac{\text{J}}{\text{g }^\circ\text{C}} \times \Delta t$

$\Delta t = \dfrac{852 \times 10^3 \text{ J}}{214 \text{ g} \times 0.8 \text{ J g}^{-1} \text{ }^\circ\text{C}^{-1}} = 5 \times 10^3 \text{ }^\circ\text{C}$

The temperature only needs to increase from 20°C to 1530°C or $\Delta t = 1510^\circ\text{C} \approx 1.5 \times 10^3$ °C. Since the actual Δt is almost twice as large as this value, the iron indeed will melt, even if a large fraction of the heat evolved is lost to the surroundings and is not retained in the products.

41. **(a)**   We first compute the heat produced by this reaction, then determine the value of ΔH in kJ/mol KOH.

$\text{heat produced} = (0.150 + 45.0) \text{ g} \times 4.184 \dfrac{\text{J}}{\text{g }^\circ\text{C}} (24.9 - 24.1) = 2 \times 10^2 \text{ J heat} = -q_{\text{absorbed}}$

$$\Delta H = -\frac{2 \times 10^2 \text{ J} \times \dfrac{1 \text{ kJ}}{1000 \text{ J}}}{0.150 \text{ g} \times \dfrac{1 \text{ mol KOH}}{56.1 \text{ g KOH}}} = -7 \times 10^1 \text{ kJ/mol}$$

(b) The temperature change in this experiment is known to but one significant figure (0.8°C). Doubling the amount of solute should give a temperature change known to two significant figures (1.6°C) and using twenty times the mass of solute should give a temperature change known to three significant figures (16.0°C). This would require but 3.000 g KOH rather than the 0.150 g KOH actually used, and would increase the precision from 1 part in 7 to 1 part in 700, an imprecision of about 0.1%

42. Let x represent the mass of NH_4Cl added to the water.

heat = mass × sp.ht. × Δt

$$x \text{ g } NH_4Cl \times \frac{1 \text{ mol } NH_4Cl}{53.49 \text{ g}} \times \frac{14.8 \text{ kJ}}{1 \text{ mol } NH_4Cl} \times \frac{1000 \text{ J}}{1 \text{ kJ}}$$

$$= -\left(\left(1400 \text{ mL} \times \frac{1.00 \text{ g}}{1 \text{ mL}}\right) + x \text{ g } NH_4Cl\right) 4.18 \frac{\text{J}}{\text{g °C}} \ (10.-25)°C$$

$$277 x = 8.8 \times 10^4 - 63 x \qquad x = \frac{8.8 \times 10^4}{277 + 63} = 2.6 \times 10^2 \text{ g } NH_4Cl$$

43. We assume that the solution volumes are additive; that is, that 200.0 mL of solution is formed. Then we compute the heat needed to warm the solution and the cup, and then ΔH for the reaction.

$$\text{heat} = \left(200.0 \text{ mL} \times \frac{1.02 \text{ g}}{1 \text{ mL}}\right) 4.02 \frac{\text{J}}{\text{g °C}} \ (27.9 - 21.1)°C + 10 \frac{\text{J}}{\text{°C}} \ (27.9 - 21.1) = 5.6 \times 10^3 \text{ J}$$

$$\Delta H_{\text{neutr.}} = \frac{-5.6 \times 10^3 \text{J}}{0.100 \text{ mol}} \times \frac{1 \text{ kJ}}{1000 \text{ J}} = -56 \text{ kJ/mol} \quad (-56.4 \text{ kJ/mol to three significant figures})$$

44. First determine the heat absorbed during the chemical reaction.

$$\text{heat of reaction} = 150.0 \text{ mL} \times \frac{1 \text{ L}}{1000 \text{ mL}} \times \frac{2.50 \text{ mol KI}}{1 \text{ L soln}} \times \frac{21.3 \text{ kJ}}{1 \text{ mol KI}} = 7.99 \text{ kJ} = q_{\text{rxn}}$$

$$-q_{\text{rxn}} = q_{\text{soln}} = \left(150.0 \text{ mL} \times \frac{1.30 \text{ g}}{1 \text{ mL}}\right) \times \frac{2.7 \text{ J}}{\text{g °C}} \times \Delta t \qquad \Delta t = \frac{-7.99 \times 10^3 \text{ J}}{150.0 \text{ mL} \times \dfrac{1.30 \text{ g}}{1 \text{ mL}} \times \dfrac{2.7 \text{ J}}{\text{g °C}}} = -15°C$$

final t = initial $t + \Delta t = 23.5°C - 15°C = 9°C$

45. $\text{heat} = 500 \text{ mL} \times \dfrac{1 \text{ L}}{1000 \text{ mL}} \times \dfrac{7.0 \text{ mol NaOH}}{1 \text{ L soln}} \times \dfrac{-42 \text{ kJ}}{1 \text{ mol NaOH}} = -1.5 \times 10^2 \text{ kJ}$

= heat of reaction = −heat absorbed by solution   or $q_{\text{rxn}} = -q_{\text{soln}}$

$$\Delta t = \frac{1.5 \times 10^5 \text{ J}}{500. \text{ mL} \times \dfrac{1.00 \text{ g}}{1 \text{ mL}} \times \dfrac{4.2 \text{ J}}{\text{g °C}}} = 71°C \qquad \text{final } t = 21°C + 71°C = 92°C$$

Our final value is approximate because of the assumed density (1.00 g/mL) and specific heat (4.2 J g⁻¹°C⁻¹). The solution's density probably is a bit larger than 1.00 g/mL. Many aqueous solutions are somewhat more dense than water.

46. Neutralization reaction: $NaOH(aq) + HCl(aq) \longrightarrow NaCl(aq) + H_2O$

Since NaOH and HCl react in a one-to-one molar ratio, and since there is twice the volume of NaOH solution as of HCl solution, but the [HCl] is not twice the [NaOH], the HCl solution is the limiting reagent.

$$\text{heat} = 25.00 \text{ mL} \times \frac{1 \text{ L}}{1000 \text{ mL}} \times \frac{1.86 \text{ mol HCl}}{1 \text{ L}} \times \frac{1 \text{ mol } H_2O}{1 \text{ mol HCl}} \times \frac{-55.84 \text{ kJ}}{1 \text{ mol } H_2O} = -2.60 \text{ kJ}$$

= heat of reaction = −heat absorbed by solution   or $q_{\text{rxn}} = -q_{\text{soln}}$

$$\Delta t = \frac{2.60 \times 10^3 \text{ J}}{75.00 \text{ mL} \times \dfrac{1.02 \text{ g}}{1 \text{ mL}} \times \dfrac{3.98 \text{ J}}{\text{g °C}}} = 8.54°C \qquad \text{final } t = 8.54°C + 24.72°C = 33.26°C$$

Bomb calorimetry

47. Note that the heat evolved is the negative of the heat absorbed.

$$\text{heat cap.} = \frac{\text{heat evolved}}{\Delta t} = \frac{1.078 \text{ g} \times \dfrac{1 \text{ mol C}_8\text{H}_6\text{O}_4}{166.1 \text{ g C}_8\text{H}_6\text{O}_4} \times \dfrac{3224 \text{ kJ}}{1 \text{ mol C}_8\text{H}_6\text{O}_4}}{(30.76 - 24.96)°\text{C}} = 3.61 \text{ kJ/°C}$$

48. The temperature should increase as the result of an exothermic combustion reaction.

$$\Delta T = 1.227 \text{ g C}_{12}\text{H}_{22}\text{O}_{11} \times \frac{1 \text{ mol C}_{12}\text{H}_{22}\text{O}_{11}}{342.2 \text{ g C}_{12}\text{H}_{22}\text{O}_{11}} \times \frac{5.65 \times 10^3 \text{ kJ}}{1 \text{ mol C}_{12}\text{H}_{22}\text{O}_{11}} \times \frac{1 \text{ °C}}{3.87 \text{ kJ}} = 5.23°\text{C}$$

49. **(a)** The heat of combustion is the negative of the heat absorbed by the calorimeter during the reaction.

$$q_{\text{comb}} = -q_{\text{caolrim}} = -\frac{(24.92 - 31.41)°\text{C} \times 4.912 \text{ kJ/°C}}{2.051 \text{ g C}_6\text{H}_{12}\text{O}_6 \times \dfrac{1 \text{ mol C}_6\text{H}_{12}\text{O}_6}{180.2 \text{ g C}_6\text{H}_{12}\text{O}_6}} = -2.80 \times 10^3 \text{ kJ/mol C}_6\text{H}_{12}\text{O}_6$$

(b) $C_6H_{12}O_6(s) + 6 O_2(g) \longrightarrow 6 CO_2(g) + 6 H_2O(l)$ $\Delta H = -2.80 \times 10^3$ kJ

50. First determine the heat capacity of the calorimeter. (The heat evolved by the reaction is the negative of the heat of combustion.) Then, use this value to determine the value of q_{comb} for thymol.

$$\text{heat cap.} = \frac{\text{heat evolved}}{\Delta t} = \frac{1.567 \text{ g C}_{10}\text{H}_8 \times \dfrac{1 \text{ mol C}_{10}\text{H}_8}{128.2 \text{ g C}_{10}\text{H}_8} \times \dfrac{5153.9 \text{ kJ}}{1 \text{ mol C}_{10}\text{H}_8}}{8.37°\text{C}} = 7.53 \text{ kJ/°C}$$

$$q_{\text{comb}} = \frac{-6.12°\text{C} \times 7.53 \text{ kJ/°C}}{1.227 \text{ g C}_{10}\text{H}_{14}\text{O} \times \dfrac{1 \text{ mol C}_{10}\text{H}_{14}\text{O}}{150.2 \text{ g C}_{10}\text{H}_{14}\text{O}}} = -5.64 \times 10^3 \text{ kJ/mol C}_{10}\text{H}_{14}\text{O}$$

Hess's Law

51.
$$\begin{array}{ll}
2 \text{ HCl}(g) + C_2H_4(g) + \tfrac{1}{2} O_2(g) \longrightarrow C_2H_4Cl_2(g) + H_2O(l) & \Delta H = -319.6 \text{ kJ} \\
Cl_2(g) + H_2O(l) \longrightarrow 2 \text{ HCl}(g) + \tfrac{1}{2} O_2(g) & \Delta H = 0.5 \times +202.5 = +101.3 \text{ kJ} \\
\hline
C_2H_4(g) + Cl_2(g) \longrightarrow C_2H_4Cl_2(g) & \Delta H = -218.3 \text{ kJ}
\end{array}$$

52.
$$\begin{array}{ll}
N_2H_4(l) + O_2(g) \longrightarrow N_2(g) + 2 H_2O(l) & \Delta H = -622.3 \text{ kJ} \\
2 H_2O_2(l) \longrightarrow 2 H_2(g) + 2 O_2(g) & \Delta H = -2 \, (-187.8 \text{ kJ}) = +375.6 \text{ kJ} \\
2 H_2(g) + O_2(g) \longrightarrow 2 H_2O(l) & \Delta H = 2 \, (-285.8 \text{ kJ}) = -571.6 \text{ kJ} \\
\hline
N_2H_4(l) + 2 H_2O_2(l) \longrightarrow N_2(l) + 4 H_2O(l) & \Delta H = -818.3 \text{ kJ}
\end{array}$$

53.
$$\begin{array}{ll}
CH_4(g) + CO_2(g) \longrightarrow 2 CO(g) + 2 H_2(g) & \Delta H = +206 \text{ kJ} \\
2 CH_4(g) + 2 H_2O(g) \longrightarrow 2 CO(g) + 6 H_2(g) & \Delta H = 2 \, (+247 \text{ kJ}) = +494 \text{ kJ} \\
CH_4(g) + 2 O_2(g) \longrightarrow CO_2(g) + 2 H_2O(g) & \Delta H = -802 \text{ kJ} \\
\hline
4 CH_4(g) + 2 O_2(g) \longrightarrow 4 CO(g) + 8 H_2(g) & \Delta H = -102 \text{ kJ} \\
\div 4 \text{ produces} \quad CH_4(g) + \tfrac{1}{2} O_2(g) \longrightarrow CO(g) + 2 H_2(g) & \Delta H = -25.5 \text{ kJ}
\end{array}$$

54.
$$\begin{array}{ll}
CO(g) + \tfrac{1}{2} O_2(g) \longrightarrow CO_2(g) & \Delta H = -283.0 \text{ kJ} \\
3 \text{ C(graphite)} + 6 H_2(g) \longrightarrow 3 CH_4(g) & \Delta H = 3 \times -74.81 = -224.43 \text{ kJ} \\
2 H_2(g) + O_2(g) \longrightarrow 2 H_2O(l) & \Delta H = 2 \times -285.81 = -571.6 \text{ kJ} \\
3 CO(g) \longrightarrow \tfrac{3}{2} O_2(g) + 3 \text{ C(graphite)} & \Delta H = 3 \times +110.54 = +331.62 \text{ kJ} \\
\hline
4 CO(g) + 8 H_2(g) \longrightarrow CO_2(g) + 3 CH_4(g) + 2 H_2O(l) & \Delta H = -747.4 \text{ kJ}
\end{array}$$

55.
$$\begin{array}{ll}
CO(g) \longrightarrow \text{C(graphite)} + \tfrac{1}{2} O_2(g) & \Delta H = +110.5 \text{ kJ} \\
CO_2(g) + 2 H_2O(l) \longrightarrow CH_3OH(l) + \tfrac{3}{2} O_2(g) & \Delta H = +726.6 \text{ kJ} \\
2 H_2(g) + O_2(g) \longrightarrow 2 H_2O(l) & \Delta H = 2 \times -285.8 = -571.6 \text{ kJ} \\
\text{C(graphite)} + O_2(g) \longrightarrow CO_2(g) & \Delta H = -393.5 \text{ kJ} \\
\hline
CO(g) + 2 H_2(g) \longrightarrow CH_3OH(l) & \Delta H = -128.0 \text{ kJ}
\end{array}$$

56.

$CS_2(l) + 3 O_2(g) \longrightarrow CO_2(g) + 2 SO_2(g)$ $\Delta H = -1077$ kJ

$2 S(s) + Cl_2(g) \longrightarrow S_2Cl_2(g)$ $\Delta H = -60.2$ kJ

$C(s) + 2 Cl_2(g) \longrightarrow CCl_4(l)$ $\Delta H = -135.4$ kJ

$2 SO_2(g) \longrightarrow 2 S(s) + 2 O_2(g)$ $\Delta H = -2 (-296.8 \text{ kJ}) = +593.6$ kJ

$CO_2(g) \longrightarrow C(s) + O_2(g)$ $\Delta H = -(-393.5 \text{ kJ}) = +393.5$ kJ

$CS_2(l) + 3 Cl_2(g) \longrightarrow CCl_4(l) + S_2Cl_2(g)$ $\Delta H = -286$ kJ

57. The thermochemical combustion reactions follow.

$C_4H_6(g) + \frac{11}{2} O_2(g) \longrightarrow 4 CO_2(g) + 3 H_2O(l)$ $\Delta H = -2543.5$ kJ

$C_4H_{10}(g) + \frac{13}{2} O_2(g) \longrightarrow 4 CO_2(g) + 5 H_2O(l)$ $\Delta H = -2878.6$ kJ

$H_2(g) + \frac{1}{2} O_2(g) \longrightarrow H_2O(l)$ $\Delta H = -285.85$ kJ

Then these equations are combined in the following manner.

$C_4H_6(g) + \frac{11}{2} O_2(g) \longrightarrow 4 CO_2(g) + 3 H_2O(l)$ $\Delta H = -2543.5$ kJ

$4 CO_2(g) + 5 H_2O(l) \longrightarrow C_4H_{10}(g) + \frac{13}{2} O_2(g)$ $\Delta H = -(-2878.6 \text{ kJ}) = +2878.6$ kJ

$2 H_2(g) + O_2(g) \longrightarrow 2 H_2O(l)$ $\Delta H = 2 (-285.85 \text{ kJ}) = -571.7$ kJ

$C_4H_6(g) + 2 H_2(g) \longrightarrow C_4H_{10}(g)$ $\Delta H = -236.6$ kJ

Enthalpies (Heats) of Formation

58. The correct answer is (2): "The molar enthalpy of formation of $CO_2(g)$ is the molar enthalpy of combustion of C(graphite)." The formation reaction for $CO_2(g)$ is the reaction in which one mol $CO_2(g)$ is formed from 1 mole of C(graphite) and 1 mol $O_2(g)$; that is: $C(graphite) + O_2(g) \longrightarrow CO_2(g)$. This is the combustion reaction of graphite. (1) This reaction does not have an enthalpy change of zero. (3) & (4) Carbon monoxide is not involved at all in the formation reaction.

59. $\Delta H^\circ = 4 \Delta H_f^\circ[HCl(g)] + \Delta H_f^\circ[O_2(g)] - 2 \Delta H_f^\circ[Cl_2(g)] - 2 \Delta H_f^\circ[H_2O(l)]$

$= 4 (-92.31) + (0.00) - 2 (0.00) - 2 (-285.8) = +202.4$ kJ

60. $\Delta H^\circ = 2 \Delta H_f^\circ[Fe(s)] + 3 \Delta H_f^\circ[CO_2(g)] - \Delta H_f^\circ[Fe_2O_3(s)] - 3 \Delta H_f^\circ[CO(g)]$

$= 2 (0.00) + 3 (-393.5) - (-824.2) - 3 (-110.5) = -24.8$ kJ

61. Balanced equation: $C_2H_5OH(l) + 3 O_2(g) \longrightarrow 2 CO_2(g) + 3 H_2O(l)$

$\Delta H^\circ = 2 \Delta H_f^\circ[CO_2(g)] + 3 \Delta H_f^\circ[H_2O(l)] - \Delta H_f^\circ[C_2H_5OH(l)] - 3\Delta H_f^\circ[O_2(g)]$

$= 2 (-393.5) + 3 (-285.8) - (-277.7) - 3 (0.00) = -1366.7$ kJ

62. $\Delta H^\circ = -397 \text{ kJ} = \Delta H_f^\circ[CCl_4(g)] + 4 \Delta H_f^\circ[HCl(g)] - \Delta H_f^\circ[CH_4(g)] - 4 \Delta H_f^\circ[Cl_2(g)]$

$= \Delta H_f^\circ[CCl_4(g)] + 4 (-92.31) - (-74.81) - 4 (0.00) = \Delta H_f^\circ[CCl_4(g)] - 294.4$

$\Delta H_f^\circ[CCl_4(g)] = -397 + 294.4 = -103$ kJ

63. $\Delta H^\circ = -8326 \text{ kJ} = -2 \Delta H_f^\circ[C_6H_{14}(l)] - 19 \Delta H_f^\circ[O_2(g)] + 12 \Delta H_f^\circ[CO_2(g)] + 14 \Delta H_f^\circ[H_2O(g)]$

$= -2 \Delta H_f^\circ[C_6H_{14}(l)] - 19 (0.00) + 12 (-393.5) + 14 (-285.8) = -2 \Delta H_f^\circ[C_6H_{14}(l)] - 8723$

$\Delta H_f^\circ[C_6H_{14}(l)] = \dfrac{+8326 - 8723}{2} = -199$ kJ/mol

64. First determine the value of ΔH_f° for $C_5H_{12}(l)$.

Balanced combustion equation: $C_5H_{12}(l) + 8 O_2(g) \longrightarrow 5 CO_2(g) + 6 H_2O(l)$

$\Delta H^\circ = -3534 \text{ kJ} = 5 \Delta H_f^\circ[CO_2(g)] + 6 \Delta H_f^\circ[H_2O(l)] - \Delta H_f^\circ[C_5H_{12}(l)] - 8 \Delta H_f^\circ[O_2(g)]$

$= 5 (-393.51) + 6 (-285.85) - \Delta H_f^\circ[C_5H_{12}(l)] - 8 (0.00) = -3683 \text{ kJ} - \Delta H_f^\circ[C_5H_{12}(l)]$

$\Delta H_f^\circ[C_5H_{12}(l)] = -3683 + 3534 = -149$ kJ

Then use this value to determine the value of ΔH° of the reaction in question.

$\Delta H^\circ = \Delta H_f^\circ[C_5H_{12}(l)] + 5 \Delta H_f^\circ[H_2O(l)] - 5 \Delta H_f^\circ[CO(g)] - 11 \Delta H_f^\circ[H_2(g)]$

$= -149 + 5 (-285.8) - 5 (-110.5) - 11 (0.00) = -1025$ kJ

65. Let us first determine the $\Delta H°$ for the combustion: $C_3H_8(g) + 5\ O_2(g) \longrightarrow 3\ CO_2(g) + 4\ H_2O(l)$

$\Delta H° = 3\ \Delta H_f°[CO_2(g)] + 4\ \Delta H_f°[H_2O(l)] - \Delta H_f°[C_3H_8(g)] - 5\ \Delta H_f°[O_2(g)]$
$= 3\ (-393.5) + 4\ (-285.8) - (-103.8) - 5\ (0.00) = -2219\ kJ$

Then the temperature change is determined by using the heat capacity of the calorimeter.

$\Delta t = 0.806\ g\ C_3H_8 \times \dfrac{1\ mol\ C_3H_8}{44.097\ g\ C_3H_8} \times \dfrac{2219\ kJ}{1\ mol\ C_3H_8} \times \dfrac{1\ °C}{4.90\ kJ} = 8.28°C$

66. **(a)** Balanced equation: $CaCO_3(s) \longrightarrow CaO(s) + CO_2(g)$

$\Delta H° = \Delta H_f°[CaO(s)] + \Delta H_f°[CO_2(g)] - \Delta H_f°[CaCO_3(s)] = -635.1 - 393.5 - (-1207) = +178\ kJ$

$heat = 1.35 \times 10^3\ kg\ CaCO_3 \times \dfrac{1000\ g}{1\ kg} \times \dfrac{1\ mol\ CaCO_3}{100.09\ g\ CaCO_3} \times \dfrac{178\ kJ}{1\ mol\ CaCO_3} = 2.40 \times 10^6\ kJ$

(b) Determine the heat of combustion: $CH_4(g) + 2\ O_2(g) \longrightarrow CO_2(g) + 2\ H_2O(l)$

$\Delta H° = \Delta H_f°[CO_2(g)] + 2\ \Delta H_f°[H_2O(l)] - \Delta H_f°[CH_4(g)] - 2\ \Delta H_f°[O_2(g)]$
$= (-393.5) + 2\ (-285.8) - (-74.81) - (0.00) = -890.3\ kJ/mol$

Now compute the volume of gas needed, with the ideal gas equation, rearranged to: $V = nRT/P$.

$volume = \dfrac{\left(2.40 \times 10^6\ kJ \times \dfrac{1\ mol\ CH_4}{890.3\ kJ}\right) 0.08206\ \dfrac{L\ atm}{mol\ K} (24.6 + 273.2)K}{756\ mmHg \times \dfrac{1\ atm}{760\ mmHg}} = 6.62 \times 10^4\ L\ CH_4$

67. **(1)** $N_2(g) + 2\ O_2(g) \longrightarrow 2\ NO_2$ $\qquad \Delta H°_{rxn} = 2\ \Delta H_f°[NO_2(g)] = 2 \times 33.18\ kJ = 66.36\ kJ$

(2) $N_2(g) + 2\ O_2(g) \longrightarrow 2\ NO_2$ $\qquad\qquad \Delta H = 66.36\ kJ$

$2\ NO_2(g) \longrightarrow N_2O_3(g) + \frac{1}{2}\ O_2(g)$ $\qquad\quad \Delta H = 16.02\ kJ$

$N_2(g) + 2\ O_2(g) \longrightarrow N_2O_3(g) + \frac{1}{2}\ O_2(g)$ $\qquad \Delta H = 82.38\ kJ$

68. $\Delta H° = \Delta H_f°[CaSO_4(s)] + \Delta H_f°[CO_2(g)] + \Delta H_f°[H_2O(l)] - \Delta H_f°[CaCO_3(s)] - \Delta H_f°[H_2SO_4(aq,1:5)]$
$= (-1434) + (-393.5) + (-285.8) - (-1207) - (-871.5) = -35\ kJ$

69. $\Delta H° = \Delta H_f°[Al(OH)_3(s)] - \Delta H_f°[Al^{3+}(aq)] - 3\ \Delta H_f°[OH^-(aq)]$
$= (-1276) - (-531) - 3\ (-230.0) = -55\ kJ$

70. $\Delta H° = \Delta H_f°[Mg^{2+}(aq)] + 2\ \Delta H_f°[NH_3(g)] + 2\ \Delta H_f°[H_2O(l)] - \Delta H_f°[Mg(OH)_2(s)] - 2\ \Delta H_f°[NH_4^+(aq)]$
$= (-466.9) + 2\ (-46.11) + 2\ (-285.8) - (-924.5) - 2\ (-132.5) = +59.0\ kJ$

71. A plausible reaction follows, which is the basis of the $\Delta H°$ calculation. $\qquad HCl(g) \xrightarrow{\text{water}} H^+(aq) + Cl^-(aq)$

$\Delta H° = \Delta H_f°[H^+(aq)] + \Delta H_f°[Cl^-(aq)] - \Delta H_f°[HCl(g)]$
$= (0.0) + (-167.2) - (-92.31) = -74.9\ kJ$

8 THE ATMOSPHERIC GASES

AND HYDROGEN

REVIEW QUESTIONS

1. **(a)** A noble gas is one of the elements in periodic table family 8A: He, Ne, Ar, Kr, Xe, and Rn. These elements are called "noble" because they form few compounds with the other elements. Like "nobility," they do not associate with the other elements.
   (b) Water gas reactions are chemical reactions in which a combustible fuel is made (usually industrially) using water as one reactant.
   (c) A chlorofluorocarbon is a compound, derived from a hydrocarbon, that contains the elements carbon, chlorine, and fluorine. These compounds are good to excellent refrigerants, but their release into the atmosphere is implicated in the destruction of the ozone layer in the upper atmosphere.
   (d) A catalytic converter is a device placed in the exhaust system of an automobile for the main purpose of converting unburned hydrocarbons and carbon monoxide into water and carbon dioxide.

2. **(a)** Allotropy refers to an element existing in more than one form, such as diamond and graphite being the two allotropes of carbon.
   (b) Electrolysis refers to the decomposition of a compound with electricity.
   (c) A nonstoichiometric compound is one in which the elements are not present in amounts that are related as small whole numbers. Furthermore, the proportions of the elements in such compounds are variable. Often such compounds form when two elements of approximately the same size and the same charge replace each other in a compound. Another possibility is when the second element fits into holes in the lattice of the first, such as metallic hydrides.
   (d) The "greenhouse" explanation of global warming is that gases such as CO_2 absorb infrared (heat) radiation coming from the Earth's surface, not allowing it to escape into space. This trapped heat will cause the temperature of the Earth to rise abnormally rapidly.

3. **(a)** The troposphere is the first 12 km of the Earth's atmosphere, the region in which we live and within which weather occurs. The stratosphere is the next 12 to 50 km above the Earth's surface.
   (b) An allotrope is a form of an element that differs from another form in physical and chemical properties. An isotope is an alternative version of an atom that differs in the neutron number.
   (c) A "fuel lean" mixture in an engine is one in which there is more air than is needed to completely burn the fuel. In a "fuel rich" mixture, there is an abundance of fuel and insufficient air.
   (d) A metallic hydride is composed of the hydride ion and an active metal cation; it is a stoichiometric compound. A nonmetallic hydride is composed of hydrogen and a less active metal. It actually consists of hydrogen atoms in the voids present in the metallic lattice. It is a nonstoichiometric compound.

4. **(a)** O_3 ozone   **(b)** N_2O dinitrogen monoxide
   (c) KO_2 potassium superoxide   **(d)** CaH_2 calcium hydride
   (e) Mg_3N_2 magnesium nitride   **(f)** K_2CO_3 potassium carbonate
   (g) $(NH_4)H_2PO_4$ ammonium dihydrogen phosphate

5. **(a)** ammonia: NH_3 **(b)** coke: almost pure C
   (c) urea: $CO(NH_2)_2$ **(d)** limestone: principally $CaCO_3$
   (e) synthesis gas: a mixture of $CO(g)$ and $H_2(g)$, produced by steam reforming a hydrocarbon.

6. **(a)** NO_2 has N in O.S. = +4.   **(b)** KO_2 has O in O.S. = $-\frac{1}{2}$. H_2O_2 has O in O.S. = –1.
   (c) N_2O has N in O.S. = +1.   **(d)** CaH_2 has H in O.S. = –1.   **(e)**   CO has C in O.S. = +2.

7. **(a)** Small quantities of $O_2(g)$ can be prepared either by the electrolysis of water, or by the gentle heating of $KClO_3(s)$ in the presence of a $MnO_2(s)$ catalyst.

$$2\ H_2O(aq) \xrightarrow{\text{electrolysis}} 2\ H_2(g) + O_2(g) \qquad\qquad 2\ KClO_3(s) \xrightarrow{\text{heat, MnO2}} 2\ KCl(s) + 3\ O_2(g)$$

(b) Small quantities of $N_2O(g)$ can be produced by the thermal decomposition of ammonium nitrate.

$$NH_4NO_3(s) \xrightarrow{200-260°C} N_2O(g) + 2\ H_2O(g)$$

(c) Small quanitites of $H_2(g)$ can be produced by the reaction of a moderately active metal with a strong acid: $Zn(s) + 2\ HCl(aq) \longrightarrow ZnCl_2(aq) + H_2(g)$

(d) Small quantities of $CO_2(g)$ can be produced by the reaction of a metal carbonate with a strong acid.

$$CaCO_3(s) + 2\ HCl(aq) \longrightarrow CaCl_2(aq) + H_2O(l) + CO_2(g)$$

8. **(a)** $LiH(s) + H_2O \longrightarrow Li^+(aq) + OH^-(aq) + H_2(g)$ **(b)** $C(s) + H_2O(g) \xrightarrow{\Delta} CO(g) + H_2(g)$
(c) $3\ NO_2(g) + H_2O(l) \longrightarrow 2\ HNO_3(aq) + NO(g)$

9. **(a)** $Mg(s) + 2\ HCl(aq) \longrightarrow MgCl_2(aq) + H_2(g)$ **(b)** $NH_3(g) + HNO_3(aq) \longrightarrow NH_4NO_3(aq)$
(c) $MgCO_3(s) + 2\ HCl(aq) \longrightarrow MgCl_2(aq) + H_2O + CO_2(g)$
(d) $NaHCO_3(s) + HC_2H_3O_2(aq) \longrightarrow NaC_2H_3O_2(aq) + H_2O + CO_2(g)$

10. Aqueous sulfuric acid is $H_2SO_4(aq)$; aqueous ammonia is $NH_3(aq)$. The equation for their complete neutralization is: $H_2SO_4(aq) + 2\ NH_3(aq) \longrightarrow (NH_4)_2SO_4(aq)$

11. We can identify oxidizing and reducing agents by changes in oxidation state. The oxidation state of one element in an oxidizing agent is lowered, while the oxidation state of one element in a reducing agent is raised.

(8.4) $4\ NH_3(g) + 5\ O_2(g) \xrightarrow{450°C,\ Pt} 4\ NO(g) + 6\ H_2O(g)$

The oxidation state of N = –3 on the left- and = +2 on the right-hand side of this equation; the O.S. of N increases. $NH_3(g)$ is the reducing agent. The oxidation state of O = 0 on the left- and = –2 on the right-hand side of this equation; the O.S. of O decreases. $O_2(g)$ is the oxidizing agent.

(8.8) $3\ NO_2(g) + H_2O(g) \longrightarrow 2\ HNO_3(aq) + NO(g)$

The oxidation state of N = +4 on the left- and = +5 in $HNO_3(aq)$ on the right-hand side of this equation; the O.S. of N increases. $NO_2(g)$ is the reducing agent. The oxidation state of N = +4 on the left- and = +2 in $NO(g)$ on the right-hand side of this equation; the O.S. of O decreases. $NO_2(g)$ is also the oxidizing agent. This is a disproportionation reaction.

(8.14) $C_8H_{18}(l) + \frac{25}{2} O_2(g) \longrightarrow 8\ CO_2(g) + 9\ H_2O(l)$

The oxidation state of C = $-\tfrac{9}{4}$ on the left- and = +4 on the right-hand side of this equation; the O.S. of C increases. $C_8H_{18}(l)$ is the reducing agent. The oxidation state of O = 0 on the left- and = –2 on the right-hand side of this equation; the O.S. of O decreases. $O_2(g)$ is the oxidizing agent.

12. The oxide of iron with an oxidation state of +3 is $Fe_2O_3(s)$: $Fe_2O_3(s) + 3\ H_2(g) \longrightarrow 2\ Fe(s) + 3\ H_2O(g)$

13. We assume that copper is oxidized to copper(II) ion. The skeleton half-equations are as follows.
$Cu(s) \longrightarrow Cu^{2+}(aq)$ $NO_3^-(aq) \longrightarrow NO(g)$
The nitrogen-containing half-equation has its elements balanced in two steps, first for O and then for H.
$NO_3^-(aq) \longrightarrow NO(g) + 2\ H_2O$ $NO_3^-(aq) + 4\ H^+(aq) \longrightarrow NO(g) + 2\ H_2O$
The two half-equations then are balanced for charge with electrons and combined to produce the net ionic equation. Finally, nitrate spectator ions are added to produce the balanced equation.

Oxidation: $Cu(s) \longrightarrow Cu^{2+}(aq) + 2\ e^-$ × 3
Reduction: $NO_3^-(aq) + 4\ H^+(aq) + 3\ e^- \longrightarrow NO(g) + 2\ H_2O$ × 2
_____
Net: $3\ Cu(s) + 2\ NO_3^-(aq) + 8\ H^+(aq) \longrightarrow 3\ Cu^{2+}(aq) + 2\ NO(g) + 4\ H_2O$

14. The gaseous thermal decomposition product is given as the last product in each of these chemical equations.

$2\ KClO_3(s) \xrightarrow{\text{heat, MnO2}} 2\ KCl(s) + 3\ O_2(g)$ Oxygen gas is produced.

$CaCO_3(s) \xrightarrow{\Delta} CaO(s) + CO_2(g)$ Carbon dioxide gas is formed.

$NH_4NO_3(s) \xrightarrow{200-260°C} 2\ H_2O(g) + N_2O(g)$ Dinitrogen monoxide gas is formed.

15. mass NO_x = 75 × 10^9 gal × $\dfrac{15 \text{ mi}}{1 \text{ gal}}$ × $\dfrac{5 \text{ g } NO_x}{1 \text{ mi}}$ × $\dfrac{1 \text{ kg}}{1000 \text{ g}}$ = 6 × 10^9 kg NO_x

16. He is found in natural gas deposits principally because alpha particles are produced during natural radioactive decay processes. These alpha particles are ^{4}He nuclei; they obtain two electrons from the surrounding material to become helium atoms. This gaseous helium then accumulates with the natural gas trapped beneath the earth. Although other noble gases are produced by radioactive decay—notably ^{40}Ar—they are not produced in the large quantities that helium is.

17. The correct answer is natural gas. H accounts for $^4/_{16}$ = 25% of the mass in CH_4, about $^2/_{18}$ = 11% of the mass of seawater (which is principally H_2O), about $^3/_{17}$ = 18% of the mass of ammonia (NH_3), and almost none of the mass of the atmosphere.

18. The reaction is $CaH_2(s) + 2\ H_2O(l) \longrightarrow Ca(OH)_2(s) + 2\ H_2(g)$
First calculate the amount of $H_2(g)$ needed and then the mass of $CaH_2(s)$ required.

mol H_2 = $\dfrac{PV}{RT}$ = $\dfrac{726 \text{ mmHg} \times \dfrac{1 \text{ atm}}{760 \text{ mmHg}} \times 215 \text{ L}}{0.08206 \text{ L atm mol}^{-1} \text{ K}^{-1} \times 294 \text{ K}}$ = 8.51 mol H_2

mass CaH_2 = 8.51 mol H_2 × $\dfrac{1 \text{ mol } CaH_2}{2 \text{ mol } H_2}$ × $\dfrac{42.09 \text{ g } CaH_2}{1 \text{ mol } CaH_2}$ = 179 g CaH_2

19. $NH_3(g) + \frac{3}{4} O_2(g) \longrightarrow \frac{1}{2} N_2(g) + \frac{3}{2} H_2O(g)$

$\Delta H^\circ = \frac{1}{2}\Delta H^\circ_f[N_2(g)] + \frac{3}{2}\Delta H^\circ_f[H_2O(l)] - \Delta H^\circ_f[NH_3(g)] + \frac{3}{4}\Delta H^\circ_f[O_2(g)]$

$= 0.5 \times (0.00) + 1.5 \times (-241.8) - (-46.11) - 0.75 \times (0.00) = -316.6$ kJ

EXERCISES

The Atmosphere

20. We assume that the gases in the atmosphere obey the ideal gas law, $PV = nRT$. Then the amount in moles of a particular gas will be given by $n = \dfrac{PV}{RT}$. Since all of the gases in the mixture are at the same temperature and subjected to the same total pressure, their amounts in moles are proportional to their volumes, with the same proportionality constant for each gas in the mixture. Therfore, volume percents and mole percents will be the same.

21. The composition of air on a mass percent basis is not the same as on a volume percent basis. The easiest way to see this is to realize that volume percents and mole percents are equal (Exercise 20), and that different gases have different mole weights. The gases with the heavier mole weights—CO_2, Ar, and O_2—will have a higher percent by mass than by volume. The calculation for the four most abundant gases in air, based on 100.00 moles of air, follows.

mass N_2 = 78.084 mol N_2 × $\dfrac{28.0135 \text{ g } N_2}{1 \text{ mol } N_2}$ = 2187.4 g N_2

mass O_2 = 20.946 mol O_2 × $\dfrac{31.9988 \text{ g } O_2}{1 \text{ mol } O_2}$ = 670.25 g O_2

mass Ar = 0.934 mol Ar × $\dfrac{39.95 \text{ g Ar}}{1 \text{ mol Ar}}$ = 37.3 g   mass CO_2 = 0.035 mol CO_2 × $\dfrac{44.01 \text{ g } CO_2}{1 \text{ mol } CO_2}$ = 1.5 g

N_2 mass% = $\dfrac{2187.4 \text{ g } N_2}{2187.4 \text{ g } N_2 + 670.25 \text{ g } O_2 + 37.3 \text{ g Ar} + 1.5 \text{ g } CO_2}$ × 100%

$= \dfrac{2187.4 \text{ g } N_2}{2896.5 \text{ g total}}$ × 100% = 75.519 % N_2 O_2 mass% = $\dfrac{670.25 \text{ g } O_2}{2896.5 \text{ g total}}$ × 100% = 23.140% O_2

Ar mass% = $\dfrac{37.3 \text{ g Ar}}{2896.5 \text{ g total}}$ × 100% = 1.29% Ar CO_2 mass% = $\dfrac{1.5 \text{ g } CO_2}{2896.5 \text{ g total}}$ × 100% = 0.052% CO_2

22. The product of the molecular sieve method—95% O_2 and 5% Ar—can be used in most of the applications that call for oxygen. Since argon is almost totally inert, it will not combine in most of the chemical reactions in which oxygen is used. Thus, other than diluting the final product, the argon impurity will have little or no effect. Consequently, a chemical manufacturer will choose the less expensive reagent when both give the same products.

23. The relative humidity compares the pressure exerted by water vapor, P_{water}, to the vapor pressure of water.

$$P_{water} = \frac{\left(0.10 \text{ mL} \times \frac{0.998 \text{ g}}{1 \text{ mL}} \times \frac{1 \text{ mol H}_2\text{O}}{18.02 \text{ g}}\right) \times \frac{0.08206 \text{ L atm}}{\text{mol K}} \times (20 + 273) \text{ K}}{25.05 \text{ L}} \times \frac{760 \text{ mmHg}}{1 \text{ atm}}$$

$$= 4.0 \text{ mmHg}$$

$$\% \text{ relative humidity} = \frac{4.0 \text{ mmHg from water}}{17.5 \text{ mmHg}} \times 100\% = 23\% \text{ relative humidity}$$

Nitrogen

24. (a) The Haber-Bosch process is the principal artificial method of fixing atmospheric nitrogen.
   $$N_2(g) + 3 \text{ H}_2(g) \rightleftharpoons 2 \text{ NH}_3(g)$$
   (b) The first step of the Ostwald process: $4 \text{ NH}_3(g) + 5 \text{ O}_2(g) \xrightarrow{450°C, Pt} 4 \text{ NO}(g) + 6 \text{ H}_2\text{O}(g)$
   (c) The second and third steps of the Ostwald process: $2 \text{ NO}(g) + O_2(g) \longrightarrow 2 \text{ NO}_2(g)$
   $$3 \text{ NO}_2(g) + H_2\text{O}(g) \longrightarrow 2 \text{ HNO}_3(aq) + \text{NO}(g)$$

25. Decomposition of $HNO_3(l)$: $\quad HNO_3(l) \longrightarrow N_2O_4(g) + H_2O(l) + O_2(g)$
Although this equation can be balanced by the half-reaction method, it also can be balanced by inspection. First notice that all of the N is present in $N_2O_4(g)$ in the products. This implies that there should be two $HNO_3(aq)$: $\quad 2 \text{ HNO}_3(l) \longrightarrow N_2O_4(g) + H_2O(l) + O_2(g)$ With this addition, N and H are balanced, and we need to only consider balancing oxygen. There are 6 O's on the left and 7 O's on the right, an imbalance that can be corrected either by making $1/2$ the coefficient of $O_2(g)$ or by doubling all of the other coefficients and leaving the $O_2(g)$ coefficient alone: $\quad 4 \text{ HNO}_3(l) \longrightarrow 2 \text{ N}_2\text{O}_4(g) + 2 \text{ H}_2\text{O}(l) + O_2(g)$

26. The two skeleton half-equations are: $\quad Zn(s) \longrightarrow Zn^{2+}(aq) \quad$ and $\quad NO_3^-(aq) \longrightarrow NH_4^+(aq)$
The nitrogen-containing half-equation has its elements balanced in two steps, first for O and then for H.
$NO_3^-(aq) \longrightarrow NH_4^+(aq) + 3 \text{ H}_2\text{O} \quad\quad\quad NO_3^-(aq) + 10 \text{ H}^+(aq) \longrightarrow NH_4^+(aq) + 3 \text{ H}_2\text{O}$
The two half-equations then are balanced for charge with electrons and combined to produce the net ionic equation. Finally, nitrate spectator ions are added to produce the balanced equation.

Oxidation: $\quad Zn(s) \longrightarrow Zn^{2+}(aq) + 2 \text{ e}^- \quad\quad\quad\quad\quad\quad\quad \times 4$
Reduction: $\quad NO_3^-(aq) + 10 \text{ H}^+(aq) + 8 \text{ e}^- \longrightarrow NH_4^+(aq) + 3 \text{ H}_2\text{O}$

Net: $\quad 4 \text{ Zn}(s) + NO_3^-(aq) + 10 \text{ H}^+(aq) \longrightarrow 4 \text{ Zn}^{2+}(aq) + NH_4^+(aq) + 3 \text{ H}_2\text{O}$
Spectators: $\quad\quad\quad\quad 9 \text{ NO}_3^-(aq) \longrightarrow 9 \text{ NO}_3^-(aq)$

Result: $\quad 4 \text{ Zn}(s) + 10 \text{ HNO}_3(aq) \longrightarrow 4 \text{ Zn}(NO_3)_2(aq) + NH_4NO_3(aq) + 3 \text{ H}_2\text{O}$

27. (a) $2 \text{ NH}_4NO_3(s) \xrightarrow{400°C} 2 \text{ N}_2(g) + O_2(g) + 4 \text{ H}_2\text{O}(g) \quad\quad$ The equation was balanced by inspection, by realizing first that each mole of $NH_4NO_3(s)$ would produce 1 mol $N_2(g)$ and 2 mol $H_2O(l)$, with 1 mol O [or $1/2$ mol $O_2(g)$] remaining. Then each coefficient was doubled.
   (b) $NaNO_3(s) \longrightarrow NaNO_2(s) + 1/2 \text{ O}_2(g) \quad$ or $\quad 2 \text{ NaNO}_3(s) \longrightarrow 2 \text{ NaNO}_2(s) + O_2(g)$
   (c) $Pb(NO_3)_2(s) \longrightarrow PbO(s) + 2 \text{ NO}_2(g) + 1/2 \text{ O}_2(g)$
   $2 \text{ Pb}(NO_3)_2(s) \longrightarrow 2 \text{ PbO}(s) + 4 \text{ NO}_2(g) + O_2(g)$

28. We recall that 1 mol of gas occupies 22.414 L at STP.

$$\text{mass N}_2 = 7.70 \times 10^{11} \text{ ft}^3 \times \left(\frac{12 \text{ in.}}{1 \text{ ft}} \times \frac{2.54 \text{ cm}}{1 \text{ in.}}\right)^3 \times \frac{1 \text{ L}}{1000 \text{ cm}^3} \times \frac{1 \text{ mol N}_2}{22.414 \text{ L}} \times \frac{28.01 \text{ g N}_2}{1 \text{ mol}}$$

$$\times \frac{1 \text{ kg}}{1000 \text{ g}} = 2.72 \times 10^{10} \text{ kg N}_2$$

29. $\text{mass\% HNO}_3 = \dfrac{15 \text{ mol HNO}_3 \times \dfrac{63.01 \text{ g HNO}_3}{1 \text{ mol HNO}_3}}{1 \text{ L soln} \times \dfrac{1000 \text{ mL}}{1 \text{ L}} \times \dfrac{1.41 \text{ g}}{1 \text{ mL soln}}} \times 100\% = 67\% \text{ HNO}_3$

Oxygen

30. (a) $2 HgO(s) \xrightarrow{\Delta} 2 Hg(l) + O_2(g)$ **(b)** $2 KClO_4(s) \xrightarrow{\Delta} 2 KClO_3(s) + O_2(g)$

31. $2 NH_4NO_3(s) \xrightarrow{400°C} 2 N_2(g) + O_2(g) + 4 H_2O(g)$ $2 H_2O_2(l) \longrightarrow 2 H_2O(l) + O_2(g)$

$2 KClO_3(s) \xrightarrow{\Delta} 2 KCl(s) + 3 O_2(g)$

(a) All three reactions have two moles of reactant, and the decomposition of potassium chlorate produces three moles of $O_2(g)$, compared to one mol $O_2(g)$ for each of the others. The potassium chlorate decomposition produces the most oxygen per mole of reactant.

(b) If each reaction produced the same amount of $O_2(g)$, then the one with the lightest mass of reactant would produce the most oxygen per gram of reactant. When the first two reactions are compared, it is clear that 2 mol H_2O_2 have less mass than 2 mol NH_4NO_3. (Notice that each mol NH_4NO_3 contains the amounts of elements—2 mol H and 2 mol O—present in each mol H_2O_2 plus some more besides.) So now we need to compare hydrogen peroxide with potassium chlorate. Notice that 6 H_2O_2 produces the same amount of $O_2(g)$ as 2 $KClO_3$. 6 mol H_2O_2 has a mass of $34.01 \times 6 = 204.1$ g, while 2 mol $KClO_3$ has a mass of $122.2 \times 2 = 245.2$ g. Thus, H_2O_2 produces the most $O_2(g)$ per gram of reactant.

32. (a) $O_3(g) + 2 I^-(aq) + 2 H^+(aq) \longrightarrow I_2(aq) + H_2O + O_2(g)$

(b) $S(s) + H_2O(l) + 3 O_3(g) \longrightarrow H_2SO_4(aq) + 3 O_2(g)$

(c) $2 [Fe(CN)_6]^{4-}(aq) + O_3(g) + H_2O \longrightarrow 2 [Fe(CN)_6]^{3-}(aq) + O_2(g) + 2 OH^-(aq)$

33. The net reaction for the system follows. $4 KO_2(s) + 2 CO_2(g) \longrightarrow 2 K_2CO_3(s) + 3 O_2(g)$

We determine the volume of $O_2(g)$ produced when 0.82 L $CO_2(g)$ is consumed, assuming that both gases are at the same temperature and pressure.

volume $O_2(g) = 0.82$ L $CO_2(g) \times \dfrac{3 \text{ L } O_2(g)}{2 \text{ L } CO_2(g)} = 1.23$ L $CO_2(g)$

The potassium superoxide system produces more than 1.00 L $O_2(g)$ for every 0.82 L $CO_2(g)$ consumed and is therefore a satisfactory atmospheric regeneration system.

34. The combustion reaction: $C_8H_{18}(l) + {}^{25}/_2 O_2(g) \longrightarrow 8 CO_2(g) + 9 H_2O(l)$

First compute the enthalpy change for combustion.

$\Delta H° = 8 \Delta H_f°[CO_2(g)] + 9 \Delta H_f°[H_2O(l)] - \Delta H_f°[C_8H_{18}(l)] - {}^{25}/_2 \Delta H_f°[O_2(g)]$

$= 8 (-393.5) + 9 (-285.8) - (-250.0) - {}^{25}/_2 (0.00) = -5470.$ kJ/mol C_8H_{18}

Then determine the heat produced from each gallon of gasoline.

heat produced $= 1.00$ gal $\times \dfrac{3.785 \text{ L}}{1 \text{ gal}} \times \dfrac{1000 \text{ mL}}{1 \text{ L}} \times \dfrac{0.703 \text{ g}}{1 \text{ mL}} \times \dfrac{1 \text{ mol } C_8H_{18}}{114.23 \text{ g } C_8H_{18}} \times \dfrac{5470 \text{ kJ}}{1 \text{ mol } C_8H_{18}}$

$= 1.27 \times 10^5$ kJ of heat $\Delta H = -1.27 \times 10^5$ kJ

35. $P = \dfrac{nRT}{V} = \dfrac{\left(5 \times 10^{12} \text{ molecules} \times \dfrac{1 \text{ mol } O_3}{6.022 \times 10^{23} \text{ molec.}}\right) \times \dfrac{0.08206 \text{ L atm}}{\text{mol K}} \times 220 \text{ K}}{1 \text{ cm}^3 \times \dfrac{1 \text{ L}}{1000 \text{ cm}^3}}$

$\times \dfrac{760 \text{ mmHg}}{1 \text{ atm}} = 1 \times 10^{-4}$ mmHg

36. If h is the thickness of the O_3 layer and r is the radius of the earth,

volume $O_3 = 4 \pi r^2 h = 4 \times 3.142 \left(4 \times 10^3 \text{ mi} \times \dfrac{5280 \text{ ft}}{1 \text{ mi}} \times \dfrac{12 \text{ in.}}{1 \text{ ft}} \times \dfrac{2.54 \text{ cm}}{1 \text{ in.}} \right)^2 0.3$ cm

$= 1._6 \times 10^{18}$ cm$^3 = 1._6 \times 10^{15}$ L

no. O_3 molecules $= 1._6 \times 10^{15}$ L $\times \dfrac{1 \text{ mol } O_3 \text{ STP}}{22.414 \text{ L}} \times \dfrac{6.022 \times 10^{23} \text{ molecules}}{1 \text{ mol } O_3} = 4._2 \times 10^{37} O_3$ molecules

37. The electrolysis reaction is $2 H_2O(l) \xrightarrow{\text{electrolysis}} 2 H_2(g) + O_2(g)$ In this reaction, 2 moles of $H_2(g)$ are produced for each mole of $O_2(g)$, by the law of combining volumes, we would expect the volume of hydrogen to be twice the volume of oxygen produced. (Actually the volumes are not exactly in the ratio of 2:1 because of the different solubilities of oxygen and hydrogen in water.)

38. The electrolysis reaction is $2 H_2O(l) \xrightarrow{\text{electrolysis}} 2 H_2(g) + O_2(g)$ In this reaction, 2 moles of water are decomposed to produce each mole of $O_2(g)$. The data in the problem are used to detetmine the amount of $O_2(g)$ produces, which then is converted to the mass of H_2O needed.

$$\text{mass } H_2O(l) = \frac{\left(735.2 \text{ mmHg} \times \frac{1 \text{ atm}}{760 \text{ mmHg}}\right) \times \left(25.08 \text{ mL} \times \frac{1 \text{ L}}{1000 \text{ mL}}\right)}{\frac{0.08206 \text{ L atm}}{\text{mol K}} \times (25.0 + 273.2) \text{ K}} \times \frac{2 \text{ mol } H_2O}{1 \text{ mol } O_2}$$

$$\times \frac{18.02 \text{ g } H_2O}{1 \text{ mol } H_2O} = 0.03573 \text{ g } H_2O = 35.73 \text{ mg } H_2O$$

39. In the stratosphere, ozone is acts to absorb ultraviolet radiation that could be harmful if it reached the surface of the Earth. Thus, ozone in the stratosphere acts rather as a shield, protecting the life below. But ozone itself is a highly reactive substance which can and does cause harm to living things when it is in contact with them in the troposphere, where its concentration should be very low.

The Noble Gases

40. First use the ideal gas law to determine the number of moles of argon.

$$n = \frac{PV}{RT} = \frac{137 \text{ atm} \times 55 \text{ L}}{0.08206 \text{ L atm mol}^{-1} \text{ K}^{-1} \times 296 \text{ K}} = 3.1 \times 10^2 \text{ mol Ar}$$

no. L air $= 310.$ mol Ar $\times \frac{22.414 \text{ L Ar STP}}{1 \text{ mol Ar}} \times \frac{100.000 \text{ L air}}{0.934 \text{ L Ar}} = 7.4 \times 10^5 \text{ L air}$

41. 1 mole of the mixture at STP occupies a volume of 22.414 L. It contains 0.79 mol He and 0.21 mol O_2.

$$\text{STP density} = \frac{\text{mass}}{\text{volume}} = \frac{0.79 \text{ mol He} \times \frac{4.003 \text{ g He}}{1 \text{ mol He}} + 0.21 \text{ mol } O_2 \times \frac{32.0 \text{ g } O_2}{1 \text{ mol } O_2}}{22.414 \text{ L}} = 0.44 \text{ g/L}$$

25°C is a temperature higher than STP and 745 mmHg is a lower pressure. Both of these conditions increase the 1.00-L volume that contains 0.44 g of the mixture at STP. We can calculate the expanded volume with the combined gas law.

$V_{\text{final}} = 1.00 \text{ L} \times \frac{760 \text{ mmHg}}{745 \text{ mmHg}} \times \frac{(25 + 273.2) \text{ K}}{273.2 \text{ K}} = 1.11 \text{ L}$ $\qquad$ final density $= \frac{0.44 \text{ g}}{1.11 \text{ L}} = 0.40 \text{ g/L}$

42. Nitrogen extracted from air gave the higher mass. Nitrogen obtained from a nitrogen compound is pure nitrogen with a molar mass of 28.01 g/mol. Nitrogen was obtained from air by removing all of the oxygen by combustion and then removing the H_2O and CO_2. This leaves almost pure nitrogen, but with a little argon (of higher mole weight, 39.948 g/mol Ar) present. We can calculate the average molar mass of this mixture, from the 78.084% N_2 and 0.934% Ar in dry air. Recall that volume percents and mole percents are equal.

$$\text{molar mass} = \frac{\left(78.084 \text{ mol } N_2 \times \frac{28.01348 \text{ g } N_2}{1 \text{ mol } N_2}\right) + \left(0.934 \text{ mol Ar} \times \frac{39.948 \text{ g Ar}}{1 \text{ mol Ar}}\right)}{78.084 \text{ mol } N_2 + 0.934 \text{ mol Ar}}$$

$$= \frac{2187.4 \text{ g } N_2 + 37.3 \text{ g Ar}}{79.018 \text{ mol total}} = 28.154 \text{ g/mol "}N_2 \text{ from air"}$$

Carbon

43. (a) $2 C_6H_{14}(l) + 19 O_2(g) \longrightarrow 12 CO_2(g) + 14 H_2O(l)$

(b) $PbO(s) + CO(g) \xrightarrow{\text{heat}} Pb(s) + CO_2(g)$

(c) $2 KOH(aq) + CO_2(g) \longrightarrow K_2CO_3(aq) + H_2O$

(d) $MgCO_3(s) + 2 HCl(aq) \longrightarrow MgCl_2(aq) + H_2O(l) + CO_2(g)$

44. (a) $HC_2H_3O_2(aq) + NaHCO_3(s) \longrightarrow NaC_2H_3O_2(aq) + H_2O + CO_2(g)$

(b) $ZnO(s) + CO(g) \longrightarrow Zn(l) + CO_2(g)$

(c) $2 CO(g) + 3 H_2(g) \longrightarrow CH_2OHCH_2OH(l)$

45. The reaction referred to is exemplified by $\qquad$ $CaCO_3(s) + 2 HCl(aq) \longrightarrow CaCl_2(aq) + CO_2(g) + H_2O$

the action of hydrochloric acid on calcium carbonate. Notice that no oxidation states change in this reaction; it is a double displacement reaction followed by the decomposition of "$H_2CO_3(aq)$." Since the carbon dioxide produced in this reaction is not formed from elemental carbon or by the oxidation of a carbon-containing compound, it cannot be partially oxidized. Thus, no $CO(g)$ will form.

46. First we write the half-equation for the reduction of permanganate ion. This begins with the skeleton half-equation $MnO_4^-(aq) \longrightarrow Mn^{2+}(aq)$ which is balanced with H_2O's and $H^+(aq)$'s

$MnO_4^-(aq) \longrightarrow Mn^{2+}(aq) + 4 H_2O$ $\qquad$ $MnO_4^-(aq) + 8 H^+(aq) \longrightarrow Mn^{2+}(aq) + 4 H_2O$

and finally charge balanced with e^-'s $\qquad$ $MnO_4^-(aq) + 8 H^+(aq) + 5 e^- \longrightarrow Mn^{2+}(aq) + 4 H_2O$

(a) The skeleton half-equation $H_2S(g) \longrightarrow SO_4^{2-}(aq)$ is balanced with H_2O's and $H^+(aq)$'s

$H_2S(g) + 4 H_2O \longrightarrow SO_4^{2-}(aq)$ $\qquad$ $H_2S(g) + 4 H_2O \longrightarrow SO_4^{2-}(aq) + 10 H^+(aq)$

and then charge balanced with e^-'s $\qquad$ $H_2S(g) + 4 H_2O \longrightarrow SO_4^{2-}(aq) + 10 H^+(aq) + 8 e^-$

The balanced oxidation and reduction half-equations are combined to form the net equation.

Oxidation: $H_2S(g) + 4 H_2O \longrightarrow SO_4^{2-}(aq) + 10 H^+(aq) + 8 e^-$ $\qquad \times 5$

Reduction: $MnO_4^-(aq) + 8 H^+(aq) + 5 e^- \longrightarrow Mn^{2+}(aq) + 4 H_2O$ $\qquad \times 8$

Net: $\qquad$ $8 MnO_4^-(aq) + 5 H_2S(g) + 14 H^+(aq) \longrightarrow 8 Mn^{2+}(aq) + 5 SO_4^{2-}(aq) + 12 H_2O$

(b) The skeleton half-equation $SO_2(g) \longrightarrow SO_4^{2-}(aq)$ is balanced with H_2O's and $H^+(aq)$'s

$SO_2(g) + 2 H_2O \longrightarrow SO_4^{2-}(aq)$ $\qquad$ $SO_2(g) + 2 H_2O \longrightarrow SO_4^{2-}(aq) + 4 H^+(aq)$

and then charge balanced with e^-'s $\qquad$ $SO_2(g) + 2 H_2O \longrightarrow SO_4^{2-}(aq) + 4 H^+(aq) + 2 e^-$

The balanced oxidation and reduction half-equations are combined to form the net equation.

Oxidation: $SO_2(g) + 2 H_2O \longrightarrow SO_4^{2-}(aq) + 4 H^+(aq) + 2 e^-$ $\qquad \times 5$

Reduction: $MnO_4^-(aq) + 8 H^+(aq) + 5 e^- \longrightarrow Mn^{2+}(aq) + 4 H_2O$ $\qquad \times 2$

Net: $\qquad$ $2 MnO_4^-(aq) + 5 SO_2(g) + 2 H_2O \longrightarrow 2 Mn^{2+}(aq) + 5 SO_4^{2-}(aq) + 4 H^+(aq)$

47. Both carbon dioxide and chlorofluorocarbons are greenhouse gases. They absorb infrared radiation, preventing this heat from leaving the Earth. But only the chlorine released by chlorofluorocarbons plays a role in the destruction of ozone in the upper atmosphere.

48. Let us determine the difference between the two reactions.

combustion: $\qquad$ $C_8H_{18}(l) + {}^{25}/_2 O_2(g) \longrightarrow 8 CO_2(g) + 9 H_2O(l)$

$-$(incomplete): $\qquad$ $7 CO_2(g) + CO(g) + 9 H_2O(l) \longrightarrow C_8H_{18}(l) + 12 O_2(g)$

difference: $\qquad$ $CO(g) + {}^1/_2 O_2(g) \longrightarrow CO_2(g)$

The heat the accompanies the "difference" reaction is the difference in the heat of the two reactions. To determine the percent of the maximum possible heat this is, first compute the enthalpy change for the combustion reaction.

$\Delta H° = 8 \Delta H_f°[CO_2(g)] + 9 \Delta H_f°[H_2O(l)] - \Delta H_f°[C_8H_{18}(l)] - {}^{25}/_2 \Delta H_f°[O_2(g)]$

$= 8 (-393.5) + 9 (-285.8) - (-250.0) - {}^{25}/_2 (0.00) = -5470.$ kJ/mol C_8H_{18}

Then compute the enthalpy change for the "difference" reaction.

$\Delta H° = \Delta H_f°[CO_2(g)] - \Delta H_f°[CO(g)] - {}^1/_2 \Delta H_f°[O_2(g)] = (-393.5) - (-110.5) - {}^1/_2 (0.00) = -504$ kJ

And finally compute the percent decrease that the difference reaction represents.

%decrease $= \dfrac{-504 \text{ kJ}}{-5470 \text{ kJ}} \times 100\% = 9.21\%$ decrease

Hydrogen

49. **(a)** $2 Al(s) + 6 HCl(aq) \longrightarrow 2 AlCl_3(aq) + 3 H_2(g)$

$\quad$ **(b)** $C_3H_8(g) + 3 H_2O(g) \longrightarrow 3 CO(g) + 7 H_2(g)$

$\quad$ **(c)** $MnO_2(s) + 2 H_2(g) \longrightarrow Mn(s) + 2 H_2O(g)$

50. **(a)** $2 H_2O(l) \xrightarrow{\text{electricity}} O_2(g) + 2 H_2(g)$

$\quad$ **(b)** $2 HI(aq) + Zn(s) \longrightarrow MgI_2(aq) + H_2(g)$ $\qquad$ Any moderately active metal is suitable.

(c) $Mg(s) + H_2SO_4(aq) \longrightarrow MgSO_4(aq) + H_2(g)$ Any strong acid is suitable.

(d) $CO(g) + H_2O(g) \longrightarrow CO_2(g) + H_2(g)$

51. $CaH_2(s) + 2\,H_2O \longrightarrow Ca(OH)_2(aq) + 2\,H_2(g)$ $Ca(s) + 2\,H_2O \longrightarrow Ca(OH)_2(aq) + H_2(g)$

$2\,Na(s) + 2\,H_2O \longrightarrow 2\,NaOH(aq) + H_2(g)$

The one of these three that produces the largest volume of $H_2(g)$ per liter of water also produces the largest amount of hydrogen gas per mole of water used. All three reactions use two moles of water and the reaction with $CaH_2(s)$ produces the greatest amount—2 moles—of hydrogen gas.

We can compare three reactions that produce the same amount of hydrogen; the one that requires the smallest mass of solid will have produced the greatest amount of hydrogen per gram of solid. The amount of hydrogen we choose is 2 moles, which meanse that we compare 1 mol CaH_2 (42.09 g) with 2 mol Ca (80.16 g) with 4 mol Na (91.96 g). Clearly CaH_2 produces the greatest amount of H_2 per gram of solid.

9 ELECTRONS IN ATOMS

REVIEW QUESTIONS

1. (a) λ is the symbol for wavelength, the distance between like features (two peaks, for instance) of successive waves.

   (b) ν is the symbol for frequency, the number of times that a particular feature of successive waves (such as a peak) passes a fixed point in a second.

   (c) h is the symbol for Planck's constant, which relates the energy and frequency of radiation: $E = h\nu$.

   (d) ψ is the symbol for the wavefunction, the function that contains all the information known about a given atomic or molecular system.

   (e) The principal quantum number, n, is related to both the distance of an electron from the nucleus and to the electron's energy.

2. (a) The atomic (line) spectrum is the result of atoms being energized and then radiating energy as they lose that energy and return to lower energy states. Since only certain energy levels are permitted for each atom, the energies radiated are definite, giving rise to light of only certain wavelengths, or lines in the spectrum.

   (b) The photoelectric effect occurs when light (*photo-*) shining on a surface causes electrons (*-electric*) to be emitted from that surface.

   (c) The wave-particle duality refers to the nature of small particles and electromagnetic radiation to have properties both of particles and of waves.

   (d) The Heisenberg uncertainty principle states that it is impossible to simultaneously determine the position and linear momentum of a particle to any desired degree of precision.

   (e) Electron spin is that property of electrons that makes them appear as if they were spinning, or rotating, on an axis. Electrons can spin in only one of two states: clockwise (spin up) or counterclockwise (spin down).

   (f) Hund's rule state that electrons will fill a subshell by entering orbitals unpaired, rather than first pairing up with two electrons in each orbital.

3. (a) The frequency of radiation is the number of oscillations that occur per second (as the radiation is observed passing a fixed point), the wavlength is the distance between two wave peaks on successive waves.

   (b) Ultraviolet light is radiation that has wavelengths just a bit shorter than those of visible light. Infrared light has wavelengths just a bit longer than visible light.

   (c) A continuous spectrum is visible light of all wavelengths that has been spread out by a prism or a grating. A discontinuous spectrum is light of only certain wavelengths, such as light emitted from an atomic species that has been excited.

   (d) A traveling wave is one whose wave peaks move past a given point as time passes, such as waves in the ocean. A standing wave is one whose nodes remain fixed, such as sound waves in an organ pipe or the waves in a violin string.

   (e) A quantum number is one of the four (principal, orbital, magnetic, spin) definite values that are used to specify the properties of an electron in an atom. An orbital is a region of space in which there is a good chance of finding an electron.

(f) *spdf* notation specifies the electron configuration of an atom by giving the principal and orbital quantum numbers of electrons, grouped into subshells. An orbital diagram, places each electron of an atom, symbolized as an arrow, into a space that represents an orbital.

4. (a) length (nm) = $1875 \, Å \times \dfrac{1 \, m}{10^{10} \, Å} \times \dfrac{10^9 \, nm}{1 \, m} = 187.5 \, nm$

   (b) length (μm) = $2350 \, Å \times \dfrac{1 \, m}{10^{10} \, Å} \times \dfrac{10^6 \, μm}{1 \, m} = 0.235 \, μm$

   (c) length (nm) = $4.67 \times 10^{-5} \, m \times \dfrac{10^9 \, nm}{1 \, m} = 4.67 \times 10^4 \, nm$

   (d) length (m) = $403 \, nm \times \dfrac{1 \, m}{10^9 \, nm} = 4.03 \times 10^{-7} \, m$

   (e) length (nm) = $1.72 \, cm \times \dfrac{1 \, m}{10^2 \, cm} \times \dfrac{10^9 \, nm}{1 \, m} = 1.72 \times 10^7 \, nm$

   (f) length (cm) = $5.2 \times 10^5 \, Å \times \dfrac{1 \, m}{10^{10} \, Å} \times \dfrac{10^2 \, cm}{1 \, m} = 5.2 \times 10^{-3} \, cm$

5. (a) $\lambda = \dfrac{c}{v} = \dfrac{3.00 \times 10^8 \, m/s}{4.5 \times 10^{13} \, s^{-1}} = 6.7 \times 10^{-6} \, m$ infrared

   (b) $\lambda = \dfrac{c}{v} = \dfrac{3.00 \times 10^8 \, m/s}{6.2 \times 10^{16} \, s^{-1}} = 4.8 \times 10^{-9} \, m$ ultraviolet

   (c) $\lambda = \dfrac{c}{v} = \dfrac{2.998 \times 10^8 \, m/s}{7.03 \times 10^5 \, s^{-1}} = 426 \, m$ radio

   (d) $\lambda = \dfrac{c}{v} = \dfrac{2.998 \times 10^8 \, m/s}{2.54 \times 10^8 \, s^{-1}} = 1.18 \, m$ radio

6. Since frequency and wavelength are inversely related ($v = c/\lambda$), the shortest wavelength will correspond to the highest frequency. Since 1 mm = 0.1 cm, 6.7×10^{-4} cm is smaller than 1.35 mm. Since 1 cm = 10^{-2} m and 1 μm = 10^{-6} m, 5.89 μm is just a bit smaller than 6.7×10^{-4} cm (6.7 μm). And since 1 Å = 10^{-10} m, 912 Å is smaller than 5.89 μm (5.89×10^4 Å).

7. The wavelength is the distance between successive peaks. Thus, $4 \times 1.04 \, nm = \lambda = 4.16 \, nm$

8. (a) $v = \dfrac{c}{\lambda} = \dfrac{3.00 \times 10^8 \, m/s}{4.16 \, nm \times \dfrac{1 \, m}{10^9 \, nm}} = 7.21 \times 10^{16} \, Hz$

   (b) $E = hv = 6.626 \times 10^{-34} \, J \, s \times 7.21 \times 10^{16} \, Hz = 4.78 \times 10^{-17} \, J$

9. (a) The velocity of electromagnetic radiation in a vacuum, the speed of light, is a universal constant.

   (b) The wavelength of electromagnetic radiation is inversely proportional to its frequency: $\lambda = c/v$

   (c) The energy per photon of electromagnetic radiation is directly proportional to its frequency: $E = hv$.

10. (a) $v = 3.2881 \times 10^{15} \, s^{-1} \left(\dfrac{1}{2^2} - \dfrac{1}{n^2} \right) = 3.2881 \times 10^{15} \, s^{-1} \left(\dfrac{1}{2^2} - \dfrac{1}{4^2} \right) = 6.1652 \times 10^{14} \, s^{-1}$

   (b) $v = 3.2881 \times 10^{15} \, s^{-1} \left(\dfrac{1}{2^2} - \dfrac{1}{6^2} \right) = 7.3069 \times 10^{14} \, s^{-1}$

   $\lambda = \dfrac{2.9979 \times 10^8 \, m/s}{7.3069 \times 10^{14} \, s^{-1}} = 4.1028 \times 10^{-7} \, m \times \dfrac{10^9 \, nm}{1 \, m} = 410.28 \, nm$

   (c) $v = \dfrac{3.00 \times 10^8 \, m}{380 \, nm \times \dfrac{1 \, m}{10^9 \, nm}} = 7.89 \times 10^{14} \, s^{-1} = 3.2881 \times 10^{15} \, s^{-1} \left(\dfrac{1}{2^2} - \dfrac{1}{n^2} \right)$

   $0.250 - \dfrac{1}{n^2} = \dfrac{7.89 \times 10^{14} \, s^{-1}}{3.2881 \times 10^{15} \, s^{-1}} = 0.240$ $\dfrac{1}{n^2} = 0.250 - 0.240 = 0.010$ $n = 10$

11. The frequencies of hydrogen emission lines in the infrared region of the spectrum other than the visible region would be predicted by replacing the constant "2" in the Balmer equation by the variable *m*, where *m* is an integer smaller than *n*: *m* = 3, 4, ... The resulting equation is $v = 3.2881 \times 10^{15} \, s^{-1} \left(\dfrac{1}{m^2} - \dfrac{1}{n^2} \right)$

12. (a) $E = hv = 6.626 \times 10^{-34} \, J \, s \times 6.75 \times 10^{15} \, s^{-1} = 4.47 \times 10^{-18} \, J/photon$

(b) $E_m = 6.626 \times 10^{-34}$ J s $\times 7.05 \times 10^{14}$ s$^{-1} \times \dfrac{6.022 \times 10^{23} \text{ photons}}{\text{mol}} \times \dfrac{1 \text{ kJ}}{1000 \text{ J}} = 281$ kJ/mol

13. **(a)** $v = \dfrac{E}{h} = \dfrac{3.54 \times 10^{-20} \text{ J}}{6.626 \times 10^{-34} \text{ J s}} = 5.34 \times 10^{13}$ s^{-1}

(b) $E = hv = \dfrac{hc}{\lambda};\ \lambda = \dfrac{hc}{E} = \dfrac{6.26 \times 10^{-34} \text{ J s} \times 3.00 \times 10^8 \text{ m/s}}{166 \text{ kJ/mol} \times \dfrac{1000 \text{ J}}{1 \text{ kJ}} \times \dfrac{1 \text{ mol}}{6.022 \times 10^{23} \text{ photons}}} = 7.21 \times 10^{-7}$ m $= 721$ nm

14. We know that $E = hv$ and that $v = c/\lambda$, thus $E = hc/\lambda$. This means that energy and wavelength are inversely proportional. The radiation of shortest wavelength will have the greatest energy per photon; that with the longest wavelength will be the least energetic. It would be nice to have all four wavelengths in the same units for comparison; we choose μm since $1\ \mu$m $= 10^{-6}$ m $= 10^{+3}$ nm $= 10^{-4}$ cm. Thus we have

(a) 735 nm = 0.735 μm   **(b)** 6.35×10^{-5} cm = 0.635 μm   most energetic

(c) 1.05 μm   **(d)** 3.5×10^{-6} m = 3.5 μm   least energetic

15. **(a)** Bohr atom radius $= r_n = n^2 a_0 = 3^2 \times 0.53$ Å $= 4.8$ Å

(b) We let $r_n = 3.00$ Å, and solve for n. Only if n is an integer does such an orbit exist.

3.00 Å $= n^2 \times 0.53$ Å $n = \sqrt{\dfrac{3.00}{0.53}} = 2.25$ There is no Bohr orbit with radius $= 3.00$ Å.

(c) $E_n = -\dfrac{R_H}{n^2} = -\dfrac{2.179 \times 18^{-18} \text{ J}}{5^2} = -8.716 \times 10^{-20}$ J

(d) We let $E_n = -4.00 \times 10^{-18}$ J, and solve for n. Only if n is an integer does such an orbit exist.

-4.00×10^{-18} J $= -\dfrac{2.179 \times 18^{-18} \text{ J}}{n^2}$ $n = \sqrt{\dfrac{-2.179 \times 10^{-18}}{-4.00 \times 10^{-18}}} = 0.738$

There is no Bohr orbit with energy $= -4.00 \times 10^{-18}$ J.

16. We can use the Balmer equation (9.2, but generalized) to determine the frequency of this line. Alternatively, we can use equation (9.6) to determine the energy of the transition, followed by the Planck equation to find the frequency of the radiation. $v = 3.2881 \times 10^{15}$ s$^{-1} \left(\dfrac{1}{3^2} - \dfrac{1}{5^2}\right) = 2.3382 \times 10^{14}$ s^{-1}

$\Delta E = 2.179 \times 10^{-18}$ J $\left(\dfrac{1}{3^2} - \dfrac{1}{5^2}\right) = 1.5495 \times 10^{-19}$ J $v = \dfrac{E}{h} = \dfrac{1.5495 \times 10^{-19} \text{ J}}{6.626 \times 10^{-34} \text{ J s}} = 2.339 \times 10^{-14}$ s^{-1}

17. **(a)** Light of the longest wavelength will be light of the smallest energy. Energy change is computed from the expression $\Delta E = B[(1/n^2) - (1/m^2)]$, where B is a constant. Also, light is emitted only when we proceed from a state with a higher value of n to one with a lower value; thus answers (b) and (c) are incorrect, since they refer to absorption of energy rather than to emission. We see that energy emitted is proportional to the difference between the inverses of the squares of the quantum numbers. Remember that we want the smaller value. The two possibilities are: **(a)** $[(1/4^2) - (1/3^2)] = 0.0489$ **(d)** $[(1/2^2) - (1/1^2)] = 0.7500$

18. **(a)** deBroglie's equation $(\lambda = h/mv)$ indicates that, if the velocity is constant, the wavelength is inversely proportional to the mass. Thus, the particle with the least mass will have the longest wavelength. Of the four particles given, the *electron* is the least massive.

19. **(a)** $m_l = 0$ Since $l = 0$ and $|m_l|$ must be $\leq l$.

(b) $l = 1$ or 2 l must be less than n (which equals 3) and also must be $\geq |m_l|$, and $m_l = -1$.

(c) $n = 2, 3,...$ n must be greater than l (which equals 1 in this case).

20. **(a)** $4s$ has $n = 4$ $l = 0$ $m_l = 0$

(b) $3p$ has $n = 3$ $l = 1$ $m_l = -1, 0, +1$

(c) $5f$ has $n = 5$ $l = 3$ $m_l = -3, -2, -1, 0, +1, +2, +3$

(d) $3d$ has $n = 3$ $l = 2$ $m_l = -2, -1, 0, +1, +2$

21. **(a)** $n = 3$ $l = 2$ $m_l = -1$ is permitted.

(b) $n = 2$ $l = 3$ $m_l = -1$ is not allowed, since it has $l > n$.

(c) $n = 4$ $l = 0$ $m_l = -1$ is not allowed, since it has $|m_l| > l$.

(d) $n = 5$ $l = 2$ $m_l = -1$ is permitted.

(e) $n = 3$ $l = 3$ $m_l = -3$ is not permitted, since it has $l > n$.

(f) $n = 5$ $l = 3$ $m_l = +2$ is permitted.

22. The number of orbitals in a subshell is determined by the value of the orbital quantum number, l. The possible m_l values range from $-l$ to $+l$ in whole number steps. Each m_l value corresponds to a different orbital. Therefore, the number of orbitals in a subshell $= 2l + 1$.

(a) Since an s orbital has $l = 0$, there is but one orbital in the $2s$ subshell.

(b) There is no $3f$ subshell. For $3f$, $n = 3$ and $l = 3$, but l must always be less than n.

(c) A p orbital has $l = 1$. There are three orbitals in the $4p$ subshell.

(d) A d orbital has $l = 2$. There are five orbitals in the $5d$ subshell.

23. The order of subshell filling is $1s\ 2s\ 2p\ 3s\ 3p\ 4s\ 3d\ 4p\ 5s\ 4d\ 5p\ 6s\ 4f\ 5d\ 6p\ 7s\ 5f\ 6d$ An s subshell can have a maximum of 2 electrons, a p subshell can have 6 electrons maximum, a d subshell can have 10, and an f subshell can have 14.

(a) F has 9 electrons. $1s^2 2s^2 2p^5$

(b) P has 15 electrons. $1s^2 2s^2 2p^6 3s^2 3p^3$

(c) Ge has 32 electrons. $1s^2 2s^2 2p^6 3s^2 3p^6 4s^2 3d^{10} 4p^2$ or $1s^2 2s^2 2p^6 3s^2 3p^6 3d^{10} 4s^2 4p^2$

(d) Te has 52 electrons. $1s^2 2s^2 2p^6 3s^2 3p^6 4s^2 3d^{10} 4p^6 5s^2 4d^{10} 5p^4$

       or $1s^2 2s^2 2p^6 3s^2 3p^6 3d^{10} 4s^2 4p^6 4d^{10} 5s^2 5p^4$

24. (a) Mg has 12 electrons, 2 more than the noble gas Ne.   [Ne] $_{3s}$ ⇅

(b) Sc has 21 electrons, 3 more than the noble gas Ar.   [Ar] $_{3d}$ [↑| | | |] $_{4s}$ ⇅

(c) Cd has 48 electrons, 12 more than the noble gas Kr.   [Kr] $_{4d}$ [⇅|⇅|⇅|⇅|⇅] $_{5s}$ ⇅

(d) Tl has 81 electrons, 27 more than Xe.   [Xe] $_{4f}$ [⇅|⇅|⇅|⇅|⇅|⇅|⇅] $_{5d}$ [⇅|⇅|⇅|⇅|⇅] $_{6s}$ ⇅ $_{6p}$ [↑| |]

25. We write the orbital diagram of each element and determine the number of unpaired electrons from that.

(a) Ca has 20 electrons, 2 more than the noble gas Ar.   [Ar] $_{4s}$ ⇅   Zero unpaired electrons.

(b) Ga has 31 electrons, 13 more than the noble gas Ar.

     [Ar] $_{3d}$ [⇅|⇅|⇅|⇅|⇅] $_{4s}$ ⇅ $_{4p}$ [↑| |]   One unpaired electron.

(c) I has 53 electrons, 17 more than the noble gas Kr.

     [Kr] $_{4d}$ [⇅|⇅|⇅|⇅|⇅] $_{5s}$ ⇅ $_{4p}$ [⇅|⇅|↑]   One unpaired electron.

(d) Pb has 82 electrons, 18 more than the noble gas Xe.

     [Xe] $_{4f}$ [⇅|⇅|⇅|⇅|⇅|⇅|⇅] $_{5d}$ [⇅|⇅|⇅|⇅|⇅] $_{6s}$ ⇅ $_{6p}$ [↑|↑|]   Two unpaired electrons.

EXERCISES

Electromagnetic Radiation

26. (a) TRUE   Since frequency and wavelength are inversely related to each other, radiation of shorter wavelength has higher frequency.

(b) FALSE   Light of wavelengths between 390 nm and 790 nm is visible to the eye.

(c) FALSE   All electromagnetic radiation has the same speed in vacuum.

(d) TRUE   The wavelength of an x ray approximates 0.1 nm.

27. The speed of light is used to convert the distance into an elapsed time.

$$\text{time} = 93 \times 10^6\ \text{mi} \times \frac{5280\ \text{ft}}{1\ \text{mi}} \times \frac{12\ \text{in.}}{1\ \text{ft}} \times \frac{2.54\ \text{cm}}{1\ \text{in}} \times \frac{1\ \text{s}}{3.00 \times 10^{10}\ \text{cm}} \times \frac{1\ \text{min}}{60\ \text{s}} = 8.3\ \text{min}$$

28. The speed of light is used to convert the time into a distance spanned by light.

$$\text{distance} = 1\ \text{y} \times \frac{365.25\ \text{d}}{1\ \text{y}} \times \frac{24\ \text{h}}{1\ \text{d}} \times \frac{3600\ \text{s}}{1\ \text{h}} \times \frac{2.9979 \times 10^8\ \text{m}}{1\ \text{s}} \times \frac{1\ \text{km}}{1000\ \text{m}} = 9.4607 \times 10^{12}\ \text{km}$$

Atomic Spectra

29. The longest wavelength component has the smallest frequency (and the highest energy).

$$\nu = 3.2881 \times 10^{15} \text{ s}^{-1} \left(\frac{1}{2^2} - \frac{1}{3^2}\right) = 4.5668 \times 10^{14} \text{ s}^{-1} \qquad \lambda = \frac{c}{\nu} = \frac{2.9979 \times 10^8 \text{ m/s}}{4.5668 \times 10^{14} \text{ s}^{-1}} = 6.5646 \times 10^{-7} \text{ m}$$

$$\nu = 3.2881 \times 10^{15} \text{ s}^{-1} \left(\frac{1}{2^2} - \frac{1}{4^2}\right) = 6.1652 \times 10^{14} \text{ s}^{-1} \qquad \lambda = \frac{c}{\nu} = \frac{2.9979 \times 10^8 \text{ m/s}}{6.1652 \times 10^{14} \text{ s}^{-1}} = 4.8626 \times 10^{-7} \text{ m}$$

$$\nu = 3.2881 \times 10^{15} \text{ s}^{-1} \left(\frac{1}{2^2} - \frac{1}{5^2}\right) = 6.9050 \times 10^{14} \text{ s}^{-1} \qquad \lambda = \frac{c}{\nu} = \frac{2.9979 \times 10^8 \text{ m/s}}{6.9050 \times 10^{14} \text{ s}^{-1}} = 4.3416 \times 10^{-7} \text{ m}$$

$$\nu = 3.2881 \times 10^{15} \text{ s}^{-1} \left(\frac{1}{2^2} - \frac{1}{6^2}\right) = 7.3069 \times 10^{14} \text{ s}^{-1} \qquad \lambda = \frac{c}{\nu} = \frac{2.9979 \times 10^8 \text{ m/s}}{7.3069 \times 10^{14} \text{ s}^{-1}} = 4.1028 \times 10^{-7} \text{ m}$$

30. First we determine the frequency of the radiation, and then match it with the Balmer equation.

$$\nu = \frac{c}{\lambda} = \frac{2.9979 \times 10^8 \text{ m s}^{-1} \times \dfrac{10^9 \text{ nm}}{1 \text{ m}}}{389 \text{ nm}} = 7.71 \times 10^{14} \text{ s}^{-1} = 3.2881 \times 10^{15} \text{ s}^{-1} \left(\frac{1}{2^2} - \frac{1}{n^2}\right)$$

$$\left(\frac{1}{2^2} - \frac{1}{n^2}\right) = \frac{7.71 \times 10^{14} \text{ s}^{-1}}{3.2881 \times 10^{15} \text{ s}^{-1}} = 0.234 = 0.2500 - \frac{1}{n^2} \qquad \frac{1}{n^2} = 0.016 \qquad n = 7.9 \approx 8$$

31. $\lambda = 1880 \text{ nm} \times \dfrac{1 \text{ m}}{10^9 \text{ nm}} = 1.88 \times 10^{-6} \text{ m}$ From Exercise 29, we see that wavelengths in the Balmer series range downward from 6.5646×10^{-7} m. Since this is less than 1.88×10^{-6} m, we conclude that light with a wavelength of 1880 nm cannot be in the Balmer series.

32. The convergence wavelength limit of the Balmer series occurs when the integer n is as large as possible. Then the fraction $1/n^2$ will be essentially equal to zero. The corresponding wavelength is computed as follows. $\nu = 3.2881 \times 10^{15} \text{ s}^{-1} \left(\frac{1}{2^2} - 0\right) = 8.2203 \times 10^{14} \text{ s}^{-1}$

$$\lambda = \frac{c}{\nu} = \frac{2.9979 \times 10^8 \text{ m/s}}{8.2203 \times 10^{14} \text{ s}^{-1}} = 3.6470 \times 10^{-7} \text{ m} \times \frac{10^9 \text{ nm}}{1 \text{ m}} = 364.70 \text{ nm}$$

33. **(a)** The maximum wavelength occurs when $n = 2$ and the minimum wavelength occurs at the series convergence limit, when n is exceedingly large.

$$\nu = 3.2881 \times 10^{15} \text{ s}^{-1} \left(\frac{1}{1^2} - \frac{1}{2^2}\right) = 2.4661 \times 10^{15} \text{ s}^{-1} \qquad \lambda = \frac{c}{\nu} = \frac{2.9979 \times 10^8 \text{ m/s}}{2.4661 \times 10^{15} \text{ s}^{-1}} = \frac{1.2156 \times 10^{-7} \text{ m}}{121.56 \text{ nm}}$$

$$\nu = 3.2881 \times 10^{15} \text{ s}^{-1} \left(\frac{1}{1^2} - 0\right) = 3.2881 \times 10^{15} \text{ s}^{-1} \qquad \lambda = \frac{c}{\nu} = \frac{2.9979 \times 10^8 \text{ m/s}}{3.2881 \times 10^{15} \text{ s}^{-1}} = \frac{9.1174 \times 10^{-8} \text{ m}}{91.174 \text{ nm}}$$

(b) First determine the frequency of this spectral line, and then the value of n to which it corresponds.

$$\nu = \frac{c}{\lambda} = \frac{2.9979 \times 10^8 \text{ m/s}}{95.0 \text{ nm} \times \dfrac{1 \text{ m}}{10^9 \text{ nm}}} = 3.16 \times 10^{15} \text{ s}^{-1} = 3.2881 \times 10^{15} \text{ s}^{-1} \left(\frac{1}{1^2} - \frac{1}{n^2}\right)$$

$$\left(\frac{1}{1^2} - \frac{1}{n^2}\right) = \frac{3.16 \times 10^{15} \text{ s}^{-1}}{3.2881 \times 10^{15} \text{ s}^{-1}} = 0.961 \qquad \frac{1}{n^2} = 1.000 - 0.961 = 0.039 \qquad n = 5$$

(c) Let us pursue the same attack as in part (c).

$$\nu = \frac{c}{\lambda} = \frac{2.9979 \times 10^8 \text{ m/s}}{108.5 \text{ nm} \times \dfrac{1 \text{ m}}{10^9 \text{ nm}}} = 2.763 \times 10^{15} \text{ s}^{-1} = 3.2881 \times 10^{15} \text{ s}^{-1} \left(\frac{1}{1^2} - \frac{1}{n^2}\right)$$

$$\left(\frac{1}{1^2} - \frac{1}{n^2}\right) = \frac{2.763 \times 10^{15} \text{ s}^{-1}}{3.2881 \times 10^{15} \text{ s}^{-1}} = 0.8403 \quad \frac{1}{n^2} = 1.000 - 0.8403 = 0.1597$$

This gives as a result $n = 2.502$. Since n is not an integer, there is no line in the Lyman spectrum with a wavelength of 108.5 nm.

Quantum Theory

34. **(a)** Combine $E = h\nu$ and $c = \nu\lambda$ to obtain $E = hc/\lambda$

$$E = \frac{6.626 \times 10^{-34} \text{ J s} \times 2.998 \times 10^8 \text{ m/s}}{335 \text{ nm} \times \frac{1 \text{ m}}{10^9 \text{ nm}}} = 5.93 \times 10^{-19} \text{ J/photon}$$

(b) $E_m = 5.93 \times 10^{-19} \dfrac{\text{J}}{\text{photon}} \times 6.022 \times 10^{23} \dfrac{\text{photons}}{\text{mol}} = 3.57 \times 10^5 \text{ J/mol} = 357 \text{ kJ/mol}$

35. First determine the energy of an individual photon, and then its wavelength in nm.

$$E = \frac{535 \frac{\text{kJ}}{\text{mol}} \times \frac{1000 \text{ J}}{1 \text{ kJ}}}{6.022 \times 10^{23} \frac{\text{photons}}{\text{mol}}} = 8.88 \times 10^{-19} \frac{\text{J}}{\text{photon}} = \frac{hc}{\lambda} \qquad \text{or} \qquad \frac{hc}{E} = \lambda$$

$$\lambda = \frac{6.626 \times 10^{-34} \text{ J s} \times 2.998 \times 10^8 \text{ m/s}}{8.88 \times 10^{-19} \text{ J}} \times \frac{10^9 \text{ nm}}{1 \text{ m}} = 224 \text{ nm} \qquad \text{This is ultraviolet radiation.}$$

36. Notice that energy and wavelength are inversely related: $E = \dfrac{hc}{\lambda}$. Therefore radiation that is 100 times as energetic as readiation with a wavelength of 1020 nm will have a wavelength one hundredth as long: 10.2 nm. The frequency of this radiation is: $\nu = \dfrac{c}{\lambda} = \dfrac{2.998 \times 10^8 \text{ m/s}}{10.2 \text{ nm} \times \frac{1 \text{ m}}{10^9 \text{ nm}}} = 2.939 \times 10^{16} \text{ s}^{-1}$ From Figure 9-3, we see that this is ultraviolet radiation.

37. $E_1 = h\nu = \dfrac{hc}{\lambda} = \dfrac{6.62608 \times 10^{-34} \text{ J·s} \times 2.99792 \times 10^8 \text{ m s}^{-1}}{589.00 \text{ nm} \times \frac{1 \text{ m}}{10^9 \text{ nm}}} = 3.3726 \times 10^{-19} \text{ J/photon}$

$E_2 = \dfrac{6.62608 \times 10^{-34} \text{ J·s} \times 2.99792 \times 10^8 \text{ m s}^{-1}}{589.59 \text{ nm} \times \frac{1 \text{ m}}{10^9 \text{ nm}}} = 3.3692 \times 10^{-19} \text{ J/photon}$

$\Delta E = E_1 - E_2 = 3.3726 \times 10^{-19} \text{ J} - 3.3692 \times 10^{-19} \text{ J} = 0.0034 \times 10^{-19} \text{ J/photon}$

The Photoelectric Effect

38. **(a)** $E = h\nu = 6.63 \times 10^{-34} \text{ J s} \times 1.3 \times 10^{15} \text{ s}^{-1} = 8.6 \times 10^{-19} \text{ J/photon}$

(b) Platinum will display a photoelectric effect when exposed to ultraviolet light since ultraviolet light has a maximum frequency of $1 \times 10^{16} \text{ s}^{-1}$, which is above the threshhold frequency of platinum. It will not display a photoelectric effect when exposed to infrared light since the maximum frequency of infrared light is $4 \times 10^{14} \text{ s}^{-1}$, which is below the threshhold frequency of platinum.

39. In his explanation of the photoelectric effect, Einstein stated that each photon strikes and is absorbed by only one electron (cannot kill two birds with one stone). He also, and very importantly, stated that no more than one photon could contribute its energy to a given electron (cannot kill one bird with two stones). It is this second principle which explains why photoelectrons are not produced in larger number when the intensity of (sub-threshhold frequency) light is increased.

The Bohr Atom

40. **(a)** radius $= n^2 a_0 = 6^2 \times 0.53 \text{ Å} \times \dfrac{1 \text{ m}}{10^{10} \text{ Å}} \times \dfrac{10^9 \text{ nm}}{1 \text{ m}} = 1.9 \text{ nm}$

(b) $E = -\dfrac{B}{n^2} = -\dfrac{2.179 \times 10^{-18} \text{ J}}{6^2} = -6.054 \times 10^{-20} \text{ J}$

41. **(a)** $\nu = 3.289 \times 10^{15} \text{ s}^{-1} \left(\dfrac{1}{4^2} - \dfrac{1}{6^2} \right) = 1.142 \times 10^{14} \text{ s}^{-1}$

(b) $\lambda = \dfrac{c}{v} = \dfrac{2.998 \times 10^8 \text{ m/s}}{1.142 \times 10^{14} \text{ s}^{-1}} = 2.625 \times 10^{-6} \text{ m} \times \dfrac{10^9 \text{ nm}}{1 \text{ m}} = 2625 \text{ nm}$

(c) This is infrared radiation.

42. The greatest quantity of energy is absorbed in the situation where the difference between the inverses of the squares of the two quantum numbers is the largest and the system begins with a lower quantum number than it finishes with. The second condition eliminates answer (d) from consideration. Now we can consider the difference of the inverses of the squares of the two quantum numbers in each case.

(a) $\left(\dfrac{1}{1^2} - \dfrac{1}{2^2}\right) = 0.75$ (b) $\left(\dfrac{1}{2^2} - \dfrac{1}{4^2}\right) = 0.1875$ (c) $\left(\dfrac{1}{3^2} - \dfrac{1}{6^2}\right) = 0.0833$

Thus, the largest amount of energy is absorbed in the transition from $n = 1$ to $n = 2$, answer (a), among the four choices given.

43. (a) $r_1 = 1^2 \times 0.53 \text{ Å} = 0.53 \text{ Å}$ $r_4 = 4^2 \times 0.53 \text{ Å} = 8.5 \text{ Å}$
   increase in distance $= r_4 - r_1 = 8.5 \text{ Å} - 0.53 \text{ Å} = 8.0 \text{ Å}$

   (b) $E_1 = \dfrac{-2.179 \times 10^{-18} \text{ J}}{1^2} = -2.179 \times 10^{-18} \text{ J}$ $E_4 = \dfrac{-2.179 \times 10^{-18} \text{ J}}{4^2} = -1.362 \times 10^{-19} \text{ J}$

   increase in energy $= -1.362 \times 10^{-19} \text{ J} - (-2.179 \times 10^{-18} \text{ J}) = 2.043 \times 10^{-18} \text{ J}$

44. If infrared light is produced, the quantum number of the final state must have a lower value (be of lower energy) than the quantum number of the initial state. First we compute the frequency of the transition being considered (from $v = c/\lambda$), and then solve for the final quantum number.

$v = \dfrac{c}{\lambda} = \dfrac{3.00 \times 10^8 \text{ m/s}}{2170 \text{ nm} \times \dfrac{1 \text{ m}}{10^9 \text{ nm}}} = 1.38 \times 10^{14} \text{ s}^{-1} = 3.2881 \times 10^{15} \text{ s}^{-1} \left(\dfrac{1}{n^2} - \dfrac{1}{7^2}\right)$

$\left(\dfrac{1}{n^2} - \dfrac{1}{7^2}\right) = \dfrac{1.38 \times 10^{14} \text{ s}^{-1}}{3.2881 \times 10^{15} \text{ s}^{-1}} = 0.04197 \quad \dfrac{1}{n^2} = 0.04197 + \dfrac{1}{7^2} = 0.06238 \qquad n = 4$

Wave-Particle Duality

45. The de Broglie equation is $\lambda = h/mv$. This means that, for a given wavelength to be produced, a lighter particle would have to be moving faster. Thus, electrons would have to move faster than protons to display matter waves of the same wavelength.

46. First, we rearrange the de Broglie equation, solving it for velocity: $v = h/m\lambda$. Then we obtain the velocity of the electron. From Table 2-1, the mass of an electron is 9.109×10^{-28} g.

$v = \dfrac{h}{m\lambda} = \dfrac{6.626 \times 10^{-34} \text{ J s}}{\left(9.109 \times 10^{-28} \text{ g} \times \dfrac{1 \text{ kg}}{1000 \text{ g}}\right)\left(1 \text{ nm} \times \dfrac{1 \text{ m}}{10^9 \text{ nm}}\right)} = 7 \times 10^5 \text{ m/s}$

47. $\lambda = \dfrac{h}{mv} = \dfrac{6.626 \times 10^{-34} \text{ J s}}{\left(145 \text{ g} \times \dfrac{1 \text{ kg}}{1000 \text{ g}}\right)\left(155 \text{ km/h} \times \dfrac{1 \text{ h}}{3600 \text{ s}} \times \dfrac{1000 \text{ m}}{1 \text{ km}}\right)} = 1.06 \times 10^{-34} \text{ m}$

The diameter of a nucleus approximates 10^{-15} m, far larger than this wavelength.

48. $\lambda = \dfrac{h}{mv} = \dfrac{6.626 \times 10^{-34} \text{ J s}}{1000 \text{ kg} \times 25 \text{ m/s}} = 2.7 \times 10^{-38} \text{ m}$

This wavelength is so much smaller than the smallest particle of matter known (nuclear diameter $\approx 10^{-15}$ m) that its experimental measurement is virtually impossible.

The Uncertainty Principle

49. The Bohr model is a determinant model of an atom. It implies that the position of the electron is exactly known at any time in the future, once that position is known at the present. The distance of the electron from the nucleus also is exactly known, as is its energy. And finally, the velocity of the electron in its orbit is exactly known. All of these exactly known quantities—position, distance from nucleus, energy, and velocity—should, according to the Heisenberg uncertainty principle, be known with some imprecision.

50. Einstein believed very strongly in the law of cause and effect, what is known as a deterministic view of the universe. He felt that the need to use probability and chance ("playing dice") in describing atomic structure resulted because a suitable theory had not yet been developed to permit accurate predictions. He believed that such a theory could be developed, and had good reason for his belief: The developments in the theory of atomic structure came very rapidly during the first thirty years of this century, and those, such as Einstein, who had lived through this period had seen the revision of a number of theories and explanations that were thought to be the final answer. Another viewpoint in this area is embodied in another famous quotation: "Nature is subtle, but not malicious." The meaning of this statement is that the causes of various effects may be quite small (subtle) but there are causes nonetheless (not malicious).

51. Bohr was stating that we should accept theories as they are revealed by experimentation and logic, rather than attempting to make these theories fit our preconceived notions of what we believe they should be. In other words, Bohr was telling Einstein to keep an open mind.

Wave Mechanics

52. A sketch of this situation is presented at right. We see that 2.50 waves span the space of the 35 cm. Thus, the length of each wave is obtained by equating: $2.50\ \lambda = 35$ cm, giving λ = 14 cm.

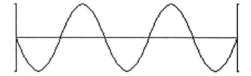

53. The differences between Bohr orbits and wave mechanical orbitals are given below.
a.   The first difference is that of shape. Bohr orbits, as originally proposed, are circular (later Sommerfeld proposed elliptical orbits). Orbitals, on the other hand, are spherical; or shaped like two tear drops or two squashed spheres; or shaped like four tear drops meeting at their points.
b.   Bohr orbits are planar pathways, while orbitals are three-dimensional regions of space in which there is a good probability of finding electrons.
c.   The electron in a Bohr orbit has a definite trajectory. Its position and velocity are known at all times. The electron in an orbital, however, does not have a well-known position or velocity. In fact, there is a small but definite probability that the electron may be found outside the boundaries generally drawn for the orbital.
Orbits and orbitals are similar in that the radii of Bohr orbits correspond to the distance from the nucleus in an orbital at which the electron is found with high probability.

54. We must be careful to distinguish between probability density—the chance of finding the electron within a definite volume of space—and the probability of finding the electron at a certain distance from the nucleus. The probability density—that is the probability of finding the electron within a small volume element (such as 1 pm²)—at the nucleus may well be high, in fact higher than the probability density at a distance 0.53 Å from the nucleus. But the probability of finding the electron at a fixed distance from the nucleus is this probability density multiplied by all of the many small volume elements that are located at this distance. (Recall the dart board analogy of Figure 9-19.)

Quantum Numbers and Electron Orbitals

55. Answer (1) is incorrect because the values of m_s may be either $+\frac{1}{2}$ or $-\frac{1}{2}$. Answer (2) is incorrect because the value of l must be $\geq |m_l|$. Since m_l cannot be larger than l, and l must be smaller than n, the only possible value is $l = 2$. Thus, answer (3) is incorrect, and answer (4) is correct.

56.

$n =$	4	4	4	4	4	4	4	4	4	4	4	4	4	4	4	4
$l =$	0	1	1	1	2	2	2	2	2	3	3	3	3	3	3	3
$m_l =$	0	–1	0	+1	–2	–1	0	+1	+2	–3	–2	–1	0	+1	+2	+3
designation	4s	4p	4p	4p	4d	4d	4d	4d	4d	4f	4f	4f	4f	4f	4f	4f
no. orbitals	1	——3——			————5————					——————7——————						

total number of orbitals = $1 + 3 + 5 + 7 = 16 = 4^2$ There are seven f orbitals in the $4f$ subshell.

57. **(a)** $n \geq 3$, $l = 2$, $m_l = 0$, $m_s = +\frac{1}{2}$ — (n must be larger than l.) This is a $3d$, $4d$, $5d$, ... orbital.

(b) $n = 2$, $l = 1$, $m_l = -1$, $m_s = -\frac{1}{2}$ — (l must be smaller than n and $\geq |m_l|$.) This is a $2p$ orbital.

(c) $n = 4$, $l = 2$, $m_l = 0$, $m_s = +\frac{1}{2}$ — (m_s can be either $+\frac{1}{2}$ or $-\frac{1}{2}$.) This is $4d$ orbital.

(d) $n = 1$, $l = 0$, $m_l = 0$, $m_s = +\frac{1}{2}$ — (The principles are stated above; n can be any positive integer, m_s could also equal $-\frac{1}{2}$.) This is a $1s$ orbital.

58. **(a)** $n = 2$ $l = 1$ $m_l = 0$ is permitted

(b) $n = 2$ $l = 2$ $m_l = -1$ is not allowable. l must be less than the value of n.

(c) $n = 3$ $l = 0$ $m_l = 0$ is permitted

(d) $n = 3$ $l = 1$ $m_l = -1$ is permitted.

(e) $n = 2$ $l = 0$ $m_l = -1$ is not allowable. $|m_l|$ cannot be greater than l.

(f) $n = 2$ $l = 3$ $m_l = -2$ is not allowable. l must be less than the value of n.

59. **(a)** $n = 2$ $l = 1$ $m_l = -1$ designates a $2p$ orbital. ($l = 1$ for all p orbitals.)

(b) $n = 4$ $l = 2$ $m_l = 0$ designates a $4d$ orbital. ($l = 2$ for all d orbitals.)

(c) $n = 5$ $l = 0$ $m_l = 0$ designates a $5s$ orbital. ($l = 0$ for all s orbitals.)

60. **(1)** TRUE The fourth principal shell has $n = 4$.

(2) TRUE A d orbital has $l = 2$. Since $n = 4$, l can be = 3, 2, 1, 0. Since $m_l = -2$, l can be = 2, 3.

(3) FALSE A p orbital has $l = 1$. But we demonstrated in part (b) that the only allowed values for l are 2 and 3.

(4) FALSE Either $+\frac{1}{2}$ or $-\frac{1}{2}$ is permitted as a value of m_s.

Electron Configurations

61.

		Incorrect	Correct	Reason for correction
(a)	B	$1s^2 2s^3$	$1s^2 2s^2 2p^1$	There can be only two electrons in an s subshell. The incorrect configuration violates the Pauli Principle.
(b)	Na	$1s^2 2s^2 2p^6 2d^1$	$1s^2 2s^2 2p^6 3s^1$	The incorrect configuration violates the permitted values of l, which always must be less than n.
(c)	K	[Ar]$3d^1$	[Ar]$4s^1$	In a multi-electron atom, the $4s$ subshell is lower in energy than the $3d$ subshell.
(d)	Ti	[Ar]$4s^2 4p^2$	[Ar]$4s^2 3d^2$	In a multi-electron atom, the $3d$ subshell is lower in energy than the $4p$ subshell.
(e)	Xe	[Kr]$5s^2 5p^6 5d^{10}$	[Kr]$4d^{10} 5s^2 5p^6$	The $4d$ subshell has not yet been filled in the [Kr] configuration.
(f)	Hg	[Xe]$4f^{14} 5d^{10} 6s^2 6p^4$	[Xe]$4f^{14} 5d^{10} 6s^2$	There should be no $6p$ electrons in a Hg atom.

62. Configuration **(b)** is correct for phosphorus. The reason why each of the other configurations is incorrect is given below.

(a) The two electrons in the $3s$ subshell must have opposed spins, that is different m_s quantum numbers.

(c) The three $3p$ orbitals must each contain one electron, before a pair of electrons is placed in any one of these orbitals.

(d) The three unpaired electrons in the $3p$ subshell must all have the same spin, either all spin up or all spin down.

63. The electron configuration of Mo is [Kr] $4d^5 5s^1$

(a) [Ar] $3d^{10} 3f^{14}$ Although this configuration has the correct number of electrons, it incorrectly assumes that there is a $3f$ subshell.

(b) [Kr] $4d^5 5s^1$ This is the correct electron configuration.

(c) [Kr] $4d^5 5s^2$ This configuration has one electron too many.

(d) [Ar] $3d^{14} 4s^2 4p^8$ This configuration has the correct number of electrons but incorrectly assumes 14 (rather than 10) electrons in a d subshell and 8 (rather than 6) electrons in a p subshell.

(e) [Ar] $3d^{10} 4s^2 4p^6 4d^6$ This configuration incorrectly assumes that electrons enter the $4d$ subshell before they enter the $5s$ subshell.

64. We write the correct electron configuration first in each case.

    (a) Si   [Ne]$3s^2 3p^2$ where $3p^2$ is ▯↑▯ ▯↑▯ ▯   There are 2 unpaired electrons in each Si atom.

    (b) S $1s^2 2s^2 2p^6 3s^2 3p^4$ There are no $3d$ electrons in an atom of S.

    (c) As   [Ar]$3d^{10} 4s^2 4p^3$ There are three $4p$ electrons in an atom of As.

    (d) Sr   [Ar]$3s^2 3p^6 3d^{10} 4s^2 4p^6 5s^2$ There are two $3s$ electrons in an atom of Sr.

    (e) Au   [Xe]$4f^{14} 5d^9 6s^2$ There are fourteen $4f$ electrons in an atom of Au.

65. **(a)** The $4p$ subshell of Br contains 5 electrons   [Ar] $_{3d}$ ↑↓↑↓↑↓↑↓↑↓ $_{4s}$ ↑↓ $_{4p}$ ↑↓↑↓↑

    (b) The $3d$ subshell of Co^{2+} contains 7 electrons   [Ar] $_{3d}$ ↑↓↑↓↑↓↑↑ $_{4s}$ ▯

    (c) The $5d$ subshell of Pb contains 10 electrons   [Xe] $_{5d}$ ↑↓↑↓↑↓↑↓↑↓ $_{6s}$ ↑↓ $_{6p}$ ↑↑▯

10 THE PERIODIC TABLE AND SOME ATOMIC PROPERTIES

1. (a) Two isoelectronic species (an atom and an ion, or two ions) have the same number of electrons.
   (b) The valence electronic shell is the outer shell of electrons. The electrons in this shell are those in the atom that have the highest principal quantum number.
   (c) A metal is an element that is relatively easily oxidized. Most elements are metals. In the periodic table, all of the transition elements are metals, as well as the lanthanides and actinides. In addition, metals are the representative elements in families 1A and 2A, as well as Al, Ga, In, Tl, Sn, Pb, and Bi.
   (d) A nonmetal is an element that is relatively easily reduced. In the periodic table, nonmetals are at the right and include H, F, Cl, Br, I, O, S, Se, N, P, C, and B.
   (e) A metalloid is an element that is fairly easy to oxidize or to reduce. That is, their behavior is between that of metals and of nonmetals. Metalloids include Si, Ge, As, Sb, Te, Po, and At.

2. (a) The periodic law is, like all natural laws, a summary of experimental observations. It states that a given property of elements varies periodically of the elements are arranged in order of increasing atomic number.
   (b) Ionization energy is the quantity of energy that, when added to a gaseous atom (or ion), will remove an electron.
   (c) Electron affinity is the quantity of energy given off by an atom (or ion) when it gains an electron.
   (d) Paramagnetism, the result of a species having one or more unpaired electrons, is the small attraction of that species into a magnetic field. It is not nearly as strong as the more familiar property of ferromagnetism.

3. (a) A group in the periodic table consists of those elements in the same vertical column, while a period consists of those elements in the same horizontal row. Elements in the same group often usually are quite similar chemically; those in the same period show a range of behaviors.
   (b) Elements in the s-block of the periodic table are those in families 1A and 2A; the last electrons added to them in the Aufbau process are s electrons; they all are metals. Elements in the p-block of the periodic table are those in families 3A, 4A, 5A, 6A, 7A, and 8A; the last electrons added to them in the Aufbau process are p electrons; some are metals, some metalloids, some nonmetals, and some noble gases.
   (c) Main-group elements are those in the s-block and the p-block; they are the elements in the "A" families. Often called representative elements, those in adjacent families differ by one valence electron and often are quite different in chemical and physical properties. Transition elements are d-block and f-block elements. All of them are metals, and adjacent elements differ by an inner shell electron, rather than a valence shell electron. Hence they are similar chemically. This similarity is particularly strong among the f-block elements.
   (d) Lanthanide elements are f-block elements are those in which the $4f$ subshell is filling: La through Lu. Actinide elements are those in which the $5f$ subshell is filling: Ac through Lr.
   (e) A covalent radius is the radius of an atom that is bonded covalently to another. For example, one half the internuclear distance in Cl_2 is the covalent radius of chlorine. A metallic radius is one half the shortest internuclear distance in a crystal of solid metal.

4. (a) The element in the fifth period and Group 3A is two spaces below Al. This is the element In.
   (b) An element similar to sulfur is any element in family 6A: O, Se, Te. (Po is not included because it is below the line that divides metals from nonmetals, and S is a nonmetal.) An element unlike sulfur is

any other element in the periodic table, particularly a metal—below and to the left of the metal-nonmetal dividing line.

(c)   A highly reactive metal in the sixth period would be the alkali metal in the sixth period: Cs.

(d)   The halogen element in the fifth period is the element in family 7A: I.

(e)   Element 18 is argon, a noble gas. An element with atomic number greater than 50 that also is a noble gas is Xe (or Rn, which is radioactive). (It is relatively easy to determine which elements are radioactive, and thus may not be suitable for chemical investigation. When the atomic weight of an element is a whole number enclosed in parentheses, that number is the mass number of the most stable—longest lived—isotope. This is a clear indication that all isotopes are radioactive.)

5.   (a)   A main-group metal is any element in families 1A or 2A, plus any of the *p*-block metals: Al, Ga, In, Tl, Sn, Pb, or Bi.

(b)   A representative nonmetal is a *p*-block element above and to the right of the metal/nonmetal dividing line, which is not a noble gas. That is, any of the following: H, F, Cl, Br, I, O, S, Se, N, P, C, B.

(c)   The noble gases are the elements in family 8A: He, Ne, Ar, Kr, Xe, and Rn.

(d)   *d*-block elements are those in the "B" families of the periodic table: Sc through Zn, Y through Cd, and Hf through Hg.

(e)   The inner transition elements are the *f*-block elements, the lanthanides (La through Lu) and the actinides (Ac through Lr).

6.   The pairs of elements that are "out of order" based on their atomic masses are presented, together with their atomic numbers below. It is clear that in each instance the elements are ordered by increasing atomic number. For one of these pairs there is a further explanation. Most of the Ar in the atmosphere is thought to result from the radioactive decay of ^{40}K, a nuclide of potassium that once was more plentiful than it is presently.

Element	Ar	K	Te	I	Co	Ni
Z	18	19	52	53	27	28
At. ms.	39.948	39.098	127.60	126.90	58.93	58.60
Element	Th	Pa	Pu	Am	Unh	Uns
Z	90	91	94	95	106	107
At. ms.	232.038	231.0359	244	243	263	262

7.   (a)   The number of protons in the nucleus of $^{79}_{34}$Se equals the atomic number: 34 protons.

(b)   The number of neutrons in the nucleus is the difference between the mass number and the atomic number of the nuclide: $79 - 34 = 45$ neutrons.

(c)   Se is in Group 6A of the fourth period of the periodic table. The $3d$ subshell completed filling with element 29 (Cu). Thus, there are ten $3d$ electrons in an atom of Se.

(d)   The $2s$ subshell is filled. Se has two $2s$ electrons.

(e)   The outer (valence) electron configuration of Se is $4s^2\,4p^4$. Se has four $4p$ electrons.

(f)   There are six electrons in the shell of highest principal quantum number ($n = 4$) in an atom of Se.

8.   Base the prediction of each electron configuration upon the configuration of the preceding noble gas.

		prediction		from the *text* Appendix
(a)	In	[Kr] $4d^{10}\,5s^2\,5p^1$	in family 3A	[Kr] $4d^{10}\,5s^2\,5p^1$
(b)	Y	[Kr] $4d^1\,5s^2$	in family 3B	[Kr] $4d^1\,5s^2$
(c)	Sb	[Kr] $4d^{10}\,5s^2\,5p^3$	in family 3A	[Kr] $4d^{10}\,5s^2\,5p^3$
(d)	Au	[Xe] $4f^{14}\,5d^{10}\,6s^1$	in family 1B	[Xe] $4f^{14}\,5d^{10}\,6s^1$

The predictions exactly match the given electron configurations for these four atoms.

9.   First write the electron configuration of the element. Then remove or add electrons in the outer shell to achieve the desired ionic charge.

(a)	Rb	[Kr] $5s^1$	Rb$^+$	[Kr] $5s^0$ = [Kr]
(b)	Br	[Ar] $3d^{10}\,4s^2\,4p^5$	Br$^-$	[Ar] $3d^{10}\,4s^2\,4p^6$ = [Kr]
(c)	O	[He] $2s^2\,2p^4$	O^{2-}	[He] $2s^2\,2p^6$ = [Ne]
(d)	Ba	[Xe] $6s^2$	Ba^{2+}	[Xe] $6s^0$ = [Xe]
(e)	Zn	[Ar] $3d^{10}\,4s^2$	Zn^{2+}	[Ar] $3d^{10}$
(f)	Ag	[Kr] $4d^{10}\,5s^1$	Ag$^+$	[Kr] $4d^{10}$
(g)	Bi	[Xe] $4f^{14}\,5d^{10}\,6s^2\,6p^3$	Bi^{3+}	[Xe] $4f^{14}\,5d^{10}\,6s^2$

10. In the literal sense, isoelectronic means having the same number of electrons. (In another sense, not used *in the text*, it means having the same electron configuration.) We determine the total number of electrons and the electron configuration for each species and make our decisions based on this information.

Fe^{2+}	24 electrons	[Ar] $3d^6 4s^0$		Sc^{3+}	18 electrons	[Ar] $3d^0 4s^0$
K^+	18 electrons	[Ar] $3d^0 4s^0$		Br^-	36 electrons	[Ar] $3d^{10} 4s^2 4p^6$
Co^{2+}	25 electrons	[Ar] $3d^7 4s^0$		Co^{3+}	24 electrons	[Ar] $3d^6 4s^0$
Sr^{2+}	36 electrons	[Ar] $3d^{10} 4s^2 4p^6$		O^{2-}	10 electrons	[He] $2s^2 2p^6$
Zn^{2+}	28 electrons	[Ar] $3d^{10} 4s^0$		Al^{3+}	10 electrons	[He] $2s^2 2p^6$

Species with the same number of electrons and the same electron configuration are the following.

Fe^{2+} and Co^{3+} $\qquad$ Sc^{3+} and K^+ $\qquad$ Br^- and Sr^{2+} $\qquad$ O^{2-} and Al^{3+}

11. Since isoelectronic species must have the same number of electrons, and each element has a different atomic number, atoms of different elements cannot be isoelectronic. Two different cations may be isoelectronic, as may two different anions, or an anion and a cation. An example would be two anions (or two cations, or an anion and a cation) that have the electron configuration of a noble gas, such as: O^{2-} and F^-, Na^+ and Mg^{2+}, or F^- and Na^+.

12. (a) K is in family 1A and in the fourth period. Elements in family 1A have an s^1 outer electron configuration; those in the fourth period are filling the 4th principal quantum level. Thus K has one $4s$ electron.

(b) I is in family 7A (s^2p^5 outer electron configuration) and in the 5th period (5th principal quantum level). I has five $5p$ electrons.

(c) Zn is in family 2B [$(n–1) d^{10} n s^2$ outer electron configuration] and in the 4th period ($n = 4$). Zn has ten $3d$ electrons.

(d) S is in family 6A [$n s^2 n p^4$] and in the 3rd period. The $2p^6$ electron configuration was complete with the preceding noble gas (Ne). S has six $2p$ electrons.

(e) Pb follows the lanthanide series in which fourteen $4f$ electrons were added. Pb has fourteen $4f$ electrons.

(f) Ni is in the d-block of elements, two from the end. It therefore has eight d electrons. It also is in the fourth period, so these eight d electrons are $3d$ electrons. Ni has eight 3d electrons.

13. (a) The most nonmetallic element is the one farthest to the right and the top (with the exception of the noble gases). This is the element fluorine.

(b) The transition elements are those with incomplete d subshells, or which form ions with incomplete d subshells. Scandium ([Ar] $3d^1 4s^2$) is the transition element with the lowest atomic number.

(c) The metalloids are defined as the elements in Groups 4A and 5A that are adjacent to the stair-step diagonal line in the periodic table, that is, Si, Ge, As, Sb, and Te; along with the elements at the bottom of Groups 6A and 7A, that is, Po and At. Of these 6 elements, only Si ($Z = 14$) has an atomic number exactly midway between those of two noble gases, Ne ($Z = 10$) and Ar ($Z = 18$).

14. In general, atomic size in the periodic table increases from top to bottom of a family and decreases from left to right of a period, as indicated in Figure 10-13. The larger element is indicated first, followed by the reason for making the choice.

(a) As $\quad$ As is to the left of Br in the 4th period.

(b) Sr $\quad$ Sr is below Mg in the periodic table.

(c) Cs $\quad$ Cs is both below and to the left of Ca in the periodic table.

(d) Xe $\quad$ Xe is below Ne in the periodic table.

(e) C $\quad$ C is to the left of O in the periodic table.

(f) Hg $\quad$ Hg is both below and to the left of Cl in the periodic table.

15. An Al atom is larger than a F atom since Al is both below and to the left of F in the periodic table. As is larger than Al, since As is below Al in the periodic table. (Even though As is to the right of Al, we would not conclude that As is smaller than Al, since increases in size down a group are more pronounced than decreases in size across—left to right—a period.) A Cs^+ ion is isoelectronic with an I^- ion, and in an isoelectronic series, anions are larger than cations; thus I^- is larger than Cs^+. I^- also is larger than As, since I is below As in the periodic table (and increases in size down a group are more pronounced than those across a period). Finally, N is larger than F, since N is to the left of F in the periodic table. Therefore, we conclude that F is the smallest species listed and I⁻ is the largest. In fact, with the exception of Cs⁺, we can rank the species in order of decreasing size.   I⁻ > As > Al > N > F and also I⁻ > Cs⁺

16. Ionization energy in the periodic table decreases from top to bottom of a family, and increases from left to right of a period, as summarized in Figure 10-13. Cs has the lowest ionization energy; it is furthest to the left and nearest to the bottom of the periodic table. Next comes Sr, then As, then S, and finally F, the most nonmetallic element in the group (and in the periodic table). Thus the elements, listed in order of increasing ionization energy are: Cs < Sr < As < S < F

17. (a) The first element in a group has the smallest atoms. In Group 4A this is C.
    (b) The atom in a period with the largest atoms is furthest to the left. In the fifth period this is Rb.
    (c) The bottom element in a group has atoms with the lowest first ionization energy. In Group 7A this is At (or I if we do not wish to consider radioactive elements).

18. First we convert the mass of Li given to an amount in moles of Li. Then we compute the energy needed to ionize this much Li.

$$\text{Energy} = 1.00 \text{ mg Li} \times \frac{1 \text{ g Li}}{1000 \text{ mg}} \times \frac{1 \text{ mol Li}}{6.941 \text{ g Li}} \times \frac{520.2 \text{ kJ}}{1 \text{ mol Li}} = 0.0750 \text{ kJ} \times \frac{1000 \text{ J}}{1 \text{ kJ}} = 74.9 \text{ J}$$

19. (a) The most metallic element is the one closest to the lower left-hand corner. This is the element Ba.
    (b) The most nonmetallic element is the one closest to the upper right-hand corner, the element S.
    (c) There are two distinct metals in this group, Ba and Ca; they have the lowest ionization energies. There is one distinct nonmetal, S; it has the highest ionization energy. The remaining two elements, As and Bi, are a metalloid and a metal, respectively. The metalloid has a higher ionization energy than does the metal. Thus Bi has the intermediate value of ionization energy of the five elements listed.

20. Metallic character decreases from left to right and from bottom to top in the periodic table. Thus, in order of decreasing metallic character the elements listed are: Rb > Ca > Sc > Fe > Te > Br > O > F The difficulty in establishing this series is in placing the elements Te and Br. First, the metal Fe is more metallic than the nonmetal Te. And Te clearly is more metallic than the halogen Br. Finally, we only need to recognize that Cl and O have approximately the same nonmetallic character, and Br clearly is more metallic than is Cl.

21. Paramagnetism indicates unpaired electrons, which in turn is often associated with partially filled subshells. First we write the electron configurations of the elements, and then those of the ions. From those electron configurations, we determine whether the species is paramagnetic or diamagnetic.

K	$[Ar] 4s^1$	K⁺	$[Ar] 4s^0$	diamagnetic, all subshells filled.
Cr	$[Ar] 3d^5 4s^1$	Cr³⁺	$[Ar] 3d^4$	paramagnetic.
Zn	$[Ar] 3d^{10} 4s^2$	Zn²⁺	$[Ar] 3d^{10}$	diamagnetic, all subshells filled.
Cd	$[Kr] 4d^{10} 5s^2$			diamagnetic, all subshells filled.
Co	$[Ar] 3d^7 4s^2$	Co³⁺	$[Ar] 3d^6$	paramagnetic
Sn	$[Kr] 4d^{10} 5s^2 5p^2$	Sn²⁺	$[Kr] 4d^{10} 5s^2 5p^0$	diamagnetic, all subshells filled.
Br	$[Ar] 3d^{10} 4s^2 4p^5$			paramagnetic, an odd number of electrons.

22. (a) 6   Tl's electron configuration $[Xe] 4f^{14} 5d^{10} 6s^2 6p^1$ has one *p* electron in its outermost shell.
    (b) 8   Z = 70 identifies the element as Yb. Yb should have the largest radius of the elements given—but *not* of all the elements—since atomic radii generally increase from top to bottom in the periodic table and from right to left. Yb also is in the *f*-block of the periodic table, an inner transition element.
    (c) 5   Ni has the electron configuration $[Ar] 4s^2 3d^8$. It also is a *d*-block element.
    (d) 1   An s^2 outer electron configuration, with the underlying configuration of a noble gas, is characteristic of elements of family 2A, the alkaline earth elements.
    (e) 2   The element in the fifth period and Group 5A is Sb, a metalloid.
    (f) 4   The element in the fourth period and Group 6A is Se, a nonmetal. (6 might also be a choice, since boron also is a nonmetal, and it has one *p* electron in the shell of highest principal quantum number, but the other elements of Group 3A are *not* nonmetals.)

The Periodic Law

23. Silver has an atomic number of $Z = 47$, and thus its atomic volume is slightly less than 10 cm³/mol. Let us assume 9.8 cm³/mol. $\text{density} = \dfrac{\text{molar mass}}{\text{atomic volume}} = \dfrac{107.87 \text{ g/mol}}{9.8 \text{ cm}^3/\text{mol}} = 11 \text{ g/cm}^3$

24. Element 114 will be a metal in the same family as Pb, element 82 (18 cm³/mol); Sn, element 50 (18 cm³/mol); and Ge, element 32 (14 cm³/mol). We note that the atomic volume of Pb and Sn are essentially equal, probably due to the lanthanide contraction. If there is also an actinide contraction, element 114 will have an atomic volume of 18 g/cm³; if not, the value is probably about 22 cm³/mol. This need to estimate atomic volume is what makes the value for density inaccurate.

$\text{density (g/cm}^3) = \dfrac{298 \text{ g/mol}}{18 \text{ cm}^3/\text{mol}} = 16 \text{ g/cm}^3$ $\qquad$ $\text{density (g/cm}^3) = \dfrac{298 \text{ g/mol}}{22 \text{ cm}^3/\text{mol}} = 14 \text{ g/cm}^3$

25. Tellurium has an atomic number of $Z = 52$, and thus its atomic volume is somewhat greater than 20 g/cm³. Let us assume 20.4 cm³/mol.

atomic mass = density × atomic volume = 6.24 g/cm³ × 20.4 cm³/mol = 127 g/mol

This compares very well with the listed value of 127.60 for Te.

26. The following data are plotted at right below.

	Z	m.p., °C
Al	13	660
Ar	18	−189
Be	4	1278
B	5	2300
C	6	3350
Cl	17	−101
F	9	−220
Li	3	179
Mg	12	651
Ne	10	−249
N	7	−210
O	8	−218
P	15	−590
Si	14	1410
Na	11	98
S	16	119

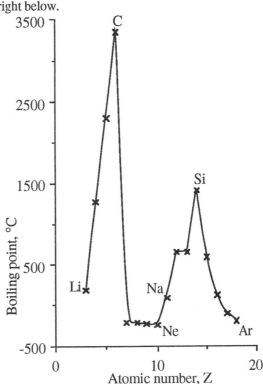

We see that melting point indeed does vary periodically with atomic number. It peaks in family 4A and has a minimum in family 8A.

The Periodic Table

27. Mendeleev arranged elements in the periodic table in order of increasing atomic weight. Of course, fractional atomic weights are permissable. Hence, there always is room for an added element between two elements that already are present in the table. On the other hand, Moseley arranged elements in order of increasing atomic number. Only integral (whole number) values of atomic number are permitted. Thus, when elements with all possible integral values in a certain range have been discovered, no new elements are possible in that range.

28. (a) The noble gas following radon ($Z = 86$) will have an atomic number of (86 + 32 =) 118.

(b) The alkali metal following francium (Z = 87) will have an atomic number of (87 + 32 =) 119.

(c) The mass number of radon (A = 222) is (222 ÷ 86 =) 2.58 times its atomic number. The mass number of Lr (A = 260) is (260 ÷ 103 =) 2.52 times its atomic number. Thus, we would expect the mass numbers, and hence approximate atomic masses of elements 118 and 119 to be about 2.52 times their atomic numbers, that is, $A_{118} \approx 297$ u and $A_{119} \approx 300$ u.

29. For there to be the same number of elements in each period of the periodic table, each shell of an electron configuration would have to contain the same number of electrons. This is not the case, the shells have 2 (*K* shell), 8 (*L* shell), 18 (*M* shell), and 32 (*N* shell) electrons each. This is because each of the periods of the periodic table begins with one *s* electron beyond a noble gas electron configuration, and the noble gas electron configuration corresponds to either s^2 or $s^2 p^6$, with various other full subshells.

30. This periodic table would have to have the present width of 18 elements plus the width of the 14 lanthanides or actinides. That is, the periodic table would be (18 + 14 =) 32 elements wide.

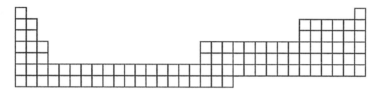

31. **(a)** The 6*d* subshell will be complete with an element in the same family as Hg. This is three elements beyond Une (Z = 109) and thus this element has an atomic number of Z = 112.

(b) The element in this period that will most closely resemble bismuth is six elements beyond Une (Z = 109) and thus has Z = 115.

(c) The element in this period that we might expect to be a metalloid would fall below At, eight elements beyond Une (Z = 109) and thus has Z = 117.

(d) The element in this period that would be a noble gas would fall below Rn, nine elements beyond Une (Z = 109) and thus has Z = 118.

Periodic Table and Electron Configurations

32. **(a)** Sb is in family 5A, with outer shell electron configuration $ns^2 np^3$. Sb has five outer-shell electrons.

(b) Pt has an atomic number of Z = 78. The fourth principal shell fills as follows: 4*s* from Z = 19 (K) to Z = 20 (Ca); 4*p* from Z = 31 (Ga) to Z = 36 (Kr); 4*d* from Z = 39 (Y) to Z = 48 (Cd); and 4*f* from Z = 58 (Ce) to Z = 71 (Lu). Since the atomic number of Pt is greater than Z = 71, the entire fourth principal shell is filled, with 32 electrons.

(c) The five elements with six outer-shell electrons are those in family 6A: O, S, Se, Te, Po.

(d) The outer-shell electron configuration is $ns^2 np^4$, giving the following as a partial orbital diagram:
s^2 ⬆⬇ p^4 ⬆⬇ ⬆ ⬆   There are two unpaired electrons in an atom of Te.

(e) The sixth period begins with Cs and ends with Rn. There are 10 outer transition elements in this period (La, and Hf through Hg) and there are 14 inner transition elements in the period (Ce through Lu). Thus, there are (10 + 14 =) 24 transition elements in the sixth period.

33. Since the periodic table is based on electron structure, two elements in the same family (Pb and element 114, Uuq) should have similar electron configurations.

(a) Pb [Xe] $4f^{14} 5d^{10} 6s^2 6p^2$ **(b)** Uuq [Rn] $5f^{14} 6d^{10} 7s^2 7p^2$

34. We first write the electron configuration for the atoms and then write the electron configuration for the ion, assuming that the appropriate number of outer shell electrons have been added (anions) or removed (cations) to produce the desired ionic charge.

(a) Sr [Kr] $5s^2$ Sr^{2+} [Kr] $5s^0$
(b) Y   [Kr] $5s^2 4d^1$ Y^{3+} [Kr] $5s^0 4d^0$
(c) Se [Ar] $3d^{10} 4s^2 4p^4$ Se^{2-} [Ar] $3d^{10} 4s^2 4p^6$ = [Kr]
(d) Cu [Ar] $3d^{10} 4s^1$ Cu^{2+} [Ar] $3d^9 4s^0$
(e) Ni [Ar] $3d^8 4s^2$ Ni^{2+} [Ar] $3d^8 4s^0$
(f) Ga [Ar] $3d^{10} 4s^2 4p^1$ Ga^{3+} [Ar] $3d^{10} 4s^0 4p^0$
(g) Ti [Ar] $3d^2 4s^2$ Ti^{2+} [Ar] $3d^2 4s^0$

35. Two ions can indeed be isoelectronic without having a noble gas electron configuration. As one example, they may have the "18 + 2" electron configuration described in Table 10-3. Specifically, Bi^{3+}, Pb^{2+}, and Tl^+ are isoelectronic, with an electron configuration of [Xe] $4f^{14}\, 5d^{10}\, 6s^2\, 6p^0$.

36. **(a)** A fifth period noble gas is the element in Group 8A and the fifth period, the element Xe.

   (b) A sixth period element whose atoms have three unpaired electrons is an element in family 5A, which has an outer electron configuration of $ns^2\, np^3$, and thus has three unpaired p electrons. This is the element Bi.

   (c) One d-block element that has one $4s$ electron is Cu: [Ar] $3d^{10}\, 4s^1$. Another is Cr: [Ar] $3d^5\, 4s^1$.

   (d) There are several p-block elements that are metalloids—elements that lie on either side of the jagged dividing line in the p-block: Si, Ge, As, Sb, Te, Po, At.

   (e) Representative metals that form the oxide M_2O_3 are found in family 3A of the periodic table: Al, Gal, In, Tl. There are also some transition elements that form an M_2O_3 oxide: Fe, Cr, and Au are examples. But oxide formulas of transition elements cannot be predicted reliably from their position in the periodic table.

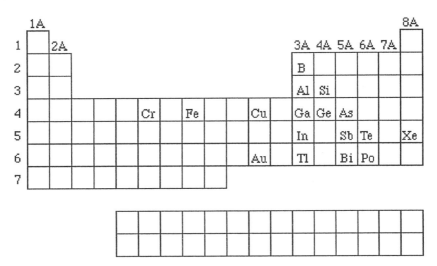

37. **(a)** N is the third element in the p-block of the second period and thus has three $2p$ electrons.

   (b) Rb is the first element in the s-block of the *fifth* period, thus Rb has two $4s$ electrons.

   (c) As is in the p-block of the fourth period. The $3d$ subshell is filled with ten electrons, but no $4d$ electrons have been added.

   (d) Au is in the d-block of the sixth period; the $4f$ subshell is filled. Au has fourteen $4f$ electrons.

   (e) Pb is the second element in the p-block of the sixth period; it has two $6p$ electrons. Since these two electrons are placed in separate $6p$ orbitals, they are unpaired. There are two unpaired electrons.

   (f) Group 4A of the periodic table is the group with the elements C, Si, Ge, Sn, and Pb. This group has five elements.

   (g) The sixth period begins with the element Cs ($Z = 55$) and ends with the element Rn ($Z = 86$). This period is thirty-two elements long.

Atomic Radii

38. The reason why the sizes of atoms do not simply increase with atomic number is because often electrons are added successively to the same subshell. These electrons do not screen each other from the nuclear charge (they do not effectively get between each other and the nucleus). Consequently, as each electron is added to a subshell and the nuclear charge increases by one unit, all of the electrons in this subshell are drawn more closely into the nucleus.

39. The reason why atomic sizes are uncertain is because the electron cloud that surrounds an atom has no specific limit. It can be pictured as gradually fading away, rather like the edge of a town. In both cases we pick an arbitrary boundary.

40. **(a)** The smallest atoms in Group 3A is the first: B

(b) Po is in the sixth period, and is larger than the others, which are rewritten in the following list from right to left in the fifth period, that is, from largest to smallest: Sr, In, Sb, Te. Thus, Te is the smallest of the elements given.

41. The hydrogen ion contains no electrons, only a nucleus. It is exceedingly tiny, much smaller than any other atom or electron-containing ion. Both He and H⁻ contain two electrons, but He has a nuclear charge of +2, while H⁻ has one of only +1. The smaller nuclear charge of H⁻ is less effective at attracting electrons than the nuclear charge of He. Thus, ordered by increasing size: $H^+ < He < H^-$.

42. In an isoelectronic series all of the species have the same number of electrons. The size is determined by the nuclear charge. Those species with the largest (positive) nuclear charge are the smallest. Those with smaller nuclear charges are smaller in size. Thus, the more positively charged an ion is in an isoelectronic series, the smaller it will be. $Y^{3+} < Sr^{2+} < Rb^+ < Kr < Br^- < Se^{2-}$

43. Li^+ is the smallest; it not only is in the second period, but also is a cation. Br^- is the largest, an anion in the 4th period. Next largest is Y in the next (the 5th) period. We expect Y to be larger than Se because it is both to the left of Se and in the next period. So finally, Se is intermediate between Li^+ and Y. $Li^+ < Se < Y < Br^-$

44. Li^+ is a cation in the first period. B is an atom in the same period. Thus, $Li^+ < B$. P is an atom in Group 5A of the 3rd period. Cl is an atom in Group 7A of the same period. Thus $Cl < P$. Br is in the fourth period, family 7A, while Br^- is its anion: $Br < Br^-$. Br would be expected to be larger than P and Cl larger than B; generally atoms are larger than atoms of the previous period. Also, Rb should be larger than Br, a nonmetal in the previous period, and probably larger than Br^- also. In the following ordering, radii in picometers (from Figure 10-9) are in parentheses. $Li^+(60) < B < Cl(99) < P(110) < Br (114) < Br^-(196) < Rb(248)$

45. Size decreases from left to right in the periodic table; on this basis I should be smaller than Al. But size increases from top to bottom in the periodic table; on this basis I should be larger than Al. There really is no good way of resolving these conflicting predictions.

Ionization Energies, Electron Affinities

46. The second ionization energy for an atom (I_2) cannot be smaller than the first ionization energy (I_1) for the same atom. The reason is that, when the first electron is removed it is being taken away from a species with a charge of +1. On the other hand, when the second electron is removed, it is being taken from a species with a charge of +2. Since the force between two charged particles is proportional to q_+q_-/r^2 (r is the distance between the particles), the higher the charge, the more difficult it will be to remove an electron.

47. In the case of a first electron affinity, a negative electron is being added to a neutral atom. This process may be either exothermic or endothermic. In the case of an ionization potential, however, a negatively charged electron is being removed from a positively charged cation, a process that must always require energy, since unlike charges attract each other.

48. The reasoning here is similar to that for the answer to Exercise 46. In both cases, an electron is being removed from a neon electron configuration. But in the case of Na^+, the electron is being removed from a species with a charge of +2, while in the case of Ne, the electron is being removed from a species with a charge of +1. The more highly charged the resulting species is, the more difficult it is to produce it by removing an electron.

49. There are five third-shell electrons in each atom of phosphorus. In Table 10-6, the values for the ionization energies of these five electrons are listed in kJ/mol. Thus, the energy needed to ionize all five electrons from a mole of phosphorus atoms is given by:
Energy = 1012 + 1903 + 2912 + 4957 + 6274 = 17058 kJ

50. Data ($I_1 = 375.7$ kJ/mol) are obtained from Table 10-5.
$$\text{no. } Cs^+ \text{ ions} = 1 \text{ J} \times \frac{1 \text{ kJ}}{1000 \text{ J}} \times \frac{1 \text{ mol } Cs^+}{375.7 \text{ kJ}} \times \frac{6.022 \times 10^{23} \text{ } Cs^+ \text{ ions}}{1 \text{ mol}} = 1.603 \times 10^{18} \text{ } Cs^+ \text{ ions}$$

51. From Figure 9-14, the ground state of the hydrogen atom is -2.179×10^{-18} J/atom below the zero of energy: the point where the electron is completely removed from the hydrogen atom.

$$I_1 = \frac{2.179 \times 10^{-18} \text{ J}}{\text{atom}} \times \frac{1 \text{ kJ}}{1000 \text{ J}} \times \frac{6.022 \times 10^{23} \text{ atom}}{1 \text{ mol H}} = 1312 \text{ kJ/mol}$$

52. The electron affinity of chlorine is –348.7 kJ/mol. We use Hess's law to determine the heat of reaction for $Cl_2(g)$ becoming $Cl^-(g)$.

$Cl_2(g) \longrightarrow 2 \, Cl(g)$ $\qquad\qquad \Delta H = + 242.8 \text{ kJ}$

$2 \, Cl(g) + 2 \, e^- \longrightarrow 2 \, Cl^-(g)$ $\qquad E.A. = 2(-348.7) \text{ kJ}$

$Cl_2(g) + 2 \, e^- \longrightarrow 2 \, Cl^-(g)$ $\qquad \Delta H = -454.6 \text{ kJ}$ $\qquad$ The overall process is *exothermic*.

53. The electron affinity of fluorine is –322.2 kJ/mol (Figure 10-11) and the first and second ionization energies of Mg (Table 10-6) are 737.7 kJ/mol and 1451 kJ/mol.

$Mg(g) \longrightarrow Mg^+(g) + e^-$ $\qquad\qquad\qquad I_1 = 737.7 \text{ kJ/mol}$

$Mg^+(g) \longrightarrow Mg^{2+}(g) + e^-$ $\qquad\qquad\quad I_2 = 1451 \text{ kJ/mol}$

$2 \, F(g) + 2 \, e^- \longrightarrow 2 \, F^-(g)$ $\qquad\qquad\quad 2 \, E.A. = 2(-322.2) \text{ kJ/mol}$

$Mg(g) + 2 \, F(g) \longrightarrow Mg^{2+}(g) + 2 \, F^-(g)$ $\quad \Delta H = +1544 \text{ kJ/mol}$ $\qquad$ This reaction is endothermic.

54. While the electron affinity of Li is –59.8 kJ/mol, the smallest ionization energy listed in Table 10-5 (and, except for Fr, displayed in Figure 10-10) is that of Cs, 375.7 kJ/mol. Thus, insufficient energy is produced by the electron affinity of Li to account for the ionization of Cs. And we would predict that compounds containing the Li^- ion will not be stable. (The other consideration is the energy released when the positive and negative ion combine to form an ion pair. This is an exothermic process, but we have no way of assessing the value of this energy.)

Magnetic Properties

55. Three of the ions have noble gas electron configurations and thus have no unpaired electrons:

$\qquad$ F^- is $1s^2 \, 2s^2 \, 2p^6$ $\qquad$ Ca^{2+} and S^{2-} are $[Ne] \, 3s^2 \, 3p^6$

Only, Fe^{2+} has unpaired electrons in its configuration: $[Ar] \, 3d^6 \, 4s^0$.

56. First we write the electron configuration of the element, then that of the ion, and then an orbital diagram. In each case, the number of unpaired electrons shown in the orbital diagram agrees with the data given in the statement of the problem.

(a) V $\quad [Ar] \, 3d^3 \, 4s^2$ $\qquad$ V^{3+} $\;[Ar] \, 3d^2 \, 4s^2$ $\qquad$ $[Ar] \, _{3d}$ ⬆⬆☐☐☐ $_{4s}$☐ two

(b) Cu $\;[Ar] \, 3d^{10} \, 4s^1$ $\qquad$ Cu^{2+} $[Ar] \, 3d^9 \, 4s^0$ $\qquad$ $[Ar] \, _{3d}$ ⬆⬇⬆⬇⬆⬇⬆⬇⬆ $_{4s}$☐ one

(c) Cr $\;[Ar] \, 3d^5 \, 4s^1$ $\qquad$ Cr^{3+} $[Ar] \, 3d^3 \, 4s^0$ $\qquad$ $[Ar] \, _{3d}$ ⬆⬆⬆☐☐ $_{4s}$☐ three

57. All atoms with an odd number of electrons must be paramagnetic. There is no way to create all pairs with an odd number of electrons. Many atoms with an even number of electrons are diamagnetic, but some are paramagnetic. The one of lowest atomic number is carbon $(Z = 6)$: $[He] \, _{2s}$⬆⬇ $_{2p}$⬆☐☐.

58. The electron configuration of each iron ion can be determined by starting with the electron configuration of the iron atom. $[Fe] = [Ar] \, 3d^6 \, 4s^2 = [Ar] \, _{3d}$ ⬆⬇⬆⬆⬆⬆ $_{4s}$⬆⬇ Then the two ions result from the loss of both $4s$ electrons, followed by the loss of a $3d$ electron in the case of Fe^{3+}.

$[Fe^{2+}] = [Ar] \, 3d^6 \, 4s^0 = [Ar] \, _{3d}$ ⬆⬇⬆⬆⬆⬆ $_{4s}$☐ $\qquad$ four unpaired electrons

$[Fe^{3+}] = [Ar] \, 3d^5 \, 4s^0 = [Ar] \, _{3d}$ ⬆⬆⬆⬆⬆ $_{4s}$☐ $\qquad$ five unpaired electrons

Predictions Based on the Periodic Table

59. (a) Elements that one would expect to exhibit the photoelectric effect with visible light should be ones that have a small value of their first ionization energy. Based on Figure 10-10, the alkali metals have the lowest first ionization potentials: Cs, Rb, and K are three suitable metals.

$\qquad$ Metals that would not exhibit the photoelectric effect with visible light are those that have high values of their first ionization energy. Again from Figure 10-10, Zn, Cd, and Hg seem to be three metals that would not exhibit the photoelectric effect with visible light.

(b) From Figure 10-1, we notice that the atomic (molar) volume does not increase significantly for the solid forms of the noble gases (the data points just before the alkali metal peaks). It probably increases less rapidly than does the molar mass. This means that the density should increase with atomic mass, and Rn should be the most dense solid. We expect densities of liquids to follow the same trend as densities of solids.

(c) To estimate the first ionization energy of fermium, we note in Figure 10-10 that the ionization energies of the lanthanides (following the Cs valley) are approximately the same. We expect similar behavior of the actinides, and estimate a first ionization energy of about +600 kJ/mol.

(d) We can estimate densities of solids from the information in Figure 10-1. Radium has $Z = 88$ and an approximate atomic volume of 40 cm^3/mol. Then we use the atomic weight of radium to determine its density:
$$\text{density} = \frac{1 \text{ mol}}{40 \text{ cm}^3} \times \frac{226 \text{ g Ra}}{1 \text{ mol}} = 5.6 \text{ g/cm}^3$$

60. (a) From Figure 10-1, the atomic (molar) volume of Al is 10. cm^3/mol and that of In is 15 cm^3/mol. Thus, we predict 12.5 cm^3/mol as the molar volume of Ga. Then we compute the density of Ga.
$$\text{density} = \frac{1 \text{ mol Ga}}{12.5 \text{ cm}^3} \times \frac{69.7 \text{ g Ga}}{1 \text{ mol Ga}} = 5.58 \text{ g/cm}^3 \text{ or } 5.6 \text{ g/cm}^3 \text{ as closely as the graph can be read.}$$

(b) Since Ga is in family 3A, the formula of its oxide should be Ga_2O_3.
mol weight = 2×69.7 g Ga + 3×16.0 g O = 187.4 g Ga_2O_3
$$\%\text{Ga} = \frac{2 \times 69.7 \text{ g Ga}}{187.4 \text{ g Ga}_2\text{O}_3} \times 100\% = 74.4\% \text{ Ga}$$

61. (a) The boiling point increases by 54 C° from CH_4 to SiH_4, and by 32 C° from SiH_4 to GeH_4. One expects another increase in boiling point from GeH_4 to SnH_4, probably by about 15 C°. Thus the predicted boiling point of SnH_4 is –75°C. (The actual boiling point of SnH_4 is –52 °C.)

(b) The boiling point decreases by 39 C° from H_2Te to H_2Se, and by 20 C° from H_2Se to H_2S. It probably will decrease by about 10 C° to reach H_2O. Therefore, one predicts a value of the boiling point of H_2O is –71°C. Of course, the actual boiling point of water is 100°C. The prediction is seriously in error because of hydrogen bonding between water molecules, a topic that is discussed in Chapter 13.

62. (a) Size increases down a group and from right to left in a period. Ba is closest to the lower left corner of the periodic table and has the largest size.

(b) Ionization energy decreases down a group and from right to left in a period. Although Pb is closest to the bottom of its group, Sr is farthest left in its period (and only one period above Pb). Sr should have the lowest first ionization energy.

(c) Electron affinity becomes more negative from left to right in a period and from bottom to top in a group. Cl is closest to the upper right in the periodic table and has the most negative (smallest) electron affinity.

(d) The number of unpaired electrons can be determined from the orbital diagram for each species.

F   [He] $_{2s}$⇅ $_{2p}$⇅↑↑   1 unpaired eln   N   [He] $_{2s}$⇅ $_{2p}$↑↑↑   3 unpaired elns
S²⁻   [Ne] $_{3s}$⇅ $_{3p}$⇅⇅⇅   0 unpaired eln   Mg²⁺ [Ne] $_{3s}$⇅ $_{3p}$⇅⇅⇅   0 unpaired eln
Sc³⁺   [Ne] $_{3s}$⇅ $_{3p}$⇅⇅⇅   0 unpaired eln   Ti³⁺ [Ar] $_{4s}$☐ $_{3d}$↑☐☐☐☐   1 unpaired eln

Thus, N has the largest number of unpaired electrons.

63. (a) $Z = 37$ 4   This is the element Rb, which has one $5s$ electron. It is easier to remove this $5s$ electron than to remove the outermost $4s$ electron of Ca, but harder than removing the outermost $6s$ electron of Cs.

(b) $Z = 9$ 3   This is the element F, with an electron affinity more negative than that of the adjacent atoms: Ne and O.

(c) $Z = 16$ 1   This is the element S. Each atom has an outer electron configuration of $ns^2 np^4$, which produces a partial (valence) orbital diagram of: $2s$ ⇅ $2p$ ⇅↑↑. Thus, there are two unpaired p electrons.

(d) $Z = 30$ 2   This is the element Zn. Since its electron configuration is [Ar] $3d^{10} 4s^2$, all electrons are paired and it is diamagnetic.

(e) $Z = 82$ 1   This is the element Pb, with an outer electron configuration of $ns^2 np^2$. Thus, Pb has two unpaired p electrons.

(f) $Z = 12$ 2   This is the element Mg. Its outer electron configuration is s^2 and thus it is diamagnetic. Its ionization potential is greater than that of Ca.

11 CHEMICAL BONDING I:
BASIC CONCEPTS

REVIEW QUESTIONS

Important Note: In this and subsequent chapters, a lone pair of electrons in a Lewis structure often is shown as a line rather than a pair of dots. Thus, the Lewis structure of Be is Be| rather than Be:

1. (a) Valence electrons are those of highest principal quantum number, those in the outermost shell, those furthest from the nucleus.
   (b) Electronegativity is a measure of the attraction that an atom in a compound has for electrons.
   (c) Lattice energy is the energy given off when a mole of a crystalline solid is formed from its constituent gaseous ions.
   (d) A double covalent bond results when two pairs of electrons are shared by two atoms.
   (e) A coordinate covalent bond is formed when both electrons of a shared pair come originally from one of the bonded atoms.

2. (a) Formal charge is a measure of how many valence electrons surround an atom in a covalently bonded structure, compared with the number of valence electrons of the isolated atom.
   (b) Resonance occurs when the bonding in a compound cannot be completely represented by one Lewis structure, but requires the "blending" of two or more.
   (c) An "expanded octet" refers to more than eight valence electrons surrounding an atom in a covalently bonded structure.

3. (a) An ionic bond is the result of the attraction between a cation and an anion. A covalent bond results when two electrons share one or more pairs of electrons.
   (b) A bonding pair of electrons is a pair that is shared between two atoms. A lone pair resides on one atom only, not shared between atoms.
   (c) Electron-pair geometry describes the orientation of electron pairs—bonding and lone—around a central atom. Molecular geometry describes the orientation of bonding pairs around the central atom.
   (d) A bond moment is the result of the unequal sharing of bonding electrons between two atoms. The resultant dipole moment of a molecule is the result of all of the bond moments and the molecule's geometry in producing a net dipole moment.
   (e) A polar molecule is one that has a net dipole moment, a slight overall separation of positive and negative charge. In a nonpolar molecule, there is no such net dipole moment.

4. (a) H·   (b) :Kr: (c) ·Sn·   (d) $[Ca]^{2+}$ (e) $[:\ddot{I}:]^-$

   (f) ·Ga·   (g) $[Sc]^{3+}$ (h) Rb·   (i) $[:\ddot{Se}:]^{2-}$

5. (a) $[K]^+ [:\ddot{I}:]^-$ (b) $[Ca]^{2+} [:\ddot{O}:]^{2-}$ (c) $[:\ddot{Br}:]^- [Ba]^{2+} [:\ddot{Br}:]^-$

   (d) $[Na]^+ [:\ddot{F}:]^-$ (e) $[K]^+ [:\ddot{S}:]^{2-} [K]^+$

6. (a) $|\bar{\text{I}}{-}\bar{\text{I}}|$ (b) $|\bar{\text{Br}}{-}\bar{\text{Cl}}|$ (c) $|\bar{\text{F}}{-}\bar{\text{O}}{-}\bar{\text{F}}|$ (d) $|\bar{\text{I}}{-}\bar{\text{N}}{-}\bar{\text{I}}|$ (e) $\text{H}{-}\bar{\text{Se}}{-}\text{H}$

                                                                           $|\bar{\text{I}}|$

7. (a) $|\bar{\text{S}}{=}\text{C}{=}\bar{\text{S}}|$ (b) $\text{H}{-}\overset{\displaystyle |\text{O}|}{\overset{\|}{\text{C}}}{-}\text{H}$ (c) $|\bar{\text{Cl}}{-}\overset{\displaystyle |\text{O}|}{\overset{\|}{\text{C}}}{-}\bar{\text{Cl}}|$ (d) $|\bar{\text{Cl}}{-}\bar{\text{N}}{=}\bar{\text{O}}|$

8. (a) $\text{H}{-}\text{H}{-}\bar{\text{N}}{-}\bar{\text{O}}{-}\text{H}$ has two bonds to (four electrons around) the second hydrogen, and only three

     bonds to (six electrons around) the nitrogen. A better Lewis structure is $\text{H}{-}\overset{\displaystyle \text{H}}{\overset{|}{\text{N}}}{-}\bar{\text{O}}{-}\text{H}$

     (b) $|\bar{\text{O}}{-}\bar{\text{Cl}}{-}\bar{\text{O}}|$ has 20 valence electrons, whereas the molecule ClO_2 has 19 valence electrons. This is

     a proper Lewis structure for the chlorite ion, although the brackets and the minus charge are missing. A

     plausible Lewis structure for the molecule ClO_2 is $|\bar{\text{O}}{-}\overset{\displaystyle \cdot}{\bar{\text{Cl}}}{-}\bar{\text{O}}|$

     (c) $[\cdot\overset{}{\text{C}}{=}\bar{\text{N}}|]^-$ has only six electrons around the C atom. $[|\text{C}{\equiv}\text{N}|]^-$ is a more plausible Lewis structure for
     the cyanide ion.

     (d) $\text{Ca}{-}\bar{\text{O}}|$ is improperly written as a covalent Lewis structure, although CaO is an ionic compound. In

     addition, there are only two electrons around the Ca atom. $[\text{Ca}]^{2+} [|\bar{\text{O}}|]^{2-}$ is a more plausible Lewis

     structure for CaO.

9. (a) The formal charge on each I is 0, computed as follows:

		number of valence electrons	=	7
		$-2 \times$ no. lone pairs	=	-6
		$-$ no. bonds	=	$\underline{-1}$
		formal charge	=	0

(b) computations for:

	$={=}\text{O}$	$-\text{O}$	$=\text{S}-$
number of valence electrons	6	6	6
$-2 \times$ no. lone pairs	-4	-6	-2
$-$no. bonds	$\underline{-2}$	$\underline{-1}$	$\underline{-3}$
formal charge	0	-1	$+1$

(c) computations for:

	$={=}\text{O}$	$-\text{O}$	C
number of valence electrons	6	6	4
$-2 \times$ no. lone pairs	-4	-6	-0
$-$no. bonds	$\underline{-2}$	$\underline{-1}$	$\underline{-4}$
formal charge	0	-1	0

(d) computations for:

	O	Cl	C
number of valence electrons	6	7	4
$-2 \times$ no. lone pairs	-4	-6	-0
$-$no. bonds	$\underline{-2}$	$\underline{-1}$	$\underline{-4}$
formal charge	0	0	0

(e) computations for:

	H	$-\text{O}-$	$-\text{O}$
number of valence electrons	1	6	6
$-2 \times$ no. lone pairs	-0	-4	-6
$-$no. bonds	$\underline{-1}$	$\underline{-2}$	$\underline{-1}$
formal charge	0	0	-1

(f) computations for:

	$=O$	$-O$	N
number of valence electrons	6	6	5
$-2 \times$ no. lone pairs	-4	-6	-1
$-$no. bonds	$\underline{-2}$	$\underline{-1}$	$\underline{-3}$
formal charge	0	-1	$+1$

10. (a) $[\bar{\text{I}}\text{O}\!-\!\text{H}]^- \; \text{Mg}^{2+} \; [|\bar{\text{O}}\!-\!\text{H}]^-$ **(b)** $\left(\begin{array}{c} \text{H} \\ | \\ \text{H}\!-\!\text{N}\!-\!\text{H} \\ | \\ \text{H} \end{array}\right)^+ [|\bar{\text{I}}\bar{\text{I}}|]^-$ **(c)** $[\bar{\text{I}}\text{O}\!-\!\bar{\text{C}}\text{l}\!-\!\text{O}\bar{\text{I}}]^- \; \text{Ca}^{2+} \; [\bar{\text{I}}\text{O}\!-\!\bar{\text{C}}\text{l}\!-\!\text{O}\bar{\text{I}}]^-$

11. Since electrons pair up if at all possible in plausible Lewis structures, we can get a very good indication if a species is paramagnetic if it has an odd number of (valence) electrons.

(a)	OH^-	$6 + 1 + 1 = 8$ valence electrons	diamagnetic
(b)	OH	$6 + 1 = 7$ valence electrons	paramagnetic
(c)	NO_3	$5 + (3 \times 6) = 23$ valence electrons	paramagnetic
(d)	SO_3	$6 + (3 \times 6) = 24$ valence electrons	diamagnetic
(e)	SO_3^{2-}	$6 + (3 \times 6) + 2 = 26$ valence electrons	diamagnetic
(f)	HO_2	$1 + (2 \times 6) = 13$ valence electrons	paramagnetic

12. (a) **(b)** $\bar{\text{I}}\text{F}\!-\!\bar{\text{P}}\!-\!\text{F}\bar{\text{I}}$ with $|\text{F}|$ below **(c)** $|\bar{\text{C}}\text{l}\!-\!\overset{\frown}{\text{I}}\!-\!\bar{\text{C}}\text{l}|$ with $|\bar{\text{C}}\text{l}|$ below **(d)** $\bar{\text{I}}\text{F}\!-\!\text{S}\!-\!\text{F}\bar{\text{I}}$ with $|\bar{\text{F}}|$ above and $|\bar{\text{F}}|$ below

13. In order to form a compound with five bond pairs to a central atom, such as PCl_5, there must be at least five electron pairs around the central atom. (Indeed, there are five bonding pairs and no lone pairs around P in PCl_5.) This requires that P be able to accomodate an expanded octet of valence electrons, which it can, since it is an element in the third period. N cannot accomodate an expanded octet, however, since it is an element in the second period of the periodic table. Neither NCl_3 nor PCl_3 requires an expanded octet. Both have 3 bonding pairs and one lone pair around the central atom.

14. We give all the Lewis structures first, from each of which we deduce the electron pair geometry and the molecular shape.

(a)	H_2Te	tetrahedral electron pair geometry, bent (angular) molecular geometry
(b)	C_2Cl_4	trigonal planar electron pair geometry around each C, planar molecule
(c)	CO_2	linear electron pair geometry, linear molecular geometry
(d)	$SbCl_6^-$	octahedral electron pair geometry, octahedral geometry
(e)	SO_4^{2-}	tetrahedral electron pair geometry, tetrahedral molecular geometry

15. (a) In SO_3, there are a total of $6 + (3 \times 6) = 24$ valence electrons, or 12 pairs. A plausible Lewis structure has three resonance forms, of which one is: $\bar{\text{O}}\!=\!\text{S}\!-\!\text{O}\bar{\text{I}}$ with $|\bar{\text{O}}|$ below. This molecule is of the type AX_3; it has a trigonal planar electron-pair geometry and a trigonal planar shape.

(b) In SO_3^{2-} the total number of valence electrons is $2 + (3 \times 6) + 6 = 26$ valence electrons, or 13 pairs. A plausible Lewis structure is $\left(\bar{\text{O}}\!=\!\bar{\text{S}}\!-\!\text{O}\bar{\text{I}}\right)^{2-}$ with $|\bar{\text{O}}|$ below. Two other resonance forms can be drawn. This molecule is of the type AX_3E; it has a tetrahedral electron-pair geometry and a trigonal pyramidal molecular shape.

(c) In SO_4^{2-}, there are a total of $2 + 6 + (4 \times 6) = 32$ valence electrons, or 16 pairs. There are several

possible resonance forms, but a simple Lewis structure is $\left(\begin{array}{c} |\overline{O}| \\ | \\ |\overline{O}-S-\overline{O}| \\ | \\ |\overline{O}| \end{array}\right)^{2-}$ This ion is of the type

AX_4; it has a tetrahedral electron-pair geometry and a tetrahedral shape.

16. In each case, a plausible Lewis structure is given first, followed by the AX_nE_m notation of each species. Then comes the electron pair geometry and, finally, the molecular geometry.

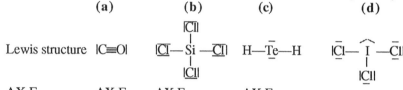

	(a)	**(b)**	**(c)**	**(d)**
Lewis structure	$|C\equiv O|$	$\overline{Cl}-Si-\overline{Cl}$ with $\overline{Cl}$ above and $|\overline{Cl}|$ below	$H-\overline{Te}-H$	$\overline{Cl}-I-\overline{Cl}$ with $|\overline{Cl}|$ below
AX_nE_m electron pair geometry	AX_1E_1 linear	AX_4E_0 tetrahedral	AX_2E_2 tetrahedral	trigonal bipyramid
molecular geometry	linear	tetrahedral	bent	T-shaped

	(e)	**(f)**	**(g)**
Lewis structure	$\overline{Cl}-Sb-\overline{Cl}$ with $|\overline{Cl}|, \overline{Cl}|$ above and $|\overline{Cl}|$ below	$\overline{O}-\overline{S}=\overline{O}$	$\left(\begin{array}{c} |\overline{F} \quad \overline{F}| \\ |\overline{F}-Al-\overline{F}| \\ |\overline{F} \quad \overline{F}| \end{array}\right)^{3-}$
AX_nE_m electron pair geometry	AX_5E_0 trigonal bipyramid	AX_2E_1 triangular planar	AX_6E_0 octahedral
molecular geometry	trigonal bipyramid	bent	octahedral

17.

bond	H—C	C=O	C—C	C—Cl	
(a) bond length	110	120	154	178	pm
(b) bond energy	414	736	347	339	kJ/mol

18. **(c)** is the longest bond. Lewis structures are of assistance, for single bonds are generally the longest. Of the two molecules with single bonds, Br_2 is expected to have longer bonds than BrCl, since Br is a larger atom than a Cl atom. (a) $\overline{O}=\overline{O}$ (b) $|N\equiv N|$ (c) $|\overline{Br}-\overline{Br}|$ (d) $|\overline{Br}-\overline{Cl}|$

19. The reaction $O_2 \longrightarrow 2\, O$ is an endothermic reaction since it requires the breaking of the bond between two oxygen atoms without the formation of any bonds. Since bond breakage is endothermic, the entire process must be endothermic.

20. **(a)** The net result of this reaction, per mole of $CH_4(g)$, is to break one mole of C—H bonds (which requires 414 kJ) and to form 1 mole of H—I bonds (which produces 297 kJ). Thus, this reaction is endothermic; a net infusion of energy is necessary.

(b) The net result of this reaction, per mole of $H_2(g)$, is to break 1 mol of H—H bonds (which requires 436 kJ) and 1 mol of I—I bonds (which requires 151 kJ), and to form two moles of H—I bonds (which produces $2 \times 297 = 594$ kJ). Thus, this reaction is just barely exothermic.

21. We draw Lewis structures of the reactants and the products to aid in determining the value of ΔH.

$$2\, \overset{\bullet}{\underset{}{N}}=\overline{O} \;+\; 5\, H-H \;\longrightarrow\; 2\, H-\overset{}{\underset{H}{\overline{N}}}-H \;+\; 2\, H-\overline{O}-H$$

ΔH = energy of bonds broken – energy of bonds formed

$$= [(2 \times N{=}O) + (5 \times H{-}H)] - [(6 \times N{-}H) + (4 \times H{-}O)]$$
$$= [(2 \times 631 \text{ kJ/mol}) + (5 \times 436 \text{ kJ/mol})] - [(6 \times 389 \text{ kJ/mol}) + (4 \times 463 \text{ kJ/mol})]$$
$$= 1.26 \times 10^3 + 2.18 \times 10^3 - 2.33 \times 10^3 - 1.85 \times 10^3$$
$$= -0.74 \times 10^3 \text{ kJ/mol} = -7.4 \times 10^2 \text{ kJ/mol}$$

22. Recall that electronegativity increases in the periodic table from lower left to upper right, and specifically that metals have lower electronegativities than do nonmetals, with metalloids having intermediate electronegativities. Based on these principles, the electronegativity of S is greater than that of As, which in turn is greater than that of Bi. The electronegativity of Ba is less than that of Mg. And finally, we would predict the electronegativity of Mg (a definite metal) to be less than that of Bi (somewhat metalloid in character). Thus, ranked in order of increasing electronegativity, these elements are: Ba < Mg < Bi < As < S, with Bi having the intermediate electronegativity of this group of five elements. The actual electronegativities are: Ba (0.9) < Mg (1.3) < Bi (2.0) < As (2.2) < S(2.6)

23. Na—Cl and K—F both are bonds between a metal and a nonmetal; they have the largest ionic character, with the ionic character of K—F being greater than that of Na—Cl, both because K is more metallic (closer to the lower left of the periodic table) than Na and F is more nonmetallic (closer to the upper right) than Cl. The remaining three bonds are covalent bonds to H. Since H and C have about the same electronegativity (a fact you need to memorize), the H—C bond is the most covalent (or the least ionic). Br is somewhat more electronegative than is C, while F is considerably more electronegative than C, making the F—H bond the most polar of the three covalent bonds. Thus, ranked in order of increasing ionic character, these five bonds are: C—H < Br—H < F—H < Na—Cl < K—F   The actual electronegativity differences follow: C(2.6)—H(2.2) < Br(3.0)—H(2.2) < F(4.0)—H(2.2) < Na(0.9)—Cl(3.2) < K(0.8)—F(4.0)
$\Delta EN = 0.4$ $\qquad = 0.8$ $\qquad = 1.8$ $\qquad = 2.3$ $\qquad = 3.2$

24. **(a)** F_2 cannot possess a dipole moment, since all of the atoms in the molecule are the same. This means that there is no electronegativity difference between atoms, and hence no polar bonds.

(b) $\overline{|O}{-}\overset{\bullet\bullet}{N}{=}\overline{O|}$ Each nitrogen-to-oxygen bond in this molecule is polar toward oxygen, the more

electronegative element. The molecule is of the AX_2E_2 category and hence is bent. Therefore the two bond dipoles do not cancel, and the molecule is polar.

(c) $|\overline{F}{-}B{-}\overline{F}|$ Although each B—F bond is polar toward F, in this trigonal planar
$\qquad\quad |\overline{F}|$ AX_3E_0 molecule these bond dipoles cancel. The molecule is nonpolar.

(d) $H{-}\overline{Br}|$ The H—Br bond is polar toward Br, and this molecule is polar as well.

(e) $H{-}\underset{|\overline{Cl}|}{\overset{H}{\underset{|}{C}}}{-}\overline{Cl}|$ The H—C bonds are not polar, but the C—Cl bonds are, toward Cl. The molecular shape is tetrahedral (AX_4) and thus these two C—Cl dipoles do not cancel each other; the molecule is polar.

(f) $|\overline{F}{-}\underset{|\overline{F}|}{\overset{|\overline{F}|}{\underset{|}{Si}}}{-}\overline{F}|$ Although each Si—F bond is polar toward F, in this tetrahedral AX_4E_0 molecule these bond dipoles oppose and cancel each other; the molecule is nonpolar.

(g) $\overline{O}{=}C{=}\overline{S}$ In this linear molecule, the two bonds from carbon both are polar away from carbon. But the C=O bond is more polar than the C=S bond and hence the molecule is polar.

25. **(a)** There is at least one instance in which one atom must bear a formal charge—a polyatomic ion—because the charge on the species equals the sum of the formal charges. There are other types of molecules that have formal charges. Sometimes this occurs when removing formal charges would require that there be more than one unpaired electron in the structure.

(b) Since three points define a plane, stating that a triatomic molecule is planar is not saying anything that is new. In fact, it is misleading, for some triatomic molecules are linear. HCN is one of them, as is

CO_2. Of course, some molecules with more than three atoms also are planar; examples are XeF_4 and $H_2C=CH_2$.

(c) This statement is incorrect, for in some molecules that contain polar bonds, the bonds are so oriented in space that there is no resulting molecular dipole moment. Examples of such molecules are CO_2, BCl_3, CCl_4, PCl_5, and SF_6.

EXERCISES

Lewis theory

26. In ionic compounds, electrons are transferred from one atom to another to form ions. Thus, in the Lewis structure of an ionic compound the resulting ions are written as separate entities, with the anion at least surrounded by pairs of electrons. In covalent compounds, these surrounding pairs of electrons are obtained by sharing electrons between atoms. Thus, the Lewis structure for a covalent compound is one complete entity, not several parts, since the atoms must be adjacent in order to share electrons.

27. Hydrogen never has an octet of electrons in any of its compounds, but only a pair (or duet, if you prefer). An example is the Lewis structure of H_2O (below). In many compounds in which the central atom is from the second period or higher, there are more than eight electrons around the central atom; an example of a compound with such an "expanded octet" is ICl_3 (below). Finally, in some compounds, there simply are not eight electrons around the central atom; such an "electron deficient" compound is BF_3.

$$H-\overline{O}-H \qquad \overline{|Cl}-\overset{\frown}{I}-\overline{Cl|} \qquad \overline{|F}-B-\overline{F|}$$
$$\qquad\qquad\quad \overline{|Cl|} \qquad\qquad\quad \overline{|F|}$$

28. (a) $Cs^+ [:\overset{\cdot\cdot}{\underset{\cdot\cdot}{Br}}:]^-$ **(b)** $H-\overline{Se}-H$ **(c)** $\overline{|Cl}-B-\overline{Cl|}$
$\qquad\qquad\qquad\qquad\qquad\qquad\qquad\qquad\qquad\qquad\qquad \overline{|Cl|}$

  CsBr, cesium bromide H_2Se, hydrogen selenide BF_3, boron trichloride

(d) $Rb^+ [:\overset{\cdot\cdot}{\underset{\cdot\cdot}{S}}:]^{2-} Rb^+$ **(e)** $Mg^{2+} [:\overset{\cdot\cdot}{\underset{\cdot\cdot}{O}}:]^{2-}$ **(f)** $\overline{|F}-\overline{O}-\overline{F|}$

  Rb_2S, rubidium sulfide   MgO, magnesium oxide OF_2, oxygen difluoride

29. (a) $Li-\overline{O}-Li$ is written as a covalent structure even though the compound is composed of a metal (Li) and a nonmetal (O). One expects an ionic Lewis structure. $[Li]^+ [|\overline{O}|]^{2-} [Li]^+$

(b) $[|\overline{Cl}|]^+ [|\overline{O}|]^{2-} [\overline{Cl}|]^+$ is written as an ionic structure, even though we expect a covalent structure between nonmetallic atoms. A more plausible structure is $|\overline{Cl}-\overline{O}-\overline{Cl}|$

(c) $|\overline{O}=N=\overline{O}|$ has too few valence electrons—8 pairs or 16 valence electrons—it should have $(2 \times 6) + 5$ = 17 valence electrons: eight pairs plus a lone electron. A plausible Lewis structure is $|\overline{O}-\overset{\cdot}{N}=\overline{O}|$, and its resonance form, both of which have +1 formal charge on N and –1 formal charge on the single-bonded oxygen. Another plausible Lewis structure, $|\overset{\cdot}{\underline{O}}-\overline{N}=\overline{O}|$, has zero formal charge on each of its atoms.

(d) In the structure $[\overline{|S}-C=\overline{N}]^-$ there is not an octet of electrons around either S nor C. In addition, there are only 7 pairs of valence electrons in this structure, 14 valence electrons. There should be $6 + 4 + 5 + 1 = 16$ valence electrons, or 8 pairs. At least two structures are possible. $[\overline{S}=C=\overline{N}]^-$ has a formal charge of –1 on N and is preferred over $[\overline{|S}-C\equiv N|]^-$, which has a formal charge of –1 on S, which is less electronegative than N.

30. **(c)** is the correct answer. **(a)** $[\overline{\text{O}}\text{—C=}\overline{\text{N}}]^-$ does not have an octet of electrons around C.

(b) $[\text{C≡Cl}]^-$ does not have an octet around the left-hand C, it has only 8 valence electrons, and should have 10, and the formal charges on the two carbons are different. **(d)** The total number of valence electrons in $|\overline{\text{N}}\text{=}\overline{\text{O}}|$ is incorrect; this odd-electron species should have 11 valence electrons, not 12.

31. **(a)** In an H_3 molecule, one of the H atoms would have to be a central atom, with two pairs of electrons, one bond to each of the other H atoms. This places more than 2 electrons around this central H atom, more than the stable pair found around H in most Lewis structures.

(b) In HHe there would either be a shared triplet of electrons between the two atoms, or three electrons around the He atom H—He· Either of these is a not particularly stable situation.

(c) He_2 would have either a double bond between two He atoms He=He and thus four electrons around each He atom, or three electrons around each He atom ·He—He· Neither situation achieves the electron configuration of the nearest noble gas.

(d)
$$\text{H—}\underset{\cdot\cdot}{\overset{\overset{\displaystyle \text{H}}{|}}{\text{O}}}\text{—H}$$
has an expanded octet (9 electrons) on oxygen; expanded octets are not found on elements of the second period. Other structures place a multiple bond between O and H. Both situations are unstable.

Ionic bonding

32. **(a)** Li· forms Li^+ cations, and $:\overset{\cdot\cdot}{\underset{\cdot\cdot}{\text{O}}}:$ forms $[:\overset{\cdot\cdot}{\underset{\cdot\cdot}{\text{O}}}:]^{2-}$ anions. Thus, lithium oxide is $Li^+ [:\overset{\cdot\cdot}{\underset{\cdot\cdot}{\text{O}}}:]^{2-} Li^+$

(b) Na· forms Na^+ cations, and $:\overset{\cdot\cdot}{\underset{\cdot\cdot}{\text{I}}}:$ forms $[:\overset{\cdot\cdot}{\underset{\cdot\cdot}{\text{I}}}:]^-$ anions. Thus, sodium iodide is $Na^+ [:\overset{\cdot\cdot}{\underset{\cdot\cdot}{\text{I}}}:]^-$

(c) ·Ba· forms Ba^{2+} cations, and $:\overset{\cdot\cdot}{\underset{\cdot\cdot}{\text{F}}}:$ forms $[:\overset{\cdot\cdot}{\underset{\cdot\cdot}{\text{F}}}:]^-$ anions. Thus, barium fluoride is $[:\overset{\cdot\cdot}{\underset{\cdot\cdot}{\text{F}}}:]^- Ba^{2+} [:\overset{\cdot\cdot}{\underset{\cdot\cdot}{\text{F}}}:]^-$

(d) ·Sc· forms Sc^{3+} cations, and $:\overset{\cdot\cdot}{\underset{\cdot\cdot}{\text{Cl}}}:$ forms $[:\overset{\cdot\cdot}{\underset{\cdot\cdot}{\text{Cl}}}:]^-$ anions. Scandium chloride is $[:\overset{\cdot\cdot}{\underset{\cdot\cdot}{\text{Cl}}}:]^- Sc^{3+} [:\overset{\cdot\cdot}{\underset{\cdot\cdot}{\text{Cl}}}:]$

The formulas of the compounds are Li_2O, NaI, BaF_2, and $ScCl_3$. $[:\overset{\cdot\cdot}{\underset{\cdot\cdot}{\text{Cl}}}:]^-$

33. The Lewis symbols are $[\text{H}|]^-$ for the hydride ion, $[|\overline{\text{N}}|]^{3-}$ for the nitride ion.

(a) $[Li]^+ [\text{H}|]^-$ **(b)** $[\text{H}|]^- [Ca]^{2+} [\text{H}|]^-$ **(c)** $[Mg]^{2+} [|\overline{\text{N}}|]^{3-} [Mg]^{2+} [|\overline{\text{N}}|]^{3-} [Mg]^{2+}$
lithium hydride   calcium hydride   magnesium nitride

34. The term "molecules" is inappropriate when describing solid potassium fluoride because potassium ions and fluoride ions are present in the crystal in an alternating pattern, with no particular potassium ion being any closer to a specific fluoride ion than are a number of other potassium ions. Furthermore, there is not a covalent bond between adjacent ions, but rather an ionic attraction. In the gas phase, there indeed can be a distinct potassium ion associated with a specific fluoride ion, but still there is no covalent bond between them. Often the term "ion pair" is used to refer to a K^+, F^- grouping.

Formal Charge

35. There are three common features of formal charge and oxidation state. Both indicate how electrons are distributed in the bonding of the compound. Second, negative formal charge (in the most plausible Lewis structure) and negative oxidation state are generally those of the most electronegative type of atom. And third, both numbers are determined by a set of rules, rather than being determined experimentally. Differences include some instances in which the same type of atom in a compound (which of course has the same oxidation state), has different formal charges, such as oxygen in ozone, O_3. Another is that formal charges are used to decide between alternative Lewis structures while oxidation state is used in balancing equations and naming compounds. The most significant difference, though, is that whereas the oxidation state of an element in its compounds is usually not zero, its formal charge often is.

36. Formal charge = number of valence electrons – 2 × number of lone pairs – number of bond pairs
The formal charge of the central atom is calculated below the Lewis structure of each species.

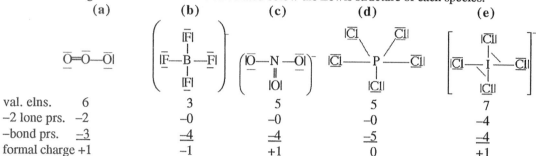

	(a)	(b)	(c)	(d)	(e)
val. elns.	6	3	5	5	7
–2 lone prs.	–2	–0	–0	–0	–4
–bond prs.	–3	–4	–4	–5	–4
formal charge	+1	–1	+1	0	+1

37. We calculate formal charge = no. valence electrons – no. bonds – 2 × no. lone pairs.

(a) H—N̄—Ō—H   f.c. of N = 5 – 3 – (1 × 2) = 0   most plausible
     | f.c. of O = 6 – 2 – (2 × 2) = 0
     H

   H—Ō—N̄—H   f.c. of N = 5 – 2 – (2 × 2) = –1   f.c. of O = 6 – 3 – (1 × 2) = +1
         |
         H

(b) S̄=C=S̄   f.c. of C = 4 – 4 – 0 = 0   f.c. of S = 6 – 2 – (2 × 2) = 0

           most plausible

   C̄=S=S̄   f.c. of C = 4 – 2 – (2 × 2) = –2   f.c. of central S = 6 – 4 – 0 = +2

           f.c. of terminal S = 6 – 2 – (2 × 2) = 0

(c) N̄=Ō—C̄l| f.c. of N = 5 – 2 – (2 × 2) = –1   f.c. of O = 6 – 3 – (1 × 2) = +1

           f.c. of Cl = 7 – 1 – (3 × 2) = 0

   Ō=N̄—C̄l| f.c. of O = 6 – 2 – (2 × 2) = 0   f.c. of N = 5 – 3 – (1 × 2) = 0

           f.c. of Cl = 7 – 1 – (3 × 2) = 0   most plausible

(d) [S̄=C=N̄]⁻   [C̄=N=S̄]⁻   [C̄=S=N̄]⁻

   f.c. of S = 6 – 2 – (2 × 2) = 0   f.c. of S = 6 – 2 – (2 × 2) = 0   f.c. of S = 6 – 4 – 0 = +2
   f.c. of C = 4 – 4 – 0 = 0   f.c. of C = 4 – 2 – (2 × 2) = –2   f.c. of C = 4 – 2 – (2 × 2) = –2
   f.c. of N = 5 – 2 – (2 × 2) = –1   f.c. of N = 5 – 4 – 0 = +1   f.c. of N = 5 – 2 – (2 × 2) = –1
   most plausible

Lewis structures

38. (a) In H_2NOH, N and O atoms are the central atoms and the terminal atoms are the H atoms. The number
of valence electrons in the molecule totals: $(3 \times 1) + 5 + 6 = 14$ valence electrons, or seven pairs. A
resonable Lewis structure is   H—N̄—Ō—H
                                  |
                                  H

(b) In N_2F_2, the F atoms are terminal atoms, and the N atoms are the central atoms. The total number of
valence electrons in the structure is $(2 \times 7) + (2 \times 5) = 24$ valence electrons, or 12 pairs. A reasonable
Lewis structure is |F̄—N̄=N̄—F̄|

(c) In HONO, the H atoms and one O atom are the terminal atoms; the other O atom and the N atom are the central atoms. The total number of valence electrons in the structure is $1 + 5 + (2 \times 6) = 18$ valence electrons, or nine (9) pairs. A plausible Lewis structure is   H—$\bar{\text{O}}$—$\bar{\text{N}}$=$\bar{\text{O}}$

(d) In H_2NNO_2, N atoms are the central atoms, and H and O atoms are terminal atoms. The total number of valence electrons are $(2 \times 1) + (2 \times 5) + (2 \times 6) = 24$ valence electrons, or 12 pairs. A plausible Lewis structure is   H—$\bar{\text{N}}$—N=$\bar{\text{O}}$
  | |
  H $|\underline{\text{O}}|$

39. The total number of valence electrons is $(2 \times 7) + (2 \times 6) = 26$ valence electrons, or 13 pairs of valence electrons. It is unlikely to have F as a central atom; that would require an expanded octet on F. One structure is $|\bar{\text{F}}$—$\bar{\text{S}}$—$\bar{\text{S}}$—$\bar{\text{F}}|$ another is $\bar{\text{S}}$=$\bar{\text{S}}$—$\bar{\text{F}}|$ Both have zero formal charge on each atom, but only the first has
      |
    $|\text{F}|$
just an octet around each atom.

40. (a) The total number of valence electrons in OCl^- is $1 + 6 + 7 = 14$, or 7 pairs. A plausible Lewis structure for the hypochlorite ion is $[|\bar{\text{Cl}}$—$\bar{\text{O}}|]^-$

(b) The total number of valence electrons in NO_2^- is $5 + (2 \times 6) + 1 = 18$, or 9 pairs. There are two resonance forms for the nitrite ion: $[\bar{\text{O}}$—$\bar{\text{N}}$=$\bar{\text{O}}]^- \leftrightarrow [\bar{\text{O}}$=$\bar{\text{N}}$—$\bar{\text{O}}|]^-$

(c) The total number of valence electrons in CO_3^{2-} is $4 + (3 \times 6) + 2 = 24$, or 12 pairs. There are three resonance forms for the carbonate ion: $\left(\bar{\text{O}}$—C=$\bar{\text{O}}\right)^{2-} \leftrightarrow \left(\bar{\text{O}}$=C—$\bar{\text{O}}|\right)^{2-} \leftrightarrow \left(\bar{\text{O}}$—C—$\bar{\text{O}}|\right)^{2-}$
        | | ‖
      $|\underline{\text{O}}|$ $|\underline{\text{O}}|$ $|\underline{\text{O}}|$

41. Each of the cations has an empty valence shell as the result of ionization. The main task is to determine the Lewis structure of each anion.

(a) The total number of valence electrons in OH^- is $6 + 1 + 1 = 8$, or 4 pairs. A plausible Lewis structure for magnesium hydroxide is $[\bar{\text{O}}$—H]$^-$ $[Mg]^{2+}$ $[|\bar{\text{O}}$—H]$^-$

(b) The total number of valence electrons in IO_3^- is $7 + (3 \times 6) + 1 = 24$, or 12 pairs. A plausible Lewis structure for potassium iodate is $[K]^+$ $\left(\begin{array}{c}|\text{O}| \\ | \\ \bar{\text{O}}\text{—I—}\bar{\text{O}}| \\ | \\ |\underline{\text{O}}|\end{array}\right)^-$

(c) The total number of valence electrons in ClO_2^- is $7 + (2 \times 6) + 1 = 20$, or 10 pairs. A plausible Lewis structure for calcium chlorite is $[\bar{\text{O}}$—$\bar{\text{Cl}}$—$\bar{\text{O}}|]^-$ $[Ca]^{2+}$ $[|\bar{\text{O}}$—$\bar{\text{Cl}}$—$\bar{\text{O}}|]^-$

(d) The total number of valence electrons in SO_4^{2-} is $6 + (4 \times 6) + 2 = 32$, or 16 pairs. A plausible structure for aluminum sulfate is $[SO_4]^{2-}$ $[Al]^{3+}$ $[SO_4]^{2-}$ $[Al]^{3+}$ $[SO_4]^{2-}$ Because of the ability of S to expand its octet, SO_4^{2-} has several resonance forms, a few of which are:

$$\left(\begin{array}{c}|\text{O}| \\ | \\ \bar{\text{O}}\text{—S—}\bar{\text{O}}| \\ | \\ |\underline{\text{O}}|\end{array}\right)^{2-} \leftrightarrow \left(\begin{array}{c}|\text{O}| \\ \| \\ \bar{\text{O}}\text{=S—}\bar{\text{O}}| \\ | \\ |\underline{\text{O}}|\end{array}\right)^{2-} \leftrightarrow \left(\begin{array}{c}|\text{O}| \\ | \\ \bar{\text{O}}\text{—S—}\bar{\text{O}}| \\ \| \\ |\underline{\text{O}}|\end{array}\right)^{2-} \leftrightarrow \left(\begin{array}{c}|\text{O}| \\ | \\ \bar{\text{O}}\text{=S=O} \\ | \\ |\underline{\text{O}}|\end{array}\right)^{2-} \leftrightarrow \left(\begin{array}{c}|\text{O}| \\ \| \\ \bar{\text{O}}\text{=S—}\bar{\text{O}}| \\ | \\ |\underline{\text{O}}|\end{array}\right)^{2-}$$

Bond lengths

42. A heteronuclear bond length (one between two different atoms) is equal to the average of two homonuclear bond lengths (one between two like atoms) of the same order (both single, both double, or both triple).

(a) I—Cl bond length   = [(I—I bond length) + (Cl—Cl bond length)] $\div$ 2
              = [266 pm + 199 pm] $\div$ 2 = 233 pm

(b) O—F bond length   = [(O—O bond length) + (F—F bond length)] $\div$ 2

$$= [145 \text{ pm} + 143 \text{ pm}] \div 2 = 144 \text{ pm}$$

(c) C—F bond length $= [(C—C \text{ bond length}) + (F—F \text{ bond length})] \div 2$

$$= [154 \text{ pm} + 143 \text{ pm}] \div 2 = 149 \text{ pm}$$

(d) N—I bond length $= [(N—N \text{ bond length}) + (I—I \text{ bond length})] \div 2$

$$= [145 \text{ pm} + 266 \text{ pm}] \div 2 = 206 \text{ pm}$$

43. In $\overline{O}=N—\overline{F}|$ The N—F bond is a single bond. Its bond length should be the average
of the N—N single bond (145 pm) and the F—F single bond (143 pm).
This average is   N—F bond length = (145 + 143) ÷ 2 = 144 pm

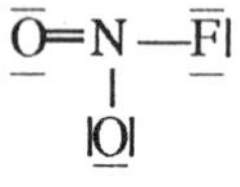

44. First we need to draw the Lewis structure of each of the compounds cited, so that we can determine the order, and thus the relative length, of each N—N bond.

(a) In N_2H_4 the total number of valence electrons is $(2 \times 5) + (4 \times 1) = 14$ valence electrons, or 9 pairs. Each N is a central atom, attached to the other N atom and to 2 H atoms. In the plausible Lewis structure given below, all atoms have zero formal charge.

(b) In N_2, the total number of valence electrons is $(2 \times 5) = 10$ valence electrons, or 5 pairs. Both N atoms have zero formal charge in the Lewis structure given below.

(c) In N_2O_4, the total number of valence electrons is $(2 \times 5) + (4 \times 6) = 34$ valence electrons or 7 pairs. Each N is attached to the other N atom, and to 2 O atoms, which are terminal atoms. The double-bonded O atoms have zero formal charge, each N has a formal charge of +1, and the single bonded O atoms have a formal charge of –1.

(d) In N_2O, the total number of valence electrons is $(2 \times 5) + 6 = 16$ valence electrons, or 8 pairs. N is the central atom. In both Lewis structures drawn below, the central N has a formal charge of +1. The double-bonded terminal N has a formal charge of –1, and the single bonded O has a formal charge of – 1. All other atoms have formal charges of zero.

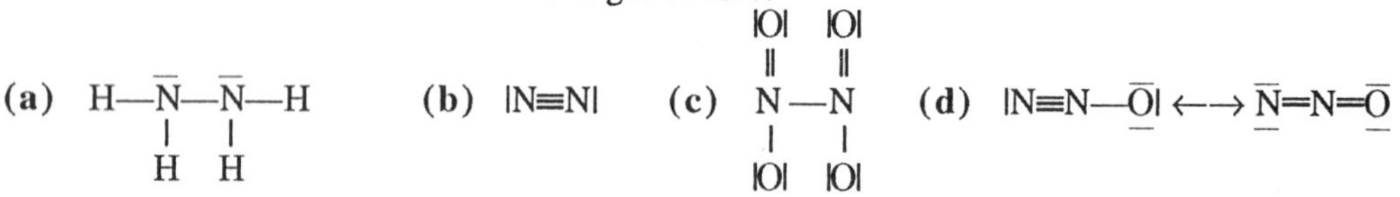

(a) H—N̄—N̄—H   **(b)** |N≡N| **(c)**   **(d)** |N≡N—Ō|⁻ ⟷ N̄=N=Ō

The shortest N—N bond of these four should be that in N_2, since it is a "pure" triple bond. That in N_2O would be next in length, being partially a double bond and partially a triple bond. Since a resonance structure for N_2O_4 can be drawn that contains a N=N bond, the bond in N_2O_4 would probably be next in length, with the N—N bond in N_2H_4 being the longest of the four.

45. In H_2NOH, there are $(3 \times 1) + 5 + 6 = 14$ valence electrons total, or 7 pairs. N and O are the two central atoms. A plausible Lewis structure has zero formal charge on each atom. H—N̄—Ō—H The N—H bond lengths are 100 pm, the O—H bond is 97 pm, and the N—O bond is 136 pm. All values are taken from Table 11-1. All bond angles approximate the tetrahedral bond angle of 109.5°, but are expected to be somewhat smaller, perhaps by 2-4° each.

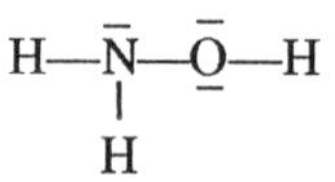

Resonance

46. In NO_2^-, the total number of valence electrons is $1 + 5 + (2 \times 6) = 18$ valence electrons, or 9 pairs. N is the central atom. There are two resonance forms. $[\overline{O}—\overline{N}=\overline{O}]^- \longleftrightarrow [\overline{O}=\overline{N}—\overline{O}]^-$

47. The only one of the four species that requires resonance forms for its adequate representation is CO_3^{2-}. In none of the others are there dissimilar electron distributions around like atoms. All four Lewis structures are drawn below.

(a) In CO_2, there are $4 + (2 \times 6) = 16$ valence electrons, or 8 pairs.

(b) In OCl^-, there are $6 + 7 + 1 = 14$ valence electrons, or 7 pairs,

(c) In CO_3^{2-}, there are $4 + (3 \times 6) + 2 = 24$ valence electrons, or 12 pairs.

(d) In OH^-, there are $6 + 1 + 1 = 8$ valence electrons, or 4 pairs.

$$|\overline{O}=C=\overline{O}| \qquad [|\overline{\underline{C}l}-\overline{\underline{O}}|]^- \qquad \left(\overset{|\overline{O}-\overset{\displaystyle C=\overline{O}}{\underset{\displaystyle |\underline{O}|}{|}}}{}\right)^{2-} \leftrightarrow \left(|\overline{O}=\overset{\displaystyle C-\overline{\underline{O}}|}{\underset{\displaystyle |\underline{O}|}{|}}\right)^{2-} \leftrightarrow \left(|\overline{O}-\overset{\displaystyle \underset{-}{C}-\overline{\underline{O}}|}{\underset{\displaystyle |\underline{O}|}{\|}}\right)^{2-} \qquad [|\overline{O}-H]^-$$

48. Bond length data from Table 11-1 follow: N≡N 110 pm   N=N 123 pm   N—N 145 pm

                                                  N=O 120 pm   N—O 136 pm

The experimental N—N bond length of 113 pm approximates that of the N≡N triple bond, which appears in structure (1). The experimental N—O bond length of 119 pm approximates that of the N=O double bond, which appears in structure (2). Structure (4) is highly unlikely because it contains no nitrogen-to-nitrogen bonds, one of which is found experimentally. Structure (3) also is unlikely, because it contains a very long (145 pm) N—N single bond, which does not agree at all well with the experimental N-to-N bond length. The molecule is probably best represented as a resonance hybrid. $|N≡N-\overline{O}| \longleftrightarrow \overline{N}=N=\overline{O}$

49. We begin by drawing all three resonance forms of HNO_3 and then analyzing their distribution of formal charge to determine which is the most plausible.

   (a) $H-\overline{O}-\underset{\underset{\displaystyle |\underline{O}|}{\|}}{N}-\overline{O}|$ **(b)** $H-\overline{O}-\underset{\underset{\displaystyle |\underline{O}|}{|}}{N}=\overline{O}$ **(c)** $H-\overline{O}=\underset{\underset{\displaystyle |\underline{\underline{O}}|}{|}}{N}-\overline{O}|$

In all three structures, the formal charges of N and H are the same:

     f.c. of H = 1 – 1 – 0 = 0   f.c. of N = 5 – 4 – 0 = +1

For an oxygen that forms two bonds (either 2 single or one double), f.c. = 6 – 2 – (2 × 2) = 0

For an oxygen that forms only one bond, f.c. = 6 – 1 – (3 × 2) = –1

For the oxygen that forms three bonds (a single and a double), f.c. = 6 – 3 – (1 × 2) = +1

Thus, structures **(a)** and **(b)** are equivalent in their distributions of formal charges, zero on all atoms except +1 on N and –1 on one O. These are resonance forms. Structure **(c)** is quite different, with formal charges of –1 on two O's, and +1 on the other, and a formal charge of +1 on N. Structure **(c)** thus is the least plausible.

Odd-electron species

50. In NO_2, there are 5 + (2 × 6) = 17 valence electrons, 8 pairs and a lone electron. N is the central atom. A plausible Lewis structure is $|\overline{O}-\overset{\displaystyle \cdot}{N}=\overline{O}|$ which has a formal charge of +1 on N and one of –1 on the single-bonded O. Another Lewis structure, with zero formal charge on each atom, is $|\dot{O}-\overline{N}=\overline{O}|$ Because of the unpaired electron we expect NO_2 to be paramagnetic.

51. Refer to the answer to Exercise 50. We would expect a bond—a pair of electrons—to form between two NO_2 molecules as a result of the pairing of of the electrons than are unpaired in the NO_2 molecules. If the second structure for NO_2 is used, the one with zero formal charge on each atom, a plausible structure for

N_2O_4 is $\overline{O}=\overline{N}-\overline{O}-\overline{O}-\overline{O}=\overline{O}$ If the first Lewis structure for NO_2 is used, a plausible Lewis

structure contains a N—N bond. $\underset{\underset{\displaystyle |\underline{O}|}{|}}{\overset{\overset{\displaystyle |O| \quad |O|}{\| \quad \|}}{N}}-\underset{\underset{\displaystyle |\underline{O}|}{|}}{N}$ Resonance structures can be drawn for this second version of

N_2O_4. A N—N bond is observed experimentally in N_2O_4. In either structure for N_2O_4, all electrons are paired; the molecule is expected to be (and is) diamagnetic.

52. **(a)** HO_2 has a total of 1 + (2 × 6) = 13 valence electrons, or 6 pairs and a lone electron. O is the central atom. A plausible Lewis structure is $H-\overline{O}-\overline{O}\cdot$

     (b) CH_3 has a total of (3 × 1) + 6 = 9 valence electrons, or 4 pairs and a lone electron. C is the central atom. A plausible Lewis structure is $H-\underset{\underset{\displaystyle H}{|}}{\overset{\displaystyle \cdot}{C}}-H$

(c) ClO_2 has a total of $(2 \times 6) + 7 = 19$ valence electrons, or 9 pairs and a lone electron. Cl is the central atom. A plausible Lewis structure is $|\overline{O}$—$\overline{Cl}$—$\overline{O}|\cdot$

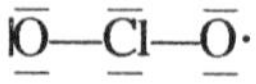

(d) NO_3 has a total of $(3 \times 6) + 5 = 23$ valence electrons, or 11 pairs, plus a lone electron. N is the central atom. A plausible Lewis structure is $\cdot\overline{O}$—N=$\overline{O}$ Resonance forms can also be drawn.
$$|\overline{O}|$$

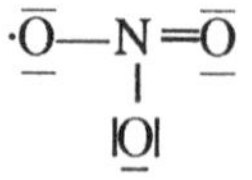

Expanded octets

53. (a) In SO_3^{2-} the total number of valence electrons is $2 + (3 \times 6) + 6 = 26$ valence electrons, or 13 pairs. S is the central atom. A plausible Lewis structure is $\left(\overline{O}{=}\overline{S}{-}\overline{O}|\right)^{2-}$ Two other resonance forms can be drawn. In each structure, the double-bonded oxygen and the S have zero formal charge, while the single-bonded O has formal charge of –1.

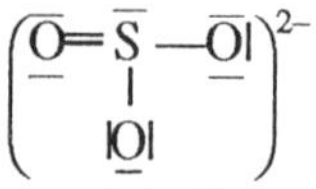

(b) In $HOClO_2$ the total number of valence electrons is $1 + 7 + (3 \times 6) = 26$ valence electrons, or 13 pairs. Cl is at the center of 3 O's, to one of which is bonded H. A plausible Lewis structure for chloric acid is 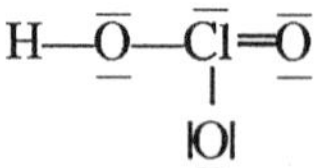
H—$\overline{O}$—$\overline{Cl}$=$\overline{O}$ In this structure, H, the O bonded to H, and the double-bonded O both have 0 formal
$$|\overline{O}|$$
charge, Cl has a formal charge of +1, and the single-bonded terminal O has formal charge of –1.

(c) In HONO, the total number of valence electrons is $1 + 5 + (2 \times 6) = 18$ valence electrons, or 9 pairs. O and N are the central atoms. The following plausible Lewis structure has zero formal charge for each atom in the molecule. H—$\overline{O}$—$\overline{N}$=$\overline{O}$

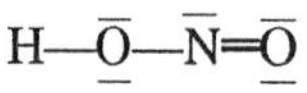

(d) In $OP(OH)_3$, the total number of valence electrons is $(3 \times 1) + 5 + (4 \times 6) = 32$ valence electrons, or 16 pairs. P is the central atom, with 4 O's surrounding it. An H is bonded to three of these O's. The following plausible Lewis structure has zero formal charge on each atom.

$$
\begin{array}{c}
|\overline{O}| \\
\| \\
\text{H—}\overline{O}\text{—P—}\overline{O}\text{—H} \\
| \\
|\overline{O}| \\
| \\
\text{H}
\end{array}
$$

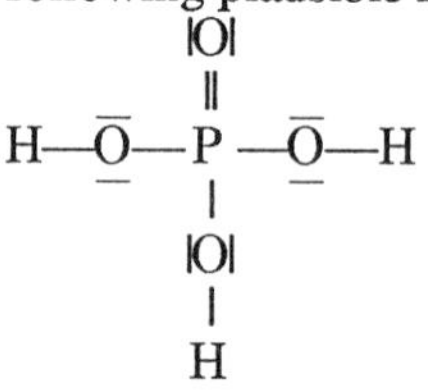

(e) In O_2SCl_2, the total number of valence electrons is $(2 \times 6) + 6 + (2 \times 7) = 32$ valence electrons, or 16 pairs. S is the central atom, surrounded by the other four atoms. There is zero formal charge on each atom in the following plausible Lewis structure.

$$
\begin{array}{c}
|\overline{Cl}| \\
| \\
\overline{O}\text{=}\text{S}\text{=}\overline{O} \\
| \\
|\overline{Cl}|
\end{array}
$$

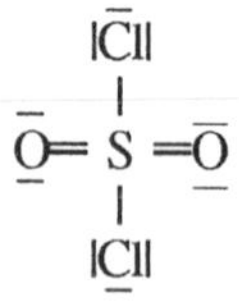

(f) In XeO_3, the total number of valence electrons is $(3 \times 6) + 8 = 26$ valence electrons, or 13 pairs. Xe is the central atom. In the plausible Lewis structure to the left below, each atom has zero formal charge. Some might object to this Lewis structure because there are seven pairs of valence electrons around the Xe (although that is the case in XeF_6). The Lewis structure at right below avoids that difficulty; it is but

one of three resonance structures. $\overline{O}$=$\overline{Xe}$=$\overline{O}$ $\overline{O}$=$\overline{Xe}$—$\overline{O}|$
$$\qquad\qquad\qquad\quad \|\qquad\qquad\quad \|$$
$$\qquad\qquad\qquad\quad |\overline{O}|\qquad\qquad\quad |\overline{O}|$$

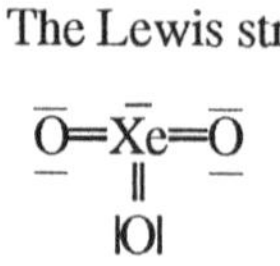

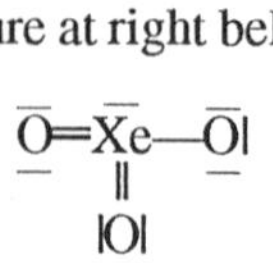

54. Let us draw the Lewis structure of H_2CSF_4. The molecule has $(2 \times 1) + 4 + 6 + (4 \times 7) = 40$ valence electrons, or 20 pairs. With only single bonds and all octets complete, there is a –1 formal charge on C, as in the structure below left. The structure below right avoids that difficulty by creating a carbon-to-sulfur double

bond.

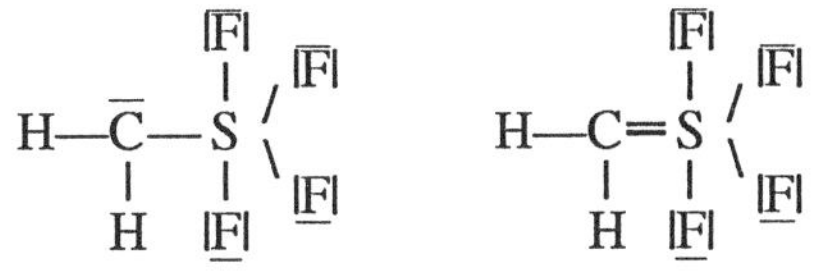

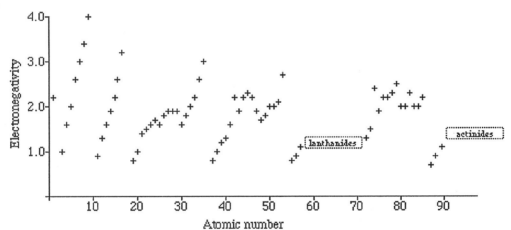

55. In SO_3, there are a total of $6 + (3 \times 6) = 24$ valence electrons, or 12 pairs. S is the central atom. In the plausible Lewis structures that follow, each single-bonded O has a formal charge of –1, and each double-bonded O has a zero formal charge. In the first three structures, the formal charge on S is +2; in the second three, the formal charge on S is +1; and in the last structure, S has zero formal charge. Thus, the last structure is the most plausible from the viewpoint of formal charge. Experimental evidence that would be valuable in distinguishing among these structures would be S—O bond lengths, as S—O single bonds would be expected to be longer than S=O bonds.

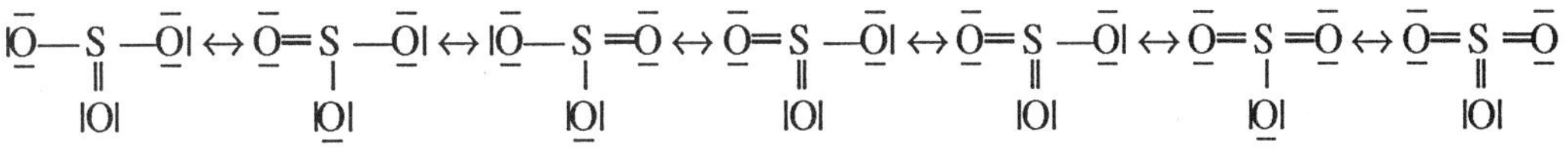

Polar Covalent Bonds

56. The percent ionic character of a bond is based on the difference in electronegativity of its constituent atoms from Figure 11-6.

	(a) Si(1.9)—H(2.2)	**(b)** O(3.4)—Cl(3.2)	**(c)** Br(3.0)—Cl(3.2)	**(d)** As(2.2)—O(3.4)
ΔEN	0.3	0.2	0.2	1.2
%ionic	≈5%	<5%	<5%	≈25%

57.

The property of electronegativity does indeed conform to the periodic law. Each of the "low points" corresponds to an alkali metal, and the end of each trend corresponds to a halogen. This is not unexpected, since electronegativity is derived from two other properties, ionization energy and electron affinity, both of which are periodic in their behavior.

Molecular shapes

58. The AX_nE_m designations that are cited below are to be found in Table 11-2 of the text, along with a sketch and a picture of a model of each type of structure.

 (a) |N≡N| N_2 is linear; two points define a line.

 (b) H—C≡N| HCN is linear. The molecule belongs to the AX_2E_0 category, and these species are linear.

(c) 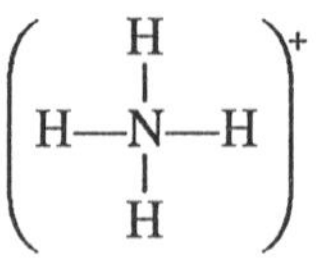 $\begin{pmatrix} H \\ | \\ H-N-H \\ | \\ H \end{pmatrix}^{+}$ NH_4^+ is tetrahedral. The ion is of the AX_4 type, which has a tetrahedral electron-pair geometry and a tetrahedral shape.

(d) 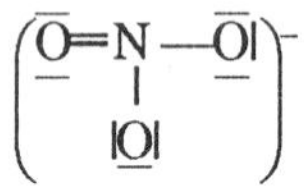 $\begin{pmatrix} \overline{O}=N-\overline{O}| \\ | \\ |\underline{O}| \end{pmatrix}^{-}$ NO_3^- is trigonal planar. The ion is of the AX_3 type, which has a trigonal planar electron-pair geometry and a trigonal planar shape. The other resonance forms are of the same type.

(e) 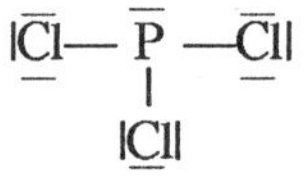 $|\overline{C}l-\overline{P}-\overline{C}l|$ PCl_3 is a trigonal pyramid. The molecule is of the AX_3E_1 type, and has a tetrahedral electron-pair geometry and a trigonal pyramid shape.
   $|\underline{C}l|$

(f) 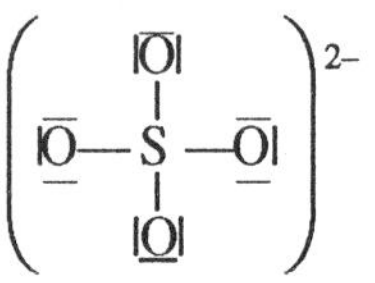 $\begin{pmatrix} |\overline{O}| \\ | \\ |\overline{O}-S-\overline{O}| \\ | \\ |\underline{O}| \end{pmatrix}^{2-}$ SO_4^{2-} has a tetrahedral shape. The ion is of the type AX_4, and has a tetrahedral electron-pair geometry and a tetrahedral shape. The other resonance forms of the sulfate ion have the same shape.

(g) 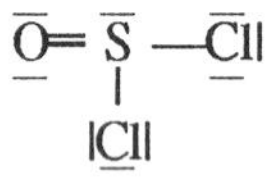 $\overline{O}=\overline{S}-\overline{C}l|$ $SOCl_2$ has a trigonal pyramidal shape. This molecule is of the AX_3E_1 type
   $|\underline{C}l|$ and has a tetrahedral electron-pair geometry and a trigonal pyramidal shape.

59. A trigonal planar shape requires that three ligands and no lone pairs be bonded to the central atom. Thus CN^- cannot have a trigonal planar shape, since only one atom is attached to the central atom. In addition, PO_4^{3-} cannot have a trigonal planar shape, since four O atoms are attached to the central P atom. We now draw the Lewis structure of each of the remaining ions, as a first step in predicting their shapes.

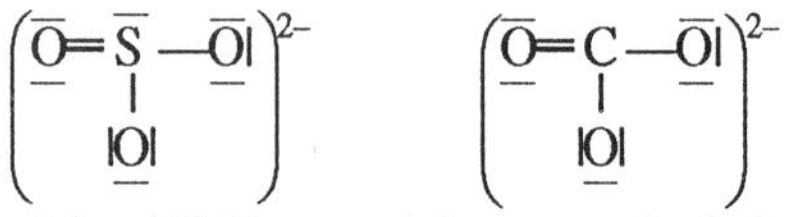

$$\begin{pmatrix} \overline{O}=\overline{S}-\overline{O}| \\ | \\ |\underline{O}| \end{pmatrix}^{2-} \qquad \begin{pmatrix} \overline{O}=C-\overline{O}| \\ | \\ |\underline{O}| \end{pmatrix}^{2-}$$

The SO_3^{2-} ion is of the AX_3E_1 type; it has a tetrahedral electron-pair geometry, and a trigonal pyramidal shape. The CO_3^{2-} ion is of the AX_3 type, and has a trigonal planar electron-pair geometry and a trigonal planar shape.

60. (a) In CO_2 there are a total of $4 + (2 \times 6) = 16$ valence electrons, or 8 pairs. The following Lewis structure is plausible. $\overline{O}=C=\overline{O}$ This is a molecule of type AX_2; CO_2 has a linear electron-pair geometry and a linear shape.

   (b) In NSF there are a total of $5 + 6 + 7 = 18$ valence electrons, or 9 pairs. The molecule can be represented by a Lewis structure with S as the central atom. $|N\equiv\overline{S}-\overline{F}|$ This molecule is of the AX_2E_1 type; it has a trigonal planar electron-pair geometry and a bent shape.

   (c) In $ClNO_2$ there are a total of $7 + 5 + (2 \times 6) = 24$ valence electrons, or 12 pairs. N is the central atom. A plausible Lewis structure is $|\overline{C}l-N=\overline{O}$ This molecule is of the AX_3 type; it has a trigonal planar
   $|\underline{O}|$
   electron-pair geometry and a trigonal planar shape.

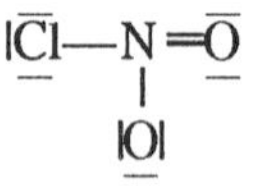

61. First we draw the Lewis structure of each species, then use these Lewis structures to predict the molecular shape.

   (a) In ClO_4^- there are $7 + 4 \times 6 + 1 = 32$ valence electrons or 16 pairs. A plausible Lewis structure is shown below. Since there are four atoms and no lone pairs bonded to the central atom, the molecular shape and the electron pair geometry are the same: tetrahedral.

   (b) In $S_2O_3^{2-}$ there are $2 \times 6 + 3 \times 6 + 2 = 32$ valence electrons or 16 pairs. A plausible Lewis structure is shown below. Since there are four atoms and no lone pairs bonded to the central atom, the electron pair geometry and the molecular shape are the same: tetrahedral.

   (c) In SbF_4^- there are $5 + 4 \times 7 + 1 = 34$ valence electrons or 17 pairs. A plausible Lewis structure is shown below. Since there are four atoms and one lone pair bonded to the central atom, the electron pair geometry is trigonal bipyramidal. The molecular shape is irregular tetrahedral, with the lone pair in an equatorial position (see Table 11-2).

(a) $\begin{pmatrix} |\overline{O}| \\ \| \\ |\overline{O}-Cl-\overline{O}| \\ | \\ |\overline{O}| \end{pmatrix}^{-}$ (b) $\begin{pmatrix} |S| \\ \| \\ |\overline{O}-S-\overline{O}| \\ | \\ |\overline{O}| \end{pmatrix}^{2-}$ (c) $\begin{pmatrix} |\overline{F}| \\ | \\ |\overline{F}-Sb-\overline{F}| \\ | \\ |\overline{F}| \end{pmatrix}^{-}$

62. **(a)** In OSF_2, there are a total of $6 + 6 + (2 \times 7) = 26$ valence electrons, or 13 pairs. S is the central atom. A

plausible Lewis structure is $\overline{O}=\overline{S}-\overline{F}|$ This molecule is of the type AX_3E_1; it has a tetrahedral
$\qquad\qquad\qquad\qquad\quad |F|$

electron-pair geometry and a trigonal pyramidal shape.

(b) In O_2SF_2, there are a total of $(2 \times 6) + 6 + (2 \times 7) = 32$ valence electrons, or 16 pairs. S is the central

atom. A plausible Lewis structure, which has zero formal charge on each atom, follows. $\overline{O}=\underset{|F|}{\overset{|F|}{S}}=\overline{O}$

This molecule is of the AX_4 type; it has a tetrahedral electron-pair geometry and a tetrahedral shape.

(c) In ClO_4^-, there are a total of $1 + 7 + (4 \times 6) = 32$ vallence electrons, or 16 pairs. Cl is the central atom.

A plausible Lewis structure is $\begin{pmatrix} |\overline{O}| \\ | \\ |\overline{O}-Cl-\overline{O}| \\ | \\ |\underline{O}| \end{pmatrix}^{-}$ This ion is of the AX_4 type; it has a tetrahedral

electron-pair geometry and a tetrahedral shape.

(d) In I_3^-, there are a total of $1 + (3 \times 7) = 22$ valence electrons, or 11 pairs. A plausible Lewis structure

follows. $[|\overline{I}-\overset{\frown}{I}-\overline{I}|]^-$ This ion is of the AX_2E_3 type; it has a trigonal bipyramidal electron-pair

geometry and a linear shape.

63. In BF_4^-, there are a total of $1 + 3 + (4 \times 7) = 32$ valence electrons, or 16 pairs. A plausible Lewis structure

has B as the central atom. $\begin{pmatrix} |\overline{F}| \\ | \\ |\overline{F}-B-\overline{F}| \\ | \\ |\underline{F}| \end{pmatrix}^{-}$ This ion is of the type AX_4; it has a tetrahedral electron-pair

geometry and a tetrahedral shape.

64. The O_3 molecule has $(3 \times 6) = 18$ valence electrons, 9 pairs. A plausible Lewis structure does indeed appear
to be linear; it has two resonance forms. $\overline{O}-\overline{O}=\overline{O} \leftrightarrow \overline{O}=\overline{O}-\overline{O}|$ But in each resonance form, there
are two bonding pairs and one lone pair on the central O atom. This indicates a trigonal planar electron pair
geometry and a bent molecular geometry.

65. Lewis structures enable us to determine molecular shapes.
   (a) N_2O_4 has $(2 \times 5) + (4 \times 6) = 34$ valence electrons, or 17 pairs.
   (b) C_2N_2 has $(2 \times 4) + (2 \times 5) = 18$ valence electrons, or 9 pairs.
   (c) C_2H_6 has $(2 \times 4) + (6 \times 1) = 14$ valence electrons, or 7 pairs.

(a) $\begin{matrix} |\overline{O}| & |\overline{O}| \\ \| & \| \\ N-N \\ | & | \\ |\underline{O}| & |\underline{O}| \end{matrix}$ (b) $|N\equiv C-C\equiv N|$ (c) $\begin{matrix} H & H \\ | & | \\ H-C-C-H \\ | & | \\ H & H \end{matrix}$ $\begin{matrix} H & H & H \\ \diagdown & & \diagup \\ & C-C & \\ \diagup & & \diagdown \\ H & & H & H \end{matrix}$

    (a) There are three atoms bound to each N in N_2O_4, making the molecule triangular planar around each N. The entire molecule does not have to be planar, however, since there is free rotation around the N—N bond. The molecule is a bit more stable if it is planar, however.

    (b) There are two atoms attached to each C, requiring a linear geometry. The entire molecule is linear.

    (c) There are four atoms attached to each C. The molecular geometry around each C is tetrahedral, as shown in the sketch above.

66. We predict shapes by beginning with the Lewis structures of the species.

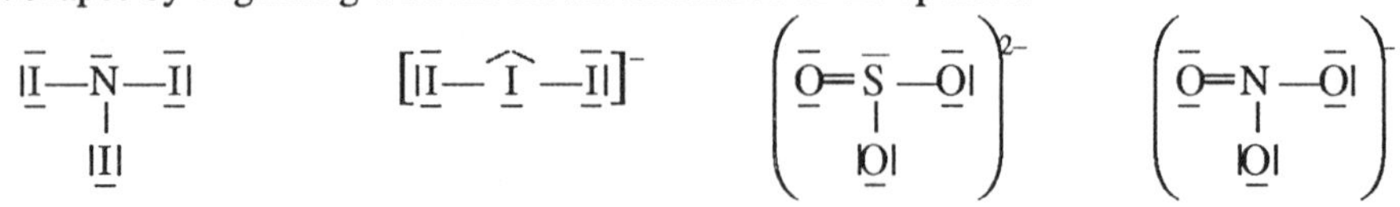

AX_nE_m	AX_3E_1	AX_2E_3	AX_3E_1	AX_3E_0
shape	trigonal pyramid	linear	trigonal pyramid	trigonal planar

Thus, SO_3^{2-} and NI_3 both have the same shape.

67. (a) TRUE   Diatomic molecules must have a linear shape; two points define a line.

    (b) FALSE   Molecules with four atoms bonded to the central atom can have many shapes: tetrahedral, irregular tetrahedral, or square planar.

    (c) FALSE   Triatomic molecules are either planar or linear. But planar molecules can have three atoms (bent, planar) four atoms (trigonal planar), or four atoms (square planar).

    (d) TRUE   An octahedral molecular shape requires the distribution of six electron pairs around the central atom. This is in turn requires a Lewis structure with an expanded octet which cannot be accomodatedby elements in the first or second period.

Polar molecules

68. For each molecule, we first draw the Lewis structure, which we use to predict the shape.

    (a) SO_2 has a total of $6 + (2 \times 6) = 18$ valence electrons, or 9 pairs. The molecule has two resonance forms. $|\overline{O}—\overline{S}=\overline{O} \leftrightarrow \overline{O}=\overline{S}—\overline{O}|$ Each of these resonance forms is of the type AX_2E_1; it has a triangular planar electron-pair geometry and a bent shape. Since each S—O bond is polar toward O, and since the bond dipoles do not point in opposite directions, the molecule has a resultant dipole moment, pointing from S through a point midway between the two O atoms; SO_2 is polar.

    (b) NH_3 has a total of $5 + (3 \times 1) = 8$ valence electrons, or 4 pairs. N is the central atom. A plausible Lewis structure is H—$\overline{N}$—H The molecule is of the AX_3E_1 type; it has a tetrahedral electron-pair geometry and a trigonal pyramidal shape. Each N—H bond is polar toward N. Since the bonds do not symmetrically oppose each other, there is a resultant molecular dipole moment, pointing from the triangular base (formed by the three H atoms) through N. The molecule is polar.

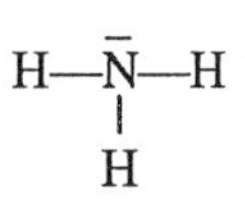

    (c) H_2S has a total of $6 + (2 \times 1) = 8$ valence electrons, or 4 pairs. S is the central atom and a plausible Lewis structure is H—$\overline{S}$—H This molecule is of the AX_2E_2 type; it has a tetrahedral electron-pair geometry and a bent shape. Each H—S bond is polar toward S. Since the bonds do not symmetrically oppose each other, the molecule has a net dipole moment, pointing through S from a point midway between the two H atoms. H_2S is polar.

    (d) C_2H_4 consists of atoms that have all about the same electronegativities. Of course, the C—C bond is not polar, but neither are the C—H bonds. Thus, the entire molecule is nonpolar.

    (e) SF_6 has a total of $6 + (6 \times 7) = 48$ valence electrons, or 24 pairs. S is the central atom. All atoms have zero formal charge in the Lewis structure. 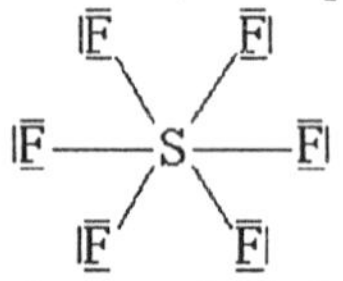 This molecule is of the AX_6E_0 type; it has an octahedral electron-pair geometry and an octahedral shape. Even though each S—F bond is polar toward F, the bonds symmetrically oppose each other resulting in a molecule that is nonpolar.

(f) CH_2Cl_2 has a total of $4 + (2 \times 1) + (2 \times 7) = 20$ valence electrons, or 10 pairs. A plausible Lewis structure is drawn below. The molecule is tetrahedral and polar, since the two polar bonds (C—Cl) do not cancel the effect of each other.

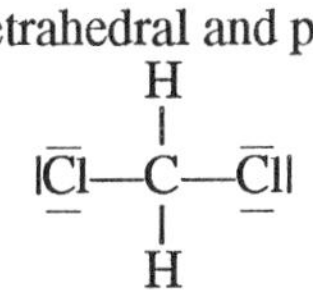

$$\overline{|Cl}—\underset{\underset{H}{|}}{\overset{\overset{H}{|}}{C}}—\overline{Cl|}$$

69. In H_2O_2, there are a total of $(2 \times 1) + (2 \times 6) = 14$ valence electrons, 7 pairs. The two O atoms are central atoms. A plausible Lewis structure has zero formal charge on each atom. $H—\overline{O}—\overline{O}—H$ In the hydrogen peroxide molecule, the O—O bond is non-polar, while the H—O bonds are polar, toward O. Since the molecule has a resultant dipole moment, it cannot be linear, for, if it were linear the two polar bonds would oppose each other and their polarities would cancel.

70. (a) HCN is a linear molecule, which can be derived from its Lewis structure. $H—C{\equiv}N|$ The $C{\equiv}N$ bond is strongly polar toward nitrogen, while the H—C bond is generally considered to be nonpolar. Thus, the molecule has a dipole moment, pointed toward N from C.

(b) SO_3 is a trigonal planar molecule, which can be derived from its Lewis structure. $\overline{O}{=}\underset{\underset{\overline{|O|}}{|}}{S}—\overline{O|}$

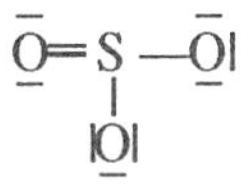

Each sulfur-oxygen bond is polar from S to O, but the three bonds are equally polar and are pointed in opposition so that they cancel. The SO_3 molecule has zero dipole moment.

(c) CS_2 is a linear molecule, which can be derived from its Lewis structure. $\overline{S}{=}C{=}\overline{S}$ Each carbon-sulfur bond is polar from C to S, but the two bonds are equally polar and are pointed in opposition to each other so that they cancel. The CS_2 molecule has zero dipole moment.

(d) OCS also is a linear molecule. Its Lewis structure is $\overline{O}{=}C{=}\overline{S}$ But the carbon-oxygen bond is more polar than the carbon-sulfur bond. Although both bond dipoles point from the central atom to the bonded atom, these two bond dipoles are unequal in strength and thus the molecule is polar in the direction from C to O.

(e) $SOCl_2$ is a trigonal pyramidal molecule. Its Lewis structure is $\overline{|Cl}—\underset{\underset{|O|}{\|}}{\overline{S}}—\overline{Cl|}$ The lone pair is at one corner of the tetrahedron. Each bond in the molecule is polar, pointing away from the central atom, but the sulfur-chlorine bond is less polar than is the sulfur-oxygen bond, making the molecule polar. The dipole moment of the molecule points from the sulfur atom to the base of the trigonal pyramid, not toward the center of the base but slightly toward the O apex of that base.

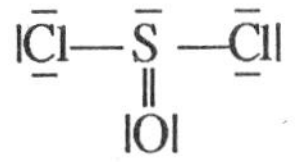

(f) SiF_4 is a tetrahedral molecule, with the following Lewis structure $\overline{F}—\underset{\underset{\overline{|F|}}{|}}{\overset{\overset{\overline{|F|}}{|}}{Si}}—\overline{F|}$ Each Si—F bond is polar, with its negative end away from the central atom toward F in each case. These four Si—F bond dipoles oppose each other in arrangement and thus cancel. SiF_4 has no dipole moment.

(g) POF_3 is a tetrahedral molecule, with the following Lewis structure $\overline{F}—\underset{\underset{\overline{|F|}}{|}}{\overset{\overset{\overline{|F|}}{|}}{P}}{=}\overline{O}$ All four bonds are polar, pointing away from the central atom, but the P—F bond polarity is stronger than that of the P=O bond. Thus, POF_3 is a polar molecule with its dipole moment pointing away from the P to the center of the triangle formed by the three F atoms.

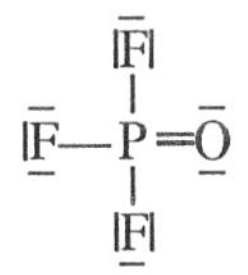

(h) XeF_2 is a linear molecule. Its Lewis structure $|\overline{F}-\widehat{Xe}-\overline{F}|$ indicates that the electron pair geometry is trigonal bipyramidal. The molecule is linear. The two Xe—F bonds are polar, but pointed in opposition to each other. The XeF_2 molecule is nonpolar.

71. (a) FNO has a total of $7 + 5 + 6 = 18$ valence electrons, or 9 pairs. N is the central atom. A plausible

Lewis structure is $|\overline{F}-\overline{N}=\overline{O}$ The formal charge on each atom in this structure is zero.

(b) FNO is of the AX_2E_1 type; it has a trigonal planar electron pair geometry and a bent shape.
(c) The N—F bond is polar toward F and the N—O bond is polar toward O. In FNO, these two bond dipoles point in the same general direction, producing a polar molecule. In FNO_2, however, the additional N—O bond dipole partially opposes the polarity of the other two bond dipoles, yielding a smaller net dipole moment.

Bond Energies

72.

$$\underset{\substack{| \\ H}}{\overset{\substack{H \\ |}}{H-C}}\underset{\substack{| \\ H}}{\overset{\substack{H \\ |}}{-C}}-H + |\overline{Cl}-\overline{Cl}| \longrightarrow \underset{\substack{| \\ H}}{\overset{\substack{H \\ |}}{H-C}}\underset{\substack{| \\ H}}{\overset{\substack{H \\ |}}{-C}}-\overline{Cl}| + H-\overline{Cl}|$$

Analysis of the Lewis structures of products and reactants indicates that a C—H bond and a Cl—Cl bond are broken, and a C—Cl and a H—Cl bond are formed.
Energy required to break bonds = C—H + Cl—Cl = 414 kJ/mol + 243 kJ/mol = 657 kJ/mol
Energy realized by forming bonds = C—Cl + H—Cl = 339 kJ/mol + 431 kJ/mol = 770 kJ/mol
$\Delta H = 657$ kJ/mol – 770 kJ/mol = –113 kJ/mol

73. The chemical equation in terms of Lewis structures is $\overline{O}=\overline{O}-\overline{O}| + \cdot\overline{O}\cdot \longrightarrow 2\ \overline{O}=\overline{O}$ The net result is

replacing a O—O single bond with a O=O double bond.
Energy required to break O—O bonds = 142 kJ/mol
Energy realized by forming O=O bonds = 498 kJ/mol
$\Delta H = 142$ kJ/mol – 498 kJ/mol = –356 kJ/mol

74. In each case we write the formation reaction, but specify reactants and products with their Lewis structures. All species are assumed to be gases.

(a) $\frac{1}{2} \overline{O}=\overline{O} + \frac{1}{2} H-H \longrightarrow \cdot\overline{O}-H$
Bonds broken = $\frac{1}{2}$ O=O + $\frac{1}{2}$ H—H = 0.5 (498 kJ + 436 kJ) = 467 kJ
Bond formed = O—H = 464 kJ
$\Delta H^\circ_f = 467$ kJ – 464 kJ = 3 kJ/mol
If the O—H bond dissociation energy of 428.0 kJ/mol from Figure 11-15 is used, $\Delta H^\circ_f = 36$ kJ/mol.

(b) $|N\equiv N| + 2\ H-H \longrightarrow \underset{\substack{| \\ H}}{\overset{\substack{H \\ |}}{H-N}}-\underset{\substack{| \\ H}}{\overset{\substack{H \\ |}}{N}}-H$

Bonds broken = N≡N + 2 H—H = 946 kJ + 2 × 436 kJ = 1818 kJ
Bonds formed = N—N + 4 N—H = 163 kJ + 4 × 389 kJ = 1719 kJ
$\Delta H^\circ_f = 1818$ kJ – 1719 kJ = 99 kJ

75. $\underset{\substack{| \\ H}}{\overset{\substack{H \\ |}}{H-C}}-\overline{O}-\underset{\substack{| \\ H}}{\overset{\substack{H \\ |}}{C}}-H + 3\ \overline{O}=\overline{O} \longrightarrow 3\ H-\overline{O}-H + 2\ \overline{O}=C=\overline{O}$

Bonds broken = 6 C—H + 2 C—O + 3 O=O = 6 × 414 kJ + 2 × 360 + 3 × 498 kJ = 4.70×10^3 kJ
Bonds formed = 6 H—O + 4 C=O = 6 × 464 kJ + 4 × 736 kJ = 5.73×10^3 kJ
$\Delta H^\circ = 4.70 \times 10^3$ kJ – 5.73×10^3 kJ = -1.03×10^3 kJ

76. The reaction of Example 11-14 is $CH_4(g) + Cl_2(g) \longrightarrow CH_3Cl(g) + HCl(g)$ $\Delta H_{rxn} = -113$ kJ. In Appendix D are the following values: $\Delta H_f^\circ[CH_4(g)] = -74.81$ kJ, $\Delta H_f^\circ[HCl(g)] = -92.31$ kJ, $\Delta H_f^\circ[Cl_2(g)] = 0$

Thus, we have $\Delta H_{rxn} = \Delta H_f^\circ[CH_3Cl(g)] + \Delta H_f^\circ[HCl(g)] - \Delta H_f^\circ[CH_4(g)] - \Delta H_f^\circ[Cl_2(g)]$

-113 kJ $= \Delta H_f^\circ[CH_3Cl(g)] - 92.31$ kJ $- (-74.81$ kJ$) - (0.00)$

$\Delta H_f^\circ[CH_3Cl(g)] = -113$ kJ $+ 92.31$ kJ -74.81 kJ $= -96$ kJ

77. First determine the value of ΔH for reaction (2).

(1) $C(s) \longrightarrow C(g)$ $\Delta H = 717$ kJ

(2) $\underline{C(g) + 2\,H_2(g) \longrightarrow CH_4(g) \quad \Delta H = ?}$

Net: $C(s) + 2\,H_2(g) \longrightarrow CH_4(g)$ $\Delta H_f = -75$ kJ

717 kJ $+ ? = -75$ kJ   or $? = -75 -717$ kJ $= -792$ kJ

To determine the energy of a C—H bond, we need to analyze reaction (2) in some detail.

$$C + 2\,H\text{—}H \longrightarrow H\text{—}\underset{\underset{H}{|}}{\overset{\overset{H}{|}}{C}}\text{—}H$$

In this reaction, 2 H—H bonds are broken and 4 C—H bonds are formed resulting in the production of 792 kJ/mol.

-792 kJ/mol   = energy of broken bonds – energy of formed bonds

= $(2 \times 436$ kJ/mol$) - (4 \times$ C—H$)$

$4 \times$ C—H   = 792 kJ/mol $+ (2 \times 436$ kJ/mol$) = 1664$ kJ/mol

C—H   = $1662 \div 4 = 416$ kJ/mol

This value compares favorably with the value of 414 kJ/mol given in Table 11-3.

Lattice Energy

78. The cycle of reactions is at right. Recall that Hess's law states that the enthalpy change is the same, whether a chemical change is produced by one reaction or several.

$$
\begin{array}{ccc}
K(s) + \tfrac{1}{2} F_2(g) & \xrightarrow{\text{formation}} & KF(s) \\
\text{sublimation}\downarrow & \downarrow\text{dissociation} & \uparrow \\
K(g) & F(g) & | \\
\text{ionization}\downarrow & \downarrow\substack{\text{electron}\\\text{affinity}} & | \\
K^+(g) & + \quad F^-(g) \xrightarrow{\text{lattice energy}} &
\end{array}
$$

Formation reaction: $K(s) + \tfrac{1}{2} F_2(g) \longrightarrow KF(s)$ $\Delta H_f^\circ = -567.3$ kJ/mol

Sublimation: $K(s) \longrightarrow K(g)$ $\Delta H_{sub} = 89.24$ kJ/mol

Ionization: $K(s) \longrightarrow K^+(g) + e^-$ $I_1 = 418.9$ kJ/mol

Dissociation: $\tfrac{1}{2} F_2(g) \longrightarrow F(g)$ D.E. $= (159/2)$ kJ/mol F

Electron Affinity: $F(g) + e^- \longrightarrow F^-(g)$ E.A. $= -328$ kJ/mol

$\Delta H_f^\circ = \Delta H_{sub} + I_1 + $ D.E. $+$ E.A. $+$ lattice energy (L.E.)

-567.3 kJ/mol $= 89.24$ kJ/mol $+ 418.9$ kJ/mol $+ (159/2)$ kJ/mol $- 328$ kJ/mol $+$ L.E.

L.E. $= -827$ kJ/mol

79. Whether or not enthalpies of sublimation of the alkali metals are approximately the same, lattice energies of a series such as LiCl(s), NaCl(s), KCl(s), RbCl(s), and CsCl(s) will vary approximately with the size of the cation. A small cation will produce a more exothermic lattice energy. Thus, the lattice energy for LiCl(s) should be the most exothermic and CsCl(s) the least in this series.

80. Second ionization energy: $Mg^+(g) \longrightarrow Mg^{2+}(g) + e^-$ $I_2 = 1451$ kJ/mol

Lattice energy: $Mg^{2+}(g) + 2\,Cl^-(g) \longrightarrow MgCl_2(s)$ L.E. $= -2526$ kJ/mol

Sublimation: $Mg(s) \longrightarrow Mg(g)$ $\Delta H_{sub} = 150.$ kJ/mol

First ionization energy: $Mg(g) \longrightarrow Mg^+(g) + e^-$ $I_1 = 738$ kJ/mol

Dissociation energy: $Cl_2(g) \longrightarrow 2\,Cl(g)$ D.E. $= 243$ kJ/mol

Electron Affinity: $2\,Cl(g) + 2\,e^- \longrightarrow Cl_2(g)$ $2 \times$ E.A. $= 2\,(-349)$ kJ/mol

$\Delta H_f^\circ = \Delta H_{sub} + I_1 + I_2 + $ D.E. $+ (2 \times$ E.A.$) +$ L.E.

= 150. kJ/mol $+ 738$ kJ/mol $+ 1451$ kJ/mol $+ 243$ kJ/mol $- 698$ kJ/mol $- 2526$ kJ/mol

$= -642$ kJ/mol

In Example 11-16, the value of ΔH_f° for MgCl is calculated as -15 kJ/mol. Therefore, $MgCl_2$ is much more stable than MgCl, since considerably more energy is released when it forms. We expect $MgCl_2(s)$ to be more stable than MgCl(s) because a $+2$ cation more strongly sttracts anions than does a $+1$ cation.

81.

Formation reaction:	$Na(s) + \frac{1}{2} H_2(g) \longrightarrow NaH(s)$	$\Delta H_f^\circ = -57$ kJ/mol
Heat of sublimation:	$Na(s) \longrightarrow Na(g)$	$\Delta H_{sub} = +108$ kJ/mol
Ionization energy:	$Na(g) \longrightarrow Na^+(g) + e^-$	$I_1 = +496$ kJ/mol
Dissociation energy:	$\frac{1}{2} H_2(g) \longrightarrow H(g)$	D.E. $= +218$ kJ/mol H
Lattice energy:	$Na^+(g) + H^-(g) \longrightarrow NaH(s)$	L.E. $= -812$ kJ/mol

$\Delta H_f^\circ = \Delta H_{sub} + I_1 + \text{D.E.} + \text{E.A.} + \text{lattice energy (L.E.)}$
-57 kJ/mol $= 108$ kJ/mol $+ 496$ kJ/mol $+$ E.A. $+ 218$ kJ/mol $- 812$ kJ/mol
E.A. $= -67$ kJ/mol

12 CHEMICAL BONDING II:

ADDITIONAL ASPECTS

REVIEW QUESTIONS

Note: In VSEPR theory the term "bond pair" is used for a single bond, a double bond, or a triple bond, even though a single bond consists of one pair of electrons, a double bond consists of two pairs, and a triple bond consists of three. To avoid any confusion between the number of electron pairs actually involved in the bonding to a central atom, and the number of atoms bonded to that central atom, we shall occasionally use the term "ligand" to indicate an atom or a group of atoms attached to the central atom.

1. (a) An sp^3 hybrid orbital is one of a set of four orbitals that are identical in shape and energy and are formed by the combination of an s orbital and three p orbitals. The four orbitals are tetrahedrally oriented.

    (b) A σ_{2p}^* molecular orbital is a high energy orbital formed by the combination of two $2p$ orbitals on adjacent atoms. The original atomic orbitals are oriented endwise to each other.

    (c) The bond order is an indication of the strength of a bond. In valence bond theory, it indicates the number of electron pairs shared by two atoms. In molecular orbital theory it is one half of the difference between the number of bonding and the number of antibonding electrons.

    (d) A σ bond is a bond that is located along the internuclear axis.

2. (a) An excited-state electron configuration is an indication of the orbitals occupies by a species's electrons after that species has absorbed energy.

    (b) Hybridization of atomic orbitals is that process in which atomic orbitals are blended together to produce new atomic orbitals that point in directions other than those of pure atomic orbitals.

    (c) The Kekulé structures of benzene are the two resonance forms of benzene in which the alternating single and double carbon-to-carbon bonds exchange positions.

    (d) The band theory of metallic bonding describes the bonding between metal atoms in terms of molecular orbitals that extend throughout the crystal.

3. (a) A σ bond is located along the internuclear axis and is the first bond formed between two atoms. A π bond is located above and below the internuclear axis and only forms after a σ bond has formed.

    (b) Localized electrons are those confined either to the region around one atom or to the region between two atoms. Delocalized electrons are free to move among three or more atoms.

    (c) Valence-bond method attempts to explain why two atoms are joined in terms of the atomic orbitals that interact to produce a bond. VSEPR theory attempts to predict the shapes of molecules based on their electronic (Lewis) structures.

    (d) A metal is a material that readily conducts electricity (due to a small or non-existent bond gap). A semiconductor conducts electricity only once it is heated or has a moderate voltage applied to it (due to an moderate band gap).

4. The Lewis structure of H_2Se H—$\overline{\underline{Se}}$—H indicates that there are two atoms and two lone pairs attached to the central atom. The bond angle of $91°$ indicates that the $1s$ orbitals on H overlap the $4p$ orbitals on Se. This would give an undistorted bond angle of $90°$. But repulsions between the H's open up the bond angle a slight bit.

5. Determining hybridization is made easier if we begin with Lewis structures. Only one resonance form is drawn for SO_2 and NO_2^-.

The structures are drawn: $\overset{|\overline{Cl}|}{\underset{|\overline{Cl}|}{\overline{Cl}—P—\overline{Cl}|}}$ with two $\overline{Cl}$ above, $\overline{O}—\overline{S}=\overline{O}$, $\overset{|\overline{Cl}|}{\underset{|\overline{Cl}|}{\overline{Cl}—C—\overline{Cl}|}}$, $|C≡O|$, $[\overline{O}—\overline{N}=\overline{O}]^-$

The P atom is attached to five ligands and no lone pairs and thus is sp^3d hybridized in PCl_5
The S atom is attached to two ligands and one lone pair and thus is sp^2 hybridized in SO_2.
The C atom is attached to four ligands and no lone pairs and thus is sp^3 hybridized in CCl_4.
Neither atom needs to be hybridized in CO.
The N atom is attached to two ligands and one lone pair in NO_2^- and thus is sp^2 hybridized.
The central atom is sp^2 hybridized in both SO_2 and NO_2^-.

6. We first draw Lewis structures as an aid to determining central atom hybridization.

$H—\overline{Se}—H$ $\cdot\overline{O}—N=\overline{O}$ $[|\overline{I}—\overset{\frown}{I}—\overline{I}|]^-$ $\overset{|\overline{Cl}|}{\underset{}{\overline{Cl}—P—\overline{Cl}|}}$ with $\overline{Cl}$ below

The central atom is sp^3 hybridized in H_2Se, sp^2 hybridized in NO_2, sp^3d hybridized in I_3^-, and sp^3 hybridized in PCl_3. d orbitals are involved in the hybridization scheme only for I_3^-.

7. Statement (2) is correct. The only consistently single bonds in C,H,O compounds are bonds to H atoms (a H atom can only form one bond), and single bonds must be σ bonds. Thus, both C—H and O—H bonds must be σ bonds, making (1) false and (2) true. Only C=C bonds (or C=O) bonds in these compounds, that is double bonds, consist of a σ bond and a π bond; (3) is false. A σ bond must form before a π bond forms to complete a double bond; (4) is false.

8. (a) HCl: H $_{1s}\boxed{\uparrow}$ Cl [Ne] $_{3s}\boxed{\uparrow\downarrow}$ $_{3p}\boxed{\uparrow\downarrow}\boxed{\uparrow\downarrow}\boxed{\uparrow}$
The $1s$ orbital of H overlaps the half-filled $3p$ orbital of Cl to produce a linear molecule.
(b) ICl: I [Kr]$4d^{10}$ $_{5s}\boxed{\uparrow\downarrow}$ $_{5p}\boxed{\uparrow\downarrow}\boxed{\uparrow\downarrow}\boxed{\uparrow}$ Cl [Ne] $_{3s}\boxed{\uparrow\downarrow}$ $_{3p}\boxed{\uparrow\downarrow}\boxed{\uparrow\downarrow}\boxed{\uparrow}$
The half-filled $3p$ orbital of Cl overlaps with the half-filled $5p$ orbital of I to produce a linear molecule.
(c) H_2Te: H $_{1s}\boxed{\uparrow}$ Te [Kr]$4d^{10}$ $_{5s}\boxed{\uparrow\downarrow}$ $_{5p}\boxed{\uparrow\downarrow}\boxed{\uparrow}\boxed{\uparrow}$
Each half-filled $5p$ orbital of Te overlaps with a half-filled $1s$ orbital of H to produce a bent molecule, with a bond angle of approximately 90°.
(d) OCl_2:   O $_{1s}\boxed{\uparrow\downarrow}$ $_{2s}\boxed{\uparrow\downarrow}$ $_{2p}\boxed{\uparrow\downarrow}\boxed{\uparrow}\boxed{\uparrow}$ Cl [Ne] $_{3s}\boxed{\uparrow\downarrow}$ $_{3p}\boxed{\uparrow\downarrow}\boxed{\uparrow\downarrow}\boxed{\uparrow}$
Each half-filled $2p$ orbital of O overlaps with a half-filled $3p$ orbital of Cl to produce a bent molecule, with a bond angle of approximately 90°.

9. The BF_3 molecule is planar with bond angles of 120°. The $2p$ orbitals are perpendicular to each other, with angles between them of 90°. Thus, three atoms bound to three $2p$ orbitals would not form a planar structure and would have the wrong bond angles. If two F's bonded to two $2p$ orbitals on B and the third F bound to the the $2s$ orbital, the resulting molecule could indeed be planar, but the bond angles would not equal 120°, and two of the bonds would be different from the other one. By hybridizing the $2s$ and two $2p$ orbitals of B, we create three equivalent orbitals that produce the observed bond angles.

10. (a) The central C atom employs four sp^3 hybrid orbitals; these have a tetrahedral symmetry. Each of the two H atoms employs a $1s$ orbital, and each of the two Cl atoms employs a half-filled $3p$ orbital.

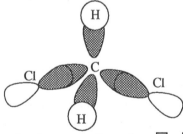

(b) The ground-state Be atom has the electron configuration $_{1s}\boxed{\uparrow\downarrow}$ $_{2s}\boxed{\uparrow\downarrow}$ $_{2p}\boxed{}\boxed{}\boxed{}$ Based in this ground-state electron configuration, we would expect no bond formation at all. To produce the two half-filled orbitals required to form two covalent Be—Cl bonds requires an excited, and hybridized electron configuration.
Be* $_{1s}\boxed{\uparrow\downarrow}$ $_{2s}\boxed{\uparrow}$ $_{2p}\boxed{\uparrow}\boxed{}\boxed{}$ $\longrightarrow$ Be*$_{hyb}$ $_{1s}\boxed{\uparrow\downarrow}$ $_{sp}\boxed{\uparrow}\boxed{\uparrow}$ $_{2p}\boxed{}\boxed{}$

The overlap of the two half-filled *sp* orbitals of Be, each with a half-filled 3*p* orbital on a Cl atom produces the linear molecule, $BeCl_2$. Recall that the ground-state electron configuration of Cl is

Cl [Ne] $3s$ [↑↓] $3p$ [↑↓][↑↓][↑]

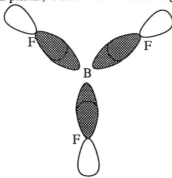

(c) The ground-state electron configuration of B suggests an ability to form one B—F bond, rather than three. $_{1s}$[↑↓] $_{2s}$[↑↓] $_{2p}$[↑][ ][ ] Again, excitation, or promotion of an electron to a higher energy orbital, followed by hybridization is required to form three equivalent half-filled B orbitals.

B* $_{1s}$[↑↓] $_{2s}$[↑] $_{2p}$[↑][↑][ ] $\longrightarrow$ Be*$_{hyb}$ $_{1s}$[↑↓] $_{sp^2}$[↑][↑][↑] $_{2p}$[ ]

Recall that the ground-state electron configuration of F is [He] $_{2s}$[↑↓] $_{2p}$[↑↓][↑↓][↑]

The BF_3 molecule is trigonal planar, with F—B—F bond angles of 120°, as shown on the next page.

11. (a) In SF_6 there are a total of $6 + (6 \times 7) = 48$ valence electrons, or 24 pairs. A plausible Lewis structure is drawn below. In order to form the six S—F bonds, the hybridization on S must be sp^3d^2.

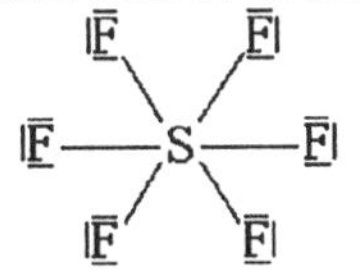

(b) In CS_2 there are a total of $4 + (2 \times 6) = 16$ valence electrons, or 8 pairs. A plausible Lewis structure follows. In order to bond two S atoms to the central C, the hybridization on that C atom must be *sp*.

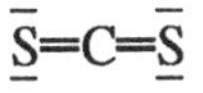

$$\bar{\underline{S}}{=}C{=}\bar{\underline{S}}$$

(c) In $SnCl_4$, there are a total of $4 + (4 \times 7) = 32$ valence electrons, or 16 pairs. A plausible Lewis structure follows. In order to form four Sn—Cl bonds, the hybridization on Sn must be sp^3.

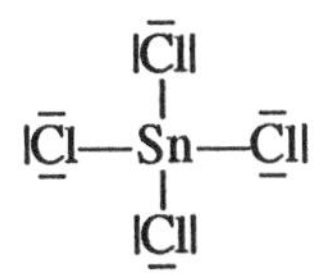

$$\begin{array}{c} |\bar{\underline{C}}l| \\ | \\ |\bar{C}l{-}Sn{-}\bar{C}l| \\ | \\ |\bar{C}l| \end{array}$$

(d) In NO_3^-, there are a total of $5 + (3 \times 6) + 1 = 24$ valence electrons, or 12 pairs. A plausible Lewis structure is drawn below. In order to bond three O atoms to the central N atom, with no lone pairs on that N atom, its hybridization must be sp^2.

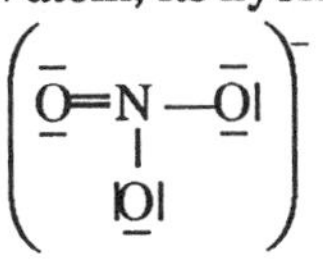

$$\left(\begin{array}{c} \bar{\underline{O}}{=}N{-}\bar{\underline{O}}| \\ | \\ |\underline{O}| \end{array} \right)^{-}$$

(e) In AsF_5, there are a total of $5 + (5 \times 7) = 40$ valence electrons, or 20 pairs. A plausible Lewis structure is drawn below. This is a molecules of the type AX_5E_0. To form five As—F bonds, the hybridization on As must be sp^3d.

12. We base each hybridization scheme on the Lewis structure for the molecule.

(a) H—C≡N| There are two atoms bonded to C and no lone pairs in this molecule. Thus, the hybridization on C must be *sp*.

(b) 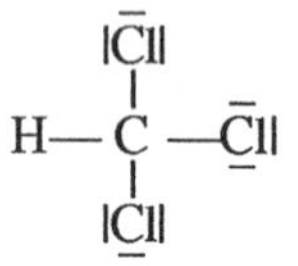
$$
\begin{array}{c}
|\bar{C}l| \\
| \\
H—C—\bar{C}l| \\
| \\
|\bar{C}l|
\end{array}
$$
There are four atoms bonded to C and no lone pairs in this molecule. Thus, the hybridization on C must be *sp³*.

(c) 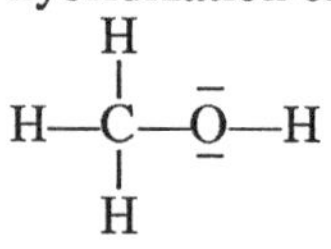
$$
\begin{array}{c}
H \\
| \\
H—C—\bar{O}—H \\
| \\
H
\end{array}
$$
There are four atoms bonded to C and no lone pairs in this molecule. Thus, the hybridization on C must be *sp³*.

(d) 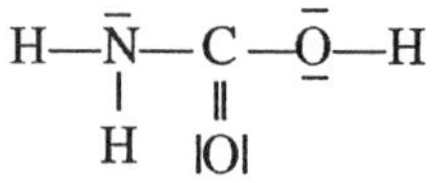
$$
\begin{array}{c}
H—\bar{N}—C—\bar{O}—H \\
| \quad \| \\
H \quad |O|
\end{array}
$$
There are three atoms bonded to C and no lone pairs in this molecule. This requires *sp²* hybridization.

13. (a) H—C≡N| is a linear molecule. The hybridization on C is *sp* (one bond to each of the two ligands).

(b) |N≡C—C≡N| is a linear molecule. The hybridization on each C is *sp* (one bond to each of the two ligands).

(c) 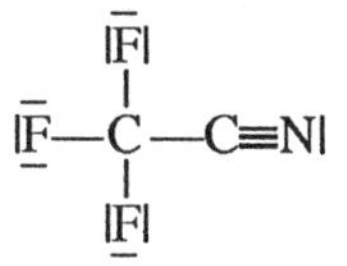
$$
\begin{array}{c}
|\bar{F}| \\
| \\
\bar{F}—C—C≡N| \\
| \\
|F|
\end{array}
$$
is neither linear nor planar. The shape around the left-hand C is tetrahedral and that C has *sp³* hybridization. The shape around the right-hand carbon is linear and that C has *sp* hybridization.

(d) 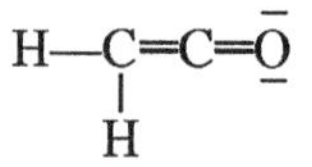
$$
\begin{array}{c}
H—C=C=\bar{O} \\
| \\
H
\end{array}
$$
is a planar molecule. The shape around the left-hand C is trigonal planar, and its hybridization is *sp²*. The shape around the right-hand C is linear, and its hybridization is *sp*.

14. (a) In HCN, there are a total of $1 + 4 + 5 = 10$ valence electrons, or 5 pairs. A plausible Lewis structure follows.   H—C≡N| The H—C bond is a σ bond, and the C≡N bond is composed of 1 σ and 2 π bonds.

(b) In C_2N_2, there are a total of $(2 \times 4) + (2 \times 5) = 18$ valence electrons, or 9 pairs. A plausible Lewis structure follows. |N≡C—C≡N| The C—C bond is a σ bond, and each C≡N bond is composed of 1 σ and 2 π bonds.

(c) In $CH_3CHCHCCl_3$, there are a total of $(4 \times 4) + (5 \times 1) + (3 \times 7) = 42$ valence electrons, or 21 pairs.

A plausible Lewis structure is 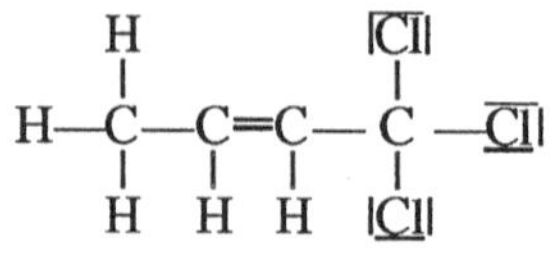
$$
\begin{array}{c}
\quad\quad H \quad\quad\quad |\bar{C}l| \\
\quad\quad | \quad\quad\quad\quad | \\
H—C—C=C—C—\bar{C}l| \\
\quad\quad | \quad | \quad | \quad | \\
\quad\quad H \quad H \quad H \quad |\bar{C}l|
\end{array}
$$
All bonds are σ bonds except one of the bonds that comprise the C=C bond. That bond is composed of one σ and one π bond.

(d) In HONO, there is a total of $1 + 5 + (2 \times 6) = 18$ valence electrons, or 9 pairs. A plausible Lewis structure is   H—$\bar{O}$—$\bar{N}$=$\bar{O}$ All single bonds in this structure are σ bonds. The double bond is composed of one σ and one π bond.

15. (a) In CO_2, there are $4 + (2 \times 6) = 16$ valence electrons, or 8 pairs. A plausible Lewis structure is

$$\overline{\underline{O}}=C=\overline{\underline{O}}$$

(b) The molecule is linear and the hybridization on the C atom is *sp*. Each C=O bond is composed of a σ bond (end-to-end overlap of a half-filled *sp* orbital on C with a half-filled 2*p* orbital on O) and a π bond (overlap of a half-filled 2*p* orbital on O with a half-filled 2*p* orbital on C).

16. In CH_3NCO, there is a total of $(3 \times 1) + (2 \times 4) + 5 + 6 = 22$ valence electrons, or 11 pairs. A plausible

Lewis structure is

$$\begin{array}{c} \quad\quad H \\ \quad\quad | \\ H-\overset{\displaystyle |}{\underset{\displaystyle |}{C}}-\overline{N}=C=\overline{\underline{O}} \\ \quad\quad H \end{array}$$

(a) Of course, all single bonds are σ bonds, and a double bond is composed of a σ bond and a π bond. Thus, each bond in the molecule consists of at least a σ bond. There are six (6) σ bonds in this molecule.

(b) Since each double bond consists of a σ bond and a π bond, there are two (2) π bonds in this molecule.

17. (a) A σ_{1s} orbital must be lower in energy than a σ_{1s}^* orbital. The bonding orbital always is lower than the antibonding orbital if they are derived from the same atomic orbitals.

(b) A σ_{2s} orbital should be lower in energy than a σ_{2p} orbital, since the 2*s* atomic orbital is lower in energy than the 2*p* atomic orbital, the orbitals from which the molecular orbitals, respectively, are derived.

(c) A σ_{1s}^* orbital should be lower than a σ_{2s} orbital, since the 1*s* atomic orbital is considerably lower in energy than is the 2*s* orbital.

(d) A σ_{2p} orbital should be lower in energy than a π_{2p}^* orbital. Both orbitals are from atomic orbitals in the same subshell but we expect a bonding orbital to be more stable than an antibonding orbital.

18. Based on Lewis theory, one would expect C_2 to have the greater bond energy, due to the formation of quadruple bond, C≡C vs. Li—Li. The molecular orbital diagrams and bond orders are as follows.

C_2 KK σ_{2s}⇅ σ_{2s}^*⇅ π_{2p}⇅⇅ σ_{2p}☐   bond order = (6 bonding e^- – 2 antibonding e^-) ÷ 2 = 2

Li_2 KK σ_{2s}⇅ σ_{2s}^*☐ π_{2p}☐☐ σ_{2p}☐   bond order = (2 bonding e^- – 0 antibonding e^-) ÷ 2 = 1

Molecular orbital theory predicts that C_2 has a greater bond energy than does Li_2.

19. (a) In each case, the total number of electrons in the species is determined first; this is followed by the molecular orbital diagram for this species.   KK σ_{2s} σ_{2s}^* π_{2p} σ_{2p} π_{2p}^* σ_{2p}^*

C_2^+ no. valence elns = $(2 \times 4) - 1 = 11$ KK   ⇅   ⇅   ⇅⥮   ☐   ☐☐   ☐

O_2^{2-} no. valence elns = $(2 \times 6) + 2 = 14$ KK   ⇅   ⇅   ⇅⇅   ⇅   ⇅⇅   ☐

F_2^+ no. valence elns = $(2 \times 7) - 1 = 17$ KK   ⇅   ⇅   ⇅⇅   ⇅   ⇅⥮   ☐

(b) Bond order = (no. bonding electrons – no. antibonding electrons) ÷ 2

C_2^+ bond order = $(5 - 2) \div 2 = 3/2$ This species is stable.

O_2^{2-} bond order = $(8 - 6) \div 2 = 1$ This species is stable.

F_2^+ bond order = $(8 - 5) \div 2 = 3/2$ This species is stable.

(c) C_2^+ has an odd number of electrons and it is paramagnetic.

O_2^{2-} has an even number of electrons and it is diamagnetic.

F_2^+ has an odd number of electrons and it is paramagnetic.

20. The largest band gap between valence and conduction band is in an insulator. There is no energy gap between the valence and conduction bands in a metal. Metalloids are semiconductors; they have a small band gap between the valence and conduction bands.

EXERCISES

Valence-Bond Theory

21. There are several ways in which valence bond theory is superior to Lewis structures in describing covalent bonds. *First,* valence bond theory clearly distinguishes between sigma and pi bonds. In Lewis theory, a double bond appears to be just two bonds and it is not clear why a double bond is not simply twice as strong as a single bond. In valence bond theory, it is clear that a sigma bond must be stronger than a pi bond, for the orbitals overlap more efficiently in a sigma bond (end-to-end) than they do in a pi bond (side-to-side). *Second,* molecular geometries are more directly obtained in valence bond theory than in Lewis theory. Although valence bond theory requires the somewhat artificial introduction of hybridization to explain these geometries, Lewis theory does not predict geometries at all; it simply provides the basis from which VSEPR theory predicts geometries. *Third,* Lewis theory does not explain hindered rotation about double bonds. With valence bond theory, one realizes that any rotation about a double bond will produce poorer overlap of the p orbitals that produce the pi bond.

22. The overlap of pure atomic orbitals gives bond angles of 90°. These bond angles are suitable only for 3- and 4-atom compound in which the central atom is an atom of the third (or higher) period of the periodic table. For central atoms of the second period of the periodic table, 3- and 4-atom compounds have bond angles closer to 180°, 120°, and 109.5° than to 90°. These other bond angles can only be explained well through hybridization.

Secondly, hybridization clearly distinguishes between the (hybrid) orbital that form σ bonds and the p orbitals that form π bonds. It places those p orbitals in their proper orientation so that they can overlap side-to-side to form π bonds.

Thirdly, the bond angles of 90° and 120° that result when the octet of the central atom is expanded cannot be produced with pure atomic orbitals. Hybrid orbitals are necessary.

Finally, the overlap of pure atomic orbitals does not usually result in all σ bonds being equivalent. Overlaps with hybrid orbitals produce equivalent bonds.

23. (a) Lewis theory really does not describe the shape of the water molecule. It does indicate that there is a single bond between each H atom and the O atom, and that there are two lone pairs attached to the O atom, but it says nothing about molecular shape.

(b) In valence bond theory using simple atomic orbitals, each H—O bond results from the overlap of a $1s$ orbital on H with a $2p$ orbital on O. Since the angle between 2p orbitals is 90°, this method initially predicts a 90° bond angle of 90°. The observed 104° bond angle is explained as being due to the repulsion between the two slightly positive H atoms.

(c) In VSEPR theory the H_2O molecule is categorized as being of the AX_2E_2 type, with two atoms and two lone pairs attached to the central oxygen atom. The lone pairs repel each other more than do the bond pairs, explaining the smaller than 109.5° bond angle.

(d) In valence bond theory using hybrid orbitals, each H—O bond results from the overlap of a $1s$ orbital on H with a sp^3 orbital on O. The angle between sp^3 orbitals is 109.5°. The observed bond angle of 104° is explained as the result of incomplete hybridization.

24. For each species, we first draw the Lewis structure, as an aid to explaining the bonding.

(a) In CO_2, there are a total of $4 + (2 \times 6) = 16$ valence electrons, or 8 pairs. C is the central atom . The

Lewis structure is $\quad \bar{O}_a\!=\!C\!=\!\bar{O}_b$ The molecule is linear and C is sp hybridized.

(b) In $ClNO_2$, there are a total of $7 + 5 + (2 \times 6) = 24$ valence electrons, or 12 pairs. N is the central atom,

and a plausible Lewis structure is $\quad |\underline{Cl}\!-\!N\!=\!\bar{O} \quad$ The molecule is trigonal planar and N is sp^2
$$\qquad\qquad\qquad\qquad\qquad\quad |\;\;$$
$$\qquad\qquad\qquad\qquad\qquad \underline{|O|}$$

hybridized.

(c) In ClO_3^-, there are a total of $7 + (3 \times 6) + 1 = 26$ valence electrons, or 13 pairs. Cl is the central atom,

and a plausible Lewis structure is
$$\left(\overline{\underline{O}} - \overline{\underline{C}}l - \overline{\underline{O}}| \atop | \atop |\underline{O}| \right)^-$$
The electron-pair geometry around Cl is

tetrahedral, indicating that Cl is sp^3 hybridized.

(d) In ICl_4^-, there are a total of $7 + (4 \times 7) + 1 = 36$ valence electrons, or 18 pairs. I is the central atom,

and a plausible Lewis structure is
$$\left(|\overline{\underline{C}}l \overset{|\underline{C}l|}{\underset{|\underline{C}l|}{\diagup~I~}} \overline{\underline{C}}l \right)^-$$
The electron-pair geometry is

octahedral, indicating that I is sp^3d^2 hybridized.

25. In CO, there is a total of $4 + 6 = 10$ valence electrons, or 5 pairs. A plausible Lewis structure is $|C{\equiv}O|$ The ground-state electron configuration of C is [He] $_{2s}\boxed{\uparrow\downarrow}$ $_{2p}\boxed{~}\boxed{\uparrow}\boxed{\uparrow}$ and that of O is [He] $_{2s}\boxed{\uparrow\downarrow}$ $_{2p}\boxed{\uparrow\downarrow}\boxed{\uparrow}\boxed{\uparrow}$ The σ bond results from end-to-end overlap of two half-filled $2p$ orbitals σ: $C(2p_y)^1$—$O(2p_y)^1$ One of the π bonds results from the side-to side overlap of two $2p$ orbitals π: $C(2p_z)^1$—$O(2p_z)^1$ while the other π bond is coordinate covalent π: $C(2p_x)^0$—$O(2p_x)^2$

An alternative bonding scheme has both the oxygen and carbon atoms sp hybridized, with some rearrangement of electrons. This gives hybridized configurations of [C] = [He] $_{sp}\boxed{\uparrow\downarrow}\boxed{\uparrow}$ $_{2p}\boxed{~}\boxed{\uparrow}$ and that of O is [He] $_{sp}\boxed{\uparrow\downarrow}\boxed{\uparrow}$ $_{2p}\boxed{\uparrow\downarrow}\boxed{\uparrow}$ The σ bond is formed by the end-to-end overlap of two half-filled sp orbitals σ: $C(sp)^1$—$O(sp)^1$ There is a lone pair in an sp hybrid orbital on each atom. The π bonds are formed as before.

26. Since a double bond corresponds to a σ bond and a π bond, it seems reasonable that the energy of a double bond equals the energy of a σ bond plus that of a π bond. The σ bond energy is that of a single bond. Therefore, the energy of a π bond = double bond energy – single bond energy = 611 kJ/mol – 347 kJ/mol = 264 kJ/mol = π bond energy. Now, a triple bond is composed of a σ bond and two π bonds; its energy should correspond to the energy of a σ bond plus that of two π bonds.

$C{\equiv}C$ bond energy = σ bond energy + (2 × π bond energy) = 347 kJ/mol + (2 × 264 kJ/mol) = 875 kJ/mol In Table 11-3, the $C{\equiv}C$ bond energy is given as 837 kJ/mol, some 5% less than the value we have calculated.

27. (a) In CCl_4, there is a total of $4 + (4 \times 7) = 32$ valence electrons, or 8 pairs. C is the central atom. A

plausible Lewis structure is
$|\overline{\underline{C}}l - \overset{|\underline{C}l|}{\underset{|\underline{C}l|}{C}} - \overline{\underline{C}}l|$
The geometry around C is tetrahedral, and C is sp^3

hybridized. Cl—C—Cl bond angles are 109.5°. Each C—Cl bond is represented by σ: $C(sp_3)^1$—$Cl(3p)^1$

(b) In ONCl, there is a total of $6 + 5 + 7 = 18$ valence electrons, or 9 pairs. N is the central atom. A

plausible Lewis structure is $\overline{\underline{O}}{=}\overline{N}{-}\overline{\underline{C}}l|$ The geometry around N is triangular planar, and N is sp^2

hybridized. The O—N—Cl bond angle is about 120°. The three bonds are represented as follows.
σ: $O(2p_y)^1$—$N(sp^2)^1$ $\qquad$ σ: $N(sp^2)^1$—$Cl(3p_z)^1$ $\qquad$ π: $O(2p_z)^1$—$N(2p_z)^1$

(c) In HONO, there are a total of $1 + (2 \times 6) + 5 = 18$ valence electrons, or 9 pairs. A plausible Lewis

structure is $H{-}\overline{\underline{O}}_a{-}\overline{N}{=}\overline{\underline{O}}_b$ The geometry around O_a is tetrahedral, O_a is sp^3 hybridized, and the H—

O_a—N bond angle is (at least close to) 109.5°. The geometry around N is C is trigonal, N is sp^2 hybridized and the O_a—N—O_b bond angle is 120°. The four bonds are represented as follows.

σ: H$(1s)^1$—O$_a(sp^3)^1$ $\qquad$ σ: O$_a(sp^3)^1$—N$(sp^2)^1$ $\qquad$ σ: N$(sp^2)^1$—O$_b(2p_y)^1$

π: N$(2p_z)^1$—O$_b(2p_z)^1$

(d) In CH$_3$CCH, there are a total of $(4 \times 1) + (3 \times 4) = 16$ valence electrons, or 8 pairs. A plausible

Lewis structure follows. $\quad$ H$_a$—C$_a$—C$_b$≡C$_c$—H$_d$ $\quad$ The geometry around C$_a$ is tetrahedral, all bond

(with H$_c$ above and H$_b$ below C$_a$)

angles around that atom is 109.5°, and the hybridization of C$_a$ is sp^3. The geometries around both C$_b$
and C$_c$ are linear, both the C$_a$—C$_b$—C$_c$ and C$_b$—C$_c$—H$_d$ bond angles are 180°, and the hybridization
of both C$_b$ and C$_c$ is sp. The eight bonds in this molecule are represented as follows.

σ: H$_a(1s)^1$—C$_a(sp^3)^1$ $\qquad$ σ: H$_b(1s)^1$—C$_a(sp^3)^1$ $\qquad$ σ: H$_c(1s)^1$—C$_a(sp^3)^1$

σ: C$_a(sp^3)^1$—C$_b(sp)^1$ $\qquad$ σ: C$_b(sp)^1$—C$_c(sp)^1$ $\qquad$ σ: C$_c(sp)^1$—H$_d(1s)^1$

π: C$_b(2p_y)^1$—C$_c(2p_y)^1$ $\qquad$ π: C$_b(2p_z)^1$—C$_c(2p_z)^1$

(e) In I$_3^-$, there are a total of $(3 \times 7) + 1 = 22$ valence electrons, or 11 pairs. A plausible Lewis structure

follows. $\quad$ $[\overline{\underline{I}}—\overset{\frown}{\underline{I}}—\underline{\overline{I}}]^-$ $\quad$ Since there are three lone pairs and two ligands attached to the central I,

the electron-pair geometry around that atom is trigonal bipyramidal; its hybridization is sp^3d. Since the
two ligand I atoms are located in the axial positions of the trigonal bipyramid, the I—I—I bond angle is
180°.

(f) In C$_3$O$_2$, there are $(3 \times 4) + (2 \times 6) = 24$ valence electrons, or 12 pairs. A plausible Lewis structure

follows. $\quad$ $\overline{O}$=C$_a$=C$_b$=C$_c$=$\overline{O}$ $\quad$ The geometry around each C atom is linear, with a bond angle of

180°; the hybridization on each C atom is sp. The eight bonds in the molecule are represented as
follows.

σ: O$(2p_y)^1$—C$_a(sp)^1$ $\qquad$ σ: C$_a(sp)^1$—C$_b(sp)^1$ $\qquad$ σ: C$_b(sp^3)^1$—C$_c(sp)^1$

σ: C$_a(sp)^1$—O$(2p_z)^1$ $\qquad$ π: O$(2p_z)^1$—C$_a(2p_z)^1$ $\qquad$ π: C$_a(2p_y)^1$—C$_b(2p_y)^1$

π: C$_b(2p_z)^1$—C$_c(2p_z)^1$ $\qquad$ π: C$_c(2p_y)^1$—O$(2p_y)^1$

(g) In C$_2$O$_4^{2-}$, there are a total of $(2 \times 4) + (4 \times 6) + 2 = 34$ valence electrons, or 17 pairs. A plausible

Lewis structure is $\left(\overline{\underline{O}}—\text{C}—\text{C}—\underline{\overline{O}}_b \right)^{2-}$ (with $\overset{||}{\underline{|O|}}$ and $\overset{||}{\underline{|O_a|}}$ below) $\quad$ As we see, the ion is symmetrical. There are three atoms

or groups of atoms attached to each C, the bond angles around each C are 120°, and the hybridization
on each C is sp^2. The bonding to one of these carbons (the right-hand one) is represented as follows.

σ: C$(sp^2)^1$—C$(sp^2)^1$ $\qquad$ σ: C$(sp^2)^1$—O$_a(2p_z)^1$ $\qquad$ σ: C$(sp^2)^1$—O$_b(2p_z)^1$

π: C$(2p_y)^1$—O$_a(2p_y)^1$

28. We begin with the Lewis structure of phosgene. $\quad$ $\underline{\overline{C}l}_a$—C—$\underline{\overline{C}l}_b$ (with $\overset{|O|}{||}$ above C) $\quad$ This structure indicates that there are

three atoms and no lone pairs attached to the central atom, thus its hybridization is sp^2. We now draw orbital
diagrams for the atoms in this molecule. $\quad$ O [He] $_{2s}\boxed{\uparrow\downarrow}$ $_{2p}\boxed{\uparrow\downarrow|\uparrow|\uparrow}$ $\quad$ Cl [Ne] $_{3s}\boxed{\uparrow\downarrow}$ $_{3p}\boxed{\uparrow\downarrow|\uparrow\downarrow|\uparrow}$ $\quad$ C$_{hyb}$
[He] $_{sp^2}\boxed{\uparrow|\uparrow|\uparrow}$ $_{2p}\boxed{\uparrow}$ $\quad$ The four bonds in the molecule are the result of the following overlaps. A sigma
bond results from the end-to-end overlap of a $3p$ orbital on Cl$_a$ with an sp^2 hybrid orbital on C. A second
sigma bond results from the end-to-end overlap of a $3p$ orbital on Cl$_b$ with a second sp^2 hybrid orbital on C.
The third sigma bond results from the overlap of one of the half-filled $2p$ orbitals on O with the third sp^2
orbital on C. The pi bond results from the side-to-side overlap of the remaining half-filled $2p$ orbital on O
with the (half-filled) $2p$ orbital on C. Since sp^2 orbitals lie in a plane with angles between them of 120°,
phosgene is trigonal planar.

29. HNCO has $1 + 4 + 5 + 6 = 16$ valence electrons, or 8 pairs. A plausible Lewis structure, in which all atoms have an octet of electrons and zero formal charge, follows.   H—N̄=C=Ō   In this Lewis structure, there are two atoms and one lone pair on the N atom; it is sp^2 hybridized. There are two atoms bonded to the C atom and no lone pairs; it is sp hybridized. The orbital diagram for each atom in the structure follows.

O [He] $_{2s}$[⇅] $_{2p}$[⇅][↑][↑]   C [He] $_{2s}$[⇅] $_{2p}$[↑][↑][ ]   ⟶   C [He] $_{sp}$[↑][↑] $_{2p}$[↑][↑]

H $_{1s}$[↑]   N [He] $_{2s}$[⇅] $_{2p}$[↑][↑][↑]   ⟶   N [He] $_{sp^2}$[⇅][↑][↑] $_{2p}$[↑]

Bonds are formed by the following overlaps.

σ: H($1s$)—N(sp^2)   σ: N(sp^2)—C(sp)   σ: C(sp)—O($2p_z$)

π: N($2p_z$)—C($2p_z$)   π: C($2p_y$)—O($2p_y$)

The N, C, and O atoms lie along a straight line; the H—N—C bond angle is approximately 120°. A sketch follows.

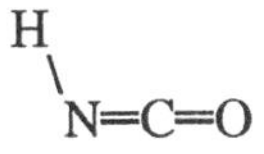

$$\begin{array}{l} \text{H} \\ \;\;\backslash \\ \;\;\;\text{N}=\text{C}=\text{O} \end{array}$$

30. The Lewis structure of ClF_3 is   F̄—Ĉl—F̄   There are three atoms and two lone pairs attached to the

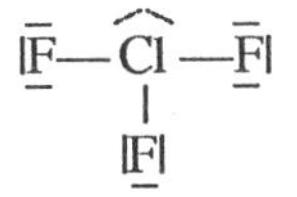

central atom, its hybridization is sp^3d, which is achieved as follows.

Cl_{unhyb} [Ne] $_{3s}$[⇅] $_{3p}$[⇅][⇅][↑] $_{3d}$[ ][ ][ ][ ][ ]   ⟶ Cl_{hyb} [Ne] $_{dsp^3}$[⇅][⇅][↑][↑][↑] $_{3d}$[ ][ ][ ][ ]

The orbital diagram of F is [He] $_{2s}$[⇅] $_{2p}$[⇅][⇅][↑]   Each of the three sigma bonds are formed by the overlap of a $2p$ orbital on F with one of the half-filled dsp^3 orbitals on Cl. Since the dsp^3 hybridization has a trigonal bipyramidal shape, and the two lone pairs occupy the axial positions in the molecule, ClF_3 is T-shaped.

31. In XeF_4, there is a total of $8 + (4 \times 7) = 36$ valence electrons, or 18 pairs. Xe is the central atom. A plausible Lewis structure is

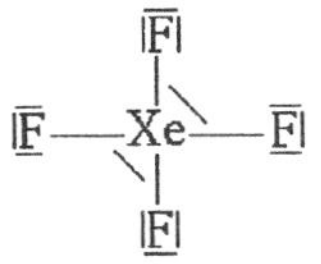

$$\begin{array}{c} \text{F̄} \\ | \\ \text{F̄}——\text{Xe}——\text{F̄} \\ | \\ \text{F̄} \end{array}$$

There are two lone pairs and four atoms attached to Xe. This AX_4E_2 type of molecule has an octahedral electron-pair geometry which requires six hybrid orbitals and thus a hybridization of sp^3d^2 for Xe. The molecular geometry is square planar; lone pairs occupy the two axial (top and bottom) positions.

32. In BrF_5 there is a total of $7 + (5 \times 7)$ $= 42$ valence electrons or 21 pairs.

In PF_5 there is a total of $5 + (5 \times 7)$ $= 40$ valence electrons or 20 pairs.

Plausible Lewis structures are drawn at right.

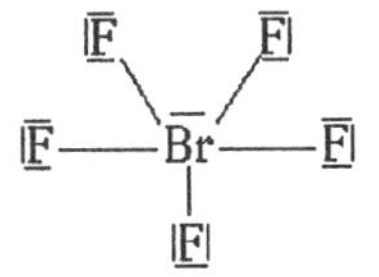

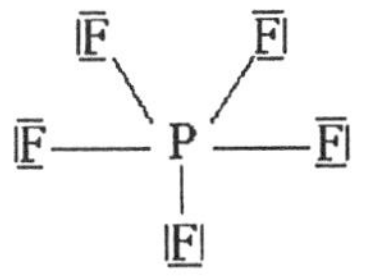

$$\begin{array}{cc} \text{F̄}\;\;\;\;\;\text{F̄} & \text{F̄}\;\;\;\;\;\text{F̄} \\ \backslash\;\;/ & \backslash\;\;/ \\ \text{F̄}——\text{Br}——\text{F̄} & \text{F̄}——\text{P}——\text{F̄} \\ | & | \\ \text{F̄} & \text{F̄} \end{array}$$

The additional electron pair in BrF_5 is a lone pair on the central atom. Thus, the electron-pair geometry of BrF_5 is octahedral, its molecular geometry is square pyramidal, and the hybridization of Br is sp^3d^2. For PF_5, however, the electron-pair geometry and the molecular shape are trigonal bipyramidal, and the hybridization of P is sp^3d.

33. We draw the Lewis structure of each species to account for the electron pairs around each central atom, which is Cl in each species. In F_2Cl^-, there is a total of $7 + (2 \times 7) + 1 = 22$ valence electrons, or 11 pairs. In F_2Cl^+, there is a total of $7 + (2 \times 7) - 1 = 20$ valence electrons, or 10 pairs. Plausible Lewis structures follow.   [F̄—Ĉl—F̄]⁻   [F̄—Ĉl—F̄]⁺   Since there are two atoms and three lone pairs attached to Cl in F_2Cl^-, the electron-pair geometry around the Cl atom is trigonal bipyramidal, the Cl atom hybridization is sp^3d, and the shape of the species is linear. There are two atoms and only two lone pairs attached to the Cl atom in F_2Cl^+, which produces a tetrahedral electron-pair geometry, a hybridization of sp^3 for Cl, and a bent shape for the species.

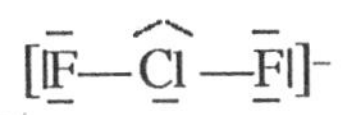

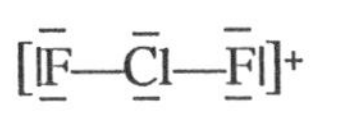

34. In acetic acid, there is a total of $(4 \times 1) + (2 \times 4) + (2 \times 6) = 24$ valence electrons, or 12 pairs. A plausible

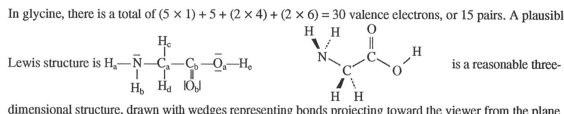

Lewis structure is $H_a—C_a—C_b—\overline{O}_a—H_d$ is a reasonable three-dimensional

structure, drawn with wedges representing bonds projecting toward the viewer from the plane of the paper and dashes representing bonds projecting away from the viewer. This sketch represents C_b as trigonal planar, C_a as tetrahedral, and O_a as bent. C_a and O_a each have sp^3 hybridization, while C_b has sp^2 hybridization. The overlap that produces each bond is as follows.

σ: $C_a(sp^3)^1$—$H_a(1s)^1$ σ: $C_a(sp^3)^1$—$H_b(1s)^1$ σ: $C_a(sp^3)^1$—$H_c(1s)^1$

σ: $C_a(sp^3)^1$—$C_b(sp^2)^1$ σ: $C_b(sp^2)^1$—$O_a(sp^3)^1$ σ: $O_a(sp^3)^1$—$H_d(1s)^1$

σ: $C_b(sp^2)^1$—$O_b(2p)^1$ π: $C_b(2p)^1$—$O_b(2p)^1$

35. In glycine, there is a total of $(5 \times 1) + 5 + (2 \times 4) + (2 \times 6) = 30$ valence electrons, or 15 pairs. A plausible

Lewis structure is $H_a—\overline{N}—C_a—C_b—\overline{O}_a—H_e$ is a reasonable three-

dimensional structure, drawn with wedges representing bonds projecting toward the viewer from the plane of the paper and dashes representing bonds projecting away from the viewer. C_b is trigonal planar, N is trigonal pyramidal, C_a is tetrahedral, and O_a is bent. C_a, O_a, and N each have sp^3 hybridization, while C_b has sp^2 hybridization. The overlap that produces each bond is as follows.

σ: $N(sp^3)^1$—$H_a(1s)^1$ σ: $N(sp^3)^1$—$H_b(1s)^1$ σ: $N(sp^3)^1$—$C_a(sp^3)^1$

σ: $C_a(sp^3)^1$—$H_c(1s)^1$ σ: $C_a(sp^3)^1$—$H_d(1s)^1$ σ: $C_a(sp^3)^1$—$C_b(sp^2)^1$

σ: $C_b(sp^2)^1$—$O_a(sp^3)^1$ σ: $O_a(sp^3)^1$—$H_e(1s)^1$ σ: $C_b(sp^2)^1$—$O_b(2p)^1$

π: $C_b(2p)^1$—$O_b(2p)^1$

36. The bond lengths are consistent with the left-hand C—C bond being a triple bond (120 pm), the remaining C—C bond being a single bond (154 pm) rather than a double bond (134 pm), and the C—O bond being a double bond (123 pm). Of course, the two C—H bonds are single bonds (110 pm). All of this is depicted in

the following Lewis structure. $H_a—C_a\equiv C_b—C_c=\overline{O}$ The overlap that produces each bond follows.
$\qquad\qquad\qquad\qquad\qquad\qquad\qquad\qquad\quad |\\ \qquad\qquad\qquad\qquad\qquad\qquad\qquad\qquad\quad H_b$

σ: $C_a(sp)^1$—$H_a(1s)^1$ σ: $C_a(sp)^1$—$C_b(sp)^1$ σ: $C_b(sp)^1$—$C_c(sp^2)^1$

σ: $C_c(sp^2)^1$—$H_b(1s)^1$ σ: $C_c(sp^2)^1$—$O(2p_y)^1$ π: $C_a(2p_y)^1$—$C_b(2p_y)^1$

π: $C_a(2p_z)^1$—$C_b(2p_z)^1$ π: $C_c(2p_z)^1$—$O(2p_z)^1$

37. All of the valence electron pairs in the molecule are indicated in the sketch; there are no lone pairs. The two end carbon atoms have three atoms attached to them; they are sp² hybridized. This means there is one half-filled 2p orbital available on each of these end carbon atoms for the formation of a pi bond. The central carbon atom has two atoms attached to it; it is sp hybridized. Thus there are two (perpendicular) half-filled 2p orbitals (such as 2p$_y$ and 2p$_z$) on this central carbon atom available for the formation of pi bonds. One pi bond is formed from the central carbon atom to each of the terminal carbon atoms. Because the two 2p orbitals on the central carbon atom are perpendicular to each other, these two pi bonds are also. Thus, the two H—C—H planes are at 90° from each other, because of the pi bonding in the molecule.

38. Again we draw the Lewis structure of each species to account for the electron pairs around each central atom.

(a) In S_2O, there is a total of $(2 \times 6) + 6 = 18$ valence electrons, or 9 pairs. A plausible Lewis structure,

with S as the central atom, follows. $\overline{S}=\overline{S}—\overline{O}| \leftrightarrow |\overline{S}—\overline{S}=\overline{O}$ In either of the resonance forms,

the hybridization on the central S is sp^2. The molecule is bent. The σ bonding is represented as σ: S(3p)—S(sp²)   s: S(sp²)—O(2p)

In the left-hand resonance structure, the π bond is formed by the overlap of two 3p orbitals, one from each S atom. In the right-hand resonance structure, the π bond is between a 3p orbital on S and a 2p orbital on O.

(b) In BrF_3, there is a total of $7 + (3 \times 7) = 28$ valence electrons, or 14 pairs. Br is the central atom. A plausible Lewis structure is |F̅—B̂r—F̅| The electron-pair geometry is trigonal bipyramidal, the
                                                             |
                                                           |F̅|

hybridization on Br is sp^3d, and the molecule is T-shaped. The three Br—F bonds result from the overlap of a hybrid sp^3d Br orbital with a 2p F orbital.

(c) In CH_3OH, there is a total of $(4 \times 1) + 4 + 6 = 14$ valence electrons, or 7 pairs. A plausible Lewis

structure follows.

```
        H
        |
  H ——— C ——— O̅ ——— H
        |
        H
```

Both C and O have tetrahedral electron-pair geometries and are sp^3 hybridized. The molecular shape around C is tetrahedral, while that around O is bent. The five bonds are represented as follows.   3 C—H bonds each of which is σ: $H(1s)^1$—$C(sp^3)^1$

σ: $C(sp^3)$—$O(sp^3)$ σ: $O(sp^3)$—$H(1s)$

Molecular-Orbital Theory

39. The valence-bond method describes a covalent bond as the result of the overlap of atomic orbitals. The more complete the overlap, the stronger the bond. Molecular orbital theory describes a bond as a region of space in a molecule where there is a good chance of finding the bonding electrons. The molecular orbital bond does not have to be created from atomic orbitals (although it often is) and the orientations of atomic orbitals do not have to be manipulated to obtain the correct geometric shape. There is little concept of the relative energies of bond in valence bond theory. In molecular orbital theory, bonds are ordered energetically. These energy orderings, in fact, provide a means of checking the predictions of the theory through spectroscopy of molecules.

40. The two molecular orbital diagrams follow.

KK σ_{2s} σ_{2s}^* π_{2p} σ_{2p} π_{2p}^* σ_{2p}^*

N_2^- no. valence elns = $(2 \times 5) + 1 = 11$ KK [↑↓] [↑↓] [↑↓][↑↓] [↑↓] [↑][ ] [ ]

N_2^{2-} no. valence elns = $(2 \times 5) + 2 = 12$ KK [↑↓] [↑↓] [↑↓][↑↓] [↑↓] [↑][↑] [ ]

N_2^- bond order = $(8 - 3) \div 2 = 2.5$ (stable) N_2^{2-} bond order = $(8 - 4) \div 2 = 2$ (stable)

41. The species with the greater bond order is expected to be the one with the shorter bond length. The molecular orbital diagrams for the three species follow.

KK σ_{2s} σ_{2s}^* π_{2p} σ_{2p} π_{2p}^* σ_{2p}^*

O_2 no. valence elns = $(2 \times 6) = 12$ KK [↑↓] [↑↓] [↑↓][↑↓] [↑↓] [↑][↑] [ ]

O_2^+ no. valence elns = $(2 \times 6) - 1 = 11$ KK [↑↓] [↑↓] [↑↓][↑↓] [↑↓] [↑][ ] [ ]

O_2^{2-} no. valence elns = $(2 \times 6) + 2 = 14$ KK [↑↓] [↑↓] [↑↓][↑↓] [↑↓] [↑↓][↑↓] [ ]

Bond orders: $O_2 = (8 - 4) \div 2 = 2$ $O_2^+ = (8 - 3) \div 2 = 2.5$ $O_2^{2-} = (8 - 6) \div 2 = 1.0$

We expect O_2^+ to have the shorter bond length, since its bond order is slightly larger than that of O_2, and substantiy greater than that of O_2^{2-}.

42. B_2 has $(2 \times 5) = 10$ electrons. We compare the number of unpaired electrons predicted if the order is π_{2p}^b

before σ_{2p}^b with the number predicted if the σ_{2p}^b orbital is before the π_{2p}^b orbital. (^{b} indicates a bonding orbital)

σ_{1s}^b σ_{1s}^* σ_{2s}^b σ_{2s}^* π_{2p}^b σ_{2p}^b π_{2p}^* σ_{2p}^*

B_2 π_{2p}^b before σ_{2p}^b [↑↓] [↑↓] [↑↓] [↑↓] [↑][↑][ ] [ ][ ][ ]

$$\sigma_{1s}^b \quad \sigma_{1s}^* \quad \sigma_{2s}^b \quad \sigma_{2s}^* \quad \sigma_{2p}^b \quad \pi_{2p}^b \quad \pi_{2p}^* \quad \sigma_{2p}^*$$

$B_2 \qquad \sigma_{2p}^b$ before π_{2p}^b 　[↑↓]　[↑↓]　[↑↓]　[↑↓]　[↑↓]　[ ][ ]　[ ][ ]　[ ]

The order of orbitals with π_{2p}^b below σ_{2p}^b in energy results in B_2 being paramagnetic, with two unpaired electrons, as observed experimentally. However, if σ_{2p}^b comes before the π_{2p}^b orbital in energy, a diamagnetic B_2 molecule is predicted, in contradiction to the experimental evidence.

43. C_2 has $(2 \times 6) = 12$ electrons total. A plausible Lewis structure, that obeys the octet rule and has zero formal charge for each C atom, incorporates a quadruple bond 　C::::C or C≡C 　Molecular orbital theory, on the other hand, distributes those same 12 electrons as shown below, resulting in a bond order of $(8 - 4) \div 2 = 2$, that is, a double bond.

$$\sigma_{1s}^b \quad \sigma_{1s}^* \quad \sigma_{2s}^b \quad \sigma_{2s}^* \quad \pi_{2p}^b \quad \sigma_{2p}^b \quad \pi_{2p}^* \quad \sigma_{2p}^*$$

(^{b} indicates a bonding orbital) 　　C_2 　[↑↓]　[↑↓]　[↑↓]　[↑↓]　[↑↓][↑↓][ ]　[ ][ ][ ]

44. In order to have a bond order higher than triple, there would have to be a region in a molecular orbital diagram where four bonding orbitals occurred together in order of increasing energy, with no intervening antibonding orbitals. No such region exists in any of the molecular orbital diagrams in Figure 12-21. Alternatively, three bonding orbitals would have to occur together energetically, following an electron configuration which, when full, results in a bond order greater than zero. This arrangement does not occur either.

45. The statement is not true, for there are many instances when the bond is strengthened when a diatomic molecule loses an electron. Specifically, whenever the electron being lost is an antibonding electron (as is the case for $F_2 \longrightarrow F_2^+$ and $O_2 \longrightarrow O_2^+$), the bond will actually be stronger in the resulting cation than it was in the starting molecule.

46. (^{b} indicates a bonding orbital) 　　$KK \quad \sigma_{2s}^b \quad \sigma_{2s}^* \quad \pi_{2p}^b \quad \sigma_{2p}^b \quad \pi_{2p}^* \quad \sigma_{2p}^*$

			KK	σ_{2s}^b	σ_{2s}^*	π_{2p}^b	σ_{2p}^b	π_{2p}^*	σ_{2p}^*
(a)	NO	$5 + 6 = 11$ valence electrons	KK	[↑↓]	[↑↓]	[↑↓][↑↓]	[↑↓]	[↑][ ]	[ ]
(b)	NO^+	$5 + 6 - 1 = 10$ valence electrons	KK	[↑↓]	[↑↓]	[↑↓][↑↓]	[↑↓]	[ ][ ]	[ ]
(c)	CO	$4 + 6 = 10$ valence electrons	KK	[↑↓]	[↑↓]	[↑↓][↑↓]	[↑↓]	[ ][ ]	[ ]
(d)	CN	$4 + 5 = 9$ valence electrons	KK	[↑↓]	[↑↓]	[↑↓][↑↓]	[↑]	[ ][ ]	[ ]
(e)	CN^-	$4 + 5 + 1 = 10$ valence electrons	KK	[↑↓]	[↑↓]	[↑↓][↑↓]	[↑↓]	[ ][ ]	[ ]
(f)	CN^+	$4 + 5 - 1 = 8$ valence electrons	KK	[↑↓]	[↑↓]	[↑↓][↑↓]	[ ]	[ ][ ]	[ ]
(g)	BN	$3 + 5 = 8$ valence electrons	KK	[↑↓]	[↑↓]	[↑↓][↑↓]	[ ]	[ ][ ]	[ ]

47. In Exercise 46, the species that are isoelectronic are:

　　with 8 electrons: 　　CN^+ 　　BN
　　with 10 electrons: 　　NO^+ 　　CO 　　CN^-

48. We first produce the molecular orbital diagram for each species.

　(a) (^{b} indicates a bonding orbital) 　　$KK \quad \sigma_{2s}^b \quad \sigma_{2s}^* \quad \pi_{2p}^b \quad \sigma_{2p}^b \quad \pi_{2p}^* \quad \sigma_{2p}^*$

		KK	σ_{2s}^b	σ_{2s}^*	π_{2p}^b	σ_{2p}^b	π_{2p}^*	σ_{2p}^*
NO^+	$5 + 6 - 1 = 14$ valence electrons	KK	[↑↓]	[↑↓]	[↑↓][↑↓]	[↑↓]	[ ][ ]	[ ]
N_2^+	$5 + 5 - 1 = 13$ valence electrons	KK	[↑↓]	[↑↓]	[↑↓][↑↓]	[↑]	[ ][ ]	[ ]

　　Since bond order = (no. bonding electrons − no. antibonding electrons)÷2
　　For NO^+ 　bond order = $(8 - 2) \div 2 = 3$ 　　　　For N_2^+ 　bond order = $(7 - 2) \div 2 = 2.5$

　(b) NO^+ is diamagnetic; it has no unpaired electrons. N_2^+ is paramagnetic; it has one unpaired electron.

　(c) The molecule with the lower bond order will also have the greater bond length. Thus, N_2^+ should have the greater bond length.

49. Refer to Figure 12-20. Both of the π_{2p} orbitals have a nodal plane. This plane passes through both nuclei and between the two lobes of the π bond. The difference between a node in the two types of orbital is that, in an antibonding orbital, that node lies between the two nuclei and is perpendicular to the internuclear axis, whereas in the bonding orbital, the two nuclei lie in the plane of the node, which plane therefor contains the internuclear axis.

Delocalized Molecular Orbitals

50. With either Lewis structures or the valence bond method, two structures must be drawn (and "averaged") to explain the π bonding in C_6H_6. The σ bonding is well explained by assuming sp^2 hybridization on each C atom. But the π bonding requires that all six C—C π bonds must be equivalent. This can be achieved by creating six π molecular orbitals—three bonding and three antibonding—into which the $6\,\pi$ electrons are placed. This creates one structure for the C_6H_6 molecule.

51. Resonance is replaced in molecular orbital theory with delocalized molecular orbitals. These orbitals extend over the entire molecule, rather than being localized between two atoms. Because Lewis theory represents all bonds as being localized, the only way to depict a situation in which a pair of electrons spreads over a larger region of the molecule is to draw several Lewis structures and envision the molecule as a combination, or average of these (localized) Lewis structures.

52. We expect to find delocalized orbitals in those species for which bonding cannot be represented thoroughly by one Lewis structure, that is, for compounds that require several resonance forms.

(a) In C_2H_4, there is a total of $(2 \times 4) + (4 \times 1) = 12$ valence electrons, or 6 pairs. The C atoms are the

central atoms
$$\begin{array}{cc} \text{H} & \text{H} \\ | & | \\ \text{C}{=}\text{C} \\ | & | \\ \text{H} & \text{H} \end{array}$$
The bonding description of C_2H_4 does not require the use of delocalized orbitals.

(b) In CO_3^{2-}, there is a total of $4 + (3 \times 6) + 2 = 18$ valence electrons, or 9 pairs. A plausible Lewis structure has three resonance forms.

$$\left(\bar{O}{=}C{-}\bar{O}|\right)^{2-} \leftrightarrow \left(|\bar{O}{-}C{-}\bar{O}|\right)^{2-} \leftrightarrow \left(|\bar{O}{-}C{=}\bar{O}\right)^{2-}$$

The bonding description of CO_3^{2-} will require the use of delocalized molecular orbitals. (See the answer to Review Problem 12.)

(c) In SO_2, there is a total of $6 + (2 \times 6) = 18$ valence electrons, or 9 pairs. N is the central atom. A plausible Lewis structure has two resonance forms. $\bar{O}{=}\bar{S}{-}\bar{O}| \leftrightarrow |\bar{O}{-}\bar{S}{=}\bar{O}$ The bonding description of SO_2 will require the use of delocalized molecular orbitals.

(d) In H_2CO, there is a total of $(2 \times 1) + 4 + 6 = 12$ valence electrons, or 6 pairs. C is the central atom. A plausible Lewis structure is $\text{H}{-}\underset{\underset{\text{H}}{|}}{\text{C}}{=}\bar{O}$ Since one Lewis structure adequately represents the bonding in the molecule, the bonding description of H_2CO does not require the use of delocalized molecular orbitals.

Metallic Bonding

53. (a) Atomic number, by itself, is not particularly important in determining whether a substance has metallic properties. However, because atomic number determines where an element appears in the periodic table, and since the location of to the left and toward to bottom of the periodic table is a region where one finds atoms of high metallic character, atomic number—by locating an element in the periodic table—has some minimal predictive value in determining metallic character.

(b) The answer for this part is much the same as the answer to part (a), since atomic mass generally parallels atomic number for the elements.

(c) The number of valence electrons (electrons in the shell of highest principal quantum number) is important in predicting metallic character. Electrons with few valence electrons—generally four or less—lose those electrons fairly readily to form cations, which is chemical behavior characteristic of metals. These elements also have few valence electrons and a larger number of valence orbitals, which aids in the formation of metallic bonding.

(d) The more vacant atomic orbitals there are in the valence shell, the more readily metallic bonding will occur.

(e) Since metals occur in every period of the periodic table—except the first—there is no particular relationship between number of electrons shells and the metallic behavior of an element. (Remember that one shell is opened at the start of each period.)

54. We first determine the ground state electron configuration of Na, Fe, and Zn

[Na] = [Ne] $3s^1$ [Fe] = [Ar] $3d^6 \, 4s^2$ [Zn] = [Ar] $3d^{10} \, 4s^2$

Based on the number of electrons in the valence shell, we would expect Na to be the softest of these three metals. The one s electron per atom simply will not contribute as strongly to bonding as will two s electrons. Based on valence electron configurations, we expect Fe and Zn to be equally hard, equally strongly bonded together. However, the incomplete $3d$ shell in Fe contributes to bonding between Fe atoms, while the full $3d$ shell in Zn does not so contribute. Thus, Fe should be harder than Zn. Melting point should also reflect the strengths of bond between atoms, with higher melting points occurring in metals that are more strongly bonded. Thus in order, melting point: Fe (1530°C) > Zn (420°C) > Na (98°C)

and hardness: Fe (4.5) > Zn (2.5) > Na (0.4). Parenthesized numbers are actual values, with hardness on a scale of 0 (talc) to 10 (diamond).

55. We first determine the number of Na atoms in the sample.

$$\text{no. Na atoms} \quad = 26.8 \text{ mg Na} \times \frac{1 \text{ g Na}}{1000 \text{ mg Na}} \times \frac{1 \text{ mol Na}}{22.99 \text{ g Na}} \times \frac{6.022 \times 10^{23} \text{ Na atoms}}{1 \text{ mol Na}}$$

$$= 7.02 \times 10^{20} \text{ Na atoms}$$

Since there is one $3s$ orbital per Na atom, and as many energy levels (molecular orbitals) are created as there are atomic orbitals initially present, there are 7.02×10^{20} energy levels present in the conduction band of this sample. Also, there is one $3s$ electron contributed by each Na atom, for a total of 7.02×10^{20} electrons. Since each energy level can hold two electrons, the conduction band is half full.

13 LIQUIDS, SOLIDS, AND INTERMOLECULAR FORCES

REVIEW QUESTIONS

1. (a) ΔH_{vap} symbolizes the molar enthalpy of vaporization, the quantity of heat needed to convert one mole of a substance from liquid to vapor.

   (b) T_c symbolizes the critical temperature, that temperature above which a gas cannot be condensed to a vapor, no matter how hgih the applied pressure.

   (c) An instantaneous dipole is a momentary imbalance of the positive and negative charges in an atom or a molecule, caused when the electron cloud moves off center in a random manner.

   (d) Crystal coordination number refers to the number of atoms of one type surrounding another atom at the same distance, usually those touching the central atom.

   (e) A unit cell is the smallest portion of a crystal that will replicate the crystal through simple translations.

2. (a) Surface tension is the work required to create a unit (such as 1 cm²) of liquid surface area.

   (b) Sublimation is the physical change of a substance being converted directly from a solid to a vapor.

   (c) Supercooling refers to the cooling of a liquid below its freezing point without the formation of any solid. The supercooled liquid is in a metastable state.

   (d) The freezing point of a liquid is that place on the cooling curve of a liquid where the temperature remains constant while heat is withdrawn from the sample. During that period, solid forms as liquid freezes.

3. (a) Adhesive forces are those between different types of substances, usually implying forces between two condensed phases. Cohesive forces are those within a substance, again usually referring to liquids or solids.

   (b) Vaporization is the physical process of transforming a liquid into a vapor. Condensentation is the reverse process: vapor to liquid.

   (c) A triple point of a substance is the point where three phases coexist, often solid, liquid, and vapor. The critical point is that temperature and pressure above which liquid and vapor cannot be distinguished; a fluid exists.

   (d) Face-centered and body-centered cubic unit cells both have atoms at the vertices of a cube. In a face-centered unit cell there also is an atom on the center of each of the six faces of the cube. In contrast, in a body-centered cubic unit cell there is an atom exactly in the center of the three-dimensional cubic cell.

   (e) A tetrahedral hole is one surrounded by four atoms at the vertices of a tetrahedron. An octahedral hole is surrounded by six atoms: above and below, in front and behind, and to the right and the left.

4. *Instantaneous dipole-induced dipole* forces occur after a chance distortion of the electron cloud of an atom or molecule creates an instantaneous imbalance of charge in that molecule: an instantaneous dipole. This imbalance creates a similar imbalance in an adjacent atom or molecule: an induced dipole. The two dipoles reinforce each other and can induce dipoles in other nearby atoms or molecules. The dipoles thus created attract each other and hold the particles together. *Dipole-dipole forces* exist between molecules that have permanent resultant dipole moments, which attract each other and hold these molecules together. *Hydrogen bonds* result from the attraction between a highly electronegative atom—usually N, O, or F—and a hydrogen atom bonded to another highly electronegative atom. *Relative strengths* of these forces are hydrogen bonds > dipole-dipole forces > instantaneous dipole-induced dipole forces. Each force in the preceding ordering is approximately ten times stronger than the one that follows it.

5. **(a & b)** It is possible for a substance not to have a normal melting point or a normal boiling point. Since normal means that the process occurs at 760 mmHg, if the solid-liquid-vapor triple point occurs at a pressure higher than 1 atm, for instance, then melting and vaporization must occur at pressures greater than 1 atm. In such a case, however, there will be a normal sublimation point.

   (c) Every substance should have a critical point. There should always be a temperature above which molecules are moving so rapidly that forces of attraction between them are ineffective in producing a condensed phase.

6. One would expect the enthalpy of sublimation **(d)** to be the largest of the four quantities cited. Molar heat capacities are quite small, on the order of fractions of a kilojoule per mole-degree. (Remember that specific heats have values of joules per gram-degree.) All of the heats of transistion are positive numbers and on the order of kilojoules per mole. Since the heat of sublimation is the sum of the heat of fusion and the heat of vaporization, ΔH_{subl} must be the largest of the three.

7. **(a)** Intermolecular forces in a liquid *do* affect its vapor pressure. The stronger these forces are, the more difficult it is for molecules to vaporize and the lower the vapor pressure is at a given temperature.

   (b) The volume of liquid present *does not* affect its vapor pressure. Realize that vaporization occurs at the surface of the liquid and thus volume of liquid should be of no consequence.

   (c) The volume of vapor present also *does not* affect the vapor pressure of a liquid, for similar resons to those given in part (b).

   (d) The size of the container also also *does not* affect a liquid's vapor pressure, again following the reasoning in part (b). Of course, both liquid and vapor must be present at the point of equilibrium.

   (e) The temperature of the liquid *does* affect that liquid's vapor pressure. Higher temperature means that more energy is present per molecule, each molecule has a larger fraction of the energy needed to overcome cohesive forces, and thus it is more likely that it will be able to overcome those cohesive forces and vaporize.

8. **(a)** mass $CHCl_3$ vaporized $= 4.23 \text{ kJ} \times \dfrac{1000 \text{ J}}{1 \text{ kJ}} \times \dfrac{1 \text{ g } CHCl_3}{247 \text{ J}} = 17.1 \text{ g } CHCl_3$

   (b) $\Delta H_{vap} = \dfrac{247 \text{ J}}{1 \text{ g } CHCl_3} \times \dfrac{1 \text{ kJ}}{1000 \text{ J}} \times \dfrac{119.38 \text{ g } CHCl_3}{1 \text{ mol } CHCl_3} = 29.5 \text{ kJ/mol chloroform}$

   (c) heat evolved $= 16.2 \text{ g } CHCl_3 \times \dfrac{247 \text{ J}}{1 \text{ g } CHCl_3} \times \dfrac{1 \text{ kJ}}{1000 \text{ J}} = 4.00 \text{ kJ}$

9. **(a)** We read up the 50°C line until we arrive at the C_6H_6 curve (b). This occurs at about 280 mmHg.

   (b) We read across the 760 mmHg line until we arrive at the $C_4H_{10}O$ curve (a). This occurs at about 35°C.

10. **(a)** The normal boiling point occurs where the vapor pressure is 760 mmHg, and thus $\ln P = 6.63$. For aniline, this occurs at about the uppermost data point (open circle) on the aniline line. This corresponds to $1/T = 2.18 \times 10^{-3} \text{ K}^{-1}$. Thus $T_{nbp} = 1 \div (2.18 \times 10^{-3} \text{ K}^{-1}) = 459 \text{ K}$

   (b) $55°C = 328 \text{ K} = T$ and thus $1/T = 3.05 \times 10^{-3} \text{ K}^{-1}$ This is the lowest point on the line and it occurs at about $\ln P = 4.50$. Thus, $P = e^{4.50} = 90 \text{ mmHg}$

11. We use the ideal gas equation, with $n = $ moles $Br_2 = 0.486 \text{ g } Br_2 \times \dfrac{1 \text{ mol } Br_2}{159.8 \text{ g } Br_2} = 3.04 \times 10^{-3} \text{ mol } Br_2$

$P = \dfrac{nRT}{V} = \dfrac{3.04 \times 10^{-3} \text{ mol } Br_2 \times 0.08206 \text{ L atm mol}^{-1} \text{ K}^{-1} \times 298.2 \text{ K}}{0.2500 \text{ L}} \times \dfrac{760 \text{ mmHg}}{1 \text{ atm}}$

$= 226 \text{ mmHg}$

12. We use the Clausius-Clapeyron equation (13.2). $T_1 = (56.0 + 273.2) \text{ K} = 329.2 \text{ K}$, $T_2 = (103.7 + 273.2) \text{ K} = 376.9 \text{ K}$

$\ln \dfrac{10.0 \text{ mmHg}}{100.0 \text{ mmHg}} = -\dfrac{\Delta H_{vap}}{8.3145 \text{ J mol}^{-1} \text{ K}^{-1}} \left(\dfrac{1}{329.2 \text{ K}} - \dfrac{1}{376.9 \text{ K}} \right) = -2.30 = -4.624 \times 10^{-5} \Delta H_{vap}$

$\Delta H_{vap} = 4.97 \times 10^4 \text{ J/mol} = 49.7 \text{ kJ/mol}$

13. We use the Clausius-Clapeyron equation (13.2). $T_1 = (5.0 + 273.2) \text{ K} = 278.2 \text{ K}$

$\ln \dfrac{760.0 \text{ mmHg}}{40.0 \text{ mmHg}} = 2.944 = -\dfrac{39.2 \times 10^3 \text{ J/mol}}{8.3145 \text{ J mol}^{-1} \text{ K}^{-1}} \left(\dfrac{1}{T_{nbp}} - \dfrac{1}{278.2 \text{ K}} \right)$

$$\left(\frac{1}{T_{nbp}} - \frac{1}{278.2 \text{ K}}\right) = -2.944 \times \frac{8.1345 \text{ K}^{-1}}{39.2 \times 10^3} = -6.24 \times 10^{-4} \text{ K}^{-1}$$

$$\frac{1}{T_{nbp}} = \frac{1}{278.2 \text{ K}} - 6.24 \times 10^{-4} \text{ K}^{-1} = 2.97 \times 10^{-3} \text{ K}^{-1} \qquad T_{nbp} = 337 \text{ K}$$

14. heat needed $= (17.0 \text{ cm})^3 \times \dfrac{0.92 \text{ g}}{1 \text{ cm}^3} \times \dfrac{1 \text{ mol}}{18.0 \text{ g}} \times \dfrac{6.01 \text{ kJ}}{1 \text{ mol}} = 1.51 \times 10^3 \text{ kJ} = 1.51 \text{ MJ}$

15. (a) heat evolved $= 4.17 \text{ kg Cu} \times \dfrac{1000 \text{ g}}{1 \text{ kg}} \times \dfrac{1 \text{ mol Cu}}{63.55 \text{ g Cu}} \times \dfrac{13.05 \text{ kJ}}{1 \text{ mol Cu}} = 856 \text{ kJ}$

    (b) heat absorbed $= (82 \text{ cm} \times 18 \text{ cm} \times 12 \text{ cm}) \times \dfrac{8.92 \text{ g}}{1 \text{ cm}^3} \times \dfrac{1 \text{ mol Cu}}{63.55 \text{ g Cu}} \times \dfrac{13.05 \text{ kJ}}{1 \text{ mol Cu}} = 3.2 \times 10^4 \text{ kJ}$

16. (a) $C_{10}H_{22}$ has a higher boiling point than C_7H_{16}. All else being equal, the substance of the higher mole weight has the higher boiling point.

    (b) CH_3—O—CH_3 ($\mathfrak{M} = 46.0$ g/mol) has a higher boiling point than C_3H_8 ($\mathfrak{M} = 44.0$ g/mol). For two substances with approximately equal mole weights, the polar substance has a higher boiling point than the nonpolar substance.

    (c) CH_3CH_2—O—H ($\mathfrak{M} = 46.0$ g/mol) has a higher boiling point than CH_3CH_2—S—H ($\mathfrak{M} = 62.1$ g/mol). Even though CH_3CH_2—S—H is the heavier molecule, hydrogen bonds form between CH_3CH_2—O—H molecules, requiring more energy to vaporize them.

17. All of these substances are homonuclear molecules, composed of the same type of atoms. Thus boiling point should increase with increasing mole weight: N_2 ($\mathfrak{M} = 28.0$ g/mol), O_3 ($\mathfrak{M} = 48.0$ g/mol), F_2 ($\mathfrak{M} = 38.0$ g/mol), Ar ($\mathfrak{M} = 39.9$ g/mol), Cl_2 ($\mathfrak{M} = 70.9$ g/mol). Thus, O_3 is seen to be out of order. The correct order is $N_2 < F_2 < Ar < O_3 < Cl_2$.

18. Ionic compounds that are composed of small ions with high charge have high melting points. In this case CaO is composed of a cation of charge +2 and an anion of charge –2. These ions also are small; CaO is the highest melting. Although the cation in MgF_2 is smaller than Ca^{2+}, the F^- anion has a smaller charge than does O^{2-}. Finally CsI is composed of a large cation and a large anion, both univalent. Order of increasing melting point: Cs I (626°C) < MgF_2 (1261°C) < CaO (2614°C) (Melting points are from a handbook.)

19. Both Ne and C_3H_8 are nonpolar; their solids are held together by relatively weak London forces, which are stronger for substances with high molecular weights. Ne has a lower melting point than C_3H_8. Next higher in melting are CH_3CH_2OH and $CH_2OHCHOHCH_2OH$, which are held together with hydrogen bonds in addition to London forces. There are more hydrogen bonds possible for a molecule of $CH_2OHCHOHCH_2OH$ than for one of CH_3CH_2OH, and $CH_2OHCHOHCH_2OH$ molecules are heavier besides; $CH_2OHCHOHCH_2OH$ has the higher melting point. KI, K_2SO_4, and MgO are ionic solids and have the highest melting points. For ionic solid the melting point is high for small, highly charged ions. Thus MgO has the highest melting point of these three, KI the lowest. Thus, arranged in order of increasing melting point, the seven substances are as follows. (Melting points, in parentheses, are from a handbook.)

    Ne   < C_3H_8 < CH_3CH_2OH < $CH_2OHCHOHCH_2OH$ <   KI   < K_2SO_4 <   MgO

    (–248.6°C < –187.7°C < –97.8°C   <   18.2°C   < 681°C < 1067°C < 2825°C)

20. The polarity of hydrazine (34.0 g/mol) might explain its high boiling point. But HCl (36.5 g/mol) also is polar, and it is a gas at room temperature. The distinction between the two substances is that hydrazine can form strong hydrogen bonds, while HCl cannot. This would raise the boiling point of hydrazine significantly above that of HCl (–114.18°C).

21. (a) Si(s) is a network covalent solid. Its structure is similar to that of diamond; both elements are in the same family of the periodic table.

    (b) CCl_4(s) is a molecular solid. Each CCl_4 formula unit is a discrete molecule. The molecules are attracted to each other by London forces.

    (c) $CaCl_2$(s) is an ionic solid. Calcium chloride consists of Ca^{2+} cations and Cl^- anions.

    (d) Ag(s) is a metallic solid. Silver is one of the metals in the periodic table.

    (e) HCl(s) is a molecular solid. Each HCl formula unit is a discrete molecule. The molecules are held together with London forces, dipole-dipole attractions, and weak hydrogen bonds.

22. Answer (c) is correct. (a) is incorrect; often there are several formula units within a unit cell, as in the case for NaCl. (b) is incorrect; the unit cell need not be cubic; that of hexagonal close packing is not. (d) is incorrect; a unit cell will not contain the same number of cations as anions if their numbers are not equal in the formula of the compound. (c) is correct; atoms at the corners, along the edges, and on the faces of a unit cell are shared with adjacent unit cells.

23. In the unit cell of Figure 13-43, there is one Cs^+ ion totally contained within the unit cell. There are eight "corner" Cl^- ions, each shared by eight other unit cells. Thus, no. Cl^- ions = $(8 \times 1/8) = 1$ Cl^- ions. Thus, the "formula of cesium chloride" is CsCl.

24. Let us use the ideal gas law to determine the final pressure in the container, assuming that all of the dry ice vaporizes. We then locate this pressure, at a temperature of 25°C, on the phase diagram of Figure 13-18.

$$P = \frac{nRT}{V} = \frac{\left(80.0 \text{ g } CO_2 \times \dfrac{1 \text{ mol } CO_2}{44.0 \text{ g } CO_2}\right) \times 0.08206 \dfrac{\text{L atm}}{\text{mol K}} \times 298 \text{ K}}{0.500 \text{ L}} = 88.9 \text{ atm}$$

This point (25°C and 88.9 atm) is most likely in the region labeled "liquid" in Figure 13-18. Thus, only liquid is present in the container.

25. (a) The atoms touch along the face diagonal, d. The length of that diagonal thus is $4r$ (r from one corner atom + $2r$ from the center atom + r from the other corner atom). The Pythagorean theorem relates the length of the unit cell, l, to the face diagonal: $d^2 = l^2 + l^2$ or $d = \sqrt{2}\, l$. Thus, two quantities equal to the face diagonal are equal to each other: $d = \sqrt{2}\, l = 4\, r = 4 \times 128$ pm $l = \dfrac{4 \times 128 \text{ pm}}{1.414} = 362$ pm

(b) The volume of the unit cell is that of a cube: $V = l^3 = (362 \text{ pm})^3 = 4.74 \times 10^7 \text{ pm}^3$

(c) In a face-centered unit cell, there are eight "corner" Cu atoms, each shared by eight other unit cells, and there are six "face" Cu atoms, each shared by two other unit cells. Thus, no. Cu atoms = $(8 \times 1/8) + (6 \times 1/2) = 4$ Cu atoms.

(d) The mass of 4 Cu atoms is determined from the atomic mass.

$$\frac{\text{mass}}{\text{unit cell}} = \frac{4 \text{ Cu atoms}}{\text{unit cell}} \times \frac{1 \text{ mol Cu atoms}}{6.022 \times 10^{23} \text{ Cu atoms}} \times \frac{63.55 \text{ g Cu}}{1 \text{ mol Cu atoms}} = \frac{4.221 \times 10^{-22} \text{ g Cu}}{1 \text{ unit cell}}$$

(e) density $= \dfrac{\text{mass}}{\text{volume}} = \dfrac{4.221 \times 10^{-212} \text{ g}}{4.74 \times 10^7 \text{ pm}^3} \times \left(\dfrac{10^{12} \text{ pm}}{1 \text{ m}} \times \dfrac{1 \text{ m}}{100 \text{ cm}}\right)^3 = 8.91 \text{ g/cm}^3$

EXERCISES

Surface Tension; Viscosity

26. Since both the silicone oil and the cloth or leather are composed of relatively nonpolar molecules, they attract each other. The oil thus adheres well to the material. Water, on the other hand is polar and adheres very poorly to the silicone oil (in fact, the water is repelled by the oil), much more poorly, in fact, than it adheres to the cloth or leather. This is because the oil is more nonpolar than is the cloth or the leather. Thus, water is repelled from the silicone-treated cloth or leather.

27. The product can lower the surface tension of water. Then the water can more easily wet a solid substance, because a greater surface area of water can be created with the same energy. (Surface tension equals the work needed to create a given quantity of surface area.) This greater water surface area means a greater area of contact with a solid object, such as a piece of fabric. More of the fabric being in contact with the water means that the water indeed is wetter.

28. Both surface tension and viscosity deal in some way with the work needed to overcome the attractions between molecules. Increasing the temperature of a liquid sample causes the molecules to move around somewhat. Some of the work has been done by adding thermal energy (heat) and less work needs to be done by the experimenter; both surface tension and viscosity decrease. The vapor pressure is a measure of the concentration of molecules that have come free of the surface. As thermal energy is added to the liquid sample, more and more molecules have enough energy to break free of the surface, and the vapor pressure increases.

Vaporization

29. The process of evaporation is an endothermic one; it requires energy. If evaporation occurs from an uninsulated container, this energy is obtained from the surroundings, through the walls of the container. However, if the evaporation occurs from an insulated container, the only source of the needed energy is the liquid that is evaporating. Therfore, the temperature of the liquid decreases.

30. $n_{benzene} = \dfrac{PV}{RT} = \dfrac{1.00 \text{ atm} \times 0.94 \text{ L}}{0.08206 \text{ L atm mol}^{-1} \text{ K}^{-1} \times 353.3 \text{ K}} = 0.032 \text{ mol benzene}$

$\Delta H_{vap} = \dfrac{1.00 \text{ kJ}}{0.032 \text{ mol}} = 31 \text{ kJ/mol benzene}$

31. heat needed $= 2.73 \text{ L H}_2\text{O} \times \dfrac{1000 \text{ cm}^3}{1 \text{ L}} \times \dfrac{0.958 \text{ g}}{1 \text{ cm}^3} \times \dfrac{1 \text{ mol H}_2\text{O}}{18.02 \text{ g H}_2\text{O}} \times \dfrac{40.7 \text{ kJ}}{1 \text{ mol H}_2\text{O}} = 5.91 \times 10^3 \text{ kJ}$

amount CH_4 needed $= 5.91 \times 10^3 \text{ kJ} \times \dfrac{1 \text{ mol CH}_4}{890 \text{ kJ}} = 6.64 \text{ mol CH}_4$

$V = \dfrac{nRT}{P} = \dfrac{6.64 \text{ mol} \times 0.08206 \text{ L atm mol}^{-1} \text{ K}^{-1} \times 296.6 \text{ K}}{756 \text{ mmHg} \times \dfrac{1 \text{ atm}}{760 \text{ mmHg}}} = 162 \text{ L methane}$

32. Determine the quantity of heat needed to vaporize 1.000 g H_2O at each temperature. At 20.°C, 2447 J of heat is needed to vaporize each 1.000 g H_2O. At 100.°C., the quantity of heat needed is

$\dfrac{10.00 \text{ kJ}}{4.430 \text{ g H}_2\text{O}} \times \dfrac{1000 \text{ J}}{1 \text{ kJ}} = 2257 \text{ J/g H}_2\text{O}$

Thus, less heat is needed to vaporize 1.000 g of H_2O at the higher temperature of 100.°C. This makes sense, for at the higher temperature the molecules of the liquid already are in rapid motion, or at least are vibrating rapidly. Some of this energy of motion or vibration will contribute to the energy needed to break the cohesive forces and vaporize the molecules.

33. (a) When the water vaporizes in the outer container, heat is required; vaporization is an endothermic process. When this vapor (steam) condenses on the outside walls of the inner container, that same heat is liberated; condensation is an exothermic process.

   (b) Liquid water—condensed on the outside wall—is in equilibriuum with the water vapor which fills the space between the two containers. This equilibrium exists at the boiling point of water. We assume that the pressure is 1.000 atm, and thus, the temperature of the equilibrium must be 373.15 K or 100.00°C. This is the maximum temperature that can be realized without pressurizing the apparatus.

Vapor Pressure and Boiling Point

34. As the can cools, the water vapor in the can condenses to liquid, and the pressure inside the can lowers to the vapor pressure of water at room temperature, around 20 to 30 mmHg. But the pressure on the outside of the can is, of course, atmospheric pressure, close to 760 mmHg. This huge difference in pressure, amounting to about 14 lb/in.², is sufficient to crush the can.

35. (a) We need the temperature at which the vapor pressure of water is 600 mmHg. This is a temperature between 93.0°C (588.6 mmHg) and 94.0°C (610.9 mmHg). Almost exactly midway between, in fact. We estimate a boiling point of 93.5°C.

   (b) If the observed boiling point is 94°C, the atmospheric pressure equals the vapor pressure of water at 94°C, that is, 611 mmHg.

36. The 25.0 L of He becomes saturated with aniline vapor, at a pressure equal to the vapor pressure of aniline.

$n_{aniline} = (6.220 \text{ g} - 6.108 \text{ g}) \times \dfrac{1 \text{ mol aniline}}{93.13 \text{ g aniline}} = 0.00120 \text{ mol aniline}$

$P = \dfrac{nRT}{V} = \dfrac{0.00120 \text{ mol} \times 0.08206 \text{ L atm mol}^{-1} \text{ K}^{-1} \times 303.2 \text{ K}}{25.0 \text{ L}} = 0.00119 \text{ atm} = 0.904 \text{ mmHg}$

37. At 27°C the vapor pressure of water is 26.7 mmHg. We use this value to determine the mass of water that could exist within the container as vapor.

$$\text{mass of water vapor} = \frac{\left(26.7 \text{ mmHg} \times \dfrac{1 \text{ atm}}{760 \text{ mmHg}}\right) \times \left(1515 \text{ mL} \times \dfrac{1 \text{ L}}{1000 \text{ mL}}\right)}{\dfrac{0.08206 \text{ L atm}}{\text{mol K}} \times (27 + 273) \text{ K}} = 0.00216 \text{ mol } H_2O(g)$$

$$\text{mass of water vapor} = 0.00216 \text{ mol } H_2O(g) \times \frac{18.02 \text{ g } H_2O}{1 \text{ mol } H_2O} = 0.0389 \text{ g } H_2O(g)$$

38. (a)

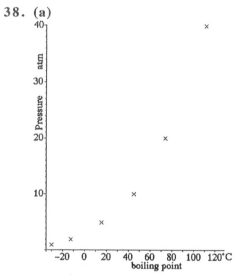

The graph of pressure *vs.* boiling point for Freon-12 is drawn at left.

(b) At a temperature of 25°C the vapor pressure is approximately 6.5 atm for Freon-12. Thus the compressor must be capable of producing pressure greater than 6.5 atm.

39. The expression cited is valid for a linear extrapolation, which would be the case for vapor pressure and temperature if a graph of vapor pressure vs. temperature were a straight line, with a zero temperature intercept of zero pressure (that is, the pressure equals zero when the temperature does). But the actual graph is a curved line that does not intercept the origin. On the other hand, the graph of ln P vs. 1/T is a straight line. It does not, however, intercept the origin. Thus, the Clausius-Clapyeron equation (or an equivalent expression, as is sometimes found in handbooks) is the correct method of determining the vapor pressure at various temperatures. The student's calculation determines how the pressure of the vapor would increase with temperature if liquid did not form. That is, the calculation is an application of Charles's law.

The Clausius-Clapeyron Equation

40. First, we need to calculate lnP and 1/T for the given data points.

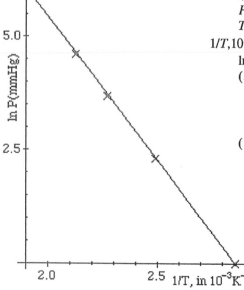

t, °C	76.6	128.0	166.7	197.3	251.0
P, mmHg	1	10	40	100	400
T, K	349.8	401.2	439.9	470.5	524.2
1/T, 10^{-3} K^{-1}	2.859	2.493	2.273	2.125	1.908
ln P	0	2.303	3.689	4.605	5.992

(a) The normal boiling point of phosphorus occurs at a pressure of 760 atm, where ln P = 6.63. Based on the graph to the left, this occurs when $1/T$ = 1.81 × 10^{-3} K^{-1}, or T = 552 K, 279°C.

(b) When ln P is plotted vs $1/T$, the slope of the line equals $-\Delta H_{vap}/R$. For the graph at left, the slope is
$\{(0 - 5.992)/[(2.859 - 1.908) \times 10^{-3}]\} = -6.30 \times 10^3$ K
Thus, $\Delta H_{vap} = 6.30 \times 10^3$ K × 8.314 J mol^{-1} K^{-1}
$$= 5.24 \times 10^4 \text{ J/mol}$$

41. $t = 56.2°C$ is $T = 329.4$ K

$$\ln\frac{760 \text{ mmHg}}{225 \text{ mmHg}} = \frac{-32.0 \times 10^3 \text{ J/mol}}{8.314 \text{ J mol}^{-1} \text{ K}^{-1}}\left(\frac{1}{329.4 \text{ K}} - \frac{1}{T}\right) = 1.217$$

$$\left(\frac{1}{329.4 \text{ K}} - \frac{1}{T}\right) = \frac{1.217 \times 8.314}{-32.0 \times 10^3} \text{ K}^{-1} = -3.16 \times 10^{-4} \text{ K}^{-1} = 3.03_6 \times 10^{-3} \text{ K}^{-1} - 1/T$$

$$1/T = (3.03_6 + 0.316) \times 10^{-3} \text{ K}^{-1} = 3.35_2 \times 10^{-3} \text{ K}^{-1} \qquad T = 298 \text{ K} = 25°C$$

42. First use the Clausius-Clapeyron equation and the data supplied to determine the value of ΔH_{vap}.

$$T_1 = (113.5 + 273.2) \text{ K} = 386.7 \text{ K} \qquad T_2 = (380 + 273) = 653 \text{ K}$$

$$\ln\frac{145.4 \text{ atm}}{1.00 \text{ atm}} = -\frac{\Delta H_{vap}}{R}\left(\frac{1}{653 \text{ K}} - \frac{1}{386.7 \text{ K}}\right) = 0.00105_5 \text{ K}^{-1}\frac{\Delta H_{vap}}{R} = 4.979$$

$$\Delta H_{vap} = \frac{4.979 \times \dfrac{8.314\overset{\cdot}{5} \text{ J}}{\text{mol K}}}{0.00105_5 \text{ K}^{-1}} \times \frac{1 \text{ kJ}}{1000 \text{ J}} = 39.3 \text{ kJ/mol}$$

Then, at $100.0°C = 373.2$ K, we use the Clausius-Clapeyron equation to find the vapor pressure.

$$\ln\frac{P}{1.00 \text{ atm}} = -\frac{39.3 \times 10^3 \text{ J/mol}}{8.3145 \text{ J mol}^{-1} \text{ K}^{-1}}\left(\frac{1}{373.2} - \frac{1}{386.7}\right) = -0.442 \qquad \frac{P}{1.00 \text{ atm}} = e^{-0.442} = 0.643$$

$$P = 0.643 \text{ atom} = 489 \text{ mmHg}$$

Critical Point

43. A substance that can exist as a liquid at room temperature (about 20°C) is one whose critical temperature is above 20°C, 293 K. Of the substances listed in Table 13-3, this includes CO_2 ($T_c = 304.2$ K), HCl ($T_c = 324.6$ K), NH_3 ($T_c = 405.7$ K), SO_2 ($T_c = 431.0$ K), and H_2O ($T_c = 647.3$ K). In fact, CO_2 exists as a liquid in CO_2 fire extinguishers.

44. The critical temperature of SO_2 is 431.0 K, which is above the temperature of 0°C, 273°K. The critical pressure of SO_2 is 77.7 atm, which is below the pressure of 100 atm. Thus, SO_2 can be maintained as a liquid at 0°C and 100 atm.

The critical temperature of methane, CH_4, is 191.1 K, which is below the temperature of 0°C, 273 K. Thus, CH_4 cannot exist as a liquid at 0°C, no matter what pressure is applied.

Fusion/Freezing

45. The heat needed is required to raise the temperature to the melting point and then to melt the solid.

heat to raise temperature = 677 g × (327.4°C – 25.0°C) × 0.134 J g^{-1} °C^{-1} = 2.74×10^4 J = 27.4 kJ

heat to melt solid = 677 g × $\dfrac{1 \text{ mol Pb}}{207.2 \text{ g Pb}}$ × 4.774 kJ/mol = 15.6 kJ

total heat needed = 27.4 kJ + 15.6 kJ = 43.0 kJ

46. The liquid in the can is supercooled. When the can is opened, gas bubbles released from the carbonated beverage serve as nuclei for the formation of ice crystals. The condition of supercooling is destroyed and the liquid reverts to the solid phase instantly.

An alternative explanation follows. The process of the gas coming out of solution is endothermic (heat is required). (We know this to be true because the reaction

solution of gas in water ⟶ gas + liquid water

proceeds to the right as the temperature is raised, a characteristic direction of an endothermic reaction.) The required heat is taken from the cooled liquid, causing it to freeze.

47. Data for cooling curves should not be collected from a system confined in an insulated container. The whole purpose of insulation is to prevent the loss of heat to the surroundings. If there is no heat loss in a nonreacting system the temperature will not change. Thus it will be impossible to collect temperature vs. time data, as required for a cooling curve.

States of Matter and Phase Diagrams

48. 0.180 g H_2O corresponds to 0.0100 mol H_2O. If the water does not vaporize completely, the pressure of the vapor in the flask equals the vapor pressure of water at the indicated temperature. However, if the water vaporizes completely, the pressure of the vapor is determined by the ideal gas law.

(a) 30.0°C, vapor pressure of H_2O = 31.8 mmHg = 0.0418 atm

$$n = \frac{PV}{RT} = \frac{0.0418 \text{ atm} \times 2.50 \text{ L}}{0.08206 \text{ L atm mol}^{-1} \text{ K}^{-1} \times 303.2 \text{ K}}$$

= 0.00420 mol H_2O vapor < 0.0100 mol H_2O; not all the H_2O vaporizes. P = 0.0418 atm

(b) 50.0°C, vapor pressure of H_2O = 92.5 mmHg = 0.122 atm

$$n = \frac{PV}{RT} = \frac{0.122 \text{ atm} \times 2.50 \text{ L}}{0.08206 \text{ L atm mol}^{-1} \text{ K}^{-1} \times 323.2 \text{ K}}$$

= 0.0115 mol H_2O vapor > 0.0100 mol H_2O; all the H_2O vaporizes. Thus,

$$P = \frac{nRT}{V} = \frac{0.0100 \text{ mol} \times 0.08206 \text{ L atm mol}^{-1} \text{ K}^{-1} \times 323.2 \text{ K}}{2.50 \text{ L}} = 0.106 \text{ atm} = 80.6 \text{ mmHg}$$

(c) 70.0°C All the H_2O must vaporize, as this temperature is higher than that of part (b). Thus,

$$P = \frac{nRT}{V} = \frac{0.0100 \text{ mol} \times 0.08206 \text{ L atm mol}^{-1} \text{ K}^{-1} \times 343.2 \text{ K}}{2.50 \text{ L}} = 0.112_6 \text{ atm} = 85.6 \text{ mmHg}$$

49. (a) If the pressure exerted by 2.50 g of H_2O(vapor) is less than the vapor pressure of water at 120.°C (393 K), then the water must exist entirely as a vapor.

$$P = \frac{nRT}{V} = \frac{\left(2.50 \text{ g} \times \dfrac{1 \text{ mol } H_2O}{18.02 \text{ g } H_2O}\right) \times 0.08206 \text{ L atm mol}^{-1} \text{ K}^{-1} \times 393 \text{ K}}{5.00 \text{ L}} = 0.895 \text{ atm}$$

Since this is less than the 1.00 atm vapor pressure at 100.°C, it must be less than the vapor pressure of water at 120°C; the water exists entirely as a vapor.

(b) 0.895 atm = 680 mmHg. From Table 13-2, we see that this corresponds to a temperature of 97.0°C (at which temperature the vapor pressure of water is 682.1 mmHg). Thus, at temperatures slightly less than 97.0°C the water will begin to condense to liquid. **But** we have forgotten that, in this constant-volume container, the pressure (of water) will decrease as the temperature decreases. Calculating the precise decrease in both involves linking the Clausius-Clapeyron equation with the expression $P = k \cdot T$, but we can estimate the final temperature. First, we determine the pressure if we lower the temperature from 393 K to 97°C (370 K). $\quad P_f = \dfrac{370 \text{ K}}{393 \text{ K}} \times 680 \text{ mmHg} = 640 \text{ mmHg}$

From Table 13-2, this pressure occurs at a temperature of about 95°C (368 K). We now determine the final pressure at this temperature. $\quad P_f = \dfrac{368 \text{ K}}{393 \text{ K}} \times 680 \text{ mmHg} = 637 \text{ mmHg}$

This is just slightly above the vapor pressure of water (633.9 mmHg) at 95°C, which we conclude must be the temperature at which liquid water appears.

50. 0.100 mol of H_2O is formed. The vapor pressure of water at 27°C is, from Table 13-2, 26.7 mmHg. This is the maximum pressure due to water vapor that can exist in the vessel. If all of the water were present as vapor, its pressure would be

$$P = \frac{nRT}{V} = \frac{0.100 \text{ mol} \times 0.08206 \text{ L atm mol}^{-1} \text{ K}^{-1} \times 300. \text{ K}}{20.0 \text{ L}} = 0.123 \text{ atm} = 93.5 \text{ mmHg}$$

Hence, the pressure inside the container is 26.7 mmHg.

51. The mass of unmelted ice is 0.22 kg. First compute the heat needed to melt this ice.

$$\text{heat to melt ice} = 0.22 \text{ kg} \times \frac{1000 \text{ g ice}}{1 \text{ kg ice}} \times \frac{1 \text{ mol ice}}{18.02 \text{ g ice}} \times \frac{6.01 \text{ kJ}}{1 \text{ mol ice}} = 73._4 \text{ kJ}$$

Then compute the heat produced by 1 mole (18.02 g) of steam condensing and then lowering in temperature from 100.0°C to 0.0°C, the melting point of ice.

$$\text{heat evolved} = \left(1 \text{ mol} \times \frac{-40.7 \text{ kJ}}{1 \text{ mol}}\right) + \left(18.02 \text{ g} \times \frac{4.18 \text{ J}}{\text{g °C}} \times \frac{1 \text{ kJ}}{1000 \text{ J}} \times (0 - 100)\text{°C}\right)$$

$$= -40.7 \text{ kJ/mol} - 7.53 \text{ kJ/mol} = -48.2 \text{ kJ/mol of steam}.$$

Use the conversion factor just obtained to determine the mass of steam needed to melt the remaining ice.

mass of steam = $73._4$ kJ $\times \dfrac{1 \text{ mol steam}}{48.2 \text{ kJ}} \times \dfrac{18.02 \text{ g steam}}{1 \text{ mol steam}}$ = 27 g steam.

52. The heat gained by the ice equals the the negative of the heat lost by the water. Let us use this fact in a step-by-step approach. We first compute the heat needed to raise the temperature of the ice to 0.0°C and then the heat given off when the temperature of the water is lowered to 0.0°C.

to heat the ice = (8.0 cm × 2.5 cm × 2.7 cm) $\times \dfrac{0.917 \text{ g}}{1 \text{ cm}^3} \times 2.01$ J g^{-1} °C^{-1} × (0°C + 25.0°C)

$\qquad\qquad = 2.5 \times 10^3$ J

to cool the water = -400.0 cm$^3 \times \dfrac{0.998 \text{ g}}{1 \text{ cm}^3} \times 4.18$ J g^{-1} °C^{-1} × (0°C − 32.0°C) = 53.4×10^3 J

Thus, at 0°C, we have 50. g ice, 399 g water, and $(53.4 - 2.5) \times 10^3$ J = 50.9 kJ of heat available

Since 50.0 g of ice is a bit less than 3 mol of ice and 50.9 kJ is enough heat to melt at least 8 mole of ice, all of the ice will melt. The heat needed to melt the ice is

50. g ice $\times \dfrac{1 \text{ mol H}_2\text{O}}{18.0 \text{ g H}_2\text{O}} \times \dfrac{6.01 \text{ kJ}}{1 \text{ mol ice}}$ = 17 kJ

We now have 50.9 kJ – 17 kJ = 34 kJ of heat, and 399 g + 50. g = 449 g of water at 0°C. We compute the temperature change that is produced by adding the heat to the water.

$\Delta t = \dfrac{34 \times 10^3 \text{ J}}{449 \text{ g} \times 4.18 \text{ J g}^{-1} \text{ °C}^{-1}} = 18$°C $\qquad$ The final temperature is 18°C

53. First we compute the mass and the amount of mercury.

mass Hg = 525 cm$^3 \times \dfrac{13.6 \text{ g}}{1 \text{ cm}^3} = 7.14 \times 10^3$ g $\qquad$ amount Hg = 7.14×10^3 g $\times \dfrac{1 \text{ mol Hg}}{200.59 \text{ g}}$ = 35.6 mol Hg

Then we calculate the heat given up by the mercury in lowering its temperature, as the sum of the following three terms.

cool liquid $\quad = 7.14 \times 10^3$ g × 0.138 J g^{-1} °C^{-1} × (−39°C − 20°C) = -5.8×10^4 J = − 58 kJ

freeze liquid $\quad = 35.6$ mol Hg × (−2.30 kJ/mol) = −81.9 kJ

cool solid $\qquad = 7.14 \times 10^3$ g × 0.126 J g^{-1} °C^{-1} × (−196 + 39)°C = -1.41×10^5 J = −141 kJ

Total heat lost by Hg = −58 kJ − 81.9 kJ − 141 kJ = −281 kJ = −heat gained by N$_2$

mass of N$_2$(l) vaporized = 281 kJ $\times \dfrac{1 \text{ mol N}_2}{5.58 \text{ kJ}} \times \dfrac{28.0 \text{ g N}_2}{1 \text{ mol N}_2} = 1.41 \times 10^3$ g N$_2$(l) = 1.41 kg N$_2$

54. Both the melting point of ice and the boiling point of water are temperatures that vary as the pressure changes, the boiling point more substantially than the melting point. The triple point, however, does not vary with pressure. Solid, liquid, and vapor coexist only at one fixed temperature and pressure.

55. (a) According to Figure 13-18, CO$_2$(s) exists at temperatures below −78.5°C when the pressure is 1 atm or less. We do not expect to find temperatures this low and partial pressures of CO$_2$(g) of 1 atm on the surface of the earth.

(b) According to Table 13-3, the critical temperature of CH$_4$—the maximum temperature at which CH$_4$(l) can exist—is 191.1 K = −82.1°C. We do not expect to find temperatures this low on the surface of the earth.

(c) Since, according to Table 13-3, the critical temperature of SO$_2$ is 431.0 K = 157.8°C, SO$_2$(g) can be found on the surface of the earth.

(d) According to Figure 13-17, I$_2$(l) can exist at pressures less than 1.00 atm between the temperatures of 114°C and 184°C. There are very few places on the surface of the earth that reach temperatures this far above the boiling point of water, places such as volcano mouths. Essentially, I$_2$(l) is not found on the surface of the earth.

(e) According to Table 13-3, the critical temperature—the maximum temperature at which O$_2$(l) exists—is 154.8 K = −118.4°C. Temperatures this low do not exist on the surface of the earth.

56. (a) The upper-right region of the phase diagram is the liquid region, while the lower-right region is the region of gas. One way to figure this out is to imagine moving from left to right (toward higher temperatures) at constant pressure. (Assume 45 atm for this case.) From experience, we known that the progression of states is solid (low temperature) ⟶ liquid (intermediate temperature) ⟶ gas (high temperature).

(b) Melting means that we convert the solid to a liquid. As the phase diagram shows, the lowest pressure at which liquid exists is at the triple point pressure, 43 atm. 1.00 atm is far below 43 atm; liquid cannot exist at this temperature.

(c) As we move from point *A* to point *B* by lowering the pressure, initially nothing happens. At a certain pressure, the solid liquefies. The pressure continues to drop, with the entire sample being liquid while it does, until another, lower pressure is reached. At this lower pressure the entire sample vaporizes. The pressure then continues to drop, with the gas becoming less dense as it does so, until point *B* is reached.

57. The final pressure specified (100. atm) is below the ice III-ice I-liquid water triple point, according to the caption of Figure 13-19; no ice III is formed. The starting point is that of $H_2O(g)$. When the pressure is increased to about 4.5 mmHg, the vapor condenses to ice I. As the pressure is raised, the melting point of ice I decreases—by 1°C per 125 atm pressure increase. Thus, somewhere between 10 and 15 atm, the ice I melts to liquid water, in which state it remains until the final pressure is reached.

58. (a) As heat is added initially, the temperature of the ice rises from –20°C to 0°C. At (or just slightly below) 0°C, ice begins to melt to liquid water. The temperature remains at 0°C until all of the ice has melted. Adding heat then warms the liquid until a temperature of about 93.5°C is reached, where the liquid begins to vaporize to steam. The temperature remains fixed until all the water is converted to steam. Adding heat then warms the steam to 200°C. (Data for this part are taken from Figure 13-19 and Table 13-2.)

(b) As the pressure is raised, initially gaseous iodine is compressed. At about 91 mmHg, liquid iodine appears; the system remains at a fixed pressure with further compression until all vapor is converted to liquid. Increasing the pressure further simply compresses the liquid until a high pressure is reached, perhaps 50 atm, where solid iodine appears. Again the pressure remains fixed with further compression until all iodine is converted to solid. After this has occurred further compression raises the pressure of the system until 100 atm is reached. (Data for this part are from Figure 13-17 and the surrounding text.)

(c) Cooling of gaseous CO_2 simply lowers the temperature until a temperature of perhaps 20°C is reached. At this point, liquid CO_2 appears. The temperature remains constant as more heat is removed until all the gas is converted to liquid. Further cooling then lowers the temperature of the liquid until a temperature of slightly higher than –56.7°C is reached, where solid CO_2 appears. At this point, further cooling simply converts liquid to solid at constant temperature, until all solid has been converted to liquid. From this point, further cooling lowers the temperature of the solid. (Data for this part are taken from Figure 13-18 and Table 13-3.)

Van der Waals Forces

59. (a) HCl is not a very heavy diatomic molecule; London forces are expected to only moderately important. Hydrogen bonding is weak in the case of H—Cl bonds; Cl is not one of the three atoms (F, O, N) that form strong hydrogen bonds. Finally, because Cl is an electronegative atom, and H is only moderately electronegative, dipole-dipole interactions should be relatively strong.

(b) In Br_2 neither hydrogen bonds nor dipole-dipole attractions are important; there are no H atoms in the molecule, and homonuclear molecules are nonpolar. London forces are more important than in HCl since Br_2 is heavier.

(c) In ICl there are no hydrogen bonds since there are no H atoms in the molecule. The London forces are as strong as in Br_2 since the two molecules have the same number of electrons. However, dipole-dipole interactions are important in ICl; the molecule is polar toward Cl.

(d) In HF London forces are not very important; the molecule has only 10 electrons. Hydrogen bonding is quite important and definitely overshadows even the strong dipole-dipole interactions.

(e) In CH_4, H bonds are not important; the H atoms are not bonded to F, O, or N. In addition the molecule is not polar, so there are no dipole-dipole interactions. Finally, London forces are quite weak since the molecule contains only 10 electrons, and this is why CH_4 is a gas with a low critical temperature.

60. The differences in boiling point between successive members of the series are 41.6, 36.6, 32.6, 29.7, and 27.2 C°. Notice that the difference between differences become smaller. We expect that the next difference might be about 2 C° smaller than 27.2 C°, that is a difference of 25 C°. Thus, we expect the boiling point of normal nonane to be 125.6 + 25.2 = about 151°C. (A handbook gives the boiling point of nonane as 150.8°C.)

61. Substituting Cl for H both makes the molecule heavier and thus increases London forces, and also makes it polar which increases dipole-dipole interactions. Both of these increases make it more difficult to disrupt the forces of attraction between molecules, increasing the boiling point. A substitution of Br for Cl increases the London forces, but makes the molecule less polar. Since London forces are relatively more important than dipole-dipole interactions, the boiling point increases yet again. Finally, substituting OH for Br decreases the London forces but increases both the dipole-dipole interactions and creates opportunities for hydrogen bonding. Since hydrogen bonds are much stronger than London forces, the boiling point increases yet once more.

Hydrogen Bonding

62. We expect CH_3OH to be a liquid from among the four substances listed. Of these four molecules, C_3H_8 has the most electrons and should have the strongest London forces. However, only CH_3OH satisfies the conditions for hydrogen bonding (H bonded to and attracted to N, O, or F) and its intermolecular attractions should be much stronger than those of the other substances.

63. (a) If we just extrapolate the straight-line portion of the graph of the boiling points of group 6A hydrides in Figure 13-25 back to a molecular weight of 18, we obtain a boiling point of 210 K, or –73°C.

    (b) The freezing point of H_2O also would be expected to be lower if hydrogen bonding were less. We might expect a freezing point similar to that of Ne, even though Ne is nonpolar, since dipole-dipole interactions are quite weak. Alternatively, since the boiling point of water is expected to decrease by some 160 K, we expect the freezing point to decrease the same amount, to perhaps 110 K, –160°C.

    (c) The liquid should have its maximum density at the freezing point.

    (d) As is the case for non-hydrogen bonded substances, the solid would be more dense than the liquid.

64. (a) Intermolecular hydrogen bonding cannot occur in C_2H_6 since the conditions for hydrogen bonding (H bonded to and also attracted to N, O, or F) are not satisfied in this molecule. There is no N, O, or F in this molecule.

    (b) Intermolecular hydrogen bonding will not occur in H_3CCH_2OH since there is not another F, O, or N in the molecule to which the H of —OH can hydrogen bond.

    (c) Intermolecular hydrogen bonding is not important in H_3CCOOH. Although there is another O atom to which the H of —OH can hydrogen bond, the resulting configurations will create a four membered ring, $\left(\begin{smallmatrix} & O\cdots H \\ & \| \quad | \\ -C&-O \end{smallmatrix}\right)$ with bond angles of 90°, quite different, and therfore strained, compared to the normal bond angles of 109.5° and 120°.

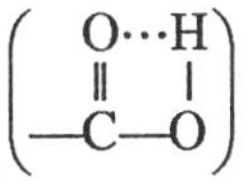

    (d) In orthophthalic acid, intermolecular hydrogen bonds can be important. The H of one —COOH group can be attracted to one of the O atoms of the other —COOH group. The resulting ring is seven atoms around and thus should not cause bond angle strain.

Network Covalent Solids

65. (a) We expect Si and C atoms to alternate in the structure, as shown at right. The C atoms are on the corners (8 × 1/8 = 1 C atom) and on the faces (6 × 1/2 = 3 C atoms), a total of four C atoms/unit cell. The Si atoms are each totally within the cell, a total of four Si atoms/unit cell.

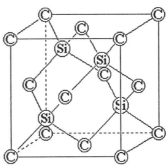

    (b) To have a graphite structure, we expect sp^2 hybridization for each atom. The hybridization schemes for B and N atoms follow.

        B   [He] $_{sp^2}$ ⊡⊡⊡ $_{2p}$ ☐   N   [He] $_{sp^2}$ ⊡⊡⊡ $_{2p}$ ⊡

The half-filled sp^2 hybrid orbitals overlap to form the σ bonding structure, and a hexagonal array of atoms. The $2p_z$ orbitals then overlap to form the π bonding orbitals. There are as many π electrons in a sample of BN as there are in a sample of graphite, assuming both samples have the same number of atoms.

(c) The probable reason why SiC does not form a graphite-like structure is the considerable difference in size between Si and C atoms. Since the Si atoms are so much larger than the C atoms, the p orbitals on adjacent atoms will not be able to overlap as efficiently as if the atoms were of identical or near identical sizes.

66. One would expect diamond to have a greater density than graphite. Although the bond distance in graphite— "one-and-a-half" bonds—would be expected to be shorter than the single bonds in diamond, the large spacing between the layers of C atoms in graphite makes its crystals much less dense than those of diamond.

67. Diamond works well in glass cutters because of its extreme hardness, a hardness that is due to the crystal being held together entirely by covalent bonds. Graphite will not function effectively in a glass cutter, since it is quite soft, soft enough to flake off in microscopic pieces when used in pencils. In fact, graphite is so soft that pure graphite is rarely used in common wooden pencils. Often clay or some other substance is mixed with the graphite to produce a mechanically strong pencil "lead".

Ionic Bonding and Properties

68. We expect forces in ionic compounds to increase as sizes of ions become smaller and as ionic charges become greater. As the forces between ions become stronger a higher temperature is required to melt the crystal. In the series of compounds NaF, NaCl, NaBr, and NaI, the anions are progressively larger, and thus the ionic forces become weaker. We expect the melting points to decrease in this series from NaF to NaI, and this is precisely what we see.

69. (a) BaF_2 should be more water soluble. Because the Ba^{2+} ion is larger than the Mg^{2+} ion, interionic forces in MgF_2 should be stronger than those in BaF_2 and more difficult to disrupt, making it more difficult to dissolve MgF_2.

(b) $MgCl_2$ should be more water soluble. Because the F^- ion is smaller than the Cl^- ion, interionic forces should be stronger and harder to disrupt in MgF_2 than those in $MgCl_2$, making it more difficult to dissolve MgF_2 than in $MgCl_2$.

70. Coulomb's law (Appendix B) states that the force between two particles of charges Q_1 and Q_2 that are separated by a distance r is given by $F = Q_1 Q_2 / \varepsilon r^2$ where ε, the dielectric constant, equals 1 for a vacuum. Since we are comparing forces, we can use +1, −1, +2, and −2 for the charges on ions. We take r to equal the sum of the cation and anion radii, which are taken from Figure 12-35.

For NaCl, $r_+ = 95$ pm, $r_- = 181$ pm $\quad F = (+1)(-1)/(95 + 181)^2 = -1.3 \times 10^{-5}$

For MgO, $r_+ = 65$ pm, $r_- = 140$ pm $\quad F = (+2)(-2)/(65 + 140)^2 = -9.5 \times 10^{-5}$

Thus, it is clear that interionic forces are about seven times stronger in MgO than in NaCl.

Crystal Structures

71. In each layer of a closest packing arrangement of spheres, there are six spheres surrounding and touching any given sphere. A second similar layer then is placed on top of this first layer so that its spheres fit into the identations in the layer below. The two different closest packing arrangements arise from two different ways of placing the third layer on top of these two, with its spheres fitting into the indentations of the layer below. In one case, one can look down into these indentations and see a sphere of the bottom (first) layer. If these indentations are used, the closest packing arrangement *abab* results (hexagonal closest packing). In the other case, no first layer sphere is visible through the indentation; the closest packing arrangement *abcabc* results (cubic closest packing).

72. Physical properties are determined by the type of bonding between atoms in a crystal or the types of interactions between ions or molecules in the crystal. Different types of interactions can produce the same geometrical relationships of unit cell components. In both Ar and CO_2, London forces hold the particles in the crystal. In NaCl, the forces are interionic attractions, while metallic bonds hold Cu atoms in their crystals. It is the type of force, not the geometric arrangement of the components, that largely determines the physical properties of the crystalline material.

73. (a) We naturally tend to look at crystal structures in right-left, up-down terms. If we do that here, we might assign a unit cell as a square, with its corners at the centers of the light-colored squares. But the crystal does not "know" right and left, or top and bottom. If we look at this crystal from the lower right corner, we see a unit cell that has its corners at the centers of dark-colored diamonds. These two types of unit cells are outlined at the top of the diagram at right.

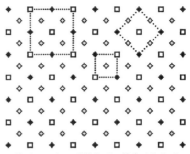

(b) The unit cell has one light-colored square fully inside it. It has four light-colored "circles" (which the computer doesn't draw as very round) on the edges, each shared with one other unit cell. So a total of $4 \times \frac{1}{2} = 2$ circles per unit cell. And the unit cell has a four dark-colored diamonds, one at each corner, and each shared with four other unit cells, for a total of $4 \times \frac{1}{4} = 1$ diamond per unit cell.

(c) One example of an erroneous unit cell is the small square outlined near the center of the figure drawn in part (a). But notice that simply repeatedly translating this unit cell toward the right, so that its left edge sits where its right edge is now, will not generate the lattice.

74. This question reduces to asking what percentage of the area of a square is covered by a circle inscribed within it. The diameter of the circle equals the side of the square. Since diameter = 2 × radius, we have

$$\frac{\text{area of circle}}{\text{area of square}} = \frac{\pi\, r^2}{(2r)^2} = \frac{\pi}{4} = \frac{3.14159}{4} = 0.7854 \qquad or \qquad 78.54\%$$

Area uncovered = $100.00\% - 78.54\% = 21.46\%$

75. In Figure 13-41 we see that the body diagonal of a cube has a length of $\sqrt{3}\, l$, where l is the length of one edge of the cube. The length of this body diagonal also equals $4r$, where r is the radius of the atom in the structure. Hence $4\,r = \sqrt{3}\, l$ or $l = 4\, r \div \sqrt{3}$ Recall that the volume of a cube is l^3, and $\sqrt{3} = 1.732$

$$\text{density} = \frac{\text{mass}}{\text{volume}} = \frac{\dfrac{2\ \text{K atoms}}{1\ \text{unit cell}} \times \dfrac{1\ \text{mol K}}{6.022 \times 10^{23}\ \text{K atoms}} \times \dfrac{39.10\ \text{g K}}{1\ \text{mol K}}}{\left(\dfrac{4 \times 227.2\ \text{pm}}{1.732} \times \dfrac{1\ \text{m}}{10^{12}\ \text{pm}} \times \dfrac{100\ \text{cm}}{1\ \text{m}} \right)^3} = 0.899\ \text{g/cm}^3$$

76. One atom lies entirely within the hcp unit cell and there are eight corner atoms, each of which are shared among eight unit cells. Thus, there are a total of two atoms per unit cell. The volume of the unit cell is given by the product of its height times the area of its base. The base is a parallelogram, which can be subdivided along its shorter diagonal into two triangles. The height of either triangle, equivalent to the perpendicular distance between the parallel sides of the parallelogram, is given by 320. pm × sin(60°) = 277 pm.

Area of base = 2 × area of triangle = 2 × (½ × base × height)

= 277 pm × 320. pm = 8.86×10^4 pm²

Volume of unit cell = 8.86×10^4 pm² × 520. pm = 4.61×10^7 pm³

$$\text{Density} = \frac{\text{mass}}{\text{volume}} = \frac{2\ \text{Mg atoms} \times \dfrac{1\ \text{mol Mg}}{6.022 \times 10^{23}\ \text{Mg atoms}} \times \dfrac{24.305\ \text{g Mg}}{1\ \text{mol Mg}}}{4.61 \times 10^7\ \text{pm}^3 \times \left(\dfrac{1\ \text{m}}{10^{12}\ \text{pm}} \times \dfrac{100\ \text{cm}}{1\ \text{m}} \right)^3} = 1.75\ \text{g/cm}^3$$

Ionic Crystal Structures

77. CaF_2   There are eight Ca^{2+} ions on the corners, each shared among eight unit cells, for a total of ($8 \times \frac{1}{8}$) one corner ion per unit cell. There are six Ca^{2+} ions on the faces, each shared between two unit cells, for a total of ($6 \times \frac{1}{2}$) three face ions per unit cell. This gives a total of four Ca^{2+} ions per unit cell. There are eight F^- ions, each wholly contained within the unit cell. The ratio of Ca^{2+} ions to F^- ions is 4 Ca^{2+} ions per 8 F^- ions: Ca_4F_8 or CaF_2.

TiO_2 There are eight Ti^{4+} ions on the corner, each shared among eight unit cells, for a total of $(8 \times 1/8)$ one Ti^{4+} corner ion per unit cell. There is one Ti^{4+} ion in the center, wholly contained within the unit cell. Thus, there are a total of two Ti^{4+} ions per unit cell. There are four O^{2-} ions on the faces of the unit cell, each shared between two unit cells, for a total of $(4 \times 1/2)$ 2 face atoms per unit cells. There are two O^{2-} ions totally contained within the unit cell. This gives a total of four O^{2-} ions per unit cell. The ratio of Ti^{4+} ions to O^{2-} ions is 2 Ti^{4+} ions per 4 O^{2-} ion: Ti_2O_4 or TiO_2.

78. The unit cell of CsCl is pictured in Figure 13-43. The length of the body diagonal equals $2\,r_+ + 2\,r_-$. From Figure 10-9, for Cs^+, $r_+ = 169$ pm and for Cl^-, $r_- = 181$ pm. Thus, the length of the body diagonal equals 2(169 pm + 181 pm) = 700. pm. From Figure 13-41, the length of this body diagonal is $\sqrt{3}\,l = 700.$ pm. The volume of the unit cell is $V = l^3$. $\sqrt{3} = 1.732$

$$V = \left(\frac{700.\ pm}{1.732} \times \frac{1\ m}{10^{12}\ pm} \times \frac{100\ cm}{1m}\right)^3 = 6.60 \times 10^{-23}\ cm^3$$

Per unit cell, there is one Cs^+ ion and one Cl^- ion (8 corner ions $\times$ $1/8$ corner ion per unit cell).

$$density = \frac{mass}{volume} = \frac{\left(1\ f.u.\ CsCl\ \times\ \dfrac{1\ mol\ CsCl}{6.022\ \times\ 10^{23}\ CsCl\ f.u.'s}\ \times\ \dfrac{168.4\ g\ CsCl}{1\ mol\ CsCl}\right)}{6.60\ \times\ 10^{-23}\ cm^3}$$

$$= 4.24\ g/cm^3$$

79. **(a)** In a sodium chloride type of lattice, there are six cations around each anion and six anions around each cation. These oppositely charged ions are arranged as follows: one above, one below, one in front, one in back, one to the right, and one to the left. Thus the crystal coordination number of Mg^{2+} is 6 and that of O^{2-} is 6 also.

(b) In the unit cell, there is an oxide ion at each of the eight corners; each of these is shared between eight unit cells. There also is an oxide ion at the center of each of the six faces; each of these oxide ions is shared between two unit cells. Thus, the total number of oxide ions is computed as follows.

$$total\ number\ of\ oxide\ ions = 8\ corners \times \frac{1\ oxide\ ion}{8\ unit\ cells} + 6\ faces \times \frac{1\ oxide\ ion}{2\ unit\ cells} = 4\ O^{2-}\ ions$$

There is a magnesium ion on each of the twelve edges; each of these is shared between four unit cells. There also is a magnesium ion in the center which is not shared with another unit cell.

$$total\ number\ of\ magnesium\ ions = 12\ edges \times \frac{1\ magnesium\ ion}{4\ unit\ cells} + 1\ central\ Mg^{2+}\ ion = 4\ Mg^{2+}\ ions$$

Thus there are four formula units per unit cell of MgO.

(c) Along the edge of the unit cell, Mg^{2+} and O^{2-} ions are in contact. The length of the edge is equal to the radius of one O^{2-}, plus the diameter of Mg^{2+}, plus the radius of another O^{2-}.

edge length = $2 \times O^{2-}$ radius + $2 \times Mg^{2+}$ radius = 2×140 pm + 2×65 pm = 410 pm

The unit cell is a cube; its volume is the cube of its length.

$$volume = (410\ pm)^3 = 6.89 \times 10^7\ pm \left(\frac{1\ m}{10^{12}\ pm} \times \frac{100\ cm}{1\ m}\right)^3 = 6.89 \times 10^{-23}\ cm^3$$

(d) $density = \dfrac{mass}{volume} = \dfrac{4\ MgO\ f.u}{6.89 \times 10^{-23}\ cm^3} \times \dfrac{1\ mol\ MgO}{6.022 \times 10^{23}\ f.u.} \times \dfrac{40.30\ g\ MgO}{1\ mol\ MgO} = 3.89\ g/cm^3$

80. From the analysis of Figure 13-42 we conclude that there are four formula units of KCl in a unit cell and the length of the edge of that unit cell is twice the internuclear distance between K^+ and Cl^- ions. (The analysis is given in more detail in the answer to Exercise 79.) The unit cell is a cube; its volume is the cube of its length.

length = 2×314.54 pm = 629.08 pm

$$volume = (629.08\ pm)^3 = 2.4895 \times 10^8\ pm \left(\frac{1\ m}{10^{12}\ pm} \times \frac{100\ cm}{1\ m}\right)^3 = 2.4895 \times 10^{-22}\ cm^3$$

unit cell mass = volume $\times$ density = $2.4895 \times 10^{-22}\ cm^3 \times 1.9893\ g/cm^3 = 4.9524 \times 10^{-22}\ g$

= mass of 4 KCl formula units

$$N_A = \frac{4\ KCl\ f.u.}{4.9524 \times 10^{-22}\ g} \times \frac{74.5510\ g\ KCl}{1\ mol} = 6.0214 \times 10^{23}\ f.u./mol$$

14 SOLUTIONS AND THEIR PHYSICAL PROPERTIES

1. (a) χ_B is the symbol for the mole fraction of component B: the amount in moles of B divided by the total amount of the solution.

   (b) P_A° represents the vapor pressure of pure substance A.

   (c) K_f is the freezing point depression constant, the distance that the freezing point of a solvent is depressed by having 1 mole of dissolved particles present in solution.

   (d) i is the symbol for the van't Hoff factor, the measured value of a colligative property divided by the value calculated assuming that each mole of solute produces one mole of particles in solution.

   (e) The activity of a solute is its effective concentration, the concentration it would have to have if solute particles did not interact with each other in solution.

2. (a) Henry's law states that the solubility of a gas in a solution is directly related to the pressure of that gas above the solution.

   (b) Freezing point depression is the change in the freezing point of a liquid due to the presence of dissolved solute.

   (c) Osmosis is the motion of solute through a membrane permeable only to it. This creates an imbalance of pressures, if the solvent and solution volumes are limited, known as osmotic pressure.

   (d) A hydrated ion is an ion surrounded by and attracted to water molecules.

   (e) Deliquescence is the absorbance by a solid of water vapor from the air to such an extent that a saturated solution is formed.

3. (a) Molarity is the amount of dissolved solute in moles per liter of solution, while molality is the amount of dissolved solute in moles per kilogram of solvent.

   (b) In an ideal solution the attractions among particles of all types are approximately the same in strength. In a nonideal solution, there is a disparity in the strengths of these attractive forces.

   (c) More solute can be dissolved in an unsaturated solution. The attempt to dissolve more solute in a supersaturated solution results in the precipitation of some solute.

   (d) Fractional crystallization is the process of first producing a concentrated and warm solution and then chilling it to a point of lower solute solubility to precipitate out some of that solute. Repetition of this process produces quite pure solute. Fractional distillation is the repeated vaporization and condensation of a liquid mixture. This also produces quite pure components.

4. $\% \text{ KI} = \dfrac{144 \text{ g KI}}{144 \text{ g KI} + 100 \text{ g H}_2\text{O}} \times 100\% = 59.0\% = 59.0 \text{ g KI}/100 \text{ g solution}$

5. (a) $\% \text{ by volume} = \dfrac{11.3 \text{ mL CH}_3\text{OH}}{75.0 \text{ mL soln}} \times 100\% = 15.1\% \text{ CH}_3\text{OH by volume}$

   (b) $\% \text{ by mass} = \dfrac{11.3 \text{ mL CH}_3\text{OH} \times \dfrac{0.793 \text{ g}}{1 \text{ mL}}}{75.0 \text{ mL} \times \dfrac{0.980 \text{ g}}{1 \text{ mL}}} \times 100\% = 12.2\% \text{ CH}_3\text{OH by mass}$

   (c) $\% \text{ (mass/volume)} = \dfrac{11.3 \text{ mL CH}_3\text{OH} \times \dfrac{0.793 \text{ g}}{1 \text{ mL}}}{75.0 \text{ mL}} \times 100\% = 11.9 \dfrac{\text{mass}}{\text{vol}} \% \text{ CH}_3\text{OH}$

6. soln. volume $= 725$ kg NaCl $\times \dfrac{1000 \text{ g NaCl}}{1 \text{ kg NaCl}} \times \dfrac{100.00 \text{ g soln}}{2.52 \text{ g NaCl}} \times \dfrac{75.0 \text{ mL soln}}{76.7 \text{ g soln}} \times \dfrac{1 \text{ L soln}}{1000 \text{ mL soln}}$

$\quad\quad\quad = 2.81 \times 10^4$ L soln $= 28.1$ kL soln

7. mass AgNO$_3$ $= 125.0$ mL soln $\times \dfrac{1 \text{ L}}{1000 \text{ mL}} \times \dfrac{0.0234 \text{ mol AgNO}_3}{1 \text{ L soln}} \times \dfrac{169.9 \text{ g}}{1 \text{ mol AgNO}_3} \times \dfrac{100.00 \text{ g mixt.}}{99.73 \text{ g AgNO}_3}$

$\quad\quad\quad = 0.498$ g mixture

8. molality $= \dfrac{3.12 \text{ g C}_6\text{H}_4\text{Cl}_2 \times \dfrac{1 \text{ mol C}_6\text{H}_4\text{Cl}_2}{147.0 \text{ g}}}{50.0 \text{ mL} \times \dfrac{0.879 \text{ g}}{1 \text{ mL}} \times \dfrac{1 \text{ kg}}{1000 \text{ g}}} = 0.483$ m

9. molarity $= \dfrac{12.00 \text{ g C}_3\text{H}_8\text{O} \times \dfrac{1 \text{ mol C}_3\text{H}_8\text{O}_2}{76.097 \text{ g C}_3\text{H}_8\text{O}_2}}{100.00 \text{ g soln} \times \dfrac{1 \text{ mL}}{1.007 \text{ g}} \times \dfrac{1 \text{ L}}{1000 \text{ mL}}} = 1.588$ M $= [\text{C}_3\text{H}_8\text{O}_2]$

10. total amount $= 1.56$ mol C$_7$H$_{16}$ $+ 3.06$ mol C$_8$H$_{18}$ $+ 2.17$ mol C$_9$H$_{20}$ $= 6.79$ moles

$\chi_7 = \dfrac{1.56 \text{ mol C}_7\text{H}_{16}}{6.79 \text{ moles total}} = 0.230 \quad\quad\quad \times 100\% = 23.0$ mol% C$_7$H$_{16}$

$\chi_8 = \dfrac{3.06 \text{ mol C}_8\text{H}_{18}}{6.79 \text{ moles total}} = 0.451 \quad\quad\quad \times 100\% = 45.1$ mol% C$_8$H$_{18}$

$\chi_9 = \dfrac{2.17 \text{ mol C}_9\text{H}_{20}}{6.79 \text{ moles total}} = 0.320 \quad\quad\quad \times 100\% = 32.0$ mol% C$_9$H$_{20}$

11. (a) $[\text{C}_3\text{H}_8\text{O}_3] = \dfrac{90.0 \text{ g C}_3\text{H}_8\text{O}_3 \times \dfrac{1 \text{ mol C}_3\text{H}_8\text{O}_3}{92.09 \text{ g C}_3\text{H}_8\text{O}_3}}{100.0 \text{ g soln} \times \dfrac{1 \text{ mL soln}}{1.235 \text{ g soln}} \times \dfrac{1 \text{ L soln}}{1000 \text{ mL soln}}} = 12.1$ M

(b) $[\text{H}_2\text{O}] = \dfrac{10.0 \text{ g H}_2\text{O} \times \dfrac{1 \text{ mol H}_2\text{O}}{18.02 \text{ g H}_2\text{O}}}{100.0 \text{ g soln} \times \dfrac{1 \text{ mL soln}}{1.235 \text{ g soln}} \times \dfrac{1 \text{ L soln}}{1000 \text{ mL soln}}} = 6.85$ M

(c) H$_2$O molality $= \dfrac{10.0 \text{ g H}_2\text{O} \times \dfrac{1 \text{ mol H}_2\text{O}}{18.02 \text{ g}}}{90.0 \text{ g C}_3\text{H}_8\text{O}_3 \times \dfrac{1 \text{ kg}}{1000 \text{ g}}} = 6.17$ m

(d) C$_3$H$_8$O$_3$ mole fraction $= \dfrac{90.0 \text{ g C}_3\text{H}_8\text{O}_3 \times \dfrac{1 \text{ mol C}_3\text{H}_8\text{O}_3}{92.09 \text{ g C}_3\text{H}_8\text{O}_3}}{90.0 \text{ g C}_3\text{H}_8\text{O}_3 \times \dfrac{1 \text{ mol C}_3\text{H}_8\text{O}_3}{92.09 \text{ g C}_3\text{H}_8\text{O}_3} + 10.0 \text{ g H}_2\text{O} \times \dfrac{1 \text{ mol H}_2\text{O}}{18.02 \text{ g H}_2\text{O}}}$

$\quad = \dfrac{0.977 \text{ mol C}_3\text{H}_8\text{O}_3}{0.977 \text{ mol C}_3\text{H}_8\text{O}_3 + 0.555 \text{ mol H}_2\text{O}} = 0.638$

(e) H$_2$O mole percent $= \dfrac{0.555 \text{ mol H}_2\text{O}}{0.977 \text{ mol C}_3\text{H}_8\text{O}_3 + 0.555 \text{ mol H}_2\text{O}} \times 100\% = 36.2$ mol % H$_2$O

12. Solubilities in Figure 14-8 are given as grams of solute per 100 g of solvent.

mass NH$_4$Cl $= 0.80$ mol NH$_4$Cl $\times \dfrac{53.5 \text{ g NH}_4\text{Cl}}{1 \text{ mol NH}_4\text{Cl}} = 43$ g NH$_4$Cl

NH$_4$Cl concentration $= \dfrac{43 \text{ g NH}_4\text{Cl}}{150.0 \text{ g H}_2\text{O}} \times 100 \text{ g H}_2\text{O} = 29$ g NH$_4$Cl

From Figure 14-8, the solubility of NH$_4$Cl in water at 25°C is 39 g/100 g water. Thus, the solution described in the problem is unsaturated.

13. Answer **(d)** is correct. An ideal solution is most likely to form when the solute and the solvent are the same type of substance, and thus the forces between solvent and solute molecules are about the same strength as those between solute or solvent molecules. In **(a)**, NaCl is an ionic solid, while H$_2$O is a polar liquid. In **(b)**, C$_2$H$_5$OH is a polar liquid, while C$_6$H$_6$(l) is a nonpolar one. In **(c)**, C$_7$H$_{16}$ a nonpolar liquid, while H$_2$O is a highly polar one. However, in **(d)** both C$_7$H$_{16}$ and C$_8$H$_{18}$ are nonpolar liquids.

14. NH$_2$OH(s) should be the most water soluble. Both C$_6$H$_6$(l) and C$_{10}$H$_8$(s) are composed of nonpolar molecules, which are poorly (if at all) soluble in water. Both NH$_2$OH(s) and CaCO$_3$(s) should be able to be attracted by water molecules. But $CaCO_3(s)$ contains ions of high charge, difficult to dissolve because of the high lattice energy.

15. Butyl alcohol should be moderately soluble in both water and benzene. A solute that is moderately soluble in both solvents will have some properties that are common to each solvent. Both naphthalene and hexane are nonpolar molecules, like benzene, but have no properties in common with water molecules. Sodium chloride consists of charged ions, similar to the charges in the polar bonds of water. But butyl alcohol possesses both a nonpolar part (C_6H_5—) like benzene, and a polar bond (—O—H) like water. In fact, water and phenol can mutually hydrogen bond.

16. We expect small, highly charged ions to form crystals with large lattice energies—a factor that decreases solubility. Based on this information, we would expect MgF_2 to be insoluble and KF to be soluble. It is probable that CaF_2 is insoluble due to a high lattice energy, but that NaF, with a smaller lattice energy, is soluble. KF is probably the most water soluble. Actual solubilities at 25°C are: 0.00020 M CaF_2 < 0.0021 M MgF_2 < 0.95 M NaF < 16 M KF.

17. We first determine the number of moles of O_2 that have dissolved.

$$\text{amount } O_2 = \frac{PV}{RT} = \frac{1.00 \text{ atm} \times 0.02831 \text{ L}}{0.08206 \text{ L atm mol}^{-1} \text{ K}^{-1} \times 298 \text{ K}} = 1.16 \times 10^{-3} \text{ mol } O_2$$

$$[O_2] = \frac{1.16 \times 10^{-3} \text{ mol } O_2}{1.00 \text{ L soln}} = 1.16 \times 10^{-3} \text{ M}$$

The oxygen concentration now is computed at the higher pressure.

$$[O_2] = \frac{1.16 \times 10^{-3}}{1 \text{ atm } O_2} \times 5.52 \text{ atm } O_2 = 6.40 \times 10^{-3} \text{ M}$$

18. First determine the number of moles of each component, then its mole fraction in the solution, then the partial pressure due to that component above the solution, and finally the total pressure.

$$\text{amount benzene} = n_b = 25.5 \text{ g } C_6H_6 \times \frac{1 \text{ mol } C_6H_6}{78.11 \text{ g } C_6H_6} = 0.326 \text{ mol } C_6H_6$$

$$\text{amount toluene} = n_t = 45.8 \text{ g } C_7H_8 \times \frac{1 \text{ mol } C_7H_8}{92.14 \text{ g } C_7H_8} = 0.497 \text{ mol } C_7H_8$$

$$\chi_b = \frac{0.326 \text{ mol } C_6H_6}{(0.326 + 0.497) \text{ total moles}} = 0.396 \qquad \chi_t = \frac{0.497 \text{ mol } C_7H_8}{(0.326 + 0.497) \text{ total moles}} = 0.604$$

$P_b = 0.396 \times 95.1 \text{ mmHg} = 37.7 \text{ mmHg}$ $\qquad\qquad$ $P_t = 0.604 \times 28.4 \text{ mmHg} = 17.2 \text{ mmHg}$

total pressure = 37.7 mmHg + 17.2 mmHg = 54.9 mmHg

19. vapor fraction of benzene $= f_b = \dfrac{37.7 \text{ mmHg}}{54.9 \text{ mmHg}} = 0.687$

   vapor fraction of toluene $= f_t = \dfrac{17.2 \text{ mmHg}}{54.9 \text{ mmHg}} = 0.313$

   These are both pressure fractions, as calculated, and also are the mole fractions in the vapor phase.

20. The solution with the lowest freezing point is 0.010 m MgI_2. The freezing point depression is proportional to the product of the solute's molality and the van't Hoff factor. For nonelectrolytes, such as C_2H_5OH (0.050 m), $i = 1$, and thus $i \cdot m = 0.050$ m. For 1:1 electrolytes, such as $MgSO_4$ and NaCl, $i = 2$, and thus $i \cdot m = 0.020$ m for (0.010 m) $MgSO_4$ and $im = 0.022$ m for (0.011 m) NaCl. And for 1:2 electrolytes, such as MgI_2 (0.010 m), $i = 3$, and thus $i \cdot m = 0.030$ m.

21. The solution with the largest freezing-point depression will have the lowest freezing point. Since C_2H_5OH is a nonelectrolyte, it will produce 1 mol of particles per mole of solute, and give the smallest freezing point depression, the highest freezing point. $HC_2H_3O_2$ is a weak electrolyte and produces slightly more than 1 mole of particles per mole of solute; its freezing point is next lower. NaCl, $MgBr_2$, and $Al_2(SO_4)_3$ all are strong electrolytes, producing 2, 3, and 5 moles of particles per mole of solute. They are listed in order of decreasing value of the freezing points of their solutions. Thus, the solutes, listed in order of decreasing freezing points of their 0.01 m aqueous solutions, are: C_2H_5OH > $HC_2H_3O_2$ > NaCl > $MgBr_2$ > $Al_2(SO_4)_3$

22. (a) $[Na^+] = \dfrac{0.90 \text{ g NaCl}}{100 \text{ mL soln}} \times \dfrac{1 \text{ mol NaCl}}{58.44 \text{ g NaCl}} \times \dfrac{1 \text{ mol Na}^+}{1 \text{ mol NaCl}} \times \dfrac{1000 \text{ mL}}{1 \text{ L soln}} = 0.15 \text{ M Na}^+$

(b) $[ions] = [Na^+] + [Cl^-] = 2 \times [Na^+] = 2 \times 0.15\ M = 0.30\ M$

(c) $\pi = CRT = \dfrac{0.30\ \text{mol ions}}{1\ L} \times \dfrac{0.08206\ \text{L atm}}{\text{mol K}} \times 310\ K = 7.6\ atm$

(d) To determine freezing point depression, we first need molality, which means we need mass of solvent.

$$\text{molality} = \dfrac{0.90\ \text{g NaCl} \times \dfrac{1\ \text{mol NaCl}}{58.44\ \text{g NaCl}} \times \dfrac{2\ \text{mol ions}}{1\ \text{mol NaCl}}}{\left(\left(100\ \text{mL soln} \times \dfrac{1.005\ g}{1\ \text{mL soln}}\right) - 0.90\ \text{g NaCl}\right) \times \dfrac{1\ \text{kg}}{1000g}} = 0.31\ m$$

$\Delta T_f = K_f \times m = 1.86\ °C/m \times 0.30\ m = 0.58°C \qquad T_f = -0.58°C$

23. First compute the molality of the benzene solution, then the number of moles of solute dissolved, and finally the mole weight of the unknown compound. $m = \dfrac{\Delta T_f}{K_f} = \dfrac{5.53°C - 4.92°C}{5.12°C/m} = 0.12\ m$

amount solute $= 0.07522\ \text{g benzene} \times \dfrac{0.12\ \text{mol solute}}{1\ \text{kg benzene}} = 9.0 \times 10^{-3}\ \text{mol solute}$

$\mathcal{M} = \dfrac{1.10\ \text{g unknown compound}}{9.0 \times 10^{-3}\ \text{mol}} = 1.2 \times 10^2\ \text{g/mol}$

24. We compute the value of the van't Hoff factor, then use this value to compute the boiling point elevation.

$i = \dfrac{\Delta T_f}{K_f\ m} = \dfrac{0.072°C}{1.86°C/m \times 0.010\ m} = 3.9 \qquad \Delta T_b = i\ K_b\ m = 3.9 \times 0.512°C\ m^{-1} \times 0.010\ m = 0.020°C$

The solution begins to boil at 100.02°C.

25. We compute the concentration of the solution, assuming that the solution volume is the same as that of the solvent. We then determine the amount of solute dissolved, and finally the mole weight.

$c = \dfrac{\pi}{RT} = \dfrac{0.79\ \text{mmHg} \times \dfrac{1\ \text{atm}}{760\ \text{mmHg}}}{0.0821\ \text{L atm mol}^{-1}\ K^{-1} \times 298\ K} = 4.2 \times 10^{-5}\ M$

amount solute $= 0.2500\ L \times \dfrac{4.2 \times 10^{-5}\ \text{mol}}{1\ L} = 1.1 \times 10^{-5}\ \text{mol}$

$\mathcal{M} = \dfrac{0.61\ g}{1.1 \times 10^{-5}\ \text{mol}} = 5.5 \times 10^4\ \text{g/mol}$

EXERCISES

Homogeneous and Heterogeneous Mixtures

26. **(b)** salicyl alcohol probably is moderately soluble in both benzene and water. The reason for this selection is that salicyl alcohol contains a benzene ring, which would make it soluble in benzene, and also can hydrogen bond to water molecules. On the other hand, **(c)** diphenyl contains only nonpolar benzene rings; it should be soluble in benzene but not in water. **(a)** *para*-dichlorobenzene contains a benzene ring, making it soluble in benzene, and two polar C—Cl bonds, which oppose each other, producing a nonpolar—and thus water-insoluble—molecule. **(d)** hydroxyacetic acid is a very polar molecule with many opportunities for hydrogen bonding. Its polar nature would make it insoluble in benzene, while the prospective hydrogen bonds will enhance aqueous solubility.

27. Vitamin C is a water-soluble vitamin. Its molecules contain a number of polar —OH groups, capable of hydrogen bonding with water, thus making these molecules soluble in water. Vitamin E is fat-soluble. Its molecules are composed of largely non-polar hydrocarbon chains, with few polar groups and thus should be soluble in nonpolar solvents.

28. **(c)** formic acid and **(f)** propylene glycol are soluble in water. They both can form hydrogen bonds with water and they both have small nonpolar portions.
(b) benzoic acid and **(d)** butyl alcohol are only slightly soluble in water. Although they both can form hydrogen bonds with water, both molecules contain reasonably large nonpolar portions, which will not dissolve well in water.
(a) iodoform and **(e)** chlorobenzene are insoluble in water. Although both molecules are polar, this is not sufficient to enable them to successfully disrupt the hydrogen bonds in water and be dissolved by water.

29. The nitrate ion is reasonably large and has a small negative charge. It thus should produce crystals with small lattice energies. In contrast, S^{2-} is highly charged and easily polarized. It should form crystals with relatively high lattice energies. In fact, partial covalent bonding occurs in many metal sulfides, as predicted from the polarizability of the anion. And these partially covalent solids are difficult to break apart.

30. Benzoic acid will react with NaOH to form a solution of sodium ions and benzoate ions. (In the equation below, ϕ represents the benzene ring with one hydrogen omitted, C_6H_5—.)

$$Na^+(aq) + OH^-(aq) + \phi\text{—COOH} \longrightarrow Na^+(aq) + \phi\text{—COO}^-(aq) + H_2O$$

The result of this reaction is that the concentration of molecular benzoic acid decreases, and thus more of the acid can be dissolved before the solution is saturated.

Percent Concentration

31. $\text{mass } HC_2H_3O_2 = 355 \text{ mL vinegar} \times \dfrac{1.01 \text{ g vinegar}}{1 \text{ mL}} \times \dfrac{5.88 \text{ g } HC_2H_3O_2}{100.00 \text{ g vinegar}} = 21.1 \text{ g } HC_2H_3O_2$

32. For water, the mass in grams and the volume in mL are about equal; the density of water is close to 1.0. For ethanol, on the other hand, the density is about 0.8. Thus, if the solution has about the same volume as its components, the percent by volume of ethanol will have to be larger than its percent by mass. This would not necessarily be true of other ethanol solutions. It would only be true in those cases where the density of the other component is greater than the density of ethanol.

33. Volume percent is not usually independent of temperature. As temperature increases, the volume of a liquid also usually increases. Unless the volume of the solution and the volume of the solute increase in the same proportion, the volume precent will be altered by these volume increases. Mass, on the other hand, is unaffected by temperature, and thus mass percent is independent of temperature.

34. $\text{mass P} = 12 \text{ fl oz} \times \dfrac{29.6 \text{ mL}}{1 \text{ fl oz}} \times \dfrac{1.00 \text{ g}}{1 \text{ mL}} \times \dfrac{0.013 \text{ g acid soln}}{100.0 \text{ g root beer}} \times \dfrac{75 \text{ g } H_3PO_4}{100 \text{ g acid soln}} \times \dfrac{1 \text{ mol } H_3PO_4}{98.00 \text{ g } H_3PO_4}$
$\times \dfrac{1 \text{ mol P}}{1 \text{ mol } H_3PO_4} \times \dfrac{30.97 \text{ g P}}{1 \text{ mol P}} \times \dfrac{1000 \text{ mg P}}{1 \text{ g P}} = 11 \text{ mg P}$

Molarity Concentration

35. $[H_2SO_4] = \dfrac{94.0 \text{ g } H_2SO_4 \times \dfrac{1 \text{ mol } H_2SO_4}{98.08 \text{ g } H_2SO_4}}{100.0 \text{ g soln} \times \dfrac{1 \text{ mL}}{1.831 \text{ g}} \times \dfrac{1 \text{ L}}{1000 \text{ mL}}} = 17.5 \text{ M}$

36. $\text{volume conc. soln} = 825 \text{ mL} \times \dfrac{0.285 \text{ mmol } C_2H_5OH}{1 \text{ mL soln}} \times \dfrac{1 \text{ mL conc. soln}}{1.71 \text{ mmol } C_2H_5OH} = 138 \text{ mL conc. soln}$

37. $[C_2H_5OH] = \dfrac{10.00 \text{ g } C_2H_5OH \times \dfrac{1 \text{ mol } C_2H_5OH}{46.07 \text{ g } C_2H_5OH}}{100.0 \text{ g soln} \times \dfrac{1 \text{ mL}}{0.9831 \text{ g}} \times \dfrac{1 \text{ L}}{1000 \text{ mL}}} = 2.134 \text{ M at } 15°C$

$[C_2H_5OH] = \dfrac{10.00 \text{ g } C_2H_5OH \times \dfrac{1 \text{ mol } C_2H_5OH}{46.07 \text{ g } C_2H_5OH}}{100.0 \text{ g soln} \times \dfrac{1 \text{ mL}}{0.9804 \text{ g}} \times \dfrac{1 \text{ L}}{1000 \text{ mL}}} = 2.128 \text{ M at } 25°C$

Molality Concentration

38. $\text{mass } I_2 = 315.0 \text{ mL } CS_2 \times \dfrac{1.261 \text{ g}}{1 \text{ mL}} \times \dfrac{1 \text{ kg}}{1000 \text{ g}} \times \dfrac{0.182 \text{ mol } I_2}{1 \text{ kg } CS_2} \times \dfrac{253.8 \text{ g } I_2}{1 \text{ mol } I_2} = 18.3 \text{ g } I_2$

39. $[HNO_3] = \dfrac{36.0 \text{ g } HNO_3 \times \dfrac{1 \text{ mol } HNO_3}{63.01 \text{ g } HNO_3}}{100.0 \text{ g soln} \times \dfrac{1 \text{ mL}}{1.2205 \text{ g}} \times \dfrac{1 \text{ L}}{1000 \text{ mL}}} = 6.97 \text{ M}$

$$\text{molality} = \frac{36.0 \text{ g HNO}_3 \times \dfrac{1 \text{ mol HNO}_3}{63.01 \text{ g HNO}_3}}{64.0 \text{ g solvent} \times \dfrac{1 \text{ kg}}{1000 \text{ g}}} = 8.93 \; m$$

40. In the original solution the mass of CH_3OH associated with each 1.00 kg = 1000 g of water is

$$\text{mass CH}_3\text{OH} = 1.00 \text{ kg H}_2\text{O} \times \frac{1.12 \text{ mol CH}_3\text{OH}}{1 \text{ kg H}_2\text{O}} \times \frac{32.04 \text{ g CH}_3\text{OH}}{1 \text{ mol CH}_3\text{OH}} = 35.9 \text{ g CH}_3\text{OH}$$

Then we compute the mass of methanol in 1.0000 kg = 1000.0 g of the original solution.

$$\text{mass CH}_3\text{OH} = 1000.0 \text{ g soln} \times \frac{35.9 \text{ g CH}_3\text{OH}}{1035.9 \text{ g soln}} = 34.7 \text{ g CH}_3\text{OH}$$

Thus, 1.0000 kg of this original solution contains (1000.0 – 34.7) 965.3 g H_2O

Now, a 1.00 m CH_3OH solution contains 32.04 g CH_3OH (one mole) for every 1000.0 g H_2O. We can compute the mass of H_2O associated with 34.7 g CH_3OH in such a solution.

$$\text{mass H}_2\text{O} = 34.7 \text{ g CH}_3\text{OH} \times \frac{1000.0 \text{ g H}_2\text{O}}{32.04 \text{ g CH}_3\text{OH}} = 1083 \text{ g H}_2\text{O}$$

(We have temporarily retained an extra significant figure in the calculation.) Since we already have 965.3 g H_2O in the original solution, the mass of water that must be added is

mass of H_2O added = 1083 g – 965.3 g = 118 g H_2O = 118 g H_2O

41. C_2H_5OH molality =

$$\frac{10.00 \text{ g C}_2\text{H}_5\text{OH} \times \dfrac{1 \text{ mol C}_2\text{H}_5\text{OH}}{46.069 \text{ g C}_2\text{H}_5\text{OH}}}{90.00 \text{ g H}_2\text{O} \times \dfrac{1 \text{ kg}}{1000 \text{ g}}} = 2.412 \; m$$

The molality of this solution does not vary with temperature. The reason why molarity varies with temperature is that the solution's density, and thus the volume of solution that contains one mole, varies with temperature. But, unlike the volume, the mass of the solution does not vary with temperature, and so its molality does not vary either.

Mole Fraction, Mole Percent

42. (a) amount C_2H_5OH = 13.2 g C_2H_5OH $\times \dfrac{1 \text{ mol C}_2\text{H}_5\text{OH}}{46.07 \text{ g C}_2\text{H}_5\text{OH}}$ = 0.287 mol C_2H_5OH

amount H_2O = 86.8 g H_2O $\times \dfrac{1 \text{ mol H}_2\text{O}}{18.02 \text{ g H}_2\text{O}}$ = 4.82 mol H_2O

$$\chi_{\text{water}} = \frac{4.82 \text{ mol H}_2\text{O}}{(0.287 + 4.82) \text{ total moles}} = 0.944 \qquad \chi_{\text{ethanol}} = \frac{0.287 \text{ mol ethanol}}{5.11 \text{ total moles}} = 0.0562$$

(b) amount H_2O = 1000. g $\times \dfrac{1 \text{ mol H}_2\text{O}}{18.02 \text{ g H}_2\text{O}}$ = 55.49 mol H_2O

$$\chi_{\text{water}} = \frac{55.49 \text{ mol H}_2\text{O}}{(55.49 + 0.512) \text{ total moles}} = 0.9909 \qquad \chi_{\text{urea}} = \frac{0.512 \text{ mol urea}}{56.00 \text{ total moles}} = 0.00914$$

(c) There is no way of precisely determining the number of moles of water present. (We need the density of the solution.) Thus, we assume that there is 1.00 L of water associated with 0.0421 mol $C_6H_{12}O_6$ and that this water has a density of 1.00 g/mL. This gives 55.49 mol H_2O and 0.0421 mol $C_6H_{12}O_6$. This is the only part of this problem for which the calculation of mole fraction is approximate.

$$\chi_{\text{water}} = \frac{55.49 \text{ mol H}_2\text{O}}{(55.49 + 0.0421) \text{ total moles}} = 0.9992 \qquad \chi_{\text{glucose}} = \frac{0.0421 \text{ mol glucose}}{55.53 \text{ total moles}} = 0.000758$$

43. Solve the following relationship for n_{ethanol}, the number of moles of ethanol, C_2H_5OH. The number of moles of water is given in the example is 5.01 moles.

$$\chi_{\text{ethanol}} = \frac{n_{\text{ethanol}}}{n_{\text{ethanol}} + 5.01 \text{ mol H}_2\text{O}} = 0.0450 \qquad n_{\text{ethanol}} = 0.0450 \, n_{\text{ethanol}} + 0.225$$

$$n_{\text{ethanol}} = \frac{0.225}{0.9550} = 0.267 \text{ mol ethanol}$$

mass added ethanol = (0.267 – 0.171) mol C_2H_5OH added $\times \dfrac{46.07 \text{ g C}_2\text{H}_5\text{OH}}{1 \text{ mol C}_2\text{H}_5\text{OH}}$ = 4.4 g C_2H_5OH

44. The amount of water present in 1 kg is n_{water} = 1000 g H_2O $\times \dfrac{1 \text{ mol H}_2\text{O}}{18.02 \text{ g H}_2\text{O}}$ = 55.49 mol H_2O

Now solve the following expression for n_{gly}, the amount of glycerol. 5.15% = 0.0515.

$$\chi_{\text{gly}} = 0.0515 = \frac{n_{\text{gly}}}{n_{\text{gly}} + 55.49} \qquad n_{\text{gly}} = 0.0515 \, n_{\text{gly}} + 2.86$$

$$n_{gly} = \frac{2.86}{(1.0000 - 0.0515)} = 3.02 \text{ mol glycerol}$$

$$\text{volume glycerol} = 3.02 \text{ mol } C_3H_8O_3 \times \frac{92.09 \text{ g } C_3H_8O_3}{1 \text{ mol } C_3H_8O_3} \times \frac{1 \text{ mL}}{1.26 \text{ g}} = 221 \text{ mL glycerol}$$

Solubility Equilibrium

45. The data in Figure 14-8 are given in mass of solute per 100 g water. 0.200 m $KClO_4$ contains 1.00 mol $KClO_4$ associated with each 1 kg = 1000 g of H_2O. The mass of $KClO_4$ dissolved in 100 g water is

$$\text{mass } KClO_4 = 100 \text{ g } H_2O \times \frac{0.200 \text{ mol } KClO_4}{1000 \text{ g } H_2O} \times \frac{138.5 \text{ g } KClO_4}{1 \text{ mol } KClO_4} = 2.77 \text{ g } KClO_4$$

This concentration is realized at a temperature of about 32°C. At this temperature a saturated solution of $KClO_4$ is 0.200 m.

46. (a) The concentration of this mixture is calculated first.

$$\frac{\text{mass solute}}{100 \text{ g } H_2O} = 100 \text{ g } H_2O \times \frac{26.0 \text{ g } KClO_4}{500.0 \text{ g water}} = 5.20 \text{ g } KClO_4$$

But at 20°C a saturated $KClO_4$ solution has a concentration of about 2.5 g $KClO_4$ dissolved in 100 g water. Thus, the solution is supersaturated

(b) The mass of $KClO_4$ that can be crystallized is the difference between the mass now present in the mixture and the mass that is dissolved in 500 g of water to produce a saturated solution.

$$\text{crystallized mass} = 26.0 \text{ g } KClO_4 - \left(500.0 \text{ g } H_2O \times \frac{2.5 \text{ g } KClO_4}{100 \text{ g } H_2O}\right) = 14 \text{ g } KClO_4$$

47. From Figure 14-8, the solubility of KNO_3 in water is 38 g KNO_3 per 100 g water at 25.0°C, while at 0.0°C its solubility is 15 g KNO_3 per 100 g water. At 25.0°C, every 138 g solution contains 38 g KNO_3 and 100.0 g water; and at 0.0°C every 115 g solution contains 15 g KNO_3 and 100.0 g water. We calculate the mass of water and of KNO_3 present at 25.0°C.

$$\text{mass water} = 335 \text{ g soln} \times \frac{100.0 \text{ g water}}{138 \text{ g soln}} = 243 \text{ g water}$$

$$\text{mass } KNO_3 = 335 \text{ g soln} - 243 \text{ g water} = 92 \text{ g } KNO_3$$

At 0.0°C the mass of water has been reduced 55 g by evaporation to (243 g − 55 g =) 188 g. The mass of KNO_3 soluble in this mass of water is

$$\text{mass } KNO_3 \text{ dissolved} = 188 \text{ g } H_2O \times \frac{15 \text{ g } KNO_3}{100.0 \text{ g } H_2O} = 28 \text{ g } KNO_3$$

Thus, of the 92 g KNO_3 in solution, the mass that recrystallizes is

mass KNO_3 recrystallized = 92 g KNO_3 originally dissolved − 28 g KNO_3 still dissolved = 64 g KNO_3

Solubility of Gases

48. $\text{mass } CH_4 = 1.00 \times 10^3 \text{ kg } H_2O \times \frac{0.02 \text{ g } CH_4}{1 \text{ kg } H_2O \cdot \text{atm}} \times 20 \text{ atm} = 4 \times 10^2 \text{ g } CH_4$

49. $\text{volume } N_2 \text{ dissolved} = 0.7808 \text{ atm } N_2 \times \frac{14.34 \text{ mL}}{1 \text{ L soln} \cdot \text{atm } N_2} = 11.20 \text{ mL } N_2/\text{L soln}$

$$n = \frac{PV}{RT} = \frac{1.00 \text{ atm} \times 11.20 \text{ mL}}{82.06 \text{ mL atm mol}^{-1} \text{ K}^{-1} \times 298 \text{ K}} = 4.58 \times 10^{-4} \text{ mol } N_2$$

$[N_2] = 4.58 \times 10^{-4} \text{ mol } N_2/\text{L soln} = 4.58 \times 10^{-4} \text{ M}$

50. We calculate the solubility as a mass percent for the purposes of comparison.

$$\text{mass } H_2S = 258 \text{ mL} \times \frac{1 \text{ mol } H_2S \text{ at STP}}{22,414 \text{ mL}} \times \frac{34.08 \text{ g } H_2S}{1 \text{ mol } H_2S} = 0.392 \text{ g } H_2S$$

$$\text{mass \% } H_2S = \frac{0.392 \text{ g } H_2S}{(100.000 + 0.392) \text{ g soln}} \times 100\% = 0.390\% \text{ H}_2S$$

Now we compute the concentration of H_2S under a pressure of 255 mmHg. Recall that STP is a pressure of 760 mmHg.

$$H_2S \text{ concentration} = 255 \text{ mmHg} \times \frac{0.390\% \text{ H}_2S}{760 \text{ mmHg}} = 0.131\% \text{ H}_2S$$

Since the mineral water's H_2S concentration (0.15%) is higher than 0.131%, some of the H_2S dissolved in the mineral water will be lost.

51. We first compute the amount of O_2 dissolved in 285 mL = 0.285 L at each temperature.

$$\text{amount } O_2 = 0.285 \text{ L} \times \frac{2.18 \times 10^{-3} \text{ mol } O_2}{1 \text{ L}} = 6.21 \times 10^{-4} \text{ mol } O_2 \text{ at } 0°C$$

$$\text{amount } O_2 = 0.285 \text{ L} \times \frac{1.26 \times 10^{-3} \text{ mol } O_2}{1 \text{ L}} = 3.59 \times 10^{-4} \text{ mol } O_2 \text{ at } 25°C$$

Determine the difference between these amounts, and the corresponding volume.

$$O_2 \text{ volume} = \frac{(6.12 - 3.59) \times 10^{-4} \text{ mol } O_2 \times \dfrac{0.08206 \text{ L atm}}{\text{mol K}} \times (25 + 273) \text{ K}}{748 \text{ mmHg} \times \dfrac{1 \text{ atm}}{760 \text{ mmHg}}}$$

$$= 6.29 \times 10^{-3} \text{ L} = 6.29 \text{ mL expelled}$$

Raoult's Law and Liquid-Vapor Equilibrium

52. We determine the mole fraction of water in this solution.

$$n_{\text{urea}} = 26.9 \text{ g CO(NH}_2)_2 \times \frac{1 \text{ mol CO(NH}_2)_2}{60.05 \text{ g CO(NH}_2)_2} = 0.448 \text{ mol CO(NH}_2)_2$$

$$n_{\text{water}} = 712 \text{ g H}_2\text{O} \times \frac{1 \text{ mol H}_2\text{O}}{18.02 \text{ g H}_2\text{O}} = 39.5 \text{ mol H}_2\text{O} \qquad \chi_{\text{water}} = \frac{39.5 \text{ mol H}_2\text{O}}{(39.5 + 0.448) \text{ total moles}} = 0.989$$

$$P_{\text{water}} = \chi_{\text{water}} \, P^*_{\text{water}} = 0.989 \times 23.8 \text{ mmHg} = 23.5 \text{ mmHg}$$

53. (a) NaCl is a nonvolatile solute. The mole fraction of water in the saturated solution is smaller than its mole fraction in the unsaturated solution. Thus, by Raoult's law, the vapor pressure of water above the unsaturated solution (1) is greater than the vapor pressure of water above the saturated solution (2).

(b) As the water evaporates from the unsaturated solution, the solution becomes more concentrated, the mole fraction of water becomes smaller, and the vapor pressure of water decreases. On the other hand, as water evaporates from the saturated solution (1), some of the solute crystallizes, the mole fraction of water remains constant, and so also does the vapor pressure of water above the solution.

(c) The solution with the lower vapor pressure will require a higher temperature before its vapor pressure reaches atmospheric pressure and thus its boiling point will be higher. The saturated solution (1) has the higher boiling point.

54. We consider a sample of 100.0 g of the solution and determine the number of moles of each component in this sample. From this information and the given vapor pressures, we determine the vapor pressure of each component.

$$\text{no. moles styrene} = n_s = 38 \text{ g styrene} \times \frac{1 \text{ mol C}_8\text{H}_8}{104 \text{ g C}_8\text{H}_8} = 0.37 \text{ mol C}_8\text{H}_8$$

$$\text{no. moles ethylbenzene} = n_e = 62 \text{ g ethylbenzene} \times \frac{1 \text{ mol C}_8\text{H}_{10}}{106 \text{ g C}_8\text{H}_{10}} = 0.58 \text{ mol C}_8\text{H}_{10}$$

$$\chi_s = \frac{n_s}{n_s + n_e} = \frac{0.37 \text{ mol}}{(0.37 + 0.58) \text{ total moles}} = 0.39 \qquad P_s = 0.39 \times 134 \text{ mmHg} = 52 \text{ mmHg for C}_8\text{H}_8$$

$$\chi_e = \frac{n_e}{n_s + n_e} = \frac{0.58 \text{ mol}}{(0.37 + 0.58) \text{ total moles}} = 0.61 \qquad P_e = 0.61 \times 182 \text{ mmHg} = 111 \text{ mmHg for C}_8\text{H}_{10}$$

Then the mole fraction in the vapor can be determined.

$$y_e = \frac{P_e}{P_e + P_s} = \frac{111 \text{ mmHg}}{(111 + 52) \text{ mmHg}} = 0.68 \qquad y_s = 1.00 - 0.68 = 0.32$$

Freezing Point Depression and Boiling Point Elevation

55. We determine the molality of the solution, then the number of moles of solute present, and then the molar masof the solute. Then we determine the compound's empirical formula, and combine this with the molar mass to determine the molecular formula.

$$m = \frac{\Delta T_f}{K_f} = \frac{5.53°C - 1.37°C}{5.12°C/m} = 0.813 \, m$$

$$\text{amount} = 50.0 \text{ mL C}_6\text{H}_6 \times \frac{0.879 \text{ g}}{1 \text{ mL}} \times \frac{1 \text{ kg}}{1000 \text{ g}} \times \frac{0.813 \text{ mol solute}}{1 \text{ kg C}_6\text{H}_6} = 3.57 \times 10^{-2} \text{ mol}$$

$$\mathfrak{M} = \frac{6.45 \text{ g}}{3.57 \times 10^{-2} \text{ mol}} = 181 \text{ g/mol}$$

$$42.4 \text{ g C} \times \frac{1 \text{ mol C}}{12.01 \text{ g C}} = 3.53 \text{ mol C} \qquad \div 1.18 \longrightarrow 2.99 \text{ mol C}$$

$$2.4 \text{ g H} \times \frac{1 \text{ mol H}}{1.01 \text{ g H}} = 2.4 \text{ mol H} \qquad \div 1.18 \longrightarrow 2.0 \text{ mol H}$$

$$16.6 \text{ g N} \times \frac{1 \text{ mol N}}{14.01 \text{ g N}} = 1.18 \text{ mol N} \qquad \div 1.18 \longrightarrow 1.00 \text{ mol N}$$

$$37.8 \text{ g O} \times \frac{1 \text{ mol O}}{16.00 \text{ g O}} = 2.36 \text{ mol O} \qquad \div 1.18 \longrightarrow 2.00 \text{ mol O}$$

The empirical formula is $C_3H_2NO_2$, with a formula mass of 84.0 g/mol. This is almost one-half the experimentally determined molar mass. Thus, the molecular formula is $C_6H_4N_2O_4$.

56. (a) First determine the molality of the solution, then the value of the freezing-point depression constant.

$$m = \frac{1.00 \text{ g } C_6H_6 \times \dfrac{1 \text{ mol } C_6H_6}{78.11 \text{ g } C_6H_6}}{80.00 \text{ g solvent} \times \dfrac{1 \text{ kg}}{1000 \text{ g}}} = 0.160 \text{ } m \qquad K_f = \frac{\Delta T_f}{m} = \frac{6.5°C - 3.3°C}{0.160 \text{ } m} = 20.°C/m$$

(b) For benzene, $K_f = 5.12$ °C m^{-1}. Cyclohexane is the better solvent for freezing point depression determinations of molar mass, since a less concentrated solution will still give a substantial freezing-point depression. For the same concentration, cyclohexane solutions will show a freezing point depression approximately four times that of benzene. (Another factor is that benzene is labeled as a carcinogen and should be avoided.)

57. We determine the molality of the benzene solution first, then the molar mass of the solute.

$$m = \frac{\Delta T_f}{K_f} = \frac{1.183°C}{5.12 \text{ °C}/m} = 0.231 \text{ } m$$

$$\text{amount solute} = 0.04456 \text{ kg benzene} \times \frac{0.231 \text{ mol solute}}{1 \text{ kg benzene}} = 0.0103 \text{ mol solute}$$

$$\mathfrak{M} = \frac{0.867 \text{ g thiophene}}{0.0103 \text{ mol thiophene}} = 84.2 \text{ g/mol}$$

Next, we determine the empirical formula from the masses of the combustion products.

$$\text{amount C} = 4.913 \text{ g } CO_2 \times \frac{1 \text{ mol } CO_2}{44.010 \text{ g } CO_2} \times \frac{1 \text{ mol C}}{1 \text{ mol } CO_2} = 0.1116 \text{ mol C} \div 0.02791 \longrightarrow 3.999 \text{ mol C}$$

$$\text{amount H} = 1.005 \text{ g } H_2O \times \frac{1 \text{ mol } H_2O}{18.015 \text{ g } H_2O} \times \frac{2 \text{ mol H}}{1 \text{ mol } H_2O} = 0.1116 \text{ mol H} \div 0.02791 \longrightarrow 3.999 \text{ mol H}$$

$$\text{amount S} = 1.788 \text{ g } SO_2 \times \frac{1 \text{ mol } SO_2}{64.065 \text{ g } SO_2} \times \frac{1 \text{ mol S}}{1 \text{ mol } SO_2} = 0.02791 \text{ mol S} \div 0.02791 \longrightarrow 1.000 \text{ mol S}$$

A reasonable empirical formula is C_4H_4S, which has an empirical mass of 84.1. Since this is the same as the experimentally determined molar mass, the molecular formula of thiophene if C_4H_4S.

58. First we determine the empirical formula of coniferin from the combustion data.

$$0.698 \text{ g } H_2O \times \frac{1 \text{ mol } H_2O}{18.02 \text{ g } H_2O} \times \frac{2 \text{ mol H}}{1 \text{ mol } H_2O} = 0.0775 \text{ mol H} \times \frac{1.01 \text{ g H}}{1 \text{ mol H}} = 0.0783 \text{ g H}$$

$$2.479 \text{ g } CO_2 \times \frac{1 \text{ mol } CO_2}{44.01 \text{ g } CO_2} \times \frac{1 \text{ mol C}}{1 \text{ mol } CO_2} = 0.05633 \text{ mol C} \times \frac{12.011 \text{ g C}}{1 \text{ mol C}} = 0.6766 \text{ g C}$$

$$(1.205 \text{ g} - 0.0783 \text{ g H} - 0.6766 \text{ g C}) \times \frac{1 \text{ mol O}}{16.00 \text{ g O}} = 0.0281 \text{ mol O}$$

$$0.0775 \text{ mol H} \div 0.0281 = 2.76 \text{ mol H} \qquad \times 4 \longrightarrow 11.04 \text{ mol H}$$

$$0.05633 \text{ mol C} \div 0.0281 = 2.00 \text{ mol C} \qquad \times 4 \longrightarrow 8.00 \text{ mol C}$$

$$0.0281 \text{ mol O} \div 0.0281 = 1.00 \text{ mol O} \qquad \times 4 \longrightarrow 4.00 \text{ mol O}$$

Empirical formula = $C_8H_{11}O_4$ with a formula weight of 171 g/mol.

Now we determine the molality of the boiling solution, the number of moles of solute present in that solution, and the molar mass of the solute. $\qquad m = \dfrac{\Delta T_b}{K_b} = \dfrac{0.068°C}{0.512°C/m} = 0.13 \text{ } m$

$$\text{amount solute} = 48.68 \text{ g } H_2O \times \frac{1 \text{ kg solvent}}{1000 \text{ g}} \times \frac{0.13 \text{ mol solute}}{1 \text{ kg solvent}} = 6.3 \times 10^{-3} \text{ mol solute}$$

$$\mathfrak{M} = \frac{2.216 \text{ g solute}}{6.3 \times 10^{-3} \text{ mol solute}} = 3.5 \times 10^2 \text{ g/mol}$$

This molar mass is twice the formula weight of the empirical formula. Thus, the molecular formula is twice the empirical formula. Molecular formula = $C_{16}H_{22}O_8$

59. First compute the molality of the solution, then the mass of sucrose that must be added to 1 kg (1000 g) H_2O, and finally the % sucrose in the solution.

$$m = \frac{\Delta T_b}{K_b} = \frac{0.15°C}{0.512°C/m} = 0.29\ m \longrightarrow \frac{0.29\ mol\ C_{12}H_{22}O_{11}}{1\ kg\ H_2O} \times \frac{342.3\ g\ C_{12}H_{22}O_{11}}{1\ mol\ C_{12}H_{22}O_{11}} = 99\ g\ C_{12}H_{22}O_{11}$$

$$\text{mass\%}\ C_{12}H_{22}O_{11} = \frac{99\ g\ C_{12}H_{22}O_{11}}{(1000.+99)\ g\ total} \times 100\% = 9.0\%\ C_{12}H_{22}O_{11}$$

Osmotic Pressure

60. Both the flowers and the cucumber contain solutions (plant sap), but both of these solutions are less concentrated than the salt solution. Thus, the solvent water in the plant material moves across the semipermeable membrane in an attempt to dilute the salt solution, leaving behind wilted flowers and shriveled pickles (little water in their tissues).

61. We first determine the molarity of the solution. Let's work the problem with three significant figures.

$$c = \frac{\pi}{RT} = \frac{1.00\ atm}{0.08206\ L\ atm\ mol^{-1}\ K^{-1} \times 273\ K} = 0.0446\ M$$

$$\text{volume} = 1\ mol \times \frac{1\ L}{0.0446\ mol\ solute} = 22.4\ L\ solution = 22.4\ L\ solvent$$

We have assumed that the solution is so dilute that its volume closely approximates the volume of the solvent constituting it. Note that this volume corresponds to the STP molar volume of an ideal gas. The osmotic pressure equation also resembles the idea gas equation.

62. First calculate the concentration of the solution, then the mass of solute in 100.0 mL of that solution.

$$[solute] = \frac{\pi}{RT} = \frac{6.15\ mmHg \times \frac{1\ atm}{760\ mmHg}}{0.08206\ L\ atm\ mol^{-1}\ K^{-1} \times 298\ K} = 3.31 \times 10^{-4}\ M$$

$$\text{mass of hemoglobin} = 100.0\ mL \times \frac{1\ L}{1000\ mL} \times \frac{3.31 \times 10^{-4}\ mol\ hemoglobin}{1\ L\ soln} \times \frac{6.86 \times 10^4\ g}{1\ mol\ hemoglobin}$$
$$= 2.27\ g\ hemoglobin$$

63. First determine the concentration of the solution from the osmotic pressure, then the amount of solute dissolved, and finally the molar mass of that solute.

$$\pi = 5.1\ mm\ soln \times \frac{0.88\ mmHg}{13.6\ mm\ soln} \times \frac{1\ atm}{760\ mmHg} = 4.3 \times 10^{-4}\ atm$$

$$c = \frac{\pi}{RT} = \frac{4.3 \times 10^{-4}\ atm}{0.0821\ L\ atm\ mol^{-1}\ K^{-1} \times 298\ K} = 1.8 \times 10^{-5}\ M$$

$$\text{amount solute} = 100.0\ mL \times \frac{1\ L}{1000\ mL} \times 1.8 \times 10^{-5}\ M = 1.8 \times 10^{-6}\ mol\ solute$$

$$\mathfrak{M} = \frac{0.50\ g}{1.8 \times 10^{-6}\ mol} = 2.8 \times 10^5\ g/mol$$

64. We need to determine the molarity of each solution. The solution with the higher molarity has the higher osmotic pressure, and water will flow from the other solution into that solution, in an attempt to create two solutions with equal concentrations.

$$[C_3H_8O_3] = \frac{14.0\ g\ C_3H_8O_3 \times \frac{1\ mol\ C_3H_8O_3}{92.09\ g\ C_3H_8O_3}}{55.2\ mL\ soln \times \frac{1\ L}{1000\ mL}} = 2.75\ M$$

$$[C_{12}H_{22}O_{11}] = \frac{17.2\ g\ C_{12}H_{22}O_{11} \times \frac{1\ mol\ C_{12}H_{22}O_{11}}{342.3\ g\ C_{12}H_{22}O_{11}}}{62.5\ mL\ soln \times \frac{1\ L}{1000\ mL}} = 0.804\ M$$

Thus, water will move from the $C_{12}H_{22}O_{11}$ solution into the $C_3H_8O_3$ solution, that is, from right to left.

65. We need the molar concentration of ions in the solution to determine its osmotic pressure. We assume that the solution has a density of 1.00 g/mL.

$$[\text{ions}] = \frac{0.9 \text{ g NaCl} \times \dfrac{1 \text{ mol NaCl}}{58.4 \text{ g NaCl}} \times \dfrac{2 \text{ mol ions}}{1 \text{ mol NaCl}}}{100.0 \text{ mL soln} \times \dfrac{1 \text{ L soln}}{1000 \text{ mL}}} = 0.3_1 \text{ M}$$

$$\pi = cRT = 0.3_1 \frac{\text{mol}}{\text{L}} \times 0.0821 \text{ L atm mol}^{-1} \text{ K}^{-1} \times (37.0 + 273.2)\text{K} = 7._9 \text{ atm} = 8 \text{ atm}$$

66. The reverse osmosis process requires a pressure equal to or slightly greater than the osmotic pressure of the solution. We assume that this solution has a density of 1.00 g/mL. First, we determine the molar concentration of ions in the solution.

$$[\text{ions}] = \frac{3.0 \text{ g NaCl} \times \dfrac{1 \text{ mol NaCl}}{58.4 \text{ g NaCl}} \times \dfrac{2 \text{ mol ions}}{1 \text{ mol NaCl}}}{100.0 \text{ g soln} \times \dfrac{1 \text{ mL soln}}{1.00 \text{ g}} \times \dfrac{1 \text{ L soln}}{1000 \text{ mL}}} = 1.0 \text{ M}$$

$$\pi = cRT = 1.0 \frac{\text{mol}}{\text{L}} \times 0.0821 \text{ L atm mol}^{-1} \text{ K}^{-1} \times (25.0 + 273.2)\text{K} = 24 \text{ atm}$$

67. The freezing point depression is given by $\Delta T_f = i\, K_f m$. Since $K_f = 1.86°C/m$ for water, $\Delta T_f = i\, 0.0186 °C$ for this group of 0.01 m solutions.

    (a) $T_f = -0.019°C$ Urea is a nonelectrolyte, and $i = 1$.

    (b) $T_f = -0.037°C$ NH_4NO_3 is a strong electrolyte, composed of two ions per formula unit; $i = 2$.

    (c) $T_f = -0.037°C$ HCl also is a strong electrolyte, composed of two ions per formula unit; $i = 2$.

    (d) $T_f = -0.056°C$ $CaCl_2$ is a strong electrolyte, composed of three ions per formula unit; $i = 3$.

    (e) $T_f = -0.037°C$ $MgSO_4$ is a strong electrolyte, composed of two ions per formula unit; $i = 2$.

    (f) $T_f = -0.019°C$ Ethanol is a nonelectrolyte; $i = 1$.

    (g) $T_f < -0.019°C$ $HC_2H_3O_2$ is a weak electrolyte; i is somewhat larger than 1.

68. **(a)** Calculate the freezing point for a 0.050 m solution of a nonelectrolyte.

    $\Delta T_{f,\text{calc}} = K_f m = 1.86°C/m \times 0.050\ m = 0.093°C$ $i = \dfrac{\Delta T_{f,\text{obs}}}{\Delta T_{f,\text{calc}}} = \dfrac{0.0986°C}{0.093°C} = 1.06$

    (b) $[H^+] = [NO_2^-] = 6.91 \times 10^{-3} \text{ M} = 0.00693 \text{ M}$ $[HNO_2] = 0.100 \text{ M} - 0.00693 \text{ M} = 0.093 \text{ M}$

    $[\text{particles}] = 0.00693 \text{ M} + 0.00693 \text{ M} + 0.093 \text{ M} = 0.107 \text{ M}$ $i = \dfrac{0.107 \text{ M}}{0.100 \text{ M}} = 1.07$

69. The mixture of $NH_3(aq)$ and $HC_2H_3O_2(aq)$, results in the formation of $NH_4C_2H_3O_2(aq)$, a solution of an ionic substance and a strong electrolyte.

$$NH_3(aq) + HC_2H_3O_2(aq) \longrightarrow NH_4C_2H_3O_2(aq) \longrightarrow NH_4^+(aq) + C_2H_3O_2^-(aq)$$

This solution of strong electrolyte conducts a current very well.

70. For two solutions to be isotonic, they must contain the same concentration of particles. For them to also have the same % mass/volume would mean that each gram of each substance produced the same number of particles. This would be true if they had the same molar mass and the same van't Hoff factor. It would also be true if the quotient of molar mass and van't Hoff factor were the same. While this condition is not impossible to meet, it is pretty unlikely.

15 CHEMICAL KINETICS

1. **(a)** $[A]_0$ symbolizes the concentration of some species A, usually a reactant, at time $= 0$.
  (b) k is the symbol for the specific rate constant of a reaction, the speed of the reaction if all rate-determining species were present at unit molarity.
  (c) $t_{1/2}$ is the symbol for the half-life of a reaction, the time during which the concentration of the reactant drops to one-half of its initial value.
  (d) A zero-order reaction is one whose rate is independent of the concentration of reactant.
  (e) A catalyst is a substance that alters the rate of a chemical reaction but that can be recovered unchanged at the end of the reaction.

2. **(a)** The method of initial rates is a means of determining the reaction order and the rate constant for a reaction by measuring the effect on the beginning reaction rate of different reactant concentrations.
  (b) An activated complex is a high-energy species that exists as a result of the collision of two molecules.
  (c) The reaction mechanism is the series of molecular processes that occounts for an overall reaction and its observed kinetics.
  (d) Heterogeneous catalysis refers to catalytic activity that occurs at the interface between two phases, usually of a liquid or a gas in contact with a solid catalytic surface.
  (e) The rate-determining step is the slowest step of a reaction mechanism.

3. **(a)** The rate of a first-order reaction depends on the first power of the concentration of a reactant; a second-order reaction's rate depends on the second power of the concentration of a reactant, or the first power of each of two reactants.
  (b) A rate equation is the relationship between the rate of a reaction and the powers of concentrations of the reactants. An integrated rate equation gives the dependence of concentration of reactant on elapsed time and initial concentration.
  (c) The activation energy is the energy that must be supplied to energize reactants to the level of the activated complex; it always is endothermic. The enthalpy of reaction is the energy absorbed as the overall reaction occurs; it may be exothermic or endothermic.
  (d) An elementary process is a description of how molecules interact with each other. The net reaction is the result of the several elementary processes that constitute it.
  (e) An enzyme is a biological catalyst. A substrate is the reactant whose reaction the catalyst promotes.

4. **(a)** $\text{Rate} = \dfrac{\Delta[A]}{\Delta t} = \dfrac{0.1867 \text{ M} - 0.1832 \text{ M}}{44 \text{ s}} = 8.0 \times 10^{-5} \text{ M s}^{-1}$

  (b) $\text{Rate} = 8.0 \times 10^{-5} \dfrac{\text{mol}}{\text{L s}} \times \dfrac{60 \text{ s}}{1 \text{ min}} = 4.8 \times 10^{-3} \text{ M min}^{-1}$

5. **(a)** $\dfrac{\Delta[B]}{\Delta t} = \dfrac{2.8 \times 10^{-3} \text{ mol A}}{\text{L s}} \times \dfrac{1 \text{ mol B}}{2 \text{ mol A}} = 1.4 \times 10^{-3} \text{ mol B L}^{-1} \text{ s}^{-1}$

  (b) 3 mol D are produced (+3 mol D) for every 2 mol A that reacts (−2 mol A).

  $\dfrac{\Delta[D]}{\Delta t} = \dfrac{-2.8 \times 10^{-3} \text{ mol A}}{\text{L s}} \times \dfrac{+3 \text{ mol D}}{-2 \text{ mol A}} = 4.2 \times 10^{-3} \text{ mol D L}^{-1} \text{ s}^{-1}$

6. In each case, we draw the tangent line to the plotted curve.
  (a) The slope of the line is $\dfrac{\Delta[H_2O_2]}{\Delta t} = \dfrac{2.00 \text{ M} - 0.000 \text{ M}}{0 \text{ s} - 2300 \text{ s}} = -8.7 \times 10^{-4} \text{ M s}^{-1}$

$$\text{Reaction rate} = -\frac{\Delta[H_2O_2]}{\Delta t} = 8.7 \times 10^{-4} \text{ M s}^{-1}$$

(b) The slope of the line is $\dfrac{\Delta[H_2O_2]}{\Delta t} = \dfrac{1.35 \text{ M} - 0.000 \text{ M}}{0 \text{ s} - 3400 \text{ s}} = -4.0 \times 10^{-4} \text{ M s}^{-1}$

$$\text{Reaction rate} = -\frac{\Delta[H_2O_2]}{\Delta t} = 4.4 \times 10^{-4} \text{ M s}^{-1}$$

7. Statement **(b)** is correct. After each half-life—that is, after each 75 s—the amount of reactant remaining is half of the amount that was present at the beginning of that half-life. Statement **(a)** is incorrect; the quantity of A remaining after 150 s is half of what was present after 75 s. Statement **(c)** is incorrect because different quantities of A are consumed in each 75 s of the reaction: $1/2$ of the original amount in the first 75 s, $1/4$ of the original amount in the second 75 s; $1/8$ of the original amount in the third 75 s, and so on. Statement **(d)** is incorrect; it implies a constant rate during the first half life. The rate of a first-order reaction actually decreases as time passes and reactant is consumed.

8. Substitute the given values into the rate equation to obtain the rate of reaction.
Rate $= k \,[A]^2 \,[B]^0 = (0.0018 \text{ M}^{-1} \text{ min}^{-1}) (0.155 \text{ M})^2 (4.68 \text{ M})^0 = 4.3 \times 10^{-5} \text{ M/min}$
Recall that (any quantity)$^0 = 1$.

9. The rate constant is determined as $k = 0.693/t_{1/2} = 0.693/13.9 \text{ min} = 0.0499 \text{ min}^{-1}$ Then the rate can be determined.   Rate $= k[A] = 0.0499 \text{ min}^{-1} \times 0.405 \text{ M} = 0.0202 \text{ M min}^{-1}$

10. (a) From Expt. 1 to Expt. 3, when [B] remains constant and [A] doubles, the rate increases by a factor of
$$\frac{2.51 \times 10^{-3} \text{ M s}^{-1}}{6.30 \times 10^{-4} \text{ M s}^{-1}} = 3.98 \approx 4 \qquad \text{Thus, the reaction is second order with respect to A.}$$
From Expt 1 to Expt. 2, when [A] remains constant and [B] doubles, the rate increases by a facor of
$$\frac{1.25 \times 10^{-3} \text{ M s}^{-1}}{6.30 \times 10^{-4} \text{ M s}^{-1}} = 1.98 \approx 2 \qquad \text{Thus, the reaction is first order with respect to B.}$$

(b) Overall rate = order with respect to A + order with respect to B = 2 + 1 = 3   The reaction is third order overall.

11. (a) A first-order reaction has a constant half-life. Thus, half of the initial concentration remains after 30.0 minutes, and at the end of another half-life—60.0 minutes total—half of the concentration present at 30.0 minutes is gone: the concentration has decreased to one-quarter of its initial value. Or, we could say that the reaction is 75% complete after two half-lives—60.0 minutes.

(b) A zero-order reaction proceeds at a constant rate. Thus, if the reaction is 50% complete in 30.0 minutes, in twice the time—60.0 minutes—the reaction will be 100% complete. And in one-fifth the time—6.0 minutes—the reaction will be 10% complete. Alternatively, we can say that the rate of

reaction is 10%/6.0 min.   time to be 75% complete $= 75\% \times \dfrac{6.0 \text{ min}}{10\%} = 45 \text{ min}$

12. (a) Although we might be tempted to apply equation (15.16) to the given data, and calculate k first, there is an easier method. During one half-life [A] decreases from 2.00 M to 1.00 M. Then, during the second half-life, [A] decreases from 1.00 M to 0.500 M. And finally, during the third half-life [A] decreases 0.500 M to 0.250 M. Consequently, three half-lives must have elapsed while [A] has decreased from 2.00 M to 0.250 M. $3\, t_{1/2} = 159 \text{ min}$ $t_{1/2} = 159 \text{ min} \div 3 = 53.0 \text{ min}$

(b) $k = \dfrac{0.693}{t_{1/2}} = \dfrac{0.693}{53.0 \text{ min}} = 0.0131 \text{ min}^{-1}$

13. (a) The mass of A has decreased to one fourth of its original value, from 1.60 g to 0.40 g. Since $\frac{1}{4} = \frac{1}{2} \times \frac{1}{2}$, we see that two half-lives have elapsed. Thus, $2 \times t_{1/2} = 22 \text{ min}$, or $t_{1/2} = 11 \text{ min}$

(b) $k = 0.693/t_{1/2} = 0.693/11 \text{ min} = 0.063 \text{ min}^{-1}$ $\ln \dfrac{[A]_t}{[A]_0} = -kt = -0.063 \text{ min}^{-1} \times 38 \text{ min} = -2.4$

$\dfrac{[A]_t}{[A]_0} = e^{-2.4} = 0.091 \quad or \quad [A]_t = [A]_0\, e^{-kt} = 1.60 \text{ g A} \times 0.091 = 0.15 \text{ g A}$

14. (a) $\ln \dfrac{[A]_t}{[A]_0} = -k\, t = \ln \dfrac{0.586 \text{ M}}{0.724 \text{ M}} = -0.211$ $k = -\dfrac{-0.211}{16.0 \text{ min}} = 0.0132 \text{ min}^{-1}$

(b) $t_{1/2} = \dfrac{0.693}{k} = \dfrac{0.693}{0.0132 \text{ min}^{-1}} = 52.5 \text{ min}$

(c) Solve the integrated rate equation for the elapsed time.

$\ln \dfrac{[A]_t}{[A]_0} = -k\,t = \ln \dfrac{0.185 \text{ M}}{0.724 \text{ M}} = -1.364 = -0.0132 \text{ min}^{-1} \times t \qquad t = \dfrac{-1.364}{-0.0132 \text{ min}^{-1}} = 103 \text{ min}$

(d) $\ln \dfrac{[A]}{[A]_0} = -kt \quad$ becomes $\quad \dfrac{[A]}{[A]_0} = e^{-kt} \quad$ which in turn becomes

$[A] = [A]_0\, e^{-kt} = 0.724 \text{ M} \exp\left(-0.0132 \text{ min}^{-1} \times 2.5 \text{ h} \times \dfrac{60 \text{ min}}{1\text{h}}\right) = 0.724 \times 0.138 = 0.10 \text{ M}$

15. In the first 500 s, [A] decreases from 2.00 M to 1.00 M; this is the first half-life. From 500 s to 1500 s, an elapsed period of 1000 s, [A] decreases by half again; from 1.00 M to 0.50 M; this is the second half-life. Since the half-life is not constant, the reaction is not first order.

During the first 500 s, $\text{Rate} = -\dfrac{1.00 \text{ M} - 2.00 \text{ M}}{500 \text{ s}} = 0.00200 \text{ M/s}$. Then during the first 1500 s,

the rate is computed as $\text{Rate} = -\dfrac{0.50 \text{ M} - 2.00 \text{ M}}{1500 \text{ s}} = 0.00100 \text{ M/s}$. Since the rate is not constant, this

reaction is not zero order.

We conclude that the reaction is second order, by the process of elimination.

We could confirm this conclusion by computing several value of the second-order rate constant with

the equation $\qquad 1/[A]_t - 1/[A]_0 = kt$

$\dfrac{1}{1.00 \text{ M}} - \dfrac{1}{2.00 \text{ M}} = 0.500 \text{ M}^{-1} = k \times 500 \text{ s} \qquad k = \dfrac{0.500 \text{ M}^{-1}}{500 \text{ s}} = 0.0010 \text{ M}^{-1} \text{ s}^{-1}$

$\dfrac{1}{0.50 \text{ M}} - \dfrac{1}{2.00 \text{ M}} = 1.50 \text{ M}^{-1} = k \times 1500 \text{ s} \qquad k = \dfrac{1.50 \text{ M}^{-1}}{1500 \text{ s}} = 0.0010 \text{ M}^{-1} \text{ s}^{-1}$

$\dfrac{1}{0.25 \text{ M}} - \dfrac{1}{2.00 \text{ M}} = 3.50 \text{ M}^{-1} = k \times 3500 \text{ s} \qquad k = \dfrac{3.50 \text{ M}^{-1}}{3500 \text{ s}} = 0.0010 \text{ M}^{-1} \text{ s}^{-1}$

The constant value of the second-order rate constant indicates that this reaction indeed is second order.

16. Statement **(d)** is correct; the activation energy of an endothermic reaction must at least equal the value of ΔH for that reaction. However, if $E_a = \Delta H_{rxn}$, the products will readily "slip back down the hill" to reactants; that is, no net reaction will occur. (There will be no activation energy for the reverse reaction, nothing to prevent that reaction from occuring regardless of the energy of its reactants.) Thus E_a should be at least a slight bit greater than ΔH_{rxn}.

17. (a) $\ln \dfrac{k_1}{k_2} = \dfrac{E_a}{R}\left(\dfrac{1}{T_2} - \dfrac{1}{T_1}\right) = \ln \dfrac{1.2 \times 10^{-4} \text{ L mol}^{-1} \text{ s}^{-1}}{3.8 \times 10^{-2} \text{ L mol}^{-1} \text{ s}^{-1}} = \dfrac{E_a}{R}\left(\dfrac{1}{666 \text{ K}} - \dfrac{1}{556 \text{ K}}\right)$

$-5.76\,R = -E_a \times 2.97 \times 10^{-4}$

$E_a = \dfrac{5.76\,R}{2.97 \times 10^{-4}} = 1.94 \times 10^4 \text{ K}^{-1} \times 8.3145 \text{ J mol}^{-1} \text{ K}^{-1} = 1.61 \times 10^5 \text{ J/mol} = 161 \text{ kJ/mol}$

(b) $\ln \dfrac{k_1}{k_2} = \dfrac{E_a}{R}\left(\dfrac{1}{T_2} - \dfrac{1}{T_1}\right) = \ln \dfrac{1.0 \times 10^{-3} \text{ L mol}^{-1} \text{ s}^{-1}}{3.8 \times 10^{-2} \text{ L mol}^{-1} \text{ s}^{-1}} = \dfrac{1.61 \times 10^5 \text{ J/mol}}{8.3145 \text{ J mol}^{-1} \text{ K}^{-1}}\left(\dfrac{1}{666 \text{ K}} - \dfrac{1}{T}\right)$

$-3.64 = 1.94 \times 10^4 \left(\dfrac{1}{666 \text{ K}} - \dfrac{1}{T}\right) \qquad \left(\dfrac{1}{666 \text{ K}} - \dfrac{1}{T}\right) = \dfrac{-3.64}{1.94 \times 10^4} = -1.88 \times 10^{-4}$

$\dfrac{1}{T} = 1.88 \times 10^{-4} + 1.50 \times 10^{-3} = 1.69 \times 10^{-3} \qquad T = 592 \text{ K}$

18. (a) We can demonstrate consistency with the stoichiometry by adding the two reactions.

Sum of the reactions: $A + B + I + B \longrightarrow I + C + D$

and then cancel the "I" that appears on both sides of the resultant equation.

$$A + 2\,B \longrightarrow C + D$$

We can determine the rate of the reaction from the mechanism by assuming that the rate of the elementary slow step is equal to the rate of the reaction. $\qquad$ Reaction rate $= k_{slow}[A][B] \qquad$ This is equivalent to the observed rate law.

(b) Again, we demonstrate consistency with the stoichiometry by adding the two reactions.

$2\,B + A + B_2 \longrightarrow B_2 + C + D$

and then cancel the "B_2" that appears on both sides of the resultant equation.

$$2 B + A \longrightarrow C + D$$

The reaction rate is the rate of the elementary slow step.   Reaction rate = $k_{slow}[A][B_2]$

But B_2 is a reaction intermediate whose concentration is difficult to determine. We can determine that concentration by assuming that the fast initial step goes equally rapidly in the forward and reverse directions. $k_f[B]^2 = k_r[B_2]$ $[B_2] = (k_f/k_r)[B]^2$

The resulting expression for $[B_2]$ is substituted into the expression for reaction rate, and we see that the experimentally determined rate law is *not* recovered.

Reaction rate = $k_{slow}[A][B_2] = k_{slow}[A](k_f/k_r)[B]^2 = k'[A][B]^2$

19. (a) Set II is data from a zero-order reaction. We know this because the rate of set II is constant. 0.25 M/25 s = 0.010 M s^{-1}. A zero-order reaction has a constant rate.

(b) A first-order reaction has a constant half-life. In set I, the first half-life is slightly less than 75 sec, since the concentration decreases by slightly more than half (from 1.00 M to 0.47 M) in 75 s. Again, from 75 s to 150 s the concentration decreases from 0.47 M to 0.22 M, again slightly more than half, in a time of 75 s. Finally, two half-lives should see the concentration decrease to one-fourth of its initial value. This is what we see: from 100 s to 250 sec, 150 s of elapsed time, the concentration decreases from 0.37 M to 0.08 M; that is also slightly less than one-fourth of its initial value. Notice that we cannot make the same statement of constancy of half-life regarding set III. The first half-life is 100 s, but it takes more than 150 s (from 100 s to 250 s) for [A] to again decrease by half.

(c) For a second-order reaction, $1/[A]_t - 1/[A]_0 = kt$ For the initial 100 s, we have

$$\frac{1}{0.50 \text{ M}} - \frac{1}{1.00 \text{ M}} = 1.0 \text{ L mol}^{-1} = k\ 100 \text{ s} \qquad k = 0.010 \text{ L mol}^{-1}\text{ s}^{-1}$$

For the initial 200 s, we have

$$\frac{1}{0.33 \text{ M}} - \frac{1}{1.00 \text{ M}} = 2.0 \text{ L mol}^{-1} = k\ 200 \text{ s} \qquad k = 0.010 \text{ L mol}^{-1}\text{ s}^{-1}$$

Since we obtain the same value of the rate constant, and have used the equation for second-order kinetics, set III must be second order.

20. Since Set I is the data for a first-order reaction, we can analyze those data to determine the half-life. In the first 75 s, the concentration decreases by a bit more than half. This implies a half-life slightly less than 75 s, perhaps 70 s. This is consistent with the other time periods noted in the answer to Review Question 19 (b) and also to the fact that in the 150 s from 50 s to 200 s, the concentration decreases from 0.61 M to 0.14 M, a bit more than a factor-of-four decrease. The factor-of-four decrease, to one-fourth of the initial value, is what we would expect of two successive half-lives. We can determine the half-life more accurately, by obtaining a value of k from the relation $\ln ([A]_t/[A]_0) = -kt$ followed by $t_{1/2} = 0.693/k$

21. We can determine an approximate initial rate by using data from the first 25 s.

$$\text{Rate} = -\frac{\Delta[A]}{\Delta t} = -\frac{0.80 \text{ M} - 1.00 \text{ M}}{25 \text{ s} - 0 \text{ s}} = 0.0080 \text{ M s}^{-1}$$

22. The approximate rate at 75 s can be taken as the rate over the time period from 50 s to 100 s.

(a) $\text{Rate}_{II} = -\dfrac{\Delta[A]}{\Delta t} = -\dfrac{0.00 \text{ M} - 0.50 \text{ M}}{100 \text{ s} - 50 \text{ s}} = 0.010 \text{ M s}^{-1}$

(b) $\text{Rate}_I = -\dfrac{\Delta[A]}{\Delta t} = -\dfrac{0.37 \text{ M} - 0.61 \text{ M}}{100 \text{ s} - 50 \text{ s}} = 0.0048 \text{ M s}^{-1}$

(c) $\text{Rate}_{III} = -\dfrac{\Delta[A]}{\Delta t} = -\dfrac{0.50 \text{ M} - 0.67 \text{ M}}{100 \text{ s} - 50 \text{ s}} = 0.0034 \text{ M s}^{-1}$

Alternatively we can use [A] at 75 s (the values given in the table) in the relationship Rate = $k\ [A]^m$, where $m = 0, 1,$ or 2.

(a) Rate = 0.010 M s^{-1} × (0.25 mol/L)^0 = 0.010 M s^{-1}

(b) Since $t_{1/2} = 70$ s, $k = 0.693/70$ s = 0.0099 s^{-1} Rate = 0.0099 s^{-1} × (0.47 mol/L)^1 = 0.0047 M s^{-1}

(c) Rate = 0.010 L mol^{-1} s^{-1} × (0.57 mol/L)^2 = 0.0032 M s^{-1}

23. We can combine the approximate rates from Review Question 22, with the fact that 10 s have elapsed, and the concentration at 100 s.

(a) $[A]_{II} = 0.00$ M   There is no reactant left after 100 s.

(b) $[A]_I = [A]_{100} - (10 \text{ s} \times \text{rate}) = 0.37 \text{ M} - (10 \text{ s} \times 0.0047 \text{ M s}^{-1}) = 0.32 \text{ M}$

(c) $[A]_{III} = [A]_{100} - (10 \text{ s} \times \text{rate}) = 0.50 \text{ M} - (10 \text{ s} \times 0.0032 \text{ M s}^{-1}) = 0.47 \text{ M}$

EXERCISES

Rates of Reactions

24. $\text{Rate} = \dfrac{-\Delta[A]}{\Delta t} = \dfrac{-(0.540 \text{ M} - 0.550 \text{ M})}{80.3 \text{ s} - 60.2 \text{ s}} = 5.0 \times 10^{-4} \text{ M s}^{-1}$

25. (a) $\text{Rate} = -\dfrac{\Delta[A]}{\Delta t} = -\dfrac{0.1455 \text{ M} - 0.1503 \text{ M}}{1.00 \text{ min} - 0.00 \text{ min}} = 0.0048 \text{ M min}^{-1}$

$\text{Rate} = -\dfrac{\Delta[A]}{\Delta t} = -\dfrac{0.1409 \text{ M} - 0.1455 \text{ M}}{2.00 \text{ min} - 1.00 \text{ min}} = 0.0046 \text{ M min}^{-1}$

(b) The rates are not equal because, in all except zero-order reactions, the rate depends on the concentration of reactant. And, of course, as the reaction proceeds reactant is consumed and its concentration decreases, therefore changing the rate of the reaction.

26. (a) $\Delta[A] = \dfrac{\Delta[A]}{\Delta t} \Delta t = -1.0 \times 10^{-2} \text{ M/min} \times (5.00 \text{ min} - 4.50 \text{ min}) = -0.0050 \text{ M}$

$[A] = [A]_i + \Delta[A] = 0.800 \text{ M} - 0.0050 \text{ M} = 0.795 \text{ M}$

(b) $\Delta[A] = 0.775 \text{ M} - 0.800 \text{ M} = -0.025 \text{ M}$

$\Delta t = \Delta[A] \dfrac{\Delta t}{\Delta[A]} = \dfrac{-0.025 \text{ M}}{-1.0 \times 10^{-2} \text{ M/min}} = 2.5 \text{ min}$　　　$\text{time} = t + \Delta t = (4.50 + 2.5) \text{ min} = 7.0 \text{ min}$

27. (a) $\text{Rate} = \dfrac{\Delta[C]}{2 \Delta t} = 1.76 \times 10^{-5} \text{ M s}^{-1}$　　　$\dfrac{\Delta[C]}{\Delta t} = 2 \times 1.76 \times 10^{-5} \text{ M s}^{-1} = 3.52 \times 10^{-5} \text{ M/s}$

(b) $\dfrac{\Delta[A]}{\Delta t} = -\dfrac{\Delta[C]}{2 \Delta t} = -1.76 \times 10^{-5} \text{ M s}^{-1}$　　　Assume this rate is constant.

$[A] = 0.4000 \text{ M} + \left(-1.76 \times 10^{-5} \text{ M s}^{-1} \times 1.00 \text{ min} \times \dfrac{60 \text{ s}}{1 \text{ min}}\right) = 0.3989 \text{ M}$

(c) $\dfrac{\Delta[A]}{\Delta t} = -1.76 \times 10^{-5} \text{ M s}^{-1}$　　　$\Delta t = \dfrac{\Delta[A]}{-1.76 \times 10^{-5} \text{ M/s}} = \dfrac{-0.0100 \text{ M}}{-1.76 \times 10^{-5} \text{ M/s}} = 568 \text{ s}$

28. (a) $\dfrac{\Delta n[O_2]}{\Delta t} = 1.00 \text{ L soln} \times \dfrac{5.7 \times 10^{-4} \text{ mol } H_2O_2}{1 \text{ L soln·s}} \times \dfrac{1 \text{ mol } O_2}{2 \text{ mol } H_2O_2} = 2.9 \times 10^{-4} \text{ mol } O_2\text{/s}$

(b) $\dfrac{\Delta n[O_2]}{\Delta t} = 2.8 \times 10^{-4} \dfrac{\text{mol } O_2}{\text{s}} \times \dfrac{60 \text{ s}}{1 \text{ min}} = 1.7 \times 10^{-2} \text{ mol } O_2\text{/min}$

(c) $\dfrac{\Delta V[O_2]}{\Delta t} = 1.7 \times 10^{-2} \dfrac{\text{mol } O_2}{\text{min}} \times \dfrac{22,414 \text{ mL } O_2 \text{ at STP}}{1 \text{ mol } O_2} = 3.8 \times 10^2 \text{ mL } O_2 \text{ at STP}$

29. Initial concentrations are $[S_2O_8^{2-}] = 0.076 \text{ M}$ and $[I^-] = 0.060 \text{ M}$. The initial rate of the reaction is 2.8×10^{-5} M s^{-1}. Of course, the reaction is $S_2O_8^{2-} + 3 \text{ I}^- \longrightarrow 2 \text{ SO}_4^{2-} + I_3^-$ The rate of reaction equals the rate of disappearance of $S_2O_8^{2-}$. Then, after 1 minute, assuming that the rate is the same as the initial rate,

(a) $[S_2O_8^{2-}] = 0.076 \text{ M} - \left(2.8 \times 10^{-5} \dfrac{\text{mol}}{\text{L·s}} \times 1 \text{ min} \times \dfrac{60 \text{ s}}{1 \text{ min}}\right) = 0.074 \text{ M}$

(b) $[I^-] = 0.060 \text{ M} - \left(2.8 \times 10^{-5} \dfrac{\text{mol } S_2O_8^{2-}}{\text{L·s}} \times \dfrac{3 \text{ mol } I^-}{1 \text{ mol } S_2O_8^{2-}} \times 1 \text{ min} \times \dfrac{60 \text{ s}}{1 \text{ min}}\right) = 0.055 \text{ M}$

30. Notice that, for every 100 mmHg drop in the pressure of A(g), there is a corresponding 200 mmHg rise in the pressure of B(g) plus a 100 mmHg rise in the pressure of C(g).

(a) We set up the calculation with three lines of information below the balanced equation: (1) the initial conditions, (2) the changes that occur, which are related to each other by reaction stoichiometry, and (3) the final conditions, which simply are initial conditions + changes.

	A(g)	$\longrightarrow$	2 B(g)	+	C(g)
Initial	1000. mmHg		0. mmHg		0. mmHg
Changes	−1000. mmHg		+2000. mmHg		+1000. mmHg
Final	0. mmHg		2000. mmHg		1000. mmHg

Total final pressure = 0. mmHg + 2000. mmHg + 1000. mmHg = 3000. mmHg

(b)

	A(g)	$\longrightarrow$	2 B(g)	+	C(g)
Initial	1000. mmHg		0. mmHg		0. mmHg
Changes	–200. mmHg		+400. mmHg		+200. mmHg
Final	800. mmHg		400. mmHg		200. mmHg

Total pressure = 800. mmHg + 400. mmHg + 200. mmHg = 1400 mmHg

31. The rate of a reaction is proportional to either the rate at which product is formed or reactant is consumed. It usually is expressed in units of mol L^{-1} s^{-1} or in mol/s. The rate constant of a reaction, on the other hand, is but one factor of the rate law of the reaction. That is, the rate of a reaction equals its rate constant multiplied by one or more concentrations or pressures (each raised to an appropriate power). Only in zero-order reactions is the rate constant equal to the rate of the reaction. (In addition, the rate constant and the rate are numerically equal when the concentrations of all species in the rate equation are 1.00 M.)

32. **(a)** The reaction is obviously not zero order, for the rate is not constant with time. That is, the concentration *vs.* time graph is not a straight line. The concentration decreases from 1.000 M to 0.500 M in 2.37 min; this is the first half-life. The concentration also decreases from 0.800 M at 0.75 min to 0.400 M at 3.07 min; this second half-life is therefore (3.07 – 0.75 =) 2.32 min. Finally, the concentration decreases from 0.400 M at 3.07 min to 0.200 M at 5.45 min; this third half-life therefore is (5.45 – 3.07 =) 2.38 min. The deviation among these three half-lives is due to the difficulty in obtaining precise information from the graph. (Even though the numbers have been obtained with a ruler, there still is considerable imprecision in them.) Since the half-life is constant, the reaction is first-order.

(b) $t_{1/2} = (2.37 + 2.32 + 2.38) \div 3 = 2.36$ min $\qquad k = \dfrac{0.693}{t_{1/2}} = \dfrac{0.693}{2.36 \text{ min}} = 0.294 \text{ min}^{-1}$

or perhaps better expressed as $k = 0.29$ min^{-1} due to imprecision.

(c) When $t = 3.5$ min, [A] = 0.352 M. Then, rate = k[A] = 0.294 min^{-1} × 0.352 M = 0.10_3 M/min

(d) Slope = $\dfrac{\Delta[\text{A}]}{\Delta t} = -$ Rate $= \dfrac{0.148 \text{ M} - 0.339 \text{ M}}{6.00 \text{ min} - 3.00 \text{ min}} = -0.0637$ M/min $\qquad$ Rate = 0.064 M/min

(e) Rate = k[A] = 0.294 min^{-1} × 1.000 M = 0.294 M/min

Method of Initial Rates

33. From Example 15-3 we have the rate law, and Example 15-4 establishes the numerical value of the rate constant as $k = 6.1 \times 10^{-3}$ L mol^{-1} s^{-1}

Rate = $k[\text{S}_2\text{O}_8{}^{2-}]^1[\text{I}^-]^1 = 6.1 \times 10^{-3}$ L mol^{-1} s^{-1} × 0.019 M × 0.015 M = 1.7×10^{-6} M min^{-1}

34. From Expt. 1 and Expt. 2 we see that [B] remains fixed while [A] triples. As a result, the initial rate increase from 4.2×10^{-3} M/min to 1.3×10^{-2} M/min, that is, the rate triples. Therefore the reaction is first order in [A]. Between Expt. 2 and Expt. 3, we see that [A] doubles, which would double the rate, and [B] doubles. As a consequence, the initial rate goes from 1.3×10^{-2} M/min to 5.2×10^{-2} M/min, that is, the rate quadruples. Since an additional doubling of the rate is due to the change in [B], the reaction is first order in [B]. Now we determine the value of the rate constant.

Rate = $k[\text{A}]^1[\text{B}]^1$ $\qquad k = \dfrac{\text{Rate}}{[\text{A}][\text{B}]} = \dfrac{5.2 \times 10^{-2} \text{ M/min}}{3.00 \text{ M} \times 3.00 \text{ M}} = 5.8 \times 10^{-3}$ L mol^{-1} min^{-1}

The rate law is $\quad$ Rate = $(5.8 \times 10^{-3}$ L mol^{-1} $\text{min}^{-1})[\text{A}]^1[\text{B}]^1$

35. **(a)** From Expt. 1 to Expt. 2, [HgCl_2] is constant at 0.105 M, but [$\text{C}_2\text{O}_4{}^{2-}$] doubles (× 2.0), from 0.15 M to 0.30 M. At the same time, the rate increases nearly four times (actually × 3.9), from 1.8×10^{-5} M min^{-1} to 7.1×10^{-5} M min^{-1}. Thus, the reaction is second order in $\text{C}_2\text{O}_4{}^{2-}$, since $2.0^x \approx 3.9$ when $x = 2$.

From Expt. 3 to Expt. 2, [$\text{C}_2\text{O}_4{}^{2-}$] is constant at 0.30 M, but [HgCl_2] slightly more than doubles (× 2.02) from 0.52 M to 1.05 M. At the same time, the rate slightly more than doubles (× 2.03) from 3.5×10^{-5} M min^{-1} to 7.1×10^{-5} M min^{-1}. Thus, the reaction is first order in HgCl_2, since $2.03 \approx 2.02^y$ when $y = 1$.

The reaction is third order overall (the sum of the orders of the two reactants).

(b) The rate constant is determined from the rate law and the data from any one of the experiments.

Rate $= k[HgCl_2]^1[C_2O_4{}^{2-}]^2 = 1.8 \times 10^{-5}$ M min^{-1} $= k$ (0.105 M)^1(0.15 M)^2

$$k = \frac{1.8 \times 10^{-5} \text{ M min}^{-1}}{(0.105 \text{ M})^1(0.15 \text{ M})^2} = 7.6 \times 10^{-3} \text{ L}^2 \text{ mol}^{-2} \text{ min}^{-1}$$

(c) Rate $= k[HgCl_2]^1[C_2O_4{}^{2-}]^2 = 7.6 \times 10^{-3}$ L^2 mol^{-2} min^{-1} (0.020 M)^1(0.22 M)^2

$= 7.4 \times 10^{-6}$ M min^{-1}

36. (a) From Expt. 1 to Expt. 2, [B] remains constant at 1.40 M and [C] remains constant at 1.00 M, but [A] is halved ($\times$ 0.50). At the same time the rate is halved ($\times$ 0.50). Thus the reaction is first order with respect to A, since $0.50^x = 0.50$ when $x = 1$.

From Expt. 2 to Expt. 3, [A] remains constant at 0.70 M and [C] remains constant at 1.00 M, but [B] is halved ($\times$ 0.50), from 1.40 M to 0.70 M. At the same time, the rate is quartered ($\times$ 0.25). Thus, the reaction is second order with respect to B, since $0.50^y = 0.25$ when $y = 2$.

From Expt. 1 to Expt. 4, [A] remains constant at 1.40 M and [B] remains constant at 1.40 M, but [C] is halved ($\times$ 0.5), from 1.00 M to 0.50 M. At the same time, the rate is increased by a factor of [$R_4 = 16 \, R_3 = 16 \times \frac{1}{4} \, R_2 = 4 \, R_2 = 4 \times \frac{1}{2} \, R_1 = 2 \times R_1$] 2.0. Thus, the order of the reaction with respect to C is -1, since $0.5^z = 2.0$ when $z = -1$.

(b) $R_5 = k$ (0.70 M)^1 (0.70 M)^2 (0.50 M)$^{-1}$ $= k \left(\dfrac{1.40 \text{ M}}{2}\right)^1\left(\dfrac{1.40 \text{ M}}{2}\right)^2\left(\dfrac{1.00 \text{ M}}{2}\right)^{-1}$

$= k \, \frac{1}{2}^1 (1.40 \text{ M})^1 \, \frac{1}{2}^2 (1.40 \text{ M})^2 \, \frac{1}{2}^{-1} (1.00 \text{ M})^{-1} = R_1 \, (\frac{1}{2})^{1+2-1} = R_1 \, (\frac{1}{2})^2 = \frac{1}{4} \, R_1$

This is based on $R_1 = k$ (1.40 M)^1 (1.40 M)^2 (1.00 M)$^{-1}$.

First-Order Reactions

37. (a) **TRUE**   The rate of the reaction does decrease as more and more of B and C are formed, but not because more and more of B and C are formed. Rather the rate decreases because the concentration of A must decrease to form more and more of B.

(b) **FALSE**   The time required for one half of substance A to react—the half-life—is independent of the quantity of "A" present.

(c) **FALSE**   A plot of ln [A] or log [A] vs. time yields a straight line. One of [A] vs. time yields a curved line.

(d) **TRUE**   The rate of formation of "C" is related to the rate of disappearance of "A" by the stoichiometry of the reaction.

38. (a) Since the half-life is 120 s, after 600 s five half-lives have elapsed, and the original quantity of A has been cut in half five times.

final quantity of A $= (0.5)^5 \times$ initial quantity of A $= 0.03125 \times$ initial quantity of A

3.125% of the original quantity of A remains unreacted after 600 s.

(b) For a first order reaction $k = \dfrac{0.693}{t_{1/2}} = \dfrac{0.693}{120 \text{ s}} = 0.00578$ s^{-1}

Rate $= k$ [A] $= 0.00578$ s$^{-1} \times 0.50$ M $= 0.00289$ M/s

39. We note that the final concentration is one-eighth of the initial concentration. Thus, three half-lives have elapsed, the first to reduce [A] from 0.800 M to 0.400 M, the second to reduce [A] from 0.400 M to 0.200 M, and the third half-life to reduce [A] from 0.200 M to 0.100 M.

$$54 \text{ min} = 3 \, t_{1/2} \qquad t_{1/2} = \frac{54 \text{ min}}{3} = 18 \text{ min}$$

To reduce the concentration yet further requires two additional half-lives: the first of these to lower [A] from 0.100 M to 0.050 M and the second of them to reduce [A] from 0.050 M to 0.025 M. These two half-lives equal 2 $\times$ 18 min = 36 min. Thus at 54 min + 36 min = 90 min from the start of the reaction [A] = 0.025 M.

40. When acetoacetic acid is 65% decomposed, its concentration has decreased to 35% of its initial value. The value of the first-order rate constant is obtained from the given value of the half-life.

$$k = \frac{0.693}{t_{1/2}} = \frac{0.693}{144 \text{ min}} = 4.81 \times 10^{-3} \text{ min}^{-1}$$

$$\ln \frac{[A]_t}{[A]_0} = -kt = \ln \frac{0.35 \times 0.135 \text{ M}}{0.135 \text{ M}} = -1.05 = -4.81 \times 10^{-3} \text{ min}^{-1}\, t$$

$$t = \frac{1.05}{4.81 \times 10^{-3} \text{ min}^{-1}} = 218 \text{ min}$$

Notice that we really did not need to know the initial concentration of acetoacetic acid.

41. We determine the value of the first-order rate constant and from that we determine the half-life. If the reactant is 99% decomposed in 137 min, then only 1% (0.010) of the initial concentration remains.

$$\ln \frac{[A]_t}{[A]_0} = -kt = \ln \frac{0.010}{1.000} = -4.61 = -k \times 137 \text{ min} \qquad k = \frac{4.61}{137 \text{ min}} = 0.0336 \text{ min}^{-1}$$

$$t_{1/2} = \frac{0.0693}{k} = \frac{0.693}{0.0336 \text{ min}^{-1}} = 20.6 \text{ min}$$

42. If 99% of the radioactivity of P-32 is lost, 1% of that radioactivity remains. First we compute the value of the rate constant from the half life. $\quad k = \frac{0.693}{t_{1/2}} = \frac{0.693}{14.3 \text{ d}} = 0.0485 \text{ d}^{-1}$

Then we use the integrated rate equation to determine the elapsed time.

$$\ln \frac{[A]_t}{[A]_0} = -kt \qquad t = -\frac{1}{k} \ln \frac{[A]_t}{[A]_0} = -\frac{1}{0.0485 \text{ d}^{-1}} \ln \frac{0.010}{1.000} = 95 \text{ days}$$

43. (a) $\ln \dfrac{[A]_t}{[A]_0} = -kt = \ln \dfrac{2.5 \text{ g}}{80.0 \text{ g}} = -3.47 = -6.2 \times 10^{-4} \text{ s}^{-1}\, t \qquad t = \dfrac{3.47}{6.2 \times 10^{-4} \text{ s}^{-1}} = 5.6 \times 10^3 \text{ s}$

(b) $\text{vol. O}_2 = 77.5 \text{ g N}_2\text{O}_5 \times \dfrac{1 \text{ mol N}_2\text{O}_5}{108.0 \text{ g N}_2\text{O}_5} \times \dfrac{1 \text{ mol O}_2}{2 \text{ mol N}_2\text{O}_5} \times \dfrac{22414 \text{ mL O}_2}{1 \text{ mol O}_2 \text{ STP}} = 8.04 \times 10^3 \text{ mL O}_2$

44. (a) If the reaction is first order, we will obtain the same value of the rate constant from several sets of data.

$$\ln \frac{[A]_t}{[A]_0} = -kt = \ln \frac{264 \text{ mmHg}}{312 \text{ mmHg}} = -k \times 390 \text{ s} = -0.167 \qquad k = \frac{0.167}{390 \text{ s}} = 4.28 \times 10^{-4} \text{ s}^{-1}$$

$$\ln \frac{[A]_t}{[A]_0} = -kt = \ln \frac{224 \text{ mmHg}}{312 \text{ mmHg}} = -k \times 777 \text{ s} = -0.331 \qquad k = \frac{0.331}{777 \text{ s}} = 4.26 \times 10^{-4} \text{ s}^{-1}$$

$$\ln \frac{[A]_t}{[A]_0} = -kt = \ln \frac{187 \text{ mmHg}}{312 \text{ mmHg}} = -k \times 1195 \text{ s} = -0.512 \qquad k = \frac{0.512}{1195 \text{ s}} = 4.28 \times 10^{-4} \text{ s}^{-1}$$

$$\ln \frac{[A]_t}{[A]_0} = -kt = \ln \frac{78.5 \text{ mmHg}}{312 \text{ mmHg}} = -k \times 3155 \text{ s} = -1.380 \qquad k = \frac{1.380}{3155 \text{ s}} = 4.37 \times 10^{-4} \text{ s}^{-1}$$

The virtual constancy of the rate constant throughout the time of the reaction confirms that the reaction is first order.

(b) The value of the rate constant was determined in part (a) as $4.3 \times 10^{-4} \text{ s}^{-1}$

(c) At 390 s, the pressure of dimethyl ether has dropped to 264 mmHg. Thus, an amount of dimethyl ether equivalent to a pressure of (312 mmHg – 264 mmHg =) 48 mmHg has decomposed. For each 1 mmHg pressure of dimethyl ether that decomposes, 3 mmHg of pressure of the products is produced. Thus, the increase in the pressure of the products is $3 \times 48 = 144$ mmHg. The total pressure at this point is 264 mmHg + 144 mmHg = 408 mmHg. Done in a more systematic fashion:

	$(CH_3)_2O(g)$	$\longrightarrow$	$CH_4(g)$ +	$H_2(g)$ +	$CO(g)$
Initial	312 mmHg		0 mmHg	0 mmHg	0 mmHg
Changes	–48 mmHg		+48 mmHg	+48 mmHg	+48 mmHg
Final	264 mmHg		48 mmHg	48 mmHg	48 mmHg

$$P_{\text{total}} = P_{\text{DME}} + P_{\text{methane}} + P_{\text{hydrogen}} + P_{\text{CO}}$$
$$= 264 \text{ mmHg} + 48 \text{ mmHg} + 48 \text{ mmHg} + 48 \text{ mmHg} = 408 \text{ mmHg}$$

(d) In the same manner as we solved part (c) we have the following.

	$(CH_3)_2O(g)$	$\longrightarrow$	$CH_4(g)$ +	$H_2(g)$ +	$CO(g)$
Initial	312 mmHg		0 mmHg	0 mmHg	0 mmHg
Changes	–312 mmHg		+312 mmHg	+312 mmHg	+312 mmHg
Final	0 mmHg		312 mmHg	312 mmHg	312 mmHg

$$P_{\text{total}} = P_{\text{DME}} + P_{\text{methane}} + P_{\text{hydrogen}} + P_{\text{CO}}$$
$$= 0 \text{ mmHg} + 312 \text{ mmHg} + 312 \text{ mmHg} + 312 \text{ mmHg} = 936 \text{ mmHg}$$

(e) We first determine P_{DME} at 1000 s. $\quad \ln \dfrac{P_{1000}}{P_0} = -kt = -4.3 \times 10^{-4} \text{ s}^{-1} \times 1000 \text{ s} = -0.43$

$$\frac{P_{1000}}{P_0} = e^{-0.43} = 0.65 \qquad\qquad P_{1000} = 312 \text{ mmHg} \times 0.65 = 203 \text{ mmHg}$$

Then we use the technique of parts (c) and (d)

	$(CH_3)_2O(g)$	$\longrightarrow$	$CH_4(g)$ +	$H_2(g)$ +	$CO(g)$
Initial	312 mmHg		0 mmHg	0 mmHg	0 mmHg
Changes	−109 mmHg		+109 mmHg	+109 mmHg	+109 mmHg
Final	203 mmHg		109 mmHg	109 mmHg	109 mmHg

$P_{total} = P_{DME} + P_{methane} + P_{hydrogen} + P_{CO}$

$\qquad = 203 \text{ mmHg} + 109 \text{ mmHg} + 109 \text{ mmHg} + 109 \text{ mmHg} = 530. \text{ mmHg}$

Second-Order Reactions

45. (a) We can either graph $1/[C_4H_6]$ vs. time and obtain a straight line, or we can determine the second-order rate constant from several data points. Then, if k indeed is a constant, the reaction is demonstrated to be second order. We shall use the second technique in this case. First we do a bit of algebra.

$$\frac{1}{[A]_t} - \frac{1}{[A]_0} = kt \qquad \frac{1}{t}\left(\frac{1}{[A]_t} - \frac{1}{[A]_0}\right) = k$$

$$k = \frac{1}{12.18 \text{ min}}\left(\frac{1}{0.0144 \text{ M}} - \frac{1}{0.0169 \text{ M}}\right) = 0.843 \text{ L mol}^{-1} \text{ min}^{-1}$$

$$k = \frac{1}{24.55 \text{ min}}\left(\frac{1}{0.0124 \text{ M}} - \frac{1}{0.0169 \text{ M}}\right) = 0.875 \text{ L mol}^{-1} \text{ min}^{-1}$$

$$k = \frac{1}{42.50 \text{ min}}\left(\frac{1}{0.0103 \text{ M}} - \frac{1}{0.0169 \text{ M}}\right) = 0.892 \text{ L mol}^{-1} \text{ min}^{-1}$$

$$k = \frac{1}{68.05 \text{ min}}\left(\frac{1}{0.0085 \text{ M}} - \frac{1}{0.0169 \text{ M}}\right) = 0.859 \text{ L mol}^{-1} \text{ min}^{-1}$$

The nearly constant value of the rate constant indicates that the reaction is second order.

(b) We use the same equation as in part (a), but solved for t, rather than k.

$$t = \frac{1}{k}\left(\frac{1}{[A]_t} - \frac{1}{[A]_0}\right) = \frac{1}{0.867 \text{ L mol}^{-1} \text{ min}^{-1}}\left(\frac{1}{0.0050 \text{ M}} - \frac{1}{0.0169 \text{ M}}\right) = 1.6 \times 10^2 \text{ min}$$

Establishing the Order of a Reaction

46. (a) $\text{Initial rate} = -\dfrac{\Delta[A]}{\Delta t} = -\dfrac{1.490 \text{ M} - 1.512 \text{ M}}{1.0 \text{ min} - 0.0 \text{ min}} = +0.022 \text{ M/min}$

$\qquad \text{Initial rate} = -\dfrac{\Delta[A]}{\Delta t} = -\dfrac{2.935 \text{ M} - 3.024 \text{ M}}{1.0 \text{ min} - 0.0 \text{ min}} = +0.089 \text{ M/min}$

(b) When the initial concentration is doubled ($\times 2.0$), from 1.512 M to 3.024 M, the initial rate quadruples ($\times 4.0$). Thus, the reaction is second order, since $2.0^x = 4.0$ when $x = 2$.

47. (a) Let us assess the possibilities. If the reaction is zero order, its rate will be constant. During the first 8 min, the rate is $-(0.60 \text{ M} - 0.80 \text{ M})/8 \text{ min} = 0.025 \text{ M/min}$. Then, during the first 24 min, the rate is $-(0.35 \text{ M} - 0.80 \text{ M})/24 \text{ min} = 0.019 \text{ M/min}$. Thus, the reaction is not zero order.

  If the reaction is first order, it will have a constant value of its rate constant, which is related to its constant half-life. The half-life can be assessed from the fact that 40 min elapse while the concentration drops from 0.80 M to 0.20 M, that is, to one-fourth of its initial value. Thus, 40 min equals two half-lives and $t_{1/2} = 20$ min. This gives $k = 0.693/t_{1/2} = 0.693/20 \text{ min} = 0.035 \text{ min}^{-1}$. Also

$$k t = -\ln\frac{[A]_t}{[A]_0} = -\ln\frac{0.35 \text{ M}}{0.80 \text{ M}} = 0.827 = k \times 24 \text{ min} \qquad k = \frac{0.827}{24 \text{ min}} = 0.034 \text{ min}^{-1}$$

The constancy of the value of k indicates that the reaction is first order.

(b) The value of rate constant is $k = 0.034 \text{ min}^{-1}$.

(c) $\text{Reaction rate} = \frac{1}{2}$ (rate of formation of B) $= k[A]^{-1}$ First we need [A] at $t = 30.$ min

$$\ln\frac{[A]}{[A]_0} = -kt = -0.034 \text{ min}^{-1} \times 30. \text{ min} = -1.0_2 \qquad \frac{[A]}{[A]_0} = e^{-1.02} = 0.36$$

$$[A] = 0.36 \times 0.80 \text{ M} = 0.29 \text{ M}$$

$$\text{rate of formation of B} = 2 \times 0.034 \text{ min}^{-1} \times 0.29 \text{ M} = 2.0 \times 10^{-2} \text{ M min}^{-1}$$

48. The data appear to indicate a first-order reaction with a constant half-life. Notice that from $t = 0$ s with $[N_2O_5] = 1.46$ M to $t = 1116$ s with $[N_2O_5] = 0.72$ M, the $[N_2O_5]$ is approximately halved and the elapsed

time in 1116 s. From $t = 1116$ s with $[N_2O_5] = 0.72$ M to $t = 2343$ s with $[N_2O_5] = 0.35$ M, the $[N_2O_5]$ is approximately halved and the elapsed time is 1227 s. And from $t = 423$ s with $[N_2O_5] = 1.09$ M to $t = 1582$ s with $[N_2O_5] = 0.54$ M, the $[N_2O_5]$ is approximately halved and the elapsed time is 1159 s. The half life is approximately constant. To make sure, we determine the value of k from several sets of data and see that it is indeed constant.
$$\ln\frac{[A]_t}{[A]_0} = -kt \qquad k = -\frac{1}{t}\ln\frac{[A]_t}{[A]_0} = -\frac{1}{423\ s}\ln\frac{1.09\ M}{1.46\ M} = 6.91 \times 10^{-4}\ s^{-1}$$

$$k = -\frac{1}{1582\ s}\ln\frac{0.54\ M}{1.46\ M} = 6.29 \times 10^{-4}\ s^{-1} \qquad k = -\frac{1}{1116\ s}\ln\frac{0.72\ M}{1.46\ M} = 6.33 \times 10^{-4}\ s^{-1}$$

$$k = -\frac{1}{753\ s}\ln\frac{0.89\ M}{1.46\ M} = 6.57 \times 10^{-4}\ s^{-1} \qquad k = -\frac{1}{1986\ s}\ln\frac{0.43\ M}{1.46\ M} = 6.16 \times 10^{-4}\ s^{-1}$$

The approximate constancy of the rate constant over a wide range of elapsed times indicates that the reaction indeed is first order, with a rate constant of $6.5 \times 10^{-4}\ s^{-1}$ (the average of the 5 values we have computed).

49. (a) In the first 22 s, [A] decreases from 0.715 M to 0.605 M, that is, $\Delta[A] = -0.110$ M. The rate in these 22 s is then determined. Rate $= \dfrac{-\Delta[A]}{\Delta t} = \dfrac{0.110\ M}{22\ s} = 5.0 \times 10^{-3}$ M/s In the first 74 s, $\Delta[A] = 0.345\ M - 0.715\ M$, and the rate is determined. Rate $= \dfrac{-\Delta[A]}{\Delta t} = \dfrac{0.370\ M}{74\ s} = 5.0 \times 10^{-3}$ M/s

Finally, in the first 132 s, $\Delta[A] = 0.055\ M - 0.715\ M$, and the rate is determined as follows. Rate $= \dfrac{-\Delta[A]}{\Delta t} = \dfrac{0.660\ M}{132\ s} = 5.0 \times 10^{-3}$ M/s. Since the rate is constant for this reaction, it must be zero order.

(b) The half-life of this reaction is the time needed for one half of the initial [A] to react. Thus, $\Delta[A] = 0.715\ M \div 2 = 0.358$ M and $\qquad t_{1/2} = \dfrac{0.358\ M}{5.0 \times 10^{-3}\ M/s} = 72$ s

50. The half-life of the reaction depends on the concentration of "A" and thus this reaction cannot be first order. For a second-order reaction, the half-life varies inversely with the rate: $t_{1/2} = 1/(k[A]_0)$ or $k = 1/(t_{1/2}[A]_0)$. Let us attempt to verify the second-order nature of this reaction by seeing if the rate constant is fixed.

$$k = \frac{1}{1.00\ M \times 50\ min} = 0.020\ L\ mol^{-1}\ min^{-1} \qquad k = \frac{1}{2.00\ M \times 25\ min} = 0.020\ L\ mol^{-1}\ min^{-1}$$

$$k = \frac{1}{0.50\ M \times 100\ min} = 0.020\ L\ mol^{-1}\ min^{-1}$$

The constancy of the rate constant demonstrates that this reaction indeed is second order. The rate equation is Rate $= k[A]^2$ and $k = 0.020\ L\ mol^{-1}\ min^{-1}$.

51. (a) The half-life depends on the initial $[NH_3]$ and thus the reaction cannot be first order. Let us attempt to verify second-order kinetics.

$$k = \frac{1}{[NH_3]_0\, t_{1/2}} \text{ for a second-order reaction} \qquad k = \frac{1}{0.0031\ M \times 7.6\ min} = 42\ L\ mol^{-1}\ min^{-1}$$

$$k = \frac{1}{0.0015\ M \times 3.7\ min} = 180\ L\ mol^{-1}\ min^{-1} \qquad k = \frac{1}{0.00068\ M \times 1.7\ min} = 865\ L\ mol^{-1}\ min^{-1}$$

The reaction is not second order. But, if the reaction is zero order, its rate will be constant.

$$\text{Rate} = \frac{[A]_0/2}{t_{1/2}} = \frac{0.0031\ M \div 2}{7.6\ min} = 2.0 \times 10^{-4}\ M/min \qquad \text{Rate} = \frac{0.0015\ M \div 2}{3.7\ min} = 2.0 \times 10^{-4}\ M/min$$

$$\text{Rate} = \frac{0.00068\ M \div 2}{1.7\ min} = 2.0 \times 10^{-4}\ M/min \qquad \text{zero-order reaction}$$

(b) The constancy of the rate indicates that the decomposition of ammonia under these conditions is zero order, and the rate constant is $k = 2.0 \times 10^{-4}$ M/min.

Collision Theory; Activation Energy

52. (a) The rate of a reaction depends on at least two factors other than the frequency of collisions. The first of these is whether each collision possesses sufficient energy to lead to a successful reaction. This depends on the activation energy of the reaction; the higher it is, the smaller will be the fraction of successful collisions. The second factor is whether the molecules in a given collision are oriented for a successful reaction. The more complicated the molecules are, the smaller the fraction of collisions that will be successfuly oriented.

(b) Although the collision frequency increases relatively slowly with temperature, the fraction of those collisions that have sufficient energy to overcome the activation energy increases much more rapidly. Therefore, the rate of reaction will increase dramatically with temperature.

(c) The addition of a catalyst has the net effect of decreasing the activation energy of the reaction, often by enabling another mechanism. The decreased activation energy means that a larger fraction of molecules has sufficient energy to lead to successful reaction. Thus the rate increases, even though the temperature does not.

53. (a) The products are 21 kJ/mol closer in energy to the constant energy activated complex than are the reactants. Thus, the activation energy for the reverse reaction is

$$84 \text{ kJ/mol} - 21 \text{ kJ/mol} = 63 \text{ kJ/mol}.$$

(b) The reaction profile for this reaction is sketched at right.

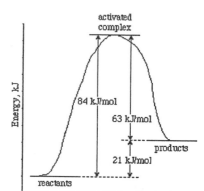

54. In an endothermic reaction, sketched at near right, we see that the value of the activation energy must be larger than the value of the enthalpy change for the reaction, ΔH. On the other hand, for an exothermic reaction (far right) there is no such requirement; we may have a small activation energy leading to a large value of ΔH.

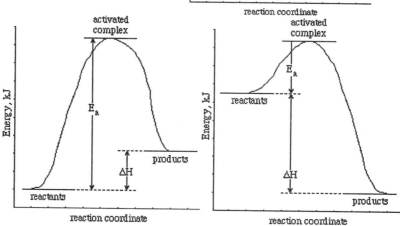

55. (a) The activation energy for the reaction of hydrogen with oxygen is quite high, too high to be supplied by the energy ordinarily available in a mixture of the two gases. However, the spark supplies a suitably concentrated form of energy to initiate the reaction of at least a few molecules. Since the reaction is highly exothermic, the reaction of these first few molecules supplies sufficient energy for yet other molecules to react and the reaction proceeds to completion.

(b) A larger spark simply means that a larger number of molecules reacts initially. But the eventual course of the reaction remains the same, with the initial reaction producing enough energy to initiate still more molecules and so on.

Effect of Temperature on Rates of Reaction

56. (a) First we need to compute values of ln k and $1/T$. Then we plot the graph.

t, °C	3°C	13°C	24°C	33°C
T, K	276 K	286 K	297 K	306 K
$1/T$, K^{-1}	0.00362	0.00350	0.00337	0.00327
k, M^{-1} s^{-1}	0.0014	0.0029	0.0062	0.0120
ln k	-6.57	-5.84	-5.08	-4.423

(b) The slope $= -E_a/R$.

$$E_a = -R \times \text{slope} = -8.3145 \frac{\text{J}}{\text{mol K}} \times \frac{-4.42 - (-6.57)}{0.00327 \text{ K}^{-1} - 0.00362 \text{ K}^{-1}} = 51 \text{ kJ/mol}$$

(c) We first apply the Arrhenius equation, with $k = 0.0120$ L mol^{-1} s^{-1} at 33°C (306 K), $k = ?$ at 40°C (313 K), and $E_a = 51 \times 10^3$ J/mol.

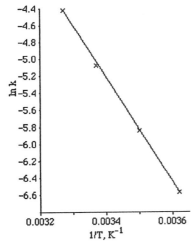

$$\ln\frac{k}{0.0120\ \text{L mol}^{-1}\ \text{s}^{-1}} = \frac{E_a}{R}\left(\frac{1}{T_1} - \frac{1}{T_2}\right) = \frac{51\times10^3\ \text{J/mol}}{8.3145\ \text{J mol}^{-1}\ \text{K}^{-1}}\left(\frac{1}{306\ \text{K}} - \frac{1}{313\ \text{K}}\right) = 0.45$$

$$e^{0.45} = 1.57 = \frac{k}{0.0120\ \text{L mol}^{-1}\ \text{s}^{-1}} \qquad k = 1.57\times0.0120\ \text{L mol}^{-1}\ \text{s}^{-1} = 0.0188\ \text{L mol}^{-1}\ \text{s}^{-1}$$

(d) We apply the Arrhenius equation, with $k = 0.0120$ L mol^{-1} s^{-1} at 33°C (306 K), $k = ?$ at 50°C (323 K), and $E_a = 51\times10^3$ J/mol, to first determine the rate constant at 50°C.

$$\ln\frac{k}{0.0120\ \text{L mol}^{-1}\ \text{s}^{-1}} = \frac{E_a}{R}\left(\frac{1}{T_1} - \frac{1}{T_2}\right) = \frac{51\times10^3\ \text{J/mol}}{8.3145\ \text{J mol}^{-1}\ \text{K}^{-1}}\left(\frac{1}{306\ \text{K}} - \frac{1}{323\ \text{K}}\right) = 1.1$$

$$e^{1.1} = 3.0 = \frac{k}{0.0120\ \text{L mol}^{-1}\ \text{s}^{-1}} \qquad k = 3.0\times0.0120\ \text{L mol}^{-1}\ \text{s}^{-1} = 0.036\ \text{L mol}^{-1}\ \text{s}^{-1}$$

Then we use the initial concentrations and the rate law determined in Example 15-3 to determine the initial rate.

Rate = $k\ [\text{S}_2\text{O}_8{}^{2-}]^1\ [\text{I}^-]^1 = 0.036$ L mol^{-1} s$^{-1} \times 0.060$ M $\times 0.120$ M $= 2.6\times10^{-4}$ M s^{-1}

57. The half-life of a first-order reaction is inversely proportional to its rate constant: $k = 0.693/t_{1/2}$. Thus we can apply a modified version of the Arrhenius equation (15.19).

(a) $\ln\dfrac{k_2}{k_1} = \ln\dfrac{(t_{1/2})_1}{(t_{1/2})_2} = \dfrac{E_a}{R}\left(\dfrac{1}{T_1} - \dfrac{1}{T_2}\right) = \ln\dfrac{46.2\ \text{min}}{2.6\ \text{min}} = \dfrac{E_a}{R}\left(\dfrac{1}{298\ \text{K}} - \dfrac{1}{(102+273)\ \text{K}}\right)$

$2.88 = \dfrac{E_a}{R}\ 6.89\times10^{-4} \qquad E_a = \dfrac{2.88\times8.3145}{6.89\times10^{-4}}\times\dfrac{1\ \text{kJ}}{1000\ \text{J}} = 34.8$ kJ/mol

(b) $\ln\dfrac{10.0\ \text{min}}{46.2\ \text{min}} = \dfrac{34.8\times10^3\ \text{J/mol}}{8.3145\ \text{J mol}^{-1}\ \text{K}^{-1}}\left(\dfrac{1}{T} - \dfrac{1}{298}\right) = -1.530 = 4.19\times10^3\left(\dfrac{1}{T} - \dfrac{1}{298}\right)$

$\left(\dfrac{1}{T} - \dfrac{1}{298}\right) = \dfrac{-1.530}{4.19\times10^3} = -3.65\times10^{-4} \qquad \dfrac{1}{T} = 2.99\times10^{-3} \qquad T = 334\ \text{K} = 61°\text{C}$

58. (a) We plot $\ln k$ vs. $1/T$. The slope of the straight line equals $-E_a/R$. First we tabulate the data to plot.

t	15.83	32.02	59.75	90.61	°C
T	288.98	305.17	332.90	363.76	K
$1/T$	0.0034604	0.0032769	0.0030039	0.0027491	K^{-1}
k	5.03×10^{-5}	3.69×10^{-4}	6.71×10^{-3}	0.119	M^{-1} s^{-1}
$\ln k$	-9.898	-7.904	-5.004	-2.129	

The slope of this graph $= -1.089\times10^4$ K $= -E_a/R$ The graph is on the next page.

$E_a = -(-1.089\times10^4\ \text{K})\times8.3145\ \text{J mol}^{-1}\ \text{K}^{-1} = 9.055\times10^4$ J/mol $= 90.55$ kJ/mol

(b) We calculate the activation energy with the Arrhenius equation from the two extreme data points.

$$\ln\frac{k_2}{k_1} = \ln\frac{0.119}{5.03\times10^{-5}} = +7.769 = \frac{E_a}{R}\left(\frac{1}{T_1} - \frac{1}{T_2}\right) = \frac{E_a}{R}\left(\frac{1}{288.98\ \text{K}} - \frac{1}{363.76\ \text{K}}\right)$$

$$= 7.1133\times10^{-4}\ \text{K}^{-1}\ \frac{E_a}{R} \qquad E_a = \frac{7.769\times8.3145\ \text{J mol}^{-1}\ \text{K}^{-1}}{7.1133\times10^{-4}\ \text{K}^{-1}} = 9.081\times10^4$$ J/mol

$E_a = 90.81$ kJ/mol   The values are in quite good agreement, with a 0.3% deviation.

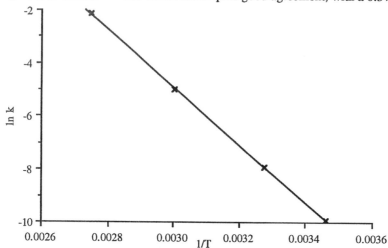

59. (a) It is the change in the value of the rate constant that causes the reaction to go faster. Let k_1 be the rate constant at room temperature, 20°C or 293 K. Then $k_2 = 2\, k_1$ is the rate constant ten degrees higher, at 30°C or 303 K.

$$\ln \frac{k_2}{k_1} = \ln \frac{2\, k_1}{k_1} = 0.693 = \frac{E_a}{R}\left(\frac{1}{T_1} - \frac{1}{T_2}\right) = \frac{E_a}{R}\left(\frac{1}{293} - \frac{1}{303\ \text{K}}\right) = 1.13 \times 10^{-4}\ \text{K}^{-1}\, \frac{E_a}{R}$$

$$E_a = \frac{0.693 \times 8.3145\ \text{J mol}^{-1}\ \text{K}^{-1}}{1.13 \times 10^{-4}\ \text{K}^{-1}} = 5.1 \times 10^4\ \text{J/mol} = 51\ \text{kJ/mol}$$

(b) Since the activation energy for the formation of HI from its elements is 171 kJ/mol, we would not expect this reaction to follow the rule of thumb.

60. Under a pressure of 2.00 atm the boiling point of water is approximately 121°C or 394 K. Under a pressure of 1 atm, the boiling point of water is 100°C or 373 K. We assume an activation energy of 5.1×10^4 J/mol and compute the ratio of the two rates.

$$\ln \frac{\text{Rate}_2}{\text{Rate}_1} = \frac{E_a}{R}\left(\frac{1}{T_1} - \frac{1}{T_2}\right) = \frac{5.1 \times 10^4\ \text{J/mol}}{8.3145\ \text{J mol}^{-1}\ \text{K}^{-1}}\left(\frac{1}{373} - \frac{1}{394\ \text{K}}\right) = 0.877$$

$\text{Rate}_2 = e^{0.877}\, \text{Rate}_1 = 2.73\, \text{Rate}_1$

Cooking will occur 2.73 times faster in the pressure cooker.

61. (a) The time required for the fixed (c) process of souring is three times as long at 20°C room temperature (293 K) as at 3°C refrigerator temperature (276 K).

$$\ln \frac{c/t_2}{c/t_1} = \ln \frac{t_1}{t_2} = \ln \frac{64\ \text{h}}{3 \times 64\ \text{h}} = -1.10 = \frac{E_a}{R}\left(\frac{1}{T_1} - \frac{1}{T_2}\right) = \frac{E_a}{R}\left(\frac{1}{293\ \text{K}} - \frac{1}{276\ \text{K}}\right) = \frac{E_a}{R}\, (-2.10 \times 10^{-4})$$

$$E_a = \frac{1.10\, R}{2.10 \times 10^{-4}\ \text{K}^{-1}} = \frac{1.10 \times 8.3145\ \text{J mol}^{-1}\ \text{K}^{-1}}{2.10 \times 10^{-4}\ \text{K}^{-1}} = 4.4 \times 10^4\ \text{J/mol} = 44\ \text{kJ/mol}$$

(b) We use the activation energy determined in part (a) to calculate the souring time at 40°C = 313 K.

$$\ln \frac{t_1}{t_2} = \frac{E_a}{R}\left(\frac{1}{T_1} - \frac{1}{T_2}\right) = \frac{4.4 \times 10^4\ \text{J/mol}}{8.3145\ \text{J mol}^{-1}\ \text{K}^{-1}}\left(\frac{1}{313\ \text{K}} - \frac{1}{293\ \text{K}}\right) = -1.15 = \ln \frac{t_1}{64\ \text{h}}$$

$$\frac{t_1}{64\ \text{h}} = e^{-1.15} = 0.317 \qquad t_1 = 0.317 \times 64\ \text{h} = 20.\ \text{h}$$

Catalysis

62. (a) Although a catalyst is *recovered unchanged from a reaction*, it does "take part in the reaction." And some catalysis actually slow down the rate of a reaction. Usually, however, these negative catalysts are called inhibitors.

(b) The function of a catalyst is to *change the mechanism of a reaction*. The new mechanism is one that has a different (lower) activation energy from the original reaction.

63. Both platinum and an enzyme can be considered heterogeneous catalysts in the sense that the catalytic activity occurs on their surfaces or near them. Even so, some would classify an enzyme as a homogeneous catalyst. The most important difference, however, is one of specificity. Platinum is a rather nonspecific catalyst, catalyzing many different reactions. An enzyme, however, is quite specific, usually catalyzing only one reaction rather than all reactions of a given class.

64. In both the enzyme and the metal surface cases, the reaction occurs in a specialized location: either with the enzyme or on the surface of the catalyst. At high concentrations of reactant, the limiting factor in determining the rate is not the concentration of reactant present but how rapidly active sites become available for reaction to occur. Thus, the rate of the reaction depends on either the quantity of enzyme present or the surface area of the catalyst, rather than on how much reactant is present; the reaction is zero order. At low concentrations or gas pressures the reaction rate depends on how rapidly molecules can reach the available active sites. Thus, the rate depends on concentration or pressure of reactant and is first order.

65. If the reaction is first order, its half-life is 100 min, for in this time period [S] decreases from 1.00 M to 0.50 M, that is, by one half. This gives a rate constant of $k = 0.693/t_{1/2} = 0.693/100$ min $= 0.00693$ min^{-1}. The rate constant also can be determined from any two of the other sets of data.

$$k\,t = \ln\frac{[A]_0}{[A]_t} = \ln\frac{1.00\text{ M}}{0.70\text{ M}} = 0.357 = k \times 60\text{ min} \qquad k = \frac{0.357}{60\text{ min}} = 0.00595\text{ min}^{-1}$$

This is not a particularly constant rate constant. Let's try zero order, where the rate should be constant.

$$\text{Rate} = -\frac{0.90\text{ M} - 1.00\text{ M}}{20\text{ min}} = 0.0050\text{ M/min} \qquad \text{Rate} = -\frac{0.50\text{ M} - 1.00\text{ M}}{100\text{ min}} = 0.0050\text{ M/min}$$

$$\text{Rate} = -\frac{0.20\text{ M} - 0.90\text{ M}}{160\text{ min} - 20\text{ min}} = 0.0050\text{ M/min} \qquad \text{Rate} = -\frac{0.50\text{ M} - 0.90\text{ M}}{100\text{ min} - 20\text{ min}} = 0.0050\text{ M/min}$$

Thus, this reaction is zero order with respect to [S].

Reaction Mechanisms

66. The molecularity of an elementary process is the number of reactant molecules in that process. This molecularity is equal to the order of the overall reaction only if the elementary process in question is the slowest and thus, the rate-determining, step of the overall reaction. In addition, the elementary process in question should be the only elementary step that influences the rate of the reaction.

67. When we use the term "bottleneck," we are referring to a situation or a location which holds up a larger process. For instance, a construction area on a high-speed highway often will become a bottleneck, reducing the speed of traffic on that highway to the speed with which vehicles can pass the construction area. In similar fashion, a rate-determining step slows down an entire reaction so that it proceeds no faster than molecules can react in the rate-determining step.

68. If the type of molecule that is expressed in the rate law as being first order collides with other molecules that are present in much larger concentrations, then the reaction will seem to depend only on the concentration of those types of molecules present in smaller concentration, since the larger concentration will be essentially unchanged during the course of the reaction. Such a situation is quite common, and has been given the name pseudo first order. Another situation is that molecules which do not participate directly in the reaction—even including product molecules—strike the reactant molecules and impart to them sufficient energy to react.

69. The three elementary steps must sum to the overall reaction. All species in the equations below are gases.

overall: $2\text{ NO} + 2\text{ H}_2 \longrightarrow \text{N}_2 + 2\text{ H}_2\text{O}$ $2\text{ NO} + 2\text{ H}_2 \longrightarrow \text{N}_2 + 2\text{ H}_2\text{O}$

–first: $-(2\text{ NO} \rightleftharpoons \text{N}_2\text{O}_2)$ $\text{N}_2\text{O}_2 \rightleftharpoons 2\text{ NO}$

–third $-(\text{N}_2\text{O} + \text{H}_2 \longrightarrow \text{N}_2 + \text{H}_2\text{O})$ *or* $\text{N}_2 + \text{H}_2\text{O} \longrightarrow \text{N}_2\text{O} + \text{H}_2$

the result is the second step, which is slow: $\text{H}_2 + \text{N}_2\text{O}_2 \longrightarrow \text{H}_2\text{O} + \text{N}_2\text{O}$

The rate of this rate-determining step is: Rate $= k_2\,[\text{H}_2]\,[\text{N}_2\text{O}_2]$

Since N_2O_2 does not appear in the overall reaction, we need to replace its concentration with the concentrations of species that do appear in the overall reaction. To do this, recall that the first step is rapid, with the forward reaction occurring at the same rate as the reverse reaction.

$k_1[\text{NO}]^2 =$ forward rate = reverse rate $= k_{-1}[\text{N}_2\text{O}_2]$

This expression is solved for $[\text{N}_2\text{O}_2]$, which then is substituted into the rate equation for the overall reaction.

$$[N_2O_2] = \frac{k_1\,[NO]^2}{k_{-1}} \qquad\qquad Rate = \frac{k_2\,k_1}{k_{-1}}\,[H_2]\,[NO]^2$$

This rate law is first-order in $[H_2]$ and second-order in $[NO]$, in agreement with the experimental determination.

70. A possible mechanism is the following.

$$O_3 \underset{k_2}{\overset{k_1}{\rightleftharpoons}} O_2 + O \quad \text{(fast)} \qquad\qquad O + O_3 \xrightarrow{\;k_3\;} 2\,O_2 \quad \text{(slow)}$$

The rate is that of the slow step. $\qquad$ Rate $= k_3[O][O_3] \qquad$ But O is a reaction intermediate, whose concentration is difficult to determine. An expression for [O] can be found by assuming that the forward and reverse "fast" steps proceed equally rapidly.

$$Rate_1 = Rate_2 \qquad k_1[O_3] = k_2[O_2][O] \qquad [O] = \frac{k_1[O_3]}{k_2[O_2]}$$

Then substitute this expression into the rate law for the reaction.

$$Rate = k_3\,\frac{k_1[O_3]}{k_2[O_2]}\,[O_3] = \frac{k_3\,k_1}{k_2}\,\frac{[O_3]^2}{[O_2]}$$

This rate equation has the same form as the experimentally determined rate law.

16 PRINCIPLES OF CHEMICAL EQUILIBRIUM

REVIEW QUESTIONS

1. **(a)** K_p is the symbol for the equilibrium constant in terms of partial pressures: the quotient of the equilibrium partial pressures of the products divided by those of the reactants, each pressure (in atmospheres) raised to a power equal to its stoichiometric coefficient.

   (b) Q_c is the symbol for the reaction quotient in terms of molarities: the quotient of the molarities of the products divided by those of the reactants, each concentration raised to a power equal to its stoichiometric coefficient. These concentrations need not be those at equilibrium.

   (c) Δn_{gas} is the difference between the sum of the stoichiometric coefficients of the gaseous products and the similar sum for gaseous reactants, using the coefficients from the balanced chemical equation.

2. **(a)** A dynamic equilibrium is a balanced situation that is created by opposing processes proceeding at such rates that the net result is no apparent change, although individual molecules may indeed be undergoing change, and probably are.

   (b) The direction of net reaction describes whether a chemical reaction produces more product (to the right) of forms more reactant (to the left), generally as the result of the alteration of reaction conditions.

   (c) Le Châtelier's principle states that if a stress is applied to a system that is at equilibrium, the system will react in such a way as to relieve the stress and attain a new equilibrium position.

   (d) The effect of a catalyst on the position of equilibrium is nonexistent. A catalyst does, however, speed up the rate at which equilibrium is established.

3. **(a)** A reaction that goes to completion is one that reacts until the amount of one of the reactants is gone. A reversible reaction is one that comes to a point of balance before all reactants are depleted.

   (b) The equilibrium constant K_c contains the equilibrium concentrations of products and reactants, while the K_p expression contains their partial pressures. The values of these two numbers are related by the expression $K_p = K_c(RT)^{\Delta n}$.

   (c) The reaction quotient, Q_c, and the equilibrium constant expression, K_c, have the same functional form, with the concentrations of products over those of reactants, each raised to a power equal to that species's stoichiometric coefficient. However, the concentrations that are used differ. In the K_c expression, these concentrations are equilibrium concentrations, while in the Q_c expression they are not necessarily the concentrations at equilibrium—often they are the initial concentrations. Similar statements can be made concerning Q_p and K_p, except that partial pressures replace concentrations.

   (d) A homogeneous equilibrium is one that takes place in a solution, either entirely gas or entirely liquid. A heterogeneous equilibrium is one that is established when two phase of matter are present, generally a solid with either a liquid or a gaseous solution.

4. **(a)** $K_c = \dfrac{[NO_2]^2}{[NO]^2[O_2]}$ **(b)** $K_c = \dfrac{[Cu^{2+}]}{[Ag^+]^2}$ **(c)** $K_c = \dfrac{[OH^-]^2}{[CO_3^{2-}]}$

5. **(a)** $K_p = \dfrac{P\{CH_4\}P\{H_2S\}^2}{P\{CS_2\}P\{H_2\}^4}$ **(b)** $K_p = P\{O_2\}^{1/2}$ **(c)** $K_p = P\{CO_2\}\{H_2O\}$

6. In each case we write the equation for the formation reaction and then the equilibrium constant expression for that reaction.

   (a) $\frac{1}{2}\,N_2(g) + \frac{1}{2}\,O_2(g) \rightleftharpoons NO(g)$ $\qquad\qquad K_c = \dfrac{[NO]}{[N_2]^{1/2}[O_2]^{1/2}}$

(b) $\frac{1}{2} H_2(g) + \frac{1}{2} Cl_2(g) \rightleftharpoons HCl(g)$ $\qquad K_c = \dfrac{[HCl]}{[H_2]^{1/2}[Cl_2]^{1/2}}$

(c) $\frac{1}{2} N_2(g) + \frac{3}{2} H_2(g) \rightleftharpoons NH_3(g)$ $\qquad K_c = \dfrac{[NH_3]}{[N_2]^{1/2}[H_2]^{3/2}}$

(d) $\frac{1}{2} Cl_2(g) + \frac{3}{2} F_2(g) \rightleftharpoons ClF_3(g)$ $\qquad K_c = \dfrac{[ClF_3]}{[Cl_2]^{1/2}[F_2]^{3/2}}$

(e) $\frac{1}{2} N_2(g) + \frac{1}{2} O_2(g) + \frac{1}{2} Cl_2(g) \rightleftharpoons NOCl(g)$ $\qquad K_c = \dfrac{[NOCl]}{[N_2]^{1/2}[O_2]^{1/2}[Cl_2]^{1/2}}$

7. **(a)** This is the reverse of the first reaction; K_c is the inverse value. $\qquad K_c = \dfrac{[CO_2][H_2]}{[CO][H_2O]} = \dfrac{1}{23.2} = 0.0431$

(b) This is twice the second reaction; K_c is its square. $\qquad K_c = \dfrac{[SO_3]^2}{[SO_2]^2[O_2]} = (56)^2 = 3.1 \times 10^3$

(c) This is half of the reverse of the third reaction; K_c is the the inverse of the square root. $\qquad K_c = \dfrac{[H_2S]}{[H_2][S_2]^{1/2}} = \dfrac{1}{\sqrt{2.3 \times 10^{-4}}} = 66$

8. We combine one-half of the reversed first reaction with the second reaction to obtain the desired reaction.

$\frac{1}{2} N_2(g) + \frac{1}{2} O_2(g) \rightleftharpoons NO(g)$ $\qquad K_c = \dfrac{1}{\sqrt{2.4 \times 10^{30}}}$

$NO(g) + \frac{1}{2} Br_2(g) \rightleftharpoons NOBr(g)$ $\qquad K_c = 1.4$

net: $\frac{1}{2} N_2(g) + \frac{1}{2} O_2(g) + \frac{1}{2} Br_2(g) \rightleftharpoons NOBr(g)$ $\qquad K_c = \dfrac{1.4}{\sqrt{2.4 \times 10^{30}}} = 9.0 \times 10^{-16}$

9. Answer **(d)** is correct. 1 mol I_2 is the limiting reagent in this reaction. If the reaction were to go to completion, 2 mol HI would be produced. Thus, the only way that 2 mol HI could be produced (answer b) would be if K_c were infinite. 1 mol HI could be produced (answer **a**), but without knowing the value of K_c for this reaction, we cannot be sure. The only certain answer is that something less than 2 mol HI are produced, answer **(d)**.

10. Simply substitute into the K_c expression. $K_c = \dfrac{[C]^2}{[A][B]} = \dfrac{(0.36 \text{ M})^2}{0.47 \text{ M} \times 0.55 \text{ M}} = 0.50$

11. The formation of 1.0×10^{-11} mol S(g) does not signifcantly change the amount of $S_2(g)$ from its original 0.0010 mol. Thus, $K_c = \dfrac{[S]^2}{[S_2]} = \dfrac{\left(\dfrac{1.0 \times 10^{-11} \text{ mol}}{0.500 \text{ L}}\right)^2}{\dfrac{0.0010 \text{ mol}}{0.500 \text{ L}}} = 2.0 \times 10^{-19}$

12. **(a)** If the amounts of SO_2 and SO_3 are equal, then $[SO_2] = [SO_3]$.

$K_c = 66.0 = \dfrac{[SO_3]^2}{[SO_2]^2[O_2]} = \dfrac{[SO_2]^2}{[SO_2]^2[O_2]} = \dfrac{1}{[O_2]}$ $\qquad [O_2] = \dfrac{1}{66.0} = 0.0152$ M

no. mol O_2 = 0.0152 M × 1.18 L = 0.0180 mol O_2

(b) If there are twice as many moles of SO_3 in the flask as SO_2, then $[SO_3] = 2 \times [SO_2]$

$K_c = 66.0 = \dfrac{[SO_3]^2}{[SO_2]^2[O_2]} = \dfrac{2^2 [SO_2]^2}{[SO_2]^2[O_2]} = \dfrac{4}{[O_2]}$ $\qquad [O_2] = \dfrac{4}{66.0} = 0.0606$ M

no. mol O_2 = 0.0606 M × 1.18 L = 0.0715 mol O_2

13. In each case we use the relationship $K_p = K_c(RT)^{\Delta n_{gas}}$

(a) For $CO_2(g) + H_2(g) \rightleftharpoons CO(g) + H_2O(g)$ $\qquad K_p = K_c = 0.0431$

(b) For $2 SO_2(g) + O_2 \rightleftharpoons 2 SO_3(g)$ $\qquad K_p = K_c(RT)^{-1} = \dfrac{3.1 \times 10^3}{0.0821 \times 900K} = 42$

(c) For $H_2(g) + \frac{1}{2} S_2(g) \rightleftharpoons H_2S(g)$ $\qquad K_p = K_c(RT)^{-1/2} = \dfrac{66}{\sqrt{0.0821 \times 1405 \text{ K}}} = 6.1$

14. We first calculate the value of K_p for the reaction $2 NO_2(g) \rightleftharpoons 2 NO(g) + O_2(g)$ $K_c = 1.8 \times 10^{-6}$ at 184°C = 457 K. For this reaction $\Delta n_{gas} = 2 + 1 - 2 = +1$.

$K_p = K_c(RT)^{\Delta n_{gas}} = 1.8 \times 10^{-6} (0.08206 \text{ L atm mol}^{-1} \text{ K}^{-1} \times 457 \text{ K})^{+1} = 6.8 \times 10^{-5}$

To obtain the final reaction $NO(g) + \frac{1}{2}O_2(g) \rightleftharpoons NO_2(g)$ from the initial reaction, that initial reaction must be reversed and then divided by two. Thus, in order to determine the value of the equilibrium constant for the final reaction, the value of K_p for the initial reaction must be inverted, and the square root taken of the result. $K_{p,final} = \sqrt{\dfrac{1}{6.8 \times 10^{-5}}} = 1.2 \times 10^2$

15. (a) $K_c = \dfrac{[CO][H_2O]}{[CO_2][H_2]} = \dfrac{\dfrac{n\{CO\}}{V} \dfrac{n\{H_2O\}}{V}}{\dfrac{n\{CO_2\}}{V} \dfrac{n\{H_2\}}{V}} = \dfrac{n\{CO\}n\{H_2O\}}{n\{CO_2\}n\{H_2\}}$ Thus, K_c does not depend on the volume of

the container for this reaction. Note also that the value of K_p does not depend on volume.

(b) Note that $K_p = K_c$ for this reaction, since $\Delta n_{gas} = 0$.

$K_c = K_p = \dfrac{0.224 \text{ mol CO} \times 0.224 \text{ mol H}_2O}{0.276 \text{ mol CO}_2 \times 0.276 \text{ mol H}_2} = 0.659$

16. (a) $K_c = \dfrac{[PCl_5]}{[PCl_3][Cl_2]} = \dfrac{\dfrac{0.105 \text{ g PCl}_5}{2.50 \text{ L}} \times \dfrac{1 \text{ mol PCl}_5}{208.2 \text{ g}}}{\left(\dfrac{0.220 \text{ g PCl}_3}{2.50 \text{ L}} \times \dfrac{1 \text{ mol PCl}_3}{137.3 \text{ g}}\right) \times \left(\dfrac{2.12 \text{ g Cl}_2}{2.50 \text{ L}} \times \dfrac{1 \text{ mol Cl}_2}{70.9 \text{ g}}\right)} = 26.3$

(b) $K_p = K_c(RT)^{\Delta n} = 26.3 \,(0.08206 \text{ L atm mol}^{-1} \text{ K}^{-1} \times 523 \text{ K})^{-1} = 0.613$

17. (a) If 5% of the 1.00 mol I_2 is dissociated into atoms, then 0.05 mol I_2 has dissociated.

Reaction:	$I_2(g)$	$\rightleftharpoons$	$2\, I(g)$
Initial:	1.00 mol		0.00 mol
Changes:	−0.05 mol		+0.10 mol
Equil:	0.95 mol		0.10 mol

$K_c = \dfrac{[I]^2}{[I_2]} = \dfrac{\left(\dfrac{0.10 \text{ mol I}}{1.00 \text{ L}}\right)^2}{\dfrac{0.95 \text{ mol I}_2}{1.00 \text{ L}}} = 0.011$

(b) $K_p = K_c(RT)^{+1} = 0.011 \,(0.0821 \times 1473 \text{ K}) = 1.3$

18. (a) We determine the concentration of each species in the gaseous solution, use these concentrations to determine the value of the reaction quotient, and compare this value of Q_c with the value of K_c.

$[SO_2] = \dfrac{0.390 \text{ mol SO}_2}{1.90 \text{ L}} = 0.205 \text{ M}$ $[O_2] = \dfrac{0.156 \text{ mol O}_2}{1.90 \text{ L}} = 0.0821 \text{ M}$

$[SO_3] = \dfrac{0.657 \text{ mol SO}_3}{1.90 \text{ L}} = 0.346 \text{ M}$ $Q_c = \dfrac{[SO_3]^2}{[SO_2]^2[O_2]} = \dfrac{(0.346)^2}{(0.205)^2 0.0821} = 34.7$

Since $Q_c = 34.7 \neq 2.8 \times 10^2 = K_c$, this mixture is not at equilibrium.

(b) Since the value of Q_c is smaller than that of K_c, the reaction will proceed to the right, forming products in reaching equilibrium.

19. Increasing the volume of an equilibrium mixture causes that mixture to shift toward the side (reactants or products) where the sum of the stoichiometric coefficients of the gaseous species is the larger. That is: shifts to the right if $\Delta n_{gas} > 0$, shifts to the left if $\Delta n_{gas} < 0$, and does not shift if $\Delta n_{gas} = 0$

(a) $C(s) + H_2O(g) \rightleftharpoons CO(g) + H_2(g)$ $\Delta n_{gas} > 0$, shift right, toward products

(b) $CO(g) + H_2O(g) \rightleftharpoons CO_2(g) + H_2(g)$ $\Delta n_{gas} = 0$, no shift, no change in equilibrium position.

(c) $4\, HCl(g) + O_2(g) \rightleftharpoons 2\, H_2O(g) + 2\, Cl_2(g)$ $\Delta n_{gas} < 0$, shift left, toward reactants

20. The equilibrium position for a reaction that is exothermic shifts to the left (favors reactants) when temperature is raised. For one that is endothermic, it shifts right (favors products) when temperature is raised.

(a) $NO(g) \rightleftharpoons \frac{1}{2}N_2(g) + \frac{1}{2}O_2(g)$ $\Delta H° = -90.2 \text{ kJ}$ shifts left, % dissociation decreases

(b) $SO_3(g) \rightleftharpoons SO_2(g) + \frac{1}{2}O_2(g)$ $\Delta H° = +98.9 \text{ kJ}$ shifts right, % dissociation increases

(c) $N_2H_4(g) \rightleftharpoons N_2(g) + 2\, H_2(g)$ $\Delta H° = -95.4 \text{ kJ}$ shifts left, % dissociation decreases

(d) $COCl_2(g) \rightleftharpoons CO(g) + Cl_2(g)$ $\Delta H° = +108.3 \text{ kJ}$ shifts right, % dissociation increases

21. $4\, HCl(g) + O_2(g) \rightleftharpoons 2\, H_2O(g) + 2\, Cl_2(g)$ $\Delta H° = -114 \text{ kJ}$

(a) Adding $O_2(g)$ to the equilibrium mixture at constant volume will cause the position of equilibrium to shift to the right, increasing the equilibrium amount of $Cl_2(g)$.

(b) Removing $HCl(g)$ from the equilibrium mixture at constant volume will cause the position of equilibrium to shift to the left, decreasing the amount of $Cl_2(g)$.

(c) Transferring the equilibrium mixture to a container of twice the volume will cause the position of equilibrium to shift to the left, decreasing the amount of $Cl_2(g)$.

(d) Adding a catalyst to the equilibrium mixture will have no effect on the equilibrium position.

(e) Raising the temperature of this exothermic reaction will cause the equilibrium position to shift to the left, decreasing the amount of $Cl_2(g)$.

22. The information for the calculation is organized around the chemical equation. Let x = mol H_2 (or I_2) that reacts. Then use stoichiometry to determine the amount of HI formed, in terms of x, and finally solve for x.

Reaction:	$H_2(g)$	+	$I_2(g)$	$\rightleftharpoons$	$2\,HI(g)$
Initial:	0.100 mol		0.100 mol		0.000 mol
Changes:	$-x$ mol		$-x$ mol		$+2x$ mol
Equil:	$0.100 - x$		$0.100 - x$		$2x$

$$K_c = \frac{[HI]^2}{[H_2][I_2]} = \frac{\left(\dfrac{2x}{1.50\ L}\right)^2}{\dfrac{0.100-x}{1.50\ L} \times \dfrac{0.100-x}{1.50\ L}}$$

Then take the square root of both sides: $\quad \sqrt{K_c} = \sqrt{50.2} = \dfrac{2x}{0.100 - x} = 7.09$

$2x = 0.709 - 7.09\,x \qquad x = \dfrac{0.709}{9.09} = 0.0780$ mol $\quad$ amount HI $= 2x = 2 \times 0.0780$ mol $= 0.156$ mol HI

amount H_2 = amount I_2 = $(0.100 - x)$ mol = $(0.100 - 0.0780)$ mol = 0.022 mol H_2 or I_2

23. This calculation is set up in a manner similar to the calculation of Review Question 22. In this case, we let x be the amount of I_2 that is formed.

Reaction:	$H_2(g)$	+	$I_2(g)$	$\rightleftharpoons$	$2\,HI(g)$
Initial:	0.000 mol		0.000 mol		0.100 mol
Changes:	$+x$ mol		$+x$ mol		$-2x$ mol
Equil:	x		x		$0.100 - 2x$

$$K_c = 50.2 = \frac{[HI]^2}{[H_2][I_2]} = \frac{\left(\dfrac{0.100-2x}{1.50\ L}\right)^2}{\dfrac{x}{1.50\ L} \times \dfrac{x}{1.50\ L}} = \frac{(0.100-2x)^2}{x^2} \qquad \sqrt{50.2} = \frac{0.100-2x}{x} = 7.09$$

$0.100 - 2x = 7.09x \qquad\qquad x = \dfrac{0.100}{2 + 7.09} = 0.0110 \qquad$ amount $I_2 = x = 0.0110$ mol I_2

24. We first determine the initial pressure of NH_3.

$$P\{NH_3(g)\} = \frac{nRT}{V} = \frac{0.170 \text{ g } NH_3 \times \dfrac{1 \text{ mol } NH_3}{17.03 \text{ g } NH_3} \times 0.08206 \text{ L atm mol}^{-1}\text{ K}^{-1} \times 298 \text{ K}}{1.60 \text{ L}}$$

$$= 0.153 \text{ atm}$$

Reaction:	$NH_4HS(s)$	$\rightleftharpoons$	$NH_3(g)$	+	$H_2S(g)$
Initial:			0.153 atm		
Changes:			$+x$ atm		$+x$ atm
Equil:			$(0.153 + x)$ atm		x atm

$K_p = P\{NH_3\}P\{H_2S\} = 0.108 = (0.153 + x)x = 0.153\,x + x^2 \qquad 0 = x^2 + 0.153\,x - 0.108$

$$x = \frac{-b \pm \sqrt{b^2 - 4ac}}{2a} = \frac{-0.153 \pm \sqrt{0.0234 + 0.432}}{2} = 0.261 \text{ atm}, -0.414 \text{ atm}$$

The negative pressure makes no physical sense. Thus, the total gas pressure is obtained as follows.

$P_{tot} = P\{NH_3\} + P\{H_2S\} = (0.153 + x) + x = 0.153 + 2\,x = 0.153 + 2 \times 0.261 = 0.675$ atm

25. Again we base the set up of the problem around the balanced chemical equation.

	$Pb(s)$	+	$2\,Cr^{3+}(aq)$	$\rightleftharpoons$	$Pb^{2+}(aq)$	+	$2\,Cr^{2+}(aq)$
initial:			0.100 M		0 M		0 M
changes:			$-2x$ M		$+x$ M		$+2x$ M
equil:			$(0.100 - 2x)$M		xM		$2x$ M

$K_c = 3.2 \times 10^{-10}$

$$K_c = \frac{[Pb^{2+}][Cr^{2+}]^2}{[Cr^{3+}]^2} = \frac{x\,(2x)^2}{(0.100 - 2x)^2}$$

$K_c \approx \dfrac{x\,(2x)^2}{(0.100)^2} = 3.2 \times 10^{-10} \qquad\qquad 4x^3 = (0.100)^2 \times 3.2 \times 10^{-10} = 3.2 \times 10^{-12}$

$$x = \sqrt[3]{\frac{3.2 \times 10^{-12}}{4}} = 9.3 \times 10^{-5} \text{ M}$$

Our assumption, that $2x << 0.100$, is valid
and thus also is our result: $[\text{Pb}^{2+}] = x = 9.3 \times 10^{-5}$ M

EXERCISES

26. (a) $4 \text{ NH}_3(\text{g}) + 3 \text{ O}_2(\text{g}) \rightleftharpoons 2 \text{ N}_2(\text{g}) + 6 \text{ H}_2\text{O}(\text{g})$ $\quad K_c = \dfrac{[\text{N}_2]^2[\text{H}_2\text{O}]^6}{[\text{NH}_3]^4[\text{O}_2]^3}$

(b) $5 \text{ H}_2(\text{g}) + 2 \text{ NO}_2(\text{g}) \rightleftharpoons 2 \text{ NH}_3(\text{g}) + 4 \text{ H}_2\text{O}(\text{g})$ $\quad K_c = \dfrac{[\text{NH}_3]^2[\text{H}_2\text{O}]^4}{[\text{H}_2]^5[\text{NO}_2]^2}$

(c) $3 \text{ Cl}_2(\text{g}) + \text{CS}_2(\text{l}) \rightleftharpoons \text{CCl}_4(\text{l}) + \text{S}_2\text{Cl}_2(\text{l})$ $\quad K_c = \dfrac{1}{[\text{Cl}_2]^3}$

(d) $\text{N}_2(\text{g}) + \text{Na}_2\text{CO}_3(\text{s}) + 4 \text{ C}(\text{s}) \rightleftharpoons 2 \text{ NaCN}(\text{s}) + 3 \text{ CO}(\text{g})$ $\quad K_c = \dfrac{[\text{CO}]^3}{[\text{N}_2]}$

27. Since $K_p = K_c(RT)^{\Delta n}$, it is also true that $K_c = K_p(RT)^{-\Delta n}$.

(a) $K_c = \dfrac{[\text{SO}_2(\text{g})][\text{Cl}_2(\text{g})]}{[\text{SO}_2\text{Cl}_2(\text{g})]} = K_p(RT)^{-(+1)} = 2.9 \times 10^{-2} (0.0821 \times 303 \text{ K})^{-1} = 0.0012$

(b) $K_c = \dfrac{[\text{NO}_2]^2}{[\text{NO}]^2[\text{O}_2]} = K_p(RT)^{-(-1)} = (1.48 \times 10^4)(0.0821 \times 457 \text{ K}) = 5.55 \times 10^5$

(c) $K_c = \dfrac{[\text{H}_2\text{S}]^3}{[\text{H}_2]^3} = K_p(RT)^0 = K_p = 0.429$

28. The equilibrium reaction is $\text{H}_2\text{O}(\text{l}) \rightleftharpoons \text{H}_2\text{O}(\text{g})$ with $\Delta n_{\text{gas}} = +1$. $K_p = K_c(RT)^{\Delta n}$ gives $K_c = K_p(RT)^{-\Delta n}$.

$$K_p = P\{\text{H}_2\text{O}\} = 23.8 \text{ mmHg} \times \frac{1 \text{ atm}}{760 \text{ mmHg}} = 0.0313$$

$$K_c = K_p(RT)^{-(+1)} = \frac{K_p}{RT} = \frac{0.0313}{0.0821 \times 298 \text{ K}} = 1.28 \times 10^{-3}$$

29. We combine the several given reactions to obtain the net reaction.

$2 \text{ N}_2\text{O}(\text{g}) \rightleftharpoons 2 \text{ N}_2(\text{g}) + \text{O}_2(\text{g})$ $\quad K_c = \dfrac{1}{(3.4 \times 10^{-18})^2}$

$4 \text{ NO}_2(\text{g}) \rightleftharpoons 2 \text{ N}_2\text{O}_4(\text{g})$ $\quad K_c = \dfrac{1}{(4.6 \times 10^{-3})^2}$

$2 \text{ N}_2(\text{g}) + 4 \text{ O}_2(\text{g}) \rightleftharpoons 4 \text{ NO}_2(\text{g})$ $\quad K_c = (4.1 \times 10^{-9})^4$

net: $2 \text{ N}_2\text{O}(\text{g}) + 3 \text{ O}_2(\text{g}) \rightleftharpoons 2 \text{ N}_2\text{O}_4(\text{g})$ $\quad K_c = \dfrac{(4.1 \times 10^{-9})^4}{(3.4 \times 10^{-18})^2 (4.6 \times 10^{-3})^2} = 1.2 \times 10^6$

30. We combine the K_c values to obtain the value of K_c for the overall reaction, and then convert this to a value of K_p.

$2 \text{ CO}_2(\text{g}) + 2 \text{ H}_2(\text{g}) \rightleftharpoons 2 \text{ CO}(\text{g}) + 2 \text{ H}_2\text{O}(\text{g})$ $\quad K_c = (1.4)^2$

$2 \text{ C}(\text{graphite}) + \text{O}_2(\text{g}) \rightleftharpoons 2 \text{ CO}(\text{g})$ $\quad K_c = (1 \times 10^8)^2$

$4 \text{ CO}(\text{g}) \rightleftharpoons 2 \text{ C}(\text{graphite}) + 2 \text{ CO}_2(\text{g})$ $\quad K_c = \dfrac{1}{(0.64)^2}$

net: $2 \text{ H}_2(\text{g}) + \text{O}_2(\text{g}) \rightleftharpoons 2 \text{ H}_2\text{O}(\text{g})$ $\quad K_c = \dfrac{(1.4)^2 (1 \times 10^8)^2}{(0.64)^2} = 5 \times 10^{16}$

$$K_p = K_c(RT)^{-\Delta n} = \frac{5 \times 10^{16}}{0.08206 \times 1200\text{K}} = 5 \times 10^{14}$$

Experimental Determination of Equilibrium Constants

31. First, we determine the concentration of PCl_5 and of Cl_2 present initially and at equilibrium, respectively. Then we use the balanced equation to help us determine the concentration of each species present at equilibrium.

$$[\text{PCl}_5] = \frac{1.00 \text{ g PCl}_5}{0.250 \text{ L}} \times \frac{1 \text{ mol PCl}_5}{208.2 \text{ g PCl}_5} = 0.0192 \text{ M} \qquad [\text{Cl}_2] = \frac{0.25 \text{ g Cl}_2}{0.250 \text{ L}} \times \frac{1 \text{ mol Cl}_2}{70.9 \text{ g Cl}_2} = 0.014 \text{ M}$$

Reaction: $PCl_5(g)$ $\rightleftharpoons$ PCl_3 + $Cl_2(g)$

Initial: 0.0192 M   0.0 M   0.0 M

Changes: −0.014 M   +0.014 M +0.014 M

Equil: 0.005$_2$ M   0.014 M   0.014 M

$$K_c = \frac{[PCl_3][Cl_2]}{[PCl_5]} = \frac{(0.014\ M)(0.014\ M)}{0.005_2\ M} = 0.03_8$$

32. (a) We determine the three concentrations. $[CO_2] = [C_6H_6O]$ since they are produced in equimolar amounts.

$$[CO_2]_{equil.} = \frac{1.50\ atm}{0.08206\ L\ atm\ mol^{-1}\ K^{-1} \times 473\ K} = 0.0386\ M = [C_6H_6O]_{equil.}$$

$$[C_7H_6O_3]_{initial} = \frac{0.300\ g \times \dfrac{1\ mol\ C_7H_6O_3}{138.1\ g\ C_7H_6O_3}}{0.0500\ L} = 0.0434\ M$$

Reaction: $C_7H_6O_3(g)$ $\rightleftharpoons$ $C_6H_6O(g)$ + $CO_2(g)$

Initial: 0.0434 M   0 M   0 M

Changes: −0.0386 M   +0.0386 M +0.0386 M

Equil: 0.0048 M   0.0386 M   0.0386 M

$$K_c = \frac{[C_6H_6O][CO_2]}{[C_7H_6O_3]} = \frac{0.0386 \times 0.0386}{0.0048} = 0.31$$

(b) For this reaction, $\Delta n_{gas} = 1 + 1 - 1 = +1$

$$K_p = K_c\,(RT)^{\Delta n} = 0.31 \times (0.08206\ L\ atm\ mol^{-1}\ K^{-1} \times 473\ K)^{+1} = 12$$

33. First we determine the partial pressure of each gas.

$$P_{initial}\{H_2(g)\} = \frac{nRT}{V} = \frac{1.00\ g\ H_2 \times \dfrac{1\ mol\ H_2}{2.016\ g\ H_2} \times \dfrac{0.08206\ L\ atm}{mol\ K} \times 1670\ K}{0.500\ L} = 136\ atm$$

$$P_{initial}\{H_2S(g)\} = \frac{nRT}{V} = \frac{1.06\ g\ H_2S \times \dfrac{1\ mol\ H_2S}{34.08\ g\ H_2S} \times \dfrac{0.08206\ L\ atm}{mol\ K} \times 1670\ K}{0.500\ L} = 8.52\ atm$$

$$P_{equil.}\{S_2(g)\} = \frac{nRT}{V} = \frac{8.00 \times 10^{-6}\ mol\ S_2 \times \dfrac{0.08206\ L\ atm}{mol\ K} \times 1670\ K}{0.500\ L} = 2.19 \times 10^{-3}\ atm$$

Reaction: 2 $H_2(g)$ + $S_2(g)$ $\rightleftharpoons$ 2 $H_2S(g)$

Initial: 136 atm   8.52 atm

Changes: +0.00438 atm +0.00219 atm −0.00438 atm

Equil: 136 atm   0.00219 atm   8.52 atm

$$K_p = \frac{P\{H_2S(g)\}^2}{P\{H_2(g)\}^2\,P\{S_2(g)\}} = \frac{(8.52)^2}{(136)^2\ 0.00219} = 1.79$$

34. We first determine the amount in moles of acetic acid in the equilibrium mixture.

$$\text{amount } CH_3CO_2H = 28.85\ mL \times \frac{1\ L}{1000\ mL} \times \frac{0.1000\ mol\ Ba(OH)_2}{1\ L} \times \frac{2\ mol\ CH_3CO_2H}{1\ mol\ Ba(OH)_2}$$
$$\times \frac{\text{complete equilibrium mixture}}{0.01\ \text{of equilibrium mixture}} = 0.5770\ mol\ CH_3CO_2H$$

Reaction: C_2H_5OH + CH_3CO_2H $\rightleftharpoons$ $CH_3CO_2C_2H_5$ + H_2O

Initial: 0.500 mol   1.000 mol   0.000 mol   0.000 mol

Changes: −0.423 mol −0.423 mol   +0.423 mol   +0.423 mol

Equil: 0.077 mol   0.577 mol   0.423 mol   0.423 mol

$$K_c = \frac{[CH_3CO_2C_2H_5][H_2O]}{[C_2H_5OH][CH_3CO_2H]} = \frac{\dfrac{0.423\ mol}{V} \times \dfrac{0.423\ mol}{V}}{\dfrac{0.077\ mol}{V} \times \dfrac{0.577\ mol}{V}} = \frac{0.423 \times 0.423}{0.077 \times 0.577} = 4.0$$

Equilibrium Relationships

35. $K_c = 375 = \dfrac{[NO_2]^2}{[NO]^2[O_2]} = \dfrac{[NO_2]^2}{[NO]^2} \times \dfrac{0.755\ L}{0.0148\ mol}$ $\dfrac{[NO]}{[NO_2]} = \sqrt{\dfrac{0.755}{0.0148 \times 375}} = 0.369$

36. $K_c = 0.011 = \dfrac{[I]^2}{[I_2]} = \dfrac{\left(\dfrac{0.50 \text{ mol I}}{V}\right)^2}{\dfrac{1.00 \text{ mol I}_2}{V}} = \dfrac{1}{V} \times 0.25$ $\qquad V = \dfrac{0.25}{0.011} = 23 \text{ L}$

37. (a) A possible equation for the oxidation of $NH_3(g)$ to $NO_2(g)$ follows.
$NH_3(g) + \frac{7}{4}O_2(g) \rightleftharpoons NO_2(g) + \frac{3}{2}H_2O(g)$

(b) We obtain K_p for the reaction in part (a) by appropriately combining the values of K_p given in the problem.

$NH_3(g) + \frac{5}{4}O_2(g) \rightleftharpoons NO_2(g) + \frac{3}{2}H_2O(g)$ $\qquad K_p = 2.11 \times 10^{19}$

$NO(g) + \frac{1}{2}O_2(g) \rightleftharpoons NO_2(g)$ $\qquad K_p = \dfrac{1}{0.524}$

net: $\quad NH_3(g) + \frac{7}{4}O_2(g) \rightleftharpoons NO_2(g) + \frac{3}{2}H_2O(g)$ $\qquad K_p = \dfrac{2.11 \times 10^{19}}{0.524} = 4.03 \times 10^{19}$

38. (a) We first determine $[H_2]$ and $[CH_4]$, and then $[C_2H_2]$. $\qquad [CH_4] = [H_2] = \dfrac{0.10 \text{ mol}}{1.00 \text{ L}} = 0.10 \text{ M}$

$K_c = \dfrac{[C_2H_2][H_2]^3}{[CH_4]^2}$ $\qquad [C_2H_2] = \dfrac{K_c [CH_4]^2}{[H_2]^3} = \dfrac{0.154 \times 0.10^2}{0.10^3} = 1.5 \text{ M}$

Since the volume of the container is 1.00 L, the concentrations are equal numerically to the amounts of each substance.

$\chi\{C_2H_2\} = \dfrac{1.5 \text{ mol C}_2\text{H}_2}{1.5 \text{ mol C}_2\text{H}_2 + 0.10 \text{ mol CH}_4 + 0.10 \text{ mol H}_2} = 0.88$

(b) The conversion of $CH_4(g)$ to $C_2H_2(g)$ is favored at low pressures, since the conversion reaction has a larger sum of the stoichiometric coefficients of gaseous products (4) than of reactants (2).

Direction and Extent of Chemical Change

39. We compute the value of Q_c for the given amounts of product and reactants.

$Q_c = \dfrac{[SO_3]^2}{[SO_2]^2[O_2]} = \dfrac{\left(\dfrac{6 \text{ mol SO}_3}{6.65 \text{ L}}\right)^2}{\left(\dfrac{2 \text{ mol SO}_2}{6.65 \text{ L}}\right)^2 \dfrac{3 \text{ mol O}_2}{6.65 \text{ L}}} = 2 \times 10^1 < K_c = 100$

Since $Q_c < K_c$, the reaction will proceed to the right, that is toward products, until equilibrium is established.

40. Use the chemical equation as a basis to organize the information we have about the reaction, and then determine the final number of moles of $Cl_2(g)$ present.

Reaction:	$CO(g)$	+	$Cl_2(g)$	$\rightleftharpoons$	$COCl_2(g)$
Initial:	1.00 mol				1.00 mol
Changes:	$+x$ mol		$+x$ mol		$-x$ mol
Equil:	$(1.00 + x)$mol		x mol		$(1.00 - x)$mol

$K_c = 1.2 \times 10^3 = \dfrac{[COCl_2]}{[CO][Cl_2]} = \dfrac{\dfrac{(1.00 - x)\text{mol}}{1.75 \text{ L}}}{\dfrac{(1.00 + x)\text{mol}}{1.75 \text{ L}} \times \dfrac{x \text{ mol}}{1.75 \text{ L}}}$ $\qquad \dfrac{1.2 \times 10^3}{1.75} = \dfrac{1.00 - x}{(1.00 + x)\, x}$

We assume that $x \ll 1.00$ This produces the following.

$\dfrac{1.2 \times 10^3}{1.75} = \dfrac{1.00}{1.00\, x}$ $\qquad x = \dfrac{1.75}{1.2 \times 10^3} = 1.5 \times 10^{-3} \text{ mol Cl}_2$ $\qquad$ Clearly $x \ll 1.00$

41. We use the balanced chemical equation as a basis to organize the information we have about the reaction.

Reaction:	$SbCl_5(g)$	$\rightleftharpoons$	$SbCl_3(g)$	+	$Cl_2(g)$
Initial:	$\dfrac{0.00 \text{ mol}}{2.50 \text{ L}}$		$\dfrac{3.00 \text{ mol}}{2.50 \text{ L}}$		$\dfrac{1.00 \text{ mol}}{2.50 \text{ L}}$
Initial:	0.000 M		1.20 M		0.400 M
Changes:	$+x$ M		$-x$ M		$-x$ M
Equil:	x M		$(1.20 - x)$M		$(0.400 - x)$M

$$K_c = 0.025 = \frac{[SbCl_3][Cl_2]}{[SbCl_5]} = \frac{(1.20 - x)(0.400 - x)}{x} = \frac{0.480 - 1.60\,x + x^2}{x}$$

$$0.025\,x = 0.480 - 1.60x + x^2 \qquad x^2 - 1.625\,x + 0.480 = 0$$

$$x = \frac{-b \pm \sqrt{b^2 - 4ac}}{2a} = \frac{1.625 \pm \sqrt{2.64 - 1.92}}{2} = 0.388 \text{ or } 1.24$$

The second of the two values for x gives a negative value of $[Cl_2]$ (= –0.84 M), and thus is physically meaningless. Thus, concentrations and amounts are the following.

$[SbCl_5] = x = 0.388$ M   amount $SbCl_5 = 2.50$ L × 0.388 M = 0.970 mol $SbCl_5$

$[SbCl_3] = 1.20 - x = 0.812$ M   amount $SbCl_3 = 2.50$ L × 0.812 M = 2.03 mol $SbCl_3$

$[Cl_2] = 0.400 - x = 0.012$ M   amount $Cl_2 = 2.50$ L × 0.012 M = 0.030 mol Cl_2

Or, in the first line labeled "Initial," we could have set y = number of moles of $SbCl_5$ produced. Then our final answer would be $y = 0.970$ mol $SbCl_5$. That is, we would have obtained the answer more directly.

42. We use the balanced chemical equation to organize the information we have about the reactants and products. Recall that the conclusion of Example 16-5 is that the reaction shifts to the left.

Reaction:	$CO(g)$	+	$H_2O(g)$	$\rightleftharpoons$	$CO_2(g)$	+	$H_2(g)$
Initital:	1.00 mol		1.00 mol		2.00 mol		2.00 mol
Changes:	+x mol		+x mol		–x mol		–x mol
Equil:	(1.00 + x) mol		(1.00 + x)mol		(2.00 – x)mol		(2.00 – x) mol

$$K_c = \frac{((2.00 - x)/V)\,((2.00 - x)/V)}{((1.00 + x)/V)\,((1.00 + x)/V)} = \left(\frac{2.00 - x}{1.00 + x}\right)^2 = 1.00 \quad \sqrt{1.00} = \frac{2.00 - x}{1.00 + x}$$

$1.00 + x = 2.00 - x$ $2x = 2.00 - 1.00 = 1.00$ $x = 0.500$

Then   amount $CO(g)$ = amount $H_2O(g) = 1.00 + x = 1.50$ mol

And   amount $CO_2(g)$ = amount $H_2(g) = 2.00 - x = 1.50$ mol

43. We first compute the initial concentration of each species present. Then we determine the equilibrium concentrations of all species. And finally, we compute the mass of CO_2 present at equilibrium.

$$[CO]_i = \frac{1.00 \text{ g}}{1.41 \text{ L}} \times \frac{1 \text{ mol CO}}{28.01 \text{ g CO}} = 0.0253 \text{ M} \qquad [H_2O]_i = \frac{1.00 \text{ g}}{1.41 \text{ L}} \times \frac{1 \text{ mol H}_2\text{O}}{18.02 \text{ g H}_2\text{O}} = 0.0394 \text{ M}$$

$$[H_2]_i = \frac{1.00 \text{ g}}{1.41 \text{ L}} \times \frac{1 \text{ mol H}_2}{2.016 \text{ g H}_2} = 0.352 \text{ M}$$

Reaction:	$CO(g)$	+	$H_2O(g)$	$\rightleftharpoons$	$CO_2(g)$	+	$H_2(g)$
Initial:	0.0253 M		0.0394 M		0.0000 M		0.352 M
Changes:	–x M		–x M		+x M		+x M
Equil:	(0.0253 – x)M		(0.0394 – x)M		x M		(0.352 + x)M

$$K_c = \frac{[CO_2][H_2]}{[CO][H_2O]} = 23.2 = \frac{x(0.352 + x)}{(0.0253 - x)(0.0394 - x)} = \frac{0.352x + x^2}{0.000997 - 0.0647x + x^2}$$

$$0.0231 - 1.50x + 23.2x^2 = 0.352x + x^2 \qquad 22.2x^2 - 1.852x + 0.0231 = 0$$

$$x = \frac{-b \pm \sqrt{b^2 - 4ac}}{2a} = \frac{1.852 \pm \sqrt{3.430 - 2.051}}{44.4} = 0.0682 \text{ M, } 0.0153 \text{ M}$$

The first value of x gives a negative $[CO]$ (= –0.0429 M) and a negative $[H_2O]$ (= –0.0298 M). Thus, x = 0.0153 M = $[CO_2]$. Now we find the mass of CO_2.

$$1.41 \text{ L} \times \frac{0.0153 \text{ mol CO}_2}{1 \text{ L mixture}} \times \frac{44.01 \text{ g CO}_2}{1 \text{ mol CO}_2} = 0.949 \text{ g CO}_2$$

44. We base each of our solutions on the balanced chemical equation.

(a)

Reaction:	$PCl_5(g)$	$\rightleftharpoons$	$PCl_3(g)$	+	$Cl_2(g)$
Initial:	$\dfrac{1.50 \text{ mol}}{2.50 \text{ L}}$		$\dfrac{1.50 \text{ mol}}{2.50 \text{ L}}$		
Changes:	$\dfrac{-x \text{ mol}}{2.50 \text{ L}}$		$\dfrac{+x \text{ mol}}{2.50 \text{ L}}$		$\dfrac{+x \text{ mol}}{2.50 \text{ L}}$
Equil:	$\dfrac{(1.50 - x) \text{ mol}}{2.50 \text{ L}}$		$\dfrac{(1.50 + x) \text{ mol}}{2.50 \text{ L}}$		$\dfrac{x \text{ mol}}{2.50 \text{ L}}$

$$K_c = \frac{[PCl_3][Cl_2]}{[PCl_5]} = 3.8 \times 10^{-2} = \frac{\dfrac{(1.50 + x)\text{ mol}}{2.50 \text{ L}} \times \dfrac{x \text{ mol}}{2.50 \text{ L}}}{\dfrac{(1.50 - x)\text{ mol}}{2.50 \text{ L}}} \qquad \frac{x(1.50 + x)}{2.50(1.50 - x)} = 3.8 \times 10^{-2}$$

$$x^2 + 1.50\,x = 0.14 - 0.095\,x \qquad x^2 \pm 1.60\,x - 0.14 = 0$$

$$x = \frac{-b \pm \sqrt{b^2 - 4ac}}{2a} = \frac{-1.60 \pm \sqrt{2.56 \pm 0.56}}{2} = 0.085 \text{ mol}, -1.51 \text{ mol}$$

The second answer gives a negative amount of Cl_2, and that makes no physical sense.

amount $PCl_5 = (1.50 - 0.085) = 1.41$ mol   amount $PCl_3 = (1.50 + 0.085) = 1.59$ mol

amount $Cl_2 = x = 0.085$ mol

(b)

Reaction:	$PCl_5(g)$	$\rightleftharpoons$	$PCl_3(g)$	+	$Cl_2(g)$
Initial:	$\dfrac{0.500 \text{ mol}}{2.50 \text{ L}}$				
Changes:	$\dfrac{-x \text{ mol}}{2.50 \text{ L}}$		$\dfrac{+x \text{ mol}}{2.50 \text{ L}}$		$\dfrac{+x \text{ mol}}{2.50 \text{ L}}$
Equil:	$\dfrac{0.500 - x \text{ mol}}{2.50 \text{ L}}$		$\dfrac{(x \text{ mol})}{2.50 \text{ L}}$		$\dfrac{(x \text{ mol})}{2.50 \text{ L}}$

$$K_c = \frac{[PCl_3][Cl_2]}{[PCl_5]} = 3.8 \times 10^{-2} = \frac{\dfrac{(x \text{ mol})}{2.50 \text{ L}} \times \dfrac{(x \text{ mol})}{2.50 \text{ L}}}{\dfrac{0.500 - x \text{ mol}}{2.50 \text{ L}}} \qquad 2.50 \times 3.8 \times 10^{-2} = \frac{x^2}{0.500 - x} = 0.095$$

$$0.048 - 0.095 x = x^2 \qquad x^2 + 0.095 x - 0.048 = 0$$

$$x = \frac{-b \pm \sqrt{b^2 - 4ac}}{2a} = \frac{-0.095 \pm \sqrt{0.0090 + 0.19}}{2} = 0.18 \text{ mol}, -0.270 \text{ mol}$$

amount $PCl_3 = 0.18$ mol = amount Cl_2 amount $PCl_5 = 0.500 - 0.18 = 0.32$ mol

45. (a) We use the balanced chemical equation as a basis to organize the information we have about the reactants and products.

Reaction:	$2 COF_2(g)$	$\rightleftharpoons$	$CO_2(g)$	+	$CF_4(g)$
Initial:	$\dfrac{0.105 \text{ mol}}{5.00 \text{ L}}$		$\dfrac{0.220 \text{ mol}}{5.00 \text{ L}}$		$\dfrac{0.055 \text{ mol}}{5.00 \text{ L}}$
Initial:	0.0210 M		0.0440 M		0.011 M

And we now compute a value of Q_c to compare with the given value of K_c.

$$Q_c = \frac{[CO_2][CF_4]}{[COF_2]^2} = \frac{(0.0440)(0.011)}{(0.0210)^2} = 1.1 < 2.00 = K_c$$

Since Q_c is not equal to K_c, the mixture is not at equilibrium.

(b) Since Q_c is smaller than K_c, the reaction will shift right, that is, products will be formed, in reaching a state of equilibrium.

(c) We continue the organization of information about reactants and products.

Reaction:	$2 COF_2(g)$	$\rightleftharpoons$	$CO_2(g)$	+	$CF_4(g)$
Initial:	0.0210 M		0.0440 M		0.011 M
Changes:	$-2x$ M		$+x$ M		$+x$ M
Equil:	$(0.0210 - 2x)$M		$(0.0440 + x)$M		$(0.011 + x)$M

$$K_c = \frac{[CO_2][CF_4]}{[COF_2]^2} = \frac{(0.0440 + x)(0.011 + x)}{(0.0210 - 2x)^2} = 2.00 = \frac{0.00048 + 0.055x + x^2}{0.000441 - 0.084x + 4x^2}$$

$$0.000882 - 0.168x + 8x^2 = 0.00048 + 0.055x + x^2 \qquad 7x^2 - 0.223x + 0.00040 = 0$$

$$x = \frac{-b \pm \sqrt{b^2 - 4ac}}{2a} = \frac{0.223 \pm \sqrt{0.0497 - 0.0112}}{14} = 0.0019 \text{ M}, 0.0299 \text{ M}$$

The second of these value for x (0.0299) gives a negative $[COF_2]$ (= −0.0388 M), clearly a nonsensical result. We now compute the concentration of each species at equilibrium, and check to ensure that the equilibrium constant is satisfied.

$[COF_2] = 0.0210 - 2x = 0.0210 - 2(0.0019) = 0.0172$ M

$[CO_2] = 0.0440 + x = 0.0440 + 0.0019 = 0.0459$ M

$[CF_4] = 0.0110 + x = 0.011 + 0.0019 = 0.013$ M

$$K_c = \frac{[CO_2][CF_4]}{[COF_2]^2} = \frac{0.0459 \text{ M} \times 0.013 \text{ M}}{(0.0172 \text{ M})^2} = 2.0$$

The agreement of this value of K_c with the cited value (2.00) indicates that this solution is correct. Now we determine the number of moles of each species at equilibrium.

no. mol $COF_2 = 5.00$ L × 0.0172 M = 0.086 mol COF_2

no. mol $CO_2 = 5.00$ L × 0.0459 M = 0.229 mol CO_2

no. mol CF_4 = 5.00 L × 0.013 M = 0.065 mol CF_4

But suppose we had *incorrectly* concluded, in part (b), that reactants would be formed in reaching equilibrium. What result would we obtain? The set up follows.

Reaction:	$2 COF_2(g)$	$\rightleftharpoons$	$CO_2(g)$	+	$CF_4(g)$
Initial:	0.0210 M		0.0440 M		0.011 M
Changes:	+ 2y M		–y M		–y M
Equil:	(0.0210 + 2y)M		(0.0440 – y)M		(0.011 – y)M

$$K_c = \frac{[CO_2][CF_4]}{[COF_2]^2} = \frac{(0.0440 - y)(0.011 - y)}{(0.0210 + 2y)^2} = 2.00 = \frac{0.00048 - 0.055y + y^2}{0.000441 + 0.084y + 4y^2}$$

$$0.000882 + 0.168y + 8y^2 = 0.00048 - 0.055y + y^2 \qquad 7y^2 + 0.223y + 0.00040 = 0$$

$$y = \frac{-b \pm \sqrt{b^2 - 4ac}}{2a} = \frac{-0.223 \pm \sqrt{0.0497 - 0.0112}}{14} = -0.0019 \text{ M}, -0.0299 \text{ M}$$

The second of these values for y (–0.0299 M) gives a negative $[COF_2]$ (= –0.0388 M), clearly a nonsensical result. We now compute the concentration of each species at equilibrium.

$[COF_2] = 0.0210 + 2y = 0.0210 + 2(-0.0019) = 0.0172$ M
$[CO_2] = 0.0440 - y = 0.0440 + 0.0019 = 0.0459$ M
$[CF_4] = 0.0110 - y = 0.011 + 0.0019 = 0.013$ M

These are the same equilibrium concentrations that we obtained by making the correct decision regarding the direction that the reaction would take. Thus, you can be assured that, if you perform the algebra correctly, it will guide you even if you make the incorrect decision about the direction of the reaction.

46. (a) We calculate the initial amount of each substance.

$$n\{C_2H_5OH\} = 15.5 \text{ g } C_2H_5OH \times \frac{1 \text{ mol } C_2H_5OH}{46.07 \text{ g } C_2H_5OH} = 0.336 \text{ mol } C_2H_5OH$$

$$n\{CH_3CO_2H\} = 25.0 \text{ g } CH_3CO_2H \times \frac{1 \text{ mol } CH_3CO_2H}{60.05 \text{ g } CH_3CO_2H} = 0.416 \text{ mol } CH_3CO_2H$$

$$n\{CH_3CO_2C_2H_5\} = 45.5 \text{ g } CH_3CO_2C_2H_5 \times \frac{1 \text{ mol } CH_3CO_2C_2H_5}{88.11 \text{ g } CH_3CO_2C_2H_5} = 0.516 \text{ mol } CH_3CO_2C_2H_5$$

$$n\{H_2O\} = 52.0 \text{ g } H_2O \times \frac{1 \text{ mol } H_2O}{18.02 \text{ g } H_2O} = 2.89 \text{ mol } H_2O$$

Since we would divide each amount by the total volume, and since there are the same numbers of product and reactant stoichiometric coefficients, we can use amounts rather than concentrations in the Q_c expression.

$$Q_c = \frac{n\{CH_3CO_2C_2H_5\} \, n\{H_2O\}}{n\{C_2H_5OH\} \, n\{CH_3CO_2H\}} = \frac{0.516 \text{ mol} \times 2.89 \text{ mol}}{0.336 \text{ mol} \times 0.416 \text{ mol}} = 10.7 > K_c = 4.0$$

Since $Q_c > K_c$ the reaction will shift to the left, forming reactants, as it attains equilibrium.

(b)

Reaction:	C_2H_5OH	+	CH_3CO_2H	$\rightleftharpoons$	$CH_3CO_2C_2H_5$	+	H_2O
Initial:	0.336 mol		0.416 mol		0.516 mol		2.89 mol
Changes:	+x mol		+x mol		–x mol		–x mol
Equil:	(0.336 + x) mol		(0.416 + x) mol		(0.516 – x) mol		(2.89 – x) mol

$$K_c = \frac{(0.516 - x)(2.89 - x)}{(0.336 + x)(0.416 + x)} = \frac{1.49 - 3.41 \, x + x^2}{0.140 + 0.752 \, x + x^2} = 4.0$$

$$x^2 - 3.41 \, x + 1.49 = 4 \, x^2 + 3.0 \, x + 0.56 \qquad 3 \, x^2 + 6.4 \, x - 0.93 = 0$$

$$x = \frac{-b \pm \sqrt{b^2 - 4ac}}{2a} = \frac{-6.4 \pm \sqrt{41 + 11}}{6} = 0.14 \text{ moles}, -2.27 \text{ moles}$$

Negative amounts make no physical sense. We compute the equilibrium amount of each substance.

$n\{C_2H_5OH\} = 0.336$ mol + 0.14 mol = 0.48 mol C_2H_5OH
$n\{CH_3CO_2H\} = 0.416$ mol + 0.14 mol = 0.56 mol CH_3CO_2H
$n\{CH_3CO_2C_2H_5\} = 0.516$ mol – 0.14 mol = 0.38 $CH_3CO_2C_2H_5$
$n\{H_2O\} = 2.89$ mol – 0.14 mol = 2.75 molH_2O

47. (a) The computed value of Q_c is compared to K_c to determine the direction in which the reaction will occur.

$$Q_c = \frac{[NOBr]^2}{[NO]^2[Br_2]} = \frac{\left(\dfrac{0.0100 \text{ mol}}{1.00L}\right)^2}{\left(\dfrac{0.100 \text{ mol}}{1.00L}\right)^2 \dfrac{0.100 \text{ mol}}{1.00L}} = 0.10 > 0.0132 = K_c$$

Thus, in reaching equilibrium, the reaction will shift to the left, forming reactants.

(b) The value of the equilibrium constant indicates that very little product is present at equilibrium. Thus, we assume that the reaction shifts to the left to establish equilibrium. (The value of $Q_c = 0.10$).

Reaction:	$2\,NO(g)$	$+$	$Br_2(g)$	$\rightleftharpoons$	$2\,NOBr(g)$
Initial:	0.100 M		0.100 M		0.0100 M
Changes:	$+2x$ M		$+x$ M		$-2x$ M
Equil:	$(0.100 + 2x)$M		$(0.100 + x)$M		$(0.0100 - 2x)$M

$$K_c = \frac{[NOBr]^2}{[NO]^2[Br_2]} = \frac{(0.0100 - 2x)^2}{(0.100 - 2x)^2(0.100 - x)} = 0.0132 \approx \frac{(0.0100 - 2x)^2}{(0.100)^2(0.100)}$$

To solve the cubic equation, we note from the numerator (which must be positive) that $2x < 0.0100$, and thus that $x < 0.00500$ M. This value is small enough that we can, to a first approximation, disregard x in the factors in the denominator. This gives the following expression.

$$0.0100 - 2x = \sqrt{0.0132\ (0.100)^2\ 0.100} = 0.00363\ M$$
$$2x = 0.0100 - \sqrt{0.0132\ (0.100)^2\ 0.100} = 0.0100 - 0.00363 = 0.0064 \qquad x = 0.0032\ M$$

Substituting this value of x back into the equation checks the consistency of the answer.

$$2x = 0.0100 - \sqrt{0.0132\ (0.100 - 0.0064)^2\ (0.100 - 0.0032)} = 0.0066 \qquad x = 0.0033$$

[NOBr] = 0.0100 − 2x = 0.0100 M − 0.0066 M = 0.0034 M

[NO] = 0.100 − 2x = 0.0934   [Br₂] = 0.100 − x = 0.0967

We verify these values by substituting into the K_c expression.

$$K_c = \frac{[NOBr]^2}{[NO]^2[Br_2]} = \frac{(0.0034)^2}{(0.0934)^2\ 0.0967} = 0.0137 \quad \text{This is fairly good agreement with } K_c = 0.132$$

$$P\{NOBr\} = \frac{nRT}{V} = [NOBr]RT = (3.4 \times 10^{-3}\ mol\ L^{-1})(0.08206\ L\ atm\ mol^{-1}\ K^{-1})(1000\ K)$$
$$= 0.279\ atm$$

48. Again we organize the known data around the balanced chemical equation.

Reaction:	$H_2(g)$	$+$	$I_2(g)$	$\rightleftharpoons$	$2\,HI(g)$
Initial:	0.250 mol		0.250 mol		0.000 mol
Changes:	$-x$ mol		$-x$ mol		$+2x$ mol
Equil:	$(0.250 - x)$mol		$(0.250 - x)$mol		$2x$ mol

$$K_c = \frac{[HI]^2}{[H_2][I_2]} = \frac{\left(\frac{2x\ mol}{4.10\ L}\right)^2}{\frac{(0.250 - x)mol}{4.10\ L} \times \frac{(0.250 - x)mol}{4.10\ L}} = \left(\frac{2x}{0.250 - x}\right)^2 = 50.2$$

$$\sqrt{50.2} = \frac{2x}{0.250 - x} = 7.09 \qquad 2x = 1.77 - 7.09\,x \qquad 9.09\,x = 1.77 \qquad x = \frac{1.77}{9.09} = 0.195\ mol$$

amount HI = 2x = 2 × 0.195 mol = 0.390 mol

amount H₂ = amount I₂ = 0.250 − x = 0.250 − 0.195 = 0.055 mol

Total amount = 0.390 mol HI + 0.055 mol H₂ + 0.055 mol I₂ = 0.500 mol

$$mol\%\ HI = \frac{0.390\ mol\ HI}{0.500\ mol\ total} \times 100\% = 78.0\%\ HI$$

49. The final volume of the mixture is 0.750 L + 2.25 L = 3.00 L. Then use the balanced chemical equation to organize the data we have concerning the reaction. The reaction should shift to the right, that is form products, in reaching a new equilibrium, since the volume is greater.

Reaction:	$N_2O_4(g)$	$\rightleftharpoons$	$2\,NO_2(g)$
Initial:	$\dfrac{0.971\ mol}{3.00\ L}$		$\dfrac{0.0580\ mol}{3.00\ L}$
Initial:	0.324 M		0.0193 M
Changes:	$-x$ M		$+2x$ M
Equil:	$(0.324 - x)$M		$(0.0193 + 2x)$M

$$K_c = \frac{[NO_2]^2}{[N_2O_4]} = \frac{(0.0193 + 2x)^2}{0.324 - x} = \frac{0.000373 + 0.0772x + 4x^2}{0.324 - x} = 4.61 \times 10^{-3}$$

$$0.000373 + 0.0772x + 4x^2 = 0.00149 - 0.00461x \qquad 4x^2 + 0.0818x - 0.00112 = 0$$

$$x = \frac{-b \pm \sqrt{b^2 - 4ac}}{2a} = \frac{-0.0818 \pm \sqrt{0.00669 + 0.0179}}{8} = 0.00938\ M$$

[NO₂] = 0.0193 + (2 × 0.00938) = 0.0381 M   amount NO₂ = 0.0381 M × 3.00 L = 0.114 mol NO₂

[N₂O₄] = 0.324 − 0.00938 = 0.314₆ M   amount N₂O₄ = 0.314₆ M × 3.00 L = 0.944 mol N₂O₄

50. $[HCONH_2]_{init} = \dfrac{0.100 \text{ mol}}{1.50 \text{ L}} = 0.0667 \text{ M}$

Reaction: $\quad HCONH_2(g) \rightleftharpoons NH_3(g) + CO(g)$

Initial: $\qquad$ 0.0667 M

Changes: $\qquad -x$ M $\qquad\qquad +x$ M $\quad +x$ M

Equil: $\qquad (0.0667 - x)$ M $\quad x$ M $\quad\;\, x$ M

$K_c = \dfrac{[NH_3][CO]}{[HCONH_2]} = \dfrac{x \cdot x}{0.0667 - x} = 4.84 \qquad x^2 = 0.323 - 4.84\,x \qquad 0 = x^2 + 4.84\,x - 0.323$

$x = \dfrac{-b \pm \sqrt{b^2 - 4ac}}{2a} = \dfrac{-4.84 \pm \sqrt{23.4 + 1.29}}{2} = 0.065 \text{ M}, -4.90 \text{ M}$

The negative concentration obviously is physically meaningless. We determine the total concentration of all species, and then the total pressure.

[total] = $[NH_3] + [CO] + [HCONH_2] = x + x + 0.0667 - x = 0.0667 + 0.065 = 0.132$ M

$P_{tot} = 0.132 \text{ mol L}^{-1} \times 0.08206 \text{ L atm mol}^{-1} \text{ K}^{-1} \times 400. \text{ K} = 4.33 \text{ atm}$

51. We first determine the initial $[H_2S]$. $\quad [H_2S] = \dfrac{1.00 \text{ mol } H_2S}{1.10 \text{ L}} = 0.909 \text{ M} \quad$ We base our solution on the

balanced chemical equation. $\qquad$ Reaction: $\quad 2\,H_2S(g) \rightleftharpoons 2\,H_2(g) + S_2(g)$

$\qquad\qquad\qquad\qquad\qquad\qquad\qquad\qquad$ Initial: $\qquad 0.909$ M

$\qquad\qquad\qquad\qquad\qquad\qquad\qquad\qquad$ Changes: $\quad -2x$ M $\qquad +2x$ M $\quad +x$ M

$\qquad\qquad\qquad\qquad\qquad\qquad\qquad\qquad$ Equil: $\quad (0.909 - 2x)$ M $\quad 2x$ M $\quad\; x$ M

$K_c = \dfrac{[H_2]^2[S_2]}{[H_2S]^2} = 1.0 \times 10^{-6} = \dfrac{(2x)^2\,x}{(0.909 - 2x)^2} \approx \dfrac{4x^3}{0.909^2} \qquad x^3 = \dfrac{1.0 \times 10^{-6}\,(0.909)^2}{4} = 2.0_7 \times 10^{-7}$

$x = 5.9 \times 10^{-3} \text{ M} \qquad$ % dissociation $= \dfrac{2x}{0.909} \times 100\% = \dfrac{2 \times 5.9 \times 10^{-3}}{0.909} \times 100\% = 1.3\%$

52. We organize the given data around the balanced chemical equation.

Reaction: $\quad Ag^+(aq) + Fe^{2+} \rightleftharpoons Fe^{3+}(aq) + Ag(s)$

Initial: $\qquad 1.00$ M $\qquad\; 1.00$ M

Changes: $\quad -x$ M $\qquad -x$ M $\qquad +x$ M

Equil: $\qquad (1.00 - x)$M $\;\; (1.00 - x)$M $\quad x$ M

$K_c = \dfrac{[Fe^{3+}]}{[Ag^+][Fe^{2+}]} = \dfrac{x}{(1.00 - x)^2} = 2.98 \qquad x = 2.98(1.00 - 2.00x + x^2) = 2.98 - 5.96x + 2.98x^2$

$2.98x^2 - 6.96x + 2.98 = 0 \qquad x = \dfrac{-b \pm \sqrt{b^2 - 4ac}}{2a} = \dfrac{6.96 \pm \sqrt{48.4 - 35.5}}{5.96} = 1.77 \text{ M}, 0.565 \text{ M}$

The first value of x (1.77 M) gives a negative $[Ag^+]$ and $[Fe^{2+}]$ (= -0.77 M). Thus, we choose $x = 0.565$ M.

$[Fe^{3+}] = 0.56 \text{ M} \qquad [Ag^+] = [Fe^{2+}] = 1.00 \text{ M} - 0.565 \text{ M} = 0.44 \text{ M}$

53. Diluting the equilibrium mixture to 2.50 times its value will decrease each concentration by a factor of 2.50. Since there are more concentrations in the denominator of the K_c expression, the denominator will become smaller than the numerator, making Q_c larger than K_c. Thus, the reaction mixture will shift left in order to regain equilibrium. Again, we organize our calculation around the balanced chemical equation.

Reaction: $\quad Ag^+(aq) + Fe^{2+}(aq) \rightleftharpoons Fe^{3+}(aq) + Ag(s) \qquad K_c = 2.98$

equil: $\qquad 0.31$ M $\qquad 0.21$ M $\qquad 0.19$ M

dilution: $\quad 0.12$ M $\qquad 0.084$ M $\qquad 0.076$ M

changes: $\quad +x$ M $\qquad +x$ M $\qquad -x$ M

new equil: $(0.12 + x)$ M $\;\; (0.084 + x)$ M $\;\; (0.076 - x)$ M

$K_c = \dfrac{[Fe^{3+}]}{[Ag^+][Fe^{2+}]} = 2.98 = \dfrac{0.076 - x}{(0.12 + x)(0.084 + x)} \qquad 2.98\,(0.12 + x)(0.084 + x) = 0.076 - x$

$0.076 - x = 0.030 + 0.61\,x + 2.98\,x^2 \qquad\qquad 2.98\,x^2 + 1.61\,x - 0.046 = 0$

$x = \dfrac{-1.61 \pm \sqrt{2.59 + 0.55}}{5.96} = 0.027, -0.57 \qquad$ Note that the negative root makes no physical sense; it

gives $[Fe^{2+}] = 0.084 - 0.57 = -0.49$ M. Thus, the new equilibrium concentrations are

$[Fe^{2+}] = 0.084 + 0.027 = 0.111$ M $\qquad\qquad [Ag^+] = 0.12 + 0.027 = 0.15$ M

$[Fe^{3+}] = 0.076 - 0.027 = 0.049$ M $\qquad\qquad$ We check by substitution.

$2.98 = K_c = \dfrac{[Fe^{3+}]}{[Ag^+][Fe^{2+}]} = \dfrac{0.049}{0.15 \times 0.111} = 2.9_7$ Agreement!

Due to rounding errors, very slightly different equilibrium concentrations are obtained if one begins with the initial mixture and dilutes it to 2.50 time its original volume and then compuites equilibrium concentrations from that point.

Partial Pressure Equilibrium Constant, K_p

54. **(a)** We first determine the initial pressure of each gas.

$$P\{SO_2\} = P\{Cl_2\} = \frac{nRT}{V} = \frac{1.00 \text{ mol} \times 0.08206 \text{ L atm mol}^{-1} \text{ K}^{-1} \times 303 \text{ K}}{1.75 \text{ L}} = 14.2 \text{ atm}$$

Then we calculate equilibrium partial pressures, organizing our calculation around the balanced chemical equation. We see that the equilibrium constant is quite small, meaning that the equilibrium position favors the reactant. Thus, we begin by forming as much reactant as possible; then proceed to equilibrium. (Significant figure rules produce a somewhat erroneous answer if we do not initially form product.)

Reaction:	$SO_2Cl_2(g)$ $\rightleftharpoons$	$SO_2(g)$	$+$	$Cl_2(g)$
Initial:	0.00 atm	14.2 atm		14.2 atm
To left:	14.2 atm			
Changes:	$-x$ atm	$+x$ atm		$+x$ atm
Equil:	$(14.2 - x)$atm	x atm		x atm

$$K_p = \frac{P\{SO_2\}\ P\{Cl_2\}}{P\{SO_2Cl_2\}} = 0.029 = \frac{x^2}{14.2 - x} \qquad x^2 = 0.41 - 0.029\ x \qquad x^2 + 0.029x - 0.41 = 0$$

$$x = \frac{-b \pm \sqrt{b^2 - 4ac}}{2a} = \frac{-0.029 \pm \sqrt{0.00084 + 1.6}}{2} = 0.62 \text{ atm}$$

$P\{SO_2\} = P\{Cl_2\} = 0.62$ atm $\qquad P\{SO_2Cl_2\} = 14.2$ atm $- 0.62$ atm $= 13.6$ atm

(b) $P_{total} = P\{SO_2\} + P\{Cl_2\} + P\{SO_2Cl_2\} = 0.62$ atm $+ 0.62$ atm $+ 13.6$ atm $= 14.8$ atm

55. Let us start with a hypothetical one mole of air, and let $2x$ be the number of moles of NO formed.

Reaction:	$N_2(g)$	$+$	$O_2(g)$	$\rightleftharpoons$	$2 NO(g)$
Initial:	0.79 mol		0.21 mol		
Changes:	$-x$ mol		$-x$ mol		$+2x$ mol
Equil:	$(0.79 - x)$mol		$(0.21 - x)$mol		$2x$ mol

$$\chi_{NO} = \frac{n\{NO\}}{n\{N_2\} + n\{O_2\} + n\{NO\}} = \frac{2x}{(0.79 - x) + (0.21 - x) + 2x} = 0.018 = \frac{2x}{1.00}$$

$x = 0.0090$ mol $\qquad 0.79 - x = 0.78$ mol $N_2 \qquad 0.21 - x = 0.20$ mol $O_2 \qquad 2x = 0.018$ mol NO

$$K_p = \frac{P\{NO\}^2}{P\{N_2\}\ P\{O_2\}} = \frac{\left(\dfrac{n\{NO\}\ RT}{P_{total}}\right)^2}{\dfrac{n\{N_2\}\ RT}{P_{total}}\dfrac{n\{O_2\}\ RT}{P_{total}}} = \frac{n\{NO\}^2}{n\{N_2\}\ n\{O_2\}} = \frac{(0.018)^2}{0.78 \times 0.20} = 2.1 \times 10^{-3}$$

56. The $I_2(s)$ maintains the presence of I_2 in the flask until it has all vaporized. Thus, if enough HI(g) is produced to deplete the $I_2(s)$, equilibrium will not be achieved.

$$P\{H_2S\} = 758.2 \text{ mmHg} \times \frac{1 \text{ atm}}{760 \text{ mmHg}} = 0.9976 \text{ atm}$$

Reaction:	$H_2S(g) + I_2(s)$ $\rightleftharpoons$	$2 HI(g) + S(s)$
Initial:	0.9976 atm	
Changes:	$-x$ atm	$+2x$ atm
Equil:	$(0.9976 - x)$atm	$2x$ atm

$$K_p = \frac{P\{HI\}^2}{P\{H_2S\}} = \frac{(2x)^2}{(0.9976 - x)} = 1.34 \times 10^{-5} \approx \frac{4x^2}{0.9976}; \quad x = \sqrt{\frac{1.34 \times 10^{-5} \times 0.9976}{4}} = 1.83 \times 10^{-3}$$
atm

Our assumption, that $0.9976 >> x$, clearly is valid. We now verify that sufficient $I_2(s)$ is present by computing the mass of I_2 needed to produce the predicted pressure of HI(g). There is 1.05 g I_2 available.

$$\text{mass } I_2 = \frac{1.83 \times 10^{-3} \text{ atm} \times 0.515 \text{ L}}{0.08206 \text{ L atm mol}^{-1} \text{ K}^{-1} \times 333 \text{ K}} \times \frac{1 \text{ mol } I_2}{2 \text{ mol HI}} \times \frac{253.8 \text{ g } I_2}{1 \text{ mol } I_2} = 0.00438 \text{ g } I_2$$

$$P_{tot} = P\{H_2S\} + P\{HI\} = (0.9976 - x) + 2x = 0.9976 + x = 0.9976 + 0.00183 = 0.9994 \text{ atm} = 759.5 \text{ mmHg}$$

57. Notice that, for this reaction, $K_p = K_n$, the equilibrium constant in terms of amount.

$$K_p = \frac{P\{HI\}^2}{P\{H_2\}\ P\{I_2\}} = \frac{\left(\dfrac{n\{HI\}RT}{P_{total}}\right)^2}{\dfrac{n\{H_2\}RT}{P_{total}} \dfrac{n\{I_2\}RT}{P_{total}}} = \frac{n\{HI\}^2}{n\{H_2\}\ n\{I_2\}} = K_n = 69$$

We start with 1.000 mol HI to determine the % dissociation. Notice that the value of K_n does not depend on either the total pressure, nor the volume of the container.

Reaction: $H_2(g) +$ $I_2(g)$ $\rightleftharpoons$ $2 HI(g)$

Initial: 1.000 mol

Changes: $+x$ mol $+x$ mol $-2x$ mol

Equil: x mol x mol $(1.000 - 2x)$mol

$$K_n = \frac{n\{HI\}^2}{n\{H_2\}\ n\{I_2\}} = \frac{(1.000 - 2x)^2}{x^2} = 69 \qquad \sqrt{69} = \frac{1.000 - 2x}{x} = 8.3$$

$$1.000 - 2x = 8.3\ x \qquad\qquad 10.3\ x = 1.000 \qquad\qquad x = \frac{1.000}{10.3} = 0.0971$$

$$\% \text{ dissociation} = \frac{2x}{1.000} \times 100\% = \frac{2 \times 0.0971}{1.000} \times 100\% = 19.4\%$$

58. Reaction: $2 SO_3(g) \rightleftharpoons 2 SO_2(g) + O_2(g)$

Initial: 1.00 atm

Changes: $-2x$ atm $+2x$ atm $+x$ atm

Equil: $(1.00 - 2x)$atm $2x$ atm x atm

Because of the small value of the equilibrium constant, the reaction does not proceed very far toward products in reaching equilibrium. Hence, we assume that $x \ll 1.00$ atm and calculate an approximate value of x.

$$K_p = \frac{P\{SO_2\}^2\ P\{O_2\}}{P\{SO_3\}^2} = \frac{(2x)^2\ x}{(1.00 - 2x)^2} = 1.6 \times 10^{-5} \approx \frac{4x^3}{(1.00)^2} \qquad x = 0.016 \text{ atm}$$

A second cycle may get closer to the true value of x. $1.6 \times 10^{-5} \approx \dfrac{4x^3}{(1.00 - 0.016)^2}$ $x = 0.016$ atm

Our initial value was sufficiently close. Now compute the total pressure at equilibrium.

$$P_{total} = P\{SO_3\} + P\{SO_2\} + P\{O_2\} = (1.00 - 2x) + 2x + x = 1.00 + x = 1.00 + 0.016 = 1.02 \text{ atm}$$

Le Châtelier's Principle

59. Continuous removal of the product, of course, has the effect of decreasing the concentration of the products below their equilibrium values. Thus, the equilibrium system is disturbed by removing the products and the system will attempt (in vain, as it turns out) to re-establish the equilibrium by shifting toward the right, that is, by producing more products.

60. **(a)** This reaction is exothermic with $\Delta H° = -150.$ kJ. Thus high temperatures favor the reverse reaction. The amount of $H_2(g)$ present at high temperatures will be less than that present at low temperatures.

     (b) $H_2O(g)$ is one of the reactants involved. Introducing more will cause the equilibrium position to shift to the right, favoring products. The amount of $H_2(g)$ will increase.

     (c) Doubling the volume of the container will favor the side of the reaction with the largest sum of gaseous stoichiometric coefficients. The sum of the stoichiometric coefficients of gaseous species is the same (4) on both sides of this reaction. Therefore, increasing the volume of the container will have no effect on the amount of $H_2(g)$ present at equilibrium.

     (d) A catalyst merely speeds up the rate at which a reaction reaches the equilibrium position. The addition of a catalyst has no effect on the amount of $H_2(g)$ present at equilibrium.

61. In reaction (1) the sum of the stoichiometric coefficients of the gaseous products is larger than the sum of the stoichiometric coefficients of the gaseous reactants. Consequently, the equilibrium position of this reaction will shift toward products when either the total pressure is lowered or, equivalently, when the total volume of the system is increased.

On the other hand, the two sums referred to in the paragraph above are equal in the case of reaction (2). Hence, changing the total pressure or the total volume will have no effect on the position of the equilibrium.

62. In a reaction of the type $I_2(g) \rightleftharpoons 2\ I(g)$ the bond between two iodine atoms, the I—I bond, must be broken. Since $I_2(g)$ is stable molecule, this bond breaking process must be endothermic. Hence, the reaction cited is an endothermic reaction. The equilibrium position of endothermic reactions is shifted toward products when the temperature is raised.

63. **(a)** The formation of NO(g) from the elements is an endothermic reaction ($\Delta H° = +181$ kJ/mol). Since the equilibrium position of endothermic reactions is shifted toward products at high temperatures, we expect the formation of NO(g) from the elements to be enhanced at high temperatures.

   (b) Reaction rates always are enhanced by high temperatures, since a larger fraction of the collisions will have an energy which surmounts the activation energy. This enhancement of rates affects both the forward and the reverse reactions. Thus, the position of equilibrium is reached more rapidly at high temperatures than at low temperatures.

64. We notice that the density of the solid ice is smaller than is that of liquid water. This means that the same mass of liquid water is present in a smaller volume than an equal mass of ice. Thus, if pressure is placed on ice, attempting to force it into a smaller volume, the ice will be transformed into the less-space-occupying water at 0°C. Thus, at 0°C ice will melt to water under pressure. This behavior is *not* expected in most cases because generally a solid is *more* dense than its liquid phase.

65. If the reaction is endothermic ($\Delta H° > 0$), the forward reaction is favored at high temperatures. If the reaction is exothermic ($\Delta H° < 0$), the forward reaction is favored at low temperatures.

   (a) $\Delta H° = \Delta H°_f[PCl_5(g)] - \Delta H°_f[PCl_3(g)] - \Delta H°_f[Cl_2(g)]$
   $= -374.9 - (-287.0) - 0.00 = -87.9$ kJ/mol   low temperatures

   (b) $\Delta H° = 2\ \Delta H°_f[H_2O(g)] + 3\ \Delta H°_f[S(rhombic)] - \Delta H°_f[SO_2(g)] - 2\ \Delta H°_f[H_2S(g)]$
   $= 2\ (-241.8) + 3\ (0.00) - (-296.8) - 2\ (-20.63) = -145.5$ kJ/mol   low temperatures

   (c) $\Delta H° = 4\ \Delta H°_f[NOCl(g)] + 2\ \Delta H°_f[H_2O(g)] - 2\ \Delta H°_f[N_2(g)] - 3\ \Delta H°_f[O_2(g)] - 4\ \Delta H°_f[HCl(g)]$
   $= 4\ (51.71) + 2\ (-241.8) - 2\ (0.00) - 3\ (0.00) - 4\ (-92.31) = +92.5$ kJ/mol   high temps

Kinetics and Equilibrium

66. Although the yield of each of these reactions is greater at lower temperatures, both reactions are faster at higher temperatures (as are all reactions). The desire is to cause these reactions to proceed as rapidly as feasible. We need to be aware that the facility in which the reaction is being carried out costs money to build and to operate. The longer this facility is occupied in producing a given batch of product, the more costly that product will be. Besides, there are other ways to form large quantities of product from a given amount of reactant. Ways other than running the reaction under conditions where the equilibrium constant has a large value include constantly removing the product as it is formed, and feeding in reactant as it is consumed. Both of these strategies will enhance the formation of product.

17 ACIDS AND BASES

REVIEW QUESTIONS

1. (a) K_w is the symbol for the ion product of water: $K_w = [H_3O^+][OH^-] = 1.0 \times 10^{-14}$ at 25°C.
    (b) pH symbolizes the negative of the logarithm of the hydrogen ion concentration: $pH = -\log [H_3O^+]$. High pH indicates an alkaline solution; low pH indicates an acidic one.
    (c) pK_a symbolizes the negative of the logarithm of the ionization constant of an acid: $pK_a = -\log K_a$.
    (d) Hydrolysis refers to the process where an ion reacts (as an acid or a base) with water to produce either hydrogen or hydroxide ion.
    (e) A Lewis acid is an electron pair acceptor, the recipient of a coordinate covalent bond.

2. (a) A conjugate base is the species produced when another species (molecule or ion) donates, and thus loses, a proton, H^+.
    (b) Percent ionization is the percentage ratio of the concentration of the ion formed by a species divided by the original concentration of that species prior to ionization.
    (c) An acid anhydride is a species, often a nonmetal oxide, whose aqueous solution is acidic.
    (d) Amphoterism refers to the ability of some substances, usually oxides, to behave as acids in the presence of strong bases, and behave as bases in the presence of strong acids.

3. (a) A Brønsted-Lowry acid is a proton (H^+) donor; a Brønsted-Lowry base is a proton acceptor..
    (b) $pH = [H_3O^+]$; the one is a logarithmic expression of the other, necessary because of the extreme range of $[H_3O^+]$ values commonly encountered.
    (c) K_a for NH_4^+ and K_b for NH_3 are inversely related to each other: $K_a \times K_b = K_w$.
    (d) The leveling effect refers the the inability of a somewhat basic solvent to distinguish between the strengths of two strong acids; they both donate proton equally well to the solvent. The electron-withdrawing effect refers to the attraction of electrons to an electronegative atom or group within a molecule. If the electrons are drawn away from an O—H bond, that bond will become polar and more easily broken, producing a stronger acid.

4. (a) HNO_2 is an acid, a proton donor. The conjugate base is NO_2^-.
    (b) OCl^- is a base, a proton acceptor. The conjugate acid is $HOCl$.
    (c) NH_2^- is a base, a proton acceptor. The conjugate acid is NH_3.
    (d) NH_4^+ is an acid, a proton donor. The conjugate base is NH_3.
    (e) $CH_3NH_3^+$ is an acid, a proton donor. The conjugate base is CH_3NH_2.

5. We write the conjugate base as the first product of the equilibrium reaction in which the acid is the first reactant.
    (a) $HIO_4(aq) + H_2O \rightleftharpoons IO_4^-(aq) + H_3O^+(aq)$
    (b) $C_6H_5COOH(aq) + H_2O \rightleftharpoons C_6H_5COO^-(aq) + H_3O^+(aq)$
    (c) $H_2PO_4^-(aq) + H_2O \rightleftharpoons HPO_4^{2-}(aq) + H_3O^+(aq)$
    (d) $C_6H_5NH_3^+(aq) + H_2O \rightleftharpoons C_6H_5NH_2(aq) + H_3O^+(aq)$

6. All of the solutes are strong acids or strong bases.
    (a) $[H_3O^+] = 0.0035 \text{ M HCl} \times \dfrac{1 \text{ mol } H_3O^+}{1 \text{ mol HCl}} = 0.0035 \text{ M}$

    $[OH^-] = \dfrac{K_w}{[H_3O^+]} = \dfrac{1.00 \times 10^{-14}}{0.0035 \text{ M}} = 2.9 \times 10^{-12} \text{ M}$

(b) $[OH^-] = 0.0182\ M\ KOH \times \dfrac{1\ mol\ OH^-}{1\ mol\ KOH} = 0.0182\ M$

$[H_3O^+] = \dfrac{K_w}{[OH^-]} = \dfrac{1.00 \times 10^{-14}}{0.0182\ M} = 5.5 \times 10^{-13}\ M$

(c) $[OH^-] = 0.0023\ M\ Sr(OH)_2 \times \dfrac{2\ mol\ OH^-}{1\ mol\ Sr(OH)_2} = 0.0046\ M$

$[H_3O^+] = \dfrac{K_w}{[OH^-]} = \dfrac{1.00 \times 10^{-14}}{0.0046\ M} = 2.2 \times 10^{-12}\ M$

(d) $[H_3O^+] = 2.6 \times 10^{-4}\ M \times \dfrac{1\ mol\ H_3O^+}{1\ mol\ HNO_3} = 2.6 \times 10^{-4}\ M$

$[OH^-] = \dfrac{K_w}{[H_3O^+]} = \dfrac{1.00 \times 10^{-14}}{2.6 \times 10^{-4}\ M} = 3.8 \times 10^{-11}\ M$

7. Again, all of the solutes are strong acids or strong bases.

(a) $[H_3O^+] = 3.0 \times 10^{-3}\ M \times \dfrac{1\ mol\ H_3O^+}{1\ mol\ HCl} = 3.0 \times 10^{-3}\ M$ $\qquad$ $pH = -\log(3.0 \times 10^{-3}) = 2.52$

(b) $[H_3O^+] = 0.000462\ M \times \dfrac{1\ mol\ H_3O^+}{1\ mol\ HNO_3} = 0.000462\ M$ $\qquad$ $pH = -\log(0.000462) = 3.335$

(c) $[OH^-] = 1.50 \times 10^{-3}\ M \times \dfrac{1\ mol\ OH^-}{1\ mol\ NaOH} = 1.50 \times 10^{-3}\ M$ $\qquad$ $pOH = -\log(1.50 \times 10^{-3}) = 2.824$

$pH = 14.000 - pOH = 14.000 - 2.824 = 11.176$

(d) $[OH^-] = 2.7 \times 10^{-3}\ M \times \dfrac{2\ mol\ OH^-}{1\ mol\ Ba(OH)_2} = 5.4 \times 10^{-3}\ M$ $\qquad$ $pOH = -\log(5.4 \times 10^{-3}) = 2.27$

$pH = 14.00 - 2.27 = 11.73$

8. Most probably, $0.11\ M = [H_3O^+]$ in $0.10\ M\ H_2SO_4$. Although H_2SO_4 is completely ionized in dilute solution, this solution is not sufficiently dilute. (Recall the paragraph before Example 17-10 on page 609 of the text.) Thus, 0.20 M is incorrect. However, the first ionization of H_2SO_4 is complete; 0.05 M cannot be correct. And the second ionization of H_2SO_4 is at least partially complete; 0.10 M incorrectly assumes that the only source of H_3O^+ is the first ionization.

9. We determine the amounts of H_3O^+ and OH^- and then the amount of the one that is in excess.

$24.10\ mL \times \dfrac{0.150\ mmol\ HNO_3}{1\ mL\ soln} \times \dfrac{1\ mmol\ H_3O^+}{1\ mmol\ HNO_3} = 3.62\ mmol\ H_3O^+$

$11.20\ mL \times \dfrac{0.412\ mmol\ KOH}{1\ mL\ soln} \times \dfrac{1\ mmol\ OH^-}{1\ mmol\ KOH} = 4.61\ mmol\ OH^-$

The net reaction is $H_3O^+(aq) + OH^-(aq) \longrightarrow 2\ H_2O$. There is an excess amount of OH^- of $(4.61 - 3.62 =)$ $0.99\ mmol\ OH^-$. This is a basic solution. The total volume of the solution is $(24.10 + 11.20 =)\ 35.30\ mL$.

$[OH^-] = \dfrac{0.99\ mmol\ OH^-}{35.30\ mL} = 0.028\ M$ $\qquad$ $pOH = -\log(0.028) = 1.55$ $\qquad$ $pH = 14.00 - 1.55 = 12.45$

10. Answer (4) is correct, $pH < 13$. If this were a strong base, its $[OH^-] = 0.10$, giving $pOH = 1.0$, and $pH = 13.0$. But since methylamine is a weak base, its $[OH^-] < 0.10$, giving $pOH > 1.0$, and $pH < 13.0$.

11. $[HC_3H_5O_2]_i = \dfrac{0.355\ mol}{715\ mL} \times \dfrac{1000\ mL}{1\ L} = 0.497\ M$ $\qquad$ $[HC_3H_5O_2]_e = 0.497\ M - 0.00273\ M = 0.494\ M$

$K_a = \dfrac{[H_3O^+][C_3H_5O_2^-]}{[HC_3H_5O_2]} = \dfrac{(0.00273)^2}{0.494} = 1.51 \times 10^{-5}$

12. (a) The set up is based on the balanced chemical equation.

Reaction:	$HC_8H_7O_2(aq) + H_2O$	$\rightleftharpoons$	$H_3O^+(aq) +$	$C_8H_7O_2^-(aq)$
Initial:	0.212 M		0 M	0 M
Changes:	$-x$ M		$+x$ M	$+x$ M
Equil:	$(0.212 - x)$ M		x M	x M

$K_a = 4.9 \times 10^{-5} = \dfrac{[H_3O^+][C_8H_7O_2^-]}{[HC_8H_7O_2]}$

$= \dfrac{x \cdot x}{0.212 - x} \approx \dfrac{x^2}{0.212}$

$x = \sqrt{0.212 \times 4.9 \times 10^{-5}} = 0.0032\ M = [H_3O^+] = [C_8H_7O_2^-]$

(b)

Reaction:	$HC_8H_7O_2 + H_2O \rightleftharpoons$	$H_3O^+ +$	$C_8H_7O_2^-$
Initial:	0.107 M		
Changes:	$-x$ M	$+x$ M	$+x$ M
Equil:	$(0.107 - x)$ M	x M	x M

$$K_a = 4.9 \times 10^{-5} = \frac{[H_3O^+][C_8H_7O_2^-]}{[HC_8H_7O_2]} = \frac{x^2}{0.107 - x} \approx \frac{x^2}{0.107} \qquad x = 0.0023 \text{ M} = [H_3O^+]$$

We have assumed $x \ll 0.107$, an assumption that clearly is correct. pH $= -\log(0.0023) = 2.64$

13. We need first to determine the molarity S of the acid needed. ([H$_3$O$^+$] $= 10^{-2.60} = 2.5 \times 10^{-3}$ M)

Reaction:	HC$_7$H$_5$O$_2$ + H$_2$O $\rightleftharpoons$	H$_3$O$^+$	+	C$_7$H$_5$O$_2^-$
Initial:	S	0 M		0 M
Changes:	-0.0025 M	$+0.0025$ M		$+0.0025$ M
Equil:	$S - 0.0025$ M	0.0025 M		0.0025 M

$$K_a = \frac{[H_3O^+][C_7H_5O_2^-]}{[HC_7H_5O_2]} = \frac{(0.0025)^2}{S - 0.0025} = 6.3 \times 10^{-5} \qquad S - 0.0025 = \frac{(0.0025)^2}{6.3 \times 10^{-5}} = 0.099$$

$S = 0.099 + 0.0025 = 0.102$ M $= [HC_7H_5O_2]$

$$250.0 \text{ mL} \times \frac{1 \text{ L}}{1000 \text{ mL}} \times \frac{0.102 \text{ mol HC}_7\text{H}_5\text{O}_2}{1 \text{ L soln}} \times \frac{122.1 \text{ g HC}_7\text{H}_5\text{O}_2}{1 \text{ mol HC}_7\text{H}_5\text{O}_2} = 3.11 \text{ g HC}_7\text{H}_5\text{O}_2$$

14. First determine [OH$^-$], and thus [(CH$_3$)$_3$NH$^+$], in the solution. Then use the K_b expression to find [(CH$_3$)$_3$N]. pOH $= 14.00 - $ pH $= 14.00 - 11.50 = 2.50$ [OH$^-$] $= 10^{-pOH} = 10^{-2.50} = 0.0032$ M

$$K_b = 6.3 \times 10^{-5} = \frac{[(CH_3)_3NH^+][OH^-]}{[(CH_3)_3N]_{equil}} = \frac{(0.0032)^2}{[(CH_3)_3N]_{equil}} \qquad [(CH_3)_3N]_{equil} = \frac{(0.0032)^2}{6.3 \times 10^{-5}} = 0.16 \text{ M}$$

Since some of the (CH$_3$)$_3$N has ionized, [(CH$_3$)$_3$N]$_{initial}$ = 0.16 M + 0.0032 M = 0.16 M

15. For H$_2$CO$_3$, $K_1 = 4.4 \times 10^{-7}$ and $K_2 = 4.7 \times 10^{-11}$

(a) The first acid ionization proceeds to a far greater extent than does the second and determines the value of [H$_3$O$^+$].

Reaction:	H$_2$CO$_3$ + H$_2$O $\rightleftharpoons$	H$_3$O$^+$	+	HCO$_3^-$
Initial:	0.025 M			
Changes:	$-x$ M	$+x$ M		$+x$ M
Equil:	$(0.025 - x)$M	x M		x M

$$K_1 = \frac{[H_3O^+][HCO_3^-]}{[H_2CO_3]} = \frac{x^2}{0.025 - x} = 4.4 \times 10^{-7} \approx \frac{x^2}{0.025} \qquad x = 1.0 \times 10^{-4} \text{ M} = [H_3O^+]$$

(b) Since the second ionization occurs to only a limited extent, [HCO$_3^-$] $= 1.0 \times 10^{-4}$ M

(c) We use the second ionization to determine [CO$_3^{2-}$].

Reaction:	HCO$_3^-$ + H$_2$O $\rightleftharpoons$	H$_3$O$^+$	+	CO$_3^{2-}$
Initial:	0.00010 M	1.0×10^{-4} M		
Changes:	$-x$ M	$+x$ M		$+x$ M
Equil:	$(0.00010 - x)$M	$(0.00010 + x)$M		x M

$$K_2 = \frac{[H_3O^+][CO_3^{2-}]}{[HCO_3^-]} = \frac{(0.00010 + x)\, x}{0.00010 - x} = 4.7 \times 10^{-11} \approx \frac{(0.00010)\, x}{0.00010}$$

$x = 4.7 \times 10^{-11}$ M $= [CO_3^{2-}]$

Again we assumed that $x \ll 0.00010$ M, which clearly is the case. We also note that our original assumption, that the second ionization is much less significant than the first, also is a valid one.

16. The species that hydrolyze are the cations of weak bases—NH$_4^+$ and C$_6$H$_5$NH$_3^+$— and the anions of weak acids—NO$_2^-$ and C$_7$H$_5$O$_2^-$.

(a) NH$_4^+$(aq) + NO$_3^-$(aq) + H$_2$O $\rightleftharpoons$ NH$_3$(aq) + NO$_3^-$(aq) + H$_3$O$^+$(aq)

(b) Na$^+$(aq) + NO$_2^-$(aq) + H$_2$O $\rightleftharpoons$ Na$^+$(aq) + HNO$_2$(aq) + OH$^-$(aq)

(c) K$^+$(aq) + C$_7$H$_5$O$_2^-$(aq) + H$_2$O $\rightleftharpoons$ K$^+$(aq) + HC$_7$H$_5$O$_2$(aq) + OH$^-$(aq)

(d) K$^+$(aq) + Cl$^-$(aq) + Na$^+$(aq) + I$^-$(aq) + H$_2$O $\longrightarrow$ no reaction

(e) C$_6$H$_5$NH$_3^+$(aq) + Cl$^-$(aq) + H$_2$O $\rightleftharpoons$ C$_6$H$_5$NH$_2$(aq) + Cl$^-$(aq) + H$_3$O$^+$(aq)

17.

Reaction:	NH$_4^+$ + H$_2$O $\rightleftharpoons$	H$_3$O$^+$ +	NH$_3$
Initial:	1.15 M	0 M	0 M
Changes:	$-x$ M	$+x$ M	$+x$ M
Equil:	$(1.15 - x)$M	x M	x M

$$K_a = \frac{[H_3O^+][NH_3]}{[NH_4^+]} = \frac{x^2}{1.15 - x} = 5.8 \times 10^{-10} \approx \frac{x^2}{1.15}$$

$x = 2.6 \times 10^{-5}$ M $= [H_3O^+]$ pH $= -\log(2.6 \times 10^{-5}) = 4.59$

18. Recall that, for a conjugate weak acid-weak base pair, $K_a \times K_b = K_w$

(a) $K_a = \dfrac{K_w}{K_b} = \dfrac{1.00 \times 10^{-14}}{1.5 \times 10^{-9}} = 6.7 \times 10^{-6}$ for $C_5H_5NH^+$

(b) $K_b = \dfrac{K_w}{K_a} = \dfrac{1.00 \times 10^{-14}}{1.8 \times 10^{-4}} = 5.6 \times 10^{-11}$ for CHO_2^-

(c) $K_b = \dfrac{K_w}{K_a} = \dfrac{1.00 \times 10^{-14}}{1.0 \times 10^{-10}} = 1.0 \times 10^{-4}$ for $C_6H_5O^-$

19. We base the calculation on the balanced chemical equation for the hydrolysis of the anion.

Reaction: $C_2H_2ClO_2^-(aq) + H_2O \rightleftharpoons HC_2H_2ClO_2(aq) + OH^-(aq)$

Initial: 2.50 M   0 M   0 M
Changes: $-x$ M $+x$ M $+x$ M
Equil: $(2.50 - x)$ M x M x M

$K_b = \dfrac{K_w}{K_a} = \dfrac{1.0 \times 10^{-14}}{1.4 \times 10^{-3}} = 7.1 \times 10^{-12} = \dfrac{[HC_2H_2ClO_2][OH^-]}{[C_2H_2ClO_2^-]} = \dfrac{x \cdot x}{2.50 - x} \approx \dfrac{x^2}{2.50}$

$x = \sqrt{2.50 \times 7.1 \times 10^{-12}} = 4.2 \times 10^{-6} = [OH^-]$ $pOH = -\log(4.2 \times 10^{-6}) = 5.38$

$pH = 14.00 - pOH = 14.00 - 5.38 = 8.62$ M

20. $HC_2Cl_3O_2$ is a stronger acid than $HC_2H_3O_2$ because in $HC_2Cl_3O_2$ there are electronegative (electron withdrawing) Cl atoms bonded to the atom adjacent to the COOH group. The electron withdrawing power of these Cl atoms polarizes the O—H bond, creating a stronger acid.

21. (a) HI is the more acidic because the H—I bond length is longer than the H—Br bond length and the H—I bond is weaker.

(b) HOClO is more acidic than HOI because there is a terminal O in HOClO but not in HOI and Cl is more electronegative than I.

(c) $H_3CCH_2CCl_2COOH$ is more acidic than $I_3CCH_2CH_2COOH$ both because Cl is more electronegative than is I and also because the Cl atoms are closer to the acidic hydrogen in the COOH group and thus can exert a stronger effect on it than can the more distant I atoms.

22. A Lewis base is an electron pair donor, while a Lewis acid is an electron pair acceptor. We draw Lewis structures to assist our interpretation.

(a) According to the following Lewis structures, SO_3 appears to be the electron pair acceptor (the Lewis acid), and H_2O is the electron pair donor (the Lewis base). Note that an additional sulfur-to-oxygen bond is formed.

(b)

Zn in $Zn(OH)_2$ accepts a pair of electrons; $Zn(OH)_2$ is a Lewis acid. OH^- donates the pair of electrons that form the covalent bond; OH^- is a Lewis base. We have assumed that Zn has sufficient covalent character to form coordinate covalent bonds with hydroxide ions.

23. (a) $NH_4Cl(aq)$ contains a cation—$NH_4^+(aq)$—that is a weak acid; it should have a pH slightly below 7.00.

(b) $NH_3(aq)$ is a weak base; it should have a reasonably high pH.

(c) $NaC_2H_3O_2(aq)$ contains an anion—$C_2H_3O_2^-(aq)$—that is a weak base; it should have a pH slightly above 7.00.

(d) Neither $K^+(aq)$ nor $Cl^-(aq)$ is an acid or a base in aqueous solution; $KCl(aq)$ should have a pH of 7.00.

In order of decreasing pH: $NH_3(aq) > NaC_2H_3O_2(aq) > KCl(aq) > NH_4Cl(aq)$

24. We might initially assume that an HCl solution of pH = 8 would simply have [HCl] = 1×10^{-8} M. But we must remember that pure water already has $[H_3O^+] = 1 \times 10^{-7}$ M. Hence, adding HCl—a strong acid which is a source of H_3O^+—will only increase $[H_3O^+]$ and thus lower the pH below the value of pH = 7, the value for pure water. In 1×10^{-8} M HCl we might expect $[H_3O^+] = 1.0 \times 10^{-7}$ M (from water) $+ 1 \times 10^{-8}$ M (from HCl) $= 1.1 \times 10^{-7}$ M. Thus pH $= -\log(1.1 \times 10^{-7}$ M$) = 6.96$, an acidic solution.

EXERCISES

Brønsted-Lowry Theory of Acids and Bases

25. The acids (proton donors) and bases (proton acceptors) are labeled below their formulas. Remember that a proton, in Brønsted-Lowry acid-base theory , is H^+.

    (a) $HOBr + H_2O \rightleftharpoons H_3O^+ + OBr^-$
          acid   base   acid   base

    (b) $HSO_4^- + H_2O \rightleftharpoons H_3O^+ + SO_4^{2-}$
          acid   base   acid   base

    (c) $HS^- + H_2O \rightleftharpoons H_2S + OH^-$
          base   acid   acid   base

    (d) $C_6H_5NH_3^+ + OH^- \rightleftharpoons C_6H_5NH_2 + H_2O$
          acid   base   base   acid

26. For each amphiprotic substance, we write its reactions with water: with the substance acting as an acid, and then with it acting as a base. Even for the substances that are not usually considered amphiprotic, both reactions are written, but one of them is labeled as unlikely. In some instances we have written an oxygen as Ø to keep track of it through the reaction.

    (a) $ØH^- + H_2O \rightleftharpoons Ø^{2-} + H_3O^+$ (unlikely) $ØH^- + H_2O \rightleftharpoons H_2Ø + OH^-$

    (b) $NH_4^+ + H_2O \rightleftharpoons NH_3 + H_3O^+$ NH_4^+ cannot add a proton without expanding octet

    (c) $H_2Ø + H_2O \rightleftharpoons ØH^- + H_3O^+$ $H_2Ø + H_2O \rightleftharpoons H_3Ø^+ + OH^-$

    (d) $HS^- + H_2O \rightleftharpoons S^{2-} + H_3O^+$ $HS^- + H_2O \rightleftharpoons H_2S + OH^-$

    (e) NO_3^- cannot act as an acid, no protons $NO_3^- + H_2O \rightleftharpoons HNO_3 + OH^-$ (unlikely)

    (f) $HCO_3^- + H_2O \rightleftharpoons CO_3^{2-} + H_3O^+$ $HCO_3^- + H_2O \rightleftharpoons H_2CO_3 + OH^-$

    (g) $HNO_3 + H_2O \rightleftharpoons H_3O^+ + NO_3^-$ $HNO_3 + H_2O \rightleftharpoons H_2NO_3^+ + OH^-$ (unlikely)

27. Although the Brønsted-Lowry theory defines an acid as a proton donor and a base as a proton acceptor, the situation is closer to one of a base actively seeking (rather than passively accepting) protons. Hence, in a solvent that is a stronger base, one that will more avidly seek protons—such as pyridine, $C_5H_5N(l)$, which is a stronger base than $H_2O(l)$—a weak acid, such as C_6H_5COOH, will become stronger, often ionizing completely to become a strong acid.

28. Answer (2), NH_3, is correct. $HC_2H_3O_2$ will react most completely with the strongest base. NO_3^- and Cl^- are very weak bases. H_2O is a weak base, but can act as an acid (donating protons) such as in the presence of NH_3. Thus, NH_3 must be the strongest base and thus the most effective in causing $HC_2H_3O_2$ to react.

29. Lewis structures are given below each equation.

    (a) $2 NH_3(l) \rightleftharpoons NH_4^+ + NH_2^-$ **(b)** $2 HF(l) \rightleftharpoons H_2F^+ + F^-$

    (c) $2 CH_3OH(l) \rightleftharpoons CH_3OH_2^+ + CH_3O^-$

    (d) $2 HC_2H_3O_2(l) \rightleftharpoons H_2C_2H_3O_2^+ + C_2H_3O_2^-$

(e) $2 H_2SO_4(l) \rightleftharpoons H_3SO_4^+ + HSO_4^-$

30. We use (ac) to represent the fact that a species is dissolved in acetic acid solution.
  (a) $C_2H_3O_2^-(ac) + HC_2H_3O_2 \rightleftharpoons HC_2H_3O_2(ac) + C_2H_3O_2^-(ac)$
  $C_2H_3O_2^-$ is a base in acetic acid solution.
  (b) $H_2O(ac) + HC_2H_3O_2 \rightleftharpoons H_3O^+(aq) + C_2H_3O_2^-(ac)$
  Since H_2O is a weaker acid than $HC_2H_3O_2$ in aqueous solution, it will also be weaker in acetic acid solution. Thus H_2O will accept a proton from the solvent acetic acid, making H_2O a base in acetic acid solution.
  (c) $HC_2H_3O_2(ac) + HC_2H_3O_2 \rightleftharpoons H_2C_2H_3O_2^+(ac) + C_2H_3O_2^-(aq)$
  Acetic acid is an acid or a base in acetic acid solution. See Exercise 29 (d).
  (d) $HClO_4(ac) + HC_2H_3O_2 \rightleftharpoons ClO_4^-(ac) + H_2C_2H_3O_2^+(ac)$
  Since $HClO_4$ is a stronger acid than $HC_2H_3O_2$ in aqueous solution, it will also be stronger in acetic acid solution. Thus, $HClO_4$ will donate a proton to the solvent acetic acid, making $HClO_4$ an acid in acetic acid solution.

31. The principle here is that the weaker acid and the weaker base will predominate at equilibrium. This is because if a substance is a strong acid it will do a good job of donating its protons and, having done so, its conjugate base will be left behind. The preferred direction is: Strong acid + strong base $\longrightarrow$ weak acid + weak base
  (a)   The reaction will favor the forward direction because OH^- (a strong base) > NH_3 as bases and NH_4^+ > H_2O as acids.
  (b)   The reaction will favor the reverse direction because HNO_3 > HSO_4^- (a weak acid in the second ionization) as acids, and SO_4^{2-} > NO_3^- as bases.
  (c)   The reaction will favor the the reverse direction because $HC_2H_3O_2$ > CH_3OH (not usually thought of as an acid) as acids, and CH_3O^- > $C_2H_3O_2^-$ as bases.
  (d)   The reaction will favor the forward direction because $HC_2H_3O_2$ (a moderate acid) > HCO_3^- (a quite weak acid) as acids and CO_3^{2-} > $C_2H_3O_2^-$ as bases.
  (e)   The reaction will favor the reverse direction because $HClO_4$ (a strong acid) > HNO_2 as acids, and NO_2^- > ClO_4^- as bases.
  (f)   The reaction will favor the forward direction, since H_2CO_3 > HCO_3^- as acids (since $K_1 > K_2$) and CO_3^{2-} > HCO_3^- as bases.

Strong Acids, Strong Bases, and pH

32. $[OH^-] = \dfrac{39 \text{ g Ba(OH)}_2 \cdot 8H_2O}{1 \text{ L soln}} \times \dfrac{1 \text{ mol Ba(OH)}_2 \cdot 8H_2O}{315.5 \text{ g Ba(OH)}_2 \cdot 8 H_2O} \times \dfrac{2 \text{ mol OH}^-}{1 \text{ mol Ba(OH)}_2 \cdot 8H_2O} = 0.25 \text{ M}$

$[H_3O^+] = \dfrac{K_w}{[OH^-]} = \dfrac{1.00 \times 10^{-14}}{0.25 \text{ M OH}^-} = 4.0 \times 10^{-14} \text{ M}$ $pH = -\log(4.0 \times 10^{-14}) = 13.40$

33. First determine the amount of HCl, and then its concentration.

amount $HCl = \dfrac{PV}{RT} = \dfrac{\left(742 \text{ mmHg} \times \dfrac{1 \text{ atm}}{760 \text{ mmHg}}\right) \times 0.187 \text{ L}}{0.08206 \text{ L atm mol}^{-1} \text{ K}^{-1} \times 295 \text{ K}} = 7.54 \times 10^{-3} \text{ mol HCl}$

$[H_3O^+] = \dfrac{7.54 \times 10^{-3} \text{ mol HCl}}{4.17 \text{ L soln}} \times \dfrac{1 \text{ mol H}_3O^+}{1 \text{ mol HCl}} = 1.81 \times 10^{-3} \text{ M}$

34. $pOH = 14.00 - 12.85 = 1.15$ $[OH^-] = 10^{-1.15} = 7.1 \times 10^{-2} \text{ M} = [NaOH]_f$

$V_i = V_f \dfrac{[NaOH]_f}{[NaOH]_i} = 1.00 \text{ L} \times \dfrac{7.1 \times 10^{-2} \text{ M}}{0.606 \text{ M}} = 0.12 \text{ L}$

35. Determine the mass of pure HCl needed, and then the volume of concentrated HCl(aq) needed.

$[H_3O^+] = 10^{-1.75} = 1.8 \times 10^{-2}$ M

$$\text{amount HCl} = 8.25 \text{ L} \times \frac{0.018 \text{ mol } H_3O^+}{1 \text{ L soln}} \times \frac{1 \text{ mol HCl}}{1 \text{ mol } H_3O^+} \times \frac{36.46 \text{ g HCl}}{1 \text{ mol HCl}} = 5.4 \text{ g HCl}$$

$$\text{vol. HCl(conc)} = 5.4 \text{ g HCl} \times \frac{100.0 \text{ g conc HCl}}{36.0 \text{ g HCl}} \times \frac{1 \text{ mL conc HCl}}{1.18 \text{ g conc HCl}} = 13 \text{ mL conc HCl soln}$$

36. The dissolved $Mg(OH)_2$ is completely dissociated into ions. $\quad$ pOH = 14.00 − pH = 14.00 − 10.53 = 3.47

$[OH^-] = 10^{-pOH} = 10^{-3.47} = 3.4 \times 10^{-4}$ M OH$^-$

$$\text{solubility} = \frac{3.4 \times 10^{-4} \text{ mol OH}^-}{1 \text{ L soln}} \times \frac{1 \text{ mol } Mg(OH)_2}{2 \text{ mol OH}^-} \times \frac{58.32 \text{ g } Mg(OH)_2}{1 \text{ mol } Mg(OH)_2} \times \frac{1000 \text{ mg}}{1 \text{ g}}$$
$$= 9.9 \text{ g } Mg(OH)_2/L$$

37. NH_3 and HCl react in a 1:1 molar ratio. Since, by Avogadro's hypothesis, equal volumes of gasses measured at the same temperature and pressure contain equal numbers of moles, we need to determine the volume at 757 mmHg and 22.0°C that is equivalent to 33.6 L at 738 mmHg and 25.0°C.

$$\text{volume } NH_3(g) = 33.6 \text{ L} \times \frac{738 \text{ mmHg}}{757 \text{ mmHg}} \times \frac{(273.2 + 22.0) \text{ K}}{(273.2 + 25.0) \text{ K}} = 32.4 \text{ L } NH_3(g)$$

We can make sure that we get the ratios correct with the approach of setting up and solving the ideal gas equation for the amount of each gas, and then equating these two expressions and solving for the volume of NH_3.

$$n\{HCl\} = \frac{738 \text{ mmHg} \cdot 33.6 \text{ L}}{R \cdot 298.2 \text{ K}} \qquad n\{NH_3\} = \frac{757 \text{ mmHg} \cdot V\{NH_3\}}{R \cdot 295.2 \text{ K}}$$

$$\frac{738 \text{ mmHg} \cdot 33.6 \text{ L}}{R \cdot 298.2 \text{ K}} = \frac{757 \text{ mmHg} \cdot V\{NH_3\}}{R \cdot 295.2 \text{ K}} \qquad \text{This yields } V\{NH_3\} = 32.4 \text{ L } NH_3(g)$$

38. The net equation is the familiar neutralization reaction: $H_3O^+ + OH^- \longrightarrow 2\, H_2O$ Then, in the solution being titrated we can determine $[OH^-]$.

$$[OH^-] = \frac{10.0 \text{ mL} \times \dfrac{0.17 \text{ g } Ca(OH)_2}{100 \text{ mL}} \times \dfrac{1 \text{ mol } Ca(OH)_2}{74.09 \text{ g } Ca(OH)_2} \times \dfrac{2 \text{ mol OH}^-}{1 \text{ mol } Ca(OH)_2}}{250.0 \text{ mL} \times \dfrac{1 \text{ L}}{1000 \text{ mL}}} = 1.8 \times 10^{-3} \text{ M}$$

$$\text{amount OH}^- = 10.0 \text{ mL} \times \frac{1.8 \times 10^{-3} \text{ mmol OH}^-}{1 \text{ mL soln}} = 0.018 \text{ mmol OH}^- \text{ titrated by HCl}$$

$$[HCl] = \frac{0.018 \text{ mmol OH}^- \times \dfrac{1 \text{ mmol } H_3O^+}{1 \text{ mmol OH}^-} \times \dfrac{1 \text{ mmol HCl}}{1 \text{ mmol } H_3O^+}}{25.1 \text{ mL HCl}} = 7.2 \times 10^{-4} \text{ M}$$

39. In each case, we need to determine the $[H_3O^+]$ or $[OH^-]$ of the solutions being mixed, and then the amount of H_3O^+ or OH^-, so that we can determine the amount in excess.

$[H_3O^+] = 10^{-2.52} = 3.0 \times 10^{-3}$ M $\qquad$ amount $H_3O^+ = 25.00$ mL $\times 3.0 \times 10^{-3}$ M $= 0.075$ mmol H_3O^+

pOH = 14.00 − 12.05 = 1.95 $\qquad\qquad [OH^-] = 10^{-1.95} = 1.1 \times 10^{-2}$ M

amount $OH^- = 25.00$ mL $\times 1.1 \times 10^{-2}$ M $= 0.28$ mmol OH^-

There is excess OH^- in the amount of 0.20 mmol OH^- (= 0.28 mmol OH^- − 0.075 mmol H_3O^+)

$[OH^-] = \dfrac{0.20 \text{ mmol OH}^-}{25.00 \text{ mL} + 25.00 \text{ mL}} = 4.0 \times 10^{-3}$ M $\qquad$ pOH $= -\log(4.0 \times 10^{-3}) = 2.40$

pH = 14.00 − 2.40 = 11.60

Weak Acids, Weak Bases, and pH

40. We base our set up on the balanced chemical equation. $[H_3O^+] = 10^{-1.46} = 3.5 \times 10^{-2}$ M

Reaction:	$CH_2FCOOH(aq) + H_2O \rightleftharpoons$	$CH_2FCOO^-(aq)$ +	$H_3O^+(aq)$
Initial:	0.500 M	0 M	0 M
Changes:	−0.035 M	+0.035 M	+0.035 M
Equil:	0.465 M	0.035 M	0.035 M

$$K_a = \frac{[H_3O^+][CH_2FCOO^-]}{[CH_2FCOOH]} = \frac{(0.035)(0.035)}{0.465} = 2.6 \times 10^{-3}$$

41. $[HC_6H_{11}O_2] = \dfrac{11\ g}{1\ L} \times \dfrac{1\ mol\ HC_6H_{11}O_2}{116.2\ g\ HC_6H_{11}O_2} = 9.5 \times 10^{-2}\ M = 0.095\ M$

$[H_3O^+] = 10^{-2.94} = 1.1 \times 10^{-3}\ M = 0.0011\ M$

The stoichiometry of the reaction, written below, indicates that $[H_3O^+] = [C_6H_{11}O_2^-]$

Reaction:	$HC_6H_{11}O_2 + H_2O \rightleftharpoons$	$C_6H_{11}O_2^-$ +	H_3O^+
Initial:	0.095 M	0 M	0 M
Changes:	−0.0011 M	+0.0011 M	+0.0011 M
Equil:	0.094 M	0.0011 M	0.0011 M

$K_a = \dfrac{[C_6H_{11}O_2^-][H_3O^+]}{[HC_6H_{11}O_2]} = \dfrac{(0.0011)^2}{0.094} = 1.3 \times 10^{-5}$

42. The balanced chemical equation indicates that $[H_3O^+] = [C_6H_5O^-]$ in a saturated aqueous solution. The value of $K_a = 1.0 \times 10^{-10}$ for phenol is from Table 17-3. The set up is based on the equilibrium constant expression. $HOC_6H_5(aq) + H_2O \rightleftharpoons C_6H_5O^-(aq) + H_3O^+(aq)$

$[H_3O^+] = 10^{-pH} = 10^{-4.90} = 1.3 \times 10^{-5}\ M = [C_6H_5O^-]$

$K_a = \dfrac{[H_3O^+][C_6H_5O^-]}{[HOC_6H_5]} = 1.0 \times 10^{-10} = \dfrac{(1.3 \times 10^{-5})^2}{[HOC_6H_5]}$ $[HOC_6H_5] = \dfrac{1.0 \times 10^{-10}}{(1.3 \times 10^{-5})^2} = 0.59\ M$

phenol solubility $= \dfrac{0.59\ mol\ HOC_6H_5}{1\ L} \times \dfrac{94.11\ g\ HOC_6H_5}{1\ mol\ HOC_6H_5} = 56\ gHOC_6H_5/L\ soln$

43. The stoichiometry of the ionization reaction indicates that $[H_3O^+] = [OC_6H_4NO_2^-]$. $K_a = 10^{-pK} = 10^{-7.23} = 5.9 \times 10^{-8}$ and $[H_3O^+] = 10^{-4.53} = 3.0 \times 10^{-5}\ M$. We let $S =$ the molar solubility of o-nitrophenol.

Reaction:	$HOC_6H_4NO_2 + H_2O \rightleftharpoons$	$OC_6H_4NO_2^-$ +	H_3O^+
Initial:	S	0 M	0 M
Changes:	$-3.0 \times 10^{-5}\ M$	$+3.0 \times 10^{-5}\ M$	$+3.0 \times 10^{-5}\ M$
Equil:	$S - 3.0 \times 10^{-5}\ M$	$3.0 \times 10^{-5}\ M$	$3.0 \times 10^{-5}\ M$

We shall initially assume that $3.0 \times 10^{-5}\ M \ll S$, to avoid the necessity of solving a quadratic equation.

$K_a = 5.9 \times 10^{-8} = \dfrac{[OC_6H_4NO_3^-][H_3O^+]}{[HOC_6H_5NO_2]} = \dfrac{(3.0 \times 10^{-5})^2}{S - 3.0 \times 10^{-5}} \approx \dfrac{(3.0 \times 10^{-5})^2}{S}$ $S = 1.5 \times 10^{-2}\ M$

We see that our initial assumption is valid, since $1.5 \times 10^{-2}\ M \gg 3.0 \times 10^{-5}\ M$

Hence solubility $= \dfrac{1.5 \times 10^{-2}\ mol\ HOC_6H_4NO_2}{1\ L\ soln} \times \dfrac{139.1\ g\ HOC_6H_4NO_2}{1\ mol\ HOC_6H_4NO_2} = 2.1\ g\ HOC_6H_4NO_2/L\ soln$

44. $[H_3O^+] = 10^{-4.86} = 1.4 \times 10^{-5}\ M = [C_6H_4ClO^-]$ The two concentrations are equal since when one ion is produced from o-chlorophenol the other is also. Then we have the following.

$K_a = 10^{-pK} = 10^{-8.49} = 3.2 \times 10^{-9} = \dfrac{[H_3O^+][C_6H_4ClO^-]}{[HOC_6H_4Cl]} = \dfrac{(1.4 \times 10^{-5})^2}{[HOC_6H_4Cl]}$ $[HOC_6H_4Cl] = 0.061\ M$

45. Since a strong acid is completely ionized, increasing its concentration has the effect of increasing $[H_3O^+]$ by an equal amount. In the case of a weak acid, however, the solution of the K_a expression gives the following equation (if we can assume that $[H_3O^+]$ is negligible compared to the original concentration of the undissociated acid, $[HA]$). $x^2 = K_a[HA]$ (where $x = [H_3O^+]$), or $x = \sqrt{K_a[HA]}$. Thus, if $[HA]$ is doubled, $[H_3O^+]$ increases by $\sqrt{2}$

46. We determine $[H_3O^+]$ which, because of the stoichiometry of the reaction, equals $[C_2H_3O_2^-]$.

$[H_3O^+] = 10^{-4.52} = 3.0 \times 10^{-5}\ M = [C_2H_3O_2^-]$

We solve for S, the concentration of $HC_2H_3O_2$ in the 0.750 L solution before it dissociates.

Reaction:	$HC_2H_3O_2 + H_2O \rightleftharpoons$	$C_2H_3O_2^-$ +	H_3O^+
Initial:	S	0 M	0 M
Changes:	$-3.0 \times 10^{-5}\ M$	$+3.0 \times 10^{-5}\ M$	$+3.0 \times 10^{-5}\ M$
Equil:	$S - 3.0 \times 10^{-5}\ M$	$3.0 \times 10^{-5}\ M$	$3.0 \times 10^{-5}\ M$

$K_a = \dfrac{[H_3O^+][C_2H_3O_2^-]}{[HC_2H_3O_2]} = 1.8 \times 10^{-5} = \dfrac{(3.0 \times 10^{-5})^2}{(S - 3.0 \times 10^{-5})}$

$(3.0 \times 10^{-5})^2 = 1.8 \times 10^{-5}(S - 3.0 \times 10^{-5}) = 9.0 \times 10^{-10} = 1.8 \times 10^{-5}S - 5.4 \times 10^{-10}$

$$S = \frac{9.0 \times 10^{-10} + 5.4 \times 10^{-10}}{1.8 \times 10^{-5}} = 2.0 \times 10^{-5} \text{ M}$$

Now we determine the mass of vinegar needed.

$$\text{mass vinegar} = 0.750 \text{ L} \times \frac{2.0 \times 10^{-5} \text{ mol HC}_2\text{H}_3\text{O}_2}{1 \text{ L soln}} \times \frac{60.05 \text{ g HC}_2\text{H}_3\text{O}_2}{1 \text{ mol HC}_2\text{H}_3\text{O}_2} \times \frac{100.0 \text{ g vinegar}}{5.7 \text{ g HC}_2\text{H}_3\text{O}_2}$$
$$= 0.016 \text{ g vinegar}$$

47. (a) We dissolve the methylamine in 1.00 L of water, which we shall assume is the volume of the final solution. First determine the number of moles of methylamine.

$$\text{mol CH}_3\text{NH}_2 = \frac{PV}{RT} = \frac{1.00 \text{ atm} \times 959 \text{ L}}{0.08206 \text{ L atm mol}^{-1} \text{ K}^{-1} \times 298 \text{ K}} = 39.2 \text{ mol CH}_3\text{NH}_2$$

And thus $[\text{CH}_3\text{NH}_2] = 39.2 \text{ mol}/1.00 \text{ L} = 39.2 \text{ M}$

Reaction:	$\text{CH}_3\text{NH}_2 + \text{H}_2\text{O} \rightleftharpoons$	$\text{CH}_3\text{NH}_3^+ +$	OH^-
Initial:	39.2 M	0 M	0 M
Changes:	$-x$ M	$+x$ M	$+x$ M
Equil:	$(39.2 - x)$M	x M	x M

We assume that $x \ll 39.2$ in order to obtain a solution.

$$K_b = \frac{[\text{CH}_3\text{NH}_3^+][\text{OH}^-]}{[\text{CH}_3\text{NH}_2]} = 4.2 \times 10^{-4} = \frac{x^2}{39.2 - x} \approx \frac{x^2}{39.2} \qquad x = 0.13 \text{ M} = [\text{OH}^-]$$

$$\text{pOH} = -\log(0.13) = 0.89 \qquad\qquad \text{pH} = 13.11$$

(b) $[\text{NaOH}] = 0.13 \text{ M OH}^- \times \frac{1 \text{ mol NaOH}}{1 \text{ mol OH}^-} = 0.13 \text{ M NaOH}$

48. $K_b = 10^{-\text{pK}} = 10^{-9.5} = 3 \times 10^{-10}$

$$[\text{C}_9\text{H}_7\text{N}] = \frac{0.6 \text{ g C}_9\text{H}_7\text{N}}{100 \text{ mL}} \times \frac{1 \text{ mol C}_9\text{H}_7\text{N}}{129.2 \text{ g C}_9\text{H}_7\text{N}} \times \frac{1000 \text{ mL}}{1 \text{ L}} = 0.05 \text{ M}$$

We base our set up on the balanced chemical equation.

Reaction:	$\text{C}_9\text{H}_7\text{N(aq)} + \text{H}_2\text{O} \rightleftharpoons$	$\text{C}_9\text{H}_7\text{NH}^+\text{(aq)} +$	$\text{OH}^-\text{(aq)}$
Initial:	0.05 M	0 M	0 M
Changes:	$-x$ M	$+x$ M	$+x$ M
Equil:	$(0.05 - x)$M	x M	x M

$$K_b = \frac{[\text{C}_9\text{H}_7\text{NH}^+][\text{OH}^-]}{[\text{C}_9\text{H}_7\text{N}]} = 3 \times 10^{-10} = \frac{x \cdot x}{0.05 - x} \approx \frac{x^2}{0.05}$$

$$x = \sqrt{0.05 \times 3 \times 10^{-10}} = 4 \times 10^{-6} \text{ M} = [\text{OH}^-] \qquad \text{pOH} = -\log(4 \times 10^{-6}) = 5.4$$
$$\text{pH} = 14.00 - 5.4 = 8.6$$

Percent Ionization

49. Let us first compute the $[\text{H}_3\text{O}^+]$ in this solution.

Reaction:	$\text{HC}_3\text{H}_5\text{O}_2 + \text{H}_2\text{O} \rightleftharpoons$	$\text{H}_3\text{O}^+ +$	$\text{C}_3\text{H}_5\text{O}_2^-$
Initial:	0.45 M	0 M	0 M
Changes:	$-x$ M	$+x$ M	$+x$ M
Equil:	$(0.45 - x)$M	x M	x M

$$K_a = \frac{[\text{H}_3\text{O}^+][\text{C}_3\text{H}_5\text{O}_2^-]}{[\text{HC}_3\text{H}_5\text{O}_2]} = \frac{x^2}{0.45 - x} = 1.3 \times 10^{-5} \approx \frac{x^2}{0.45} \qquad x = 2.4 \times 10^{-3} \text{ M}$$

We have assumed that $x \ll 0.45$ M, an assumption that clearly is correct.

(a) $\alpha = \frac{[\text{H}_3\text{O}^+]_{\text{eq}}}{[\text{HC}_3\text{H}_5\text{O}_2]_i} = \frac{2.4 \times 10^{-3} \text{ M}}{0.45 \text{ M}} = 0.0053$

(b) % ionization $= \alpha \times 100\% = 0.0053 \times 100\% = 0.53\%$

50. We first determine $[\text{H}_3\text{O}^+]$ in this solution. $\qquad K_a = \text{pK}_a = 10^{-0.70} = 0.20$

Reaction:	$\text{HC}_2\text{Cl}_3\text{O}_2 + \text{H}_2\text{O} \rightleftharpoons$	$\text{C}_2\text{Cl}_3\text{O}_2^- +$	H_3O^+
Initial:	0.050 M	0 M	0 M
Changes:	$-x$ M	$+x$ M	$+x$ M
Equil:	$(0.050 - x)$M	x M	x M

$$K_a = \frac{[C_2Cl_3O_2^-][H_3O^+]}{[HC_2Cl_3O_2]} = 0.20 = \frac{x^2}{0.050 - x} \qquad x^2 = 0.010 - 0.20\,x \qquad x^2 + 0.20\,x - 0.010 = 0$$

$$x = \frac{-b \pm \sqrt{b^2 - 4ac}}{2a} = \frac{-0.20 \pm \sqrt{0.040 + 0.040}}{2} = 0.04\ \text{M} = [C_2Cl_3O_2^-]$$

$$\alpha = \frac{[C_2Cl_3O_2^-]_{eq}}{[HC_2Cl_3O_2]_i} = \frac{0.04\ \text{M}}{0.050\ \text{M}} = 0.8 \qquad \%\ \text{ionization} = 8 \times 10^1\%$$

51. The fact that the NH_3 is 2.0% ionized means that $[NH_4^+] = 0.020\,[NH_3]_{initial} = [OH^-]$. Expressed another way, this is $[NH_3]_{initial} = 50\,[OH^-]$. Of course, we also know that $[NH_3]_{equil} = [NH_3]_{initial} - [OH^-]$. We use the base dissociation constant expression for NH_3 to organize our information.

$$K_b = 1.8 \times 10^{-5} = \frac{[NH_4^+][OH^-]}{[NH_3]_{equil}} = \frac{[OH^-]^2}{5.\times[OH^-] - [OH^-]} = \frac{[OH^-]}{49}$$

$$[OH^-] = 49 \times 1.8 \times 10^{-5} = 8.8 \times 10^{-4}\ \text{M} \qquad [NH_3]_{initial} = 50 \times 8.8 \times 10^{-4} = 0.044\ \text{M}$$

52. No, the degree of ionization of $HC_2H_3O_2$ is not expected to be 13% in 0.0010 M $HC_2H_3O_2$ or 42% in 0.00010 M $HC_2H_3O_2$. The degrees of ionization at lower values of $[HC_2H_3O_2]$ are based on the assumption that $[HC_2H_3O_2]_i \approx [HC_2H_3O_2]_i - [HC_2H_3O_2]_{eq}$. As long as $[HC_2H_3O_2]_{eq}$ is less than 5% of $[HC_2H_3O_2]_i$ this is a reasonable assumption. Of course, it has broken down at 13% ionization.

Polyprotic Acids

53. Since H_3PO_4 is a weak acid, there is little HPO_4^{2-} (produced in the second ionization) compared to the H_3O^+ (produced in the first ionization). In turn, there is little PO_4^{3-} (produced in the third ionization) compared to the HPO_4^-, and very little compared to the H_3O^+.

54. **(a)**

Reaction:	$H_2S + H_2O$	$\rightleftharpoons$	HS^-	$+$	H_3O^+
Initial:	0.075 M		0 M		0 M
Changes:	$-x$ M		$+x$ M		$+x$ M
Equil:	$(0.075 - x)$M		x M		$+x$ M

$$K_a = \frac{[HS^-][H_3O^+]}{[H_2S]} = 1.0 \times 10^{-7} = \frac{x^2}{0.075 - x} \approx \frac{x^2}{0.075} \qquad x = 8.7 \times 10^{-5}\ \text{M} = [H_3O^+]$$

$$[HS^-] = 9.1 \times 10^{-5}\ \text{M} \qquad [S^{2-}] = K_2 = 1 \times 10^{-19}\ \text{M}$$

(b) The set up for this problem is the same as for part (a), with the substitution of 0.0050 M for 0.075 M as the initial value of $[H_2S]$.

$$K_1 = \frac{[HS^-][H_3O^+]}{[H_2S]} = 1.0 \times 10^{-7} = \frac{x^2}{0.0050 - x} \approx \frac{x^2}{0.0050} \qquad x = 2.2 \times 10^{-5}\ \text{M} = [H_3O^+]$$

$$[HS^-] = 2.2 \times 10^{-5}\ \text{M} \qquad [S^{2-}] = K_2 = 1 \times 10^{-19}\ \text{M}$$

(c) The set up for this part is the same as for part (a), with the substitution of 1.0×10^{-5} M for 0.075 M as the initial value of $[H_2S]$. The solution differs in that we cannot assume $x << 1.0 \times 10^{-5}$. Rather than solve the quadratic equation, however, we shall use the method of successive approximations. This involves making the unwarranted assumption referred to above to obtain a first approximate value of x. This approximate value of x then is used in the denominator of the K_1 expression (but not as a substitute in x^2) to obtain a second approximate value of x. The cycle continues until two successive values of x are in agreement.

$$K_1 = \frac{[HS^-][H_3O^+]}{[H_2S]} = 1.0 \times 10^{-7} = \frac{x^2}{1.0 \times 10^{-5} - x} \approx \frac{x^2}{1.0 \times 10^{-5}} \qquad x = 1.0 \times 10^{-6}\ \text{M} = [H_3O^+]$$

$$K_1 = \frac{[HS^-][H_3O^+]}{[H_2S]} = 1.0 \times 10^{-7} = \frac{x^2}{1.0 \times 10^{-5} - 1.0 \times 10^{-6}} = \frac{x^2}{9.0 \times 10^{-6}}$$

$$x = 9.5 \times 10^{-7}\ \text{M} = [H_3O^+] \qquad [HS^-] = 9.5 \times 10^{-7}\ \text{M} \qquad [S^{2-}] = K_2 = 1 \times 10^{-19}\ \text{M}$$

55. The main estimate involves assuming that the mass percents can be expressed as 0.057 g 75% H_3PO_4 per 100. mL of solution and 0.084 g 75% H_3PO_4 per 100. mL of solution. That is, that the density of this aqueous solution is essentially 1.00 g/mL. Based on this assumption, we can compute the initial concentrations of H_3PO_4.

$$[H_3PO_4] = \dfrac{0.057 \text{ g impure } H_3PO_4 \times \dfrac{75 \text{ g } H_3PO_4}{100 \text{ g impure } H_3PO_4} \times \dfrac{1 \text{ mol } H_3PO_4}{98.00 \text{ g } H_3PO_4}}{100. \text{ mL soln} \times \dfrac{1 \text{ L}}{1000 \text{ mL}}} = 0.0044 \text{ M}$$

$$[H_3PO_4] = \dfrac{0.084 \text{ g impure } H_3PO_4 \times \dfrac{75 \text{ g } H_3PO_4}{100 \text{ g impure } H_3PO_4} \times \dfrac{1 \text{ mol } H_3PO_4}{98.00 \text{ g } H_3PO_4}}{100. \text{ mL soln} \times \dfrac{1 \text{ L}}{1000 \text{ mL}}} = 0.0064 \text{ M}$$

Reaction:	$H_3PO_4 + H_2O \rightleftharpoons$	$H_2PO_4^- +$	H_3O^+
Initial:	0.0044 M	0 M	0 M
Changes:	$-x$ M	$+x$ M	$+x$ M
Equil:	$(0.0044 - x)$M	x M	x M

$$K_1 = \dfrac{[H_2PO_4^-][H_3O^+]}{[H_2PO_4]} = \dfrac{x^2}{0.0044 - x} = 7.1 \times 10^{-3} \qquad x^2 + 0.0071\,x - 3.1 \times 10^{-5} = 0$$

$$x = \dfrac{-b \pm \sqrt{b^2 - 4ac}}{2a} = \dfrac{-0.0071 \pm \sqrt{5.0 \times 10^{-5} + 1.2 \times 10^{-4}}}{2} = 3 \times 10^{-3} \text{ M} = [H_3O^+]$$

The set up for the second concentration is the same as for the first, with the exception of substitution 0.0064 M for 0.0044 M.

$$K_1 = \dfrac{[H_2PO_4^-][H_3O^+]}{[H_2PO_4]} = \dfrac{x^2}{0.0064 - x} = 7.1 \times 10^{-3} \qquad x^2 + 0.0071\,x - 4.6 \times 10^{-5} = 0$$

$$x = \dfrac{-b \pm \sqrt{b^2 - 4ac}}{2a} = \dfrac{-0.0071 \pm \sqrt{5.0 \times 10^{-5} + 1.8 \times 10^{-4}}}{2} = 4 \times 10^{-3} \text{ M} = [H_3O^+]$$

The two values of pH now are determined, representing the pH range in a cola drink.

$$pH = -\log(3 \times 10^{-3}) = 2.5 \qquad\qquad pH = -\log(4 \times 10^{-3}) = 2.4$$

56. In all cases, of course, the first ionization of H_2SO_4 is complete, and establishes the initial values of $[H_3O^+]$ and $[HSO_4^-]$. We shall deal with the second ionization in each case.

(a)

Reaction:	$HSO_4^- + H_2O \rightleftharpoons$	$SO_4^{2-} +$	H_3O^+
Initial:	0.75 M	0 M	0.75 M
Changes:	$-x$ M	$+x$ M	$+x$ M
Equil:	$(0.75 - x)$M	x M	$(0.75 + x)$M

$$K_2 = \dfrac{[SO_4^{2-}][H_3O^+]}{[HSO_4^-]} = 0.011 = \dfrac{x\,(0.75 + x)}{0.75 - x} \approx \dfrac{0.75\,x}{0.75} \qquad x = 0.011 \text{ M} = [SO_4^{2-}]$$

We have assumed that $x \ll 0.75$ M, an assumption that clearly is correct.

$[HSO_4^-] = 0.75 - 0.011 = 0.74$ M $\qquad\qquad [H_3O^+] = 0.75 + 0.011 \text{ M} = 0.76$ M

(b) The set up for this part is similar to part (a), with the exception of substituting 0.075 M for 0.75 M.

$$K_2 = \dfrac{[SO_4^{2-}][H_3O^+]}{[HSO_4^-]} = 0.011 = \dfrac{x\,(0.075 + x)}{0.075 - x} \qquad 0.011(0.075 - x) = 0.075x + x^2$$

$$x^2 + 0.076\,x - 8.3 \times 10^{-4} = 0 \qquad x = \dfrac{-b \pm \sqrt{b^2 - 4ac}}{2a} = \dfrac{-0.076 \pm \sqrt{0.0058 + 0.0033}}{2} = 0.0097 \text{ M}$$

$x = 0.0097$ M $= [SO_4^{2-}]$

$[HSO_4^-] = 0.075 - 0.0099 = 0.065$ M $\qquad\qquad [H_3O^+] = 0.075 + 0.0099 \text{ M} = 0.085$ M

(c) Again, the set up is the same as for part (a), with the exception of substituting 0.00075 M for 0.75 M.

$$K_2 = \dfrac{[SO_4^{2-}][H_3O^+]}{[HSO_4^-]} = 0.011 = \dfrac{x\,(0.00075 + x)}{0.00075 - x}$$

$$0.011(0.00075 - x) = 0.00075\,x + x^2 \qquad x^2 + 0.0118x - 8.3 \times 10^{-6} = 0$$

$$x = \dfrac{-b \pm \sqrt{b^2 - 4ac}}{2a} = \dfrac{-0.0118 \pm \sqrt{1.39 \times 10^{-4} + 3.3 \times 10^{-5}}}{2} = 6.6 \times 10^{-4}$$

$x = 6.6 \times 10^{-4}$ M $= [SO_4^{2-}]$ $\qquad\qquad [HSO_4^-] = 0.00075 - 0.00066 = 9 \times 10^{-5}$ M

$[H_3O^+] = 0.00075 + 0.00068 \text{ M} = 1.4 \times 10^{-3}$ M $\qquad [H_3O^+]$ is almost twice the initial value of $[H_2SO_4]$.

57. We first determine $[H_3O^+]$, with our calculation, as usual, based on the balanced chemical equation.

Reaction:	$HOOC(CH_2)_4COOH(aq) + H_2O \rightleftharpoons$	$HOOC(CH_2)_4COO^-(aq) +$	$H_3O^+(aq)$
Initial:	0.10 M		
Changes:	$-x$ M	$+x$ M	$+x$ M
Equil:	$(0.10 - x)$ M	x M	x M

$$K_{a_1} = \frac{[H_3O^+][HOOC(CH_2)_4COO^-]}{[HOOC(CH_2)_4COOH]} = 3.9 \times 10^{-5} = \frac{x \cdot x}{0.10 - x} \approx \frac{x^2}{0.10}$$

$$x = \sqrt{0.10 \times 3.9 \times 10^{-5}} = 2.0 \times 10^{-3} \, M = [H_3O^+]$$

We see that our simplifying assumption, that $x \ll 0.10$ M, is indeed valid. Now we consider the second ionization. We shall see that very little H_3O^+ is produced in this ionization because of the small size of two terms: K_{a_2} and $[HOOC(CH_2)_4COO^-]$. Again we base our calculation on the balanced chemical equation.

Reaction:	$HOOC(CH_2)_4COO^-(aq) + H_2O \rightleftharpoons$	$^-OOC(CH_2)_4COO^-(aq) \; +$	$H_3O^+(aq)$
Initial:	2.0×10^{-3} M		2.0×10^{-3} M
Changes:	$-y$ M	$+y$ M	$+y$ M
Equil:	$(0.0020 - y)$ M	y M	$(0.0020 + y)$ M

$$K_{a_2} = \frac{[H_3O^+][^-OOC(CH_2)_4COO^-]}{[HOOC(CH_2)_4COO^-]} = 3.9 \times 10^{-6} = \frac{y(0.0020 + y)}{0.0020 - y} \approx \frac{0.0020 \, y}{0.0020}$$

$$y = 3.9 \times 10^{-6} \, M = [^-OOC(CH_2)_4COO^-]$$

Again, we see that our assumption, that $y \ll 0.0020$ M, is valid. In addition, we also see that virtually no H_3O^+ is created in this second ionization. The concentrations of all species have been calculated above, with the exception of $[OH^-]$.

$$[OH^-] = \frac{K_w}{[H_3O^+]} = \frac{1.00 \times 10^{-14}}{2.0 \times 10^{-3}} = 5.0 \times 10^{-12} \, M \qquad\qquad [HOOC(CH_2)_4COOH] = 0.10 \, M$$

$$[H_3O^+] = [HOOC(CH_2)_4COO^-] = 2.0 \times 10^{-3} \, M \qquad\qquad [^-OOC(CH_2)_4COO^-] = 3.9 \times 10^{-6} \, M$$

58. **(a)** First ionization: $\qquad\qquad C_{20}H_{24}O_2N_2 + H_2O \rightleftharpoons C_{20}H_{24}O_2N_2H^+ + OH^-$

Second ionization: $\qquad\quad C_{20}H_{24}O_2N_2H^+ + H_2O \rightleftharpoons C_{20}H_{24}O_2N_2H_2^{2+} + OH^-$

(b) $[C_{20}H_{24}O_2N_2] = \dfrac{1.00 \text{ g quinine} \times \dfrac{1 \text{ mol quinine}}{324.4 \text{ g quinine}}}{1900. \text{ mL} \times \dfrac{1 \text{ L}}{1000 \text{ mL}}} = 1.62 \times 10^{-3} \, M \qquad K_1 = 10^{-6.0} = 1 \times 10^{-6}$

Reaction:	$C_{20}H_{24}O_2N_2 + H_2O$	$\rightleftharpoons$	$C_{20}H_{24}O_2N_2H^+$	$+$	OH^-
Initial:	0.00162 M				
Changes:	$-x$ M		$+x$ M		$+x$ M
Equil:	$(0.00162 - x)$M		x M		x M

$$K_1 = \frac{[C_{20}H_{24}O_2N_2H^+][OH^-]}{[C_{20}H_{24}O_2N_2]} = 1 \times 10^{-6} = \frac{x^2}{0.00162 - x} \approx \frac{x^2}{0.00162} \qquad x = 4 \times 10^{-5} \, M$$

Our assumption, that $x \ll 0.00162$, clearly is valid.

$$pOH = -\log(4 \times 10^{-5}) = 4.4 \qquad\qquad pH = 14.00 - 4.4 = 9.6$$

Ions as Acids and Bases (Hydrolysis)

59. **(a)** KCl forms a neutral solution, being composed of the cation of a strong base, and the anion of a strong acid; no hydrolysis occurs.

(b) KF forms an alkaline solution, being composed of the cation of a strong base and the anion of a weak acid; the fluoride ion hydrolyzes. $\qquad F^- + H_2O \rightleftharpoons HF + OH^-$

(c) $NaNO_3$ forms a neutral solution, being composed of the cation of a strong base and the anion of a strong acid. No hydrolysis occurs.

(d) $Ca(OCl)_2$ forms an alkaline solution, being composed of the cation of a strong base, and the anion of a weak acid; the hypochlorite ion hydrolyzes. $\; OCl^- + H_2O \rightleftharpoons HOCl + OH^-$

(e) NH_4NO_2 forms an acidic solution, being composed of the cation of a weak base and the anion of a weak acid; the ammonium ion hydrolyzes: $\quad NH_4^+ + H_2O \rightleftharpoons NH_3(aq) + H_3O^+ \qquad K_a = 5.6 \times 10^{-10}$

and so does the nitrite ion: $\; NO_2^- + H_2O \rightleftharpoons HNO_2 + OH^- \quad K_b = 1.2 \times 10^{-11}$ Since NH_4^+ is a stronger acid than NO_2^- is a base, the solution will be acidic. The ionization constants were computed from data in Table 17-3 and with the relationship $K_w = K_a \times K_b$.

$$\text{For } NH_4^+, K_a = \frac{1.0 \times 10^{-14}}{1.8 \times 10^{-5}} = 5.6 \times 10^{-10} \qquad\qquad \text{For } NO_2^-, K_a = \frac{1.0 \times 10^{-14}}{7.2 \times 10^{-4}} = 1.2 \times 10^{-11}$$

Where $\qquad 1.8 \times 10^{-5} = K_b$ of $NH_3 \qquad\qquad$ and $\qquad\qquad 7.2 \times 10^{-4} = K_a$ of HNO_2

60.

Reaction:	$C_6H_7O_2^- + H_2O$	$\rightleftharpoons$	$HC_6H_7O_2$ +	OH^-
Initial:	0.55 M		0 M	0 M
Changes:	$-x$ M		$+x$ M	$+x$ M
Equil:	$(0.55 - x)$M		x M	x M

$$K_b = \frac{K_w}{K_a} = \frac{1.00 \times 10^{-14}}{1.7 \times 10^{-5}} = 5.9 \times 10^{-10} = \frac{[HC_6H_7O_2][OH^-]}{[C_6H_7O_2^-]} = \frac{x^2}{0.55 - x} \approx \frac{x^2}{0.55}$$

$x = 1.8 \times 10^{-5}$ M = $[OH^-]$ $pOH = -\log(1.8 \times 10^{-5}) = 4.74$ $pH = 14.00 - 4.74 = 9.26$

61. We determine $[H_3O^+]$ (which equals the concentration of the +2 aluminum cation) from the value of pH and then use the acid ionization constant expression to determine the concentration of hydrated aluminum(III) ion. $[H_3O^+] = 10^{-pH} = 10^{-3.2} = 6.3 \times 10^{-4} = [[Al(H_2O)_5OH]^{2+}]$

$$K_a = \frac{[H_3O^+][[Al(H_2O)_5OH]^{2+}]}{[[Al(H_2O)_6]^{3+}]} = 1.1 \times 10^{-5} = \frac{(6.3 \times 10^{-5})^2}{[[Al(H_2O)_6]^{3+}]}$$

$$[[Al(H_2O)_6]^{3+}] = \frac{(6.3 \times 10^{-5})^2}{1.1 \times 10^{-5}} = 3.6 \times 10^{-4} \text{ M at equilibrium}$$

Of course some of the hydrated aluminum cation initially dissolved in the solution has hydrolyzed. That initial concentration was $[[Al(H_2O)_6]^{3+}] = 3.6 \times 10^{-4}$ M $+ 6.3 \times 10^{-5}$ M $= 4.2 \times 10^{-4}$ M

62. pH = 8.75 is a basic solution. We need the salt of the cation of a strong base, and the anion of a weak acid. This must be **(c)** KNO_2. (a) NH_4Cl is the salt of the cation of a weak base and the anion of a strong acid. (b) $KHSO_4$ and (d) $NaNO_3$ both are the salts of cations of strong bases with anions of strong acids.

pOH = 14.00 - 8.75 = 5.25 $[OH^-] = 10^{-5.25} = 5.6 \times 10^{-6}$ M

Reaction:	$NO_2^+ + H_2O$	$\rightleftharpoons$	HNO_2	+	OH^-
Initial:	S		0 M		0 M
Changes:	-5.6×10^{-6} M		$+5.6 \times 10^{-6}$ M		$+5.6 \times 10^{-6}$ M
Equil:	$(S - 5.6 \times 10^{-6}$ M$)$		5.6×10^{-6} M		5.6×10^{-6} M

$$K_b = \frac{K_w}{K_a} = \frac{1.00 \times 10^{-14}}{7.2 \times 10^{-4}} = 1.4 \times 10^{-11} = \frac{[HNO_2][OH^-]}{[NO_2^-]} = \frac{(5.6 \times 10^{-6})^2}{S - 5.6 \times 10^{-6}}$$

$$S - 5.6 \times 10^{-6} = \frac{(5.6 \times 10^{-6})^2}{1.4 \times 10^{-11}} = 2.2 \text{ M}$$ $S = 2.2$ M $= [KNO_2]$

63.

Reaction:	$C_5H_5NH^+ + H_2O$	$\rightleftharpoons$	C_5H_5N	+	H_3O^+
Initial:	0.0482 M		0 M		0 M
Changes:	$-x$ M		$+x$ M		$+x$ M
Equil:	$(0.0482 - x)$M		x M		x M

$$K_a = \frac{K_w}{K_b} = \frac{1.00 \times 10^{-14}}{1.5 \times 10^{-9}} = 6.7 \times 10^{-6} = \frac{[C_5H_5N][H_3O^+]}{[C_5H_5NH^+]} = \frac{x^2}{0.0482 - x} = \frac{x^2}{0.0482}$$

$x = 5.7 \times 10^{-4}$ M $= [H_3O^+]$ $pH = -\log(5.7 \times 10^{-4}) = 3.24$

64. **(a)** $HSO_3^- + H_2O \rightleftharpoons H_3O^+ + SO_3^{2-}$ $K_2 = 6.2 \times 10^{-8}$

$HSO_3^- + H_2O \rightleftharpoons OH^- + H_2SO_3$ $K_b = \dfrac{K_w}{K_1} = \dfrac{1.00 \times 10^{-14}}{1.3 \times 10^{-2}} = 7.7 \times 10^{-13}$

Since $K_2 > K_b$, a solution of HSO_3^- is acidic.

(b) $HS^- + H_2O \rightleftharpoons H_3O^+ + S^{2-}$ $K_2 = 1 \times 10^{-19}$

$HS^- + H_2O \rightleftharpoons OH^- + H_2S$ $K_b = \dfrac{K_w}{K_1} = \dfrac{1.0 \times 10^{-14}}{1.0 \times 10^{-7}} = 1.0 \times 10^{-7}$

Since $K_2 < K_b$, a solution of HS^- is alkaline.

(c) $HPO_4^{2-} + H_2O \rightleftharpoons H_3O^+ + PO_4^{3-}$ $K_3 = 4.2 \times 10^{-13}$

$HPO_4^{2-} + H_2O \rightleftharpoons OH^- + H_2PO_4^-$ $K_b = \dfrac{K_w}{K_2} = \dfrac{1.00 \times 10^{-14}}{3.7 \times 10^{-8}} = 2.7 \times 10^{-7}$

Since $K_3 < K_b$, a solution of HPO_4^{2-} is alkaline.

65. First we look for the strong acids; they are H_2SO_4 and HNO_3. Since H_2SO_4 is diprotic, with the first ionization being strong, the second somewhat strong in dilute solution, it yields the greater $[H_3O^+]$ of these two acids. Next we look for the weak acids; there is only one, $HC_2H_3O_2$. Next in order of decreasing acidity come salts with cations from weak bases and anions from strong acids; NH_4ClO_4 is in this category. Then come salts in both ions hydrolyze to the same degree; $NH_4C_2H_3O_2$ is an example, forming a pH-neutral solution. Next come salts that have the cation of a strong base and the anion of a weak acid; $NaNO_2$ is in this category. Then come weak bases, of which $NH_3(aq)$ is an example. And finally, strong bases: $NaOH$ and $Ba(OH)_2$, of which $Ba(OH)_2$ contributes two OH^- ions per formula unit, and thus forms the more alkaline solution.

$$H_2SO_4 < HNO_3 < HC_2H_3O_2 < NH_4ClO_4 < NH_4C_2H_3O_2 < NaNO_2 < NH_3 < NaOH < Ba(OH)_2$$

66. **(a)** $HClO_3$ should be a stronger acid than is $HClO_2$. In each acid there is an H—O—Cl grouping. The remaining oxygen atoms are attached directly to Cl as terminal O atoms. Thus, there are two terminal O atoms in $HClO_3$ and one terminal O atom in $HClO_2$. Of oxoacids of the same element, the one with the higher number of terminal oxygen atoms is the stronger. $K_a = 5 \times 10^2$ for $HClO_3$ and $K_a = 1.2 \times 10^{-2}$ for $HClO_2$

(b) HNO_2 and H_2CO_3 have the same number (one) of terminal oxygen atoms. They differ in N being more electronegative than C. HNO_2 ($K_a = 7.2 \times 10^{-4}$) is stronger than H_2CO_3 ($K_a = 4.4 \times 10^{-7}$).

(c) H_3PO_4 and H_2SiO_3 have the same number (one) of terminal oxygen atoms. They differ in P being more electronegative than Si. H_3PO_4 ($K_a = 7.1 \times 10^{-3}$) is stronger than H_2SiO_3 ($K_a = 1.7 \times 10^{-10}$).

67. The weakest of the five acids is CH_3CH_2COOH. The reasoning is as follows. HBr is a strong acid, stronger than the carboxylic acids. A carboxylic acid—such as $CH_2ClCOOH$ and CH_2FCH_2COOH—with a strongly electronegative atom attached to the hydrocarbon chain will be stronger than one in which no such group is attached. But the I atom is so weakly electronegative that it influences acid strength hardly at all. Acid strengths follow.

68. For the acids given, we determine values of m and n in the formula $EO_m(OH)_n$.
(a) The formula of H_3AsO_4 can be rewritten $AsO(OH)_3$, which has $m = 1$ and $n = 3$. The expected value of $K_a = 10^{-2}$.
(b) $HOClO$ or $ClO(OH)$ has $m = 1$ and $n = 1$. We expect $K_a \approx 10^{-2}$ or $pK_a \approx 2$. The actual value is $pK_a = 1.92$.
(c) Iodic acid is $HOIO_2$ or $IO_2(OH)$ has $m = 2$ or $n = 1$. We expect K_a to be large. The accepted value is $K_a = 0.16$, not terribly large (but not small either).

Lewis Theory of Acids and Bases

69. **(a)** $[\bar{O}\text{—}H]^-$ is Lewis base; it has lone-pair electrons to donate.

(b) $|\bar{Cl}\text{—}Al\text{—}\bar{Cl}|$ is a Lewis acid; there is a site for accepting an electron pair.

(c) $H\text{—}C\text{—}\bar{N}\text{—}H$ is a Lewis base; the lone pair of electrons on N can be donated.

70. **(a)**

$$H\text{—}\bar{O}\text{—}B\text{—}\bar{O}\text{—}H + [\bar{O}\text{—}H]^- \longrightarrow \left(H\text{—}\bar{O}\text{—}B\text{—}\bar{O}\text{—}H \right)^-$$

acid: eln pr acceptor   base: eln pr donor

(b)

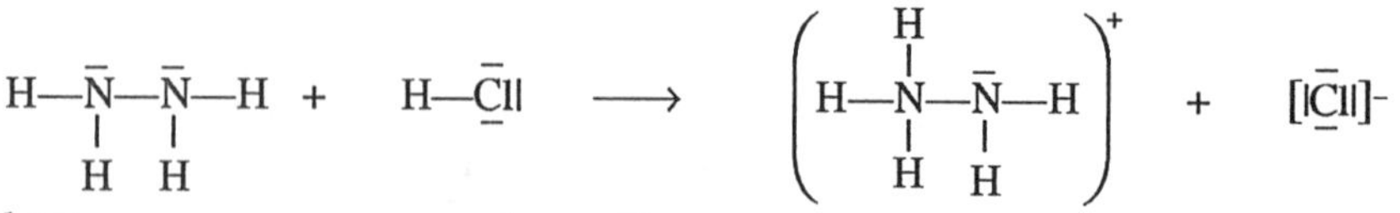

$$H{-}\overline{N}{-}\overline{N}{-}H \; + \; H{-}\overline{\underline{C}}l \; \longrightarrow \; \left(H{-}\overset{H}{\underset{H}{N}}{-}\overset{}{\underset{H}{N}}{-}H \right)^{+} \; + \; [\overline{\underline{C}}l]^{-}$$

base   secondary acid

The actual Lewis acid is H$^+$, which is supplied by HCl.

(c)

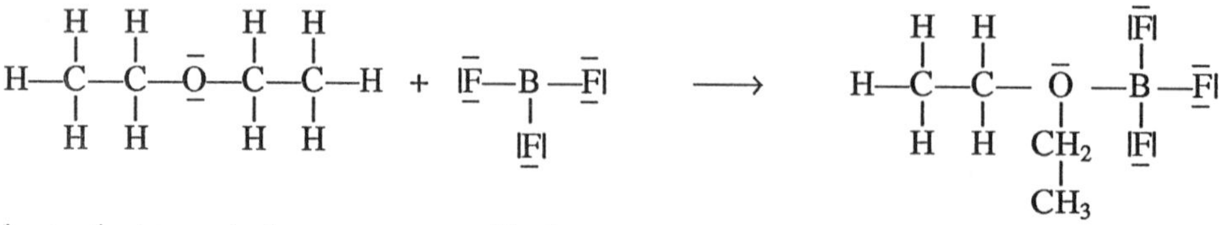

$$H{-}\overset{H}{\underset{H}{C}}{-}\overset{H}{\underset{H}{C}}{-}\overline{\underline{O}}{-}\overset{H}{\underset{H}{C}}{-}\overset{H}{\underset{H}{C}}{-}H \; + \; \overline{F}{-}B{-}\overline{F}l \; \longrightarrow \; H{-}\overset{H}{\underset{H}{C}}{-}\overset{H}{\underset{H}{C}}{-}\overline{O}{-}B{-}\overline{F}l$$

base: electron pair donor   acid: eln pr acceptor

71. The C in a CO_2 molecule has location available for the acceptance of a pair of electrons; it is the Lewis acid. The hydroxide anion is a Lewis base, with pairs of electrons to donate.

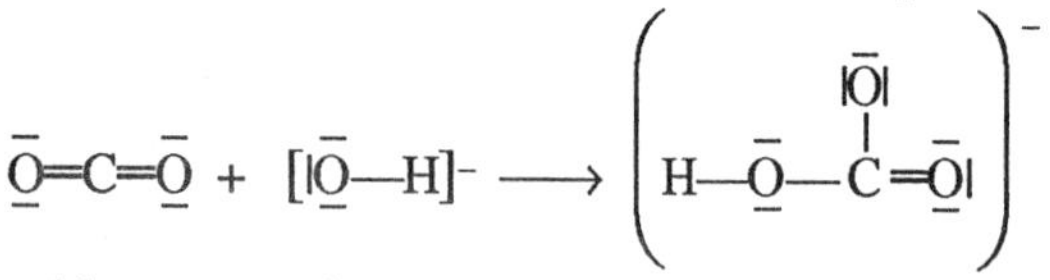

$$\overline{O}{=}C{=}\overline{O} \; + \; [l\overline{O}{-}H]^{-} \; \longrightarrow \; \left(H{-}\overline{O}{-}\overset{l\overline{O}l}{C}{=}\overline{O}l \right)^{-}$$

acid   base

18 ADDITIONAL ASPECTS OF ACID–BASE EQUILIBRIA

REVIEW QUESTIONS

1. **(a)** The abbreviation "mmol" stands for millimole, one thousandth of a mole. This amount of material often is a convenient one to use in calculations involving solutions where volumes are measured in mL.
   (b) HIn is a generalized symbol for the formula of the acid form of an indicator.
   (c) The equivalence point of a titration is that point when sufficient added titrant is present to react with all of the substance being titrated, with no titrant left over.
   (d) A titration curve is a plot of the pH of solution versus the volume of added titrant (or sometimes versus the percentage of substance that has been titrated).

2. **(a)** The common-ion effect refers to the suppression of an ionization equilibrium caused by the presence (or addition) of some of the ions produced by the ionization.
   (b) A buffer solution maintains constant pH by consuming added strong acid or added strong base. The buffer solution contains components that react with each.
   (c) The value of pK_a can be determined from the titration curve of a monoprotic weak acid since the pH at the half equivalence point is equal to the pK_a of the weak acid.
   (d) pH is measured with a series of acid-base indicator that all change color at different pH values. By determining the color of each indicator in a sample of the solution being tested, the pH of that solution can be placed within a narrow range.

3. **(a)** Buffer capacity is the amount of strong acid or base that can be added to a specified volume of a buffer before a significant change in pH occurs. Buffer range is the range of pH values within which a buffer solution will resist changes in pH.
   (b) Hydrolysis refers to the reaction of an ion with water to produce either $H_3O^+(aq)$ or $OH^-(aq)$. Neutralization refers to the reaction of an acid with a base to produce a situation with pH closer to 7.0.
   (c) The first and second equivalence points of a weak acid are the points where, respectively, the first proton has been completely ionized, and the second proton has been completely ionized.
   (d) The equivalence point of a titration is the point where stoichiometric amounts of acid and base have been combined. The end point is when the indicator changes color. Careful selectron of the indicator can ensure that these two points coincide.

4. **(a)** Note that HI is a strong acid and the initial $[H_3O^+] = [HI] = 0.0852$ M

Reaction:	$HC_3H_5O_2 + H_2O$	$\rightleftharpoons$	$C_3H_5O_2^-$	$+$	H_3O^+
Initial:	0.315 M				0.0664 M
Changes:	$-x$ M		$+x$ M		$+x$ M
Equil:	$(0.315 - x)$M		x M		$(0.0664 + x)$M

   $$K_a = \frac{[C_3H_5O_2^-][H_3O^+]}{[HC_3H_5O_2]} = 1.3 \times 10^{-5} = \frac{x(0.0664 + x)}{0.315 - x} \approx \frac{0.0664\, x}{0.315} \qquad x = 6.2 \times 10^{-5} \text{ M}$$

   We have assumed that $x \ll 0.0664$ M, an assumption that clearly is correct. $[H_3O^+] = 0.0664$ M

   (b) $[OH^-] = \dfrac{K_w}{[H_3O^+]} = \dfrac{1.00 \times 10^{-14}}{0.0664} = 1.51 \times 10^{-13}$ M

   (c) $[C_3H_5O_2^-] = x = 6.2 \times 10^{-5}$ M

   (d) $[I^-] = [HI]_i = 0.0664$ M

5. **(a)** The NH_4Cl dissociates completely and thus $[NH_4^+]_i = [Cl^-]_i = 0.0818$ M

Reaction:	$NH_3 + H_2O$	$\rightleftharpoons$	NH_4^+	$+$	OH^-
Initial:	0.212 M		0.0818 M		
Changes:	$-x$ M		$+x$ M		$+x$ M
Equil:	$(0.212 - x)$M		$(0.0818 + x)$M		x M

$$K_b = \frac{[NH_4^+][OH^-]}{[NH_3]} = \frac{(0.0818 + x)\,x}{0.212 - x} = 1.8 \times 10^{-5} \approx \frac{0.0818x}{0.212} \qquad x = 4.7 \times 10^{-5}\ M$$

In solving this problem, we have assumed $x \ll 0.0818$ M, a clearly valid assumption.

$[OH^-] = x = 4.7 \times 10^{-5}$ M   **(b)** $[NH_4^+] = 0.0818 + x = 0.0818$ M

(c) $[Cl^-] = 0.0818$ M   **(d)** $[H_3O^+] = \dfrac{1.00 \times 10^{-14}}{4.7 \times 10^{-5}} = 2.1 \times 10^{-10}$ M

6. Adding $H_3O^+(aq)$ represents reaction with strong acid; adding $OH^-(aq)$ represents reaction with strong base.

(a) $CHO_2^-(aq) + H_3O^+(aq) \longrightarrow HCHO_2(aq) + H_2O$

   $HCHO_2(aq) + OH^-(aq) \longrightarrow CHO_2^-(aq) + H_2O$

(b) $C_6H_5NH_2(aq) + H_3O^+(aq) \longrightarrow C_6H_5NH_3^+(aq) + H_2O$

   $C_6H_5NH_3^+(aq) + OH^-(aq) \longrightarrow C_6H_5NH_2(aq) + H_2O$

(c) $HPO_4^{2-}(aq) + H_3O^+(aq) \longrightarrow H_2PO_4^-(aq) + H_2O$

   $H_2PO_4^-(aq) + OH^-(aq) \longrightarrow HPO_4^{2-}(aq) + H_2O$

7. **(a)**

Reaction:	$HC_7H_5O_2(aq) + H_2O$	$\rightleftharpoons$	$C_7H_5O_2^-$	$+$	H_3O^+
Initial:	0.0782 M		0.116 M		
Changes:	$-x$ M		$+x$ M		$+x$ M
Equil:	$(0.0782 - x)$M		$(0.116 + x)$M		x M

$$K_a = \frac{[H_3O^+][C_7H_5O_2^-]}{[HC_7H_5O_2]} = 6.3 \times 10^{-5} = \frac{x(0.116 + x)}{0.0782 - x} \approx \frac{0.116\,x}{0.0782} \qquad x = 4.2 \times 10^{-5}\ M$$

To determine the value of x, we assumed $x \ll 0.0782$ M, an assumption that clearly is correct.

$[H_3O^+] = 4.2 \times 10^{-5}$ M $pH = -\log(4.2 \times 10^{-5}) = 4.38$

(b)

Reaction:	$NH_3(aq) + H_2O$	$\rightleftharpoons$	$NH_4^+(aq)$	$+$	$OH^-(aq)$
Initial:	0.352 M		0.144 M		
Changes:	$-x$ M		$+x$ M		$+x$ M
Equil:	$(0.352 - x)$M		$(0.144 + x)$M		x M

$$K_b = \frac{[NH_4^+][OH^-]}{[NH_3]} = 1.8 \times 10^{-5} = \frac{x(0.144 + x)}{0.352 - x} \approx \frac{0.144\,x}{0.352} \qquad x = 4.4 \times 10^{-5}\ M$$

To determine the value of x, we assumed $x \ll 0.144$, a clearly valid assumption.

$[OH^-] = 4.4 \times 10^{-5}$ M $pOH = -\log(4.4 \times 10^{-5}) = 4.36$ $pH = 14.00 - 4.36 = 9.64$

8. $[H_3O^+] = 10^{-4.06} = 8.7 \times 10^{-5}$ M. We let $S =$ initial $[CHO_2^-]$

Reaction:	$HCHO_2(aq) + H_2O$	$\rightleftharpoons$	$CHO_2^-(aq)$	$+$	$H_3O^+(aq)$
Initial:	0.515 M		S M		
Changes:	-8.7×10^{-5} M		$+8.7 \times 10^{-5}$ M		$+8.7 \times 10^{-5}$M
Equil:	0.515 M		$(S + 8.7 \times 10^{-5})$		8.7×10^{-5} M

$$K_a = \frac{[H_3O^+][CHO_2^-]}{[HCHO_2]} = 1.8 \times 10^{-4} = \frac{(S + 8.7 \times 10^{-5})\,8.7 \times 10^{-5}}{0.515} \approx \frac{8.7 \times 10^{-5}\,S}{0.515}$$

$S = 1.06$ M   To determine S, we assumed $S \gg 8.7 \times 10^{-5}$ M, clearly a valid assumption.

Or, we could have used the Henderson-Hasselbalch equation. $pK_a = -\log(1.8 \times 10^{-4}) = 3.74$

$$4.06 = 3.74 + \log \frac{[CHO_2^-]}{[HCHO_2]} \qquad \frac{[CHO_2^-]}{[HCHO_2]} = 2.1 \qquad [CHO_2^-] = 2.1 \times 0.515 = 1.08\ M$$

The difference in the two answers is due simply to rounding.

9. We use the Henderson-Hasselbalch equation. $pK_b = -\log(1.8 \times 10^{-5}) = 4.74$

$pK_a = 14.00 - pK_b = 14.00 - 4.74 = 9.26$ $pH = 9.12 = 9.26 + \log \dfrac{[NH_3]}{[NH_4^+]}$

$$\frac{[NH_3]}{[NH_4{}^+]} = 10^{-0.14} = 0.72 \qquad [NH_3] = 0.72 \times [NH_4{}^+] = 0.72 \times 0.884 \text{ M} = 0.64 \text{ M}$$

10. 0.60 mol $NaC_2H_3O_2$ will raise the pH of 1.00 L of 0.50 M HCl to the greatest extent. Both OH^- and $C_2H_3O_2{}^-$ are bases that will react with the strong acid (H_3O^+) in HCl(aq), resulting in an increase in the pH of the solution. Since there is more acetate ion available than hydroxide ion, 0.60 mol $NaC_2H_3O_2$ will consume more H_3O^+ than will 0.40 mol NaOH, causing a greater increase in the pH of the solution. In fact, 0.60 mol $NaC_2H_3O_2$ will consume all of the H_3O^+ in 1.00 L of 0.50 M HCl, leaving only a solution of acetic acid and sodium acetate, while 0.40 mol NaOH will leave a solution of HCl (with a concentration of 0.40 M) and NaCl. With regard to the other two possibilities, 0.70 mol NaCl will not affect the pH and 0.50 mol $HC_2H_3O_2$ will lower the pH since $HC_2H_3O_2$ is an acid.

11. We use the Henderson-Hasselbalch equation to determine K_a of lactic acid.

$$[C_3H_5O_3{}^-] = \frac{1.00 \text{ g } NaC_3H_5O_3}{100.0 \text{ mL soln}} \times \frac{1000 \text{ mL}}{1 \text{ L soln}} \times \frac{1 \text{ mol } NaC_3H_5O_3}{112.0 \text{ g } NaC_3H_5O_3} \times \frac{1 \text{ mol } C_3H_5O_3{}^-}{1 \text{ mol } NaC_3H_5O_3} = 0.0893 \text{ M}$$

$$pH = 4.11 = pK_a + \log\frac{[C_3H_5O_3{}^-]}{[HC_3H_5O_3]} = pK_a + \log\frac{0.0893 \text{ M}}{0.0500 \text{ M}} = pK_a + 0.252$$

$$pK_a = 4.11 - 0.252 = 3.86 \qquad K_a = 10^{-3.86} = 1.4 \times 10^{-4}$$

12. (a) Use the Henderson-Hasselbalch equation to determine $[CHO_2{}^-]$ in the buffer solution.

$$pH = pK_a + \log\frac{[CHO_2{}^-]}{[HCHO_2]} = 3.82 = 3.74 + \log\frac{[CHO_2{}^-]}{[HCHO_2]} \qquad \log\frac{[CHO_2{}^-]}{[HCHO_2]} = 3.82 - 3.74 = 0.08$$

$$\frac{[CHO_2{}^-]}{[HCHO_2]} = 1.2 \qquad [CHO_2{}^-] = 1.2 \ [HCHO_2] = 1.2 \times 0.505 \text{ M} = 0.61 \text{ M}$$

$$\text{mass } NaCHO_2 = 0.250 \text{ L} \times \frac{0.61 \text{ mol } CHO_2{}^-}{1 \text{ L soln}} \times \frac{1 \text{ mol } NaCHO_2}{1 \text{ mol } CHO_2{}^-} \times \frac{68.0 \text{ g } NaCHO_2}{1 \text{ mol } NaCHO_2}$$

$$= 10. \text{ g } NaCHO_2 \qquad [\text{Note that } pK_a = -\log(1.8\times 10^{-4}) = 3.74]$$

(b) $[OH^-]_i = \dfrac{0.20 \text{ g NaOH}}{0.250 \text{ L}} \times \dfrac{1 \text{ mol NaOH}}{40.0 \text{ g NaOH}} \times \dfrac{1 \text{ mol } OH^-}{1 \text{ mol NaOH}} = 0.020 \text{ M } OH^-$

Thus, $[HCHO_2]$ will decrease by 0.020 M and $[CHO_2{}^-]$ will increase by 0.020 M because of the reaction with the added hydroxide ion, as a result of: $\qquad HCHO_2 + OH^- \longrightarrow CHO_2{}^- + H_2O$

$$pH = 3.74 + \log\frac{0.61 + 0.02}{0.505 - 0.02} = 3.85$$

13. (a) The pH color change range is 1.00 pH unit on either side of pK_{HIn}. If the pH color change range is below pH = 7.00, the indicator changes color in acidic solution. If it is above pH = 7.00, the indicator changes color in alkaline solution. If pH = 7.00 falls within the pH color change ranges, the indicator changes color near the neutral point.

indicator	K_{HIn}	pK_{HIn}	pH color change range	changes color in?
bromophenol blue	1.4×10^{-4}	3.85	2.9 (yellow) to 4.9 (blue)	acidic solution
bromocresol green	2.1×10^{-5}	4.68	3.7 (yellow) to 5.7 (blue)	acidic solution
bromothymol blue	7.9×10^{-8}	7.10	6.1 (yellow) to 8.1 (blue)	neutral soln
2,4-dinitrophenol	1.3×10^{-4}	3.89	2.9 (clrless) to 4.9 (yellow)	acidic solution
chlorophenol red	1.0×10^{-6}	6.00	5.0 (yellow) to 7.0 (red)	sl. acidic soln
thymolphthalein	1.0×10^{-10}	10.00	9.0 (clrless) to 11.0 (blue)	basic solution

(b) If bromcresol green is green, the pH is between 3.7 and 5.7, probably about pH = 4.7.
If chlorophenol red is orange, the pH is between 5.0 and 7.0, probably about pH = 6.0.

14. We need to determine the pH of each solution, and then use the table developed in Exercise 13a to predict the color of the indicator. (The weakly acidic or basic character of the indicator does not affect the pH of the solution, since the indicator is added in very small quantity.)

(a) $[H_3O^+] = 0.100 \text{ M HCl} \times \dfrac{1 \text{ mol } H_3O^+}{1 \text{ mol HCl}} = 0.100 \text{ M} \qquad pH = -\log(0.100 \text{ M}) = 1.000$

2,4-dinitrophenol assumes its acid color in a solution with pH = 1.000. The solution is colorless.

(b) Solutions of NaCl(aq) are pH neutral, with pH = 7.000. Chlorophenol red assumes its basic color in such a solution; the solution is red.

(c) Reaction: $\qquad NH_3 + H_2O \rightleftharpoons NH_4{}^+ + OH^-$

Initial: $\qquad$ 1.00 M

Changes: $-x$ M $+x$ M $+x$ M
Equil: $(1.00 - x)$M x M x M

$$K_b = \frac{[NH_4^+][OH^-]}{[NH_3]} = 1.8 \times 10^{-5} = \frac{x^2}{1.00 - x} \approx \frac{x^2}{1.00} \qquad x = 4.2 \times 10^{-3} \text{ M} = [OH^-]$$

$$pOH = -\log(4.2 \times 10^{-3}) = 2.38 \qquad pH = 14.00 - 2.38 = 11.62$$

Thymolphthalein assumes its basic color in a solution with pH = 11.62; the solution is blue.

(d) From Figure 17-3, seawater has pH = 8.50 to 10.0. Bromcresol green assumes its basic color in this solution; the solution is blue.

15. (a) The titration reaction is $KOH(aq) + HI(aq) \longrightarrow KI(aq) + H_2O$

vol. KOH soln = 25.00 mL $\times \dfrac{0.182 \text{ mmol HI}}{1 \text{ mL soln}} \times \dfrac{1 \text{ mmol KOH}}{1 \text{ mmol HI}} \times \dfrac{1 \text{ mL soln}}{0.122 \text{ mmol KOH}}$

= 37.3 mL KOH soln

(b) The titration reaction is $2 KOH(aq) + H_2SO_4(aq) \longrightarrow K_2SO_4(aq) + 2 H_2O$

vol. KOH soln = 20.00 mL $\times \dfrac{0.0648 \text{ mol } H_2SO_4}{1 \text{ L soln}} \times \dfrac{2 \text{ mol KOH}}{1 \text{ mol } H_2SO_4} \times \dfrac{1 \text{ mL soln}}{0.122 \text{ mmol KOH}}$

= 21.2 mL KOH soln

16. All we can do is sketch approximate titration curves of pH vs. percent of titration, since we do not have the concentration of the acid or the base, or the volume of solution being titrated. We can, however, precisely determine the pH at the half-equivalence point [mid-way between untitrated and completely titrated for a weak acid (or base)]; this is equal to the pK_a of the weak acid (or pOH = pK_b of the weak base). Further, if we assume all solutions are 1.00 M, we can determine the pH at each equivalence point. It equals 14.00 + $\log(\sqrt{K_w/K_a})$ for the titration of a weak acid. [Indicator choices are given in square brackets.]

(a) In this titration (assuming 1.00 M KOH and 1.00 M HNO_3), the initial pH = 14.00 and that at the equivalence point is pH = 7.00. The pH drops rapidly after the equivalence point, rapidly reaching pH = 1.00. [Bromothymol blue changes from blue at pH = 8 to yellow at pH = 6.]

(b) The initial pH is that of 1.00 M NH_3, pH = 11.6. That of the half-equivalence point is pOH = pK_b = 4.76 and thus pH = 9.24. That of the equivalence point is $- \log (\sqrt{K_w/K_b})$ = pH = 4.62. The pH then drops rapidly with added strong acid to pH = 1.00. [Methyl red changes from yellow at pH = 6.2 to red at pH = 4.5.]

(c) The initial pH is that of 1.00 M $HC_2H_3O_2$, pH = 2.38. That of the half-equivalence point is pH = pK_a = 4.76. The pH at the equivalence point is determined from pOH = $-\log(\sqrt{K_w/K_a})$ = 4.62 or pH = 9.38. The pH then rises rapidly with the addition of strong base. [Phenolphthalein changes from colorless at pH = 8 to red at pH = 10.]

(d) The initial pH is that of 1.00 M $H_2PO_4^-$, pH = 3.60. That of the first half-equivalence point is pH = pK_2 = 7.20. That of the first equivalence point results from the hydrolysis of HPO_4^{2-} ion, about pH = 10.6. That of the second half equivalence point would be pH = pK_3 = 12.37, but this will be somewhat hard to reach; the solution is becoming pretty dilute. Finally, the pH of the second equivalence point will be that of the hydrolysis of the phosphate ion, about pH = 13.2, which again will be hard to reach without adding excessively concentrated base. [Alizarin yellow R changes from yellow at pH = 10 to violet at pH = 12. These is no suitable indicator given in Figure 18-6 for the second equivalence point.]

These curves are sketched below.

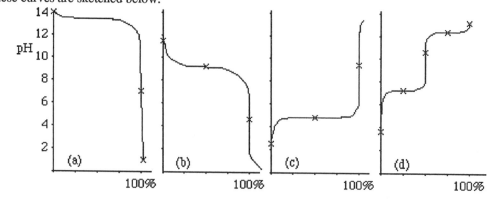

17. First we calculate the amount of HCl. The titration reaction is $HCl(aq) + KOH(aq) \longrightarrow KCl(aq) + H_2O$

amount HCl $= 25.00 \text{ mL} \times \dfrac{0.180 \text{ mmol HCl}}{1 \text{ mL soln}} = 4.50 \text{ mmol HCl} = 4.50 \text{ mmol } H_3O^+$ present

Then, in each case, we calculate the amount of OH^- that has been added, determine which ion—$OH^-(aq)$ or $H_3O^+(aq)$—is in excess, compute the concentration of that ion, and determine the pH.

(a) amount $OH^- = 10.00 \text{ mL} \times \dfrac{0.224 \text{ mmol } OH^-}{1 \text{ mL soln}} = 2.24 \text{ mmol } OH^-$ H_3O^+ is in excess.

$$[H_3O^+] = \dfrac{4.50 \text{ mmol } H_3O^+ - 2.24 \text{ mmol } OH^- \times \dfrac{1 \text{ mmol } H_3O^+}{1 \text{ mmol } OH^-}}{25.00 \text{ mL originally} + 10.00 \text{ mL titrant}} = 0.0646 \text{ M}$$

pH $= -\log(0.0646) = 1.190$

(b) amount $OH^- = 15.00 \text{ mL} \times \dfrac{0.224 \text{ mmol } OH^-}{1 \text{ mL soln}} = 3.36 \text{ mmol } OH^-$ H_3O^+ is in excess.

$$[H_3O^+] = \dfrac{4.50 \text{ mmol } H_3O^+ - 3.36 \text{ mmol } OH^- \times \dfrac{1 \text{ mmol } H_3O^+}{1 \text{ mmol } OH^-}}{25.00 \text{ mL originally} + 15.00 \text{ mL titrant}} = 0.0285 \text{ M}$$

pH $= -\log(0.0285) = 1.545$

18. The titration reaction is $KOH(aq) + HCl(aq) \longrightarrow KCl(aq) + H_2O$

no. mmol KOH $= 20.00 \text{ mL} \times \dfrac{0.350 \text{ mmol KOH}}{1 \text{ mL soln}} = 7.00 \text{ mmol KOH}$

(a) The total volume of the solution is $V = 20.00 \text{ mL} + 15.00 \text{ mL} = 35.00 \text{ mL}$

no. mmol HCl $= 15.00 \text{ mL} \times \dfrac{0.425 \text{ mmol HCl}}{1 \text{ mL soln}} = 6.38 \text{ mmol HCl}$

no. mmol excess $OH^- = (7.00 \text{ mmol KOH} - 6.38 \text{ mmol HCl}) \times \dfrac{1 \text{ mmol } OH^-}{1 \text{ mmol KOH}} = 0.62 \text{ mmol } OH^-$

$[OH^-] = \dfrac{0.62 \text{ mmol } OH^-}{35.00 \text{ mL soln}} = 0.018 \text{ M}$ pOH $= -\log(0.018) = 1.74$ pH $= 14.00 - 1.74 = 12.26$

(b) The total volume of solution is $V = 20.00 \text{ mL} + 20.00 \text{ mL} = 40.00 \text{ mL}$

no. mmol HCl $= 20.00 \text{ mL} \times \dfrac{0.425 \text{ mmol HCl}}{1 \text{ mL soln}} = 8.50 \text{ mmol HCl}$

no. mmol excess $H_3O^+ = (8.50 \text{ mmol HCl} - 7.00 \text{ mmol KOH}) \times \dfrac{1 \text{ mmol } H_3O^+}{1 \text{ mmol HCl}} = 1.50 \text{ mmol } H_3O^+$

$[H_3O^+] = \dfrac{1.50 \text{ mmol } H_3O^+}{40.00 \text{ mL}} = 0.0375 \text{ M}$ pH $= -\log(0.0375) = 1.426$

19. The titration reaction is $HNO_2(aq) + NaOH(aq) \longrightarrow NaNO_2(aq) + H_2O$

amount $HNO_2 = 25.00 \text{ mL} \times \dfrac{0.146 \text{ mmol } HNO_2}{1 \text{ mL soln}} = 3.65 \text{ mmol } HNO_2$

(a) The volume of the solution is $25.00 \text{ mL} + 10.00 \text{ mL} = 35.00 \text{ mL}$

amount NaOH $= 10.00 \text{ mL} \times \dfrac{0.116 \text{ mmol NaOH}}{1 \text{ mL soln}} = 1.16 \text{ mmol NaOH}$

1.16 mmol $NaNO_2$ are formed in this reaction and there is an excess of (3.65 mmol HNO_2 – 1.16 mmol NaOH =) 2.49 mmol HNO_2. We can use the Henderson-Hasselbalch equation to determine the pH of the solution. $pK_a = -\log(7.2 \times 10^{-4}) = 3.14$

$$pH = pK_a + \log \dfrac{[NO_2^-]}{[HNO_2]} = 3.14 + \log \dfrac{1.16 \text{ mmol } NO_2^-/35.00 \text{ mL}}{2.49 \text{ mmol } HNO_2/35.00 \text{ mL}} = 2.81$$

(b) The volume of the solution is $25.00 \text{ mL} + 20.00 \text{ mL} = 45.00 \text{ mL}$

amount NaOH $= 20.00 \text{ mL} \times \dfrac{0.116 \text{ mmol NaOH}}{1 \text{ mL soln}} = 2.32 \text{ mmol NaOH}$

2.32 mmol $NaNO_2$ are formed in this reaction and there is an excess of (3.65 mmol HNO_2 – 2.32 mmol NaOH =) 1.33 mmol HNO_2.

$$pH = pK_a + \log \dfrac{[NO_2^-]}{[HNO_2]} = 3.14 + \log \dfrac{2.32 \text{ mmol } NO_2^-/45.00 \text{ mL}}{1.33 \text{ mmol } HNO_2/45.00 \text{ mL}} = 3.38$$

20. The calculation is very similar to that of Review Question 19. In this case, however, the titration reaction is

$NH_3(aq) + HCl(aq) \longrightarrow NH_4Cl(aq) + H_2O$

amount $NH_3 = 20.00 \text{ mL} \times \dfrac{0.306 \text{ mmol } NH_3}{1 \text{ mL soln}} = 6.12 \text{ mmol } NH_3$

(a) The volume of the solution is 20.00 mL + 10.00 mL = 30.00 mL

amount HCl = $10.00 \text{ mL} \times \dfrac{0.505 \text{ mmol NaOH}}{1 \text{ mL soln}}$ = 5.05 mmol HCl

5.05 mmol NH_4Cl is formed in this reaction and there is an excess of (6.12 mmol NH_3 – 5.05 mmol HCl =) 1.07 mmol NH_3. We can use the Henderson-Hasselbalch equation to determine the pH of the solution. $pK_b = -\log(1.8 \times 10^{-5}) = 4.74$ $pK_a = 14.00 - pK_b = 14.00 - 4.74 = 9.26$

$pH = pK_a + \log \dfrac{[NH_3]}{[NH_4^+]} = 9.26 + \log \dfrac{1.07 \text{ mmol } NH_3/30.00 \text{ mL}}{5.05 \text{ mmol } NH_4^+/30.00 \text{ mL}} = 8.59$

(b) The volume of the solution is 20.00 mL + 20.00 mL = 40.00 mL

amount HCl = $20.00 \text{ mL} \times \dfrac{0.505 \text{ mmol NaOH}}{1 \text{ mL soln}}$ = 10.1 mmol HCl

6.12 mmol NH_4Cl is formed in this reaction and there is an excess of (10.1 mmol HCl – 6.12 mmol NH_3 =) 4.0 mmol HCl, which latter species determines the pH of the solution.

$[H_3O^+] = \dfrac{4.0 \text{ mmol HCl}}{40.00 \text{ mL soln}} \times \dfrac{1 \text{ mmol } H_3O^+}{1 \text{ mmol HCl}} = 0.10 \text{ M}$ $pH = -\log(0.10) = 1.0$

21. (a) This is the pH of a 0.01000 M $HC_7H_5O_2$ solution.

Reaction:	$HC_7H_5O_2 + H_2O$	$\rightleftharpoons$	$C_7H_5O_2^- +$	H_3O^+
Initial:	0.01000 M			
Changes:	$-x$ M		$+x$ M	$+x$ M
Equil:	$(0.01000 - x)$M		x M	x M

$K_a = \dfrac{[C_7H_5O_2^-][H_3O^+]}{[HC_7H_5O_2]} = 6.3 \times 10^{-5} = \dfrac{x^2}{0.0100 - x} = \dfrac{x^2}{0.0100}$ $x = 7.9 \times 10^{-4} \text{ M} = [H_3O^+]$

$pH = -\log(7.9 \times 10^{-4}) = 3.10$

(b) amount of $HC_7H_5O_2 = 25.0 \text{ mL} \times 0.01000 \text{ M} = 0.250 \text{ mmol } HC_7H_5O_2$

We determine the volume of 0.01000 M $Ba(OH)_2$ to reach the equivalence point.

The titration reaction is $Ba(OH)_2(aq) + 2 HC_7H_5O_2(aq) \longrightarrow Ba(C_7H_5O_2)_2 + 2 H_2O$

vol. base = $0.250 \text{ mmol } HC_7H_5O_2 \times \dfrac{1 \text{ mmol } Ba(OH)_2}{2 \text{ mmol } HC_7H_5O_2} \times \dfrac{1 \text{ mL soln}}{0.01000 \text{ mmol } Ba(OH)_2} = 12.5 \text{ mL}$

Thus, the addition of 6.25 mL of 0.01000 M $Ba(OH)_2$ brings us to the half-equivalence point, where $pH = pK_a = 4.20$

(c) At the equivalence point, there is 0.250 mmol $C_7H_5O_2^-$ in 25.00 + 12.50 = 37.50 mL solution. It is the hydrolysis of this anion that determines the pH of the solution.

$[C_7H_5O_2^-] = \dfrac{0.250 \text{ mmol } C_7H_5O_2^-}{37.50 \text{ mL soln}} = 6.67 \times 10^{-3} \text{ M}$

Reaction:	$C_7H_5O_2^- + H_2O$	$\rightleftharpoons$	$HC_7H_5O_2 +$	OH^-
Initial:	0.00667 M			
Changes:	$-x$ M		$+x$ M	$+x$ M
Equil:	$(0.00667 - x)$M		x M	x M

$K_b = \dfrac{K_w}{K_a} = \dfrac{1.00 \times 10^{-14}}{6.3 \times 10^{-5}} = \dfrac{x^2}{0.00667 - x} \approx \dfrac{x^2}{0.00667}$ $x = 1.0 \times 10^{-6} \text{ M} = [OH^-]$

$pOH = -\log(1.0 \times 10^{-6}) = 6.00$ $pH = 14.00 - 6.00 = 8.00$

(d) The excess added base determines the pH of the solution.

excess amount $OH^- = 2.5 \text{ mL} \times \dfrac{0.0100 \text{ mmol } Ba(OH)_2}{1 \text{ mL } Ba(OH)_2} \times \dfrac{2 \text{ mmol } OH^-}{1 \text{ mmol } Ba(OH)_2} = 0.050 \text{ mmol } OH^-$

$[OH^-] = \dfrac{0.050 \text{ mmol } OH^-}{25.0 \text{ mL} + 15.0 \text{ mL}} = 1.3 \times 10^{-3} \text{ M}$ $pOH = -\log(1.3 \times 10^{-3}) = 2.89$

$pH = 14.00 - 2.89 = 11.11$

22. 0.10 M solutions of $NaHSO_4$ are the most acidic. The Na^+ cation does not contribute to the acidity of the solution; it does not hydrolyze. Of the anions involved, S^{2-} hydrolyzes to produce an alkaline solution. Each of the other anions can ionize further to yield H_3O^+, but HSO_4^- does so to a far greater extent than do the other two; it is quite strong for a weak acid.

EXERCISES

The Common Ion Effect

23. (a) Since $NaNO_2$ contributes to the solution an ion—nitrite ion, NO_2^-—that is produced by the ionization of HNO_2, increasing the concentration of nitrite ion in this way suppresses the ionization of HNO_2 and decreases $[H_3O^+]$ of this solution. The pH of the solution increases.

(b) Although $NaNO_3$ contributes to the solution an ion—nitrate ion, NO_3^-—that is produced by the ionization of HNO_3, HNO_3 is a strong acid. There is no $HNO_3(aq)$ in equilibrium with hydrogen and nitrate ions. Thus, there is no equilibrium to be shifted by the addition of one of its products. The $[H_3O^+]$ and the pH are unaffected by the addition of $NaNO_3$ to a solution of nitric acid.

24. (a) The strong acid HCl suppresses the ionization of the weak acid HOCl so much that a negligible concentration of H_3O^+ is contributed to the solution by HOCl. Thus $[H_3O^+] = [HCl] = 0.045$ M

(b) This is a buffer solution. We can use the Henderson-Hasselbalch equation to determine its pH.

$$pK_a = -\log(7.2 \times 10^{-4}) = 3.14 \qquad pH = pK_a + \log\frac{[NO_2^-]}{[HNO_2]} = 3.14 + \log\frac{0.100\ M}{0.0650\ M} = 3.33$$

$$[H_3O^+] = 10^{-3.33} = 4.7 \times 10^{-4}\ M$$

(c) This also is a buffer solution, as we see by an analysis of the reaction between the components.

Reaction:	H_3O^+(aq, from HCl) +	$C_2H_3O_2^-$(aq, from $NaC_2H_3O_2$) $\longrightarrow$	$HC_2H_3O_2$(aq) + H_2O
In soln:	0.0416 M	0.0626 M	
Produce HAc:	−0.0416 M	−0.0416 M	+0.0416 M
Initial:	≈0 M	0.0210 M	0.0416 M

Now the Henderson-Hasselbalch equation can be used. $pK_a = -\log(1.8 \times 10^{-5}) = 4.74$

$$pH = pK_a + \log\frac{[C_2H_3O_2^-]}{[HC_2H_3O_2]} = 4.74 + \log\frac{0.0210\ M}{0.0416\ M} = 4.44 \qquad [H_3O^+] = 10^{-4.44} = 3.6 \times 10^{-5}\ M$$

25. (a) Neither Ba^{2+}(aq) nor Cl^-(aq) hydrolyzes or ionizes to affect the acidity of the solution. $[OH^-]$ is determined entirely by the $Ba(OH)_2$ solute.

$$[OH^-] = \frac{0.0088\ mol\ Ba(OH)_2}{1\ L\ soln} \times \frac{2\ mol\ OH^-}{1\ mol\ Ba(OH)_2} = 0.017_6\ M$$

(b) We use the Henderson-Hasselbalch equation for this buffer solution.

$$[NH_4^+] = 0.200\ M\ (NH_4)_2SO_4 \times \frac{2\ mol\ NH_4^+}{1\ mol\ (NH_4)_2SO_4} = 0.400\ M \qquad pK_a = 9.26\ for\ NH_4^+.$$

$$pH = pK_a + \log\frac{[NH_3]}{[NH_4^+]} = 9.26 + \log\frac{0.500\ M}{0.400\ M} = 9.36 \qquad pOH = 14.00 - 9.36 = 4.64$$

$$[OH^-] = 10^{-4.64} = 2.3 \times 10^{-5}\ M$$

(c) This solution also is a buffer solution, as analysis of the reaction between its components shows.

Reaction:	NH_4^+(aq, from NH_4Cl) +	OH^-(aq, from NaOH) $\longrightarrow$	NH_3(aq) + H_2O
In soln:	0.286 M	0.202 M	
Form NH_3:	−0.202 M	−0.202 M	+0.202 M
Initial:	0.084 M	≈0 M	0.202 M

$$pH = pK_a + \log\frac{[NH_3]}{[NH_4^+]} = 9.26 + \log\frac{0.202\ M}{0.804\ M} = 9.64 \qquad pOH = 14.00 - 9.64 = 4.36$$

$$[OH^-] = 10^{-4.36} = 4.4 \times 10^{-5}\ M$$

26. The explanation for the different result is that each of these solutions has present an ion—acetate ion, $C_2H_3O_2^-$—that is produced in the ionization of acetic acid. The presence of this ion suppresses the ionization of acetic acid, thus minimizing the increase in $[H_3O^+]$. All three solutions are buffer solutions and their pH can be found with the aid of the Henderson-Hasselbalch equation.

(a) $pH = pK_a + \log\frac{[C_2H_3O_2^-]}{[HC_2H_3O_2]} = 4.76 + \log\frac{0.10}{1.0} = 3.76 \qquad [H_3O^+] = 10^{-3.76} = 1.7 \times 10^{-4}\ M$

$$\%\ ionization = \frac{[H_3O^+]}{[HC_2H_3O_2]_i} \times 100\% = \frac{1.7 \times 10^{-4}\ M}{1.0\ M} \times 100\% = 0.017\%$$

(b) $pH = pK_a + \log\frac{[C_2H_3O_2^-]}{[HC_2H_3O_2]} = 4.76 + \log\frac{0.10}{0.10} = 4.76 \qquad [H_3O^+] = 10^{-4.76} = 1.7 \times 10^{-5}\ M$

$$\% \text{ ionization} = \frac{[H_3O^+]}{[HC_2H_3O_2]_i} \times 100\% = \frac{1.7 \times 10^{-5} \text{ M}}{0.10 \text{ M}} \times 100\% = 0.017\%$$

(c) $\quad pH = pK_a + \log\frac{[C_2H_3O_2^-]}{[HC_2H_3O_2]} = 4.76 + \log\frac{0.10}{0.010} = 5.76 \qquad [H_3O^+] = 10^{-5.76} = 1.7 \times 10^{-6} \text{ M}$

$$\% \text{ ionization} = \frac{[H_3O^+]}{[HC_2H_3O_2]_i} \times 100\% = \frac{1.7 \times 10^{-6} \text{ M}}{0.010 \text{ M}} \times 100\% = 0.017\%$$

27. amount of solute $= 1.15 \text{ mg} \times \dfrac{1 \text{ g}}{1000 \text{ mg}} \times \dfrac{1 \text{ mol } C_6H_5NH_3^+Cl^-}{129.6 \text{ g}} \times \dfrac{1 \text{ mol } C_6H_5NH_3^+}{1 \text{ mol } C_6H_5NH_3^+Cl^-}$

$\qquad\qquad\qquad = 8.87 \times 10^{-6} \text{ mol } C_6H_5NH_3^+$

$[C_6H_5NH_3^+] = \dfrac{8.87 \times 10^{-6} \text{ mol } C_6H_5NH_3^+}{3.18 \text{ L soln}} = 2.79 \times 10^{-6} \text{ M}$

Reaction: $\quad C_6H_5NH_2(aq) + H_2O \rightleftharpoons C_6H_5NH_3^+(aq) \quad + \quad OH^-(aq)$

Initial: $\qquad 0.105 \text{ M} \qquad\qquad\qquad 2.79 \times 10^{-6} \text{ M}$

Changes: $\quad -x \text{ M} \qquad\qquad\qquad\quad +x \text{ M} \qquad\qquad\qquad +x \text{ M}$

Equil: $\qquad (0.105 - x)\text{M} \qquad\quad (2.79 \times 10^{-6} + x)\text{M} \qquad x \text{ M}$

$K_b = \dfrac{[C_6H_5NH_3^+][OH^-]}{[C_6H_5NH_2]} = 7.4 \times 10^{-10} = \dfrac{(2.79 \times 10^{-6} + x)\, x}{0.105 - x}$

$7.4 \times 10^{-10}(0.105 - x) = (2.79 \times 10^{-6} + x)x = 7.8 \times 10^{-11} - 7.4 \times 10^{-10}x = 2.79 \times 10^{-6}\, x + x^2$

$x^2 + (2.79 \times 10^{-6} + 7.4 \times 10^{-10})x - 7.8 \times 10^{-11} = 0 = x^2 + 2.79 \times 10^{-6}x - 7.8 \times 10^{-11}$

$x = \dfrac{-b \pm \sqrt{b^2 - 4ac}}{2a} = \dfrac{-2.79 \times 10^{-6} \pm \sqrt{7.78 \times 10^{-12} + 1.3 \times 10^{-10}}}{2} = 7.5 \times 10^{-6}\text{M} = [OH^-]$

$pOH = -\log(7.5 \times 10^{-6}) = 5.12 \qquad pH = 14.00 - 5.12 = 8.88$

Buffer Solutions

28. (a) 0.100 M NaCl is not a buffer solution. There is neither a weak acid nor a weak base present.

(b) 0.100 M NaCl—0.100 M NH_4Cl is not a buffer solution. Although a weak acid, NH_4^+, is present, its conjugate base is not.

(c) 0.100 M CH_3NH_2—0.150 M $CH_3NH_3^+Cl$ is a buffer solution. Both a weak base, CH_3NH_2, and its conjugate acid, $CH_3NH_3^+$, are present in approximately equal concentrations.

(d) 0.100 M HCl—0.050 M $NaNO_2$ is not a buffer solution. All the NO_2^- is converted to HNO_2 and thus the solution is a mixture of a strong acid and a weak acid.

(e) 0.100 M HCl—0.200 M $NaC_2H_3O_2$ is a buffer solution. All of the HCl reacts with half of the $C_2H_3O_2^-$ to form a solution with 0.100 M $HC_2H_3O_2$, a weak acid, and 0.100 M $C_2H_3O_2^-$, its conjugate base.

(f) 0.100 M $HC_2H_3O_2$—0.125 M $NaC_3H_5O_2$ is not a buffer in the strict sense because it does not contain a weak acid and its conjugate base, but rather the conjugate base of another weak acid. These two weak acids (acetic, with $K_a = 1.74 \times 10^{-5}$ and propionic, with $K_a = 1.35 \times 10^{-5}$) have approximately the same strength, however, so this solution would resist changes in its pH on the addition of strong acid or strong base, although its pH would not be as narrowly controlled as in a "classic" buffer solution.

29. (a) Reaction with added acid: $\quad HPO_4^{2-} + H_3O^+ \longrightarrow H_2PO_4^- + H_2O$

$\qquad\quad$ Reaction with added base: $\quad H_2PO_4^- + OH^- \longrightarrow HPO_4^{2-} + H_2O$

(b) We assume initially that the buffer has equal concentrations of the two ions, $[H_2PO_4^-] = [HPO_4^{2-}]$

$\qquad pH = pK_2 + \log\dfrac{[HPO_4^{2-}]}{[H_2PO_4^-]} = 7.20 + 0.00 = 7.20$ is the pH where the buffer is most effective.

(c) $pH = 7.20 + \log\dfrac{[HPO_4^{2-}]}{[H_2PO_4^-]} = 7.20 + \log\dfrac{0.150 \text{ M}}{0.050 \text{ M}} = 7.20 + 0.48 = 7.68$

30. First use the Henderson-Hasselbalch equation $[pK_b = -\log(1.8 \times 10^{-5}) = 4.74, pK_a = 14.00 - 4.74 = 9.26]$ to determine $[NH_4^+]$ in the buffer solution.

$pH = 9.48 = pK_a + \log\dfrac{[NH_3]}{[NH_4^+]} = 9.26 + \log\dfrac{[NH_3]}{[NH_4^+]} \qquad \log\dfrac{[NH_3]}{[NH_4^+]} = 9.48 - 9.26 = +0.22$

$\dfrac{[NH_4^+]}{[NH_3]} = 10^{-0.22} = 0.60$ $\qquad$ $[NH_4^+] = 0.60 \times [NH_3] = 0.60 \times 0.216 \text{ M} = 0.13 \text{ M}$

We now assume that the volume of the solution does not change significantly when the solid is added.

$$\text{mass (NH}_4)_2\text{SO}_4 = 425.0 \text{ mL} \times \dfrac{1 \text{ L soln}}{1000 \text{ mL}} \times \dfrac{0.13 \text{ mol NH}_4^+}{1 \text{ L soln}} \times \dfrac{1 \text{ mol (NH}_4)_2\text{SO}_4}{2 \text{ mol NH}_4^+}$$

$$\times \dfrac{132.2 \text{ g (NH}_4)_2\text{SO}_4}{1 \text{ mol (NH}_4)_2\text{SO}_4} = 3.7 \text{ g (NH}_4)_2\text{SO}_4$$

31. (a) amount $HC_7H_5O_2 = 1.50 \text{ g HC}_7\text{H}_5\text{O}_2 \times \dfrac{1 \text{ mol HC}_7\text{H}_5\text{O}_2}{122.1 \text{ g HC}_7\text{H}_5\text{O}_2} = 0.0123 \text{ mol HC}_7\text{H}_5\text{O}_2$

amount $C_7H_5O_2^- = 1.50 \text{ g NaC}_7\text{H}_5\text{O}_2 \times \dfrac{1 \text{ mol NaC}_7\text{H}_5\text{O}_2}{144.1 \text{ g NaC}_7\text{H}_5\text{O}_2} \times \dfrac{1 \text{ mol C}_7\text{H}_5\text{O}_2^-}{1 \text{ mol NaC}_7\text{H}_5\text{O}_2}$

$= 0.0104 \text{ mol C}_7\text{H}_5\text{O}_2^-$

$pH = pK_a + \log \dfrac{[C_7H_5O_2^-]}{[HC_7H_5O_2]} = -\log(6.3 \times 10^{-5}) + \log \dfrac{0.0104 \text{ mol C}_7\text{H}_5\text{O}_2^-/0.1500 \text{ L}}{0.0123 \text{ mol HC}_7\text{H}_5\text{O}_2/0.1500 \text{ L}}$

$= 4.20 - 0.0729 = 4.13$

(b) To lower the pH of this buffer solution, that is, to make it more acidic, benzoic acid must be added. The quantity is determined as follows.

$4.00 = 4.20 + \log \dfrac{0.0104 \text{ mol C}_7\text{H}_5\text{O}_2^-}{x \text{ mol HC}_7\text{H}_5\text{O}_2}$ $\qquad$ $\log \dfrac{0.0104 \text{ mol C}_7\text{H}_5\text{O}_2^-}{x \text{ mol HC}_7\text{H}_5\text{O}_2} = -0.20$

$\dfrac{0.0104 \text{ mol C}_7\text{H}_5\text{O}_2^-}{x \text{ mol HC}_7\text{H}_5\text{O}_2} = 10^{-0.20} = 0.63$ $\qquad$ $x = \dfrac{0.0104}{0.63} = 0.017 \text{ mol HC}_7\text{H}_5\text{O}_2 \text{ total}$

added $HC_7H_5O_2 = 0.017 \text{ mol HC}_7\text{H}_5\text{O}_2 - 0.0123 \text{ mol HC}_7\text{H}_5\text{O}_2 = 0.005 \text{ mol HC}_7\text{H}_5\text{O}_2$

added mass $HC_7H_5O_2 = 0.005 \text{ mol HC}_7\text{H}_5\text{O}_2 \times \dfrac{122.1 \text{ g HC}_7\text{H}_5\text{O}_2}{1 \text{ mol HC}_7\text{H}_5\text{O}_2} = 0.6 \text{ g HC}_7\text{H}_5\text{O}_2$

32. $pH = 3.71 = pK_a + \log \dfrac{[CHO_2^-]}{[HCHO_2]} = -\log(1.8 \times 10^{-4}) + \log \dfrac{[CHO_2^-]}{[HCHO_2]} = 3.74 + \log \dfrac{[CHO_2^-]}{[HCHO_2]}$

$\log \dfrac{[CHO_2^-]}{[HCHO_2]} = 3.71 - 3.74 = -0.03$ $\qquad$ $\dfrac{[CHO_2^-]}{[HCHO_2]} = 10^{-0.03} = 0.93$

$[CHO_2^-] = 0.93 \times 0.312 = 0.29 \text{ M}$

$\text{mass NaCHO}_2 = 0.325 \text{ L} \times \dfrac{0.29 \text{ mol CHO}_2^-}{1 \text{ L soln}} \times \dfrac{1 \text{ mol NaCHO}_2}{1 \text{ mol CHO}_2^-} \times \dfrac{68.01 \text{ g NaCHO}_2}{1 \text{ mol NaCHO}_2} = 6.4 \text{ g NaCHO}_2$

33. The added NH_3 will react with the formic acid, and the pH of the buffer solution will increase. The original buffer solution has $[HCHO_2] = 0.312 \text{ M}$ and $[CHO_2^-] = 0.29 \text{ M}$. We first calculate the $[NH_3]$ in solution, reduced from 15 M because of dilution. $\qquad$ $[NH_3]$ added $= 15 \text{ M} \times \dfrac{0.35 \text{ mL}}{100.4 \text{ mL}} = 0.052 \text{ M}$

Reaction:	$NH_3(aq)$	$+$	$HCHO_2(aq)$	$\longrightarrow$	$NH_4^+(aq)$	$+$	$CHO_2^-(aq)$
Buffer:			0.312 M				0.29 M
Added:	0.052 M						
Changes:	−0.052 M		−0.052 M				+0.052 M
Final:	0.000 M		0.260 M				0.34 M

$pH = pK_a + \log \dfrac{[CHO_2^-]}{[HCHO_2]} = 3.74 + \log \dfrac{0.34}{0.260} = 3.86$

34. For NH_3, $pK_b = -\log(1.8 \times 10^{-5})$ $\qquad$ For NH_4^+, $pK_a = 14.00 - pK_b = 14.00 - 4.74 = 9.76$

(a) $[NH_3] = \dfrac{1.51 \text{ g NH}_3}{0.500 \text{ L}} \times \dfrac{1 \text{ mol NH}_3}{17.03 \text{ g NH}_3} = 0.177 \text{ M}$

$[NH_4^+] = \dfrac{3.85 \text{ g (NH}_4)_2\text{SO}_4}{0.500 \text{ L}} \times \dfrac{1 \text{ mol (NH}_4)_2\text{SO}_4}{132.1 \text{ g (NH}_4)_2\text{SO}_4} \times \dfrac{2 \text{ mol NH}_4^+}{1 \text{ mol (NH}_4)_2\text{SO}_4} = 0.117 \text{ M}$

$pH = pK_a + \log \dfrac{[NH_3]}{[NH_4^+]} = 9.26 + \log \dfrac{0.177 \text{ M}}{0.117 \text{ M}} = 9.44$

(b) The $OH^-(aq)$ reacts with the $NH_4^+(aq)$ to produce an equivalent amount of $NH_3(aq)$.

$[OH^-]_i = \dfrac{0.88 \text{ g NaOH}}{0.500 \text{ L}} \times \dfrac{1 \text{ mol NaOH}}{40.00 \text{ g NaOH}} \times \dfrac{1 \text{ mol OH}^-}{1 \text{ mol NaOH}} = 0.044 \text{ M}$

Reaction:	$NH_4^+(aq)$	$+$	$OH^-(aq)$	$\rightleftharpoons$	$NH_3(aq) + H_2O$
Initial:	0.117 M				0.177 M
Add NaOH:			0.044 M		
React:	−0.044 M		−0.044 M		+0.044 M
Final:	0.073 M		0.0000 M		0.221 M

$$pH = pK_a + \log\frac{[NH_3]}{[NH_4^+]} = 9.26 + \log\frac{0.221\ M}{0.073\ M} = 9.74$$

(c) Reaction: $NH_3(aq)$ + $H_3O^+(aq) \rightleftharpoons NH_4^+(aq) + H_2O$

Initial: 0.177 M   0.117 M

Add HCl: $+x$ M

React: $-x$ M $-x$ M $+x$ M

Final: (0.177 $-x$) M 10^{-9} M   (0.117 $+ x$)M

$$pH = 9.00 = pK_a + \log\frac{[NH_3]}{[NH_4^+]} = 9.26 + \log\frac{(0.177 - x)\ M}{(0.117 + x)\ M}$$

$$\log\frac{(0.177 - x)\ M}{(0.117 + x)\ M} = 9.00 - 9.26 = -0.26 \qquad \frac{(0.177 - x)\ M}{(0.117 + x)\ M} = 10^{-0.26} = 0.55$$

$$0.177 - x = 0.55(0.117 + x) = 0.064 - 0.55\ x \qquad 1.55\ x = 0.177 - 0.064 = 0.113$$

$$x = \frac{0.113}{1.55} = 0.0729\ M$$

$$\text{volume HCl} = 0.500\ L \times\frac{0.0729\ mol\ H_3O^+}{1\ L\ soln} \times\frac{1\ mol\ HCl}{1\ mol\ H_3O^+} \times\frac{1000\ mL}{12\ mol\ HCl} = 3.0\ mL$$

35. The pK_a's of the three acids help us chose the one to be used in the buffer. It is the acid with pK_a within 1.00 pH units of 3.50. $pK_a = 3.74$ for $HCHO_2$, $pK_a = 4.74$ for $HC_2H_3O_2$, and $pK_1 = 2.15$ for H_3PO_4. Thus we choose $HCHO_2$ and $NaCHO_2$ to prepare a buffer with pH = 3.50. The Henderson-Hasselbalch equation is used to determine the relative amounts of each component present in the buffer solution.

$$pH = 3.50 = 3.74 + \log\frac{[CHO_2^-]}{[HCHO_2]} \qquad \log\frac{[CHO_2^-]}{[HCHO_2]} = 3.50 - 3.74 = -0.24 \qquad \frac{[CHO_2^-]}{[HCHO_2]} = 10^{-0.24} = 0.58$$

This ratio of concentrations is also the ratio of the number of moles of each component in the buffer solution, since both concentrations are a number of moles in a certain volume, and the volumes are the same (the two solutes are in the same solution). This ratio also is the ratio of the volumes of the two solutions, since both solutions being mixed contain the same concentration of solute. If we assume 100. ml of acid solution, $V_{acid} = 100.$ mL. Then the volume of salt solution is $V_{salt} = 0.58 \times 100.$ mL = 58 mL 0.100 M $NaCHO_2$

36. $[H_3O^+]_i = 10^{-7.4} = 4._0 \times 10^{-8}$ M $[H_3O^+]_f = 10^{-7.3} = 5._0 \times 10^{-8}$ M

$$\% \text{ increase} = \frac{[H_3O^+]_f - [H_3O^+]_i}{[H_3O^+]_i} \times 100\% = \frac{5._0 \times 10^{-8}\ M - 4._0 \times 10^{-8}\ M}{4._0 \times 10^{-8}\ M} \times 100\% = 2_5\%$$

37. A buffer is most effective within one pH unit of pK_a for a weak acid or within one pOH unit of pK_b for a weak base.

(a) $pK_a = 3.14$ for HNO_2. A $HNO_2/NaNO_2$ buffer is most effective in the pH range from pH = 2.14 to pH = 4.14.

(b) $pK_b = 4.74$ for NH_3. A $NH_3/(NH_4)_2SO_4$ buffer is most effective in the pOH range from pOH = 3.74 to pOH = 5.74, or a pH range from pH = 8.26 to pH = 10.26.

(c) $pK_b = 3.38$ for CH_3NH_2. A $CH_3NH_2/CH_3NH_3^+Cl^-$ buffer is most effective in the pOH range from pOH = 2.38 to pOH = 4.38, or a pH range from pH = 9.62 to pH = 11.62.

38. (a) The pH of each buffer is determined from the Henderson-Hasselbalch equation.

$$pH = pK_a + \log\frac{[C_2H_3O_2^-]}{[HC_2H_3O_2]} = 4.74 + \log\frac{0.010\ M}{0.010\ M} = 4.74$$

$$pH = pK_a + \log\frac{[C_2H_3O_2^-]}{[HC_2H_3O_2]} = 4.74 + \log\frac{0.50\ M}{0.100\ M} = 5.44$$

But the effective pH range is the same for each of these acetate buffers: from pH = 3.74 to pH = 5.74, one pH unit on either side of pK_a for acetic acid.

(b) We determine the buffer capacity per liter of each solution.

To each liter of 0.010 M $HC_2H_3O_2$—0.010 M $NaC_2H_3O_2$ we can add 0.010 mol OH^- before all of the $HC_2H_3O_2$ is consumed, and we can add 0.010 mol H_3O^+ before all of the $C_2H_3O_2^-$ is consumed.

To each liter of 0.100 M $HC_2H_3O_2$—0.50 M $KC_2H_3O_2$ we can add 0.100 mol OH^- before all the $HC_2H_3O_2$ is consumed, and we can add 0.50 mol H_3O^+ before all of the $C_2H_3O_2^-$ is consumed.

39. (a) We can lower the pH of the 0.250 M $HC_2H_3O_2$—0.55 M $C_2H_3O_2^-$ buffer solution by increasing $[HC_2H_3O_2]$ or lowering $[C_2H_3O_2^-]$. NaCl solutions will have no effect, and the addition of NaOH(aq) or $NaC_2H_3O_2(aq)$ will raise the pH. A solution of $HC_2H_3O_2$ will lower the pH; the addition of 0.050 M $HC_2H_3O_2$ to a 0.250 M $HC_2H_3O_2$ actually will lower $[HC_2H_3O_2]$, but $[C_2H_3O_2^-]$ will be lowered even more by dilution, and thus the ratio $[C_2H_3O_2^-]/[HC_2H_3O_2]$ decreases, lowering pH by the Henderson-Hasselbalch equation. The addition of 0.150 M HCl will raise $[HC_2H_3O_2]$ and lower $[C_2H_3O_2^-]$ through the reaction $\quad H_3O^+ + C_2H_3O_2^- \longrightarrow HC_2H_3O_2 + H_2O$

(b) We first use the Henderson-Hasselbalch equation to determine the ratio of the concentration of acetate ion and acetic acid. $\qquad pH = 5.00 = 4.74 + \log \dfrac{[C_2H_3O_2^-]}{[HC_2H_3O_2]}$

$\log \dfrac{[C_2H_3O_2^-]}{[HC_2H_3O_2]} = 5.00 - 4.74 = 0.26 \qquad \dfrac{[C_2H_3O_2^-]}{[HC_2H_3O_2]} = 10^{0.26} = 1.8$

Now we compute the amount of each component in the original buffer solution.

amount of $C_2H_3O_2^- = 300.\ mL \times \dfrac{0.55\ mmol\ C_2H_3O_2^-}{1\ mL\ soln} = 165\ mmol\ C_2H_3O_2^-$

amount of $HC_2H_3O_2 = 300.\ mL \times \dfrac{0.250\ mmol\ HC_2H_3O_2}{1\ mL\ soln} = 75.0\ mmol\ HC_2H_3O_2$

$\longrightarrow$ Now let x represent the amount of H_3O^+ added in mmol.

$1.8 = \dfrac{165 - x}{75.0 + x} \qquad 165 - x = 1.8(75 + x) = 13_5 + 1.8x \qquad 165 - 13_5 = 2.7\ x$

$x = \dfrac{165 - 13_5}{2.7} = 11\ mmol\ H_3O^+$

no. mL 0.150 M HCl $= 11\ mmol\ H_3O^+ \times \dfrac{1\ mmol\ HCl}{1\ mmol\ H_3O^+} \times \dfrac{1\ mL\ soln}{0.150\ mmol\ HCl}$

$\qquad\qquad\qquad\qquad = 74\ mL\ 0.150\ M\ HCl\ solution$

$\longrightarrow$ For adding 0.050 M $HC_2H_3O_2$, we let x = volume in mL of added weak acid. Then, the number of mmols of added $HC_2H_3O_2$ is 0.050 x.

$[C_2H_3O_2^-] = \dfrac{165}{300 + x} \qquad\qquad [HC_2H_3O_2] = \dfrac{75.0 + 0.050x}{300 + x}$

$1.8 = \dfrac{[C_2H_3O_2^-]}{[HC_2H_3O_2]} = \dfrac{165}{75.0 + 0.050\ x} \qquad 13_5 + 0.080\ x = 165 \qquad x = \dfrac{165 - 13_5}{0.080} = 4 \times 10^2\ mL$

40. (a) We use the Henserson-Hasselbalch equation to determine the pH of the solution. The total solution volume is 36.00 mL + 64.00 mL = 100.00 mL. $pK_a = 14.00 - pK_b = 14.00 + \log(1.8 \times 10^{-5}) = 9.26$

$[NH_3] = \dfrac{36.00\ mL \times 0.200\ M\ NH_3}{100.00\ mL} = \dfrac{7.20\ mmol\ NH_3}{100.0\ mL} = 0.0720\ M$

$[NH_4^+] = \dfrac{64.00\ mL \times 0.200\ M\ NH_4^+}{100.00\ mL} = \dfrac{12.8\ mmol\ NH_4^+}{100.0\ mL} = 0.128\ M$

$pH = pK_a + \log \dfrac{[NH_3]}{[NH_4^+]} = 9.26 + \log \dfrac{0.0720}{0.128\ M} = 9.01$

(b) The solution has $[OH^-] = 10^{-4.99} = 1.0 \times 10^{-5}\ M$

The Henderson-Hasselbalch equation depends on the assumption: $[NH_3] \gg 1.0 \times 10^{-5}\ M \ll [NH_4^+]$

If the solution is diluted to 1.00 L, $[NH_3] = 7.20 \times 10^{-3}\ M$, and $[NH_4^+] = 1.28 \times 10^{-2}\ M$. These two concentrations are consistent with the assumption.

However, if the solution is diluted to 1000. L, $[NH_3] = 7.2 \times 10^{-6}\ M$, and $[NH_3] = 1.28 \times 10^{-6}$ M, and these two concentrations are not consistent with the assumption. Thus, in 1000. L of solution, the given quantities of NH_3 and NH_4^+ will not produce a solution with pH = 9.00. With sufficient dilution, the solution will become indistinguishable from pure water; its pH will equal 7.00.

(c) The 0.10 mL of added 1.00 M HCl does not significantly affect the volume of the solution, but it does add 0.10 mL $\times$ 1.00 M HCl = 0.10 mmol H_3O^+ This added H_3O^+ reacts with NH_3, decreasing its amount from 7.20 mmol NH_3 to 7.10 mmol NH_3, and increasing the amount of NH_4^+ from 12.8 mmol NH_4^+ to 12.9 mmol NH_4^+, through the reaction: $\quad NH_3 + H_3O^+ \longrightarrow NH_4^+ + H_2O$

$pH = 9.26 + \log \dfrac{7.10\ mmol\ NH_3/100.10\ mL}{12.9\ mmol\ NH_4^+/100.10\ mL} = 9.00$

(d) We see in the calculation of part (c) that the total volume of the solution does not affect the pOH of the solution, at least as long as the Henderson-Hasselbalch equation is obeyed. We let x represent the number of millimoles of H_3O^+ added, through 1.00 M HCl. This increases the amount of NH_4^+ and decreases the amount of NH_3, through the reaction $\qquad NH_3 + H_3O^+ \longrightarrow NH_4^+ + H_2O$

$$pH = 8.90 = 9.26 + \log \frac{7.20 - x}{12.8 + x} \qquad \log \frac{7.20 - x}{12.8 + x} = 8.90 - 9.26 = -0.36$$

Inverting, we have: $\dfrac{12.8 + x}{7.20 - x} = 10^{0.36} = 2.29 \qquad 12.8 + x = 2.29(7.20 - x) = 16.5 - 2.29\,x$

$$x = \frac{16.5 - 12.8}{1.00 + 2.29} = 1.1 \text{ mmol } H_3O^+$$

$$\text{vol } 1.00 \text{ M HCl} = 1.1 \text{ mmol } H_3O^+ \times \frac{1 \text{ mmol HCl}}{1 \text{ mmol } H_3O^+} \times \frac{1 \text{ mL soln}}{1.00 \text{ mmol HCl}} = 1.1 \text{ mL } 1.00 \text{ M HCl}$$

41. (a) $[C_2H_3O_2^-] = \dfrac{10.0 \text{ g } NaC_2H_3O_2}{0.300 \text{ L soln}} \times \dfrac{1 \text{ mol } NaC_2H_3O_2}{82.03 \text{ g } NaC_2H_3O_2} \times \dfrac{1 \text{ mol } C_2H_3O_2^-}{1 \text{ mol } NaC_2H_3O_2} = 0.406 \text{ M } C_2H_3O_2^-$

Reaction:	$HC_2H_3O_2 + H_2O$	$\rightleftharpoons$	$C_2H_3O_2^-$ +	H_3O^+
Initial:				0.200 M
Add $NaC_2H_3O_2$			0.406 M	
Consume H_3O^+	0.200 M		0.206 M	

Then use the Henderson-Hasselbalch equation.

$$pH = pK_a + \log \frac{[C_2H_3O_2^-]}{[HC_2H_3O_2]} = 4.74 + \log \frac{0.206 \text{ M}}{0.200 \text{ M}} = 4.76 + 0.01 = 4.75$$

(b) We can calculate the initial $[OH^-]$ due to the $Ba(OH)_2$.

$$[OH^-] = \frac{1.00 \text{ g } Ba(OH)_2}{0.300 \text{ L}} \times \frac{1 \text{ mol } Ba(OH)_2}{171.3 \text{ g } Ba(OH)_2} \times \frac{2 \text{ mol } OH^-}{1 \text{ mol } Ba(OH)_2} = 0.0389 \text{ M}$$

Then $HC_2H_3O_2$ is consumed.

Reaction:	$HC_2H_3O_2 +$	OH^-	$\rightleftharpoons$	$C_2H_3O_2^- +$	H_2O
Initial:	0.200 M	0.0389 M		0.206 M	
Consume OH^-	0.161 M			0.245 M	

Then use the Henderson-Hasselbalch equation.

$$pH = pK_a + \log \frac{[C_2H_3O_2^-]}{[HC_2H_3O_2]} = 4.74 + \log \frac{0.245 \text{ M}}{0.161 \text{ M}} = 4.74 + 0.18 = 4.92$$

(c) $Ba(OH)_2$ can be added until all of the $HC_2H_3O_2$ is consumed.

$$Ba(OH)_2 + 2\,HC_2H_3O_2 \longrightarrow Ba(C_2H_3O_2)_2 + 2\,H_2O$$

$$\text{quantity of } Ba(OH)_2 = 0.300 \text{ L} \times \frac{0.200 \text{ mol } HC_2H_3O_2}{1 \text{ L soln}} \times \frac{1 \text{ mol } Ba(OH)_2}{2 \text{ mol } HC_2H_3O_2} = 0.0300 \text{ mol } Ba(OH)_2$$

$$\times \frac{171.3 \text{ g } Ba(OH)_2}{1 \text{ mol } Ba(OH)_2} = 5.14 \text{ g } Ba(OH)_2$$

(d) This is an excess of 0.06 g $Ba(OH)_2$ and it is this excess that determines the pOH of the solution.

$$[OH^-] = \frac{0.06 \text{ g } Ba(OH)_2}{0.300 \text{ L soln}} \times \frac{1 \text{ mol } Ba(OH)_2}{171.3 \text{ g } Ba(OH)_2} \times \frac{2 \text{ mol } OH^-}{1 \text{ mol } Ba(OH)_2} = 2 \times 10^{-3} \text{ M } OH^-$$

$$pOH = -\log(2 \times 10^{-3}) = 2.7 \qquad pH = 14.00 - 2.7 = 11.3$$

Acid-Base Indicators

42. (a) In an acid-base titration, the pH of the solution changes sharply at a definite pH that is known in advance at the beginning of the titration. (This pH change occurs during the addition of a very small volume of titrant.) In determining the pH of a solution, on the other hand, that pH is not known in advance. Since each indicator only serves to fix the pH over a quite small region, often less than 2.0 pH units, several indicators—carefully chosen to span the entire range of 14 pH units—must be employed to even narrow the pH to ± 1 pH unit.

(b) An indicator is, after all, a weak acid. Its addition to a solution will affect the acidity of that solution. Thus, one adds only enough indicator to show a color change and not enough to affect solution acidity.

43. (a) 0.10 M KOH is an alkaline solution and phenol red will display its basic color in such a solution; the solution will be red.

(b) 0.10 M $HC_2H_3O_2$ is an acidic solution—although that of a weak acid—and phenol red will display its acidic color in such a solution; the solution will be yellow.

(c) 0.10 M NH_4NO_3 is an acidic solution due to the hydrolysis of the ammonium ion. Phenol red will display its acidic solor—that is, yellow—in this solution.

(d) 0.10 M HBr is an acidic solution, the aqueous solution of a strong acid. Phenol red will display its acidic color in this solution; the solution will be yellow.

(e) 0.10 M NaCN is an alkaline solution because of the hydrolysis of the cyanide ion. Phenol red will display its basic color—red—in this solution.

44. (a) A 0.10 M $HC_2H_3O_2$—0.10 M $NaC_2H_3O_2$ solution—having equal concentrations of acetic acid and acetate ion—has a pH equal to the pK_a of acetic acid, pH = 4.74. We use an equation similar to the Henderson-Hasselbalch equation to determine the relative concentrations of indicator, HIn, and its anion, In^-, in this solution.

$$pH = pK_{HIn} + \log \frac{[In^-]}{[HIn]} \qquad 4.74 = 4.95 + \log \frac{[In^-]}{[HIn]} \qquad \log \frac{[In^-]}{[HIn]} = 4.74 - 4.95 = -0.21$$

$$\frac{[In^-]}{[HIn]} = 10^{-0.21} = 0.62 = \frac{x}{100-x} \qquad x = 62 - 0.62\,x \qquad x = \frac{62}{1.62} = 38\% \; In^- \text{ and } 62\% \; HIn$$

(b) When the indicator is in a solution whose pH equals its pK_a (4.95), the ratio $[In^-]/[HIn] = 1.00$. However, at the midpoint of its color change range (about pH = 5.3), the ratio $[In^-]/[HIn]$ is greater than 1.00. Thus, even though $[HIn] < [In^-]$ at this midpoint, the contribution of HIn to establishing the color of the solution is about the same as the contribution of In^-. This must mean that HIn (red) is more strongly colored than In^- (yellow).

45. (a) pH = $-\log$ (0.205) = 0.688 The indicator is red in this solution.

(b) The total volume of the solution is 600.0 mL. We compute the amount of each solute.

amount H_3O^+ = 350.0 mL × 0.205 M = 71.8 mmol H_3O^+

amount NO_2^- = 250.0 mL × 0.500 M = 125 mmol NO_2^-

$$[H_3O^+] = \frac{71.8 \text{ mmol}}{600.0 \text{ mL}} = 0.120 \text{ M} \qquad [NO_2^-] = \frac{125 \text{ mmol}}{600.0 \text{ mL}} = 0.208 \text{ M}$$

The H_3O^+ and NO_2^- react to produce a buffer solution in which $[HNO_2] = 0.120$ M and $[NO_2^-] = 0.208 - 0.120 = 0.088$ M. We use the Henderson-Hasselbalch equation to determine the pH of this solution. $pK_a = -\log (7.2 \times 10^{-4}) = 3.14$

$$pH = pK_a + \log \frac{[NO_2^-]}{[HNO_2]} = 3.14 + \log \frac{0.088 \text{ M}}{0.120 \text{ M}} = 3.01 \quad \text{The indicator is yellow in this solution.}$$

(c) The total volume of the solution is 750. mL. We compute the amount and then the concentration of each solute. amount OH^- = 150 mL × 0.100 M = 15.0 mmol OH^-

This OH^- reacts with HNO_2 in the buffer solution to neutralize some of it and leave 56.8 mmol (= 71.8 − 15.0) unneutralized.

$$[HNO_2] = \frac{56.8 \text{ mmol}}{750. \text{ mL}} = 0.0757 \text{ M} \qquad [NO_2^-] = \frac{(125 + 15) \text{ mmol}}{750. \text{ mL}} = 0.186 \text{ M}$$

We use the Henderson Hasselbalch equation to determine the pH of this solution.

$$pH = pK_a + \log \frac{[NO_2^-]}{[HNO_2]} = 3.14 + \log \frac{0.186 \text{ M}}{0.0757 \text{ M}} = 3.53 \quad \text{The indicator is yellow in this solution.}$$

(d) We determine the $[OH^-]$ due to the added $Ba(OH)_2$.

$$[OH^-] = \frac{5.00 \text{ g Ba(OH)}_2}{0.750 \text{ L}} \times \frac{1 \text{ mol Ba(OH)}_2}{171.34 \text{ g Ba(OH)}_2} \times \frac{2 \text{ mol OH}^-}{1 \text{ mol Ba(OH)}_2} = 0.0778 \text{ M}$$

This is sufficient $[OH^-]$ to react with the existing $[HNO_2]$ and leave an excess $[OH^-] = 0.0778$ M − 0.0757 M = 0.0021 M. [pOH = $-\log(0.0021)$ = 2.68. pH = 14.00 − 2.68 = 11.32.] The indicator is blue in this solution.

Neutralization Reactions

46. The titration reaction is $\qquad Ca(OH)_2(aq) + 2\,HCl(aq) \longrightarrow CaCl_2(aq) + 2\,H_2O$

$$\frac{\text{g Ca(OH)}_2}{\text{L soln}}$$

$$= \frac{10.7 \text{ mL HCl} \times \dfrac{1 \text{ L}}{1000 \text{ mL}} \times \dfrac{0.1032 \text{ mol HCl}}{1 \text{ L}} \times \dfrac{1 \text{ mol Ca(OH)}_2}{2 \text{ mol HCl}} \times \dfrac{74.10 \text{ g Ca(OH)}_2}{1 \text{ mol Ca(OH)}_2}}{50.00 \text{ mL Ca(OH)}_2 \text{ soln} \times \dfrac{1 \text{ L}}{1000 \text{ mL}}}$$

$$= 0.818 \text{ g Ca(OH)}_2/\text{L soln}$$

47. To the second equivalence point the reaction is: $\qquad H_3PO_4(aq) + 2\,KOH(aq) \longrightarrow K_2PO_4(aq) + 2\,H_2O$

The molarity of the H_3PO_4 solution is determined as follows.

$$H_3PO_4 \text{ molarity} = \frac{31.15 \text{ mL KOH soln} \times \dfrac{0.242 \text{ mmol KOH}}{1 \text{ mL KOH soln}} \times \dfrac{1 \text{ mmol H}_3\text{PO}_4}{2 \text{ mmol KOH}}}{25.00 \text{ mL H}_3\text{PO}_4 \text{ soln}} = 0.151 \text{ M}$$

48. (a) $NaHSO_4(aq) + NaOH(aq) \longrightarrow Na_2SO_4(aq) + H_2O$

$HSO_4^-(aq) + OH^-(aq) \longrightarrow SO_4^{2-}(aq) + H_2O$

(b) We first determine the mass of $NaHSO_4$.

$$\text{mass NaHSO}_4 = 38.56 \text{ mL} \times \frac{1 \text{ L}}{1000 \text{ mL}} \times \frac{0.215 \text{ mol NaOH}}{1 \text{ L}} \times \frac{1 \text{ mol NaHSO}_4}{1 \text{ mol NaOH}}$$

$$\times \frac{120.06 \text{ g NaHSO}_4}{1 \text{ mol NaHSO}_4} = 0.995 \text{ g NaHSO}_4$$

$$\% \text{ NaCl} = \frac{1.028 \text{ g sample} - 0.995 \text{ g NaHSO}_4}{1.028 \text{ g sample}} \times 100\% = 3.2\% \text{ NaCl}$$

(c) At the endpoint of this titration the solution is one of SO_4^{2-}, from which the pH is determined by hydrolysis. Since K_a for HSO_4^- is relatively large (1.1×10^{-2}), hydrolysis of SO_4^{2-} should not occur to a very great extent. The pH of a neutralized solution should be very nearly 7, and most of the indicators represented in Figure 18-7 would be suitable. A more exact solution follows.

$$[SO_4^{2-}] = \frac{0.995 \text{ g NaHSO}_4}{0.03856 \text{ L}} \times \frac{1 \text{ mol NaHSO}_4}{120.06 \text{ g NaHSO}_4} \times \frac{1 \text{ mol SO}_4^{-2}}{1 \text{ mol NaHSO}_4} = 0.215 \text{ M}$$

Reaction:	$SO_4^{2-}(aq) + H_2O$	$\rightleftharpoons$	$HSO_4^-(aq)$	+	$OH^-(aq)$
Initial:	0.2150 M				
Changes:	$-x$ M		$+x$ M		$+x$ M
Equil:	$(0.2150 - x)$ M		x M		x M

$$K_b = \frac{[HSO_4^-][OH^-]}{[SO_4^{2-}]} = \frac{K_w}{K_{a_2}} = \frac{1.00 \times 10^{-14}}{1.1 \times 10^{-2}} = 9.1 \times 10^{-13} = \frac{x \cdot x}{0.2150 - x} \approx \frac{x^2}{0.2150}$$

$[OH^-] = \sqrt{9.1 \times 10^{-13} \times 0.2150} = 4.4 \times 10^{-7}$ M $pOH = -\log(4.4 \times 10^{-7}) = 6.35$

$pH = 14.00 - 6.36 = 7.64$

Thus, either bromothymol blue (pH color change range from pH = 6.1 to pH = 7.9) or phenol red (pH color change range from pH = 6.4 to pH = 8.0) would be a suitable indicator, since either changes color at pH = 7.64.

49. We determine the amount of solute in each solution. And then the amount of the reagent in excess.

A. $[H_3O^+] = 10^{-2.50} = 0.0032$ M

$$\text{mmol HCl} = 100.0 \text{ mL} \times \frac{0.0032 \text{ mmol H}_3\text{O}^+}{1 \text{ mL soln}} \times \frac{1 \text{ mmol HCl}}{1 \text{ mmol H}_3\text{O}^+} = 0.32 \text{ mmol HCl}$$

B. $pOH = 14.00 - 11.00 = 3.00$ $[OH^-] = 10^{-3.00} = 1.0 \times 10^{-3}$ M

$$\text{mmol NaOH} = 100.0 \text{ mL} \times \frac{0.0010 \text{ mmol OH}^-}{1 \text{ mL soln}} \times \frac{1 \text{ mmol NaOH}}{1 \text{ mmol OH}^-} = 0.10 \text{ mmol NaOH}$$

Result:

Titration reaction:	$NaOH(aq)$	+	$HCl(aq)$	$\rightleftharpoons$	$NaCl(aq) + H_2O$
Initial amounts:	0.10 mmol		0.32 mmol		
After reaction:	0.00 mmol		0.22 mmol		

$$[H_3O^+] = \frac{0.22 \text{ mmol HCl}}{200.0 \text{ mL soln}} \times \frac{1 \text{ mmol H}_3\text{O}^+}{1 \text{ mmol HCl}} = 1.1 \times 10^{-3} \text{ M} \quad pH = -\log(1.1 \times 10^{-3}) = 2.96$$

Titration Curves

50. In each case, the volume of acid and its molarity are the same. Thus, also the amount of acid is the same in each case. The volume of titrant needed to reach the equivalence point will also be the same in both cases, since the titrant has the same concentration in each case, and it is the same amount of base that reacts with a given amount (in moles) of acid. Realize that, as the titration of a weak acid proceeds, the weak acid will ionize—replenishing the H_3O^+ in solution. This will occur until all of the weak acid has ionized, and has reacted with the strong base.

51. (a) This equivalence point is the result of the titration of a weak acid with a strong base. The species present in solution is CO_3^{2-} which, through its hydrolysis, will form an alkaline, or basic solution. The other ionic species in solution—Na^+—will not hydrolyze. Thus, pH > 7.0

(b) This is the titration of a strong acid with a weak base. The species present in solution is NH_4^+ which hydrolyzes to form an acidic solution. Cl^- does not hydrolyze. Thus pH < 7.0

(c) This is the titration of a strong acid with a strong base. Two ions are present in the solution at the equivalence point—K^+ and Cl^-—and neither of these hydrolyze. The solution will have a pH of 7.00.

52. (a) Initial $[OH^-] = 0.100$ M OH^- pOH = $-\log(0.100) = 1.000$ pH = 13.000

Since this is the titration of a strong base with a strong acid, KI is the solute present at the equivalence point and pH = 7.000.   The titration reaction is $KOH(aq) + HI(aq) \longrightarrow KI(aq) + H_2O$

no. mL HI = 25.0 mL KOH soln $\times \dfrac{0.100 \text{ mmol KOH soln}}{1 \text{ mL soln}} \times \dfrac{1 \text{ mmol HI}}{1 \text{ mmol KOH}} \times \dfrac{1 \text{ mL HI soln}}{0.200 \text{ mmol HI}}$

= 12.5 mL HI soln

Initial amount of KOH present = 25.0 mL KOH soln $\times$ 0.100 M = 2.50 mmol KOH

At the 40% titration point: 5.00 mL HI soln $\times$ 0.200 M HI = 1.00 mmol HI

excess KOH = 2.50 mmol KOH – 1.00 mmol HI = 1.50 mmol KOH

$[OH^-] = \dfrac{1.50 \text{ mmol KOH}}{30.0 \text{ mL total}} \times \dfrac{1 \text{ mmol } OH^-}{1 \text{ mmol KOH}} = 0.0500$ M   pOH = $-\log(0.0500) = 1.30$

pH = 14.00 – 1.30 = 12.70

At the 80% titration point: 10.00 mL HI soln $\times$ 0.200 M HI = 2.00 mmol HI

excess KOH = 2.50 mmol KOH – 2.00 mmol HI = 0.50 mmol KOH

$[OH^-] = \dfrac{0.50 \text{ mmol KOH}}{35.0 \text{ mL total}} \times \dfrac{1 \text{ mmol } OH^-}{1 \text{ mmol KOH}} = 0.0143$ M   pOH = $-\log(0.0143) = 1.84$

pH = 14.00 – 1.84 = 12.16

At the 110% titration point: 13.75 mL HI soln $\times$ 0.200 M HI = 2.75 mmol HI

excess HI = 2.75 mmol KOH – 2.50 mmol HI = 0.75 mmol KOH

$[H_3O^+] = \dfrac{0.75 \text{ mmol KOH}}{38.8 \text{ mL total}} \times \dfrac{1 \text{ mmol } OH^-}{1 \text{ mmol KOH}} = 0.0193$ M   pH = $-\log(0.0193) = 1.71$

Since the pH changes very rapidly at the equivalence point, from about pH = 10 to about pH = 4, most of the indicators in Figure 18-6 can be used. The main exceptions are alizarin yellow R, bromophenol blue, thymol blue (in its acid range), and methyl violet.

(b) *Initial pH:*

Reaction: $NH_3(aq) + H_2O \rightleftharpoons NH_4^+(aq) + OH^-(aq)$

Initial: 1.00 M

Changes: $-x$ M $+x$ M $+x$ M

Equil: $(1.00 - x)$M x M x M

$K_b = \dfrac{[NH_4^+][OH^-]}{[NH_3]} = 1.8 \times 10^{-5} = \dfrac{x^2}{1.00 - x} \approx \dfrac{x^2}{1.00}$ $x = 4.2 \times 10^{-3}$ M = $[OH^-]$

pOH = $-\log(4.2 \times 10^{-3}) = 2.38$ pH = 14.00 – 2.38 = 11.62 = initial pH

Volume of titrant: $NH_3 + HCl \longrightarrow NH_4Cl + H_2O$

no. mL HCl = 10.0 mL $\times \dfrac{1.00 \text{ mmol } NH_3}{1 \text{ mL soln}} \times \dfrac{1 \text{ mmol HCl}}{1 \text{ mmol } NH_3} \times \dfrac{1 \text{ mL HCl soln}}{0.250 \text{ mmol HCl}} = 40.0$ mL soln

pH at equivalence point:

The total volume of the solution at the equivalence point is 10.0 + 40.0 = 50.0 mL

Also at the equivalence point, all of the NH_3 has reacted to form NH_4^+. It is this NH_4^+ that hydrolyzes to determine the pH of the solution.

$[NH_4^+] = \dfrac{10.0 \text{ mL} \times \dfrac{1.00 \text{ mmol } NH_3}{1 \text{ mL soln}} \times \dfrac{1 \text{ mmol } NH_4^+}{1 \text{ mmol } NH_3}}{50.0 \text{ mL total solution}} = 0.200$ M

Reaction: $NH_4^+(aq) + H_2O \rightleftharpoons NH_3(aq) + H_3O^+(aq)$

Initial: 0.200 M

Changes: $-x$ M $+x$ M $+x$ M

Equil: $(0.200 - x)$M x M x M

$K_a = \dfrac{K_w}{K_b} = \dfrac{1.00 \times 10^{-14}}{1.8 \times 10^{-5}} = \dfrac{[NH_3][H_3O^+]}{[NH_3]} = \dfrac{x^2}{0.200 - x} \approx \dfrac{x^2}{0.200}$ $x = 1.1 \times 10^{-5}$ M

$[H_3O^+] = 1.1 \times 10^{-5}$ M   pH = $-\log(1.1 \times 10^{-5}) = 4.96$

Of the indicators in Figure 18-6, the only one that has the pH of the equivalence point within its pH color change range is methyl red (yellow at pH = 6.2 and red at pH = 4.5)

At the 50% titration point, $[NH_3] = [NH_4^+]$ and $pOH = pK_b = 4.74$ pH = 14.000 − 4.74 = 9.26

The titration curves for parts (a) and (b) are sketched below.

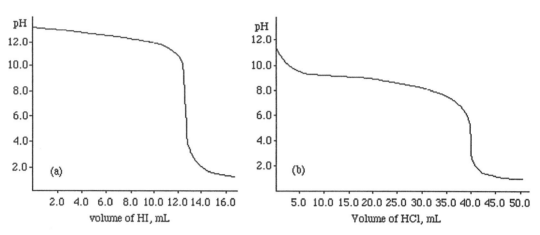

53. **(a)** Titration reaction: $HCl(aq) + NaOH(aq) \longrightarrow NaCl(aq) + H_2O$

$$100\% \text{ vol. NaOH} = 25.00 \text{ mL} \times \frac{0.1000 \text{ mmol HCl}}{1 \text{ mL HCl soln}} \times \frac{1 \text{ mmol NaOH}}{1 \text{ mmol HCl}} \times \frac{1 \text{ mL NaOH soln}}{0.1000 \text{ mmol NaOH}}$$

$$= 25.00 \text{ mL}$$

90% volume of NaOH soln = 0.9000×25.00 mL = 22.5 mL

no. mmol HCl = 25.00 mL × 0.1000 M = 2.500 mmol HCl

no. mmol NaOH = 22.50 mL × 0.1000 M = 2.250 mmol NaOH

no. mmol excess HCl = 2.500 mmol − 2.250 mmol = 0.250 mmol HCl = 0.250 mmol H_3O^+

It is this excess HCl that determines the pH of the solution.

$$[H_3O^+] = \frac{0.250 \text{ mmol } H_3O^+}{25.00 \text{ mL} + 22.50 \text{ mL}} = 5.26 \times 10^{-3} \text{ M} \qquad pH = -\log(5.26 \times 10^{-3}) = 2.28$$

(b) We have already noted in the Henderson-Hasselbalch equation that the total volume of the buffer does not affect the pH of the solution, because both the anion and the weak acid are present in the same volume of the solution. Thus, the ratio of concentrations in that equation, $[A^-]/[HA]$, equals the ratio of moles of these two components, n_A/n_{HA}. At any given percent point in the titration, the number of moles of A^- is determined by that percent: n_A = fraction × original number of moles of HA, since, for example, if the titration is 80% complete, then 0.80 of the original number of moles of HA has been converted to A^-. But this also means that 0.20 of the original number of moles of HA are present in solution as the undissociated acid. Following this line of reasoning, the Henderson-Hasselbalch equation becomes.

$$pH = pK_a + \log\frac{\% \text{ titrated}}{\% \text{ untitrated}} \qquad \text{For this case:} \qquad pH = 4.74 + \log\frac{90\%}{10\%} = 4.74 + 0.95 = 5.69$$

54. **(a)** We begin with 10.0 mL of a solution of 1.00 M $NH_3(aq)$, and determine its pH.

Reaction: $NH_3(aq) + H_2O \rightleftharpoons NH_4^+(aq) + OH^-(aq)$

Initial: 1.00 M

Changes: −x M   +x M   +x M

Equil: (1.00 − x)M   x M   x M

$$K_b = \frac{[NH_4^+][OH^-]}{[NH_3]} = 1.8 \times 10^{-5} = \frac{x^2}{1.00 - x} \approx \frac{x^2}{1.00} \qquad x = 4.2 \times 10^{-3} \text{ M} = [OH^-]$$

$$pOH = -\log(4.2 \times 10^{-3}) = 2.38 \qquad pH = 14.00 - 2.38 = 11.62 = \text{initial pH}$$

(b) The volume of 0.350 M HCl needed to reach the end point is 28.6 mL. At each point in the buffer region, we can calculate the amount of H_3O^+ added, and use that to calculate the amounts of NH_3 unreacted and NH_4^+ produced. From the Summarizing Example, we know that the initial amount of NH_3 is 10.0 mmol

amount of added H_3O^+ = 10.0 mL × 0.350 M = 3.50 mmol $H_3O^+ = y$

We organize the information around the titration reaction. V = solution volume.

Reaction: $\quad NH_3(aq) \quad + \quad H_3O^+(aq) \longrightarrow NH_4^+(aq) + H_2O$

In soln: $\qquad$ 10.0 mmol $\qquad$ y mmol

React: $\qquad -y$ mmol $\qquad -y$ mmol $\qquad +y$ mmol

Initial: $\qquad (10.0 - y)$ mmol ≈ 0 mmol $\qquad y$ mmol

$pH = pK_a + \log \dfrac{[NH_3]}{[NH_4^+]} = 9.26 + \log \dfrac{(10.0 - y)/V}{y/V} = 9.26 + \log \dfrac{10.0 - y}{y} = 9.26 + \log \dfrac{6.50}{3.50} = 9.53$

(c) amount of added H_3O^+ = 20.0 mL × 0.350 M = 7.00 mmol H_3O^+ = y

$pH = 9.26 + \log \dfrac{10.0 - y}{y} = 9.26 + \log \dfrac{3.00}{7.00} = 8.89$

(d) amount of added H_3O^+ = 30.0 mL × 0.350 M = 10.5 mmol H_3O^+

amount of excess H_3O^+ = 10.5 mmol H_3O^+ – 10.0 mmol $NH_3 \times \dfrac{1 \text{ mmol } H_3O^+}{1 \text{ mmol } NH_3} = 0.5$ mmol H_3O^+

$[H_3O^+] = \dfrac{0.5 \text{ mmol } H_3O^+}{10.0 \text{ mL original soln} + 30.0 \text{ mL added titrant}} = 0.01_{25}$ M

$pH = -\log (0.0125) = -1.9$

55. (a) This is the pH of 0.250 M NH_3.

Reaction: $\quad NH_3(aq) + H_2O \rightleftharpoons NH_4^+(aq) \quad + \quad OH^-(aq)$

Initial: $\qquad$ 0.250 M

Changes: $\qquad -x$ M $\qquad\qquad +x$ M $\qquad\qquad +x$ M

Equil: $\qquad (0.250 - x)$M $\qquad x$ M $\qquad\qquad x$ M

$K_b = \dfrac{[NH_4^+][OH^-]}{[NH_3]} = 1.8 \times 10^{-5} = \dfrac{x^2}{0.250 - x} \approx \dfrac{x^2}{0.250} \qquad x = 2.1 \times 10^{-3}$ M = $[OH^-]$

$pOH = -\log(2.1 \times 10^{-3}) = 2.68 \qquad\qquad pH = 11.32$ M

(b) This is the volume of titrant needed to reach the equivalence point.

The titration reaction is $\qquad NH_3(aq) + HI(aq) \longrightarrow NH_4I(aq)$

vol. HI = 20.00 mL NH_3(aq) $\times \dfrac{0.250 \text{ mmol } NH_3}{1 \text{ mL } NH_3 \text{ soln}} \times \dfrac{1 \text{ mmol HI}}{1 \text{ mmol } NH_3} \times \dfrac{1 \text{ mL HI soln}}{0.350 \text{ mmol HI}}$

$\qquad = 14.3$ mL HI soln

(c) The pOH at the half-equivalence point of the titration of a weak base with a strong acid is equal to the pK_b of the weak base. $\qquad pOH = pK_b = 4.74 \qquad\qquad pH = 14.00 - 4.74 = 9.26$

(d) NH_4^+ is formed during the titration, and its hydrolysis determines the pH of the solution.

total volume of solution = 20.00 mL + 14.3 mL = 34.3 mL

mmol NH_4^+ = 20.00 mL NH_3(aq) $\times \dfrac{0.250 \text{ mmol } NH_3}{1 \text{ mL } NH_3 \text{ soln}} \times \dfrac{1 \text{ mmol } NH_4^+}{1 \text{ mmol } NH_3} = 5.00$ mmol NH_4^+

$[NH_4^+] = \dfrac{5.00 \text{ mmol } NH_4^+}{34.3 \text{ mL soln}} = 0.146$ M

Reaction: $\quad NH_4^+(aq) + H_2O \rightleftharpoons NH_3(aq) \quad + \quad H_3O^+(aq)$

Initial: $\qquad$ 0.146 M

Changes: $\qquad -x$ M $\qquad\qquad +x$ M $\qquad\qquad +x$ M

Equil: $\qquad (0.146 - x)$M $\qquad x$ M $\qquad\qquad x$ M

$K_a = \dfrac{K_w}{K_b} = \dfrac{1.00 \times 10^{-14}}{1.8 \times 10^{-5}} = \dfrac{[NH_3][H_3O^+]}{[NH_4^+]} = \dfrac{x^2}{0.146 - x} \approx \dfrac{x^2}{0.146} \qquad x = 9.0 \times 10^{-6}$ M = $[H_3O^+]$

$pH = -\log(9.0 \times 10^{-6}) = 5.05$

56. (a) At a pH of 3.00, $[H_3O^+] = 10^{-3.00} = 1.0 \times 10^{-3}$ M. Since the initial pH of the 0.100 M HCl solution is 1.00, this must mean that the solution is nearly titrated. If we let V equal the volume of 0.100 M NaOH added, then the total volume at this point is 25.00 + V. The original amount of H_3O^+ in the solution is 25.00 mL × 0.100 M = 2.50 mmol H_3O^+. The amount in mmol of NaOH added is given by 0.100 M × V. That means the amount of H_3O^+ remaining is: 2.50 mmol – (0.100 × V). Therefore, the $[H_3O^+]$ is:

$[H_3O^+] = \dfrac{2.50 \text{ mmol} - (0.100 \times V) \text{ mmol}}{25.00 \text{ mL} + V \text{ mL}} = 0.0010$ M $\qquad 2.50 - 0.100\,V = 0.025 + 0.0010\,V$

$2.50 - 0.025 = (0.100 + 0.0010)\,V = 0.101\,V \qquad V = \dfrac{2.50 - 0.025}{0.101} = 24.5$ mL

(b) Since $pK_a = 4.74$ for acetic acid, a pH = 5.25 is within the buffer region, and the Henderson-Hasselbalch equation can be used. We shall let x be the number of millimoles of strong base that must be added. As in part (a), there are 2.50 mmol of acid at the beginning of the titration.

$$pH = pK_a + \log \frac{[C_2H_3O_2^-]}{[HC_2H_3O_2]} = 5.25 = 4.74 + \log \frac{x}{2.50 - x} \qquad \log \frac{x}{2.50 - x} = 0.51$$

$$\frac{x}{2.50 - x} = 10^{0.51} = 3.2 \qquad x = 3.2(2.50 - x) = 8.0 - 3.2x \qquad x = \frac{8.0}{4.2} = 1.9 \text{ mmol OH}^-$$

$$\text{volume of titrant} = 1.9 \text{ mmol OH}^- \times \frac{1 \text{ mL titrant}}{0.100 \text{ mmol OH}^-} = 19 \text{ mL}$$

(c) pH = 2.50 is before the first equivalence point in the titration of H_3PO_4. We use the Henderson-Hasselbalch equation to determine the amount of OH$^-$ that must be added to reach pH = 2.50.

$$pH = pK_a + \log \frac{[H_2PO_4^-]}{[H_3PO_4]} = 2.50 = 2.15 + \log \frac{[H_2PO_4^-]}{[H_3PO^-]} \qquad \log \frac{[H_2PO_4^-]}{[H_3PO_4]} = 0.35$$

$$\frac{[H_2PO_4^-]}{[H_3PO_4]} = 10^{+0.35} = 2.2 = \frac{x}{1.00 - x} \qquad 2.2 - 2.2x = x \qquad x = \frac{2.2}{3.2} = 0.69$$

In the expression above, we have realized that the titration begins with 1.00 mmol (= 10.00 mL × 0.100 M) H_3PO_4, and x mmol of $H_2PO_4^-$ is formed by reaction with NaOH. Thus, the volume of titrant solurion needed is   0.69 mmol NaOH × (1 mL/0.100 mmol NaOH) = 6.9 mL.

57. We calculate below many points on the titration curve. Part of the reason for this is that the computer cannot sketch, but can only plot the points given to it. The information that you should obtain is summarized in a table just prior to the titration curve. There should be two equivalence points in this titation, as in Figure 18-10. These points are not evenly spaced, since the H_3PO_4 is, in a sense, partially titrated. Initially we have a buffer of H_3PO_4(aq) and $H_2PO_4^-$(aq) for which we can determine the pH.

$$pH = pK_{a_1} + \frac{[H_2PO_4^-]}{[H_3PO_4]} = 2.15 + \frac{0.0150 \text{ M}}{0.0400 \text{ M}} = 1.72$$

The 10.00 mL 0.0400 M H_3PO_4 contains 0.400 mmol H_3PO_4, which will react first with the 0.0200 M NaOH in reaching the first equivalence point. Each mL of 0.0200 M NaOH contains 0.0200 mmol OH$^-$, which will convert 0.0200 mmol H_3PO_4 into $H_2PO_4^-$. Since we start with 10.00 mL × 0.0150 M NaH_2PO_4 = 0.150 mmol $H_2PO_4^-$, we can determine the amounts of H_3PO_4 and $H_2PO_4^-$ after x mL of 0.0200 M NaOH have been added.

amount H_3PO_4 = 0.400 mmol H_3PO_4 − x mL × 0.0200 mmol H_3PO_4 = (0.400 − 0.0200 x) mmol H_3PO_4

amount H_2PO_4 = 0.150 mmol H_3PO_4 + x mL × 0.0200 mmol $H_2PO_4^-$ = (0.150 + 0.0200 x) mmol $H_2PO_4^-$

These amounts are then used, with the Henderson-Hasselbalch equation, to determine the pH of the solution.

After the addition of 1.00 mL: $pH = pK_{a_1} + \frac{[H_2PO_4^-]}{[H_3PO_4]} = 2.15 + \frac{(0.150 + 0.0200) \text{ M}}{(0.400 - 0.0200) \text{ M}} = 1.80$

2.00 mL: $pH = 2.15 + \frac{(0.150 + 0.0400) \text{ M}}{(0.400 - 0.0400) \text{ M}} = 1.87$ The remaining points are calculated similarly.

titrant, mL	4.00	6.00	8.00	10.00	12.00	14.00	16.00	18.00
pH	2.01	2.13	2.26	2.39	2.53	2.70	2.92	3.25

At the first equivalence point, we have a solution of dihydrogen phosphate ion, for which we calculate the pH as follows. $pH = \frac{1}{2}(pK_{a_1} + pK_{a_2}) = \frac{1}{2}(2.15 + 7.20) = 4.68$

$$\text{volume titrant} = 0.400 \text{ mmol } H_3PO_4 \times \frac{1 \text{ mmol NaOH}}{1 \text{ mmol } H_3PO_4} \times \frac{1.00 \text{ mL titrant}}{0.0200 \text{ mmol NaOH}} = 20.0 \text{ mL}$$

At the second equivalence point, we have a solution of monohydrogen phosphate ion, for which we calculate the pH as follows. $pH = \frac{1}{2}(pK_{a_2} + pK_{a_3}) = \frac{1}{2}(7.20 + 12.37) = 9.78$

$$\text{volume titrant} = (0.400 + 0.150) \text{ mmol } H_2PO_4^- \times \frac{1 \text{ mmol NaOH}}{1 \text{ mmol } H_2PO_4^-} \times \frac{1.00 \text{ mL titrant}}{0.0200 \text{ mmol NaOH}} = 20.0 \text{ mL}$$

total volume of titrant to second equivalence point = 20.0 mL + 27.5 mL = 47.5 mL

In the region between the first and the second equivalence points we are titrating a solution that has 0.400 mmol $H_2PO_4^-$ (from H_3PO_4) and 0.150 mmol $H_2PO_4^-$ (from NaH_2PO_4), a total of 0.550 mmol $H_2PO_4^-$. This requires an additional volume of 0.0200 M NaOH.

Again we use the Henderson-Hasselbalch equation to determine the pH at various points between the two titration points. We let y be the volume of NaOH in excess of the 20.00 mL needed to reach the first equivalence point.

amount HPO_4^{2-} = y mL NaOH × 0.0200 M   amount $H_2PO_4^-$ = 0.550 − y mL NaOH × 0.0200 M

$$pH = pK_{a_2} + \log \frac{\text{amount HPO}_4{}^{2-}}{\text{amount H}_2\text{PO}_4{}^-} = 7.20 + \log \frac{0.0200\,y}{0.550 - 0.0200\,y}$$

When $y = 2.00$ mL $\quad pH = 7.20 + \log \dfrac{0.0200 \times 2.00}{0.550 - 0.0200 \times 2.00} = 7.20 + \log \dfrac{0.0400}{0.510} = 6.09$

y, mL	4.00	8.00	12.00	16.00	20.00	24.00	26.00
pH	6.43	6.81	7.09	7.34	7.63	8.04	8.44

Beyond the second equivalence point, we have a buffer solution of $HPO_4{}^{2-}$ and $PO_4{}^{3-}$, for which we might determine the pH in a similar fashion to the determination we made in the region between the first and the second equivalence point. We let z be the volume of 0.0200 M NaOH added beyond the second equivalence

point. $\qquad pH = pK_{a_3} + \log \dfrac{\text{amount PO}_4{}^{3-}}{\text{amount HPO}_4{}^{2-}} = 12.38 + \log \dfrac{0.0200\,z}{0.550 - 0.0200\,z}$

However, $[OH^-]$ in the titrant is so low that this expression does not apply; the pH is determined by the amount of OH^- in the excess titrant. For example, at 48.00 mL total titrant added, 0.50 mL beyond the second equivalence point, we have the following.

$$[OH^-] = 0.0200\text{ M} \times \frac{0.50\text{ mL}}{48.00\text{ mL}} = 2.1 \times 10^{-4}\text{ M} \qquad pOH = 3.68 \qquad pH = 14.00 - 3.68 = 10.32$$

To go half-way from the first to the second equivalence point, volume of titrant $= 27.5$ mL $\div 2 = 13.8$ mL
total volume of titrant needed to this half-way point $= 20.00$ mL $+ 13.8$ mL $= 33.8$ mL
At this half-way point: $pH = pK_{a_2} = 7.20$ $\qquad$ The data you should have obtained and plotted are

titrant volume	0.00 mL	20.0 mL	33.8 mL	47.5 mL
pH	1.72	4.68	7.20	9.78

All of these data are plotted in the graph below.

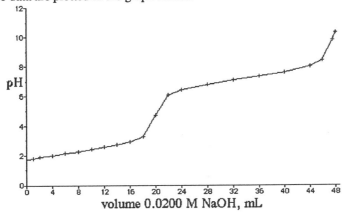

58. For each of the titrations, the pH at the half-equivalence point equals the pK_a of the acid.
The initial pH is that of 0.1000 M weak acid, determined as follows.

$$K_a = \frac{x^2}{0.1000} \qquad x = \sqrt{0.1000 \times K_a} = [H_3O^+]$$

The pH at the equivalence point is that of 0.05000 M anion of the weak acid, for which the $[OH^-]$ is

determined as follows. $\qquad K_b = \dfrac{K_w}{K_a} = \dfrac{x^2}{0.05000} \qquad x = \sqrt{\dfrac{K_w}{K_a} 0.0500} = [OH^-]$

And finally when 0.100 mL of base has been added beyond the equivalence point, the pH is determined by the excess added base, as follows.

$$[OH^-] = \frac{0.100\text{ mL} \times \dfrac{0.1000\text{ mmol NaOH}}{1\text{ mL NaOH soln}} \times \dfrac{1\text{ mmol OH}^-}{1\text{ mmol NaOH}}}{20.1\text{ mL soln total}} = 4.98 \times 10^{-4}\text{ M}$$

$pOH = -\log(4.98 \times 10^{-4}) = 3.303 \qquad pH = 14.000 - 3.303 = 10.697$

(a) Initial: $\quad [H_3O^+] = \sqrt{0.1000 \times 7.0 \times 10^{-3}} = 0.027$ M $\qquad pH = 1.58$

$\qquad$ Equiv: $\quad [OH^-] = \sqrt{\dfrac{1.00 \times 10^{-14}}{7.0 \times 10^{-3}} \times 0.0500} = 2.7 \times 10^{-7}$

$\qquad pOH = 6.57 \qquad pH = 14.00 - 6.57 = 7.43$

$\qquad$ Indicator: bromothymol blue, yellow at pH = 6.2 and blue at pH = 7.8

(b) Initial: $\quad [H_3O^+] = \sqrt{0.1000 \times 3.0 \times 10^{-4}} = 0.0055$ M $\qquad pH = 2.26$

Equiv: $[OH^-] = \sqrt{\dfrac{1.00 \times 10^{-14}}{3.0 \times 10^{-4}} \times 0.0500} = 1.3 \times 10^{-6}$

pOH = 5.89   pH = 14.00 − 5.89 = 8.11

Indicator: thymol blue, yellow at pH = 8.0 and blue at pH =10.0

(c) Initial: $[H_3O^+] = \sqrt{0.1000 \times 2.0 \times 10^{-8}} = 0.000045$ M   pH = 4.35

Equiv: $[OH^-] = \sqrt{\dfrac{1.00 \times 10^{-14}}{2.0 \times 10^{-8}} \times 0.0500} = 1.6 \times 10^{-4}$

pOH = 3.80   pH = 14.00 − 3.80 = 10.20

Indicator: alizarin yellow R, yellow at pH = 10.0 and violet at pH = 12.0

The three titration curves are sketched on the same axes below.

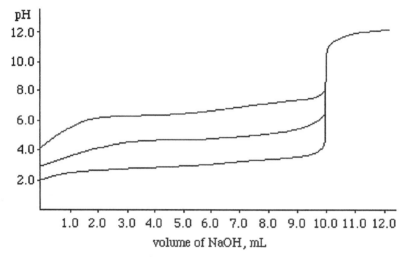

59. **(a)** At a pH = 2.00, in Figure 18-7 the pH is changing gradually with added NaOH. There would be no sudden change in color with the addition of a small volume of NaOH.

    (b) At pH = 2.0 in Figure 18-7, approximately 20.5 mL have been added. Since, equivalence required the addition of 25.0 mL, there are 4.5 mL left to add.   % HCl unneutralized $= \dfrac{4.5}{25.0} \times 100\% = 18\%$

pH of Salts of Polyprotic Acids

60. We expect a solution of Na_2S to be alkaline, or basic. This alkalinity is created by the hydrolysis of the sulfide ion, the anion of a very weak acid ($K_2 = 1 \times 10^{-19}$ for H_2S). $S^{2-}(aq) + H_2O \rightleftharpoons HS^-(aq) + OH^-(aq)$

61. **(a)** The hydrolysis of carbonate ion determines the pH of this solution. $K_2 = 4.7 \times 10^{-11}$ for H_2CO_3.

Reaction:	$CO_3^{2-}(aq) + H_2O$	$\rightleftharpoons$	$HCO_3^-(aq)$	+	$OH^-(aq)$
Initial:	1.0 M				
Changes:	−x M		+x M		+x M
Equil:	(1.0 − x)M		x M		x M

$K_b = \dfrac{K_w}{K_2} = \dfrac{1.00 \times 10^{-14}}{4.7 \times 10^{-11}} = \dfrac{[HCO_3^-][OH^-]}{[CO_3^{2-}]} = \dfrac{x^2}{1.0 - x} \approx \dfrac{x^2}{1.0}$ $x = 1.5 \times 10^{-2}$ M = $[OH^-]$

We see that our assumption, that $x \ll 1.0$ M, is valid.

pOH = −log(1.5 × 10⁻²) = 1.82   pH = 14.00 − 1.82 = 12.18

    (b) The set up is the same as for the previous part, except that $[CO_3^{2-}]_i = 0.010$ M

$K_b = \dfrac{K_w}{K_2} = \dfrac{1.00 \times 10^{-14}}{4.7 \times 10^{-11}} = \dfrac{[HCO_3^-][OH^-]}{[CO_3^{2-}]} = \dfrac{x^2}{0.010 - x} = 2.1 \times 10^{-4}$

We obtain the solution from the quadratic equation.

$x^2 = 2.1 \times 10^{-6} - 2.1 \times 10^{-4}x$ $x^2 + 2.1 \times 10^{-4}x - 2.1 \times 10^{-6} = 0$

$$x = \frac{-b \pm \sqrt{b^2 - 4ac}}{2a} = \frac{-2.1 \times 10^{-4} \pm \sqrt{4.4 \times 10^{-8} + 8.4 \times 10^{-6}}}{2} = 2.7 \times 10^{-3} = [OH^-]$$

$$pOH = -\log(2.7 \times 10^{-3}) = 2.57 \qquad pH = 14.00 - 2.57 = 11.43$$

62. We expect the pH of a solution of sodium dihydrogen citrate, NaH_2Cit, to be the average of pK_1 and pK_2, or equal to $(3.13 + 4.76) \div 2 = 3.95$. This is an acidic solution.

63. $HC_8H_4O_4^-(aq)$ is capable of reacting with H_2O in two ways.

ionization: $HC_8H_4O_4^-(aq) + H_2O \rightleftharpoons C_8H_4O_4^{-}(aq) + H_3O^+(aq)$

hydrolysis: $HC_8H_4O_4^-(aq) + H_2O \rightleftharpoons H_2C_8H_4O_4(aq) + OH^-(aq)$

For a solution of such an anion, the pH is computed as follows.

$$pH = \tfrac{1}{2}(pK_{a_1} + pK_{a_2}) = \tfrac{1}{2}[(-\log 1.1 \times 10^{-3}) + (-\log 3.7 \times 10^{-6})] = 4.20$$

64. **(a)** $H_3PO_4(aq) + CO_3^{2-}(aq) \longrightarrow H_2PO_4^-(aq) + HCO_3^-(aq)$

$H_2PO_4^-(aq) + CO_3^{2-}(aq) \longrightarrow HPO_4^{2-}(aq) + HCO_3^-(aq)$

$HPO_4^{2-}(aq) + OH^-(aq) \longrightarrow PO_4^{3-}(aq) + H_2O$

(b) The pH values of 1.00 M solutions of the three ions are as follows. (That for 1.0 M CO_3^{2-} was computed in Exercise 61).

1.0 M OH^- has pH = 14.00   1.0 M CO_3^{2-} has pH = 12.11   1.0 M PO_4^{3-} has pH = 13.19

Thus, we see that CO_3^{2-} is not a strong enough base to remove the third proton from H_3PO_4.

As an alternative method of solving this problem, we can compute the equilibrium constant for each of the reactions of carbonate ion with the various protonated species of H_3PO_4.

$$H_3PO_4 + CO_3^{2-} \longrightarrow H_2PO_4^- + HCO_3^- \qquad K = \frac{K_1\{H_3PO_4\}}{K_2\{H_2CO_3\}} = \frac{7.1 \times 10^{-3}}{4.7 \times 10^{-11}} = 1.5 \times 10^8$$

$$H_2PO_4^- + CO_3^{2-} \longrightarrow HPO_4^{2-} + HCO_3^- \qquad K = \frac{K_2\{H_3PO_4\}}{K_2\{H_2CO_3\}} = \frac{6.3 \times 10^{-8}}{4.7 \times 10^{-11}} = 1.3 \times 10^3$$

$$HPO_4^{2-} + CO_3^{2-} \longrightarrow PO_4^{3-} + HCO_3^- \qquad K = \frac{K_3\{H_3PO_4\}}{K_2\{H_2CO_3\}} = \frac{4.2 \times 10^{-13}}{4.7 \times 10^{-11}} = 8.9 \times 10^{-3}$$

Since the equilibrium constant for the third reaction is much smaller than 1.00, we conclude that it proceeds to the right to only a negligible extent and thus is not a practical method of producing PO_4^{3-}. The other two reactions have reasonably large equilibrium constants, and would be expected to form product. They have the advantage of using an inexpensive base and, even if they do not go to completion they will be drawn to completion by reaction with OH^- in the last step of the process.

65. **(a)** $Ba(OH)_2$ is a strong base. $pOH = 14.00 - 12.22 = 1.78$ $[OH^-] = 10^{-1.78} = 0.016_6$ M

$$[Ba(OH)_2] = \frac{0.016_6 \text{ mol } OH^-}{1 \text{ L}} \times \frac{1 \text{ mol } Ba(OH)_2}{2 \text{ mol } OH^-} = 0.0083 \text{ M}$$

(b) $pOH = 14.00 - 8.91 = 5.09$ $[OH^-] = 10^{-5.09} = 8.1 \times 10^{-6}$ M

Reaction: $C_6H_5NH_2(aq) + H_2O \rightleftharpoons C_6H_5NH_3^+(aq) + OH^-(aq)$

Initial: x M

Changes: -8.1×10^{-6} M $+8.1 \times 10^{-6}$ M $+8.1 \times 10^{-6}$ M

Equil: $(x - 8.1 \times 10^{-6})$M 8.1×10^{-6} M 8.1×10^{-6} M

$$K_b = \frac{[C_6H_5NH_3^+][OH^-]}{[C_6H_5NH_2]} = 7.4 \times 10^{-10} = \frac{(8.1 \times 10^{-6})^2}{x - 8.1 \times 10^{-6}}$$

$$x - 8.1 \times 10^{-6} = \frac{(8.1 \times 10^{-6})^2}{7.4 \times 10^{-10}} = 0.089 \text{ M} \qquad x = 0.089 \text{ M}$$

(c) $pH = 4.48 = pK_a + \log\frac{[C_2H_3O_2^-]}{[HC_2H_3O_2]} = 4.74 + \log\frac{0.294 \text{ M}}{[HC_2H_3O_2]}$ $\log\frac{0.294 \text{ M}}{[HC_2H_3O_2]} = 4.48 - 4.74$

$$\frac{0.294 \text{ M}}{[HC_2H_3O_2]} = 10^{-0.26} = 0.55 \qquad [HC_2H_3O_2] = \frac{0.294 \text{ M}}{0.55} = 0.53 \text{ M}$$

(d) $[H_3O^+] = 10^{-5.05} = 8.9 \times 10^{-6}$ M

Reaction: $NH_4^+(aq) + H_2O \rightleftharpoons NH_3(aq) + H_3O^+(aq)$

Initial: x M

Changes: -8.9×10^{-6} M $\qquad$ $+8.9 \times 10^{-6}$ M $\quad$ $+8.9 \times 10^{-6}$ M

Equil: $(x - 8.9 \times 10^{-6})$ M $\qquad$ 8.9×10^{-6} M $\quad$ 8.9×10^{-6} M

$$K_a = \frac{[NH_3][H_3O^+]}{[NH_4^+]} = \frac{K_w}{K_b \text{ for } NH_3} = \frac{1.00 \times 10^{-14}}{1.8 \times 10^{-5}} = 5.6 \times 10^{-10} = \frac{(8.9 \times 10^{-6})^2}{x - 8.9 \times 10^{-6}}$$

$$x - 8.9 \times 10^{-6} = \frac{(8.9 \times 10^{-6})^2}{5.6 \times 10^{-10}} = 0.14 \text{ M} \qquad x = [NH_4^+] = [NH_4Cl] = 0.14 \text{ M}$$

66. (a) pOH $= 14.00 - 6.07 = 7.93$ $\qquad$ $[OH^-] = 10^{-7.93} = 1.2 \times 10^{-8}$ M

$$Q_b = \frac{[NH_4^+][OH^-]}{[NH_3]} = \frac{(0.10 \text{ M})(1.2 \times 10^{-8} \text{ M})}{0.10} = 1.2 \times 10^{-8}$$

Since the value of Q_b does not equal the tabulated value of $K_b = 1.8 \times 10^{-5}$, the solution described cannot exist.

(b) These solutes can be added to the same solution, but the final solution will have an appreciable $[HC_2H_3O_2]$ because of the reaction of $H_3O^+(aq)$ with $C_2H_3O_2^-(aq)$.

Reaction:	$H_3O^+(aq)$	$+$	$C_2H_3O_2^-(aq)$	$\longrightarrow$	$HC_2H_3O_2(aq) + H_2O$
Initial:	0.058 M		0.10 M		
Changes:	−0.058 M		−0.058 M		+0.058 M
Final:	0.000 M		0.04 M		0.058 M

Of course, some H_3O^+ will exist in the final solution, but not equivalent to 0.058 M HI.

(c) Both 0.10 M KNO_2 and 0.25 M KNO_3 can exist together. Some hydrolysis of the $NO_2^-(aq)$ ion will occur, forming $HNO_2(aq)$.

67. (a) When $[H_3O^+]$ and $[HC_2H_3O_2]$ are high and $[C_2H_3O_2^-]$ is very low, a common ion—H_3O^+—has been added to a solution of acetic acid, suppressing its ionization.

(b) When $[C_2H_3O_2^-]$ is high and $[H_3O^+]$ and $[HC_2H_3O_2]$ are very low, we are dealing with a solution of acetate ion, which hydrolyzes to produce a limited concentration of $HC_2H_3O_2$.

(c) When $[HC_2H_3O_2]$ is high and both $[H_3O^+]$ and $[C_2H_3O_2^-]$ are low, the solution is an acetic acid solution, in which the solute is partially ionized.

(d) When both $[HC_2H_3O_2]$ and $[C_2H_3O_2^-]$ are high while $[H_3O^+]$ is low, the solution is a buffer solution, in which the presence of acetate ion suppresses the ionization of acetic acid.

19 SOLUBILITY AND COMPLEX-ION EQUILIBRIA

REVIEW QUESTIONS

1. **(a)** K_{sp} is the symbol for the solubility product constant, representing the equilibrium constant for the dissolving of a sparingly soluble (ionic) compound.

(b) K_f is the symbol for the formation contant of a complex ion: one complex ion is formed (is the product) from the reactants: a simple cation and the appropriate ligands.

(c) A ligand is a species that attaches to a central metal ion by forming a (coordinate) covalent bond; it is a Lewis base.

(d) The coordination number in a complex ion is the number of locations where ligands can attach.

2. **(a)** The common-ion effect in the dissolving of a sparingly soluble ionic compound is that less of that compound will dissolve because of the presence of one of its constituent ions in solution.

(b) Fractional precipitation refers to adding a precipitating agent (such as an anion) to a solution containing two other ions (such as cations) gradually so that only one compound precipitates.

(c) Ion-pair formation refers to the reseult of the attraction between solvated ions of unlike charges in solution. These solvated ions cluster together and behave to some extent as one particle.

(d) Qualitative cation analysis is a scheme of alternating selective precipitation and dissolving, with the goal of isolating each type of cation in its own sample where a definitive test for it can be performed.

3. **(a)** The solubility of a compound refers to the concentration of that compound in solution, either as a molarity or as a mass per unit volume. The solubility product constant is the equilibrium constant in terms of concentrations of ions, for the dissolving equilibrium.

(b) The common-ion effect describes the lowering of the solubility of a compound in a solution due to the presence of one of the ions of that compound in the solution. The salt effect describes the enhancement of the solubility of a compound in a solution due to the presence of a different type of ion in solution.

(c) A complex ion is a cation with its attached ligands. A coordination compound is a compound that contains one or more types of complex ions.

4. **(a)** $Ag_2SO_4(s) \rightleftharpoons 2\,Ag^+(aq) + SO_4^{2-}(aq)$ $K_{sp} = [Ag^+]^2[SO_4^{2-}]$

(b) $Ra(IO_3)_2(s) \rightleftharpoons Ra^{2+}(aq) + 2\,IO_3^-(aq)$ $K_{sp} = [Ra^{2+}][IO_3^-]^2$

(c) $Ni_3(PO_4)_2(s) \rightleftharpoons 3\,Ni^{2+}(aq) + 2\,PO_4^{3-}(aq)$ $K_{sp} = [Ni^{2+}]^3[PO_4^{3-}]^2$

(d) $PuO_2CO_3(s) \rightleftharpoons PuO_2^{2+}(aq) + CO_3^{2-}(aq)$ $K_{sp} = [PuO_2^{2+}][CO_3^{2-}]$

5. **(a)** $K_{sp} = [Fe^{3+}][OH^-]^3$ $Fe(OH)_3(s) \rightleftharpoons Fe^{3+}(aq) + 3\,OH^-(aq)$

(b) $K_{sp} = [BiO^+][OH^-]$ $BiOOH(s) \rightleftharpoons BiO^+(aq) + OH^-(aq)$

(c) $K_{sp} = [Hg_2^{2+}][I^-]^2$ $Hg_2I_2(s) \rightleftharpoons Hg_2^{2+}(aq) + 2\,I^-(aq)$

(d) $K_{sp} = [Pb^{2+}]^3[AsO_4^{3-}]^2$ $Pb_3(AsO_4)_2(s) \rightleftharpoons 3\,Pb^{2+}(aq) + 2\,AsO_4^{3-}(aq)$

6. **(a)** $CrF_3(s) \rightleftharpoons Cr^{3+}(aq) + 3\,F^-(aq)$ $K_{sp} = [Cr^{3+}][F^-]^3 = 6.6 \times 10^{-11}$

(b) $Au_2(C_2O_4)_3(s) \rightleftharpoons 2\,Au^{3+}(aq) + 3\,C_2O_4^{2-}(aq)$ $K_{sp} = [Au^{3+}]^2[C_2O_4^{2-}]^3 = 1 \times 10^{-10}$

(c) $PbOHCl(s) \rightleftharpoons Pb^{2+}(aq) + OH^-(aq) + Cl^-(aq)$ $K_{sp} = [Pb^{2+}][OH^-][Cl^-] = 2 \times 10^{-14}$

(d) $Ag_3[Co(NO_2)_6](s) \rightleftharpoons 3\,Ag^+(aq) + [Co(NO_2)_6]^{3-}(aq)$ $K_{sp} = [Ag^+]^3[[Co(NO_2)_6]^{3-}] = 8.5 \times 10^{-21}$

7. Let s = solubility of each compound in moles of compound per liter of solution.

(a) $K_{sp} = [Ba^{2+}][CrO_4^{2-}] = (s)(s) = s^2 = 1.2 \times 10^{-10}$ $s = 1.1 \times 10^{-5}\ M$

(b) $K_{sp} = [Pb^{2+}][Br^-]^2 = (s)(2s)^2 = 4s^3 = 4.0 \times 10^{-5}$ $s = 2.2 \times 10^{-2}$ M

(c) $K_{sp} = [Ce^{3+}][F^-]^3 = (s)(3s)^3 = 27s^4 = 8 \times 10^{-16}$ $s = 7 \times 10^{-5}$ M

(d) $K_{sp} = [Mg^{2+}]^3[AsO_4^{3-}]^2 = (3s)^3(2s)^2 = 108s^5 = 2.1 \times 10^{-20}$ $s = 4.5 \times 10^{-5}$ M

8. Again, let s = solubility of each compound in moles of solute per liter of solution.

    (a) $K_{sp} = [Cs^+][MnO_4^-] = (s)(s) = s^2 = (3.8 \times 10^{-3})^2 = 1.4 \times 10^{-5}$

    (b) $K_{sp} = [Pb^{2+}][ClO_2^-]^2 = (s)(2s)^2 = 4s^3 = 4(2.8 \times 10^{-3})^3 = 8.8 \times 10^{-8}$

    (c) $K_{sp} = [Li^+]^3[PO_4^{3-}] = (3s)^3(s) = 27s^4 = 27(2.9 \times 10^{-3})^4 = 1.9 \times 10^{-9}$

9. Statement **(d)** is correct. Consider the solubility equation : $PbI_2(s) \rightleftharpoons Pb^{2+}(aq) + 2 \; I^-(aq)$

The stoichiometry of the dissolving reaction indicates that two I- ions are formed for each Pb^{2+} ion. Thus

$[Pb^{2+}] = 0.5 \; [I^-]$. The relationship between K_{sp} and $[Pb^{2+}]$ is $[Pb^{2+}] = \sqrt[3]{K_{sp} / 4}$

10. We let s = molar solubility of $Mg(OH)_2$ in moles solute per liter of solution.

    (a) $K_{sp} = [Mg^{2+}][OH^-]^2 = (s)(2s)^2 = 4s^3 = 1.8 \times 10^{-11}$ $s = 1.7 \times 10^{-4}$ M

    (b)

Reaction:	$Mg(OH)_2(s)$	$\rightleftharpoons$	$Mg^{2+}(aq)$	+	$2\,OH^-(aq)$
Initial:			0.0315 M		
Changes:			+s M		+2s M
Equil:			(0.0315 + s)M		2s M

$K_{sp} = (0.0315 + s)(2s)^2 = 1.8 \times 10^{-11} \approx (0.0315)(2s)^2 = 0.13 \; s^2$ $s = 1.2 \times 10^{-5}$ M

    (c) $[OH^-] = [KOH] = 0.0822$ M

Reaction:	$Mg(OH)_2(s)$	$\rightleftharpoons$	$Mg^{2+}(aq)$	+	$2\,OH^-(aq)$
Initial:					0.0822 M
Changes:			+s M		+2s M
Equil:			s M		(0.0822 + 2s)M

$K_{sp} = (s)(0.0822 + 2s)^2 = 1.8 \times 10^{-11} \approx (s)(0.0822)^2 = 0.0068 \; s$ $s = 2.6 \times 10^{-9}$ M

11. The solubility equilibrium is $CaCO_3(s) \rightleftharpoons Ca^{2+}(aq) + CO_3^{2-}(aq)$

    (a) The addition of $Na_2CO_3(aq)$ produces $CO_3^{2-}(aq)$ in solution. This common ion will suppress the solubility of $CaCO_3(s)$.

    (b) HCl(aq) is a strong acid that will react with carbonate ion: $CO_3^{2-}(aq) + 2\,H_3O^+(aq) \longrightarrow CO_2(g) + 3$ H_2O. This will decrease $[CO_3^{2-}]$ in the solution and more $CaCO_3(s)$ will dissolve.

    (c) $HSO_4^-(aq)$ is a moderately weak acid. It is strong enough to protonate carbonate ion, decreasing $[CO_3^{2-}]$ and enhancing the solubility of $CaCO_3(s)$, as the value of K_c indicates.

$HSO_4^-(aq) + CO_3^{2-}(aq) \rightleftharpoons SO_4^{2-}(aq) + HCO_3^-(aq)$ $K_c = \dfrac{K(HSO_4^-)}{K(HCO_3^-)} = \dfrac{0.011}{4.7 \times 10^{-11}} = 2.3 \times 10^8$

12. In each case, compute Q and compare its value with the value of K_{sp}. If $Q > K_{sp}$, a precipitate should form.

    (a) $Q = [Mg^{2+}][CO_3^{2-}] = (0.017)(0.0068) = 1.2 \times 10^{-4} > 3.5 \times 10^{-8} = K_{sp}$

    Precipitation should occur.

    (b) $Q = [Ag^+]^2[SO_4^{2-}] = (0.0034)^2(0.0112) = 1.3 \times 10^{-7} < 1.4 \times 10^{-5} = K_{sp}$

    Precipitation is not expected to occur.

    (c) $pOH = 14.00 - 2.80 = 11.20$ $[OH^-] = 10^{-11.20} = 6.3 \times 10^{-12}$ M

    $Q = [Cr^{3+}][OH^-]^3 = (0.041)(6.3 \times 10^{-12})^3 = 1.0 \times 10^{-35} < 6.3 \times 10^{-31} = K_{sp}$

    Precipitation is not expected to occur.

13. We use the solubility product expression to determine $[Ca^{2+}]$ that can coexist with $[SO_4^{2-}] = 0.750$ M.

$K_{sp} = 9.1 \times 10^{-6} = [Ca^{2+}][SO_4^{2-}] = [Ca^{2+}](0.750 \text{ M})$ $[Ca^{2+}] = \dfrac{9.1 \times 10^{-6}}{0.750 \text{ M}} = 1.2 \times 10^{-5}$ M

% unprecipitated $= \dfrac{1.2 \times 10^{-5} \text{ M}}{0.0655 \text{ M}} \times 100\% = 0.018\%$ unprecipitated.

Of course, $[SO_4^{2-}]$ actually does decrease because it is consumed in the precipitation. In this case $[SO_4^{2-}] = 0.750$ M $- 0.0655$ M $= 0.681$ M. But this just changes the final $[Ca^{2+}]$ slightly, to 1.3×10^{-5} M.

14. (a) Determine [I⁻] when AgI just begins to precipitate, and [I⁻] when PbI_2 just begins to precipitate.

$$K_{sp} = [Ag^+][I^-] = 8.5 \times 10^{-17} = (0.10)[I^-] \qquad\qquad [I^-] = 8.5 \times 10^{-16} \text{ M}$$

$$K_{sp} = [Pb^{2+}][I^-]^2 = 7.1 \times 10^{-9} = (0.10)[I^-]^2 \qquad [I^-] = \sqrt{\frac{7.1 \times 10^{-9}}{0.10}} = 2.7 \times 10^{-4} \text{ M}$$

Since 8.5×10^{-16} M is less than 2.7×10^{-4} M, AgI will precipitate before PbI_2.

(b) [I⁻] = 2.7×10^{-4} M before the second cation—Pb^{2+}—begins to precipitate.

(c) $K_{sp} = [Ag^+][I^-] = 8.5 \times 10^{-17} = [Ag^+](2.7 \times 10^{-4})$ $[Ag^+] = 3.1 \times 10^{-13}$ M

(d) Since [Ag⁺] has decreased to less than 0.1% of its initial value before PbI_2 begins to precipitate, we conclude that Ag^+ and Pb^{2+} can be separated by precipitation with iodide ion.

15. $Mg(OH)_2(s)$ will be the most soluble in a solution of $NaHSO_4(aq)$. $NaHSO_4$ will form an acidic solution, via the reaction $HSO_4^-(aq) + H_2O \rightleftharpoons H_3O^+(aq) + SO_4^{2-}(aq)$ and this H_3O^+ will react with OH⁻ ion: $OH^-(aq) + H_3O^+(aq) \rightleftharpoons 2\ H_2O$ causing the solubility equilibrium to shift right $Mg(OH)_2(s) \rightleftharpoons Mg^{2+}(aq) + 2\ OH^-(aq)$ Since NaOH has an ion in common with $Mg(OH)_2$, its use will actually decrease the solubility of $Mg(OH)_2$. Addition of Na_2CO_3 will also decrease $Mg(OH)_2$ solubility, through hydrolysis of the carbonate ion: $H_2O + CO_3^{2-} \rightleftharpoons HCO_3^- + OH^-$ followed by the common ion effect of OH⁻.

16. (a) $Ag^+(aq) + NO_3^-(aq) + Na^+(aq) + Br^-(aq) \longrightarrow AgBr(s) + Na^+(aq) + NO_3^-(aq)$

(b) $Cu^{2+}(aq) + NO_3^-(aq) + H_3O^+(aq) + Cl^-(aq) \longrightarrow$ no reaction

(c) $Fe^{2+}(aq) + H_2S(aq,\ in\ 0.3\ M\ HCl) \longrightarrow$ no reaction

(d) $Cu(OH)_2(s) + 4\ NH_3(aq) \longrightarrow [Cu(NH_3)_4]^{2+}(aq) + 2\ OH^-(aq)$

(e) $Fe^{3+}(aq) + NH_4^+(aq) + 3\ OH^-(aq) \longrightarrow Fe(OH)_3(s) + NH_4^+(aq)$

(f) $Ag_2SO_4(s) + 4\ NH_3(aq) \longrightarrow 2\ [Ag(NH_3)_2]^+(aq) + SO_4^{2-}(aq)$

(g) $CaSO_3(s) + 2\ H_3O^+(aq) \longrightarrow Ca^{2+}(aq) + 3\ H_2O + SO_2(g)$

17. $Cu(OH)_2$ dissolves readily in acidic solutions [HCl(aq) and $HNO_3(aq)$] and in ammoniacal solutions [$NH_3(aq)$]. The net ionic equations for these two reactions are, respectively, the following.

$$Cu(OH)_2(s) + 2\ H^+(aq) \longrightarrow Cu^{2+}(aq) + 2\ H_2O$$

$$Cu(OH)_2(s) + 4\ NH_3(aq) \longrightarrow [Cu(NH_3)_4]^{2+}(aq) + 2\ OH^-(aq)$$

18. The best choice is HCl(aq). AgCl is insoluble in water, while $CuCl_2$ is water soluble. On the other hand, both Cu^{2+} and Ag^+ form insoluble hydroxides and carbonates, so NaOH(aq) and $(NH_4)_2CO_3(aq)$ would not serve to separate these two cations.

19. First we determine [OH⁻] in this buffer solution.

$$pH = pK_a + \log\frac{[C_2H_3O_2^-]}{[HC_2H_3O_2]} = 4.74 + \log\frac{0.25 \text{ M}}{0.50 \text{ M}} = 4.44 \qquad pOH = 14.00 - 4.44 = 9.56$$

$[OH^-] = 10^{-9.56} = 2.8 \times 10^{-10}$ M Now, we compute the value of Q_{sp} for $Al(OH)_3$.

$Q = [Al^{3+}][OH^-]^3 = (0.225)(2.8 \times 10^{-10})^3 = 5.0 \times 10^{-30} > 1.3 \times 10^{-33} = K_{sp}$

Precipitation should occur from this solution.

20. We first find the concentration of free metal ion. Then we determine the value of Q_{sp} for the precipitation reaction, and compare that value with the value of K_{sp} to determine whether precipitation will occur.

Reaction:	$Ag^+(aq) +$	$2\ CN^-(aq) \rightleftharpoons$	$[Ag(CN)_2]^-(aq)$
Initial:	1.05 M	0.012 M	
Changes:	+x M	+2x M	−x M
Equil:	x M	(1.05 + 2x)M	(0.012 − x)M

$$K_f = \frac{[[Ag(S_2O_3)_2]^{3-}]}{[Ag^+][S_2O_3^{2-}]^2} = 5.6 \times 10^{18} = \frac{0.012 - x}{x\ (1.05 + 2x)^2} \approx \frac{0.012}{1.05^2\ x} \qquad x = 1.9 \times 10^{-21} \text{ M} = [Ag^+]$$

$Q = [Ag^+][I^-] = (1.9 \times 10^{-21})(2.0) = 3.8 \times 10^{-21} < 8.5 \times 10^{-17} = K_{sp}$ Precipitation should not occur.

21. We use the formation constant expression to determine the concentration of free silver ion, [Ag⁺].

$$K_f = 1.6 \times 10^7 = \frac{[[Ag(NH_3)_2]^+]}{[Ag^+][NH_3]^2} = \frac{1.8 \text{ M}}{[Ag^+](1.50 \text{ M})^2} \qquad [Ag^+] = \frac{1.8}{1.6 \times 10^7\ (1.50)^2} = 5.0 \times 10^{-8} \text{ M}$$

The K_{sp} expression is used to determine the $[Cl^-]$ that can coexist with this $[Ag^+]$.

$$K_{sp} = 1.8 \times 10^{-10} = [Ag^+][Cl^-] = (5.0 \times 10^{-8} \text{ M})[Cl^-] \qquad [Cl^-] = \frac{1.8 \times 10^{-10}}{5.0 \times 10^{-8}} = 3.6 \times 10^{-3} \text{ M}$$

EXERCISES

K_{sp} and Solubility

22. We first determine the molar solubility, s, of $Cd(OH)_2$.

$$s = \frac{0.00026 \text{ g Cd(OH)}_2}{100 \text{ mL}} \times \frac{1000 \text{ mL}}{1 \text{ L}} \times \frac{1 \text{ mol Cd(OH)}_2}{146.4 \text{ g Cd(OH)}_2} = 1.8 \times 10^{-5} \text{ M}$$

$$K_{sp} = [Cd^{2+}][OH^-]^2 = (s)(2s)^2 = 4s^3 = 4(1.8 \times 10^{-5})^3 = 2.3 \times 10^{-14}$$

23. We use the value of K_{sp} for each compound in determining $[Mg^{2+}]$ in its saturated solution. In each case, s represents the molar solubility of the compound.

    (a) $MgCO_3$ $K_{sp} = [Mg^{2+}][CO_3^{2-}] = (s)(s) = s^2 = 3.5 \times 10^{-8}$

         $s = 1.9 \times 10^{-4}$ M $[Mg^{2+}] = 1.9 \times 10^{-4}$ M

    (b) MgF_2 $K_{sp} = [Mg^{2+}][F^-]^2 = (s)(2s)^2 = 4s^3 = 3.7 \times 10^{-8}$

         $s = 2.1 \times 10^{-3}$ M $[Mg^{2+}] = 2.1 \times 10^{-3}$ M

    (c) $Mg_3(PO_4)_2$ $K_{sp} = [Mg^{2+}]^3[PO_4^{3-}]^2 = (3s)^3(2s)^2 = 108s^5 = 1 \times 10^{-25}$

         $s = 4 \times 10^{-6}$ M $[Mg^{2+}] = 1 \times 10^{-5}$ M

A saturated solution of MgF_2 has the highest $[Mg^{2+}]$.

24. We determine $[F^-]$ in saturated CaF_2, and from that value the concentration of fluoride ion in ppm.

For CaF_2 $K_{sp} = [Ca^{2+}][F^-]^2 = (s)(2s)^2 = 4s^3 = 5.3 \times 10^{-9}$ $s = 1.1 \times 10^{-3}$ M

The solubility in ppm is the number of grams of CaF_2 in 10^6 g solution. We assume a solution density of

1.00 g/mL.   mass of $F^- = 10^6$ g soln $\times \dfrac{1 \text{ mL}}{1.00 \text{ g soln}} \times \dfrac{1 \text{ L soln}}{1000 \text{ mL}} \times \dfrac{1.1 \times 10^{-3} \text{ mol CaF}_2}{1 \text{ L soln}}$

$$\times \frac{2 \text{ mol F}^-}{1 \text{ mol CaF}_2} \times \frac{19.0 \text{ g F}^-}{1 \text{ mol F}^-} = 42 \text{ g F}^-$$

This is 42 times more concentrated than the optimum concentration of fluoride ion. CaF_2 is, in fact, more soluble than is necessary. Its use might lead to excessive F^- in solution.

25. We determine $[OH^-]$ in a saturated solution. From this $[OH^-]$ we determine pH.

$K_{sp} = [BiO^+][OH^-] = 4 \times 10^{-10} = s^2$ $s = 2 \times 10^{-5}$ M $= [OH^-]$

$pOH = -\log(2 \times 10^{-5}) = 4.7$ $pH = 9.3$

26. We first assume that the volume of the solution does not change appreciably when its temperature is lowered. Then we determine the mass of $PbSO_4$ dissolved in each solution, recognizing that the molar solubility of $PbSO_4$ equals the square root of its solubility product constant, since it is the only solute in the solution.

At 50°C: $s = \sqrt{2.3 \times 10^{-8}} = 1.5 \times 10^{-4}$ M   At 25°C: $s = \sqrt{1.6 \times 10^{-8}} = 1.3 \times 10^{-4}$ M

mass of $PbSO_4$ at 50°C $= 0.815$ L $\times \dfrac{1.5 \times 10^{-4} \text{ mol PbSO}_4}{1 \text{ L soln}} \times \dfrac{303.3 \text{ g PbSO}_4}{1 \text{ mol PbSO}_4} = 0.037 \text{ g PbSO}_4$

mass of $PbSO_4$ at 25°C $= 0.815$ L $\times \dfrac{1.3 \times 10^{-4} \text{ mol PbSO}_4}{1 \text{ L soln}} \times \dfrac{303.3 \text{ g PbSO}_4}{1 \text{ mol PbSO}_4} = 0.032 \text{ g PbSO}_4$

mass $PbSO_4$ precipitated $= (0.037 \text{ g} - 0.032 \text{ g}) \times \dfrac{1000 \text{ mg}}{1 \text{ g}} = 5 \text{ mg}$

27. First we determine $[I^-]$ in the saturated solution.

$K_{sp} = [Pb^{2+}][I^-]^2 = 7.1 \times 10^{-9} = (s)(2s)^2 = 4s^3$ $s = 1.2 \times 10^{-3}$ M

The $AgNO_3$ reacts with the I^- in this saturated solution in the titration. $Ag^+(aq) + I^-(aq) \longrightarrow AgI(s)$

We determine the amount of Ag^+ needed for this titration, and then $[AgNO_3]$ in the titrant.

$$\text{amount Ag}^+ = 0.02500 \text{ L} \times \frac{1.2 \times 10^{-3} \text{ mol PbI}_2}{1 \text{ L soln}} \times \frac{2 \text{ mol I}^-}{1 \text{ mol PbI}_2} \times \frac{1 \text{ mol Ag}^+}{1 \text{ mol I}^-} = 6.0 \times 10^{-5} \text{ mol Ag}^+$$

$$[\text{AgNO}_3] = \frac{6.0 \times 10^{-5} \text{ mol Ag}^+}{0.0133 \text{ L soln}} \times \frac{1 \text{ mol AgNO}_3}{1 \text{ mol Ag}^+} = 4.5 \times 10^{-3} \text{ M}$$

28. We determine $[\text{C}_2\text{O}_4^{2-}] = s$, the solubility of the saturated solution.

$$[\text{C}_2\text{O}_4^{2-}] = \frac{6.3 \text{ mL} \times \dfrac{0.00102 \text{ mmol KMnO}_4}{1 \text{ mL soln}} \times \dfrac{5 \text{ mmol C}_2\text{O}_4^{2-}}{2 \text{ mmol MnO}_4^-}}{250.0 \text{ mL}} = 6.4 \times 10^{-5} \text{ M} = s = [\text{Ca}^{2+}]$$

$$K_{sp} = [\text{Ca}^{2+}][\text{C}_2\text{O}_4^{2-}] = (s)(s) = s^2 = (6.4 \times 10^{-5})^2 = 4.1 \times 10^{-9}$$

29. We use the ideal gas law to determine the amount in moles of H_2S gas used.

$$n = \frac{PV}{RT} = \frac{\left(748 \text{ mmHg} \times \dfrac{1 \text{ atm}}{760 \text{ mmHg}}\right) \times \left(30.4 \text{ mL} \times \dfrac{1 \text{ L}}{1000 \text{ mL}}\right)}{0.08206 \text{ L atm mol}^{-1} \text{ K}^{-1} \times (23 + 273)\text{K}} = 1.23 \times 10^{-3} \text{ moles}$$

If we assume that all of the H_2S is consumed in forming Ag_2S, we can compute the $[\text{Ag}^+]$ in the $AgBrO_3$ solution. This assumption is valid if the equilibrium constant for the cited reaction is large, which we see that it is, as follows.

$$2 \text{ Ag}^+(aq) + \text{S}^{2-}(aq) \rightleftharpoons \text{Ag}_2\text{S}(s) \qquad 1/K_{sp} = 1/6 \times 10^{-50}$$

$$\text{H}_2\text{S}(aq) + 2 \text{ H}_2\text{O} \rightleftharpoons \text{S}^{2-}(aq) + 2 \text{ H}_3\text{O}^+(aq) \qquad K_1 \times K_2 = 1.0 \times 10^{-7} \times 1 \times 10^{-19}$$

$$2 \text{ Ag}^+(aq) + \text{H}_2\text{S}(aq) \rightleftharpoons \text{Ag}_2\text{S}(s) + 2 \text{ H}_3\text{O}^+ \qquad K = \frac{1.0 \times 10^{-7} \times 1 \times 10^{-19}}{6 \times 10^{-50}} = 2 \times 10^{23}$$

$$[\text{Ag}^+] = \frac{1.23 \times 10^{-3} \text{ mol H}_2\text{S}}{338 \text{ mL soln}} \times \frac{1000 \text{ mL}}{1 \text{ L soln}} \times \frac{2 \text{ mol Ag}^+}{1 \text{ mol H}_2\text{S}} = 7.28 \times 10^{-3} \text{ M}$$

Then, for $AgBrO_3$ $\qquad K_{sp} = [\text{Ag}^+][\text{BrO}_3^-] = (7.28 \times 10^{-3})^2 = 5.30 \times 10^{-5}$

The Common-Ion Effect

30. The presence of KI in a solution produces a significant $[\text{I}^-]$ in that solution. Through the solubility product principle, not as much AgI can dissolve in such a solution as in pure water, since the ion product, $[\text{Ag}^+][\text{I}^-]$, cannot exceed the value of K_{sp}. If the solution contains KNO_3, however, more AgI can dissolve than in pure water, since the activity of each ion is less than its molarity. On an ionic level, the reason for this is that ion pairs—such as $AgNO_3(aq)$ and KI(aq)—form in the solution, preventing Ag^+ and I^- ions from coming together and precipitating.

31. Reaction: $\qquad \text{Ag}_2\text{SO}_4(s) \rightleftharpoons \quad 2 \text{ Ag}^+(aq) \quad + \quad \text{SO}_4^{2-}(aq)$

Original:		0.200 M
Add solid:	$+x$ M	$+x/2$ M
Equil:	x M	$(0.200 + x/2)$M

$x = [\text{Ag}^+] = 9.2 \times 10^{-3}$ M

$K_{sp} = [\text{Ag}^+]^2[\text{SO}_4^{2-}] = (9.2 \times 10^{-3})^2(0.200 + 0.0046) = 1.7 \times 10^{-5}$

32. Even though $BaCO_3$ is more soluble than $BaSO_4$, it will still precipitate when 0.50 M Na_2CO_3(aq) is added to a saturated solution of $BaSO_4$, because there is a sufficient $[\text{Ba}^{2+}]$ in such a solution for the product $[\text{Ba}^{2+}][\text{CO}_3^{2-}]$ to exceed the value of K_{sp} for the compound. An example will demonstrate this phenomenon. Let us assume that the two solutions being mixed are of equal volume and—to make the situation even more unfavorable—that the saturated $BaSO_4$ solution is not in contact with solid $BaSO_4$, meaning that it does not maintain its saturation when it is diluted. First we determine $[\text{Ba}^{2+}]$ in saturated $BaSO_4$(aq).

$$K_{sp} = [\text{Ba}^{2+}][\text{SO}_4^{2-}] = 1.1 \times 10^{-10} = s^2 \qquad s = \sqrt{1.1 \times 10^{-10}} = 1.0 \times 10^{-5} \text{ M}$$

Mixing solutions of equal volumes means that the concentrations of solutes not common to the two solutions are halved by dilution.

$$[\text{Ba}^{2+}] = \frac{1}{2} \times \frac{1.0 \times 10^{-5} \text{ mol BaSO}_4}{1 \text{ L}} \times \frac{1 \text{ mol Ba}^{2+}}{1 \text{ mol BaSO}_4} = 5.0 \times 10^{-6} \text{ M}$$

$$[\text{CO}_3^{2-}] = \frac{1}{2} \times \frac{0.50 \text{ mol Na}_2\text{CO}_3}{1 \text{ L}} \times \frac{1 \text{ mol CO}_3^{2-}}{1 \text{ mol Na}_2\text{CO}_3} = 0.25 \text{ M}$$

$Q\{BaCO_3\} = [Ba^{2+}][CO_3^{2-}] = (5.0 \times 10^{-6})(0.25) = 1.2 \times 10^{-6} > 5.0 \times 10^{-9} = K_{sp}\{BaCO_3\}$
Thus, precipitation of $BaCO_3$ indeed should occur under the conditions described.

33. For PbI_2, $K_{sp} = 7.1 \times 10^{-9} = [Pb^{2+}][I^-]^2$

    (a) In a solution with $[PbI_2] = 1.0 \times 10^{-4}$ M, $[Pb^{2+}] = 1.0 \times 10^{-4}$ M, and $[I^-] = 2.0 \times 10^{-4}$ M

Reaction:	$PbI_2(s)$	$\rightleftharpoons$	$Pb^{2+}(aq)$	$+$	$2\,I^-(aq)$
Initial:			1.0×10^{-4} M		2.0×10^{-4} M
Add lead(II):			$+x$ M		
Equil:			$(0.00010 + x)$M		0.00020

$K_{sp} = 7.1 \times 10^{-9} = (0.00010 + x)(0.00020)^2$ $(0.00010 + x) = 0.18$
$x = 0.18$ M $= [Pb^{2+}]$

    (b) In a solution with $[PbI_2] = 1.0 \times 10^{-5}$ M, $[Pb^{2+}] = 1.0 \times 10^{-5}$ M, and $[I^-] = 2.0 \times 10^{-5}$ M

Reaction:	$PbI_2(s)$	$\rightleftharpoons$	$Pb^{2+}(aq)$	$+$	$2\,I^-(aq)$
Initial:			1.0×10^{-5} M		2.0×10^{-5} M
Add KI:					$+x$ M
Equil:			1.0×10^{-5} M		$(2.0 \times 10^{-5} + x)$M

$K_{sp} = 7.1 \times 10^{-9} = (1.0 \times 10^{-5})(2.0 \times 10^{-5} + x)^2$ $(2.0 \times 10^{-5} + x) = \sqrt{\dfrac{7.1 \times 10^{-9}}{1.0 \times 10^{-5}}} = 2.7 \times 10^{-2}$

$x = 2.7 \times 10^{-2} - 2.0 \times 10^{-5} = 2.7 \times 10^{-2}$ M $= [I^-]$

    (c) The set up here is similar to that of part (a). $K_{sp} = 7.1 \times 10^{-9} = (1.0 \times 10^{-7} + x)(2.0 \times 10^{-7})^2$

$(1.0 \times 10^{-7} + x) = \dfrac{7.1 \times 10^{-9}}{(2.0 \times 10^{-7})^2} = 71$ M $x = [Pb^{2+}] = 1.8 \times 10^5$ M

Since this concentration is so huge, $[Pb^{2+}]$ cannot be raised high enough to lower the solubility of PbI_2 to 1.0×10^{-7} M.

34. We calculate a few more points than required by a figure such as Figure 19-2. The set up for each case is as follows, with s representing the solubility of $Pb(IO_3)_2$, and c the concentration of KIO_3. We assume in each case that $c \gg s$.

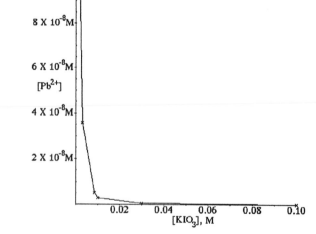

$Pb(IO_3)_2(s)$	$\rightleftharpoons$	$Pb^{2+}(aq) + 2\,IO_3^-(aq)$
from KIO_3		c
From $Pb(IO_3)_2$	s	$2s$
Equil:	s	$2s + c$

$K_{sp} = (s)(2s + c)^2 = 3.2 \times 10^{-13} \approx (s)(c)^2$

$$s = \frac{3.2 \times 10^{-13}}{c^2}$$

$c = 0.0010$ M	$s = 3.2 \times 10^{-7}$ M
$c = 0.0030$ M	$s = 3.6 \times 10^{-8}$ M
$c = 0.0080$ M	$s = 5.0 \times 10^{-9}$ M
$c = 0.010$ M	$s = 3.2 \times 10^{-9}$ M
$c = 0.030$ M	$s = 3.6 \times 10^{-10}$ M
$c = 0.060$ M	$s = 8.9 \times 10^{-11}$ M
$c = 0.080$ M	$s = 5.0 \times 10^{-11}$ M
$c = 0.10$ M	$s = 3.2 \times 10^{-11}$ M

35. $[Ca^{2+}] = \dfrac{115\text{ g }Ca^{2+}}{10^6\text{ g soln}} \times \dfrac{1\text{ mol }Ca^{2+}}{40.08\text{ g }Ca^{2+}} \times \dfrac{1000\text{ g soln}}{1\text{ L soln}} = 2.87 \times 10^{-3}$ M

$[Ca^{2+}][F^-]^2 = K_{sp} = 5.3 \times 10^{-9} = (2.87 \times 10^{-3})[F^-]^2$ $[F^-] = 1.4 \times 10^{-3}$ M

ppm $F^- = \dfrac{1.4 \times 10^{-3}\text{ mol }F^-}{1\text{ L soln}} \times \dfrac{19.00\text{ g }F^-}{1\text{ mol }F^-} \times \dfrac{1\text{ L soln}}{1000\text{ g}} \times 10^6\text{ g soln} = 27$ ppm

36. Reaction: $CaSO_4(s) \rightleftharpoons Ca^{2+}(aq) + SO_4^{2-}(aq)$

Soln: 　　　　　　　　　　　　　　　0.0010 M

+ $CaSO_4(s)$ 　　　　　　　　$+x$ M 　　　$+x$ M

Equil: 　　　　　　　　　　　x M 　　　$(0.0010 + x)$M

$K_{sp} = [Ca^{2+}][SO_4^{2-}] = 9.1 \times 10^{-6} = x(0.0010 + x) = 0.0010x + x^2$ 　$x^2 + 0.0010\,x - 9.1 \times 10^{-6} = 0$

$$x = \frac{-b \pm \sqrt{b^2 - 4ac}}{2a} = \frac{-0.0010 \pm \sqrt{1.0 \times 10^{-6} + 3.6 \times 10^{-5}}}{2} = 2.5 \times 10^{-3}\ M = [CaSO_4]$$

$$\text{mass } CaSO_4 = 100.0\ mL \times \frac{1\ L}{1000\ mL} \times \frac{2.5 \times 10^{-3}\ mol\ CaSO_4}{1\ L\ soln} \times \frac{136.1\ g\ CaSO_4}{1\ mol\ CaSO_4} = 0.034\ g\ CaSO_4$$

37. We first calculate the $[Ag^+]$ and the $[Cl^-]$ in the saturated solution.

$K_{sp} = [Ag^+][Cl^-] = 1.8 \times 10^{-10} = (s)(s) = s^2$ 　　　　$s = 1.3 \times 10^{-5}\ M = [Ag^+] = [Cl^-]$

Both of these concentrations are marginally diluted by the addition of 1 mL of NaCl(aq)

$$[Ag^+] = [Cl^-] = 1.3 \times 10^{-5}\ M \times \frac{100.0\ mL}{100.0\ mL + 1.0\ mL} = 1.3 \times 10^{-5}\ M$$

The $[Cl^-]$ in the NaCl(aq) also is diluted. 　　　$[Cl^-] = 1.0\ M \times \frac{1.0\ mL}{100.0\ mL + 1.0\ mL} = 9.9 \times 10^{-3}\ M$

Let us use this $[Cl^-]$ to determine the $[Ag^+]$ that can exist in this solution.

$[Ag^+][Cl^-] = 1.8 \times 10^{-10} = [Ag^+](9.9 \times 10^{-3}\ M)$ 　　　　$[Ag^+] = \dfrac{1.8 \times 10^{-10}}{9.9 \times 10^{-3}} = 1.8 \times 10^{-8}\ M$

We compute the amount of AgCl in this final solution, and in the initial solution.

$$\text{mmol AgCl final} = 101.0\ mL \times \frac{1.8 \times 10^{-8}\ mol\ Ag^+}{1\ L\ soln} \times \frac{1\ mmol\ AgCl}{1\ mmol\ Ag^+} = 1.8 \times 10^{-6}\ mmol\ AgCl$$

$$\text{mmol AgCl final} = 101.0\ mL \times \frac{1.3 \times 10^{-5}\ mol\ Ag^+}{1\ L\ soln} \times \frac{1\ mmol\ AgCl}{1\ mmol\ Ag^+} = 1.3 \times 10^{-3}\ mmol\ AgCl$$

The difference between these two amounts is the amount of AgCl that precipitates. We compute its mass.

$$\text{mass AgCl} = (1.3 \times 10^{-3} - 1.8 \times 10^{-6})\ mmol\ AgCl \times \frac{143.3\ mg\ AgCl}{1\ mmol\ AgCl} = 0.19\ mg$$

We conclude that the precipitate will not be visible to the unaided eye, since it weighs less than 1 mg.

Criterion for Precipitation from Solution

38. We first determine $[Mg^{2+}]$, then determine the value of Q_{sp} and compare it to the value of K_{sp}

$$[Mg^{2+}] = \frac{17.5\ mg\ MgCl_2}{325\ mL\ soln} \times \frac{1\ mmol\ MgCl_2 \cdot 6H_2O}{203.3\ mg\ MgCl_2 \cdot 6H_2O} \times \frac{1\ mmol\ Mg^{2+}}{1\ mmol\ MgCl_2} = 2.65 \times 10^{-4}\ M$$

$Q_{sp} = [Mg^{2+}][F^-]^2 = (2.65 \times 10^{-4})(0.045)^2 = 5.4 \times 10^{-7} > 3.7 \times 10^{-8} = K_{sp}$

Thus, precipitation of $MgF_2(s)$ should occur from this solution.

39. The solutions mutually dilute each other.

$$[Cl^-] = 0.016\ M \times \frac{155\ mL}{155\ mL + 245\ mL} = 6.2 \times 10^{-3}\ M$$

$$[Pb^{2+}] = 0.175\ M \times \frac{245\ mL}{245\ mL + 155\ mL} = 0.107\ M$$

Then we compute the value of the ion product and compare it to the solubility product constant value.

$Q_{sp} = [Pb^{2+}][Cl^-]^2 = (0.107)(6.2 \times 10^{-3})^2 = 4.2 \times 10^{-6} < 1.6 \times 10^{-5} = K_{sp}$

Thus, precipitation of $PbCl_2(s)$ will not occur from these mixed solutions.

40. (a) First we determine $[Cl^-]$ due to the added NaCl.

$$[Cl^-] = \frac{1.0\ mg\ NaCl}{1.0\ L\ soln} \times \frac{1\ g}{1000\ mg} \times \frac{1\ mol\ NaCl}{58.4\ g\ NaCl} \times \frac{1\ mol\ Cl^-}{1\ mol\ NaCl} = 1.7 \times 10^{-5}\ M$$

Then we determine the value of the ion product and compare it to the solubility product constant value.

$Q = [Ag^+][Cl^-] = (0.10)(1.7 \times 10^{-5}) = 1.7 \times 10^{-6} > 1.8 \times 10^{-10} = K_{sp}$ for AgCl

Precipitation of AgCl(s) should occur.

(b) The KBr(aq) is diluted on mixing, but the $[Ag^+]$ and $[Cl^-]$ are barely affected by dilution.

$$[Br^-] = 0.20\ M \times \frac{0.05\ mL}{0.05\ mL + 200\ mL} = 5 \times 10^{-5}\ M$$

Now we determine $[Ag^+]$ in a saturated AgCl solution.

$K_{sp} = [Ag^+][Cl^-] = (s)(s) = s^2 = 1.8 \times 10^{-10}$ $\qquad\qquad$ $s = 1.3 \times 10^{-5}$ M

Then we determine the value of the ion product and compare it to the solubility product constant value.

$Q = [Ag^+][Br^-] = (1.3 \times 10^{-5})(5 \times 10^{-5}) = 6 \times 10^{-10} > 5.0 \times 10^{-13} = K_{sp}$ for AgBr

Precipitation of AgBr(s) should occur.

(c) The hydroxide ion is diluted by mixing the two solutions.

$$[OH^-] = 0.0150 \text{ M} \times \frac{0.05 \text{ mL}}{0.05 \text{ mL} + 5000 \text{ mL}} = 2 \times 10^{-7} \text{ M}$$

But the $[Mg^{2+}]$ does not change significantly.

$$[Mg^{2+}] = \frac{2.0 \text{ mg Mg}^{2+}}{1.0 \text{ L soln}} \times \frac{1 \text{ g}}{1000 \text{ mg}} \times \frac{1 \text{ mol Mg}^{2+}}{24.3 \text{ g Mg}} = 8.2 \times 10^{-5} \text{ M}$$

Then we determine the value of the ion product and compare it to the solubility product constant value.

$Q = [Mg^{2+}][OH^-]^2 = (2 \times 10^{-7})(8.2 \times 10^{-5})^2 = 1 \times 10^{-15} < 1.8 \times 10^{-11} = K_{sp}$ for $Mg(OH)_2$

Thus, precipitation of $Mg(OH)_2(s)$ will not occur.

41. We determine $[C_2O_4^{2-}]$ in this solution. From assumption 3 in Section 17-6, $[C_2O_4^{2-}] = K_{a_2} = 5.4 \times 10^{-5}$

$Q_{sp} = [Ca^{2+}][C_2O_4^{2-}] = (0.150)(5.4 \times 10^{-5}) = 8.1 \times 10^{-6} > 1.3 \times 10^{-9} = K_{sp}$

Thus CaC_2O_4 should precipitate from this solution. (The quantity of $H_2C_2O_4$ used is immaterial since $[C_2O_4^{2-}]$ is independent of the oxalic acid concentration.)

42. We determine the $[OH^-]$ needed to just initiate precipitation of $Fe(OH)_3$

$K_{sp} = [Fe^{3+}][OH^-]^3 = 4 \times 10^{-38} = (0.086 \text{ M})[OH^-]^3$ $\qquad$ $[OH^-] = \sqrt[3]{\dfrac{4 \times 10^{-38}}{0.086}} = 8 \times 10^{-13}$ M

$pOH = -\log(8 \times 10^{-13}) = 12.1$ $\qquad$ $pH = 14.00 - 12.1 = 1.9$

$Fe(OH)_3$ will precipitate from a solution with pH > 1.9.

43. We determine the amount of H_2 produced during the electrolysis, and then determine $[OH^-]$.

$$\text{amount H}_2 = \frac{PV}{RT} = \frac{748 \text{ mmHg} \times \dfrac{1 \text{ atm}}{760 \text{ mmHg}} \times 1.04 \text{ L}}{0.08206 \text{ L atm mol}^{-1} \text{ K}^{-1} \times 296 \text{ K}} = 0.0421 \text{ mol H}_2$$

$$[OH^-] = \frac{0.0421 \text{ mol H}_2 \times \dfrac{2 \text{ mol OH}^-}{1 \text{ mol H}_2}}{0.315 \text{ L sample}} = 0.267 \text{ M}$$

$Q_{sp} = [Mg^{2+}][OH^-]^2 = (0.220)(0.267)^2 = 1.57 \times 10^{-2} > 1.8 \times 10^{-11} = K_{sp}$

Yes, precipitation of $Mg(OH)_2(s)$ should occur during the electrolysis.

44. (a) The solutions mutually dilute each other. We first determine the solubility of each compound in its saturated solution and then its concentration after dilution.

$K_{sp} = [Ag^+]^2[SO_4^{2-}] = 1.4 \times 10^{-5} = (2s)^2 s = 4s^3$ $\qquad$ $s = \sqrt[3]{\dfrac{1.4 \times 10^{-5}}{4}} = 1.5 \times 10^{-2}$ M

$[SO_4^{2-}] = 0.015 \text{ M} \times \dfrac{100.0 \text{ mL}}{100.0 \text{ mL} + 250.0 \text{ mL}} = 0.0043 \text{ M}$ $\qquad$ $[Ag^+] = 0.0086$ M

$K_{sp} = [Pb^{2+}][CrO_4^{2-}] = 2.8 \times 10^{-13} = (s)(s) = s^2$ $\qquad$ $s = \sqrt{2.8 \times 10^{-13}} = 5.3 \times 10^{-7}$ M

$[Pb^{2+}] = [CrO_4^{2-}] = 5.3 \times 10^{-7} \times \dfrac{250.0 \text{ mL}}{250.0 \text{ mL} + 100.0 \text{ mL}} = 3.8 \times 10^{-7}$ M

From the balanced chemical equation, we see that the two possible precipitates are $PbSO_4$ and Ag_2CrO_4. (Neither $PbCrO_4$ nor Ag_2SO_4 can precipitate because they have been diluted below their saturated concentrations.) $\qquad$ $PbCrO_4 + Ag_2SO_4 \longrightarrow PbSO_4 + Ag_2CrO_4$ $\qquad$ Thus, we compute the value of Q for each of these compounds and compare those values with the solubility constant product value.

$Q = [Pb^{2+}][SO_4^{2-}] = (3.8 \times 10^{-7})(0.0043) = 1.6 \times 10^{-9} < 1.6 \times 10^{-8} = K_{sp}$ for $PbSO_4$

Thus, $PbSO_4(s)$ will not precipitate.

$Q = [Ag^+]^2[CrO_4^{2-}] = (0.0086)^2(3.8 \times 10^{-7}) = 2.8 \times 10^{-11} > 1.1 \times 10^{-12} = K_{sp}$ for Ag_2CrO_4

Thus, $Ag_2CrO_4(s)$ should precipitate.

Completeness of Precipitation

45. First determine that a precipitate forms. The solutions mutually dilute each other.

$$[CrO_4{}^{2-}] = 0.350 \text{ M} \times \frac{200.0 \text{ mL}}{200.0 \text{ mL} + 200.0 \text{ mL}} = 0.175 \text{ M}$$

$$[Ag^+] = 0.0100 \text{ M} \times \frac{200.0 \text{ mL}}{200.0 \text{ mL} + 200.0 \text{ mL}} = 0.00500 \text{ M}$$

We determine the value of the ion product and compare it to the solubility product constant value.

$$Q = [Ag^+]^2[CrO_4{}^{2-}] = (0.00500)^2(0.175) = 4.4 \times 10^{-6} > 2.4 \times 10^{-12} = K_{sp} \text{ for Ag}_2CrO_4$$

Ag_2CrO_4 should precipitate.

Now, we assume that as much solid forms as possible, and then we approach equilibrium by dissolving that solid in a solution that contains the ion in excess.

Reaction:	$Ag_2CrO_4(s) \rightleftharpoons$	$2 Ag^+(aq) +$	$CrO_4{}^{2-}(aq)$
Orig. soln:		0.00500	0.175 M
Form solid:		0 M	0.1725 M
Changes:		$+2x$ M	$+x$ M
Equil:		$2x$ M	$(0.1725 + x)$M

$$K_{sp} = [Ag^+]^2[CrO_4{}^{2-}] = 1.1 \times 10^{-12} = (2x)^2(0.1725 + x) \approx (4x^2)(0.1725)$$

$$x = \sqrt{\frac{1.1 \times 10^{-12}}{4 \times 0.1725}} = 1.3 \times 10^{-6} \text{ M} \qquad [Ag^+] = 2x = 2.6 \times 10^{-6} \text{ M}$$

46. (a) We use the solubility product constant expression to determine $[Pb^{2+}]$ in a solution with 0.100 M Cl^-.

$$K_{sp} = [Pb^{2+}][Cl^-]^2 = 1.6 \times 10^{-5} = [Pb^{2+}](0.100)^2 \qquad [Pb^{2+}] = \frac{1.6 \times 10^{-5}}{(0.100)^2} = 1.6 \times 10^{-3} \text{ M}$$

$$\% \text{ unprecipitated} = \frac{1.6 \times 10^{-3} \text{ M}}{0.050 \text{ M}} \times 100\% = 3.2\%$$

(b) Now we want $[Pb^{2+}]_f = 1\%$ $[Pb^{2+}]_i = 0.010 \times 0.050$ M $= 5.0 \times 10^{-4}$ M

$$K_{sp} = [Pb^{2+}][Cl^-]^2 = 1.6 \times 10^{-5} = (5.0 \times 10^{-4})[Cl^-]^2 \qquad [Cl^-] = \sqrt{\frac{1.6 \times 10^{-5}}{5.0 \times 10^{-4}}} = 0.18 \text{ M}$$

47. (a) We determine $[OH^-]$ in a saturated $Ca(OH)_2$ solution.

$$K_{sp} = [Ca^{2+}][OH^-]^2 = 5.5 \times 10^{-6} = (s)(2s)^2 = 4s^3 \qquad s = \sqrt[3]{\frac{5.5 \times 10^{-6}}{4}} = 1.1 \times 10^{-2} \text{ M}$$

$[OH^-] = 2s = 2.2 \times 10^{-2}$ M   This is higher than 2.0×10^{-3} M

(b) 99.9% precipitated means that $[Mg^{2+}]_f = 0.1\%$ $[Mg^{2+}]_i = 0.001 \times 0.059$ M $= 5.9 \times 10^{-5}$ M

$$K_{sp} = [Mg^{2+}][OH^-]^2 = 1.8 \times 10^{-11} = (5.9 \times 10^{-5})[OH^-]^2$$

$$[OH^-] = \sqrt{\frac{1.8 \times 10^{-11}}{5.9 \times 10^{-5}}} = 5.5 \times 10^{-4} \text{ M}$$

This is the minimum $[OH^-]$ necessary to produce complete precipitation.

Fractional Precipitation

48. The concentrations of silver ion that are cited in Example 19-7 range from 5.0×10^{-11} M to 1.5×10^{-5} M. These are incredibly small concentrations, especially the first. Virtually any $AgNO_3(aq)$ solution that we would prepare by usual means would have at least these concentrations. However, there is is matter of dilution to be considered. If one drop (0.05 mL) of $AgNO_3(aq)$ is added to 500.0 mL of solution, the $[Ag^+]$ will decrease by a factor of 10^4. Thus we would have to begin with 0.15 M $AgNO_3$ for this dilution to produce 1.5×10^{-5} M. So we cannot be too careless and use extremely dilute $AgNO_3(aq)$.

49. Normally we would worry about the mutual dilution of the two solutions, but the values of the solubility product constants are so small that only a very small volume of 0.10 M $Pb(NO_3)_2$ solution needs to be added, as we shall see.

(a) Since the two anions are present at the same concentration and they have the same type of formula (one anion per cation), the one forming the compound with the smallest K_{sp} value will precipitate first. Thus, CrO_4^{2-} is the first ion to precipitate.

(b) At the point where SO_4^{2-} begins to precipitate, we have

$$K_{sp} = [Pb^{2+}][SO_4^{2-}] = 1.6 \times 10^{-8} = [Pb^{2+}](0.010 \text{ M}) \qquad [Pb^{2+}] = \frac{1.6 \times 10^{-8}}{0.010} = 1.6 \times 10^{-6} \text{ M}$$

Now we can test our original assumption, that only a very small volume of 0.10 M $Pb(NO_3)_2$ solution has been added. We assume that we have 1.00 L of the original solution, the one with the two anions dissolved in it, and compute the volume of 0.10 M $Pb(NO_3)_2$ that has to be added to achieve $[Pb^{2+}] = 1.6 \times 10^{-6}$ M

$$\text{added soln volume} = 1.00 \text{ L} \times \frac{1.6 \times 10^{-6} \text{ mol Pb}^{2+}}{1 \text{ L soln}} \times \frac{1 \text{ mol Pb(NO}_3)_2}{1 \text{ mol Pb}^{2+}} \times \frac{1 \text{ L Pb}^{2+}\text{soln}}{0.10 \text{ mol Pb(NO}_3)_2}$$
$$\times \frac{1000 \text{ mL}}{1 \text{ L}} = 1.6 \times 10^{-2} \text{ mL Pb}^{2+} \text{ soln} = 0.016 \text{ mL Pb}^{2+} \text{ soln}$$

This is less than one drop (0.05 mL) of the Pb^{2+} solution, clearly a very small volume.

(c) The two anions are effectively separated if $[Pb^{2+}]$ has not reached 1.6×10^{-6} M when $[CrO_4^{2-}]$ is reduced to 0.1% of its original value, that is, to 0.010×10^{-3} M $= 1.0 \times 10^{-5}$ M $= [CrO_4^{2-}]$

$$K_{sp} = [Pb^{2+}][CrO_4^{2-}] = 2.8 \times 10^{-13} = [Pb^{2+}](1.0 \times 10^{-5})$$

$$[Pb^{2+}] = \frac{2.8 \times 10^{-13}}{1.0 \times 10^{-5}} = 2.8 \times 10^{-8} \text{ M}$$

Thus, the two anions can be effectively separated by fractional precipitation.

50. (a) We answer this question by determining the $[Ag^+]$ needed to initiate precipitation of each compound.

AgCl: $\quad K_{sp} = [Ag^+][Cl^-] = 1.8 \times 10^{-10} = [Ag^+](0.250) \qquad [Ag^+] = \dfrac{1.8 \times 10^{-10}}{0.250} = 7.2 \times 10^{-10} \text{ M}$

AgBr: $\quad K_{sp} = [Ag^+][Br^-] = 5.0 \times 10^{-13} = [Ag^+](0.0022) \qquad [Ag^+] = \dfrac{5.0 \times 10^{-13}}{0.0022} = 2.3 \times 10^{-10} \text{ M}$

AgBr precipitates first, since it requires a lower $[Ag^+]$.

(b) Cl^- and Br^- cannot be separated by this fractional precipitation. $[Ag^+]$ will have to rise to 1000 times its initial value, to 2.3×10^{-7} M, before AgBr is completely precipitated. But as soon as $[Ag^+]$ reaches 7.2×10^{-10} M, AgCl will begin to precipitate.

51. (a) 0.10 M NaCl will not work at all, since both $BaCl_2$ and $CaCl_2$ are soluble in water.

(b) $K_{sp} = 1.1 \times 10^{-10}$ for $BaSO_4$ and $K_{sp} = 9.1 \times 10^{-6}$ for $CaSO_4$. Since these values differ by more than 1000, 0.50 M Na_2SO_4 would effectively separate Ba^{2+} from Ca^{2+}.
We first compute $[SO_4^{2-}]$ when $BaSO_4$ begins to precipitate.

$[Ba^{2+}][SO_4^{2-}] = (0.05)[SO_4^{2-}] = 1.1 \times 10^{-10} \qquad [SO_4^{2-}] = \dfrac{1.1 \times 10^{-10}}{0.05} = 2 \times 10^{-9} \text{ M}$

And then $[SO_4^{2-}]$ when $[Ba^{2+}]$ has decreased to 0.1% of its initial value, that is, to 5×10^{-5} M

$[Ba^{2+}][SO_4^{2-}] = (5 \times 10^{-5})[SO_4^{2-}] = 1.1 \times 10^{-10} \qquad [SO_4^{2-}] = \dfrac{1.1 \times 10^{-10}}{5 \times 10^{-5}} = 2 \times 10^{-6} \text{ M}$

And finally $[SO_4^{2-}]$ when $CaSO_4$ begins to precipitate.

$[Ca^{2+}][SO_4^{2-}] = (0.05)[SO_4^{2-}] = 9.1 \times 10^{-6} \qquad [SO_4^{2-}] = \dfrac{9.1 \times 10^{-6}}{0.05} = 2 \times 10^{-4} \text{ M}$

(c) Now, $K_{sp} = 5 \times 10^{-3}$ for $Ba(OH)_2$ and $K_{sp} = 5.5 \times 10^{-6}$ for $Ca(OH)_2$. The fact that these two K_{sp} values differ by almost a factor of 1000 does not tell the entire story, because $[OH^-]$ appears squared in both K_{sp} expressions. We compute $[OH^-]$ when $Ca(OH)_2$ begins to precipitate.

$[Ca^{2+}][OH^-]^2 = 5.5 \times 10^{-6} = (0.05)[OH^-]^2 \qquad [OH^-] = \sqrt{\dfrac{5.5 \times 10^{-6}}{0.05}} = 1 \times 10^{-2} \text{ M}$

Precipitation will not proceed this far; we only have 0.001 M NaOH.

(d) Finally, $K_{sp} = 5.1 \times 10^{-9}$ for $BaCO_3$ and $K_{sp} = 2.8 \times 10^{-9}$ for $CaCO_3$. Since these two values differ by less than a factor of 2, 0.50 M Na_2CO_3 would not effectively separate Ba^{2+} from Ca^{2+}.

Solubility and pH

52. (a) Since HI is a strong acid, $[I^-] = 1.05 \times 10^{-3} \text{ M} + 1.05 \times 10^{-3} \text{ M} = 2.10 \times 10^{-3} \text{ M}$

We determine the value of the ion product and compare it to the solubility product constant value.

$Q_{sp} = [Pb^{2+}][I^-]^2 = (1.1 \times 10^{-3})(2.10 \times 10^{-3})^2 = 4.9 \times 10^{-9} < 7.1 \times 10^{-9} = K_{sp}$ for PbI_2

A precipitate of PbI_2 will not form under these conditions.

(b) We compute the $[OH^-]$ needed for precipitation.

$$K_{sp} = [Mg^{2+}][OH^-]^2 = 1.8 \times 10^{-11} = (0.0150)[OH^-]^2 \qquad [OH^-] = \sqrt{\frac{1.8 \times 10^{-11}}{0.0150}} = 3.5 \times 10^{-5} \text{ M}$$

Then we compute $[OH^-]$ in this solution, resulting from the ionization of NH_3.

$$[NH_3] = 1.00 \text{ M} \times \frac{0.05 \times 10^{-3} \text{ L}}{2.50 \text{ L}} = 2 \times 10^{-5} \text{ M}$$

Since NH_3 is a weak base, we know that $[OH^-]$ produced from NH_3 will be less than 2×10^{-5} M. And since $[OH^-] = 3.5 \times 10^{-5}$ M is needed for precipitation to occur, we conclude that $Mg(OH)_2$ will not precipitate from this solution.

(c) 0.010 M $HC_2H_3O_2$ and 0.01 M $NaC_2H_3O_2$ is a buffer solution with pH = pK_a of acetic acid, since the acid and its anion are present in equal concentrations. (If they were not, we would have used the Henderson-Hasselbalch equation to determine the pH.) From this, we determine the $[OH^-]$.

pH = 4.74   pOH = 14.00 − 4.74 = 9.26 $[OH^-] = 10^{-9.26} = 5.5 \times 10^{-10}$

$Q = [Al^{3+}][OH^-]^3 = (0.010)(5.5 \times 10^{-10})^3 = 1.7 \times 10^{-30} > 1.3 \times 10^{-33} = K_{sp}$ of $Al(OH)_3$

Thus, $Al(OH)_3(s)$ should precipitate from this solution.

53. We determine $[Mg^{2+}]$ in the solution.

$$[Mg^{2+}] = \frac{0.95 \text{ g } Mg(OH)_2}{1 \text{ L soln}} \times \frac{1 \text{ mol } Mg(OH)_2}{58.3 \text{ g } Mg(OH)_2} \times \frac{1 \text{ mol } Mg^{2+}}{1 \text{ mol } Mg(OH)_2} = 0.016 \text{ M}$$

Then we determine $[OH^-]$ in the solution, and its pH.

$$K_{sp} = [Mg^{2+}][OH^-]^2 = 1.8 \times 10^{-11} = (0.016)[OH^-]^2 \qquad [OH^-] = \sqrt{\frac{1.8 \times 10^{-11}}{0.016}} = 3.4 \times 10^{-5} \text{ M}$$

$pOH = -\log(3.4 \times 10^{-5}) = 4.47$ $pH = 14.00 - 4.47 = 9.53$

54. In each case we indicate whether the compound is more soluble in acid or base, or neither. And we write the net ionic equation for the reaction in which the solid dissolves in acid or base. Substances are soluble in acid or base if either (1) an acid-base reaction occurs [as in (a) and (f)] or (2) a gas is produced, since escape of the gas from the reaction mixture causes the reaction to shift to the right.

(a) Base: $H_2C_2O_4(s) + 2\,OH^-(aq) \longrightarrow C_2O_4^{2-}(aq) + 2\,H_2O$

(b) Acid: $MgCO_3(s) + 2\,H^+(aq) \longrightarrow Mg^{2+}(aq) + H_2O + CO_2(g)$

(c) Acid: $CdS(s) + 2\,H^+(aq) \longrightarrow Cd^{2+}(aq) + H_2S(g)$

(d) Neither: KCl

(e) Neither: $NaNO_3$

(f) Acid: $Ca(OH)_2(s) + 2\,H^+(aq) \longrightarrow Ca^{2+}(aq) + 2\,H_2O$

55. First determine the $[Mg^{2+}]$ and $[NH_3]$ that result from dilution to a total volume of 0.500 L.

$$[Mg^{2+}] = 0.100 \text{ M} \times \frac{0.150 \text{ L}_{initial}}{0.500 \text{ L}_{final}} = 0.0300 \text{ M} \qquad [NH_3] = 0.100 \text{ M} \times \frac{0.350 \text{ L}_{initial}}{0.500 \text{ L}_{final}} = 0.0700 \text{ M}$$

Then determine the $[OH^-]$ that will allow $[Mg^{2+}] = 0.0300$ M in this solution.

$$K_{sp} = [Mg^{2+}][OH^-]^2 = (0.0300)[OH^-]^2 \qquad [OH^-] = \sqrt{\frac{1.8 \times 10^{-11}}{0.0300}} = 2.4 \times 10^{-5} \text{ M}$$

This $[OH^-]$ is maintained by the $NH_3-NH_4^+$ buffer, for which we use the Henserson-Hasselbalch equation.

$$pH = 14.00 - pOH = 14.00 + \log(2.4 \times 10^{-5}) = 9.38 = pK_a + \log\frac{[NH_3]}{[NH_4^+]} = 9.26 + \frac{[NH_3]}{[NH_4^+]}$$

$$\log\frac{[NH_3]}{[NH_4^+]} = 9.38 - 9.26 = +0.12 \qquad \frac{[NH_3]}{[NH_4^+]} = 10^{+0.12} = 1.3 \qquad [NH_4^+] = \frac{0.0700 \text{ M } NH_3}{1.3} = 0.054 \text{ M}$$

$$\text{mass } (NH_4)_2SO_4 = 0.500 \text{ L} \times \frac{0.054 \text{ mol } NH_4^+}{L \text{ soln}} \times \frac{1 \text{ mol } (NH_4)_2SO_4}{2 \text{ mol } NH_4^+} \times \frac{132.1 \text{ g } (NH_4)_2SO_4}{1 \text{ mol } (NH_4)_2SO_4} = 1.8 \text{ g}$$

56. (a)   We calculate [OH⁻] needed for precipitation.

$$K_{sp} = [Al^{3+}][OH^-]^3 = 1.3 \times 10^{-33} = (0.050 \text{ M})[OH^-]^3 \qquad [OH^-] = \sqrt[3]{\frac{1.3 \times 10^{-33}}{0.050}} = 3.0 \times 10^{-11}$$

$$pOH = -\log (3.0 \times 10^{-11}) = 10.52 \qquad\qquad pH = 14.00 - 10.52 = 3.48$$

(b)   We use the Henderson-Hasselbalch equation to determine [C₂H₃O₂⁻].

$$pH = 3.48 = pK_a + \log \frac{[C_2H_3O_2^-]}{[HC_2H_3O_2]} = 4.74 + \log \frac{[C_2H_3O_2^-]}{1.00 \text{ M}}$$

$$\log \frac{[C_2H_3O_2^-]}{1.00 \text{ M}} = 3.48 - 4.74 = -1.26 \qquad \frac{[C_2H_3O_2^-]}{1.00 \text{ M}} = 10^{-1.26} = 0.055 \qquad [C_2H_3O_2^-] = 0.055 \text{ M}$$

This situation does not quite obey the guideline that the ratio of concentrations fall in the range 0.10 to 10.0, but the resulting error is a small one in this circumstance.

$$\text{mass } NaC_2H_3O_2 = 0.2000 \text{ L} \times \frac{0.055 \text{ mol } C_2H_3O_2^-}{1 \text{ L soln}} \times \frac{1 \text{ mol } NaC_2H_3O_2}{1 \text{ mol } C_2H_3O_2^-} \times \frac{82.03 \text{ g } NaC_2H_3O_2}{1 \text{ mol } NaC_2H_3O_2}$$

$$= 0.90 \text{ g } NaC_2H_3O_2$$

Complex-Ion Equilibria

57.   Lead(II) ion forms a complex ion with chloride ion. It forms no such complex ion with nitrate ion. The formation of this complex ion decreases the concentrations of free Pb²⁺(aq) and free Cl⁻(aq). Thus, more PbCl₂ can dissolve before the value of the solubility product is satisfied.
$$Pb^{2+}(aq) + 3 \text{ Cl}^-(aq) \rightleftharpoons [PbCl_3]^-(aq)$$

58.   HCl(aq): $Zn(OH)_2(s) + 2 H_3O^+(aq) \rightleftharpoons Zn^{2+}(aq) + 4 H_2O$
HC₂H₃O₂(aq):   as in HCl(aq) *or* $Zn(OH)_2(s) + 2 HC_2H_3O_2(aq) \rightleftharpoons Zn^{2+}(aq) + 2 C_2H_3O_2^-(aq) + 2 H_2O$
NH₃(aq): $Zn(OH)_2(s) + 4 NH_3(aq) \rightleftharpoons [Zn(NH_3)_4]^{2+}(aq) + 2 OH^-(aq)$
OH⁻(aq): $Zn(OH)_2(s) + 2 OH^-(aq) \rightleftharpoons [Zn(OH)_4]^{2-}(aq)$

59. $Zn^{2+}(aq) + 4 NH_3(aq) \rightleftharpoons [Zn(NH_3)_4]^{2+}(aq) \qquad K_f = 4.1 \times 10^8$
NH₃(aq) will be least effective in reducing the concentration of the complex ion. In fact, the addition of NH₃(aq) will increase the concentration of the complex ion by favoring a shift of the equilibrium to the right. NH₄⁺(aq) will have a similar effect, but it is not as direct. NH₃(aq) is formed by the hydrolysis of NH₄⁺(aq) and thus increasing [NH₄⁻] will eventually increase NH₃(aq): $NH_4^+(aq) + H_2O \rightleftharpoons NH_3(aq) + H_3O^+(aq)$. The addition of HCl(aq) actually will decrease the concentration of complex ion. HCl(aq) will react with NH₃(aq) to decrease its concentration and that will cause the complex ion equilibrium to shift left.

60. (a)   We substitute directly into the K_f expression. $K_f = \dfrac{[[Cu(CN)_4^{3-}]]}{[Cu^{2+}][CN^-]^4} = \dfrac{0.0500}{(6.1 \times 10^{-32})(0.80)^4} = 2.0 \times 10^{30}$

(b)   Reaction: $Cu^{2+}(aq) \quad + \quad 4 NH_3(aq) \rightleftharpoons [Cu(NH_3)_4]^{2+}(aq)$
Initial: 0.10 M   6.0 M
Changes: $-x$ M $+x$ M
Equil: $(0.10 - x)$ M   6.0 M x M

$$K_f = \frac{[[Cu(NH_3)_4]^{2+}]}{[Cu^{2+}][NH_3]^4} = 1.1 \times 10^{13} = \frac{x}{(0.10 - x) \ 6.0^4} \qquad \frac{x}{0.10 - x} = 1.4 \times 10^{16}$$

We assume that most of the copper(II) in solution is present as the complex ion, or that $x \approx 0.10$ M,

then $0.10 - x = \dfrac{x}{1.4 \times 10^{16}} = \dfrac{0.10}{1.4 \times 10^{16}} = 7.1 \times 10^{-17}$ M = [Cu²⁺]

61.   We first find the concentration of free metal ion. Then we determine the value of Q_{sp} for the precipitation reaction, and compare its value with the value of K_{sp} to determine whether precipitation should occur.
Reaction: $Ag^+(aq) + \ 2 S_2O_3^{2-}(aq) \rightleftharpoons \quad [Ag(S_2O_3)_2]^{3-}(aq)$
Initial: 1.05 M   0.012 M
Changes: $+x$ M $+2x$ M $-x$ M
Equil: x M $(1.05 + 2x)$M $(0.012 - x)$M

$$K_f = \frac{[[Ag(S_2O_3)_2]^{3-}]}{[Ag^+][S_2O_3^{2-}]^2} = 1.7 \times 10^{13} = \frac{0.012 - x}{x \ (1.05 + 2x)^2} \approx \frac{0.012}{1.05^2 \ x} \qquad x = 6.4 \times 10^{-16} \text{ M} = [Ag^+]$$

$$Q = [Ag^+][I^-] = (6.4 \times 10^{-16})(2.0) = 1.3 \times 10^{-15} > 8.5 \times 10^{-17} = K_{sp}$$

Since $Q > K_{sp}$, precipitation of AgI(s) should occur.

62. We determine [OH⁻] in this solution, and also the free [Cu²⁺].

$$pH = pK_a + \log \frac{[NH_3]}{[NH_3^+]} = 9.26 + \log \frac{0.10 \text{ M}}{0.10 \text{ M}} = 9.26 \qquad pH = 14.00 - 9.26 = 4.74$$
$$[OH^-] = 10^{-4.74} = 1.8 \times 10^{-5} \text{ M}$$

$$Cu^{2+}(aq) + 4 NH_3(aq) \rightleftharpoons [Cu(NH_3)_4]^{2+}(aq)$$
$$K_f = \frac{[[Cu(NH_3)_4]^{2+}]}{[Cu^{2+}][NH_3]^4} = 1.1 \times 10^{13} = \frac{0.015}{[Cu^{2+}]\,0.10^4} \quad [Cu^{2+}] = \frac{0.015}{1.1 \times 10^{13} \times 0.10^4} = 1.4 \times 10^{-11} \text{ M}$$

Now we determine the value of Q and compare it with the value of K_{sp} for Cu(OH)₂.
$$Q = [Cu^{2+}][OH^-]^2 = (1.4 \times 10^{-11})(1.8 \times 10^{-5})^2 = 4.5 \times 10^{-21} < 2.2 \times 10^{-20} = K_{sp} \text{ for Cu(OH)}_2$$
Precipitation of Cu(OH)₂(s) from this solution should not occur.

63. We first compute the free [Ag⁺] in the original solution. The size of the equilibrium constant indicates that the reaction lies to the right.

Reaction: $Ag^+(aq) + 2 NH_3(aq) \rightleftharpoons [Ag(NH_3)_2]^+(aq)$
In soln: 0.10 M   1.00 M
Form complex: –0.10 M   –0.20 M   +0.10 M
                  0 M   0.80 M   0.10 M
Changes: +x M   +2x M   –x M
Equil: x M   (0.80 + 2x) M   (0.10 – x) M

$$K_f = 1.6 \times 10^7 = \frac{[[Ag(NH_3)_2]^+]}{[Ag^+][NH_3]^2} = \frac{0.10 - x}{x\,(0.80 + 2x)^2} \approx \frac{0.10}{x\,(0.80)^2} \quad x = \frac{0.10}{1.6 \times 10^7\,(0.80)^2} = 9.8 \times 10^{-9} \text{ M}$$

Thus, free [Ag⁺] = 9.8×10^{-9}. We determine the [I⁻] that can coexist in this solution without precipitation.

$$K_{sp} = [Ag^+][I^-] = 8.5 \times 10^{-17} = (9.8 \times 10^{-9})[I^-] \qquad [I^-] = \frac{8.5 \times 10^{-17}}{9.8 \times 10^{-9}} = 8.7 \times 10^{-9} \text{ M}$$

And now we determine the mass of KI needed to produce this [I⁻]

$$\text{mass KI} = 1.00 \text{ L soln} \times \frac{8.7 \times 10^{-9} \text{ mol I}^-}{1 \text{ L soln}} \times \frac{1 \text{ mol KI}}{1 \text{ mol I}^-} \times \frac{166.0 \text{ g KI}}{1 \text{ mol KI}} = 1.4 \times 10^{-6} \text{ g KI}$$

64. (a) CuS is in the hydrogen sulfide group of qualitative analysis. Its precipitation occurs when 0.3 M HCl is saturated with H₂S. It will certainly precipitate from a (non-acidic) saturated solution of H₂S, which has a much higher [S²⁻].
$$Cu^{2+}(aq) + H_2S(satd \ aq) \longrightarrow CuS(s) + 2 H^+(aq)$$
This reaction proceeds to a significant extent in the forward direction.

(b) MgS is soluble, according to the solubility rules listed in Chapter 5.
$$Mg^{2+}(aq) + H_2S(satd \ aq) \xrightarrow{\ 0.3 \text{ M HCl}\ } \text{no reaction}$$

(c) As in part (a), PbS is in the qualitative analysis hydrogen sulfide gourp, which precipitates from a 0.3 M HCl solution saturated with H₂S. Therefore, PbS does not dissolve appreciably in 0.3 M HCl.
$$PbS(s) + HCl \ (0.3 \text{ M HCl}) \longrightarrow \text{no reaction}$$

(d) Since ZnS(s) does not precipitate in the Hydrogen Sulfide group, we conclude that it is soluble in acidic solution. $ZnS(s) + 2 HNO_3(aq) \longrightarrow Zn(NO_3)_2(aq) + H_2S(g)$

65. Since the cation concentrations are identical, the value of Q_{spa} is the same for each one. It is this value of Q_{spa} that we compare with K_{spa} to determine if precipitation occurs.
$$Q_{spa} = \frac{[M^{2+}][H_2S]}{[H_3O^+]^2} = \frac{0.05 \text{ M} \times 0.10 \text{ M}}{(0.010 \text{ M})^2} = 5 \times 10^1$$
If $Q_{spa} > K_{spa}$ precipitation of the metal sulfide should occur. But if $Q_{spa} < K_{spa}$, precipitation will not occur.
For CuS, $K_{spa} = 6 \times 10^{-16} < Q_{spa} = 5 \times 10^{-1}$ Precipitation of CuS(s) should occur.
For HgS, $K_{spa} = 2 \times 10^{-32} < Q_{spa} = 5 \times 10^{-1}$ Precipitation of HgS(s) should occur.
For MnS, $K_{spa} = 3 \times 10^7 > Q_{spa} = 5 \times 10^{-1}$ Precipitation of MnS(s)will not occur.

66. (a) First we calculate [H₃O⁺] in the buffer solution with the Henderson-Hasselbalch equation.
$$pH = pK_a + \log \frac{[C_2H_3O_2^-]}{[HC_2H_3O_2]} = 4.74 + \log \frac{0.15 \text{ M}}{0.25 \text{ M}} = 4.52 \qquad [H_3O^+] = 10^{-4.52} = 3.0 \times 10^{-5} \text{ M}$$
We use this information to calculate a value of Q_{spa} for MnS in this solution and compare that with K_{spa}

$$Q_{spa} = \frac{[Mn^{2+}][H_2S]}{[H_3O^+]^2} = \frac{(0.15)(0.10)}{(3.0 \times 10^{-5})^2} = 1.7 \times 10^7 < 3 \times 10^7 = K_{spa} \text{ for MnS}$$

Precipitation of MnS(s) will not occur.

(b) We need to change $[H_3O^+]$ so that $Q_{spa} = 3 \times 10^7 = \dfrac{(0.15)(0.10)}{[H_3O^+]^2}$ $[H_3O^+] = \sqrt{\dfrac{(0.15)(0.10)}{3 \times 10^7}}$

$[H_3O^+] = 2.2 \times 10^{-5}$ M   pH = 4.66   This is a more basic solution, which we can achieve by increasing the basic component of the buffer solution, the acetate ion. We find out the new acetate ion concentration with the Henderson-Hasselbalch equation.

$$pH = pK_a + \log \frac{[C_2H_3O_2^-]}{[HC_2H_3O_2]} = 4.66 = 4.74 + \log \frac{[C_2H_3O_2^-]}{0.25 \text{ M}} \qquad \log \frac{[C_2H_3O_2^-]}{0.25 \text{ M}} = 4.66 - 4.74 = -0.08$$

$\dfrac{[C_2H_3O_2^-]}{0.25 \text{ M}} = 10^{-0.08} = 0.83$ $[C_2H_3O_2^-] = 0.83 \times 0.25$ M = 0.21 M

67. We know that $K_{spa} = 3 \times 10^7$ for MnS and $K_{spa} = 6 \times 10^2$ for FeS. The metal sulfide will begin to precipitate when $Q_{spa} = K_{spa}$. Let us determine $[H_3O^+]$ just necessary to form each precipitate. We assume the the solution is saturated with H_2S, $[H_2S] = 0.10$ M.

$$K_{spa} = \frac{[M^{2+}][H_2S]}{[H_3O^+]^2} \qquad [H_3O^+] = \sqrt{\frac{[M^{2+}][H_2S]}{K_{spa}}} = \sqrt{\frac{(0.10 \text{ M})(0.10 \text{ M})}{3 \times 10^7}} = 1.8 \times 10^{-5} \text{ M for MnS}$$

$$[H_3O^+] = \sqrt{\frac{(0.10 \text{ M})(0.10 \text{ M})}{6 \times 10^2}} = 4.1 \times 10^{-3} \text{ M for FeS}$$

Thus, if the solution is maintained at an acidity just a bit higher than 1.8×10^{-5} M = $[H_3O^+]$, FeS will precipitate and Mn^{2+}(aq) will remain in solution. To determine if the separation is complete, we see whether $[Fe^{2+}]$ has decreased to 0.1% or less of its original value when the solution is held at the aforementioned acidity. Let $[H_3O^+] = 2.0 \times 10^{-5}$ M and calculate $[Fe^{2+}]$.

$$K_{spa} = \frac{[Fe^{2+}][H_2S]}{[H_3O^+]^2} = 6 \times 10^2 = \frac{[Fe^{2+}](0.10 \text{ M})}{(2.0 \times 10^{-5} \text{ M})^2} \qquad [Fe^{2+}] = \frac{(6 \times 10^2)(2.0 \times 10^{-5})^2}{0.10} = 2.4 \times 10^{-6} \text{ M}$$

% Fe^{2+}(aq) remaining $= \dfrac{2.4 \times 10^{-6} \text{ M}}{0.10 \text{ M}} \times 100\% = 0.0024\%$ Separation is complete.

Qualitative Analysis

68. **(a)** Ag^+ and/or Hg_2^{2+} are probably present. Both of these cations form precipitates from an acidic solution of chloride ion.

(b) We cannot tell whether Mg^{2+} is present or not. Both MgS and $MgCl_2$ are water soluble.

(c) Pb^{2+} possibly is absent; it is the only cation of those given which forms a precipitate in an acidic solution that is treated with H_2S, and no sulfide precipitate was formed.

(d) We cannot tell whether Fe^{2+} is present or not. FeS will not precipitate from an acidic solution that is treated with H_2S; the solution must be alkaline for a FeS precipitate to form.

(a) and **(c)** are the valid conclusions.

69. The purpose of adding hot water is to separate Pb^{2+} from AgCl and Hg_2Cl_2. Thus, the most important consequence of the omission is that there not longer is a vlid test for the presence or absence of Pb^{2+}. In addition, if we add NH_3 first, $PbCl_2$ may form $Pb(OH)_2$. If $Pb(OH)_2$ does form, it will be present with Hg_2Cl_2 in the solid, although $Pb(OH)_2$ will not darken with added NH_3. Thus, we might falsely conclude that Ag^+ is present, but not falsely conclude that Hg_2^{2+} is present.

70. For $PbCl_2$(aq), 2 $[Pb^{2+}] = [Cl^-]$ and we let s = molar solubility of $PbCl_2$. Thus $s = [Pb^{2+}]$.

$K_{sp} = [Pb^{2+}][Cl^-]^2 = (s)(2s)^2 = 4s^3 = 1.6 \times 10^{-5}$ $s = \sqrt[3]{1.6 \times 10^{-5} \div 4} = 1.6 \times 10^{-2}$ M = $[Pb^{2+}]$

Both $[Pb^{2+}]$ and $[CrO_4^{2-}]$ are diluted by mixing the two solutions.

$[Pb^{2+}] = 0.016 \text{ M} \times \dfrac{1.00 \text{ mL}}{1.05 \text{ mL}} = 0.015$ M $[CrO_4^{2-}] = 1.0 \text{ M} \times \dfrac{0.05 \text{ mL}}{1.05 \text{ mL}} = 0.048$ M

$Q = [Pb^{2+}][CrO_4^{2-}] = (0.015 \text{ M})(0.050 \text{ M}) = 7.2 \times 10^{-4} > 2.8 \times 10^{-13} = K_{sp}$

Thus, precipitation should occur from the solution described.

71. (a) $Pb^{2+}(aq) + 2\ Cl^{-}(aq) \longrightarrow PbCl_2(s)$

(b) $Zn(OH)_2(s) + 2\ OH^{-}(aq) \longrightarrow [Zn(OH)_4]^{2-}(aq)$

(c) $Fe(OH)_3(s) + 3\ H_3O^{+}(aq) \longrightarrow Fe^{3+}(aq) + 6\ H_2O$ *or* $[Fe(H_2O)_6]^{3+}(aq)$

(d) $Cu^{2+}(aq) + H_2S(aq) \longrightarrow CuS(s) + 2\ H^{+}(aq)$

(e) $2\ [SbCl_4]^{-}(aq) + 3\ H_2S(aq) \longrightarrow Sb_2S_3(s) + 8\ Cl^{-}(aq) + 6\ H^{+}(aq)$

20 SPONTANEOUS CHANGE: ENTROPY

AND FREE ENERGY

REVIEW QUESTIONS

1. **(a)** ΔS_{univ} is the symbol for the entropy change of the universe. If this quantity is positive, the accompanying process is spontaneous.

   (b) ΔG_f° is the symbol for the standard free energy of formation, the free energy change of the standard state reaction when one mole of substance is produced from stable forms of its elements.

   (c) K_{eq} is the symbol for the thermodynamic equilibrium constant, in whose expression gasses are represented by pressures and solutes in aqueous solution by molarities.

2. **(a)** *Absolute* molar entropy refers to the entropy of a substance referred to pure perfect crystals at 0 K, which have zero entropy.

   (b) A reversible process is one occurring in a state of constant equilibrium; a small opposing stress would reverse the direction of the reaction.

   (c) Trouton's rule states that the entropy change on vaporization approximates 88 J mol^{-1} K^{-1}. It works well for nonpolar liquids.

   (d) An equilibrium constant is evaluated from tabulated thermodynamic data by combining values of ΔG_f° to obtain ΔG_{rxn}° and then using $\Delta G_{rxn}^{\circ} = -RT \ln K_{eq}$.

3. **(a)** A spontaneous process is one that occurs without (or sometimes in spite of) external intervention. A nonspontaneous process will only occur when some external agency operates on it.

   (b) The second law of thermodyanics places limitations on converting heat into work and on the spontaneous directions of processes. The third law of thermodynamics establishes a zero point for entropy.

   (c) ΔG refers to the free energy change for a process, while ΔG° requires that both the initial and final states of the process are standard ones: gases at 1 atm pressure, aqueous solutes at 1 molar concentration, and pure solids and liquids.

4. **(a)** Increase in entropy because a gas has been created from a liquid, a condensed phase.

   (b) Decrease in entropy because a condensed phase, a solid, is created from a solid and a gas.

   (c) No change in entropy that can be easily detected, since the number of moles of gas produced is the same as the number that reacted.

   (d) $2\,H_2S(g) + 3\,O_2(g) \longrightarrow 2\,H_2O(g) + 2\,SO_2(g)$ Decrease in entropy since five moles of gas with high entropy become only four moles of gas, with about the same quantity of entropy per mole.

5. **(a)** At 75°C, 1 mol H_2O (g, 1 atm) has a greater entropy than 1 mol H_2O (l, 1 atm) since a gas is much more disordered than a liquid.

   (b) $50.0\,g\,Fe \times \dfrac{1\,mol\,Fe}{55.8\,g\,Fe} = 0.896$ mol Fe has a higher entropy than 0.80 mol Fe, both (s) at 1 atm and 5°C, because entropy is an extensive property that depends on the amount of substance present.

   (c) 1 mol Br_2(l, 1 atm, 8°C) has a higher entropy than 1 mol Br_2(s, 1atm, 2°C) because solids are more ordered substances than are liquids.

   (d) 0.312 mol SO_2 (g, 0.110 atm, 32.5°C) has a higher entropy than 0.284 mol O_2 (g, 15.0 atm, 22.3°C) for at least three reasons. First, entropy is an extensive property that depends on the amount of substance present. Second, entropy increases with temperature. Third, entropy is greater at lower pressures. Furthermore, entropy generally is higher per mole for more complicated molecules.

6. We predict the sign of ΔS based on the number of moles of gas in the balanced chemical equation; if $\Delta n_{gas} <$ 0, ΔS will be smaller than zero. If $\Delta n_{gas} > 0$, ΔS will be greater than zero.

(a) $\Delta S > 0$ and $\Delta H > 0$ This is case 3 in Table 20-1.

(b) $\Delta S < 0$ and $\Delta H < 0$ This is case 2 in Table 20-1.

(c) $\Delta S > 0$ and $\Delta H < 0$ This is case 1 in Table 20-1.

(d) $\Delta S < 0$ and $\Delta H > 0$ This is case 4 in Table 20-1.

7. Answer (b) is correct. Br—Br bonds are broken in this reaction, meaning that it is endothermic, with $\Delta H >$ 0. Since the number of moles of gas increases during the reaction, $\Delta S > 0$. And, because $\Delta G = \Delta H - T\,\Delta S$, this reaction is nonspontaneous at low temperatures where the ΔH term predominates, with $\Delta G > 0$, and spontaneous at high temperatures where the $T\,\Delta S$ term predominates, with $\Delta G < 0$.

8. Answer (d) is correct. A reaction that proceeds only through electrolysis is a reaction that is nonspontaneous. Such a reaction has $\Delta G > 0$.

9. Answer (d) is correct. Since $\Delta G^\circ = -RT \ln K_{eq}$, if $\Delta G^\circ = 0$, $\ln K_{eq} = 0$, which means that $K_{eq} = 1.00$

10. $\Delta H^\circ = \Delta H_f^\circ[NH_4Cl(s)] - \Delta H_f^\circ[NH_3(g)] - \Delta H_f^\circ[HCl(g)]$

$= -314.4 \text{ kJ/mol} - (-46.1 \text{ kJ/mol} - 92.3 \text{ kJ/mol}) = -176.0 \text{ kJ/mol}$

$\Delta G^\circ = \Delta G_f^\circ[NH_4Cl(s)] - \Delta G_f^\circ[NH_3(g)] - \Delta G_f^\circ[HCl(g)]$

$= -203.0 \text{ kJ/mol} - (-16.5 \text{ kJ/mol} - 95.3 \text{ kJ/mol}) = -91.2 \text{ kJ/mol}$

$\Delta G^\circ = \Delta H^\circ - T\Delta S^\circ \qquad \Delta S^\circ = \dfrac{\Delta H^\circ - \Delta G^\circ}{T} = \dfrac{-176.0 \text{ kJ/mol} + 91.2 \text{ kJ/mol}}{298.15 \text{ K}} \times \dfrac{1000 \text{ J}}{1 \text{ kJ}} = -284 \text{ J mol}^{-1} \text{ K}^{-1}$

11. (a) $\Delta G^\circ = \Delta G_f^\circ[C_2H_6(g)] - \Delta G_f^\circ[C_2H_2(g)] - 2\,\Delta G_f^\circ[H_2(g)]$

$= -32.89 \text{ kJ/mol} - 209.2 \text{ kJ/mol} - 2(0.00 \text{ kJ/mol}) = -242.1 \text{ kJ/mol}$

(b) $\Delta G^\circ = 2\,\Delta G_f^\circ[SO_2(g)] + \Delta G_f^\circ[O_2(g)] - 2\,\Delta G_f^\circ[SO_3(g)]$

$= 2(-300.2 \text{ kJ/mol}) + 0.00 \text{ kJ/mol} - 2(-371.1 \text{ kJ/mol}) = +141.8 \text{ kJ/mol}$

(c) $\Delta G^\circ = 3\,\Delta G_f^\circ[Fe(s)] + 4\,\Delta G_f^\circ[H_2O(g)] - \Delta G_f^\circ[Fe_3O_4(s)] - 4\,\Delta G_f^\circ[H_2(g)]$

$= 3(0.00 \text{ kJ/mol}) + 4(-228.6 \text{ kJ/mol}) - (-1015 \text{ kJ/mol}) - 4(0.00 \text{ kJ/mol}) = 101 \text{ kJ/mol}$

(d) $\Delta G^\circ = 2\,\Delta G_f^\circ[Al^{3+}(aq)] + 3\,\Delta G_f^\circ[H_2(g)] - 2\,\Delta G_f^\circ[Al(s)] - 6\,\Delta G_f^\circ[H^+(aq)]$

$= 2(-485 \text{ kJ/mol}) + 3(0.00 \text{ kJ/mol}) - 2(0.00 \text{ kJ/mol}) - 6(0.00 \text{ kJ/mol}) = -970 \text{ kJ/mol}$

12. (a) This temperature is the melting point of $I_2(s)$ at 1 atm pressure. From the phase diagram for iodine, we see that this process occurs at 114°C.

(b) $\Delta G^\circ = 0$ for $I_2(s, 1 \text{ atm}) \rightleftharpoons I_2(l, 1 \text{ atm})$. The point of the crossing of the lines in Figure 20-9 is the point where $\Delta G^\circ = 0$. In addition the melting point of a substance is the temperature where liquid and solid have the same free energy.

13. (a) $\Delta S_{vap}^\circ = \dfrac{\Delta H_{vap}^\circ}{T_{vap}} = \dfrac{3.86 \text{ kcal/mol}}{-85.05°C + 273.15 \text{ K}} \times \dfrac{1000 \text{ cal}}{1 \text{ kcal}} \times \dfrac{4.184 \text{ J}}{1 \text{ cal}} = 85.9 \text{ J mol}^{-1} \text{ K}^{-1}$

(b) $\Delta S_{fus}^\circ = \dfrac{\Delta H_{fus}^\circ}{T_{fus}} = \dfrac{27.05 \text{ cal/g}}{97.82°C + 273.15 \text{ K}} \times \dfrac{22.99 \text{ g Na}}{1 \text{ mol Na}} \times \dfrac{4.184 \text{ J}}{1 \text{ cal}} = 7.014 \text{ J mol}^{-1} \text{ K}^{-1}$

14. (a) $K_{eq} = \dfrac{P\{NO_2(g)\}^2}{P\{NO(g)\}^2\,P\{O_2\}} = K_p$ 　　(b) $K_{eq} = P\{SO_2(g)\} = K_p$

(c) $K_{eq} = \dfrac{[H_3O^+][C_2H_3O_2^-]}{[HC_2H_3O_2]} = K_c$ 　　(d) $K_{eq} = P\{H_2O(g)\}\,P\{CO_2(g)\} = K_p$

(e) $K_{eq} = \dfrac{[Mn^{2+}(aq)]\,P\{Cl_2(g)\}}{[H^+]^4\,[Cl^-]^2}$ 　　neither K_p nor K_c

15. $\Delta G^\circ = -RT \ln K_p = -(8.3145 \text{ J mol}^{-1} \text{ K}^{-1})(1000. \text{ K})\left(\dfrac{1 \text{ kJ}}{1000 \text{ J}}\right) \ln(2.45 \times 10^{-7}) = 126.6 \text{ kJ/mol}$

16. $\Delta G^\circ = 2\,\Delta G_f^\circ[NO(g)] - \Delta G_f^\circ[N_2O(g)] - 0.5\,\Delta G_f^\circ[O_2(g)]$

$= 2(86.57 \text{ kJ/mol}) - (104.2 \text{ kJ/mol}) - 0.5(0.00 \text{ kJ/mol}) = 68.9 \text{ kJ/mol}$

$= -RT \ln K_p = -(8.3145 \times 10^{-3} \text{ kJ mol}^{-1} \text{ K}^{-1})(298 \text{ K}) \ln K_p$

$\ln K_p = -\dfrac{68.9 \text{ kJ/mol}}{8.3145 \times 10^{-3} \text{ kJ mol}^{-1} \text{ K}^{-1} \times 298.15 \text{ K}} = -27.8$

$K_p = 8.4 \times 10^{-13}$

17. We first balance the chemical equation in each case, and then calculate the value of $\Delta G°$ with data from Appendix D, and finally calculate the value of K_{eq} with the use of $\Delta G° = -RT \ln K$.

(a) $4\,HCl(g) + O_2(g) \rightleftharpoons 2\,H_2O(g) + 2\,Cl_2(s)$

$\Delta G° = 2\,\Delta G_f°[H_2O(g)] + 2\,\Delta G_f°[Cl_2(g)] - 4\,\Delta G_f°[HCl(g)] - \Delta G_f°[O_2(g)]$

$= 2 \times (-228.6\ kJ/mol) + 2 \times 0.00\ kJ/mol - 4 \times (-95.30\ kJ/mol) - 0.00\ kJ/mol = -76.0\ kJ/mol$

$\ln K_{eq} = \dfrac{-\Delta G°}{RT} = \dfrac{+76.0 \times 10^3\ J/mol}{8.3145\ J\ mol^{-1}\ K^{-1} \times 298\ K} = +30.7 \qquad K_{eq} = 2.2 \times 10^{13}$

(b) $3\,Fe_2O_3(s) + H_2(g) \rightleftharpoons 2\,Fe_3O_4(s) + H_2O(g)$

$\Delta G° = 2\,\Delta G_f°[Fe_3O_4(s)] + \Delta G_f°[H_2O(g)] - 3\,\Delta G_f°[Fe_2O_3(s)] - \Delta G_f°[H_2(g)]$

$= 2 \times (-1015\ kJ/mol) - 228.6\ kJ/mol - 3 \times (-742.2\ kJ/mol) - 0.00\ kJ/mol = -32\ kJ/mol$

$\ln K_{eq} = \dfrac{-\Delta G°}{RT} = \dfrac{32 \times 10^3\ J/mol}{8.3145\ J\ mol^{-1}\ K^{-1} \times 298\ K} = 13 \qquad K_{eq} = 4 \times 10^5$

(c) $2\,Ag^+(aq) + SO_4{}^{2-}(aq) \rightleftharpoons Ag_2SO_4(s)$

$\Delta G° = \Delta G_f°[Ag_2SO_4(s)] - 2\,\Delta G_f°[Ag^+(aq)] - \Delta G_f°[SO_4{}^{2-}(aq)]$

$= -618.5\ kJ/mol - 2 \times 77.12\ kJ/mol - (-744.6\ kJ/mol) = -28.1\ kJ/mol$

$\ln K_{eq} = \dfrac{-\Delta G°}{RT} = \dfrac{28.1 \times 10^3\ J/mol}{8.3145\ J\ mol^{-1}\ K^{-1} \times 298\ K} = 11.3 \qquad K_{eq} = 8.1 \times 10^4$

18. $\Delta S° = S°\{CO_2(g)\} + S°\{H_2(g)\} - S°\{CO(g)\} - S°\{H_2O(g)\}$

$= 213.6\ J\ mol^{-1}\ K^{-1} + 130.6\ J\ mol^{-1}\ K^{-1} - 197.6\ J\ mol^{-1}\ K^{-1} - 188.7\ J\ mol^{-1}\ K^{-1}$

$= -42.1\ J\ mol^{-1}\ K^{-1}$

19. We only compute a value of $\Delta G°$ for each reaction. If this value is significantly less than zero, we conclude that the reaction is expected to occur to some extent at 298.15 K.

(a) $\Delta G° = 2\,\Delta G_f°[O_3(g)] - 3\,\Delta G_f°[O_2(g)] = 2\,(163.2\ kJ/mol) - 3\,(0.00\ kJ/mol)$

$= 326.4\ kJ/mol \qquad$ Does not occur to a significant extent.

(b) $\Delta G° = 2\,\Delta G_f°[NO_2(g)] - \Delta G_f°[N_2O_4(g)] = 2\,(51.30\ kJ/mol) - 97.82\ kJ/mol$

$= 4.78\ kJ/mol \qquad$ Occurs only to a very small extent.

(c) $\Delta G° = 2\,\Delta G_f°[BrCl(g)] - \Delta G_f°[Cl_2(g)] - \Delta G_f°[Br_2(l)]$

$= 2\,(-0.96\ kJ/mol) - 0.00\ kJ/mol - 0.00\ kJ/mol$

$= -1.9\ kJ/mol \qquad$ Occurs to some extent.

We could have predicted the first result by simply looking at the free energy of formation of ozone. The small size of the result in the other two cases indicates that some calculation is necessary.

20. $\Delta G = \Delta H - T\,\Delta S \qquad T\,\Delta S = \Delta H - \Delta G \qquad T = \dfrac{\Delta H - \Delta G}{\Delta S}$

$T = \dfrac{-843.7 \times 10^3\ J - (-777.8 \times 10^3\ J)}{-165\ J/K} = 399\ K$

21. (a) $\Delta G° = 2\,\Delta G_f°[NO_2(g)] - 2\,\Delta G_f°[NO(g)] - \Delta G_f°[O_2(g)]$

$= 2\,(51.30\ kJ/mol) - 2\,(86.57\ kJ/mol) - (0.00\ kJ/mol) = -70.54\ kJ/mol$

(b) $\Delta G° = -R\,T \ln K_p \qquad \ln K_p = -\dfrac{\Delta G°}{R\,T} = -\dfrac{-70.54 \times 10^3\ J/mol}{8.3145\ J\ mol^{-1}\ K^{-1} \times 298.15\ K} = 28.46$

$K_p = e^{28.46} = 2.3 \times 10^{12}$

22. (a) $\Delta S° = S°[Na_2CO_3(s)] + S°[H_2O(l)] + S°[CO_2(g)] - 2\,S°[NaHCO_3(s)]$

$= 135.0\ J\ mol^{-1}\ K^{-1} + 69.91\ J\ mol^{-1}\ K^{-1} + 213.6\ J\ mol^{-1}\ K^{-1} - 2\,(102\ J\ mol^{-1}\ K^{-1})$

$= 215\ J\ mol^{-1}\ K^{-1}$

(b) $\Delta H° = \Delta H_f°[Na_2CO_3(s)] + \Delta H_f°[H_2O(l)] + \Delta H_f°[CO_2(g)] - 2\,\Delta H_f°[NaHCO_3(s)]$

$= -1131\ kJ/mol - 285.8\ kJ/mol - 393.5\ kJ/mol - 2\,(-950.8\ kJ/mol)$

$= +91\ kJ/mol$

(c) $\Delta G° = \Delta H° - T\,\Delta S° = 91\ kJ/mol - (298\ K)(215 \times 10^{-3}\ kJ\ mol^{-1}\ K^{-1})$

$= 91\ kJ/mol - 64.1\ kJ/mol = 27\ kJ/mol$

(d) $\Delta G^\circ = -RT \ln K_{eq}$ $\ln K_{eq} = -\dfrac{\Delta G^\circ}{RT} = -\dfrac{27 \times 10^3 \text{ J/mol}}{8.3145 \text{ J mol}^{-1} \text{ K}^{-1} \times 298 \text{ K}} = -10._9$

$K_{eq} = e^{-10.9} = 2 \times 10^{-5}$

23. (a) $\Delta S^\circ = S^\circ[CH_3CH_2OH(g)] + S^\circ[H_2O(g)] - S^\circ[CO(g)] - 2\,S^\circ[H_2(g)] - S^\circ[CH_3OH(g)]$
   $= 282.6 \text{ J mol}^{-1} \text{ K}^{-1} + 188.7 \text{ J mol}^{-1} \text{ K}^{-1}$
   $\qquad\qquad - 197.6 \text{ J mol}^{-1} \text{ K}^{-1} - 2\,(130.6 \text{ J mol}^{-1} \text{ K}^{-1}) - 239.7 \text{ J mol}^{-1} \text{ K}^{-1}$
   $= -227.2 \text{ J mol}^{-1} \text{ K}^{-1}$

$\Delta H^\circ = \Delta H_f^\circ[CH_3CH_2OH(g)] + \Delta H_f^\circ[H_2O(g)] - \Delta H_f^\circ[CO(g)] - 2\,\Delta H_f^\circ[H_2(g)] - \Delta H_f^\circ[CH_3OH(g)]$
   $= -234.4 \text{ kJ/mol} - 241.8 \text{ kJ/mol} - (-110.5 \text{ kJ/mol}) - 2\,(0.00 \text{ kJ/mol}) - (-200.7 \text{ kJ/mol})$
   $= -165.0 \text{ kJ/mol}$

$\Delta G^\circ = -165.0 \text{ kJ/mol} - (298.15 \text{ K})(-227.2 \times 10^{-3} \text{ kJ mol}^{-1} \text{ K}^{-1}) = -165.0 \text{ kJ/mol} + 67.7 \text{ kJ/mol}$
   $= -97.3 \text{ kJ/mol}$

(b) Since $\Delta H^\circ < 0$ for this reaction, it is favored at low temperatures. And since $\Delta n_{gas} = +2 - 4 = -2$, which is less than zero, the product side of the reaction is favored at high pressures.

(c) We assume that neither ΔS° nor ΔH° varies significantly with temperature. Then we compute a value of ΔG° at 750. K From this value of ΔG°, we compute a value of K_p.

$\Delta G^\circ = \Delta H^\circ - T\Delta S^\circ = -165.0 \text{ kJ/mol} - (750.\text{ K})(-227.2 \times 10^{-3} \text{ kJ mol}^{-1} \text{ K}^{-1})$
   $= -165.0 \text{ kJ/mol} + 170.\text{ kJ/mol} = +5 \text{ kJ/mol} = -RT \ln K_p$

$\ln K_p = -\dfrac{\Delta G^\circ}{RT} = -\dfrac{5 \times 10^3 \text{ J/mol}}{8.3145 \text{ J mol}^{-1} \text{ K}^{-1} \times 750.\text{ K}} = -0.8$ $\qquad K_p = e^{-0.8} = 0.45$

24. We use the van't Hoff equation with $\Delta H^\circ = -1.8 \times 10^5$ J/mol, $T_1 = 800.$ K, $T_2 = 100.°C = 373$ K, and $K_1 = 9.1 \times 10^2$.

$\ln \dfrac{K_2}{K_1} = \dfrac{\Delta H^\circ}{R}\left(\dfrac{1}{T_1} - \dfrac{1}{T_2}\right) = \dfrac{-1.8 \times 10^5 \text{ J/mol}}{8.3145 \text{ J mol}^{-1} \text{ K}^{-1}}\left(\dfrac{1}{800 \text{ K}} - \dfrac{1}{373 \text{ K}}\right) = 31$

$\dfrac{K_2}{K_1} = e^{31} = 3 \times 10^{13} = \dfrac{K_2}{9.1 \times 10^2}$ $\qquad K_2 = (3 \times 10^{13})(9.1 \times 10^2) = 3 \times 10^{16}$

25. Answer **(b)** is incorrect. For a reaction, such as this one, that involves only gases, $K_{eq} = K_p$. And this equation combined with $\Delta G^\circ = -RT \ln K_{eq}$ produces $K_p = \text{antiln}(-\Delta G^\circ/RT) = e^{-\Delta G/RT}$. Finally, $\Delta G = \Delta G^\circ + RT \ln Q$ is a general quation, valid for all reactions. But $K_{eq} \neq K_c$ for a reaction involving gases (unless $\Delta n_{gas} = 0$ for the reaction); $K_{eq} = K_c$ is true only for those reactions in which only the concentrations of solutes in aqueous solutions appear in the equilibrium constant expression.

EXERCISES

Spontaneous Change, Entropy, and Disorder

26. (a) The freezing of ethanol involves a *decrease* in the entropy of the system, since solids are more ordered than liquids, in general.

(b) The sublimation of dry ice involves converting a quite ordered solid into a very disordered vapor. Thus the entropy of the system *increases* substantially.

(c) The burning of rocket fuel involves converting a somewhat ordered liquid fuel into the highly disordered mixture of the gaseous combustion products. The entropy of the system *increases* substantially.

27. (a) Positive   High entropy gas is produced from two solids.

(b) Negative   A low entropy solid is produced from moderate entropy liquid and a high entropy gas.

(c) Positive   Three moles of high entropy gas are produced from two moles of gas.

(d) Uncertain   There are the same number of moles of gaseous products as of gaseous reactants.

(e) Negative   A liquid of small entropy is produced from a high entropy gas.

28. Although there is a substantial change in entropy involved in (a) changing $H_2O(l, 1 \text{ atm})$ to $H_2O(g, 1 \text{ atm})$, it is not as large as (c) converting the liquid to a gas at 10 mmHg. The gas is more ordered at higher pressures. In turn, (b) if we start with a solid and convert it to a gas at the lower pressure, the entropy change should be even larger, since a solid is more ordered than a liquid. Thus, in order of increasing ΔS, the processes are: (a) < (c) < (b).

29. The first law of thermodynamics states that energy is neither created nor destroyed (thus, "The energy of the world is constant.") A consequence of the second law of thermodynamics is that entropy increases for all spontaneous—that is, naturally occurring—processes (and therefore, "the entropy of the world increases toward a maximum.")

30. When environmental pollutants are produced they are dispersed throughout the environment. These pollutants thus start in a relatively compact form and end up spread throughout a large volume. In this large volume, because they are mixed with many other substances, they are highly disordered and thus have a high entropy. Returning them to their original compact form requires reducing this entropy, a highly nonspontaneous process. If we have had enough foresight to retain these pollutants in a reasonably compact form—such as disposing of them in a *secure* landfill, rather than dispersing them in the atmosphere or in rivers and seas—the task of permanently removing them from the environment, and perhaps even converting them to useful forms, will be considerably easier.

31. The entropy of formation of a compound would be the difference between the absolute entropy of one mole of the compound and the sum of the absolute entropies of the appropriate amounts of the elements constituting the compound, each in their most stable forms. It seems as though $CS_2(l)$ would have the highest molar entropy of formation of the compounds listed, since it is the only substance whose formation does not involve the consumption of high entropy gaseous reactants.

(a) $C(\text{graphite}) + 2\,H_2(g) \rightleftharpoons CH_4(g)$

$\Delta S_f^\circ[CH_4(g)] = S^\circ[CH_4(g)] - S^\circ[C(\text{graphite})] - 2\,S^\circ[H_2(g)]$

$\qquad = 186.2\ \text{J mol}^{-1}\,\text{K}^{-1} - 5.74\ \text{J mol}^{-1}\,\text{K}^{-1} - 2 \times 130.6\ \text{J mol}^{-1}\,\text{K}^{-1} = -80.7\ \text{J mol}^{-1}\,\text{K}^{-1}$

(b) $2\,C(\text{graphite}) + 3\,H_2(g) + \frac{1}{2}O_2(g) \rightleftharpoons C_2H_5OH(l)$

$\Delta S_f^\circ[C_2H_5OH(l)] = S^\circ[C_2H_5OH(l)] - 2\,S^\circ[C(\text{graphite})] - 3\,S^\circ[H_2(g)] - \frac{1}{2}S^\circ[O_2(g)]$

$\qquad = 160.7\ \text{J mol}^{-1}\,\text{K}^{-1} - 2 \times 5.74\ \text{J mol}^{-1}\,\text{K}^{-1} - 3 \times 130.6\ \text{J mol}^{-1}\,\text{K}^{-1} - \frac{1}{2} \times 205.1\ \text{J mol}^{-1}\,\text{K}^{-1}$

$\qquad = -345.1\ \text{J mol}^{-1}\,\text{K}^{-1}$

(c) $C(\text{graphite}) + 2\,S(\text{rhombic}) \rightleftharpoons CS_2(l)$

$\Delta S_f^\circ[CS_2(g)] = S^\circ[CS_2(l)] - S^\circ[C(\text{graphite})] - 2\,S^\circ[S(\text{rhombic})]$

$\qquad = 151.3\ \text{J mol}^{-1}\,\text{K}^{-1} - 5.74\ \text{J mol}^{-1}\,\text{K}^{-1} - 2 \times 31.80\ \text{J mol}^{-1}\,\text{K}^{-1} = 82.0\ \text{J mol}^{-1}\,\text{K}^{-1}$

Phase Transitions

32. (a) $\Delta H_{\text{vap}}^\circ = \Delta H_f^\circ[H_2O(g)] - \Delta H_f^\circ[H_2O(l)] = -241.8\ \text{kJ/mol} - (-285.8\ \text{kJ/mol}) = +44.0\ \text{kJ/mol}$

$\Delta S_{\text{vap}}^\circ = S^\circ[H_2O(g)] - S^\circ[H_2O(l)] = 188.7\ \text{J mol}^{-1}\,\text{K}^{-1} - 69.9\ \text{J mol}^{-1}\,\text{K}^{-1} = 118.8\ \text{J mol}^{-1}\,\text{K}^{-1}$

There is an alternate, but incorrect, method of obtaining $\Delta S_{\text{vap}}^\circ$.

$$\Delta S_{\text{vap}}^\circ = \frac{\Delta H_{\text{vap}}^\circ}{T} = \frac{44.03 \times 10^3\ \text{J/mol}}{298.15\ \text{K}} = 147.7\ \text{J mol}^{-1}\,\text{K}^{-1}$$

This method is invalid because the temperature in the denominator of the equation must be the temperature at which the transition is an equilibrium. Liquid water and water vapor at 1 atm pressure (standard state, indicated by °) are in equilibrium only at 100° C = 373 K.

(b) The reason why $\Delta H_{\text{vap}}^\circ$ is different from its value at 100°C has to do with the heat required to bring the reactants and products down to 298 K from 373 K. The specific heat of liquid water is higher than the heat capacity of steam. Thus, more heat is given off (this is negative heat, an exothermic process) by lowering the temperature of the liquid water from 100°C to 25°C than is given off by lowering the temperature of the same amount of steam. Another way to think of this is that hydrogen bonding is more disrupted in water at 100° than at 25° (because the molecules are in rapid—thermal—motion), and hence there is not as much energy required to convert liquid to vapor.

The reason why $\Delta S_{\text{vap}}^\circ$ has a larger value at 25°C than at 100°C has to do with disorder. A vapor at 1 atm pressure (the case at both temperatures) has about the same entropy. On the other hand, liquid water is more disordered at higher temperatures since more of the hydrogen bonds are disrupted by thermal motion. (The hydrogen bonds are totally disrupted in the two vapors).

33. Trouton's rule is obeyed most closely for liquids in which there is not a high degree of order within the liquid. In both HF and CH_3OH hydrogen bonds create considerable order within the liquid. In $C_6H_5CH_3$ the only attractive forces are non-directional London forces, which cause the molecules to attract each other, but have no preferred orientation as hydrogen bonds do. Thus, liquid $C_6H_5CH_3$ is not a particularly ordered liquid.

34. $\Delta H_{vap}^\circ = \Delta H_f^\circ[Br_2(g)] - \Delta H_f^\circ[Br_2(l)] = 30.91 \text{ kJ/mol} - 0.00 \text{ kJ/mol} = 30.91 \text{ kJ/mol}$

$\Delta S_{vap}^\circ = \dfrac{\Delta H_{vap}^\circ}{T_{vap}} \approx 88 \text{ J mol}^{-1} \text{ K}^{-1}$ or $T_{vap} = \dfrac{\Delta H_{vap}^\circ}{\Delta S_{vap}^\circ} \approx \dfrac{30.91 \times 10^3 \text{ J/mol}}{88 \text{ J mol}^{-1} \text{ K}^{-1}} = 3.5 \times 10^2 \text{ K}$

The accepted value of the boiling point of bromine is 58.8°C = 332 K = 3.32×10^2 K. Thus, our estimate is in reasonable agreement with the measured value.

35. (a) We use Trouton's rule ($\Delta S_{vap}^\circ \approx 88 \text{ J mol}^{-1} \text{ K}^{-1}$) to estimate the normal boiling point.

$T_{nbp} = \dfrac{\Delta H_{vap}^\circ}{\Delta S_{vap}^\circ} = \dfrac{[-77.2 \text{ kJ/mol} - (-105.9)] \times \dfrac{1000 \text{ J}}{1 \text{ kJ}}}{88 \text{ J mol}^{-1} \text{ K}^{-1}} = 326 \text{ K}$

(b) $\Delta G_{vap}^\circ = \Delta H_{vap}^\circ - T \Delta S_{vap}^\circ$

$= [-77.2 - (-105.9)] \text{ kJ/mol} - \dfrac{88 \text{ J mol}^{-1} \text{ K}^{-1} \times 298 \text{ K}}{\dfrac{1000 \text{ J}}{1 \text{ kJ}}} = 28.7 \text{ kJ/mol} - 26.2 \text{ kJ/mol}$

$= 2.5 \text{ kJ/mol}$

(c) A positive value of ΔG_{vap}° indicates that normal boiling (having a vapor pressure of 1.00 atm) is nonspontaneous (will not occur) at 298 K for cyclopentane. The vapor pressure of cyclopentane at 298 K is less than 1.00 atm.

Free Energy and Spontaneous Change

36. (a) $\Delta H^\circ < 0$ and $\Delta S^\circ < 0$ (since $\Delta n_{gas} < 0$) for this reaction. Thus, this reaction is case 2 of Table 20-1. It is spontaneous at low temperatures and nonspontaneous at high temperatures.

(b) We are unable to predict the sign of ΔS° for this reaction, since $\Delta n_{gas} = 0$. Thus, no prediction as to the temperature behavior of this reaction can be made.

(c) $\Delta H^\circ > 0$ and $\Delta S^\circ > 0$ (since $\Delta n_{gas} > 0$) for this reaction. This is case 3 of Table 20-1. It is nonspontaneous at low temperatures, but spontaneous at high temperatures.

(d) $\Delta H^\circ > 0$ and $\Delta S^\circ < 0$ (since $\Delta n_{gas} < 0$) for this reaction. This is case 4 of Table 20-1. It is nonspontaneous at all temperatures.

(e) $\Delta H^\circ < 0$ and $\Delta S^\circ > 0$ (since $\Delta n_{gas} > 0$) for this reaction. This is case 1 of Table 20-1. It is spontaneous at all temperatures.

37. First of all, the process is clearly spontaneous, and therefore $\Delta G < 0$. In addition, the gases are more disordered when they are at a lower pressure and therefore $\Delta S > 0$. We also conclude that $\Delta H = 0$, since the gases are ideal and thus there are no forces of attraction or repulsion between them.

38. Since an ideal solution forms spontaneously, $\Delta G < 0$. Also, the molecules of solvent and solvent are mixed together in the solution, a more disordered state than the separated solvent and solute. Therefore, $\Delta S > 0$. However, in an ideal solution, the attractive forces between solvent and solute molecules equals those forces between solvent molecules and those between solute molecules. Thus, $\Delta H = 0$.

39. (a) An exothermic reaction (one that gives off heat) may not occur spontaneously if, at the same time, the system becomes more ordered, that is , $\Delta S^\circ < 0$. This is particularly true at a high temp[erature, where the $T\Delta S$ term dominates the ΔG expression. An example of such a process is the freezing of water (clearly exothermic since the reverse process, melting ice, is endothermic) at temperatures above 0°C.

(b) A reaction in which $\Delta S > 0$ need not be spontaneous if that process also is endothermic. This is particularly true at low temperatures, where the ΔH term dominates the ΔG expression. An example is the vaporization of water (clearly an endothermic process, one that requires heat, and one that produces a gas, so $\Delta S > 0$) at low temperatures, that is, below 100°C.

40. Since this reaction produces more moles of gas than it consumes, $\Delta S > 0$. The reaction also is endothermic, since energy is required to break the A—B bond. Hence $\Delta H > 0$. Therefore, this reaction is of case 3 in

Table 20-1. It is nonspontaneous at low temperatures, but eventually becomes spontaneous as the temperature is raised.

41. $\Delta G°$ is directly related to the equilibrium constant of a chemical reaction through $\Delta G° = -RT \ln K$. And the equilibrium constant can, of course, be used to determine the relative proportions of reactants and products at equilibrium. At equilibrium ΔG (without the °) is equal to zero (that is one definition of equilibrium). The size of $\Delta G°$ is an indication of how far the equilibrium conditions differ from standard conditions.

Standard Free Energy Change

42. $\Delta S° = 2\, S°[POCl_3(l)] - 2\, S°[PCl_3(g)] - S°[O_2(g)]$
  $= 2\,(222 \text{ J/K}) - 2\,(312 \text{ J/K}) - 205 \text{ J/K} = -385 \text{ J/K}$

$\Delta G° = \Delta H° - T\,\Delta S° = -555 \times 10^3 \text{ J} - (298 \text{ K})(-385 \text{ J/K}) = -440. \times 10^3 \text{ J} = -440.\text{ kJ}$

43. **(a)** $\Delta G° = \Delta G_f°[N_2H_4(g)] + \Delta G_f°[H_2(g)] - 2\,\Delta G_f°[NH_3(g)]$
  $= 159.3 \text{ kJ/mol} + 0.00 \text{ kJ/mol} - 2\,(-16.48 \text{ kJ/mol}) = +192.3 \text{ kJ/mol}$

 (b) We calculate the enthalpy change for the reaction.
  $\Delta H° = \Delta H_f°[N_2H_4(g)] + \Delta H_f°[H_2(g)] - 2\,\Delta H_f°[NH_3(g)]$
  $= 95.4 \text{ kJ/mol} + 0.00 \text{ kJ/mol} - 2\,(-46.11 \text{ kJ/mol}) = +187.6 \text{ kJ/mol}$

  Since this is an endothermic reaction, it will be favored at higher temperatures.

44. We combine the reactions in the same way as for Hess's law calculations.

 (a) $N_2O(g) \longrightarrow N_2(g) + \frac{1}{2}O_2(g)$ $\Delta G° = \frac{1}{2}(-208.4 \text{ kJ}) = -104.2 \text{ kJ}$

   $N_2(g) + 2\,O_2(g) \longrightarrow 2\,NO_2(g)$ $\Delta G° = +102.6 \text{ kJ}$

  Net: $N_2O(g) + \frac{3}{2}O_2(g) \longrightarrow 2\,NO_2(g)$ $\Delta G° = -104.2 + 102.6 = -1.6 \text{ kJ}$

  This reaction reaches an equilibrium condition, a conclusion we reach based on the relatively small absolute value of $\Delta G°$.

 (b) $2\,N_2(g) + 6\,H_2(g) \longrightarrow 4\,NH_3(g)$ $\Delta G° = 2\,(-33.0 \text{ kJ}) = -66.0 \text{ kJ}$

   $4\,NH_3(g) + 5\,O_2(g) \longrightarrow 4\,NO(g) + 6\,H_2O(l)$ $\Delta G° = -1011 \text{ kJ}$

   $4\,NO(g) \longrightarrow 2\,N_2(g) + 2\,O_2(g)$ $\Delta G° = -2\,(+173.1 \text{ kJ}) = -346.2 \text{ kJ}$

  Net: $6\,H_2(g) + 3\,O_2(g) \longrightarrow 6\,H_2O(l)$ $\Delta G° = -66.0 \text{ kJ} - 1011 \text{ kJ} - 346.2 \text{ kJ} = -1423 \text{ kJ}$

  This reaction is three times the desired reaction, which therefore has $\Delta G° = -1423 \text{ kJ} \div 3 = -474.3 \text{ kJ}$ The quite large negative value of this value of $\Delta G°$ indicates that this reaction would tend to go to completion at 25°C.

 (c) $4\,NH_3(g) + 5\,O_2(g) \longrightarrow 4\,NO(g) + 6\,H_2O(l)$ $\Delta G° = -1011 \text{ kJ}$

   $4\,NO(g) \longrightarrow 2\,N_2(g) + 2\,O_2(g)$ $\Delta G° = -2\,(+173.1 \text{ kJ}) = -346.2 \text{ kJ}$

   $2\,N_2(g) + O_2(g) \longrightarrow 2\,N_2O(g)$ $\Delta G° = +208.4 \text{ kJ}$

  Net: $4\,NH_3(g) + 4\,O_2(g) \longrightarrow 2\,N_2O(g) + 6\,H_2O(l)$ $\Delta G° = -1011 \text{ kJ} - 346.2 \text{ kJ} + 208.4 \text{ kJ} = -1149 \text{ kJ}$

  This reaction is twice the desired reaction, which therefore has $\Delta G° = -1149 \text{ kJ} \div 2 = -574.5 \text{ kJ}$ The very large negative value of $\Delta G°$ for this reaction indicates that it will go to completion.

45. For the vaporization of water: $H_2O(l) \longrightarrow H_2O(g)$

$\Delta G° = \Delta G_f°[H_2O(g)] - \Delta G_f°[H_2O(l)] = -228.6 \text{ kJ} - (-237.2 \text{ kJ}) = +8.6 \text{ kJ}$

  $C_8H_{18}(l) + \frac{25}{2}O_2(g) \longrightarrow 8\,CO_2(g) + 9\,H_2O(l)$ $\Delta G° = -5.28 \times 10^3 \text{ kJ}$

  $9\,H_2O(l) \longrightarrow 9\,H_2O(g)$ $\Delta G° = 9\,(8.6 \text{ kJ}) = +77.4 \text{ kJ}$

Net: $C_8H_{18}(l) + \frac{25}{2}O_2(g) \longrightarrow 8\,CO_2(g) + 9\,H_2O(l)$ $\Delta G° = -5.28 \times 10^3 \text{ kJ} + 77.4 \text{ kJ} = -5.20 \times 10^3 \text{ kJ}$

46. **(a)** We compute $\Delta G°$ for the given reaction.

  $\Delta H° = \Delta H_f°[TiCl_4(l)] + \Delta H_f°[O_2(g)] - \Delta H_f°[TiO_2(s)] - 2\,\Delta H_f°[Cl_2(g)]$
  $= -804.2 \text{ kJ/mol} + 0.00 \text{ kJ/mol} - (-944.7 \text{ kJ/mol}) - 2\,(0.00 \text{ kJ/mol}) = +140.5 \text{ kJ/mol}$

  $\Delta S° = S°[TiCl_4(l)] + S°[O_2(g)] - S°[TiO_2(s)] - 2\,S°[Cl_2(g)]$
  $= 252.3 \text{ J mol}^{-1} \text{ K}^{-1} + 205.1 \text{ J mol}^{-1} \text{ K}^{-1} - (50.3 \text{ J mol}^{-1} \text{ K}^{-1}) - 2\,(223.0 \text{ J mol}^{-1} \text{ K}^{-1})$
  $= -38.9 \text{ J mol}^{-1} \text{ K}^{-1}$

$\Delta G° = \Delta H° - T \Delta S° = +140.5$ kJ/mol $- (298.2$ K$)(-38.9 \times 10^{-3}$ kJ mol^{-1} K$^{-1})$
$= +140.5$ kJ/mol $+ 11.6$ kJ/mol $= +152.1$ kJ/mol
This reaction is nonspontaneous at 25°C.

(b) For the cited reaction
$\Delta G° = 2 \Delta G_f°[CO_2(g)] - 2 \Delta G_f°[CO(g)] - \Delta G_f°[O_2(g)]$
$= 2 (-394.4$ kJ/mol$) - 2 (-137.2$ kJ/mol$) - 0.00$ kJ/mol $= -514.4$ kJ/mol
Then we couple the two reactions.

$TiO_2(s) + 2 Cl_2(g) \longrightarrow TiCl_4(l) + O_2(g)$ $\qquad \Delta G° = +152.1$ kJ/mol
$2 CO(g) + O_2(g) \longrightarrow 2 CO_2(g)$ $\qquad \Delta G° = -514.4$ kJ/mol

Net: $TiO_2(s) + 2 Cl_2(g) + 2 CO(g) \longrightarrow TiCl_4(l) + 2 CO_2(g)$ $\;\; \Delta G° = +152.1$ kJ/mol $- 514.4$ kJ/mol
$= -362.3$ kJ/mol

The coupled reaction has $\Delta G° < 0$ and therefore is spontaneous.

47. (a) $NiO(s) \longrightarrow Ni(s) + \frac{1}{2} O_2(g)$ $\qquad \Delta G° = +115$ kJ
$C(s) + \frac{1}{2} O_2(g) \longrightarrow CO(g)$ $\qquad \Delta G° = -250$ kJ
Net: $NiO(s) + C(s) \longrightarrow Ni(s) + CO(g)$ $\qquad \Delta G° = +115$ kJ $- 250$ kJ $= -135$ kJ $\qquad$ Spontaneous

(b) $MnO(s) \longrightarrow Mn(s) + \frac{1}{2} O_2(g)$ $\qquad \Delta G° = +280$ kJ
$C(s) + \frac{1}{2} O_2(g) \longrightarrow CO(g)$ $\qquad \Delta G° = -250$ kJ
Net: $MnO(s) + C(s) \longrightarrow Mn(s) + CO(g)$ $\qquad \Delta G° = +280$ kJ $- 250$ kJ $= +30$ kJ $\qquad$ Nonspontaneous

(c) $TiO_2(s) \longrightarrow Ti(s) + O_2(g)$ $\qquad \Delta G° = +630$ kJ
$2 C(s) + O_2(g) \longrightarrow 2 CO(g)$ $\qquad \Delta G° = 2 (-250$ kJ$) = -500$ kJ
Net: $TiO_2(s) + 2 C(s) \longrightarrow Ti(s) + 2 CO(g)$ $\;\; \Delta G° = +630$ kJ $- 500$ kJ $= +130$ kJ $\qquad$ Nonspontaneous

Free Energy Change and Equilibrium

48. The equilibrium between liquid water and water vapor under 0.50 atm pressure can be established at only one temperature: the point where the 0.50 atm line crosses the liquid-vapor curve in the phase diagram of water. We can determine this temperature by solving the Clausius-Clapeyron equation for T_2, with $T_1 = 373.15$ K, $P_1 = 1.00$ atm, $P_2 = 0.50$ atm, and $\Delta H_{vap} = 44.0 \times 10^3$ J/mol (the last value from Example 12-4).

$\ln \frac{P_2}{P_1} = \frac{\Delta H_{vap}}{R} \left(\frac{1}{T_1} - \frac{1}{T_2} \right) = \ln \frac{0.50 \text{ atm}}{1.00 \text{ atm}} = -0.693 = \frac{44.0 \times 10^3 \text{ J/mol}}{8.3145 \text{ J mol}^{-1} \text{ K}^{-1}} \left(\frac{1}{373.15} - \frac{1}{T_2} \right)$

$-0.69 = 5.29 \times 10^3 \text{ K} \left(\frac{1}{373.15} - \frac{1}{T_2} \right)$ $\qquad \left(\frac{1}{373.15 \text{ K}} - \frac{1}{T_2} \right) = \frac{-0.693}{5.29 \times 10^3 \text{ K}} = -1.31 \times 10^{-4} \text{ K}^{-1}$

$\frac{1}{T_2} = \frac{1}{373.15 \text{ K}} + 1.31 \times 10^{-4} \text{ K}^{-1} = 2.811 \times 10^{-3} \text{ K}^{-1}$ $\qquad T_2 = 355.8 \text{ K} = 82.6°C$

An easier way to estimate this temperature is to consult Table 13-2, recognizing that 0.50 atm = 360 mmHg. We see that the temperature where this equilibrium is established is approximately 80.°C

49. The state point of 110°C and 1 atm pressure is in the region labeled "solid" in the phase diagram of iodine. Thus the state of I_2 that exists at 110°C and 1 atm pressure is solid iodine. Thus, $I_2(s)$ has a lower free energy under these conditions than does $I_2(l)$, since the reaction $I_2(s) \longrightarrow I_2(l)$ is nonspontaneous; $\Delta G < 0$.

50. The state point of 1 atm and –60°C is in the region labeled "gas" in the phase diagram of carbon dioxide. Thus the state of CO_2 that exists at 1 atm and –60°C is gaseous CO_2. Thus, of the three states of carbon dioxide $CO_2(g)$ has the lowest free energy under these conditions.

The Thermodynamic Equilibrium Constant

51. The thermodynamic equilibrium constant is used in the expression $\Delta G° = - R T \ln K_{eq}$ rather than K_c because only the thermodynamic equilibrium constant expresses the deviation of the equilibrium conditions from the standard state (designated by °). For substances dissolved in aqueous solution this standard state is very close to 1.00 M concentration. However, for gases, the standard state is not 1.00 M concentration but rather 1.00 atm pressure. (This is approximately 0.0409 M for an ideal gas at 298.15 K). K_c may be used in place of K_{eq} when $K_c = K_{eq}$, that is, when all of the factors in the thermodynamic equilibrium constant expression are molar concentrations. Likewise K_p may be used in place of K_{eq} when all of the factors in the thermodyanic equilibrium constant expression are pressures.

52. (a) $K_{eq} = \dfrac{P\{H_2(g)\}^4}{P\{H_2O(g)\}^4}$

(b) Both solids—Fe(s) and Fe_3O_4(s)—are properly excluded from the thermodynamic equilibrium constant expression. (Actually, each solid has an activity of 1.00.) Thus the equilibrium partial pressures of both H_2(g) and H_2O(g) do not depend on the amounts of the two solids present, as long as some of each solid is present. One way to understand this, is that any chemical reaction occurs on the surface of the solids, and thus is unaffected by their volume.

(c) We can produce H_2(g) from H_2O(g) without regard to the proportions of Fe(s) and Fe_3O_4(s) with one qualification, of course. There must always be some Fe(s) present for the production of H_2(g) to continue.

Relationships Involving ΔG, $\Delta G°$, Q, and K

53. (a) To determine K_p we need the equilibrium partial pressures. For the ideal gas law, for each partial pressure, $P = nRT/V$. Since R, T, and V are the same for each gas, and since there are the same number of partial pressure factors in the numerator as in the denominator of the K_p expression, we can use the ratio of amounts to determine K_p.

$$K_p = \frac{P[CO(g)]\,P[H_2O(g)]}{P[CO_2(g)]\,P[H_2(g)]} = \frac{n[CO(g)]\,n[H_2O(g)]}{n[CO_2(g)]\,n[H_2(g)]} = \frac{0.224\ \text{mol CO} \times 0.224\ \text{mol } H_2O}{0.276\ \text{mol } H_2 \times 0.276\ \text{mol } CO_2} = 0.659$$

(b) $\Delta G° = -RT \ln K_p = -8.3145\ \text{J mol}^{-1}\ \text{K}^{-1} \times 1000.\ \text{K} \times \ln(0.659) = 3.47 \times 10^3\ \text{J/mol} = 3.47\ \text{kJ/mol}$

(c) $Q_p = \dfrac{0.0400\ \text{mol CO} \times 0.0850\ \text{mol } H_2O}{0.0500\ \text{mol } CO_2 \times 0.070\ \text{mol } H_2} = 0.971 > 0.659 = K_p$

Since Q_p is larger than K_p, the reaction will proceed to the left, forming reactants, in attaining equilibrium.

54. (a) We know that $K_p = K_c\,(RT)^{\Delta n}$ For the reaction $2\ SO_2(g) + O_2(g) \rightleftharpoons 2\ SO_3(g)$, $\Delta n_{gas} = 2 - (2 + 1) = -1$, and therefore a value of K_p can be obtained.

$$K_p = K_c\,(RT)^{-1} = \frac{2.8 \times 10^2}{\dfrac{0.08206\ \text{L atm}}{\text{mol K}} \times 1000\ \text{K}} = 3.41 = K_{eq}$$

We recognize that $K_{eq} = K_p$ since all of the substances involved in the reaction are gases. We can now evaluate $\Delta G°$.

$$\Delta G° = -RT \ln K_{eq} = -\frac{8.3145\ \text{J}}{\text{mol K}} \times 1000\ \text{K} \times \ln(3.41) = -10.2 \times 10^3\ \text{J/mol}$$

(b) We can evaluate Q_c for this situation and compare the value with that of $K_c = 2.8 \times 10^2$.

$$Q_c = \frac{[SO_3]^2}{[SO_2]^2[O_2]} = \frac{\left(\dfrac{0.72\ \text{mol } SO_3}{2.50\ \text{L}}\right)^2}{\left(\dfrac{0.40\ \text{mol } SO_2}{2.50\ \text{L}}\right)^2 \times \dfrac{0.18\ \text{mol } O_2}{2.50\ \text{L}}} = 45 < 2.8 \times 10^2 = K_c$$

Since Q_c is smaller than K_c the reaction will shift right, products will form, until the two values are equal.

55. (a) $K_{eq} = K_c$ $\Delta G° = -RT \ln K_{eq} = -(8.3145 \times 10^{-3}\ \text{kJ mol}^{-1}\ \text{K}^{-1})(445 + 273) \ln 50.2 = -23.4\ \text{kJ}$

(b) $K_{eq} = K_p = K_c(RT)^{\Delta n} = 1.7 \times 10^{-13}\,(0.0821 \times 298)^{1/2} = 8.4 \times 10^{-13}$

$\Delta G° = -RT \ln K_p = -(8.3145 \times 10^{-3}\ \text{kJ mol}^{-1}\ \text{K}^{-1})(298.15\ \text{K}) \ln(8.4 \times 10^{-13}) = +68.92\ \text{kJ/mol}$

(c) $K_{eq} = K_p = K_c(RT)^{\Delta n} = 4.61 \times 10^{-3}\,(0.08206 \times 298.15)^{+1} = 0.113$

$\Delta G° = -RT \ln K_p = -(8.3145 \times 10^{-3}\ \text{kJ mol}^{-1}\ \text{K}^{-1})(298.15\ \text{K}) \ln(0.113) = +5.40\ \text{kJ/mol}$

(d) $K_{eq} = K_c = 9.14 \times 10^{-6}$

$\Delta G° = -RT \ln K_c = -(8.3145 \times 10^{-3}\ \text{kJ mol}^{-1}\ \text{K}^{-1})(298.15\ \text{K}) \ln(9.14 \times 10^{-6}) = +28.75\ \text{kJ/mol}$

56. $\Delta G° = -RT \ln K_p = -(8.3145 \times 10^{-3}\ \text{kJ mol}^{-1}\ \text{K}^{-1})(298.15\ \text{K}) \ln(1.6 \times 10^{12}) = -69.7\ \text{kJ/mol}$

$$CO(g) + Cl_2(g) \longrightarrow COCl_2(g) \qquad\qquad \Delta G° = -69.7 \text{ kJ/mol}$$

$$\tfrac{1}{2}\, C(\text{graphite}) + \tfrac{1}{2}\, O_2(g) \longrightarrow CO(g) \qquad\qquad \Delta G_f° = -137.2 \text{ kJ/mol}$$

Net: $\tfrac{1}{2}\, C(\text{graphite}) + \tfrac{1}{2}\, O_2(g) + Cl_2(g) \longrightarrow COCl_2(g) \qquad \Delta G_f° = -206.9 \text{ kJ/mol}$

The standard free energy of formation of $COCl_2(g)$ given in Appendix D is -206.8 kJ/mol, pretty good agreement.

57. (a) The first equation involves the formation of one mole of $Mg^{2+}(aq)$ from $Mg(OH)_2(s)$ and 2 $H^+(aq)$, while the second equation involves the formation of only one-half mole of $Mg^{2+}(aq)$. We would expect only half as large a free energy change if only half the product is formed.

(b) The value of K_{eq} for the first reaction will be the square of the value of K_{eq} for the second reaction. This is the same way in which the equilibrium constant expressions are related.

$$K_1 = \frac{[Mg^{2+}]}{[H^+]^2} = \left(\frac{[Mg^{2+}]^{1/2}}{[H^+]}\right)^2 = (K_2)^2$$

(c) The equilibrium solubilities will be the same no matter which of the two expressions is used. The equilibrium conditions (solubilities in this instance) are the same no matter how we choose to express them in an equilibrium constant expression.

58. (a) We can determine the equilibrium partial pressure from the value of the equilibrium constant.

$$\Delta G° = -RT \ln K_p \qquad \ln K_p = -\frac{\Delta G°}{RT} = -\frac{58.56 \times 10^3 \text{ J/mol}}{8.3145 \text{ J mol}^{-1}\text{ K}^{-1} \times 298 \text{ K}} = -23.64$$

$$K_p = P\{O_2(g)\}^{1/2} = e^{-23.64} = 5.4 \times 10^{-11} \qquad P\{O_2(g)\} = (5.4 \times 10^{-11})^2 = 2.9 \times 10^{-21} \text{ atm}$$

(b) Lavoisier did two things to increase the quantity of oxygen that he obtained. First, he ran the reaction at a high temperature, which shifts the equilibrium to the right for this endothermic reaction which also has $\Delta S > 0$. Second, the oxygen was continuously removed from the vicinity of the $Hg(l)$ as it was formed, promoting formation of product (Le Châtelier's principle).

59. In each case, we determine the value of $\Delta G°$ for the solubility reaction. From that, we determine the value of the equilibrium constant, K_{sp}, for that solubility reaction.

(a) $AgCl(s) \rightleftharpoons Ag^+(aq) + Cl^-(aq)$

$\Delta G° = \Delta G_f°[Ag^+(aq)] + \Delta G_f°[Cl^-(aq)] - \Delta G_f°[AgCl(s)]$

$\quad = 77.12 \text{ kJ/mol} - 132.3 \text{ kJ/mol} - (-109.8 \text{ kJ/mol}) = +220.1 \text{ kJ/mol}$

$$\ln K_{eq} = \frac{-\Delta G°}{RT} = \frac{-54.6 \times 10^3 \text{ J/mol}}{8.3145 \text{ J mol}^{-1}\text{ K}^{-1} \times 298.15 \text{ K}} = -22.0 \qquad K_{sp} = e^{-22.0} = 2.8 \times 10^{-10}$$

(b) $Ag_2SO_4(s) \rightleftharpoons 2\, Ag^+(aq) + SO_4^{2-}(aq)$

$\Delta G° = 2\, \Delta G_f°[Ag^+(aq)] + \Delta G_f°[SO_4^{2-}(aq)] - \Delta G_f°[Ag_2SO_4^{2-}(s)]$

$\quad = 2 \times 77.12 \text{ kJ/mol} - 744.6 \text{ kJ/mol} - (-618.5 \text{ kJ/mol}) = +28.14 \text{ kJ/mol}$

$$\ln K_{eq} = \frac{-\Delta G°}{RT} = \frac{-28.14 \times 10^3 \text{ J/mol}}{8.3145 \text{ J mol}^{-1}\text{ K}^{-1} \times 298.15 \text{ K}} = -11.35 \qquad K_{sp} = e^{-11.35} = 1.2 \times 10^{-5}$$

(c) $Fe(OH)_3(s) \rightleftharpoons Fe^{3+}(aq) + 3\, OH^-(aq)$

$\Delta G° = \Delta G_f°[Fe^{3+}(aq)] + 3\, \Delta G_f°[OH^-(aq)] - \Delta G_f°[Fe(OH)_3(s)]$

$\quad = -4.6 \text{ kJ/mol} + 3 \times (-157.3 \text{ kJ/mol}) - (-696.6 \text{ kJ/mol}) = +54.66 \text{ kJ/mol}$

$$\ln K_{eq} = \frac{-\Delta G°}{RT} = \frac{-220.1 \times 10^3 \text{ J/mol}}{8.3145 \text{ J mol}^{-1}\text{ K}^{-1} \times 298.15 \text{ K}} = -88.79 \qquad K_{sp} = e^{-88.79} = 2.7 \times 10^{-39}$$

60. (a) We determine the values of $\Delta H°$ and $\Delta S°$ from the data in Appendix D, and then the value of $\Delta G°$ at $25°C = 298$ K.

$\Delta H° = \Delta H_f°[C_2H_2(g)] + 4\, \Delta H_f°[H_2O(g)] - 2\, \Delta H_f°[CO_2(g)] - 5\, \Delta H_f°[H_2(g)]$

$\quad = 226.7 \text{ kJ/mol} + 4\,(-241.8 \text{ kJ/mol}) - 2\,(-393.5 \text{ kJ/mol}) - 5\,(0.00 \text{ kJ/mol}) = +46.5 \text{ kJ/mol}$

$\Delta S° = S°[C_2H_2(g)] + 4\, S°[H_2O(g)] - 2\, S°[CO_2(g)] - 5\, S°[H_2(g)]$

$\quad = 200.8 + 4 \times 188.7 - 2 \times 213.6 - 5 \times 130.6 = -124.6 \text{ J mol}^{-1}\text{ K}^{-1}$

$\Delta G° = \Delta H° - T\Delta S° = +46.5 \text{ kJ/mol} - 298 \text{ K}\,(-0.1246 \text{ kJ mol}^{-1}\text{ K}^{-1}) = 83.6 \text{ kJ/mol}$

Since the value of $\Delta G°$ is quite positive, this reaction is not favored to any significant extent at $25°C$.

(b) Because the value of $\Delta H°$ is positive and that of $\Delta S°$ is negative, the reaction is *nonspontaneous* at all temperatures, if reactants and products are *in their standard states*. The reaction will proceed slightly in the forward direction, however, to produce an equilibrium mixture with small quantities of $C_2H_2(g)$ and $H_2O(g)$. And, because the forward reaction is endothermic, this reaction is favored by raising the temperature. That is, the value of K_{eq} increases with temperature.

(c) $\Delta G°_{1000} = \Delta H° - T\,\Delta S° = +46.5 \text{ kJ/mol} - 1000.\text{ K }(-0.1246 \text{ kJ mol}^{-1}\text{ K}^{-1}) = 171.1 \text{ kJ/mol}$

$$= 171.1 \times 10^3 \text{ J/mol} = -RT \ln K_p$$

$$\ln K_p = \frac{-\Delta G°}{RT} = \frac{-171.1 \times 10^3 \text{ J/mol}}{8.3145 \text{ J mol}^{-1}\text{ K}^{-1} \times 1000.\text{ K}} = -20.58 \qquad K_p = 1.15 \times 10^{-9}$$

(d) Reaction: $2\,CO_2(g)\quad+\quad 5\,H_2(g)\ \rightleftharpoons\ C_2H_2(g)\quad+\quad 4\,H_2O(g)$

Initial:	1.00 atm	1.00 atm		
Changes:	$-2x$ atm	$-5x$ atm	$+x$ atm	$+4x$ atm
Equil:	$(1.00 - 2x)$ atm	$(1.00 - 5x)$ atm	x atm	$4x$ atm

$$K_p = 1.15 \times 10^{-9} = \frac{P[C_2H_2]\,P[H_2O]^4}{P[CO_2]^2\,P[H_2]^5} = \frac{x\,(4x)^4}{(1.00 - 2x)^2\,(1.00 - 5x)^5} \approx x\,(4x)^4 = 256\,x^5$$

$$x = \sqrt[5]{\frac{1.15 \times 10^{-9}}{256}} = 5.38 \times 10^{-3} \text{ atm} = P[C_2H_2]$$

Our assumption, that $5x \ll 1.00$ atm, is a valid one.

$\Delta G°$ as a Function of Temperature

61. We first determine the value of $\Delta G°$ at $100° \text{ C}$, from the values of $\Delta H°$ and $\Delta S°$, which are determined from information listed in Appendix D.

$\Delta H° = \frac{1}{2}\,\Delta H°_f[N_2(g)] + \frac{1}{2}\,\Delta H°_f[O_2(g)] + \frac{1}{2}\,\Delta H°_f[Cl_2(g)] - \Delta H°_f[NOCl(g)]$
$\quad = 0.500\,[0.00 + 0.00 + 0.00] - 51.71 = -51.71 \text{ kJ/mol}$

$\Delta S° = \frac{1}{2}\,S°[N_2(g)] + \frac{1}{2}\,S°[O_2(g)] + \frac{1}{2}\,S°[Cl_2(g)] - S°[NOCl(g)]$
$\quad = 0.500\,[191.5 + 205.0 + 223.0] - 261.6 = 48.2 \text{ J mol}^{-1}\text{ K}^{-1}$

$\Delta G° = \Delta H° - T\,\Delta S° = -51.71 \text{ kJ/mol} - 373 \text{ K} \times 0.0482 \text{ kJ mol}^{-1}\text{ K}^{-1} = -69.7 \text{ kJ/mol} = -RT \ln K_p$

$$\ln K_p = \frac{-\Delta G°}{RT} = \frac{+69.7 \times 10^3 \text{ J/mol}}{8.3145 \text{ J mol}^{-1}\text{ K}^{-1} \times 373 \text{ K}} = 22.5 \qquad K_p = e^{22.5} = 6 \times 10^9$$

62. (a) $\Delta H° = \Delta H°_f[CO_2(g)] + \Delta H°_f[H_2(g)] - \Delta H°_f[CO(g)] - \Delta H°_f[H_2O(g)]$
$\quad = -393.5 \text{ kJ/mol} - 0.00 \text{ kJ/mol} - (-110.5 \text{ kJ/mol}) - (-241.8 \text{ kJ/mol})$
$\quad = -41.2 \text{ kJ/mol}$

$\Delta S° = S°[CO_2(g)] + S°[H_2(g)] - S°[CO(g)] - S°[H_2O(g)]$
$\quad = 213.6 \text{ J mol}^{-1}\text{ K}^{-1} + 130.6 \text{ J mol}^{-1}\text{ K}^{-1} - 197.6 \text{ J mol}^{-1}\text{ K}^{-1} - 188.7 \text{ J mol}^{-1}\text{ K}^{-1}$
$\quad = -42.1 \text{ J mol}^{-1}\text{ K}^{-1}$

$\Delta G° = \Delta H° - T\,\Delta S° = -41.2 \text{ kJ/mol} - (298.15 \text{ K})(-42.1 \times 10^{-3} \text{ kJ mol}^{-1}\text{ K}^{-1})$
$\quad = -41.2 \text{ kJ/mol} + 12.6 \text{ kJ/mol} = -28.6 \text{ kJ/mol}$

(b) $\Delta G° = \Delta H° - T\,\Delta S° = -41.2 \text{ kJ/mol} - (1025 \text{ K})(-42.1 \times 10^{-3} \text{ kJ mol}^{-1}\text{ K}^{-1})$
$\quad = -41.2 \text{ kJ/mol} + 43.2 \text{ kJ/mol} = 2.0 \text{ kJ/mol} = -RT \ln K_p$

$$\ln K_p = -\frac{\Delta G°}{RT} = -\frac{2.0 \times 10^3 \text{ J/mol}}{8.3145 \text{ J mol}^{-1}\text{ K}^{-1} \times 1025 \text{ K}} = -0.23 \qquad K_p = e^{-0.23} = 0.79$$

63. We assume that both $\Delta H°$ and $\Delta S°$ are constant with temperature.

$\Delta H° = 2\,\Delta H°_f[SO_3(g)] - 2\,\Delta H°_f[SO_2(g)] - \Delta H°_f[O_2(g)]$
$\quad = 2\,(-395.7 \text{ kJ/mol}) - 2\,(-296.8 \text{ kJ/mol}) - (0.00 \text{ kJ/mol}) = -197.8 \text{ kJ/mol}$

$\Delta S° = 2\,S°[SO_3(g)] - 2\,S°[SO_2(g)] - S°[O_2(g)]$
$\quad = 2\,(256.6 \text{ J mol}^{-1}\text{ K}^{-1}) - 2\,(248.1 \text{ J mol}^{-1}\text{ K}^{-1}) - (205.0 \text{ J mol}^{-1}\text{ K}^{-1}) = -188.0 \text{ J mol}^{-1}\text{ K}^{-1}$

$$\Delta G° = \Delta H° - T\,\Delta S° = -RT \ln K_{eq} \qquad\qquad \Delta H° = T\,\Delta S° - RT \ln K_{eq} \qquad T = \frac{\Delta H°}{\Delta S° - R \ln K_{eq}}$$

$$T = \frac{-197.8 \times 10^3 \text{ J/mol}}{-188.0 \text{ J mol}^{-1}\text{ K}^{-1} - 8.3145 \text{ J mol}^{-1}\text{ K}^{-1} \ln(1.0 \times 10^6)} = 653.1 \text{ K}$$

This value compares very favorably with the value of $T = 6.37 \times 10^2$ that was obtained in Example 20-10.

64. We use the van't Hoff equation to determine the value of ΔH°. $448°C = 721$ K and $350°C = 623$ K.

$$\ln \frac{K_1}{K_2} = \frac{\Delta H^\circ}{R}\left(\frac{1}{T_2} - \frac{1}{T_1}\right) = \ln \frac{50.0}{66.9} = -0.291 = \frac{\Delta H^\circ}{R}\left(\frac{1}{623} - \frac{1}{721}\right) = \frac{\Delta H^\circ}{R}(2.2 \times 10^{-4})$$

$$\frac{\Delta H^\circ}{R} = \frac{-0.291}{2.2 \times 10^{-4}} = 1.3 \times 10^3 \qquad \Delta H^\circ = -1.3 \times 10^3 \times 8.3145 = -11 \times 10^3 \text{ J/mol} = -11 \text{ kJ/mol}$$

ΔG° as a Function of Temperature; The van't Hoff Equation

65. (a) $\ln \dfrac{K_2}{K_1} = \dfrac{\Delta H^\circ}{R}\left(\dfrac{1}{T_1} - \dfrac{1}{T_2}\right) = \dfrac{57.2 \times 10^3 \text{J/mol}}{8.3145 \text{ J mol}^{-1} \text{ K}^{-1}}\left(\dfrac{1}{298 \text{ K}} - \dfrac{1}{273 \text{ K}}\right) = -2.1$

$\dfrac{K_2}{K_1} = e^{-2.1} = 0.12 \qquad K_2 = 0.12 \times 0.113 = 0.014$ at 273 K

(b) $\ln \dfrac{K_2}{K_1} = \dfrac{\Delta H^\circ}{R}\left(\dfrac{1}{T_1} - \dfrac{1}{T_2}\right) = \dfrac{57.2 \times 10^3 \text{J/mol}}{8.3145 \text{ J mol}^{-1} \text{ K}^{-1}}\left(\dfrac{1}{T_1} - \dfrac{1}{298 \text{ K}}\right) = \ln \dfrac{0.113}{1.00} = -2.180$

$\left(\dfrac{1}{T_1} - \dfrac{1}{298 \text{ K}}\right) = \dfrac{-2.180 \times 8.3145}{57.2 \times 10^3} \text{ K}^{-1} = -3.17 \times 10^{-4} \text{ K}^{-1}$

$\dfrac{1}{T_1} = \dfrac{1}{298} - 3.17 \times 10^{-4} = 3.36 \times 10^{-3} - 3.17 \times 10^{-4} = 3.04 \times 10^{-3} \text{ K}^{-1} \quad T_1 = 329$ K

66. First we calculate ΔG° at 298 K to obtain a value of K_{eq} at that temperature.

$\Delta G^\circ = 2 \Delta G_f^\circ[NO_2(g)] - 2 \Delta G_f^\circ[NO(g)] - \Delta G_f^\circ[O_2(g)]$
$\quad = 2 (51.30 \text{ kJ/mol}) - 2 (86.57 \text{ kJ/mol}) - 0.00 \text{ kJ/mol} = -70.54 \text{ kJ/mol}$

$$\ln K_{eq} = \frac{-\Delta G^\circ}{RT} = -\frac{-70.54 \times 10^3 \text{ J/mol K}}{\dfrac{8.3145 \text{ J}}{\text{mol K}} \times 298.15 \text{ K}} = 28.45_5 \qquad K_{eq} = e^{28.455} = 2.3 \times 10^{12}$$

Now we calculate ΔH° for the reaction so that we can apply the van't Hoff equation.

$\Delta H^\circ = 2 \Delta H_f^\circ[NO_2(g)] - 2 \Delta H_f^\circ[NO(g)] - \Delta H_f^\circ[O_2(g)]$
$\quad = 2 (33.18 \text{ kJ/mol}) - 2 (90.25 \text{ kJ/mol}) - 0.00 \text{ kJ/mol} = -114.14 \text{ kJ/mol}$

$$\ln \frac{K_2}{K_1} = \frac{\Delta H^\circ}{R}\left(\frac{1}{T_1} - \frac{1}{T_2}\right) = \frac{-114.14 \times 10^3 \text{J/mol}}{8.3145 \text{ J mol}^{-1} \text{ K}^{-1}}\left(\frac{1}{298 \text{ K}} - \frac{1}{373 \text{ K}}\right) = -9.3$$

$\dfrac{K_2}{K_1} = e^{-9.3} = 9 \times 10^{-5} \qquad K_2 = 9 \times 10^{-5} \times 2.3 \times 10^{12} = 2 \times 10^8$

67. (a)

t, °C	T, K	$1/T$, K^{-1}	K_p	$\ln K_p$
30.	303	3.30×10^{-3}	1.66×10^{-5}	-11.006
50.	323	3.10×10^{-3}	3.90×10^{-4}	-7.849
70.	343	2.92×10^{-3}	6.27×10^{-3}	-5.072
100.	373	2.68×10^{-3}	2.31×10^{-1}	-1.465

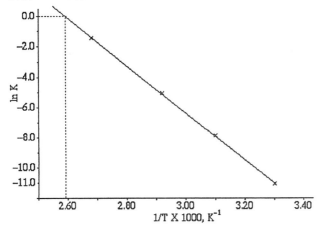

The slope of this graph is $-\Delta H^\circ/R = -1.54 \times 10^4$ K

$\Delta H^\circ = -(8.3145 \text{ J mol}^{-1} \text{ K}^{-1})(-1.54 \times 10^4 \text{ K}) = 128 \times 10^3 \text{ J/mol} = 128 \text{ kJ/mol}$

(b) When the total pressure is 2.00 atm, and both gases have been produced from $NaHCO_3(s)$,

$P\{H_2O(g)\} = P\{CO_2(g)\} = 1.00 \text{ atm}$ $K_p = P\{H_2O(g)\}P\{CO_2(g)\} = (1.00)(1.00) = 1.00$

Thus, $\ln K_p = \ln (1.00) = 0.000$ The point is shown on the graph of part (a) and corresponds to

$1/T = 2.59 \times 10^{-3} \text{ K}^{-1}$ $T = 386$ K

We can compute the same temperature from the van't Hoff equation.

$$\ln \frac{K_2}{K_1} = \frac{\Delta H^\circ}{R} \left(\frac{1}{T_1} - \frac{1}{T_2}\right) = \frac{128 \times 10^3 \text{J/mol}}{8.3145 \text{ J mol}^{-1} \text{ K}^{-1}} \left(\frac{1}{T_1} - \frac{1}{303 \text{ K}}\right) = \ln \frac{1.66 \times 10^{-5}}{1.00} = -11.006$$

$$\left(\frac{1}{T_1} - \frac{1}{303 \text{ K}}\right) = \frac{-11.006 \times 8.3145}{128 \times 10^3} \text{ K}^{-1} = -7.15 \times 10^{-4} \text{ K}^{-1}$$

$$\frac{1}{T_1} = \frac{1}{303} - 7.15 \times 10^{-4} = 3.30 \times 10^{-3} - 7.15 \times 10^{-4} = 2.58 \times 10^{-3} \text{ K}^{-1} \quad T_1 = 388 \text{ K}$$

This result agrees well with the result obtained from the graph.

68. First, the van't Hoff equation is used to obtain a value of ΔH°. 200°C = 473 K and 260°C = 533 K.

$$\ln \frac{K_2}{K_1} = \frac{\Delta H^\circ}{R} \left(\frac{1}{T_1} - \frac{1}{T_2}\right) = \ln \frac{2.15 \times 10^{11}}{4.56 \times 10^8} = 6.156 = \frac{\Delta H^\circ}{8.3145 \text{ J mol}^{-1} \text{ K}^{-1}} \left(\frac{1}{533} - \frac{1}{473 \text{ K}}\right)$$

$6.156 = -2.8625 \times 10^{-5} \Delta H^\circ$ $\Delta H^\circ = \dfrac{6.156}{-2.8625 \times 10^{-5}} = -215.1 \times 10^3 \text{ J/mol} = -215.1 \text{ kJ/mol}$

Another route to ΔH° is the combination of standard enthalpies of formation.

$CO(g) + 3 H_2(g) \rightleftharpoons CH_4(g) + H_2O(g)$

$\Delta H^\circ = \Delta H_f^\circ[CH_4(g)] + \Delta H_f^\circ[H_2O(g)] - \Delta H_f^\circ[CO(g)] - 3 \Delta H_f^\circ[H_2(g)]$

$= -74.81 \text{ kJ/mol} - 241.8 \text{ kJ/mol} - (-110.5) - 3 \times 0.00 \text{ kJ/mol} = -206.1 \text{ kJ/mol}$

The values disagree by about 4.4%. Obviously ΔH° for this reaction must show some temperature variation over this range.

21 ELECTROCHEMISTRY

1. (a) E° is the symbol for the standard cell potential, the voltage measured when no current is flowing and all cell reagents are in their standard states.
   (b) $\mathscr{F}$ is the symbol for Faraday's constant, the charge on one mole of electrons, 96,485 coulombs.
   (c) The anode is the electrode where oxidation occurs and toward which anions move.
   (d) The cathode is the electrode where reduction occurs and toward which cations move.

2. (a) A salt bridge is a tube filled with electrolyte that is used to join two half-cells in such a way that the electrochemical circuit is completed but the contents of the half-cells do not mix.
   (b) The standard hydrogen electrode is based on the reduction of hydrogen ion, at 1.000 M concentration, to hydrogen gas at 1.000 atm pressure. It is assigned a half-cell potential of 0.000 V.
   (c) Cathodic protection is achieved by electrically joining a more active metal to a less active one that is to be protected. The more active metal is oxidized, protecting the less active one from corrosion.
   (d) A fuel cell is a voltaic cell in which a reaction that normally occurs as a combustion reaction serves as the cell reaction. Reactants are continually supplied and products removed from such a cell.

3. (a) A half-reaction is either the oxidation reaction or the reduction reaction. On the other hand, the net reaction is an oxidation-reduction reaction, the combination of two half-reactions.
   (b) In a voltaic or galvanic cell, a chemical change produces electricity. This type of cell has $E > 0$. In an electrolytic cell, the passage of electric current produces a chemical change. This type of cell has $E < 0$.
   (c) A primary battery is nonreversible, like a dry cell. A secondary battery can be recharged and reused.
   (d) E°_{cell} is the cell potential when all reactants and all products are in their standard states. E_{cell} is the cell potential when reactants and products are not necessaily in their standard states.

4. The correct statement is (d). Electrons are produced at the anode and move toward the cathode, regardless of the electrode material. The electrons do not move through the salt bridge; ions do. Reduction occurs at the cathode in both galvanic and electrolytic cells—in all types of electrochemical cells.

5. (a) Oxidation: $Fe(s) \longrightarrow Fe^{2+}(aq) + 2\ e^-$
   Reduction: $Cu^{2+}(aq) + 2\ e^- \longrightarrow Cu(s)$

   Net: $Fe(s) + Cu^{2+}(aq) \longrightarrow Fe^{2+}(aq) + Cu(s)$

   (b) Oxidation: $2\ Br^-(aq) \longrightarrow Br_2(aq) + 2\ e^-$
   Reduction: $Cl_2(aq) + 2\ e^- \longrightarrow 2\ Cl^-(aq)$

   Net: $2\ Br^-(aq) + Cl_2(aq) \longrightarrow Br_2(aq) + 2\ Cl^-(aq)$

   (c) Oxidation: $Al(s) \longrightarrow Al^{3+}(aq) + 3\ e^-$
   Reduction: $\{Fe^{3+}(aq) + e^- \longrightarrow Fe^{2+}(aq)\}\quad \times 3$

   Net: $Al(s) + 3\ Fe^{3+}(aq) \longrightarrow Al^{3+}(aq) + 3\ Fe^{2+}(aq)$

   (d) Oxidation: $\{Cl^-(aq) + 3\ H_2O \longrightarrow ClO_3^-(aq) + 6\ H^+(aq) + 6\ e^-\}\quad \times 5$
   Reduction: $\{MnO_4^-(aq) + 8\ H^+(aq) + 5\ e^- \longrightarrow Mn^{2+}(aq)\ 4\ H_2O\}\quad \times 6$

   Net: $5\ Cl^-(aq) + 6\ MnO_4^-(aq) + 18\ H^+(aq) \longrightarrow 5\ ClO_3^-(aq) + 6\ Mn^{2+}(aq) + 9\ H_2O$

(e) Oxidation: $S^{2-}(aq) + 8\ OH^-(aq) \longrightarrow SO_4^{2-}(aq) + 4\ H_2O + 8\ e^-$

Reduction: $\{O_2(g) + 2\ H_2O + 4\ e^- \longrightarrow 4\ OH^-(aq)\} \qquad \times 2$

Net: $S^{2-}(aq) + 2\ O_2(g) \longrightarrow SO_4^{2-}(aq)$

6. Since $Ni^{2+}(aq)$ undergoes reduction, and the other metal (M) undergoes oxidation,

$E^\circ_{cell} = E^\circ_{Ni} - E^\circ_M \qquad$ or $\qquad E^\circ_M = E^\circ_{Ni} - E^\circ_{cell}$

(a) $E^\circ_{Mn} = E^\circ_{Ni} - E^\circ_{cell} = -0.257\ V - 0.923\ V = -1.180\ V$

(b) $E^\circ_{Pd} = E^\circ_{Ni} - E^\circ_{cell} = -0.257\ V + 1.172\ V = +0.915\ V$

(c) $E^\circ_{Ti} = E^\circ_{Ni} - E^\circ_{cell} = -0.257\ V - 1.373\ V = -1.630\ V$

(d) $E^\circ_V = E^\circ_{Ni} - E^\circ_{cell} = -0.257\ V - 0.873\ V = -1.130\ V$

7. (a) Oxidation: $Zn(s) \longrightarrow Zn^{2+}(aq) + 2\ e^- \qquad\qquad -E^\circ = +0.763\ V$

Reduction: $Sn^{2+}(aq) + 2\ e^- \longrightarrow Sn(s) \qquad\qquad E^\circ = -0.137\ V$

Net: $Zn(s) + Sn^{2+}(aq) \longrightarrow Zn^{2+}(aq) + Sn(s) \qquad E^\circ_{cell} = +0.626\ V$

(b) Oxidation: $\{Fe^{2+}(aq) \longrightarrow Fe^{3+}(aq) + e^-\} \qquad \times 2 \qquad -E^\circ = -0.771\ V$

Reduction: $Sn^{4+}(aq) + 2\ e^- \longrightarrow Sn^{2+}(aq) \qquad\qquad E^\circ = +0.154\ V$

Net: $2\ Fe^{2+}(aq) + Sn^{4+}(aq) \longrightarrow 2\ Fe^{3+}(aq) + Sn^{2+}(aq) \qquad E^\circ_{cell} = -0.617\ V$

(c) Oxidation: $Cu(s) \longrightarrow Cu^{2+}(aq) + 2\ e^- \qquad\qquad -E^\circ = -0.337\ V$

Reduction: $Cl_2(g) + 2\ e^- \longrightarrow 2\ Cl^-(aq) \qquad\qquad E^\circ = +1.358\ V$

Net: $Cu(s) + Cl_2(g) \longrightarrow Cu^{2+}(aq) + 2\ Cl^-(aq) \qquad E^\circ_{cell} = +1.021\ V$

8. (a) Oxidation: $2\ Cl^-(aq) \longrightarrow Cl_2(g) + 2\ e^- \qquad\qquad -E^\circ = -1.358\ V$

Reduction: $PbO_2(s) + 4\ H^+(aq) + 2\ e^- \longrightarrow Pb^{2+}(aq) + 2\ H_2O \qquad E^\circ = +1.455\ V$

Net: $2\ Cl^-(aq) + PbO_2(s) + 4\ H^+(aq) \longrightarrow Cl_2(g) + Pb^{2+}(aq) + H_2O \qquad E^\circ_{cell} = 0.097\ V$

(b) Oxidation: $\{Mg(s) \longrightarrow Mg^{2+}(aq) + 2\ e^-\} \qquad \times 3 \qquad -E^\circ = +2.356\ V$

Reduction: $\{Sc^{3+}(aq) + 3\ e^- \longrightarrow Sc(s)\} \qquad \times 2 \qquad E^\circ\{Sc^{3+}/Sc\}$

Net: $3\ Mg(s) + 2\ Sc^{3+}(aq) \longrightarrow 3\ Mg^{2+}(aq) + 2\ Sc(s) \qquad E^\circ_{cell} = +0.33\ V$

$E^\circ\{Sc^{3+}/Sc\} = +0.33\ V - 2.356\ V = -2.03\ V$

(c) Oxidation: $Cu^+(aq) \longrightarrow Cu^{2+}(aq) + e^- \qquad\qquad -E^\circ\{Cu^{2+}/Cu^+\}$

Reduction: $Ag^+(aq) + e^- \longrightarrow Ag(s) \qquad\qquad E^\circ = +0.800\ V$

Net: $Cu^+(aq) + Ag^+(aq) \longrightarrow Cu^{2+}(aq) + Ag(s) \qquad E^\circ_{cell} = +0.641\ V$

$-E^\circ\{Cu^{2+}/Cu^+\} = +0.641 - 0.800\ V = -0.159\ V \qquad E^\circ\{Cu^{2+}/Cu^+\} = +0.159\ V$

9. (a) Oxidation: $Sn(s) \longrightarrow Sn^{2+}(aq) + 2\ e^- \qquad\qquad -E^\circ = +0.137\ V$

Reduction: $Zn^{2+}(aq) + 2\ e^- \longrightarrow Zn(s) \qquad\qquad E^\circ = -0.763\ V$

Net: $Sn(s) + Zn^{2+}(aq) \longrightarrow Sn^{2+}(aq) + Zn(s) \qquad E^\circ_{cell} = -0.626\ V \qquad$ Nonspontaneous

(b) Oxidation: $2\ I^-(aq) \longrightarrow I_2(s) + 2\ e^- \qquad\qquad -E^\circ = -0.535\ V$

Reduction: $\{Fe^{3+}(aq) + e^- \longrightarrow Fe^{2+}(aq)\} \qquad \times 2 \qquad E^\circ = +0.771\ V$

Net: $2\ I^-(aq) + 2\ Fe^{3+}(aq) \longrightarrow 2\ Fe^{2+}(aq) + I_2(s) \qquad E^\circ_{cell} = +0.236\ V \qquad$ Spontaneous

(c) Oxidation: $\{2\ H_2O \longrightarrow O_2(g) + 4\ H^+(aq) + 4\ e^-\} \qquad \times 3 \qquad -E^\circ = -1.229\ V$

Reduction: $\{NO_3^- + 4\ H^+(aq) + 3\ e^- \longrightarrow NO(g) + 2\ H_2O\} \qquad \times 4 \qquad E^\circ = +0.956\ V$

Net: $4\ NO_3^-(aq) + 4\ H^+(aq) \longrightarrow 3\ O_2(g) + 4\ NO(g) + 2\ H_2O \qquad E^\circ_{cell} = -0.273\ V$

This is a nonspontaneous cell reaction.

(d) Oxidation: $\{Cl^-(aq) + 2\ OH^-(aq) \longrightarrow OCl^-(aq) + H_2O + 2\ e^-\} \qquad \times 2 \quad -E^\circ = -0.893\ V$

Reduction: $O_2(g) + 2\ H_2O + 4\ e^- \longrightarrow 4\ OH^-(aq) \qquad\qquad E^\circ = +0.401\ V$

Net: $2\ Cl^-(aq) + O_2(g) \longrightarrow 2\ OCl^-(aq) \qquad E^\circ_{cell} = -0.489\ V \qquad$ Nonspontaneous

10. Hg is more difficult to oxidize (–0.854 V) than H_2 (0.000 V), and thus Hg(l) will not dissolve in 1 M HCl. But we know that the reduction of nitrate ion in acidic solution is a quite spontaneous half-reaction (+0.956 V). This is sufficient to overcome the reluctance of Hg to be oxidized. Hg(l) will react with and dissolve in $HNO_3(aq)$.

11. **(a)** Oxidation: $Mg(s) \longrightarrow Mg^{2+}(aq) + 2 e^-$ $\qquad\qquad -E^\circ = +2.356$ V

Reduction: $Sn^{2+}(aq) + 2 e^- \longrightarrow Sn(s)$ $\qquad\qquad E^\circ = -0.137$ V

Net: $Mg(s) + Sn^{2+}(aq) \longrightarrow Mg^{2+}(aq) + Sn(s)$ $\quad E^\circ_{cell} = + 2.219$ V

This reaction will occur to a significant extent.

(b) Oxidation: $Sn(s) \longrightarrow Sn^{2+}(aq) + 2 e^-$ $\qquad\qquad -E^\circ = +0.137$ V

Reduction: $2 H^+(aq) \longrightarrow H_2(g)$ $\qquad\qquad E^\circ = 0.000$ V

Net: $Sn(s) + 2 H^+(aq) \longrightarrow Sn^{2+}(aq) + H_2(g)$ $\quad E^\circ_{cell} = +0.137$ V

This reaction will occur to a significant extent.

(c) Oxidation: $\{Fe^{2+}(aq) \longrightarrow Fe^{3+}(aq) + e^-\} \quad \times 2$ $\qquad -E^\circ = -0.771$ V

Reduction: $SO_4^{2-}(aq) + 4 H^+(aq) + 2 e^- \longrightarrow SO_2(g) + 2 H_2O$ $\qquad E^\circ = +0.16$ V

Net: $2 Fe^{2+}(aq) + SO_4^{2-}(aq) + 4 H^+(aq) \longrightarrow 2 Fe^{3+}(aq) + SO_2(g) + 2 H_2O$ $\quad E^\circ_{cell} = -0.61$ V

This reaction will not occur to a significant extent.

(d) Oxidation: $\{H_2O_2(aq) \longrightarrow O_2(g) + 2 H^+(aq) + 2 e^-\}$ $\qquad \times 5 \qquad -E^\circ = -0.695$ V

Reduction: $\{MnO_4^-(aq) + 8 H^+(aq) + 5 e^- \longrightarrow Mn^{2+}(aq) + 4 H_2O\}$ $\quad \times 2 \qquad E^\circ = +1.51$ V

Net: $5 H_2O_2(aq) + 2 MnO_4^-(aq) + 6 H^+(aq) \longrightarrow 5 O_2(g) + 2 Mn^{2+}(aq) + 8 H_2O$ $\quad E^\circ_{cell} = +0.82$ V

This reaction will occur to a significant extent.

(e) Oxidation: $2 Cl^-(aq) \longrightarrow Cl_2(aq) + 2 e^-$ $\qquad\qquad -E^\circ = -1.358$ V

Reduction: $I_2(s) + 2 e^- \longrightarrow 2 I^-(aq)$ $\qquad\qquad E^\circ = +0.535$ V

Net: $2 Cl^-(aq) + I_2(s) \longrightarrow Cl_2(aq) + 2 I^-(aq)$ $\quad E^\circ_{cell} = -0.823$ V

This reaction will not occur to a significant extent.

12. The relatively small value of E°_{cell} for a reaction indicates that the reaction will proceed in the forward reaction, but will stop short of completion. A much larger value of E°_{cell} would be necessary before we would conclude that the reaction goes to completion. For example, we can compute the value of the equilibrium constant for this reaction. A value of 1000 or more is needed for a reaction that would go to completion.

$$E^\circ_{cell} = \frac{0.0592}{n} \log K_{eq} \qquad \log K_{eq} = \frac{n \times E^\circ_{cell}}{0.0592} = \frac{2 \times 0.02}{0.0592} = 0.7 \qquad K_{eq} = e^{0.7} = 2$$

13. If E°_{cell} is positive, the reaction will occur. For the reduction of dicromate ion to $Cr^{3+}(aq)$ we have the following.

$Cr_2O_7^{2-}(aq) + 14 H^+(aq) + 6 e^- \longrightarrow 2 Cr^{3+}(aq) + 7 H_2O$ $\qquad E^\circ = +1.33$ V

Thus, if an oxidation has $-E^\circ$ that is smaller (more negative) than –1.33 V, the oxidation will not occur.

(a) $Sn^{2+}(aq) \longrightarrow Sn^{4+}(aq) + 2 e^- \qquad -E^\circ = -0.154$ V

$Sn^{2+}(aq)$ can be oxidized to $Sn^{4+}(aq)$ by $Cr_2O_7^{2-}(aq)$.

(b) $I_2(s) + 6 H_2O \longrightarrow 2 IO_3^-(aq) + 12 H^+(aq) + 10 e^-$ $\qquad -E^\circ = -1.20$ V

$I_2(s)$ can be oxidized to $IO_3^-(aq)$ by $Cr_2O_7^{2-}(aq)$.

(c) $Mn^{2+}(aq) + 4 H_2O \longrightarrow MnO_4^-(aq) + 8 H^+(aq) + 5 e^-$ $\qquad -E^\circ = -1.51$ V

$Mn^{2+}(aq)$ cannot be oxidized to $MnO_4^-(aq)$ by $Cr_2O_7^{2-}(aq)$.

14. **(a)** Oxidation: $\{Al(s) \longrightarrow Al^{3+}(aq) + 3 e^-\} \quad \times 2$ $\qquad -E^\circ = +1.676$ V

Reduction: $\{Cu^{2+}(aq) + 2 e^- \longrightarrow Cu(s)\} \quad \times 3$ $\qquad E^\circ = +0.337$ V

Net: $2 Al(s) + 3 Cu^{2+}(aq) \longrightarrow 2 Al^{3+}(aq) + 3 Cu(s)$ $\quad E^\circ_{cell} = +2.013$ V

$\Delta G^\circ = -n \, \mathscr{F} \, E^\circ_{cell} = -$ (6 mol elns)(96,485 C/mol elns)(2.013 V) $= -1.165 \times 10^6$ J $= -1.165 \times 10^3$ kJ

(b) Oxidation: $\{2\,I^-(aq) \longrightarrow I_2(s) + 2\,e^-\}$ $\times 2$ $\quad\quad -E^\circ = -0.535$ V

Reduction: $O_2(g) + 4\,H^+(aq) + 4\,e^- \longrightarrow 2\,H_2O$ $\quad\quad E^\circ = +1.229$ V

Net: $4\,I^-(aq) + O_2(g) + 4\,H^+(aq) \longrightarrow 2\,I_2(s) + 2\,H_2O$ $\quad E^\circ_{cell} = +0.694$ V

$\Delta G^\circ = -n\,\mathcal{F}\,E^\circ_{cell} = -(4 \text{ mol elns})(96{,}485 \text{ C/mol elns})(0.694 \text{ V}) = -2.68 \times 10^5 \text{ J} = -268 \text{ kJ}$

(c) Oxidation: $\{Ag(s) \longrightarrow Ag^+(aq) + e^-\}$ $\times 6$ $\quad\quad -E^\circ = -0.800$ V

Reduction: $Cr_2O_7^{2-}(aq) + 14\,H^+(aq) + 6\,e^- \longrightarrow 2\,Cr^{3+}(aq) + 7\,H_2O$ $\quad E^\circ = +1.33$ V

Net: $6\,Ag(s) + Cr_2O_7^{2-}(aq) + 14\,H^+(aq) \longrightarrow 6\,Ag^+(aq) + 2\,Cr^{3+}(aq) + 7\,H_2O$

$E^\circ_{cell} = -0.800 \text{ V} + 1.33 \text{ V} = +0.53$ V

$\Delta G^\circ = -n\,\mathcal{F}\,E^\circ_{cell} = -(6 \text{ mol elns})(96{,}485 \text{ C/mol elns})(0.53 \text{ V}) = -3.1 \times 10^5 \text{ J} = -3.1 \times 10^2 \text{ kJ}$

15. Since $\Delta G^\circ = -n\,\mathcal{F}\,E^\circ_{cell} = -RT \ln K$ $\quad \ln K = \dfrac{n\,\mathcal{F}\,E^\circ_{cell}}{R\,T}$ This simplifies to $\log K = \dfrac{n}{0.0592}\,E^\circ_{cell}$

(a) Oxidation: $Ag(s) \longrightarrow Ag^+(aq) + e^-$ $\quad\quad -E^\circ = -0.800$ V

Reduction: $Fe^{3+}(aq) + e^- \longrightarrow Fe^{2+}(aq)$ $\quad\quad E^\circ = +0.771$ V

Net: $Ag(s) + Fe^{3+}(aq) \longrightarrow Ag^+(aq) + Fe^{2+}(aq)$ $\quad E^\circ_{cell} = -0.029$ V

$\log K_{eq} = \dfrac{n}{0.0592}\,E^\circ_{cell} = \dfrac{1 \text{ mol elns} \times (-0.029 \text{ V})}{0.0592} = -0.49; \quad K_{eq} = 10^{-0.49} = 0.32 = \dfrac{[Fe^{2+}][Ag^+]}{[Fe^{3+}]}$

(b) Oxidation: $2\,Cl^-(aq) \longrightarrow Cl_2(g) + 2\,e^-$ $\quad\quad -E^\circ = -1.360$ V

Reduction: $MnO_2(s) + 4\,H^+(aq) + 2\,e^- \longrightarrow Mn^{2+}(aq) + 2\,H_2O$ $\quad E^\circ = +1.224$ V

Net: $2\,Cl^-(aq) + MnO_2(s) + 4\,H^+(aq) \longrightarrow Mn^{2+}(aq) + Cl_2(g) + 2\,H_2O$ $\quad E^\circ_{cell} = -0.136$ V

$\log K_{eq} = \dfrac{2 \text{ mol elns} \times (-0.13 \text{ V})}{0.0592} = -4.4 \quad\quad K_{eq} = 10^{-4.4} = 4 \times 10^{-5} = \dfrac{[Mn^{2+}]\,P\{Cl_2(g)\}}{[Cl^-]^2[H^+]^4}$

(c) Oxidation: $4\,OH^-(aq) \longrightarrow O_2(g) + 2\,H_2O + 4\,e^-$ $\quad\quad -E^\circ = -0.401$ V

Reduction: $\{OCl^-(aq) + H_2O + e^- \longrightarrow Cl^-(aq) + 2\,OH^-\} \times 2$ $\quad E^\circ = +0.890$ V

Net: $2\,OCl^-(aq) \longrightarrow 2\,Cl^-(aq) + O_2(g)$ $\quad\quad E^\circ_{cell} = +0.489$ V

$\log K_{eq} = \dfrac{n}{0.0592}\,E^\circ_{cell} = \dfrac{4 \text{ mol elns} \times (0.489 \text{ V})}{0.0592} = 33.0 \quad\quad K_{eq} = 1 \times 10^{33} = \dfrac{[Cl^-]^2\,P\{O_2(g)\}}{[OCl^-]^2}$

16. (a) Oxidation: $Fe(s) \longrightarrow Fe^{2+}(aq) + 2\,e^-$ $\quad\quad -E^\circ = +0.440$ V

Reduction: $Cu^{2+}(aq) + 2\,e^- \longrightarrow Cu(s)$ $\quad\quad E^\circ = +0.337$ V

Net: $Fe(s) + Cu^{2+}(aq) \longrightarrow Fe^{2+}(aq) + Cu(s)$ $\quad E^\circ_{cell} = +0.777$ V

Electrons flow from the Fe electrode (electrode B) to the Cu electrode (electrode A) with a potential of +0.777 V

(b) Oxidation: $Sn^{2+}(aq) \longrightarrow Sn^{4+}(aq) + 2\,e^-$ $\quad\quad -E^\circ = -0.154$ V

Reduction: $\{Ag^+(aq) + e^- \longrightarrow Ag(s)\} \times 2$ $\quad\quad E^\circ = +0.800$ V

Net: $Sn^{2+}(aq) + 2\,Ag^+(aq) \longrightarrow Sn^{4+}(aq) + 2\,Ag(s)$ $\quad E^\circ_{cell} = +0.646$ V

Electrons flow from the Pt electrode (electrode A) to the Ag electrode (electrode B) with a potential of +0.646 V.

(c) Oxidation: $Zn(s) \longrightarrow Zn^{2+}(aq) + 2\,e^-$ $\quad\quad -E^\circ = +0.763$ V

Reduction: $Fe^{2+}(aq) + 2\,e^- \longrightarrow Fe(s)$ $\quad\quad E^\circ = -0.440$ V

Net: $Zn(s) + Fe^{2+}(aq) \longrightarrow Zn^{2+}(aq) + Fe(s)$ $\quad E^\circ_{cell} = +0.323$ V

$E = E^\circ_{cell} - \dfrac{0.0592}{n} \log \dfrac{[Zn^{2+}]}{[Fe^{2+}]} = +0.323 \text{ V} - \dfrac{0.0592}{2} \log \dfrac{0.10 \text{ M}}{1.0 \times 10^{-3} \text{ M}}$

$= +0.323 \text{ V} - 0.0592 \text{ V} = +0.264 \text{V}$

Electrons flow from the Zn electrode (electrode A) to the Fe electrode (electrode B) with a potential of +0.264 V.

17. Oxidation: $Zn(s) \longrightarrow Zn^{2+}(aq) + 2\ e^-$ $\quad\quad\quad\quad -E° = +0.763$ V

Reduction: $\{Ag^+(aq) + e^- \longrightarrow Ag(s)\} \times 2$ $\quad\quad E° = +0.800$ V

$\rule{8cm}{0.4pt}$

Net: $\quad Zn(s) + 2\ Ag^+(aq) \longrightarrow Zn^{2+}(aq) + 2\ Ag(s)$ $\quad E°_{cell} = +1.563$ V

$E = E°_{cell} - \dfrac{0.0592}{n} \log \dfrac{[Zn^{2+}]}{[Ag^+]^2} = +1.563\ V - \dfrac{0.0592}{2} \log \dfrac{1.00\ M}{x^2} = +1.000\ V$

$\log \dfrac{1.00\ M}{x^2} = \dfrac{-2 \times (1.000 - 1.563)}{0.0592} = 19.0 \quad\quad x = \sqrt{1.0 \times 10^{-19}} = 3.2 \times 10^{-10}\ M$

18. In each case, we employ the equation 0.0592 pH.

    (a) $E_{cell} = 0.0592\ pH = 0.0592 \times 4.87 = 0.288$ V

    (b) $pH = -\log (0.00267) = 2.573$ $\quad\quad\quad\quad E_{cell} = 0.0592\ pH = 0.0592 \times 2.573 = 0.152$ V

    (c) $K_a = \dfrac{[H^+][C_2H_3O_2^-]}{[HC_2H_3O_2]} = 1.8 \times 10^{-5} = \dfrac{x^2}{0.486 - x} \approx \dfrac{x^2}{0.486}$

         $x = \sqrt{0.486 \times 1.8 \times 10^{-5}} = 3.0 \times 10^{-3}\ M \quad\quad pH = -\log (3.0 \times 10^{-3}) = 2.52$

         $E_{cell} = 0.0592\ pH = 0.0592 \times 2.52 = 0.149$ V

19. We predict the possible products at the anode and at the cathode. Then we choose the oxidation and the reduction which, when combined, yield the least negative cell potential.

    (a) Possible products (of oxidation) at the anode are the following.

         $2\ H_2O \longrightarrow O_2(g) + 4\ H^+(aq) + 4\ e^- \quad -1.229$ V

         $2\ Cl^-(aq) \longrightarrow Cl_2(g) + 2\ e^- \quad\quad\quad -1.358$ V

    Possible products (of reduction) at the cathode are the following.

         $2\ H^+(aq) + 2\ e^- \longrightarrow H_2(g) \quad\quad\quad 0.000$ V

         $Cu^{2+}(aq) + 2\ e^- \longrightarrow Cu(s) \quad\quad +0.337$ V

    Because of the high overpotential for the production of $O_2(g)$ the products of electrolysis of $CuCl_2(aq)$ will be $Cl_2(g)$ at the anode and $Cu(s)$ at the cathode.

    (b) Possible products (of oxidation) at the anode are the following.

         $2\ H_2O \longrightarrow O_2(g) + 4\ H^+(aq) + 4\ e^- \quad -1.229$ V

         $2\ SO_4^{2-}(aq) \longrightarrow S_2O_8^{2-}(aq) + 2\ e^- \quad -2.01$ V

    Possible products (of reduction) at the cathode are the following.

         $2\ H^+(aq) + 2\ e^- \longrightarrow H_2(g) \quad\quad\quad 0.000$ V

         $Na^+(aq) + e^- \longrightarrow Na(s) \quad\quad\quad -2.713$ V

    The products of electrolysis of $Na_2SO_4(aq)$ will be $O_2(g)$ at the anode and $H_2(g)$ at the cathode.

    (c) The only possible products (of oxidation) at the anode is the following.

         $2\ Cl^-(l) \longrightarrow Cl_2(g) + 2\ e^-$

    The only possible product (of reduction) at the cathode is the following.

         $Ba^{2+}(l) + 2\ e^- \longrightarrow Ba(l)$

    The products of electrolysis of $BaCl_2(l)$ will be $Cl_2(g)$ at the anode and $Ba(l)$ at the cathode.

    (d) The only possible product (of oxidation) at the anode is the following.

         $4\ OH^- \longrightarrow O_2(g) + 2\ H_2O + 4\ e^- \quad\quad -0.401$ V

    Possible products (of reduction) at the cathode are the following.

         $2\ H_2O + 2\ e^- \longrightarrow H_2(g) + 2\ OH^-(aq) \quad\quad -0.828$ V

         $K^+(aq) + e^- \longrightarrow K(s) \quad\quad\quad\quad\quad\quad\quad -2.924$ V

    The products of electrolysis of $KOH(aq)$ will be $O_2(g)$ at the anode and $H_2(g)$ at the cathode.

20. When $MgCl_2(l)$ is electrolyzed, $Mg(l)$ is produced at the cathode, with a half-cell voltage of -2.356 V. On the other hand, when $MgCl_2(aq)$ is electrolyzed, $H_2(g)$ is produced at the cathode, with a standard half-cell voltage of 0.000 V.

21. We calculate the total amount of charge passed and the number of moles of electrons.

     amount $e^- = 48\ min \times \dfrac{60\ s}{1\ min} \times \dfrac{1.87\ C}{1\ s} \times \dfrac{1\ mol\ e^-}{96485\ C} = 0.056\ mol\ e^-$

(a) mass Zn = 0.056 mol e$^-$ $\times \dfrac{1 \text{ mol Zn}^{2+}}{2 \text{ mol e}^-} \times \dfrac{1 \text{ mol Zn}}{1 \text{ mol Zn}^{2+}} \times \dfrac{65.39 \text{ g Zn}}{1 \text{ mol Zn}} = 1.8$ g Zn

(b) mass Al = 0.056 mol e$^-$ $\times \dfrac{1 \text{ mol Al}^{3+}}{3 \text{ mol e}^-} \times \dfrac{1 \text{ mol Al}}{1 \text{ mol Al}^{3+}} \times \dfrac{26.98 \text{ g Al}}{1 \text{ mol Al}} = 0.50$ g Al

(c) mass Ag = 0.056 mol e$^-$ $\times \dfrac{1 \text{ mol Ag}^+}{1 \text{ mol e}^-} \times \dfrac{1 \text{ mol Ag}}{1 \text{ mol Ag}^+} \times \dfrac{107.9 \text{ g Ag}}{1 \text{ mol Ag}} = 6.0$ g Ag

(d) mass Ni = 0.056 mol e$^-$ $\times \dfrac{1 \text{ mol Ni}^{2+}}{2 \text{ mol e}^-} \times \dfrac{1 \text{ mol Ni}}{1 \text{ mol Ni}^{2+}} \times \dfrac{58.69 \text{ g Ni}}{1 \text{ mol Ni}} = 1.6$ g Ni

22. The two half reactions follow: $Al^{3+}(aq) + 3\ e^- \longrightarrow Al(s)$ and $2\ H^+(aq) + 2\ e^- \longrightarrow H_2(g)$ Thus, three moles of electrons are needed to produce each mole of Al(s) while 2 moles of electrons are needed to produce each mole of $H_2(g)$. With this information, we compute the amount of $H_2(g)$ that will be produced.

amount $H_2(g)$ = 4.57 g Al $\times \dfrac{1 \text{ mol Al}}{26.98 \text{ g Al}} \times \dfrac{3 \text{ mol e}^-}{1 \text{ mol Al}} \times \dfrac{1 \text{ mol H}_2(g)}{2 \text{ mol e}^-} = 0.254$ mol $H_2(g)$

Then we use the ideal gas equation to find the volume of $H_2(g)$.

$$\text{volume of } H_2(g) = \dfrac{0.254 \text{ mol } H_2 \times \dfrac{0.08206 \text{ L atm}}{\text{mol K}} \times (273.2 + 26.5) \text{ K}}{763 \text{ mmHg} \times \dfrac{1 \text{ atm}}{760 \text{ mmHg}}} = 6.22 \text{ L}$$

EXERCISES

23. (a) If the metal dissolves in HNO_3, it has a reduction potential that is smaller than $E°[NO_3^-(aq)/NO(g)] = 0.956$ V. If it also does not dissolve in HCl, it has a reduction potential that is larger than $E°[H^+(aq)/H_2(g)] = 0.000$ V. If it displaces $Ag^+(aq)$ from solution, then it has a reduction potential that is smaller than $E°[Ag^+(aq)/Ag(s)] = 0.800$ V. But if it does not displace $Cu^{2+}(aq)$ from solution, then its reduction potential is larger than $E°[Cu^{2+}(aq)/Cu(s)] = 0.337$ V   0.337 V $< E° <$ 0.800 V

(b) If the metal dissolves in HCl, it has a reduction potential that is smaller than $E°[H^+(aq)/H_2(g)] = 0.000$ V. If it does not displace $Zn^{2+}(aq)$ from solution, its reduction potential is larger than $E°[Zn^{2+}(aq)/Zn(s)] = -0.763$ V. If it also does not displace $Fe^{2+}(aq)$ from solution, its reduction potential is larger than $E°[Fe^{2+}(aq)/Fe(s)] = -0.440$ V. -0.440 V $< E° <$ 0.000 V

24. We place a strip of solid indium in each of the metal ion solutions and see if that metal plates out on the indium strip. Eventually, we find a pair of metals which, first, are adjacent to each other in Table 21-1 and, second, for one of which indium displaces that metal from solution and for the other no such displacement occurs. (In fact, for the second metal, the solid metal displaces indium metal from a solution of In^{3+}.) The standard electrode potential lies between the standard electrode potentials of these two metals. This technique will work only if indium metal does not react with water, that is, if the standard reduction potential of $In^{3+}(aq)/In(s)$ is greater than about -1.8 V. The inaccuracy inherent in this technique is due to overpotentials, which can be as much as 0.200 V. Its imprecision is limited by the closeness of the reduction potentials of the other two metals. Probably the metals furthest apart are U and Mg, which are separated by 0.7 V.

25. We separate the given equation into its two half-equations. One of them is the reduction of nitrate ion in acidic solution, whose standard half-cell potential we retrieve from Table 21-1 and use to solve the problem.

Oxidation: $U^{4+}(aq) + 2\ H_2O(l) \longrightarrow UO_2^{2+}(aq) + 4\ H^+(aq) + 2\ e^-$ $-E°[UO_2^{2+}(aq)|U^{4+}(aq)]$

Reduction: $NO_3^-(aq) + 4\ H^+(aq) + 3\ e^- \longrightarrow NO(g) + 2\ H_2O$ $E° = +0.956$ V

Net: $3\ U^{4+}(aq) + 2\ NO_3^-(aq) + 2\ H_2O(l) \longrightarrow 3\ UO_2^{2+}(aq) + 2\ NO(g) + 4\ H^+(aq)$ $E°_{cell} = 0.63$ V

$E°_{cell} = 0.63$ V = + 0.956 V $- E°[UO_2^{2+}(aq)|U^{4+}(aq)]$

$E°[UO_2^{2+}(aq)|U^{4+}(aq)] = 0.956$ V $- 0.63$ V = $+0.33$ V

26. We separate the given equation into its two half-equations. One of them is the reduction of $Cl_2(g)$ to $Cl^-(aq)$ whose standard half-cell potential we obtain from Table 21-1 and use to solve the problem.

Oxidation: Na(in Hg) $\longrightarrow Na^+(aq) + e^-$ $\times 2$ $-E°[Na^+(aq)|Na(in Hg)]$

Reduction: $Cl_2(g) + 2\ e^- \longrightarrow 2\ Cl^-(aq)$ $E° = +1.358$ V

Net: 2 Na(in Hg) $+ Cl_2(g) \longrightarrow 2\ Na^+(aq) + 2\ Cl^-(aq)$ $E°_{cell} = 3.20$ V

$E°_{cell} = 3.20$ V = $+1.358$ V $- E°[Na^+(aq)|Na(in Hg)]$

$E°[Na^+(aq)|Na(in Hg)] = 1.358$ V $- 3.20$ V = -1.84 V

27. We divide the net cell equation into two half-equations.

Oxidation: $Al(s) + 4\ OH^-(aq) \longrightarrow [Al(OH)_4]^-(aq) + 3\ e^-$ $\times 4$ $\quad -E°[Al(s)|[Al(OH)_4]^-(aq)]$

Reduction: $O_2(g) + 2\ H_2O + 4\ e^- \longrightarrow 4\ OH^-(aq)$ $\times 3$ $\quad E° = +0.401\ V$

Net: $\quad 4\ Al(s) + 3\ O_2(g) + 6\ H_2O + 4\ OH^-(aq) \longrightarrow 4\ [Al(OH)_4]^-(aq)$ $\quad E°_{cell} = 2.71\ V$

$E°_{cell} = 2.71\ V = +0.401\ V - E°[Al(s)|[Al(OH)_4]^-(aq)]$

$E°[Al(s)|[Al(OH)_4]^-(aq)] = 0.401\ V - 2.71\ V = -2.31\ V$

Predicting Oxidation-Reduction Reactions

28. $Na(s)$ does not displace $Mg^{2+}(aq)$ from aqueous solution since solid sodium reacts with the solvent instead.

$2\ Na(s) + H_2O \longrightarrow 2\ Na^+(aq) + H_2(g) + 2\ OH^-(aq)$

29. (a) The copper does not react with the $HCl(aq)$ even under these circumstances, but it does provide a better surface for the formation of $H_2(g)$, a surface with a lower overpotential.

(b) The reaction that occurs is $Zn(s) \longrightarrow Zn^{2+}(aq) + 2\ e^-$ Some of the electrons produced in this reaction move to the copper surface, where the following reaction occurs. $2\ H^+(aq) + 2\ e^- \longrightarrow H_2(g)$.

(c) As we indicated in part (a), the copper surface is one on which bubbles of hydrogen form more readily than they do on a zinc metal surface. We say that the copper surface has a lower overpotential. Not as high a voltage is required to produce $H_2(g)$ because of the arrangement of atoms on the copper surface.

30. (a) Oxidation: $\{Ag(s) \longrightarrow Ag^+(aq) + e^-\} \times 3$ $\qquad -E° = -0.800\ V$

Reduction: $NO_3^-(aq) + 4\ H^+(aq) + 3\ e^- \longrightarrow NO(g) + 2\ H_2O$ $\qquad E° = +0.956\ V$

Net: $\quad 3\ Ag(s) + NO_3^-(aq) + 4\ H^+(aq) \longrightarrow 3\ Ag^+(aq) + NO(g) + 2\ H_2O$ $\quad E°_{cell} = +0.16\ V$

(b) Oxidation: $Zn(s) \longrightarrow Zn^{2+}(aq) + 2\ e^-$ $\qquad -E° = +0.763\ V$

Reduction: $2\ H^+(aq) + 2\ e^- \longrightarrow H_2(g)$ $\qquad E° = 0.000\ V$

Net: $\quad Zn(s) + 2\ H^+(aq) \longrightarrow Zn^{2+}(aq) + H_2(g)$ $\quad E°_{cell} = +0.763\ V$

(c) Oxidation: $Au(s) \longrightarrow Au^{3+}(aq) + 3\ e^-$ $\qquad -E° = -1.52\ V$

Reduction: $NO_3^-(aq) + 4\ H^+(aq) + 3\ e^- \longrightarrow NO(g) + 2\ H_2O$ $\qquad E° = +0.956\ V$

Net: $\quad Au(s) + NO_3^-(aq) + 4\ H^+(aq) \longrightarrow Au^{3+}(aq) + NO(g) + 2\ H_2O$ $\quad E°_{cell} = -0.52\ V$

$Au(s)$ does not dissolve in $1.00\ M\ HNO_3(aq)$

31. In each case, we determine whether $E°_{cell}$ is greater than zero; if so, the reaction will occur.

(a) Oxidation: $\{Zn(s) \longrightarrow Zn^{2+}(aq) + 2\ e^-\}$ $\times 3$ $\qquad -E° = +0.763\ V$

Reduction: $\{Al^{3+}(aq) + 3\ e^- \longrightarrow Al(s)\}$ $\times 2$ $\qquad E° = -1.676\ V$ $\qquad$ This reaction

Net: $\quad 3\ Zn(s) + 2\ Al^{3+}(aq) \longrightarrow 3\ Zn^{2+}(aq) + 2\ Al(s)$ $\quad E°_{cell} = -0.913\ V$ $\quad$ does not occur.

(b) Oxidation: $\{2\ Cl^-(aq) \longrightarrow Cl_2(g) + 2\ e^-\}$ $\times 5$ $\qquad -E° = -1.358\ V$

Reduction: $\{MnO_4^-(aq) + 8\ H^+(aq) + 5\ e^- \longrightarrow Mn^{2+}(aq) + 4\ H_2O\} \times 2$ $\qquad E° = 1.51\ V$

Net: $10\ Cl^-(aq) + 2\ MnO_4^-(aq) + 16\ H^+(aq) \longrightarrow 5\ Cl_2(g) + 2\ Mn^{2+}(aq) + 8\ H_2O$ $\quad E°_{cell} = +0.15\ V$

This reaction occurs as written.

(c) Oxidation: $\{Ag(s) \longrightarrow Ag^+(aq) + e^-\}$ $\times 2$ $\qquad -E° = -0.800\ V$

Reduction: $2\ H^+(aq) + 2\ e^- \longrightarrow H_2(g)$ $\qquad E° = +0.000\ V$

Net: $\quad 2\ Ag(s) + 2\ H^+(aq) \longrightarrow 2\ Ag^+(aq) + H_2(g)$ $\quad E°_{cell} = -0.800\ V$

This reaction will not occur as written; $Ag(s)$ is insoluble in $HCl(aq)$.

(d) Oxidation: $Mn^{2+}(aq) + 2\ H_2O \longrightarrow MnO_2(s) + 4\ H^+(aq) + 2\ e^-$ $\qquad -E° = -1.23\ V$

Reduction: $O_2(g) + 2\ H^+(aq) + 2\ e^- \longrightarrow H_2O_2(aq)$ $\qquad E° = +0.695\ V$

Net: $\quad Mn^{2+}(aq) + 2\ H_2O + O_2(g) \longrightarrow MnO_2(s) + 2\ H^+(aq) + H_2O_2(aq)$ $\quad E°_{cell} = -0.54\ V$

This reaction will not occur as written.

Voltaic Cells

32. (a) Oxidation: $Cu(s) \longrightarrow Cu^{2+}(aq) + 2 e^-$ $\qquad -E° = -0.337$ V

Reduction: $\{Fe^{3+}(aq) + e^- \longrightarrow Fe^{2+}(aq)\} \quad \times 2$ $\qquad E° = +0.771$ V

Net: $Cu(s) + 2 Fe^{3+}(aq) \longrightarrow Cu^{2+}(aq) + 2 Fe^{2+}(aq)$ $\quad E°_{cell} = +0.434$ V

(b) Oxidation: $\{Al(s) \longrightarrow Al^{3+}(aq) + 3 e^-\} \quad \times 2$ $\qquad -E° = +1.676$ V

Reduction: $\{Pb^{2+}(aq) + 2 e^- \longrightarrow Pb(s)\} \quad \times 3$ $\qquad E° = -0.125$ V

Net: $2 Al(s) + 3 Pb^{2+}(aq) \longrightarrow 2 Al^{3+}(aq) + 3 Pb(s)$ $\quad E°_{cell} = +1.551$ V

(c) Oxidation: $2 H_2O \longrightarrow O_2(g) + 4 H^+(aq) + 4 e^-$ $\qquad -E° = -1.229$ V

Reduction: $\{Cl_2(g) + 2 e^- \longrightarrow 2 Cl^-(aq)\} \quad \times 2$ $\qquad E° = +1.360$ V

Net: $2 H_2O + 2 Cl_2(g) \longrightarrow O_2(g) + 4 H^+(aq) + 4 Cl^-(aq)$ $\quad E°_{cell} = +0.131$ V

The two cells for parts (a) and (c) are sketched below. The anode is on the left in each case.

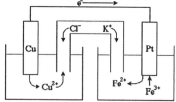 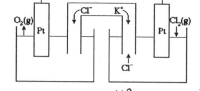

(d) Anode, Oxdn: $\{Zn(s) \longrightarrow Zn^{2+}(aq) + 2 e^-\} \qquad\qquad \times 3$ $\qquad -E° = +0.763$ V

Cathode, Redn: $\{NO_3^-(aq) + 4 H^+(aq) + 3 e^- \longrightarrow NO(g) + 2 H_2O\} \quad \times 2$ $\qquad E° = +0.956$ V

Net rxn: $3 Zn(s) + 2 NO_3^-(aq) + 8 H^+(aq) \longrightarrow 3 Zn^{2+}(aq) + 2 NO(g) + 4 H_2O$ $\quad E°_{cell} = +1.719$ V

Cell diagram: $Zn(s)|Zn^{2+}(aq)||H^+(aq),NO_3^-(aq)|NO(g)|Pt(s)$

33. (a) $Fe(s)|Fe^{2+}(aq)||Cl^-(aq)|Cl_2(g)|Pt(s)$

Oxidation: $Fe(s) \longrightarrow Fe^{2+}(aq) + 2 e^-$ $\qquad -E° = +0.440$ V

Reduction: $Cl_2(g) + 2 e^- \longrightarrow 2 Cl^-(aq)$ $\qquad E° = +1.358$ V

Net: $Fe(s) + Cl_2(g) \longrightarrow Fe^{2+}(aq) + 2 Cl^-(aq)$ $\quad E°_{cell} = +1.798$ V

(b) $Mg(s)|Mg^{2+}(aq)||Cu^{2+}(aq)|Cu(s)$

Oxidation: $Mg(s) \longrightarrow Mg^{2+}(aq) + 3 e^-$ $\qquad -E° = +2.356$ V

Reduction: $Cu^{2+}(aq) + 2 e^- \longrightarrow Cu(s)$ $\qquad E° = +0.337$ V

Net: $Mg(s) + Cu^{2+}(aq) \longrightarrow Mg^{2+}(aq) + Cu(s)$ $\quad E°_{cell} = +2.693$ V

(c) $Pt(s)|Cu^+(aq)|Cu^{2+}(aq)||Cu^+(aq)|Cu(s)$

Oxidation: $Cu^+(aq) \longrightarrow Cu^{2+}(aq) + e^-$ $\qquad -E° = -0.159$ V

Reduction: $Cu^+(aq) + e^- \longrightarrow Cu(s)$ $\qquad E° = +0.520$ V

Net: $2 Cu^+(aq) \longrightarrow Cu^{2+}(aq) + Cu(s)$ $\quad E°_{cell} = -0.361$ V

$\Delta G°$, $E°_{cell}$, and K

34. (a) Oxidation: $\{Al(s) \longrightarrow Al^{3+}(aq) + 3 e^-\} \quad \times 2$ $\qquad -E° = +1.676$ V

Reduction: $\{Zn^{2+}(aq) + 2 e^- \longrightarrow Zn(s)\} \quad \times 3$ $\qquad E° = -0.763$ V

Net: $2 Al(s) + 3 Zn^{2+}(aq) \longrightarrow 2 Al^{3+}(aq) + 3 Zn(s)$ $\quad E°_{cell} = +0.913$ V

$\Delta G° = -n \mathcal{F} E°_{cell} = -(6 \text{ mol } e^-)(96,485 \text{ C/mol } e^-)(0.913 \text{ V}) = -5.3 \times 10^5 \text{ J} = -5.3 \times 10^2 \text{ kJ}$

(b) Oxidation: $\{Pb^{2+} + 2 H_2O \longrightarrow PbO_2(s) + 4 H^+ + 4 e^-\} \quad \times 5$ $\qquad -E° = -1.455$ V

Reduction: $\{MnO_4^- + 8 H^+ + 5 e^- \longrightarrow Mn^{2+} + 4 H_2O\} \quad \times 4$ $\qquad E° = +1.51$ V

Net: $5 Pb^{2+} + 4 MnO_4^- + 12 H^+ \longrightarrow 5 PbO_2 + 4 Mn^{2+} + 6 H_2O$ $\quad E°_{cell} = -0.06$ V

$$\Delta G^\circ = -n\,\mathfrak{F}\,E^\circ_{cell} = -(20 \text{ mol e}^-)(96{,}485 \text{ C/mol e}^-)(-0.06 \text{ V}) = 1 \times 10^5 \text{ J} = 1 \times 10^2 \text{ kJ}$$

(c) Oxidation: $2 \text{ Cl}^-(\text{aq}) \longrightarrow \text{Cl}_2(\text{g}) + 2 \text{ e}^-$ $\qquad\qquad -E^\circ = -1.358 \text{ V}$

Reduction: $\text{MnO}_2(\text{s}) + 4 \text{ H}^+(\text{aq}) + 2 \text{ e}^- \longrightarrow \text{Mn}^{2+}(\text{aq}) + 2 \text{ H}_2\text{O}$ $\qquad E^\circ = +1.23 \text{ V}$

Net: $\qquad 2 \text{ Cl}^- + \text{MnO}_2(\text{s}) + 4 \text{ H}^+ \longrightarrow \text{Cl}_2(\text{g}) + \text{Mn}^{2+} + 2 \text{ H}_2\text{O} \qquad E^\circ_{cell} = -0.13 \text{ V}$

$$\Delta G^\circ = -n\,\mathfrak{F}\,E^\circ_{cell} = -(2 \text{ mol e}^-)(96{,}485 \text{ C/mol e}^-)(-0.13 \text{ V}) = 2.5 \times 10^4 \text{ J} = +25 \text{ kJ}$$

35. First we determine the value of the equilibrium constant, then we solve the resulting equilibrium expression for $[\text{Zn}^{2+}]_{eq}$.

$$\ln K = -\frac{\Delta G^\circ}{R\,T} = -\frac{-5.3 \times 10^5}{8.3145 \text{ J mol}^{-1} \text{ K}^{-1} \times 298 \text{ K}} = 2.1 \times 10^2 \qquad\qquad K = 2 \times 10^{91}$$

Because of the extremely large value of the equilibrium constant, we assume that the reaction goes to completion and then find the equilibrium $[\text{Zn}^{2+}]$ from this point.

Reaction: $\quad 2 \text{ Al}(\text{s}) + \quad 3 \text{ Zn}^{2+}(\text{aq}) \rightleftharpoons 2 \text{ Al}^{3+}(\text{aq}) + 3 \text{ Zn}(\text{s})$

Initial: $\qquad\qquad\qquad 1.00 \text{ M}$

To completion: $\qquad\qquad 0.00 \text{ M} \qquad\qquad 0.667 \text{ M}$

Changes: $\qquad\qquad\quad +3x \text{ M} \qquad\qquad -2x \text{ M}$

Equil: $\qquad\qquad\qquad 3x \text{ M} \qquad\qquad (0.667 - 2x)\text{M}$

$$K = \frac{[\text{Al}^{3+}]^2}{[\text{Zn}^{2+}]^3} = 2 \times 10^{91} = \frac{(0.667 - 2x)^2}{(3x)^3} \approx \frac{(0.667)^2}{27\,x^3} \qquad x = \sqrt[3]{\frac{0.445}{54 \times 10^{91}}} = 9 \times 10^{-32} \text{ M}$$

$[\text{Zn}^{2+}] = 3x = 3 \times 10^{-31} \text{ M}$ Clearly, the reaction goes to completion.

36. (a) A negative value of E°_{cell} indicates that $\Delta G^\circ = -n\,\mathfrak{F}\,E^\circ_{cell}$ is positive which in turn indicates that K is less than one; $\Delta G^\circ = -R\,T \ln K$. Now $K = \frac{[\text{Cu}^{2+}]^2[\text{Sn}^{2+}]}{[\text{Cu}^+]^2[\text{Sn}^{4+}]}$ Thus, when all concentrations are equal, the ion product, Q, equals 1.00. Thus, all the concentrations cannot be 0.500 M at the same time.

(b) In order to establish equilibrium, that is, to have the ion product become less than 1.00, and equal the equilibrium constant, the concentrations of the products must decrease and those of the reactants must increase. A net reaction to the left will occur.

37. Cell reaction: $\quad \text{Zn}(\text{s}) + \text{Ag}_2\text{O}(\text{s}) \longrightarrow \text{ZnO}(\text{s}) + 2 \text{ Ag}(\text{s}) \qquad$ We assume that the cell operates at 298 K.

$\Delta G^\circ = \Delta G^\circ_f[\text{ZnO}(\text{s})] + 2 \Delta G^\circ_f[\text{Ag}(\text{s})] - \Delta G^\circ_f[\text{Zn}(\text{s})] - \Delta G^\circ_f[\text{Ag}_2\text{O}(\text{s})]$

$\qquad = -318.3 \text{ kJ/mol} + 2 (0.00 \text{ kJ/mol}) - 0.00 \text{ kJ/mol} - (-11.2 \text{ kJ/mol}) = -307.1 \text{ kJ/mol} = -n\,\mathfrak{F}\,E^\circ_{cell}$

$$E^\circ_{cell} = -\frac{\Delta G^\circ}{n\,\mathfrak{F}} = -\frac{-307.1 \times 10^3 \text{ J/mol}}{2 \text{ mol e}^-\text{/mol rxn} \times 96{,}485 \text{ C/mol e}^-} = 1.591 \text{ V}$$

38. Oxidation: $\text{N}_2\text{H}_4(\text{aq}) \longrightarrow \text{N}_2(\text{g}) + 4 \text{ H}^+(\text{aq}) + 4 \text{ e}^-$

Reduction: $\text{O}_2(\text{g}) + 4 \text{ H}^+(\text{aq}) + 4 \text{ e}^- \longrightarrow 2 \text{ H}_2\text{O}(\text{l})$

Net: $\qquad \text{N}_2\text{H}_4(\text{aq}) + \text{O}_2(\text{g}) \longrightarrow \text{N}_2(\text{g}) + 2 \text{ H}_2\text{O}(\text{l})$

Thus, $n = 4$. $\quad \Delta G^\circ = -n\,\mathfrak{F}\,E^\circ_{cell} = -(4 \text{ mol e}^-)(96485 \text{ C/mol e}^-)(1.559 \text{ V}) \times \frac{1 \text{ kJ}}{1000 \text{ J}} = -601.7 \text{ kJ/mol}$

$\Delta G^\circ = -601.7 \text{ kJ/mol} = \Delta G^\circ_f[\text{N}_2(\text{g})] + 2 \Delta G^\circ_f[\text{H}_2\text{O}(\text{l})] - \Delta G^\circ_f[\text{N}_2\text{H}_4(\text{aq})] - \Delta G^\circ_f[\text{N}_2(\text{g})]$

$\qquad\qquad\qquad\qquad\quad = 0.00 \text{ kJ/mol} + 2 (-237.2 \text{ kJ/mol}) - \Delta G^\circ_f[\text{N}_2\text{H}_4(\text{aq})] - 0.00 \text{ kJ/mol}$

$\Delta G^\circ_f[\text{N}_2\text{H}_4(\text{aq})] = 2 (-237.2 \text{ kJ/mol}) + 601.7 \text{ kJ/mol} = +127.3 \text{ kJ/mol}$

39. From equation (21.28) we know that $n = 12$ and the cell equation. First we compute the value of ΔG°.

$$\Delta G^\circ = -n\mathfrak{F}E^\circ_{cell} = -12 \text{ mol e}^- \times \frac{96485 \text{ C}}{1 \text{ mol e}^-} \times 2.73 \text{ V} = -3.16 \times 10^6 \text{ J} = -3.16 \text{ MJ}$$

Then we use this value with the balanced equation and values of ΔG°_f to calculate $\Delta G^\circ_f[[\text{Al}(\text{OH})_4]^-]$.

$4 \text{ Al}(\text{s}) + 3 \text{ O}_2(\text{g}) + 6 \text{ H}_2\text{O} + 4 \text{ OH}^-(\text{aq}) \longrightarrow 4 [\text{Al}(\text{OH})_4]^-(\text{aq})$

$\Delta G^\circ = 4 \Delta G^\circ_f[[\text{Al}(\text{OH})_4]^-] - 4 \Delta G^\circ_f[\text{Al}(\text{s})] - 3 \Delta G^\circ_f[\text{O}_2(\text{g})] - 6 \Delta G^\circ_f[\text{H}_2\text{O}(\text{l})] - 4 \Delta G^\circ_f[\text{OH}^-(\text{aq})]$

$-3.16 \times 10^3 \text{ kJ} = 4 \Delta G^\circ_f[[\text{Al}(\text{OH})_4]^-] - 4 \times 0.00 \text{ kJ} - 3 \times 0.00 \text{ kJ} - 6 \times (-237.2 \text{ kJ}) - 4 \times (-157.3)$

$\qquad\qquad\qquad = 4 \Delta G^\circ_f[[\text{Al}(\text{OH})_4]^-] + 2052.4 \text{ kJ}$

$\Delta G^\circ_f[[\text{Al}(\text{OH})_4]^-] = (-3.16 \times 10^3 \text{ kJ} - 2052.4 \text{ kJ}) \div 4 = -1.30 \times 10^3 \text{ kJ} = -1.30 \text{ MJ}$

Concentration Dependence of E_{cell}—the Nernst Equation

40. We first calculate E_{cell}° for each reaction and then use the Nernst equation to calculate E_{cell}.

    (a) Oxidation: $Fe(s) \longrightarrow Fe^{2+}(0.35\ M) + 2\ e^-$ $-E^\circ = +0.440\ V$

             Reduction: $Sn^{2+}(0.070\ M) + 2\ e^- \longrightarrow Sn(s)$ $E^\circ = -0.137\ V$

             Net: $Fe(s) + Sn^{2+}(0.070\ M) \longrightarrow Fe^{2+}(0.35\ M) + Sn(s)$ $E_{cell}^\circ = +0.303\ V$

$$E_{cell} = E_{cell}^\circ - \frac{0.0592}{n} \log \frac{[Fe^{2+}]}{[Sn^{2+}]} = 0.303\ V - \frac{0.0592}{2} \log \frac{0.35\ M}{0.070\ M} = 0.282\ V$$

    (b) Oxidation: $2\ Cl^-(1.2\ M) \longrightarrow Cl_2(0.25\ atm) + 2\ e^-$ $-E^\circ = -1.358\ V$

             Reduction: $\{Ag^+(0.46\ M) + e^- \longrightarrow Ag(s)\}\ \times 2$ $E^\circ = +0.800\ V$

             Net: $2\ Cl^-(1.2\ M) + 2\ Ag^+(0.46\ M) \longrightarrow Cl_2(0.25\ atm) + 2\ Ag(s)$ $E_{cell}^\circ = -0.558\ V$

$$E_{cell} = E_{cell}^\circ - \frac{0.0592}{n} \log \frac{P\{Cl_2(g)\}}{[Cl^-]^2[Ag^+]^2} = -0.558 - \frac{0.0592}{2} \log \frac{0.25}{(0.46)^2(1.2)^2} = -0.555\ V$$

    (c) Oxidation: $Mg(s) \longrightarrow Mg^{2+}(0.016\ M) + 2\ e^-$ $-E^\circ = +2.356\ V$

             Reduction: $2\ H_2O + 2\ e^- \longrightarrow 2\ OH^-(0.65\ M) + H_2(0.75\ atm)$ $E^\circ = -0.828\ V$

             Net: $Mg(s) + 2\ H_2O \longrightarrow Mg^{2+}(0.016\ M) + 2\ OH^-(0.65\ M) + H_2(0.75\ atm)$ $E_{cell}^\circ = +1.528\ V$

$$E_{cell} = E_{cell}^\circ - \frac{0.0592}{2} \log ([Mg^{2+}][OH^-]^2 P\{H_2\}) = +1.528 - 0.0296 \log [(0.016)(0.65)^2(0.75)]$$
$$= 1.596\ V$$

41. Oxidation: $\{I_2(s) + 6\ H_2O \longrightarrow 2\ IO_3^-(aq) + 12\ H^+(aq) + 10\ e^-\}\ \times 2$ $-E^\circ = -1.20\ V$

     Reduction: $\{O_2(g) + 4\ H^+(aq) + 4\ e^- \longrightarrow 2\ H_2O\}$ $\times 5$ $E^\circ = +1.229\ V$

     Net: $2\ I_2(s) + 5\ O_2(g) + 2\ H_2O \longrightarrow 4\ IO_3^-(aq) + 4\ H^+(aq)$ $E_{cell}^\circ = +0.03\ V$

$$E_{cell} = E_{cell}^\circ - \frac{0.0592}{20} \log \frac{[H^+]^4[IO_3^-]^4}{P\{O_2\}} = +0.03 - 0.00296 \log \frac{[H^+]^4(1.00)^4}{(1.00\ atm)}$$

         $= +0.03 - 4 \times 0.00296 \log[H^+] = +0.03 - 0.0118 \log [H^+] = +0.005 + 0.0118\ pH$

    (a) $E_{cell} = +0.03 - 0.0118 \log(6.6) = +0.02\ V$ Some reaction occurs.

    (b) $E_{cell} = +0.03 - 0.0118 \log(1.0) = +0.03\ V$ Yes, some reaction occurs.

    (c) $E_{cell} = +0.03 - 0.0118 \log(0.060) = +0.04\ V$ Yes, some reaction occurs.

    (d) $E_{cell} = +0.03 + 0.0118 \times 9.22 = +0.14\ V$ A reaction occurs.

Reaction occurs foir all these conditions, but becomes more spontaneous as pH increases.

42. Oxidation: $\{2\ Cl^-(aq) \longrightarrow Cl_2(g) + 2\ e^-\}$ $\times 3$ $-E^\circ = -1.360\ V$

     Reduction: $Cr_2O_7^{2-}(aq) + 14\ H^+(aq) + 6\ e^- \longrightarrow 2\ Cr^{3+}(aq) + 7\ H_2O$ $E^\circ = +1.33\ V$

     Net: $Cr_2O_7^{2-}(aq) + 6\ Cl^-(aq) + 14\ H^+(aq) \longrightarrow 3\ Cl_2(g) + 2\ Cr^{3+}(aq) + 7\ H_2O$ $E_{cell}^\circ = -0.03\ V$

We notice that, under standard conditions, the oxidation of $Cl^-(aq)$ to Cl_2 by $Cr_2O_7^{2-}(aq)$ in acidic solution is not spontaneous. However, in the laboratory both $[Cl^-]$ and $[H^+]$ would normally be larger than 1.00 M when one is attempting to produce $Cl_2(g)$. In addition, $Cl_2(g)$ would be conducted away from the reaction mixture and would not be present at 1.00 atm pressure (nor , of course, would $[Cr^{3+}]$ be equal to 1.00 M, particularly not at the beginning of the reaction).

43. We write the half-equation for the reduction of each of these oxidizing agents. Those in which neither $H^+(aq)$ nor $OH^-(aq)$ are present, are not dependent on the pH of the solution for their oxidizing strength.

     $Cl_2(g) + 2\ e^- \longrightarrow 2\ Cl^-(aq)$ $O_2(g) + 4\ H^+(aq) + 4\ e^- \longrightarrow 2\ H_2O$

     $MnO_4^-(aq) + 8\ H^+(aq) + 5\ e^- \longrightarrow Mn^{2+} + 4\ H_2O$ $H_2O_2(aq) + 2\ H^+(aq) + 2\ e^- \longrightarrow 2\ H_2O$

     $F_2(g) + 2\ e^- \longrightarrow 2\ F^-(aq)$

    (a) For $O_2(g)$, $MnO_4^-(aq)$, and $H_2O_2(aq)$ oxidizing power is dependent on pH.

    (b) The oxidizing power of both $Cl_2(g)$ and $F_2(g)$ are independent of pH.

    (c) All three of the three oxidizing agents listed in part (a) are more effective in acidic solution, since in all three cases, $H^+(aq)$ is a reactant in the half-reaction and increasing its concentration makes the half-reaction more spontaneous.

44. This is a concentration cell, in which $E^{\circ}_{\text{cell}} = 0.000$ V. $n = 2$.

Cell reaction: $\quad H_2(g, 1atm) + 2\,H^+(0.10\,M) \longrightarrow H_2(g, 1\,atm) + 2\,H^+(x\,M)$

$E_{\text{cell}} = E^{\circ}_{\text{cell}} - \dfrac{0.0592}{2} \log \dfrac{[H^+(x)]^2}{[H^+(0.10)]^2} = 0.108\,V \qquad \log \dfrac{[H^+(x)]^2}{[H^+(0.10)]^2} = -\dfrac{0.108 \times 2}{0.0592} = -3.65$

$\log \dfrac{[H^+(x)]}{[H^+(0.10)]} = \dfrac{-3.65}{2} = -1.82 \qquad\qquad \dfrac{[H^+(x)]}{[H^+(0.10)]} = 10^{-1.82} = 0.015$

$x = 0.015 \times 0.10 = 1.5 \times 10^{-3}\,M = [H^+] \qquad\qquad pH = 2.82$

45. Oxidation: $\quad Zn(s) \longrightarrow Zn^{2+}(aq) + 2\,e^- \qquad\qquad -E^{\circ} = +0.763\,V$

Reduction: $\quad Cu^{2+}(aq) + 2\,e^- \longrightarrow Cu(s) \qquad\qquad E^{\circ} = +0.337\,V$

Net: $\qquad Zn(s) + Cu^{2+}(aq) \longrightarrow Cu(s) + Zn^{2+}(aq) \qquad E^{\circ}_{\text{cell}} = +1.100\,V$

(a) We set $E = 0.000$ V and solve for $[Cu^{2+}]$ in the Nernst equation.

$E_{\text{cell}} = E^{\circ}_{\text{cell}} - \dfrac{0.0592}{2} \log \dfrac{[Zn^{2+}]}{[Cu^{2+}]} = 0.000 = 1.100 - 0.0296 \log \dfrac{1.0\,M}{[Cu^{2+}]}$

$\log \dfrac{1.0\,M}{[Cu^{2+}]} = \dfrac{0.000 - 1.100}{-0.0296} = 37.2 \qquad [Cu^{2+}] = 10^{-37.2} = 6 \times 10^{-38}\,M$

(b) If we work the problem the other way—by assuming that $[Cu^{2+}]_i = 1.0\,M$ and $[Zn^{2+}]_i = 0.0\,M$—we obtain $[Cu^{2+}]_f = 6 \times 10^{-38}\,M$ and $[Zn^{2+}]_f = 1.0\,M$. We would say that this reaction goes to completion.

46. Oxidation: $\quad Sn(s) \longrightarrow Sn^{2+}(aq) + 2\,e^- \qquad\qquad -E^{\circ} = +0.137\,V$

Reduction: $\quad Pb^{2+}(aq) + 2\,e^- \longrightarrow Pb(s) \qquad\qquad E^{\circ} = -0.125\,V$

Net: $\quad Sn(s) + Pb^{2+}(aq) \longrightarrow Sn^{2+}(aq) + Pb(s) \qquad E^{\circ}_{\text{cell}} = +0.012\,V$

Now we wish to find out if $Pb^{2+}(aq)$ will be completely displaced—that is, will $[Pb^{2+}]$ reach 0.0010 M—if $[Sn^{2+}]$ is fixed at 1.00 M? We use the Nernst equation to determine if the cell voltage still is positive under these conditions.

$E_{\text{cell}} = E^{\circ}_{\text{cell}} - \dfrac{0.0592}{2} \log \dfrac{[Sn^{2+}]}{[Pb^{2+}]} = +0.012 - \dfrac{0.0592}{2} \log \dfrac{1.0}{0.0010} = +0.012 - 0.089 = -0.077\,V$

No, this reaction will not go to completion under the conditions stated. We can work this the other way as well: assume that $[Pb^{2+}] = (1.0 - x)\,M$ and calculating $[Sn^{2+}] = x\,M$ at equilibrium, that is, where $E_{\text{cell}} = 0$.

$E_{\text{cell}} = 0.00 = E^{\circ}_{\text{cell}} - \dfrac{0.0592}{2} \log \dfrac{[Sn^{2+}]}{[Pb^{2+}]} = +0.012 - \dfrac{0.0592}{2} \log \dfrac{[Sn^{2+}]}{1.0}$

$\log \dfrac{x}{1.0 - x} = \dfrac{2 \times 0.012}{0.0592} = 0.41 \qquad x = 10^{0.41}\,(1.0 - x) = 2.6 - 2.6\,x \qquad x = \dfrac{2.6}{3.6} = 0.72\,M$

We would expect the final $[Sn^{2+}]$ to equal 1.0 M if the reaction went to completion. Instead it equals 0.72 M.

47. (a) The two half-equations and the cell equation are given below. $E^{\circ}_{\text{cell}} = 0.000$ V

Oxidation: $\quad H_2(g) \longrightarrow 2\,H^+(0.85\,M\;KOH) + 2\,e^-$

Reduction: $\quad 2\,H^+(1.0\,M) + 2\,e^- \longrightarrow H_2(g)$

Net: $\qquad 2\,H^+(1.0\,M) \longrightarrow 2\,H^+(0.85\,M\;KOH)$

$[H^+]_{\text{base}} = \dfrac{1.00 \times 10^{-14}}{0.85} = 1.2 \times 10^{-14}\,M$

$E_{\text{cell}} = E^{\circ}_{\text{cell}} - \dfrac{0.0592}{2} \log \dfrac{[H^+]^2_{\text{base}}}{[H^+]^2_{\text{acid}}} = 0.000 - \dfrac{0.0592}{2} \log \dfrac{(1.2 \times 10^{-14})^2}{(1.0)^2} = +0.824\,V$

(b) For the reduction of H_2O to $H_2(g)$ in basic solution, $E^{\circ} = -0.828$ V. This reduction is the reverse of the reaction that occurs in the cathode of the cell described, with one small difference: in the standard half-cell $[OH^-] = 1.00$ M, while in the anode half-cell in the case $[OH^-] = 0.85$ M. Or, viewed in another way, in 1.00 M KOH $[H^+]$ is smaller still than in 0.85 M KOH. The forward reaction (dilution of H^+) should occur to an even greater extent with 1.00 M KOH than with 0.85 M KOH. This means that E°_{cell} (0.828 V) should be slightly larger than E_{cell} (0.824 V)

48. (a) Since $NH_3(aq)$ is a weaker base than $KOH(aq)$, $[OH^-]$ will be smaller than it is in the previous problem. Therefore the $[H^+]$ will be higher. Its logarithm will be less negative, and the cell voltage will be smaller. Or, viewed as in Exercise 47(b), the difference in $[H^+]$ between 1.0 M HCl and 0.85

KOH is greater than the difference in $[H^+]$ between 1.0 M HCl and 0.85 M NH_3. The forward reaction is "less spontaneous" and $E°_{cell}$ is smaller.

(b) Reaction: $NH_3(aq) + H_2O \rightleftharpoons NH_4^+(aq) + OH^-(aq)$

Initial: $\quad$ 0.85 M

Changes: $\quad -x$ M $\qquad\qquad$ $+x$ M $\qquad$ $+x$ M

Equil: $\quad$ (0.85 − x) M $\qquad$ x M $\qquad$ x M

$$K_b = \frac{[NH_4^+][OH^-]}{[NH_3]} = 1.74 \times 10^{-5} = \frac{x \cdot x}{0.85 - x} \approx \frac{x^2}{0.85}$$

$$x = [OH^-] = \sqrt{0.85 \times 1.74 \times 10^{-5}} = 3.8 \times 10^{-3} \text{ M} \qquad [H_3O^+] = \frac{1.00 \times 10^{-14}}{3.8 \times 10^{-3}} = 2.6 \times 10^{-12} \text{ M}$$

$$E_{cell} = E°_{cell} - \frac{0.0592}{2} \log \frac{[H^+]^2_{base}}{[H^+]^2_{acid}} = 0.000 - \frac{0.0592}{2} \log \frac{(2.6 \times 10^{-12})^2}{(1.0)^2} = +0.686 \text{ V}$$

49. First we need $[Ag^+]$ in a saturated solution of Ag_2SO_4.

$$K_{sp} = [Ag^+]^2[SO_4^{2-}] = (2s)^2(s) = 4s^3 = 1.4 \times 10^{-5} \qquad s = \sqrt[3]{\frac{1.4 \times 10^{-5}}{4}} = 1.5 \times 10^{-2} \text{ M}$$

The cell diagrammed is a concentration cell, for which $E°_{cell} = 0.000$ V. $n = 1$. $[Ag^+] = 2S = 0.030$ M

Cell reaction: $\quad Ag(s) + Ag^+(0.125 \text{ M}) \longrightarrow Ag(s) + Ag^+(0.030 \text{ M})$

$$E_{cell} = E°_{cell} - \frac{0.0592}{1} \log \frac{0.030 \text{ M}}{0.125 \text{ M}} = 0.000 + 0.037 \text{ V} = 0.037 \text{ V}$$

50. (a) Oxidation: $Sn(s) \longrightarrow Sn^{2+}(0.150 \text{ M}) + 2 e^- \qquad\qquad -E° = +0.137$ V

Reduction: $\underline{Pb^{2+}(0.550 \text{ M}) + 2 e^- \longrightarrow Pb(s) \qquad\qquad\qquad E° = -0.125 \text{ V}}$

Net: $\qquad Sn(s) + Pb^{2+}(0.550 \text{ M}) \longrightarrow Pb(s) + Sn^{2+}(0.150 \text{ M}) \qquad E°_{cell} = +0.012$ V

$$E_{cell} = E°_{cell} - \frac{0.0592}{2} \log \frac{[Sn^{2+}]}{[Pb^{2+}]} = 0.012 - 0.0296 \log \frac{0.150}{0.550} = 0.012 + 0.017 = 0.029 \text{ V}$$

(b) E_{cell} will decrease with time because, as the reaction proceeds, $[Sn^{2+}]$ will increase and $[Pb^{2+}]$ decrease.

(c) When $[Pb^{2+}] = 0.500$ M $= 0.550$ M $- 0.050$ M, $[Sn^{2+}] = 0.150$ M $+ 0.050$ M, because, by the stoichiometry of the reaction, a mole of Sn^{2+} is produced for every mole of Pb^{2+} that reacts.

$$E_{cell} = E°_{cell} - \frac{0.0592}{2} \log \frac{[Sn^{2+}]}{[Pb^{2+}]} = 0.012 - 0.0296 \log \frac{0.200}{0.500} = 0.012 + 0.012 = 0.024 \text{ V}$$

(d) Reaction: $\quad Sn(s) + Pb^{2+}(aq) \longrightarrow Pb(s) + Sn^{2+}(aq)$

Initial: $\qquad\qquad$ 0.550 M $\qquad\qquad\qquad$ 0.150 M

Changes: $\qquad\qquad -x$ M $\qquad\qquad\qquad$ $+x$ M

Final: $\qquad\qquad$ (0.550 − x)M $\qquad\qquad$ (0.150 + x)M

$$E_{cell} = E°_{cell} - \frac{0.0592}{2} \log \frac{[Sn^{2+}]}{[Pb^{2+}]} = 0.020 = 0.012 - 0.0296 \log \frac{0.150 + x}{0.550 - x} = 0.012 + 0.017$$

$$= 0.029 \text{ V}$$

$$\log \frac{0.150 + x}{0.550 - x} = \frac{E_{cell} - 0.012}{-0.0296} = \frac{0.020 - 0.012}{-0.0296} = -0.27$$

$$\frac{0.150 + x}{0.550 - x} = 10^{-0.27} = 0.54 \qquad 0.150 + x = 0.54(0.550 - x) = 0.30 - 0.54 x$$

$$x = \frac{0.30 - 0.150}{1.54} = 0.097 \text{ M} \qquad [Sn^{2+}] = 0.150 + 0.097 = 0.247 \text{ M}$$

(e) We use the expression that we developed in part (d).

$$\log \frac{0.150 + x}{0.550 - x} = \frac{E_{cell} - 0.012}{-0.0296} = \frac{0.000 - 0.012}{-0.0296} = +0.41$$

$$\frac{0.150 + x}{0.550 - x} = 10^{+0.41} = 2.6; \quad 0.150 + x = 2.6(0.550 - x) = 1.4 - 2.6 x \qquad x = \frac{1.4 - 0.15}{3.6} = 0.35 \text{ M}$$

$$[Sn^{2+}] = 0.150 + 0.35 = 0.50 \text{ M} \qquad\qquad [Pb^{2+}] = 0.550 - 0.35 = 0.20 \text{ M}$$

51. (a) A voltaic cell with a voltage of 0.1000 V would be possible by using two half-cells whose standard reduction potentials differ by approximately 0.10 V, such as the following pair.

Oxidation: $\quad 2 Cr^{3+}(aq) + 7 H_2O \longrightarrow Cr_2O_7^{2-}(aq) + 14 H^+(aq) + 6 e^- \qquad -E° = -1.33$ V

Reduction: $\quad \{PbO_2(s) + 4 H^+(aq) + 2 e^- \longrightarrow Pb^{2+}(aq) + 2 H_2O\} \qquad \times 3 \quad E° = +1.455$ V

Net: $\quad 2 Cr^{3+}(aq) + 3 PbO_2(s) + H_2O \longrightarrow Cr_2O_7^{2-}(aq) + 3 Pb^{2+}(aq) + 2 H^+(aq) \qquad E°_{cell} = 0.125$ V

The voltage can be adjusted to 0.1000 V by suitable alteration of the concentrations. $[Pb^{2+}]$ or $[H^+]$ could be increased or $[Cr^{3+}]$ could be decreased, or any combination of the three of these.

(b) To produce a cell with a voltage of 2.500 V requires that one start with two half-cells whose reduction potentials differ by about that much. An interesting pair follows.

Oxidation:	$Al(s) \longrightarrow Al^{3+}(aq) + 3\ e^-$	$-E^\circ = +1.676$ V
Reduction:	$\{Ag^+(aq) + e^- \longrightarrow Ag(s)\} \times 3$	$E^\circ = +0.800$ V
Net:	$Al(s) + 3\ Ag^+(aq) \longrightarrow Al^{3+}(aq) + 3\ Ag(s)$	$E^\circ_{cell} = +2.476$ V

Again, the desired voltage can be obtained by adjusting the concentrations: in this case increasing $[Ag^+]$ and/or decreasing $[Al^+]$.

(c) Since no pair of half-cells has a potential difference larger than about 6 volts, we conclude that producing a single cell with a potential of 10.00 V is impossible. It is possible, however, to join several cells together into a battery that delivers a voltage of 10.00 V. Four of the cells from part (b) would do quite nicely.

Batteries and Fuel Cells

52. (a) The cell diagram begins with the anode and ends with the cathode.

Cell diagram: $Cr(s) \mid Cr^{2+}(aq), Cr^{3+}(aq) \parallel Fe^{2+}(aq), Fe^{3+}(aq) \mid Fe(s)$

(b)

Oxidation:	$Cr^{2+}(aq) \longrightarrow Cr^{3+}(aq) + e^-$	$-E^\circ = +0.424$ V
Reduction:	$Fe^{3+}(aq) + e^- \longrightarrow Fe^{2+}(aq)$	$E^\circ = +0.771$ V
Net:	$Cr^{2+}(aq) + Fe^{3+}(aq) \longrightarrow Cr^{3+}(aq) + Fe^{2+}(aq)$	$E^\circ_{cell} = +1.194$ V

53. (a)

Anode, Oxdn:	$Zn(s) \longrightarrow Zn^{2+}(aq) + 2\ e^-$	$-E^\circ = +0.763$ V
Cathode, Redn:	$Br_2(l) + 2\ e^- \longrightarrow 2\ Br^-(aq)$	$E^\circ = +1.065$ V
Net:	$Zn(s) + Br_2(l) \longrightarrow Zn^{2+}(aq) + 2\ Br^-(aq)$	$E^\circ_{cell} = +1.828$ V

(b)

Anode, Oxdn:	$\{Li(s) \longrightarrow Li^+(aq) + e^-\} \times 2$	$-E^\circ = +3.040$ V
Cathode, Redn:	$F_2(g) + 2\ e^- \longrightarrow 2\ F^-(aq)$	$E^\circ = +2.866$ V
Net:	$2\ Li(s) + F_2(g) \longrightarrow 2\ Li^+(aq) + 2\ F^-(aq)$	$E^\circ_{cell} = +5.906$ V

(c)

Anode, Oxdn:	$Mg(s) \longrightarrow Mg^{2+}(aq) + 2\ e^-$	$-E^\circ = +2.356$ V
Cathode, Redn:	$I_2(s) + 2\ e^- \longrightarrow 2\ I^-(aq)$	$E^\circ = +0.535$ V
Net:	$Mg(s) + I_2(s) \longrightarrow Mg^{2+}(aq) + 2\ I^-(aq)$	$E^\circ_{cell} = +2.891$ V

(d)

Anode, Oxdn:	$\{Fe(s) \longrightarrow Fe^{2+}(aq) + 2\ e^-\} \times 2$	$-E^\circ = +0.440$ V
Cathode, Redn:	$O_2(g) + 4\ H^+(aq) + 4\ e^- \longrightarrow 2\ H_2O$	$E^\circ = +1.229$ V
Net:	$2\ Fe(s) + O_2(g) + 4\ H^+(aq) \longrightarrow 2\ Fe^{2+}(aq) + 2\ H_2O$	$E^\circ_{cell} = +1.669$ V

54. (a)

Oxidation:	$Zn(s) \longrightarrow Zn^{2+}(aq) + 2\ e^-$
Reduction:	$2\ MnO_2(s) + H_2O + 2\ e^- \longrightarrow Mn_2O_3(s) + 2\ OH^-(aq)$
Acid-base:	$\{NH_4^+(aq) + OH^-(aq) \longrightarrow NH_3(g) + H_2O(l)\} \times 2$
Complex:	$Zn^{2+}(aq) + 2\ NH_3(aq) + 2\ Cl^-(aq) \longrightarrow [Zn(NH_3)_2]Cl_2(s)$

with $-E^\circ = +0.763$ V for the oxidation.

Net: $Zn(s) + 2\ MnO_2(s) + 2\ NH_4^+(aq) + 2\ Cl^-(aq) \longrightarrow Mn_2O_3(s) + H_2O(l) + [Zn(NH_3)_2]Cl_2(s)$

(b) $\Delta G^\circ = -n\ \mathscr{F}\ E^\circ_{cell} = -(2\ \text{mol e}^-)(96485\ \text{C/mol e}^-)(1.55\ \text{V}) = -2.99 \times 10^5$ J/mol

This is the standard free energy change for the entire reaction, which is composed of the four reactions in part (a). We can determine the values of ΔG° for the acid-base and complex formation reactions with data from Table 17-2 and $pK_f = -4.81$.

$\Delta G^\circ_{a-b} = -RT \ln K_b = -(8.3145\ \text{J mol}^{-1}\ \text{K}^{-1})(298.15\ \text{K}) \ln(1.8 \times 10^{-5}) = 2.708 \times 10^4$ J/mol

$\Delta G^\circ_{cmplx} = -RT \ln K_f = -(8.3145\ \text{J mol}^{-1}\ \text{K}^{-1})(298.15\ \text{K}) \ln(10^{4.81}) = -2.745 \times 10^4$ J/mol

Then $\Delta G^\circ_{total} = \Delta G^\circ_{redox} + \Delta G^\circ_{a-b} + \Delta G^\circ_{cmplx}$

$\Delta G^\circ_{redox} = \Delta G^\circ_{total} - \Delta G^\circ_{a-b} - \Delta G^\circ_{cmplx}$

$$= -2.99 \times 10^5 \text{ J/mol} - 2.708 \times 10^4 + 2.745 \times 10^4 \text{ J/mol} = -2.99 \times 10^5 \text{ J/mol}$$

Note that the acid-base (2.708×10^4 J/mol) and complex ion formation (-2.745×10^4 J/mol) reactions in essence "cancel each other out" in this case. Thus, the voltage of the redox reactions alone is

$$1.55 \text{ V} = +0.763 \text{ V} + E°(MnO_2/Mn_2O_3)$$
$$E°(MnO_2/Mn_2O_3) = 1.55 \text{ V} - 0.763 \text{ V} = +0.79 \text{ V}$$

55. Cell reaction: $\quad 2 H_2(g) + O_2(g) \longrightarrow 2 H_2O(l)$

$$\Delta G_{rxn}^° = 2 \Delta G_f^°[H_2O(l)] = 2 (-237.2 \text{ kJ/mol}) = -474.4 \text{ kJ/mol}$$
$$\Delta H_{rxn}^° = 2 \Delta H_f^°[H_2O(l)] = 2 (-285.8 \text{ kJ/mol}) = -571.6 \text{ kJ/mol}$$

$$E_{cell}^° = -\frac{\Delta G°}{n \, \mathfrak{F}} = -\frac{-474.4 \times 10^3 \text{ J/mol}}{4 \text{ mol e}^- \times 96485 \text{ C/mol e}^-} = 1.229 \text{ V}$$

$$\varepsilon = \frac{\Delta G°}{\Delta H°} = \frac{-474.4 \text{ kJ/mol}}{-571.6 \text{ kJ/mol}} = 0.8300 = \text{efficiency value}$$

56. We already know that $\Delta G° = -2108$ kJ/mol for the propane-oxygen reaction. We calculate a value of $\Delta H°$ from data in Appendix D.

$$\Delta H° = 3 \Delta H_f^°[CO_2(g)] + 4 \Delta H_f^°[H_2O(l)] - \Delta H_f^°[C_3H_8(g)] - 5 \Delta H_f^°[O_2(g)]$$
$$= 3 \times (-393.5 \text{ kJ}) + 4 \times (-285.8 \text{ kJ}) - (-103.8 \text{ kJ}) - 5 \times 0.00 \text{ kJ}$$
$$= -2219.9 \text{ kJ/mol}$$

$$\varepsilon = \frac{\Delta G°}{\Delta H°} = \frac{-2108 \text{ kJ/mol}}{-2219.9 \text{ kJ/mol}} = 0.9496 = \text{efficiency value} = 94.96\%$$

Electrochemical Mechanism of Corrosion

57. During corrosion, the metal that corrodes produces electrons. Hence, this metal is negatively charged. Thus, we wish to make the pipe the cathode, where reduction occurs, in an electrochemical cell, since corrosion proceeds by oxidation. Thus, the inert electrode will be the anode, and any possible oxidation will occur there. This means that the inert electode will be the positive electrode and the pipe will be the negative electrode.

58. As soon as the iron and the copper came into contact, and electrochemical cell was created, in which the more electrochemically active metal (Fe) oxidized. In this way the iron behaved as a sacrificial anode, protecting the copper from corrosion. The two half-reactions and the net cell reaction follow.

Anode, Oxidation:	$Fe(s) \longrightarrow Fe^{2+} + 2 e^-$	$-E° = +0.440 \text{ V}$
Cathode, Reduction:	$Cu^{2+}(aq) + 2 e^- \longrightarrow Cu(s)$	$E° = +0.337 \text{ V}$
Net cell reaction:	$Fe(s) + Cu^{2+}(aq) \longrightarrow Cu(s) + Fe^{2+}(aq)$	$E_{cell}^° = +0.777 \text{ V}$

Note that, because of the presence of the iron, and its electrical contact with the copper, any copper that does corrode will be reduced back to the metal.

59. The anode reaction—that of oxidation—is the formation of $Fe^{2+}(aq)$. This occurs far below the water line. The cathode reaction—that of reduction—is the formation of $OH^-(aq)$ from $O_2(g)$. It is logical that this reaction would occur at or near the water line. This reduction reaction requires $O_2(g)$ from the atmosphere and H_2O from the water. The oxidation reaction, on the other hand simply requires iron from the pipe and also an aqueous solution into which the $Fe^{2+}(aq)$ can disperse and not build up to such a high concentration that corrosion is inhibited.

Anode, Oxidation:	$Fe(s) \longrightarrow Fe^{2+}(aq) + 2 e^-$
Cathode, Reduction:	$O_2(g) + 2 H_2O + 4 e^- \longrightarrow 4 OH^-(aq)$

Electrolysis Reactions

60. We determine the standard cell voltage of each chemical reaction. Those voltages that are negative are those of chemical reactions that require electrolysis.

(a)

Oxidation:	$2 H_2O \longrightarrow 4 H^+(aq) + O_2(g) + 4 e^-$	$-E° = -1.229 \text{ V}$
Reduction:	$\{2 H^+(aq) + 2 e^- \longrightarrow H_2(g)\} \quad \times 2$	$E° = 0.000 \text{ V}$
Net:	$2 H_2O \longrightarrow 2 H_2(g) + O_2(g)$	$E_{cell}^° = -1.229 \text{ V}$

This reaction requires electrolysis, with an applied voltage of at least +1.229 V.

(b) Oxidation: $Zn(s) \longrightarrow Zn^{2+}(aq) + 2\ e^-$ $\qquad -E° = +0.763\ V$

Reduction: $Fe^{2+}(aq) + 2\ e^- \longrightarrow Fe(s)$ $\qquad E° = -0.440\ V$

Net: $Zn(s) + Fe^{2+}(aq) \longrightarrow Fe(s) + Zn^{2+}(aq)$ $\qquad E°_{cell} = +0.323\ V$

This is a spontaneous reaction.

(c) Oxidation: $2\ Br^-(aq) \longrightarrow Br_2(l) + 2\ e^-$ $\qquad -E° = -1.065\ V$

Reduction: $O_2(g) + 2\ H^+(aq) + 2\ e^- \longrightarrow H_2O_2(aq)$ $\qquad E° = +0.695\ V$

Net: $2\ Br^-(aq) + O_2(g) + 2\ H^+(aq) \longrightarrow Br_2(l) + H_2O_2(aq)$ $\qquad E°_{cell} = -0.370\ V$

This reaction requires electrolysis, with an applied voltage of at least +0.370 V.

(d) Oxidation: $4\ OH^-(aq) \longrightarrow O_2(g) + 2\ H_2O + 4\ e^-$ $\qquad -E° = -0.401\ V$

Reduction: $\{2\ H_2O + 2\ e^- \longrightarrow H_2(g) + 2\ OH^-(aq)\}\quad \times 2$ $\qquad E° = -0.828\ V$

Net: $2\ H_2O \longrightarrow 2\ H_2(g) + O_2(g)$ $\qquad E°_{cell} = -1.229\ V$

This reaction requires electrolysis, with an applied voltage of at least +1.229 V.

61. Since oxidation occurs at the anode, we know that the product cannot be H_2 (since it is produced from H_2O in a reduction reaction), SO_2 (which is a reduction product of SO_4^{2-}), or SO_3 (which is produced from SO_4^{2-} without a change of oxidation state; it is the dehydration product of H_2SO_4). But O_2 indeed is the result of the oxidation of H_2O.

62. (a) The two gases that are produced are $H_2(g)$ and $O_2(g)$.

(b) At the anode: $2\ H_2O \longrightarrow 4\ H^+(aq) + O_2(g) + 4\ e^-$ $\qquad -E° = -1.229\ V$

At the cathode: $\{2\ H^+(aq) + 2\ e^- \longrightarrow H_2(g)\}\quad \times 2$ $\qquad E° = 0.000\ V$

Net cell reaction: $2\ H_2O \longrightarrow 2\ H_2(g) + O_2(g)$ $\qquad E°_{cell} = -1.229\ V$

63. (a) $Zn^{2+}(aq) + 2\ e^- \longrightarrow Zn(s)$

mass of Zn = $23.0\ min \times \dfrac{60\ s}{1\ min} \times \dfrac{2.33\ C}{1\ s} \times \dfrac{1\ mol\ e^-}{96485\ C} \times \dfrac{1\ mol\ Zn}{2\ mol\ e^-} \times \dfrac{65.39\ g\ Zn}{1\ mol\ Zn} = 1.09\ g\ Zn$

(b) $2\ I^-(aq) \longrightarrow I_2(s) + 2\ e^-$

time needed = $3.44\ g\ I_2 \times \dfrac{1\ mol\ I_2}{253.8\ g\ I_2} \times \dfrac{2\ mol\ e^-}{1\ mol\ I_2} \times \dfrac{96485\ C}{1\ mol\ e^-} \times \dfrac{1\ s}{1.56\ C} \times \dfrac{1\ min}{60\ s} = 27.9\ min$

(c) $Cu^{2+}(aq) + 2\ e^- \longrightarrow Cu(s)$

mmol Cu^{2+} removed $= 235\ s \times \dfrac{2.17\ C}{1\ s} \times \dfrac{1\ mol\ e^-}{96485\ C} \times \dfrac{1\ mol\ Cu^{2+}}{2\ mol\ e^-} \times \dfrac{1000\ mmol}{1\ mol}$

$= 2.64\ mmol\ Cu^{2+}$

decrease in $[Cu^{2+}] = \dfrac{2.64\ mmol\ Cu^{2+}}{335\ mL} = 0.00788\ M$

final $[Cu^{2+}] = 0.215\ M - 0.00788\ M = 0.207\ M$

(d) mmol Ag^+ removed = $255\ mL\ (0.185\ M - 0.175\ M) = 2.55\ mmol\ Ag^+$

time needed = $2.55\ mmol\ Ag^+ \times \dfrac{1\ mol\ Ag^+}{1000\ mmol\ Ag^+} \times \dfrac{1\ mol\ e^-}{1\ mol\ Ag^+} \times \dfrac{96485\ C}{1\ mol\ e^-} \times \dfrac{1\ s}{1.92\ C} = 128\ s$

64. (a) charge = $0.918\ g\ Ag \times \dfrac{1\ mol\ Ag}{107.87\ g\ Ag} \times \dfrac{1\ mol\ e^-}{1\ mol\ Ag} \times \dfrac{96,485\ C}{1\ mol\ e^-} = 821\ C$

(b) current = $\dfrac{821\ C}{1214\ s} = 0.676\ A$

65. (a) Anode, Oxdn: $2\ H_2O \longrightarrow 4\ H^+(aq) + 4\ e^- + O_2(g)$ $\qquad -E° = -1.229\ V$

Cathode, Redn: $\{Ag^+(aq) + e^- \longrightarrow Ag(s)\}\ \times 4$ $\qquad E° = +0.800\ V$

Net: $2\ H_2O + 4\ Ag^+(aq) \longrightarrow 4\ H^+(aq) + O_2(g) + 4\ Ag(s)$ $\qquad E°_{cell} = -0.429\ V$

(b) charge = $(25.2794 - 25.0782)g\ Ag \times \dfrac{1\ mol\ Ag}{107.87\ g\ Ag} \times \dfrac{1\ mol\ e^-}{1\ mol\ Ag} \times \dfrac{96485\ C}{1\ mol\ e^-} = 180.0\ C$

current = $\dfrac{180.0\ C}{2.00\ h} \times \dfrac{1\ h}{3600\ s} = 0.0250\ A$

(c) The gas is oxygen.

$$V = \frac{nRT}{P} = \frac{\left(180.0 \text{ C} \times \frac{1 \text{ mol e}^-}{96485 \text{ C}} \times \frac{1 \text{ mol O}_2}{4 \text{ mol e}^-}\right) 0.08206 \frac{\text{L atm}}{\text{mol K}} (23°\text{C} + 273)\text{K}}{755 \text{ mmHg} \times \frac{1 \text{ atm}}{760 \text{ mmHg}}}$$

$$= 0.0114 \text{ L O}_2 \times \frac{1000 \text{ mL}}{1 \text{ L}} = 11.4 \text{ mL}$$

66. The product of the electrolysis of Na_2SO_4(aq) at the anode is oxygen, $-E°[O_2(g)|H_2O] = -1.229$ V. The other possible product is $S_2O_8{}^{2-}$(aq), which will not form, since it has a considerably less favorable half-cell potential, $-E°[S_2O_8{}^{2-}(aq)|SO_4{}^{2-}(aq)] = -2.01$

$$\text{mol O}_2 = 2.11 \text{ h} \times \frac{3600 \text{ s}}{1 \text{ h}} \times \frac{3.16 \text{ C}}{1 \text{ s}} \times \frac{1 \text{ mol e}^-}{96485 \text{ C}} \times \frac{1 \text{ mol O}_2}{4 \text{ mol e}^-} = 0.0622 \text{ mol O}_2$$

The vapor pressure of water at 25°C, from Table 12-2, is 23.8 mmHg.

$$V = \frac{nRT}{P} = \frac{0.0622 \text{ mol} \times 0.08206 \text{ L atm mol}^{-1} \text{ K}^{-1} \times 298 \text{ K}}{(753 - 23.8) \text{ mmHg} \times \frac{1 \text{ atm}}{760 \text{ mmHg}}} = 1.59 \text{ L O}_2(g)$$

22 CHEMISTRY OF THE REPRESENTATIVE (MAIN GROUP) ELEMENTS I: METALS

REVIEW QUESTIONS

1. **(a)** A dimer is a molecule that is formed by the joining together of two identical simpler molecules (called monomers). For instance, N_2O_4 is a dimer of NO_2.
   (b) An adduct is formed when two simple molecules come together and a covalent bond, often a coordinate covalent bond, forms between them. $NH_3 \cdot BF_3$ is an adduct of NH_3 and BF_3.
   (c) Calcination is the process of heating a carbonate strongly to form an oxide.
   (d) An amphoteric axide is one that will react with either strong acid or with strong base.

2. **(a)** A diagonal relationship refers to the similarity between two elements that are diagonally related to each other in the periodic table, such as Li and Mg, or Be and Al. Even though these elements are in different periodic families, they have some similarities in behavior.
   (b) Deionized water is prepared by ion exchange by passing it through material in which the ions H^+ or OH^- are present. These ions substitute or exchange for the ions in the water, in a two-step process: first the cations in the water are replaced by H^+, then the anions by OH^-. The result is water nearly free of ionic contaminants.
   (c) The thermite reaction refers to the reduction of a metal oxide with another, more active, metal in a highly exothermic reaction, for instance: $Fe_2O_3(s) + 2\ Al(s) \longrightarrow 2\ Fe(liquid!) + Al_2O_3(s)$
   (d) The "inert pair" effect refers to the tendency of heavier representative metals to have oxidation states in which they have lost their np electrons, but not their ns^2 electrons. Thus, they have an oxidation state two units less than their periodic table family number. Examples include Pb^{2+}, Sn^{2+}, Bi^{3+}, Sb^{3+}, Tl^+.

3. **(a)** The peroxide ion is O_2^{2-}; the superoxide ion is O_2^-.
   (b) Quicklime is the common name for $CaO(s)$; slaked lime is the common name for $Ca(OH)_2(s)$.
   (c) Temporary hard water contains divalent cations, such as Ca^{2+}, Mg^{2+}, and Fe^{2+} and the bicarbonate anion, HCO_3^-. Heating produces, H_2O, $CO_2(g)$, and a carbonate precipitate. Permanent hard water does not form a precipitate upon heating, since the anion is one such as SO_4^{2-} that is thermally stable.
   (d) A soap is a potassium or sodium salt of a natural or slightly altered carboxylic acid (—COOH) that has a long hydrocarbon chain. A detergent is a synthetic sodium or potassium salt of a long hydrocarbon chain sulfonic acid (—OSO_3H).

4. **(a)** PbO_2 lead(IV) oxide   **(b)** SnF_2 tin(II) fluoride
   (c) $CaCl_2 \cdot 6H_2O$ calcium chloride hexahydrate   **(d)** Mg_3N_2 magnesium nitride
   (e) $Ca(OH)_2$ calcium hydroxide
   (f) $Mg(HCO_3)_2$ magnesium hydrogen carbonate   **(g)** KO_2 potassium superoxide

5. **(a)** $MgCO_3(s) \xrightarrow{\Delta} MgO(s) + CO_2(g)$
   (b) $CaO(s) + 2\ HCl(aq) \longrightarrow CaCl_2(aq) + H_2O$
   (c) $2\ Al(s) + 2\ K^+(aq) + 2\ OH^-(aq) + 6\ H_2O \longrightarrow 2\ K^+(aq) + 2\ [Al(OH)_4]^-(aq) + 3\ H_2(g)$
   (d) $CaO(s) + H_2O \longrightarrow Ca(OH)_2(s, \text{in limited water})$
   $\longrightarrow Ca^{2+}(aq) + 2\ OH^-(aq)$ [in abundant water]
   (e) $2\ Na_2O_2(s) + 2\ CO_2(g) \longrightarrow 2\ Na_2CO_3(s) + O_2(g)$
   (f) $K_2CO_3(s) \xrightarrow{\Delta} \text{no reaction}$

6. **(a)** $MgCO_3(s) + 2\,HCl(aq) \longrightarrow MgCl_2(aq) + CO_2(g) + H_2O$

(b) $2\,Na(s) + 2\,H_2O \longrightarrow 2\,NaOH(aq) + H_2(g)$

$2\,Al(s) + 2\,NaOH(aq) + 6\,H_2O \longrightarrow 2\,Na[Al(OH)]_4(aq) + 3\,H_2(g)$

(c) $2\,NaCl(s) + H_2SO_4(\text{conc., aq}) \longrightarrow 2\,HCl(g) + Na_2SO_4(s)$

7. **(a)** $K_2CO_3(aq) + Ba(OH)_2(aq) \longrightarrow BaCO_3(s) + 2\,KOH(aq)$

(b) $Mg(HCO_3)_2(s) \xrightarrow{\Delta} MgCO_3(s) + CO_2(g) + H_2O(g)$

(c) $SnO(s) + C(s) \xrightarrow{\Delta} Sn(l) + CO(g)$

(d) $2\,NaF(s) + H_2SO_4(\text{conc., aq}) \longrightarrow 2\,HF(g) + Na_2SO_4(s)$

(e) $CaCO_3(s) + 2\,HCl(aq) \longrightarrow CaCl_2(aq) + H_2O + CO_2(g)$

(f) $PbO_2(s) + 4\,HI(aq) \longrightarrow PbI_2(s) + I_2(s) + 2\,H_2O$

8. Replace the names with chemical formulas and balance the result.

$CaSO_4 \cdot 2H_2O(s) + (NH_4)_2CO_3(aq) \longrightarrow (NH_4)_2SO_4(aq) + CaCO_3(s) + 2\,H_2O$

9. Temporary hard water is softened by the addition of an alkaline (basic) material. All of the substances listed form alkaline solutions except NH_4Cl. NH_4^+ hydrolyzes to form an acidic solution, and thus could not be used to soften temporary hard water.

10. If the water is 185 ppm Ca^{2+}, 1.00×10^6 g soln will contain 185 g Ca.

$$[Na^+] = \frac{185\text{ g }Ca^{2+}}{1.00 \times 10^6\text{ g soln}} \times \frac{1.00 \times 10^6\text{ g soln}}{1000\text{ L}} \times \frac{1\text{ mol }Ca^{2+}}{40.08\text{ g }Ca^{2+}} \times \frac{2\text{ mol }Na^+}{1\text{ mol }Ca^{2+}} = 9.23 \times 10^{-3}\text{ M}$$

11. **(a)** $BaCO_3(s) \xrightarrow{\Delta} BaO(s) + CO_2(g)$

(b) $MgO(s) \xrightarrow{\Delta}$ no reaction

(c) $SnO_2(s) + 2\,CO(g) \xrightarrow{\Delta} Sn(l) + 2\,CO_2(g)$

(d) $2\,Na^+(aq) + 2\,Cl^-(aq) + 2\,H_2O \xrightarrow{electrolysis} 2\,Na^+(aq) + OH^-(aq) + Cl_2(g) + H_2(g)$

12. **(a)** $Li_2CO_3(s) + 2\,HBr(aq) \longrightarrow 2\,LiBr(aq) + H_2O + CO_2(g)$

(b) $Li_2O(s) + (NH_4)_2CO_3(aq) \longrightarrow Li_2CO_3(s) + 2\,NH_3(g) + H_2O$

$NH_4^+(aq)$ is present in very limited amount because the solution is strongly basic; Li_2O is the anhydride of a strong base.

(c) $Mg(OH)_2(s) + H_2SO_3(aq) \longrightarrow MgSO_3(aq) + 2\,H_2O$

(d) $PbO(s) + OCl^-(aq) \longrightarrow PbO_2(s) + Cl^-(aq)$

13. To answer this one, your really do not need to formally study chemistry. You just have to observe the world around you. Neither aluminum foil nor aluminum cookware reacts with water. But to explain that answer, you have to know that of these four active metals, only aluminum forms an oxide that adheres tightly to its surface and prevents further reaction.

14. The correct answer is **(a)**. Balanced equations for the four pairs follow.

$Ca(s) + 2\,H_2O \longrightarrow Ca(OH)_2(s) + H_2(g)$ $\qquad$ $CaH_2(s) + 2\,H_2O \longrightarrow Ca(OH)_2(s) + 2\,H_2(g)$

$2\,Na(s) + 2\,H_2O \longrightarrow 2\,NaOH(aq) + H_2(g)$ $\qquad$ $2\,Na_2O_2(s) + 2\,H_2O \longrightarrow 4\,NaOH(aq) + O_2(g)$

$2\,K(s) + 2\,H_2O \longrightarrow 2\,KOH(aq) + H_2(g)$ $\qquad$ $4\,KO_2(s) + 2\,H_2O \longrightarrow 4\,KOH(aq) + 3\,O_2(g)$

$Li_3N(s) + 3\,H_2O \longrightarrow 3\,LiOH(aq) + NH_3(g)$ $\qquad$ $LiH(s) + H_2O \longrightarrow LiOH(aq) + H_2(g)$

15. The symbols of the elements are given following each compound's formula.

(a) Limestone is slightly impure $CaCO_3(s)$—Ca, C, and O

(b) Gypsum is slightly impure $CaSO_4 \cdot 2H_2O$—Ca, S, O, and H

(c) Bauxite is the primary ore of aluminum: $Al_2O_3 \cdot (1\text{-}3)H_2O$—Al, O, and H

(d) Bronze is an alloy of approximately 90% copper and 10% tin—Cu and Sn

(e) Slaked lime and hydrated lime both are $Ca(OH)_2(s)$—Ca, O, and H

16. (a) Stalactites are $CaCO_3(s)$ **(b)** Plaster of Paris is $CaSO_4 \cdot \frac{1}{2}H_2O$

(c) "bathtub ring" is a salt of Ca^{2+} and a long-carbon chain carboxylate anion. An example would be calcium palmitate: $Ca[CH_3(CH_2)_{14}COO]_2(s)$

(d) barium "milkshake" is an aqueous temporary suspension of $BaSO_4(s)$

(e) rubies are Al_2O_3 with Cr^{3+} ions replacing some Al^{3+} ions.

EXERCISES

Alkali (Group 1A) Metals

17. (a) $2\,Li(s) + Cl_2(g) \longrightarrow 2\,LiCl(s)$ **(b)** $2\,Na(s) + O_2(g) \longrightarrow Na_2O_2(s)$

(c) $Li_2CO_3(s) \overset{\Delta}{\longrightarrow} Li_2O(s) + CO_2(g)$ **(d)** $Na_2SO_4(s) + 4\,C(s) \longrightarrow Na_2S(s) + 4\,CO(g)$

(e) $K(s) + O_2(g) \longrightarrow KO_2(s)$

(f) $Na^+(aq) + Cl^-(aq) + NH_3(aq) + CO_2(g) + H_2O \xrightarrow{\text{Solvay process}} NaHCO_3(s) + NH_4Cl(aq)$

18. Both LiCl and KCl are soluble in water, but Li_3PO_4 is not very soluble. Hence the addition of $Na_3PO_4(aq)$ to a solution of the white solid will produce a precipitate if the white solid is LiCl, but no precipitate if the white solid is KCl. Another possibility is a flame test; lithium gives a red color to a flame, while the potassium flame test is violet.

19. (a) $2\,Li(s) + 2\,H_2O(l) \longrightarrow 2\,LiOH(aq) + H_2(g)$ **(b)** $LiH(s) + H_2O(l) \longrightarrow LiOH(aq) + H_2(g)$

(c) $Li_2O(s) + H_2O \longrightarrow 2\,LiOH(aq)$

20. In addition to $OH^-(aq)$, the other expected product is $O_2(g)$ in the case of peroxide and superoxide. The resulting equations are readily balanced by inspection if one pays attention to charge balance.

Oxide: $O^{2-} + H_2O \longrightarrow 2\,OH^-(aq)$ Peroxide: $2\,O_2^{2-} + 2\,H_2O \longrightarrow 4\,OH^-(aq) + O_2(g)$

Superoxide: $4\,O_2^- + 2\,H_2O \longrightarrow 4\,OH^-(aq) + 3\,O_2$

21. We know that sodium metal was produced at the cathode from the reduction of sodium ion, Na^+. Thus, hydroxide must have been involved in an oxidation at the anode. The hydrogen in hydroxide ion already is in its highest oxidation state and thus could not be oxidized. This leaves oxidation of the oxygen to elementary oxygen.

Cathode, reduction: $\{Na^+ + e^- \longrightarrow Na(l)\} \times 4$

Anode, oxidation: $4\,OH^- \longrightarrow O_2(g) + 2\,H_2O(g) + 4\,e^-$

Net: $4\,Na^+ + 4\,OH^- \longrightarrow 4\,Na(l) + O_2(g) + 2\,H_2O(g)$

22. There are two principal reasons why the electrolysis of NaCl(l) is used to produce sodium commercially rather than the electrolysis of NaOH(l). First, NaCl is readily available whereas NaOH is produced from NaCl(aq)—by electrolysis, in fact. Thus the raw material NaCl is much cheaper than is NaOH. Second, but less important, when sodium is produced from the electrolysis of NaCl(l), a by-product is $Cl_2(g)$, whereas $O_2(g)$ is a by-product of the production of NaOH(l). Since $O_2(g)$ can be produced more cheaply by the fractional distillation of liquid air, $Cl_2(g)$ can be sold for a higher price than $O_2(g)$. Thus, the two reasons for producing sodium from NaCl(l) rather than NaOH(l), are a much cheaper raw material and a more profitable by-product.

23. (a) $\text{total energy} = 3.0\,V \times 0.50\,A\,h \times \dfrac{3600\,s}{1\,hr} \times \dfrac{1\,C/s}{1\,A} \times \dfrac{1\,J}{1\,V \cdot C} = 5.4 \times 10^3\,J$

$\text{time} = 5.4 \times 10^3\,J \times \dfrac{1\,s}{5 \times 10^{-6}\,J} = 1._1 \times 10^9\,s \times \dfrac{1\,hr}{3600\,s} \times \dfrac{1\,day}{24\,hr} \times \dfrac{1\,y}{365\,day} = 34\,y$

We obtained the first conversion factor for time as follows.

$5.0\,\mu W \times \dfrac{1 \times 10^{-6}\,W}{1\,\mu W} \times \dfrac{1\,J/s}{1\,W} = \dfrac{5.0 \times 10^{-6}\,J}{1\,s}$

(b) The capacity of the battery is determined by the mass of Li present.

$\text{mass Li} = 0.50\,A\,h \times \dfrac{1\,C/s}{1\,A} \times \dfrac{3600\,s}{1\,h} \times \dfrac{1\,mol\,e^-}{96500\,C} \times \dfrac{1\,mol\,Li}{1\,mol\,e^-} \times \dfrac{6.941\,g\,Li}{1\,mol\,Li} = 0.13\,g\,Li$

24. **(a)** We first compute the mass of $NaHCO_3$ that should be produced from 1.00 ton NaCl, assuming that all of the Na in the NaCl ends up in the $NaHCO_3$. We use the unit, ton-mole, to simplify the calculations.

$$mass\ NaHCO_3 = 1.00\ ton\ NaCl \times \frac{1\ ton\text{-}mol\ NaCl}{58.4\ ton\ NaCl} \times \frac{1\ ton\text{-}mol\ Na}{1\ ton\text{-}mol\ NaCl} \times \frac{1\ ton\text{-}mol\ NaHCO_3}{1\ ton\text{-}mol\ Na}$$

$$\times \frac{84.0\ ton\ NaHCO_3}{1\ ton\text{-}mol\ NaHCO_3} = 1.44\ ton\ NaHCO_3$$

$$\%\ yield = \frac{1.03\ ton\ NaHCO_3\ produced}{1.44\ ton\ NaHCO_3\ expected} \times 100\% = 71.5\%\ yield$$

(b) NH_3 is used in the principal step of the Solvay proces to produce a medium in which $NaHCO_3$ is formed and from which it will precipitate. The solution remaining from the precipitation contains NH_4Cl, from which NH_3 is recovered by treatment with $Ca(OH)_2$. Thus NH_3 is simply used during the Solvay process to produce the proper conditions for the desired reactions. Any net consumption of NH_3 is the result of unavoidable losses during production.

25. $H_2(g)$ and $Cl_2(g)$ are produced during the electrolysis of NaCl(aq), which is summarized in equation (22.7). The electrode reactions are the following.

Anode, Oxdn: $\quad\quad 2\ Cl^-(aq) \longrightarrow Cl_2(g) + 2\ e^-$

Cathode, Redn: $\quad\quad 2\ H_2O + 2\ e^- \longrightarrow H_2(g) + 2\ OH^-(aq)$

We can compute the amount of OH^- produced at the cathode.

$$mol\ OH^- = 137\ s \times \frac{1.08\ C}{1\ s} \times \frac{1\ mol\ e^-}{96500\ C} \times \frac{2\ mol\ OH^-}{2\ mol\ e^-} = 1.53 \times 10^{-3}\ mol\ OH^-$$

Then we compute the $[OH^-]$ and, from that, the pH of the solution.

$$[OH^-] = \frac{1.53 \times 10^{-3}\ mol\ OH^-}{0.445\ L\ soln} = 3.44 \times 10^{-3}\ M \quad\quad pOH = -\log(3.44 \times 10^{-3}) = 2.463$$

$$pH = 14.00 - 2.463 = 11.54$$

The final pH of the solution might, however, depend on the initial [NaCl] of the solution. Once the $[Cl^-]$ drops to a quite low value, another reaction can occur at the anode.

Anode, Oxdn: $\quad\quad 2\ H_2O \longrightarrow 4\ H^+(aq) + 4\ e^- + O_2(g)$

Note that this reaction produces 1 mol H^+ for every mole of electrons. Thus it precisely counters the cathode reaction and, once $Cl_2(g)$ is no longer produced at the anode, the pH of the solution will cease changing. Since the number of moles of OH^- produced at the cathode (1.53×10^{-3} mol OH^-) equals the number of moles of Cl^- consumed at the anode, and this is quite a small number, it is unlikely that the second anode reaction will occur.

26. **(a)** $Ca(OH)_2(s) + SO_4^{2-}(aq) \rightleftharpoons CaSO_4(s) + 2\ OH^-(aq)$

(b) We sum two solubility reactions and combine their values of K_{sp}.

$$Ca(OH)_2(s) \rightleftharpoons Ca^{2+}(aq) + 2\ OH^-(aq) \quad\quad\quad\quad K_{sp} = 5.5 \times 10^{-6}$$

$$Ca^{2+}(aq) + SO_4^{2-}(aq) \rightleftharpoons CaSO_4(s) \quad\quad\quad\quad 1/K_{sp} = 1/9.1 \times 10^{-6}$$

$$\overline{Ca(OH)_2(s) + SO_4^{2-}(aq) \rightleftharpoons CaSO_4(s) + 2\ OH^-(aq) \quad\quad K = \frac{5.5 \times 10^{-6}}{9.1 \times 10^{-6}} = 0.60}$$

Because this value of K is not significantly different from 1.00, we conclude that the reaction lies neither very far to the right (it does not go to completion) nor to the left.

(c) Reaction: $\quad Ca(OH)_2(s) + SO_4^{2-}(aq) \rightleftharpoons CaSO_4(s) + 2\ OH^-(aq)$

Initial: $\quad\quad\quad\quad\quad\quad\quad\quad 1.00\ M$

Changes: $\quad\quad\quad\quad\quad\quad\quad -x\ M \quad\quad\quad\quad\quad\quad\quad\quad +2x\ M$

Equil: $\quad\quad\quad\quad\quad\quad\quad (1.00-x)M \quad\quad\quad\quad\quad\quad 2x\ M$

$$K = \frac{[OH^-]^2}{[SO_4^{2-}]} = 0.60 = \frac{4x^2}{1.00-x} \quad\quad\quad 4x^2 = 0.60 - 0.60\ x \quad\quad\quad 4x^2 + 0.60\ x - 0.60 = 0$$

$$x = \frac{-b \pm \sqrt{b^2 - 4ac}}{2a} = \frac{-0.60 \pm \sqrt{0.36 + 9.60}}{8} = 0.32\ M$$

$$[SO_4^{2-}] = 1.00 - x = 0.68\ M \quad\quad\quad\quad [OH^-] = 2x = 0.64$$

Alkaline earth (group 2A) metals

27. $CaO \xleftarrow{\Delta} CaCO_3 \xleftarrow{CO_2} Ca(OH)_2 \xrightarrow{HCl} CaCl_2 \xrightarrow{electrolysis} Ca$

$CaHPO_4 \xleftarrow{H_3PO_4} \quad | \quad H_2SO_4 \rightarrow CaSO_4$

The reactions are as follows.

$Ca(OH)_2(s) + 2\ HCl(aq) \longrightarrow CaCl_2(aq) + 2\ H_2O$

$CaCl_2(l) \xrightarrow{\Delta,\ electrolysis} Ca(l) + Cl_2(g)$

$Ca(OH)_2(s) + CO_2(g) \longrightarrow CaCO_3(s) + H_2O(g)$

$CaCO_3(s) \xrightarrow{\Delta} CaO(s) + CO_2(g)$

$Ca(OH)_2(s) + H_2SO_4(aq) \longrightarrow CaSO_4(s) + 2\ H_2O$

$Ca(OH)_2(s) + H_3PO_4(aq) \longrightarrow CaHPO_4(aq) + 2\ H_2O$ Actually $CaCO_3(s)$ is the industrial starting material from which $Ca(OH)_2$ is made by slaking lime. $CaO(s) + H_2O \longrightarrow Ca(OH)_2(s)$

28. **(a)** $BeF_2 + 2\ Na \longrightarrow Be + 2\ NaF(s)$ **(c)** $UO_2(s) + 2\ Ca(s) \longrightarrow U(s) + 2\ CaO(s)$

(b) $Ca(s) + Cl_2(g) \longrightarrow CaCl_2(s)$ **(d)** $MgCO_3 \cdot CaCO_3(s) \xrightarrow{\Delta} MgO(s) + CaO(s) + 2\ CO_2(g)$

(e) $H_2SO_4(aq) + Ca(OH)_2(s) \longrightarrow CaSO_4(s) + 2\ H_2O(l)$

29. **(a)** $Mg(HCO_3)_2(s) \xrightarrow{heat} MgO(s) + 2\ CO_2(g) + H_2O$

(b) $CaCl_2(l) \xrightarrow{electricity} Ca(l) + Cl_2(g)$ **(c)** $Ca(s) + 2\ HCl(aq) \longrightarrow CaCl_2(aq) + H_2(g)$

(d) $H_2SO_4(aq) + Ba(OH)_2(s) \longrightarrow BaSO_4(s) + 2\ H_2O(l)$

30. The production of Mg from Mg^{2+} in the Dow process does not violate the principle of conservation of electric charge. Notice in Figure 22-7 that magnesium is present as Mg^{2+} until the final step in the process—the electrolysis of $MgCl_2(l)$. At that point, two electrons are transferred from chloride ions to reduce Mg^{2+} to Mg^0, at the same time forming $Cl_2(g)$.

31. mass Mg $= 4\ km^3 \times \left(\dfrac{1000\ m}{1\ km} \times \dfrac{100\ cm}{1\ m}\right)^3 \times \dfrac{1.03\ g}{1\ cm^3\ seawater} \times \dfrac{1\ lb}{454\ g} \times \dfrac{1\ ton}{2000\ lb} \times \dfrac{1272\ g\ Mg}{1\ ton\ seawater}$

$= 6 \times 10^{12}\ g\ Mg = 6 \times 10^6\ metric\ tons\ Mg = 6 \times 10^6\ Mg\ Mg$ (megagrams of magnesium)

32. 1.00 M NH_4Cl is a somewhat acidic solution due to the hydrolysis of $NH_4^+(aq)$; $MgCO_3$ should be most soluble in this solution. The remaining two solutions are buffer solutions, the 0.100 M NH_3–1.00 M NH_4Cl buffer being the more acidic of the two; in this solution $MgCO_3$ has intermediate solubility. $MgCO_3$ is least soluble in the most alkaline solution, 1.00 M NH_3–1.00 M NH_4Cl.

33. Let us compute the value of the equilibrium constant for each reaction by combining two solubility product constant values. Large values of equilibrium constants indicate that the reaction is displaced far to the right. Values of K that are quite a bit smaller than 1 indicate that the reaction is displaced far to the left.

(a) $Ba(OH)_2(s) \rightleftharpoons Ba^{2+}(aq) + 2\ OH^-(aq)$ $K_{sp} = 5 \times 10^{-3}$

$Ba^{2+}(aq) + SO_4^{2-}(aq) \rightleftharpoons BaSO_4(s)$ $1/K_{sp} = 1/1.1 \times 10^{-10}$

$Ba(OH)_2(s) + SO_4^{2-}(aq) \rightleftharpoons BaSO_4(s) + 2\ OH^-(aq)$ $K = \dfrac{5 \times 10^{-3}}{1.1 \times 10^{-10}} = 5 \times 10^7$

Equilibrium lies far to the right.

(b) $Mg(OH)_2(s) \rightleftharpoons Mg^{2+}(aq) + 2\ OH^-(aq)$ $K_{sp} = 1.8 \times 10^{-11}$

$Mg^{2+}(aq) + CO_3^{2-}(aq) \rightleftharpoons MgCO_3(s)$ $1/K_{sp} = 1/3.5 \times 10^{-8}$

$Mg(OH)_2(s) + CO_3^{2-}(aq) \rightleftharpoons MgCO_3(s) + 2\ OH^-(aq)$ $K = \dfrac{1.8 \times 10^{-11}}{3.5 \times 10^{-8}} = 5.1 \times 10^{-4}$

Equilibrium lies far to the left.

(c) $Ca(OH)_2(s) \rightleftharpoons Ca^{2+}(aq) + 2\ OH^-(aq)$ $K_{sp} = 5.5 \times 10^{-6}$

$Ca^{2+}(aq) + 2\ F^-(aq) \rightleftharpoons CaF_2(s)$ $1/K_{sp} = 1/5.3 \times 10^{-9}$

$Ca(OH)_2(s) + 2\ F^-(aq) \rightleftharpoons CaF_2(s) + 2\ OH^-(aq)$ $K = \dfrac{5.5 \times 10^{-6}}{5.3 \times 10^{-9}} = 1.0 \times 10^3$

Equilibrium lies far to the right.

34. We expect the reaction to occur to a significant extent if its equilibrium constant has a large value.

(a) $SrCO_3(s) \rightleftharpoons Sr^{2+}(aq) + CO_3^{2-}(aq)$ $K_{sp} = 1.1 \times 10^{-10}$

$2\ HC_2H_3O_2(aq) \rightleftharpoons 2\ H^+(aq) + 2\ C_2H_3O_2^-(aq)$ $1/K_a^2 = 1/(1.8 \times 10^{-5})^2$

$H^+(aq) + CO_3^{2-}(aq) \rightleftharpoons HCO_3^-(aq)$ $1/K_{a_2} = 1/4.7 \times 10^{-11}$

$H^+(aq) + HCO_3^-(aq) \rightleftharpoons H_2O + CO_2(g)$ $1/K_{a_1} = 4.2 \times 10^{-7}$

$SrCO_3(s) + 2\ HC_2H_3O_2(aq) \rightleftharpoons Sr(C_2H_3O_2)_2(aq) + H_2O + CO_2(g)$

$$K = \frac{1.1 \times 10^{-10}}{(1.8 \times 10^{-5})^2\ 4.7 \times 10^{-11} \times 4.4 \times 10^{-7}} = 1.6 \times 10^{16}$$

(b) $Ba(OH)_2(s) \rightleftharpoons Ba^{2+}(aq) + 2\ OH^-(aq)$ $K_{sp} = 5 \times 10^{-3}$

$2\ NH_4^+(aq) + 2\ OH^-(aq) \rightleftharpoons 2\ NH_3(aq) + 2\ H_2O$ $1/K_b^2 = 1/(1.8 \times 10^{-5})^2$

$Ba(OH)_2(s) + 2\ NH_4^+(aq) \rightleftharpoons Ba^{2+}(aq) + 2\ NH_3(aq) + 2\ H_2O$ $K = \dfrac{5 \times 10^{-3}}{(1.8 \times 10^{-5})^2} = 2 \times 10^7$

(c) $ZnS(s) + H_3O^+(aq) \rightleftharpoons Zn^{2+}(aq) + S^{2-}(aq)$ $K_{sp} = 2 \times 10^{-24}$

$H^+(aq) + OH^-(aq) \rightleftharpoons H_2O$ $1/K_w = 1/1.0 \times 10^{-14}$

$H^+(aq) + HS^-(aq) \rightleftharpoons H_2S(aq)$ $1/K_{a_1} = 1/1.0 \times 10^{-7}$

$H_2S(aq) \rightleftharpoons H_2S(g)$ $K' = 1.00\ atm/0.10\ M = 10.$

$ZnS(s) + 2\ H^+(conc.\ aq) \rightleftharpoons Zn^{2+}(aq) + H_2S(g)$ $K = \dfrac{2 \times 10^{-24} \times 10}{1.0 \times 10^{-14} \times 1.0 \times 10^{-7}} = 0.02$

This reaction will occur to a significant extent because of the concentrated acid as reactant and because the escape of $H_2S(g)$ prevents the reverse reaction from being significant.

Hard Water

35. First we combine equations representing the neutralization of bicarbonate ion with hydroxide ion and the formation of a generalized carbonate precipitate $[MCO_3(s)]$ to determine the overall stoichiometry of the reaction. We also add to the combination the slaking of lime, $CaO(s)$. Remember that every two HCO_3^- ions are associated with an ion of $M^{2+}(aq)$.

$CaO(s) + H_2O \longrightarrow Ca(OH)_2(s) \xrightarrow{\ H_2O\ } Ca^{2+}(aq) + 2\ OH^-(aq)$

$2\ \{HCO_3^-(aq) + OH^-(aq) \longrightarrow H_2O + CO_3^{2-}(aq)\}$

$CO_3^{2-}(aq) + M^{2+}(aq) \longrightarrow MCO_3(s)$

$CO_3^{2-}(aq) + Ca^{2+}(aq) \longrightarrow CaCO_3(s)$

$Ca(OH)_2(s) + M^{2+}(aq) + 2\ HCO_3^-(aq) \longrightarrow CaCO_3(s) + MCO_3(s) + 2\ H_2O$

We determine the mass of HCO_3^- per gallon of hard water.

$$\frac{mass\ HCO_3^-}{1\ gal\ water} = \frac{120.0\ g\ HCO_3^-}{10^6\ g\ water} \times \frac{1.00\ g\ water}{1\ mL\ water} \times \frac{1000\ mL\ water}{1\ L\ water} \times \frac{1\ L\ water}{1.0567\ qt\ water} \times \frac{4\ qt}{1\ gal}$$

$$= 0.4542\ g\ HCO_3^-/gal\ water$$

Then we determine the mass of $Ca(OH)_2$ needed.

$$mass\ Ca(OH)_2 = 1.00 \times 10^6\ gal\ water \times \frac{0.4542\ g\ HCO_3^-}{1\ gal\ water} \times \frac{1\ mol\ HCO_3^-}{61.02\ g\ HCO_3^-} \times \frac{1\ mol\ Ca(OH)_2}{2\ mol\ HCO_3^-}$$

$$\times \frac{78.09\ g\ CaO}{1\ mol\ CaO} = 2.76 \times 10^5\ g\ Ca(OH)_2 \times \frac{1\ kg\ Ca(OH)_2}{1 \times 10^3\ g\ Ca(OH)_2} = 276\ kg\ CaO$$

36. (a) $mass\ CaCO_3 = 276\ kg\ Ca(OH)_2 \times \dfrac{1\ kmol\ Ca(OH)_2}{78.09\ kg\ Ca(OH)_2} \times \dfrac{2\ kmol\ CaCO_3}{1\ kmol\ Ca(OH)_2} \times \dfrac{100.1\ kg\ CaCO_3}{1\ kmol\ CaCO_3}$

$$= 746\ kg\ CaCO_3$$

(b) It is clear from the equations that are summed in the answer to Exercise 35, and especially the net equation, that half of the Ca^{2+} is derived from the CaO and half from the water itself.

37. H^+ is the counterion in the cation exchange resin, and OH^- is the counterion in the anion exchange resin. We first pass the solution through the cation exchange resin, R, then the anion exchange resin, R'.

$$3\ H_2R + 2\ Fe^{3+}(aq) \longrightarrow Fe_2R_3 + 6\ H^+(aq) \qquad\qquad H_2R + Ca^{2+}(aq) \longrightarrow CaR + 2\ H^+(aq)$$

$$H_2R + Mg^{2+}(aq) \longrightarrow MgR + 2\ H^+(aq)$$

$$R'(OH) + HCO_3^-(aq) \longrightarrow R'HCO_3 + OH^-(aq) \qquad\qquad R'(OH)_2 + SO_4^{2-}(aq) \longrightarrow R'SO_4 + 2\ OH^-(aq)$$

Of course, the total charge on the removed anions is identical to the total charge on the removed cations, meaning that the number of moles of $H^+(aq)$ and $OH^-(aq)$ are identical. Then, the following reaction occurs.

$$H^+(aq) + OH^-(aq) \longrightarrow H_2O$$

38. The $[H^+]$ is computed from the data supplied in the problem.

$$[H^+] = \frac{77.5\ g\ Ca^{2+} \times \dfrac{1\ mol\ Ca^{2+}}{40.08\ g\ Ca^{2+}} \times \dfrac{2\ mol\ H^+}{1\ mol\ Ca^{2+}}}{1000\ L\ water} = 3.87 \times 10^{-3}\ M$$

$$pH = -\log(3.87 \times 10^{-3}) = 2.412$$

39. Let us determine the $[OH^-]$ in this solution first. $\quad 2\ OH^-(aq) + H_2SO_4(aq) \longrightarrow 2\ H_2O + SO_4^{2-}(aq)$

$$[OH^-] = \frac{21.58\ mL\ acid \times \dfrac{1.00 \times 10^{-3}\ mmol\ H_2SO_4}{1\ mL\ acid} \times \dfrac{2\ mmol\ OH^-}{1\ mmol\ H_2SO_4}}{25.00\ mL\ base} = 1.726 \times 10^{-3}\ M$$

Now we compute the number of grams of $CaSO_4$ in 10^6 g of hard water. This is the ppm hardness. The anion exchange reaction is $\quad R(OH)_2 + SO_4^{2-}(aq) \longrightarrow RSO_4 + 2\ OH^-(aq)$

$$mass\ CaSO_4 = 10^6\ g\ water \times \frac{1\ mL}{1.00\ g} \times \frac{1\ L}{1000\ mL} \times \frac{1.726 \times 10^{-3}\ mol\ OH^-}{1\ L} \times \frac{1\ mol\ SO_4^{2-}}{2\ mol\ OH^-}$$
$$\times \frac{1\ mol\ CaSO_4}{1\ mol\ SO_4^{2-}} \times \frac{136.1\ g\ CaSO_4}{1\ mol\ CaSO_4} = 117\ g\ CaSO_4 \qquad\qquad (117\ ppm\ CaSO_4)$$

40. We first write the balanced equation for the formation of bathtub ring.

$$Ca^{2+}(aq) + 2\ CH_3(CH_2)_{16}COO-K^+(aq) \longrightarrow Ca^{2+}[CH_3(CH_2)_{16}COO^-]_2(s,\ "ring") + 2\ K^+(aq)$$

$$Mass\ "ring" = 25.5\ L \times \frac{88\ g\ Ca^{2+}}{1000\ L\ water} \times \frac{1\ mol\ Ca^{2+}}{40.08\ g\ Ca^{2+}} \times \frac{1\ mol\ "ring"}{1\ mol\ Ca^{2+}} \times \frac{607.0\ g\ "ring"}{1\ mol\ "ring"}$$
$$= 34\ g\ bathtub\ ring$$

Aluminum

41. **(a)** $2\ Al(s) + 6\ HI(aq) \longrightarrow 2\ AlI_3(aq) + 3\ H_2(g)$ **(b)** $2\ Al(s) + 3\ Cl_2(g) \longrightarrow 2\ AlCl_3(s)$

(c) $2\ KOH(aq) + 2\ Al(s) + 6\ H_2O(l) \longrightarrow 2\ K^+(aq) + 2\ [Al(OH)_4]^-(aq) + 3\ H_2(g)$

(d) Oxidation: $\{Al(s) \longrightarrow Al^{3+}(aq) + 3\ e^-\} \qquad\qquad\qquad \times 2$

Reduction: $\{SO_4^{2-}(aq) + 4\ H^+(aq) + 2\ e^- \longrightarrow SO_2(g) + 2\ H_2O\} \quad \times 3$

Net: $\quad 2\ Al(s) + 3\ SO_4^{2-}(aq) + 12\ H^+(aq) \longrightarrow 2\ Al^{3+}(aq) + 3\ SO_2(aq) + 6\ H_2O$

(e) $2\ Al(s) + Cr_2O_3(s) \xrightarrow{\ heat\ } 2\ Cr(l) + Al_2O_3(s)$

(f) $Fe_2O_3(s) + OH^-(aq) \longrightarrow$ no reaction $\qquad Al_2O_3(s) + 2\ OH^-(aq) + 3\ H_2O \longrightarrow 2\ [Al(OH)_4]^-(aq)$

42. Oersted: $\quad 2\ Al_2O_3(s) + 3\ C(s) + 6\ Cl_2(g) \xrightarrow{\ \Delta\ } 4\ AlCl_3(s) + 3\ CO_2(g)$

Wöhler: $\quad AlCl_3(s) + 3\ K(s) \xrightarrow{\ \Delta\ } Al(s) + 3\ KCl(s)$

43. **(a)** $Fe_2O_3(s) + 2\ Al(s) \longrightarrow Al_2O_3(s) + 2\ Fe(s)$

$\Delta H^\circ_{rxn} = \Delta H^\circ_f[Al_2O_3(s)] + 2\ \Delta H^\circ_f[Fe(s)] - \Delta H^\circ_f[Fe_2O_3(s)] - 2\ \Delta H^\circ_f[Al(s)]$

$= -1676\ kJ/mol + 2\ (0.00\ kJ/mol) - (-824.2\ kJ/mol) - 2\ (0.00\ kJ/mol) = -852\ kJ/mol$

(b) $3\ MnO_2(s) + 4\ Al(s) \longrightarrow 2\ Al_2O_3(s) + 4\ Mn(s)$

$\Delta H^\circ_{rxn} = 2\ \Delta H^\circ_f[Al_2O_3(s)] + 4\ \Delta H^\circ_f[Mn(s)] - 3\ \Delta H^\circ_f[MnO_2(s)] - 4\ \Delta H^\circ_f[Al(s)]$

$= 2\ (-1676\ kJ/mol) + 4\ (0.00\ kJ/mol) - 3\ (-520.0\ kJ/mol) - 4\ (0.00\ kJ/mol) = -1792\ kJ/mol$

(c) $3\ MgO(s) + 2\ Al(s) \longrightarrow 3\ Mg(s) + Al_2O_3(s)$

$\Delta H^\circ_{rxn} = \Delta H^\circ_f[Al_2O_3(s)] + 3\ \Delta H^\circ_f[Mg(s)] - 3\ \Delta H^\circ_f[MgO(s)] - 2\ \Delta H^\circ_f[Al(s)]$

$= -1676\ kJ/mol + 2\ (0.00\ kJ/mol) - 3\ (-601.7\ kJ/mol) - 2\ (0.00\ kJ/mol)$

$= +129\ kJ/mol \qquad\qquad$ An endothermic reaction

44. A method of analyzing this reaction is to envision the HCO_3^- ion as a combination of CO_2 and OH^-. Then the OH^- reacts with Al^{3+} and forms $Al(OH)_3$. [This method of envisioning HCO_3^- is somewhat of a trick, but it does have its basis in reality. After all, H_2CO_3 (= $H_2O + CO_2$) + $OH^- \longrightarrow HCO_3^- + H_2O$.]

$Al^{3+}(aq) + 3\ HCO_3^-(aq) \longrightarrow Al(OH)_3(s) + 3\ CO_2(g)$

Another possibility is to consider the reaction as, first, the hydrolysis of hydrated aluminum ion to produce $Al(OH)_3(s)$ and an acidic solution, followed by the reaction of the acid with bicarbonate ion.

$[Al(H_2O)_6]^{3+}(aq) + 3\ H_2O \longrightarrow Al(OH)_3(s) + 3\ H_3O^+(aq) + 3\ H_2O$

$3\ H_3O^+(aq) + 3\ HCO_3^-(aq) \longrightarrow 6\ H_2O + 3\ CO_2(g)$

45. Aluminum and its oxide are soluble in both acid and in base.

$2\ Al(s) + 6\ H^+(aq) \longrightarrow 2\ Al^{3+}(aq) + 3\ H_2(g)$ $Al_2O_3(s) + 6\ H^+(aq) \longrightarrow 2\ Al^{3+}(aq) + 3\ H_2O$

$2\ Al(s) + 2\ OH^-(aq) + 6\ H_2O \longrightarrow 2\ [Al(OH)_4]^-(aq) + 3\ H_2(g)$

$Al_2O_3(s) + 2\ OH^-(aq) + 3\ H_2O \longrightarrow 2\ [Al(OH)_4]^-(aq)$

The resistance of $Al(s)$ to corrosion in the pH range 4.5 to 8.5 (and its susceptibility to corrosion outside that range) is a reflection of the ability of either acid or base to attack aluminum or its protective oxide coating. Thus, aluminum is nonreactive only when the medium to which it is exposed is not highly acidic or highly basic.

46. Both Al and Mg are attacked by acid and their ions both are precipitated by hydroxide ion.

$2\ Al(s) + 6\ H^+(aq) \longrightarrow 2\ Al^{3+}(aq) + 3\ H_2(g)$ $Mg(s) + 2\ H^+(aq) \longrightarrow Mg^{2+}(aq) + H_2(g)$

$Al^{3+}(aq) + 3\ OH^-(aq) \longrightarrow Al(OH)_3(s)$ $Mg^{2+}(aq) + 2\ OH^- \longrightarrow Mg(OH)_2(s)$

But, of these two solid hydroxides, only $Al(OH)_3(s)$ redissolves in excess $OH^-(aq)$.

$Al(OH)_3(s) + OH^-(aq) \longrightarrow [Al(OH)_4]^-(aq)$

Thus, the procedure for analysis consists of dissolving the sample in $HCl(aq)$ and then treating the resulting solution with $NaOH(aq)$ until a precipitate forms. If this precipitate dissolves completely when excess $NaOH(aq)$ is added, the sample was Aluminum 2S. If at least some of the precipitate does not dissolve, the sample was magnalium.

47. $CO_2(g)$ is, of course, the anhydride of an acid. The reaction here is an acid-base reaction.

$[Al(OH)_4]^-(aq) + CO_2(aq) \longrightarrow Al(OH)_3(s) + HCO_3^-(aq)$

48. The reaction described in Exercise 47 is: $[Al(OH)_4]^-(aq) + CO_2(aq) \longrightarrow Al(OH)_3(s) + HCO_3^-(aq)$

Substituting $HCl(aq)$ for $CO_2(aq)$ gives: $[Al(OH)_4]^-(aq) + HCl(aq) \longrightarrow Al(OH)_3(s) + H_2O + Cl^-(aq)$

Thus $HCl(aq)$ could be used in place of $CO_2(aq)$, but it has two disadvantages. First, $HCl(aq)$ is a strong acid; and the addition of too much $HCl(aq)$ will dissolve the $Al(OH)_3(s)$; this will not happen if excess $CO_2(aq)$ is added. Second, $CO_2(aq)$ is cheaper and more readily available (also safer to use) than $HCl(aq)$. Recall that $CO_2(g)$ is produced during the electrolytic production of aluminum.

49. The $Al^{3+}(aq)$ ion hydrolyzes. $Al^{3+}(aq) + 3\ H_2O \longrightarrow Al(OH)_3(s) + 3\ H^+(aq)$

And the hydrogen ion that is produced reacts with bicarbonate ion to liberate $CO_2(g)$.

$H^+(aq) + HCO_3^-(aq) \longrightarrow H_2O + CO_2(g)$

50. **(a)** $PbO(s) + 2\ HNO_3(aq) \longrightarrow Pb(NO_3)_2(aq) + H_2O$ **(b)** $2\ PbS(s) + 3\ O_2 \overset{\Delta}{\longrightarrow} 2\ PbO(s) + 2\ SO_2(g)$

   (c) $SnO_2(s) + 2\ C(s) \overset{\Delta}{\longrightarrow} Sn(l) + 2\ CO(g)$

   (d) $2\ Fe^{3+}(aq) + Sn^{2+}(aq) \longrightarrow 2\ Fe^{2+}(aq) + Sn^{4+}(aq)$

   (e) $2\ PbS(s) + 3\ O_2(g) \overset{\Delta}{\longrightarrow} 2\ PbO(s) + 2\ SO_2(g)$ $2\ SO_2(g) + O_2(g) \longrightarrow 2\ SO_3(g)$

    $SO_3(g) + PbO(s) \longrightarrow PbSO_4(s)$ Or perhaps simply: $PbS(s) + 2\ O_2(g) \longrightarrow PbSO_4(s)$

51. **(a)** Treat tin(II) oxide with hydrochloric acid. $SnO(s) + 2\ HCl(aq) \longrightarrow SnCl_2(aq) + H_2O$

   (b) Attack lead with dilute nitric acid. $3\ Pb(s) + 8\ HNO_3(aq) \longrightarrow 3\ Pb(NO_3)_2(aq) + 4\ H_2O + 2\ NO(g)$

   (c) First we dissolve $PbO_2(s)$ in $HNO_3(aq)$ and then treat the resulting solution with $K_2CrO_4(aq)$.

    $2\ PbO_2(s) + 4\ HNO_3(aq) \longrightarrow 2\ Pb(NO_3)_2(aq) + O_2(g) + 2\ H_2O$

    $Pb(NO_3)_2(aq) + K_2CrO_4(aq) \longrightarrow PbCrO_4(aq) + 2\ KNO_3(aq)$

52. We use the Nernst equation to determine whether the cell voltage still is positive when the reaction has gone to completion.

    (a) Oxidation: $\{Fe^{2+}(aq) \longrightarrow Fe^{3+}(aq) + e^-\}$ $\times 2$ $-E^\circ = -0.771$ V

         Reduction: $PbO_2(s) + 4\ H^+(aq) + 2\ e^- \longrightarrow Pb^{2+}(aq) + 2\ H_2O$ $E^\circ = +1.455$ V

         Net: $2\ Fe^{2+}(aq) + PbO_2(s) + 4\ H^+ \longrightarrow 2\ Fe^{3+}(aq) + Pb^{2+}(aq) + 2\ H_2O$

         $E^\circ_{cell} = -0.771$ V $+ 1.455$ V $= +0.684$ V

         In this case, when the reaction has gone to completion, $[Fe^{2+}] = 0.001$ M, $[Fe^{3+}] = 0.999$ M, and $[Pb^{2+}] = 0.500$ M.

$$E_{cell} = E^\circ_{cell} - \frac{0.0592}{2} \log \frac{[Fe^{3+}]^2\ [Pb^{2+}]}{[Fe^{2+}]^2} = 0.684 - \frac{0.0592}{2} \log \frac{(0.999)^2\ (0.500)}{(0.001)^2}$$

                $= 0.684 - 0.169 = +0.515$ V   Yes, this reaction will go to completion.

    (b) Oxidation: $2\ SO_4^{2-}(aq) \longrightarrow S_2O_8^{2-}(aq) + 2\ e^-$ $-E^\circ = -2.01$ V

         Reduction: $PbO_2(s) + 4\ H^+(aq) + 2\ e^- \longrightarrow Pb^{2+}(aq) + 2\ H_2O$ $E^\circ = +1.455$ V

         Net: $2\ SO_4^{2-}(aq) + PbO_2(s) + 4\ H^+ \longrightarrow S_2O_8^{2-}(aq) + Pb^{2+}(aq) + 2\ H_2O$

         $E^\circ_{cell} = -2.01 + 1.455 = -0.56$ V   This reaction is not even spontaneous initially.

    (c) Oxdn: $\{Mn^{2+}(0.0001\ M) + 4\ H_2O \longrightarrow MnO_4^-(aq) + 8\ H^+(aq) + 5\ e^-\} \times 2$ $-E^\circ = -1.51$ V

         Redn: $\{PbO_2(s) + 4\ H^+(aq) + 2\ e^- \longrightarrow Pb^{2+}(aq) + 2\ H_2O\}$ $\times 5$ $E^\circ = +1.455$ V

         Net: $2\ Mn^{2+}(0.0001\ M) + 5\ PbO_2(s) + 4\ H^+ \longrightarrow 2\ MnO_4^-(aq) + 5\ Pb^{2+}(aq) + 2\ H_2O$

         $E^\circ_{cell} = -1.51 + 1.455 = -0.06$ V   The standard cell potential indicates that this reaction is not spontaneous when all concentrations are 1 M. Since the concentration of a reactant (Mn^{2+}) is lower than 1.00 M, this reaction is even less spontaneous than the standard cell potential indicates.

53. A positive value of E°_{cell} indicates that a reaction should occur.

    (a) Oxidation: $Sn^{2+}(aq) \longrightarrow Sn^{4+}(aq) + 2\ e^-$ $-E^\circ = -0.154$ V

         Reduction: $I_2(s) + 2\ e^- \longrightarrow 2\ I^-(aq)$ $E^\circ = +0.535$ V

         Net: $Sn^{2+}(aq) + I_2(s) \longrightarrow Sn^{4+}(aq) + 2\ I^-(aq)$ $E^\circ_{cell} = +0.381$ V

         Yes, $Sn^{2+}(aq)$ will reduce I_2 to I^-.

    (b) Oxidation: $Sn^{2+}(aq) \longrightarrow Sn^{4+}(aq) + 2\ e^-$ $-E^\circ = -0.154$ V

         Reduction: $Fe^{2+}(aq) + 2\ e^- \longrightarrow Fe(s)$ $E^\circ = -0.440$ V

         Net: $Sn^{2+}(aq) + Fe^{2+}(aq) \longrightarrow Sn^{4+}(aq) + Fe(s)$ $E^\circ_{cell} = -0.594$ V

         No, $Sn^{2+}(aq)$ will not reduce $Fe^{2+}(aq)$ to $Fe(s)$.

    (c) Oxidation: $Sn^{2+}(aq) \longrightarrow Sn^{4+}(aq) + 2\ e^-$ $-E^\circ = -0.154$ V

         Reduction: $Cu^{2+}(aq) + 2\ e^- \longrightarrow Cu(s)$ $E^\circ = +0.337$ V

         Net: $Sn^{2+}(aq) + Cu^{2+}(aq) \longrightarrow Sn^{4+}(aq) + Cu(s)$ $E^\circ_{cell} = +0.183$ V

         Yes, $Sn^{2+}(aq)$ will reduce $Cu^{2+}(aq)$ to $Cu(s)$.

    (d) Oxidation: $Sn^{2+}(aq) \longrightarrow Sn^{4+}(aq) + 2\ e^-$ $-E^\circ = -0.154$ V

         Reduction: $\{Fe^{3+}(aq) + e^- \longrightarrow 2\ Fe^{2+}(aq)$ $E^\circ = +0.771$ V

         Net: $Sn^{2+}(aq) + 2\ Fe^{3+}(aq) \longrightarrow Sn^{4+}(aq) + 2\ Fe^{2+}(aq)$ $E^\circ_{cell} = +0.617$ V

         Yes, $Sn^{2+}(aq)$ will reduce $Fe^{3+}(aq)$ to $Fe^{2+}(aq)$.

23 CHEMISTRY OF THE REPRESENTATIVE (MAIN GROUP) ELEMENTS II: NONMETALS

REVIEW QUESTIONS

1. **(a)** Plastic sulfur is a metastable form of sulfur that results when liquid sulfur is cooled quickly. It is amorphous and stretchy. It slowly reverts to stable rhombic sulfur.
   (b) A poly phosphate is a condensation product of several ortho phosphate ions. $P_2O_7^{4-}$ is diphosphate, $P_3O_{10}^{5-}$ is triphosphate, etc.
   (c) Eutrophication is the result of an excess of nutrients in bodies of water, such as ponds, which causes rapid and excessive growth of plant growth in the water. This depletes the oxygen content of the water and marine life dies off. The decaying plant and animal life fill the pond from the bottom, eventually turning it into a marsh.
   (d) An interhaolgen is a compound formed between two halogens. Examples include ClF_3 and ICl.

2. **(a)** The Frasch process is used to mine underground deposits of sulfur. Superheated water and compressed air are pumped down, melting the sulfur and forcing it to the surface, where it cools and dries to form solid sulfur.
   (b) The contact process refers to oxidizing $SO_2(g)$ with $O_2(g)$ to $SO_3(g)$ using a $V_2O_5(s)$ catalyst. It is an essential step in the modern industrial production of sulfuric acid.
   (c) An acid anhydride is an oxide that dissolves in water to form an acidic solution.

3. **(a)** A polyhalide is an anion composed of two or more like halogen atoms, such as I_3^-. An interhalogen compound is composed of two different kinds of halogens.
   (b) A silane is a compound of silicon and hydrogen. A silicone has a —Si—O—Si— backbone terminated by —OH groups, with hydrocarbon groups attached to each of the silicon atoms.
   (c) A colloidal disperson of a solid in a liquid that flows readily is known as a sol. One that has had sufficient liquid removed that it does not flow readily is called a gel.

4. **(a)** $KBrO_3$ potassium bromate   **(b)** BrF_3 bromine trifluoride
   (c) NaIO   sodium hypoiodite   **(d)** NaH_2PO_4 sodium dihydrogen phosphate
   (e) Li_3N lithium nitride   **(f)** $Na_2S_2O_6$ sodium dithionate

5. **(a)** $CaCl_2(s) + H_2SO_4(\text{conc. aq}) \xrightarrow{\Delta} CaSO_4(s) + 2\ HCl(g)$
   (b) $I_2(s) + Cl^-(aq) \longrightarrow$ no reaction   **(c)** $NH_3(aq) + HClO_4(aq) \longrightarrow NH_4ClO_4(aq)$

6. **(a)** $Cl_2(aq) + 2\ NaOH(aq) + H_2O \longrightarrow NaCl(aq) + NaOCl(aq) + H_2O$
   (b) Oxidation: $2\ I^-(aq) \longrightarrow I_2(s) + 2\ e^-$
       Reduction: $SO_4^{2-}(aq) + 2\ e^- + 4\ H^+(aq) \longrightarrow SO_2(g) + 2\ H_2O$

       Net: $2\ I^-(aq) + SO_4^{2-}(aq) + 4\ H^+(aq) \longrightarrow I_2(g) + SO_2(g) + 2\ H_2O$
                $2\ NaI(s) + 2\ H_2SO_4(aq) \longrightarrow I_2(g) + 2\ SO_2(g) + H_2O + Na_2SO_4(aq)$
   (c) $Cl_2(g) + 2\ Br^-(aq) \longrightarrow 2\ Cl^-(aq) + Br_2(l)$

7.

Acidic Solution	Basic Solution
$H_5IO_6 + H^+(aq) + 2\ e^- \rightarrow IO_3^-(aq) + 3\ H_2O$	$OCl^-(aq) + H_2O + 2\ e^- \rightarrow Cl^-(aq) + 2\ OH^-(aq)$
$4\ H_3PO_2(aq) + 4\ H^+(aq) + 4\ e^- \rightarrow P_4(s) + 8\ H_2O$	$B_2H_6 + 2\ H_2O + 4\ e^- \rightarrow 2\ BH_4^-(aq) + 2\ OH^-(aq)$
$Sb_2O_5(s) + 6\ H^+(aq) + 4\ e^- \rightarrow 2\ SbO^+(aq) + 3\ H_2O$	$HXeO_4^-(aq) + 3\ H_2O + 6\ e^- \rightarrow Xe(g) + 7\ OH^-(aq)$

8. First compute the theoretical yield of P from 8.00 ton phosphate rock, then the percent yield in this case.

no. ton P = 8.00 ton rock $\times \dfrac{31 \text{ ton P}_2\text{O}_5}{100.0 \text{ ton rock}} \times \dfrac{1 \text{ ton-mol P}_2\text{O}_5}{141.9 \text{ ton P}_2\text{O}_5} \times \dfrac{2 \text{ ton-mol P}}{1 \text{ ton-mol P}_2\text{O}_5}$

$\times \dfrac{31.0 \text{ ton P}}{1 \text{ ton-mol P}_2\text{O}_5} = 1.1 \text{ ton P}$ The mass of a ton-mole in tons is numerically equal to the mass of a (gram-)mole in grams.

%yield $= \dfrac{1.00 \text{ ton P produced}}{1.1 \text{ ton P calculated}} \times 100\% = 91\%$ yield

9. $3 \text{ Cl}_2(g) + \text{I}^-(aq) + 3 \text{ H}_2\text{O} \longrightarrow 6 \text{ Cl}^-(aq) + \text{IO}_3^-(aq) + 6 \text{ H}^+(aq)$

$\text{Cl}_2(g) + 2 \text{ Br}^-(aq) \longrightarrow 2 \text{ Cl}^-(aq) + \text{Br}_2(aq)$

10. $\dfrac{\text{seawater}}{\text{volume}} = 38 \times 10^6 \text{ ton H}_2\text{SO}_4 \times \dfrac{2000 \text{ lb}}{1 \text{ ton}} \times \dfrac{454 \text{ g}}{1 \text{ lb}} \times \dfrac{1 \text{ mol H}_2\text{SO}_4}{98.1 \text{ g H}_2\text{SO}_4} \times \dfrac{1 \text{ mol SO}_4^{2-}}{1 \text{ mol H}_2\text{SO}_4} \times \dfrac{96.1 \text{ g SO}_4^{2-}}{1 \text{ mol SO}_4^{2-}}$

$\times \dfrac{1000 \text{ mg}}{1 \text{ g}} \times \dfrac{1 \text{ L seawater}}{2650 \text{ mg SO}_4^{2-}} \times \dfrac{1 \text{ m}^3}{1000 \text{ L}} \times \dfrac{1 \text{ km}^3}{10^9 \text{ m}^3} = 13 \text{ km}^3$ seawater

11. (a) Reduction: $\text{Cl}_2(g) + 2 \text{ e}^- \longrightarrow 2 \text{ Cl}^-(aq)$ $E° = +1.358 \text{ V}$

Oxidations: $2 \text{ I}^-(aq) \longrightarrow \text{I}_2(s) + 2 \text{ e}^-$ $-E° = -0.535 \text{ V}$

$2 \text{ H}_2\text{O} \longrightarrow \text{O}_2(g) + 4 \text{ H}^+(aq) + 4 \text{ e}^-$ $-E° = -1.229 \text{ V}$

The production of $\text{I}_2(s)$ is more likely; it results in the larger standard cell potential.

(b) NH_4^+ has nitrogen in its lowest common oxidation state. NH_4^+ cannot be reduced further. Thus, NH_4^+ is oxidized and H_2O_2 must be reduced to H_2O.

12. We draw the Lewis structures of the products and reactants of each reaction.

(a) $\frac{1}{2}|\bar{\text{C}}\text{l}\text{—}\bar{\text{C}}\text{l}| + \frac{1}{2}|\bar{\text{F}}\text{—}\bar{\text{F}}| \longrightarrow |\bar{\text{C}}\text{l}\text{—}\bar{\text{F}}|$

Break: 0.5 Cl—Cl + 0.5 F—F = 0.5(243 + 159) = 201 kJ/mol
Form: Cl—F   = 251 kJ/mol
$\Delta H° = 201 \text{ kJ/mol} - 251 \text{ kJ/mol} = -50. \text{ kJ/mol}$

(b) $|\bar{\text{F}}\text{—}\bar{\text{F}}| + \frac{1}{2}|\bar{\text{O}}\text{=}\bar{\text{O}}| \longrightarrow |\bar{\text{F}}\text{—}\bar{\text{O}}\text{—}\bar{\text{F}}|$

Break: 0.5 O=O + F—F = (0.5 × 498) + 159 = 408 kJ/mol
Form: 2 O—F   = 2 × 213   = 426 kJ/mol
$\Delta H° = 408 \text{ kJ/mol} - 426 \text{ kJ/mol} = -18 \text{ kJ/mol}$

(c) $|\bar{\text{C}}\text{l}\text{—}\bar{\text{C}}\text{l}| + \frac{1}{2}|\bar{\text{O}}\text{=}\bar{\text{O}}| \longrightarrow |\bar{\text{C}}\text{l}\text{—}\bar{\text{O}}\text{—}\bar{\text{C}}\text{l}|$

Break: 0.5 O=O + Cl—Cl = (0.5 × 498) + 243 = 498 kJ/mol
Form: 2 O—Cl   = 2 × 205   = 410 kJ/mol
$\Delta H° = 492 \text{ kJ/mol} - 410 \text{ kJ/mol} = +82 \text{ kJ/mol}$

(d) $\frac{3}{2}|\bar{\text{F}}\text{—}\bar{\text{F}}| + \frac{1}{2}|\text{N}\text{≡}\text{N}| \longrightarrow |\bar{\text{F}}\text{—}\bar{\text{N}}\text{—}\bar{\text{F}}|$
$\qquad\qquad\qquad\qquad\qquad\qquad\quad |\bar{\text{F}}|$

Break: 0.5 N≡N + 1.5 F—F = (0.5 × 946) + (1.5 × 159) = 712 kJ/mol
Form: 3 N—F   = 3 × 280   = 840 kJ/mol
$\Delta H° = 712 \text{ kJ/mol} - 840 \text{ kJ/mol} = -128 \text{ kJ/mol}$

13. (a) $\bar{\text{O}}\text{=}\overset{\displaystyle |\text{O}|}{\underset{\displaystyle |\text{O}|}{\overset{\|}{\underset{\|}{\bar{\text{X}}\text{e}}}}}\text{=}\bar{\text{O}}$ **(b)** $\bar{\text{O}}\text{=}\overset{}{\underset{\displaystyle |\text{O}|}{\overset{}{\bar{\text{X}}\text{e}}}}\text{=}\bar{\text{O}}$ **(c)** $|\bar{\text{F}}\text{—}\overset{\displaystyle |\bar{\text{F}} \quad \bar{\text{F}}|}{\underset{\displaystyle |\text{O}|}{\text{X}\text{e}}}\text{—}\bar{\text{F}}|$

(a) The Lewis structure has 3 ligands and one lone pair on Xe. XeO_3 has a trigonal pyramid shape.
(b) The Lewis structure has four ligands on Xe. XeO_4 has a tetrahedral chape.
(c) The Lewis structure has five ligands and one lone pair on Xe. OXeF_4 has a square pyramidal shape.

14. **(a)** AgAt silver astatide **(b)** CSe_2 carbon diselenide **(c)** MgPo magnesium polonide
(d) H_2TeO_3 tellurous acid **(e)** K_2SeSO_3 potassium thioselenate **(f)** $KAtO_4$ perastatic acid

15. A tetrahedral shape requires only four atoms and no lone pairs attached to the central atom. XeF_4 has six—four ligands and two lone pairs, as the following Lewis structures demonstrate. XeF_4 has a square planar shape. SO_4^{2-}, ClO_4^-, and XeO_4 all are tetrahedral.

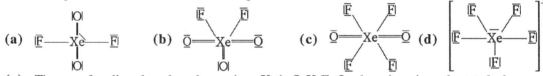

16. I_3^- is a polyhalide ion. An interhalogen compound contains atoms from two or more halogen atoms. An oxoanion is an anion that contains oxygen.

EXERCISES

Noble Gas Compounds

17. We use VSEPR theory to predict the shapes of the species involved.

(a) O_2XeF_2 has a total of $2 \times 6 + 8 + 2 \times 7 = 34$ valence electrons = 17 pairs.

(b) O_3XeF_2 has a total of $3 \times 6 + 8 + 2 \times 7 = 40$ valence electrons = 20 pairs.

(c) O_2XeF_4 has a total of $2 \times 6 + 8 + 4 \times 7 = 48$ valence electrons = 24 pairs.

(d) XeF_5^+ has a total of $5 \times 7 + 8 - 1 = 42$ valence electrons = 21 pairs.

We draw a plausible Lewis structure for each species below.

(a) **(b)** **(c)** **(d)**

(a) There are four ligands and one lone pair on Xe in O_2XeF_2. Its shape is an irregular tetrahedron.
(b) There are five ligands and no lone pairs on Xe in O_3XeF_2. Its shape is trigonal bipyramidal.
(c) There are six ligands and no lone pairs on Xe in O_2XeF_4. Its shape is octahedral.
(d) There are five ligands and one lone pair on Xe in XeF_5^+. Its shape is square pyramidal.

The Halogens

18. In general, elemental forms of the lighter elements in the halogen family will displace the ions of heavier ions in the family from their aqueous solutions. That is, Cl_2 will displace Br^- and I^-; and Br_2 will displace I^-.

19. **(a)**

$2 HOCl(aq) + 2 H^+(aq) + 2 e^- \longrightarrow Cl_2(g) + 2 H_2O$	$\Delta G°_a = -2 \mathcal{F} (+1.62 V)$
$Cl_2(aq) + 2 e^- \longrightarrow 2 Cl^-(aq)$	$\Delta G°_b = -2 \mathcal{F} (+1.358 V)$

$2 HOCl(aq) + 2 H^+(aq) + 4 e^- \longrightarrow 2 Cl^-(aq) + 2 H_2O$ $\Delta G° = -4 \mathcal{F} E°$

$E° = \dfrac{2(1.62 + 1.358) V}{4} = +1.49 V$

(b) Oxidation: $HOCl(aq) + H_2O \longrightarrow HClO_2(aq) + 2 H^+(aq) + 2 e^-$ $-E° = -1.63 V$
Reduction: $HOCl(aq) + H^+(aq) + 2 e^- \longrightarrow Cl^-(aq) + H_2O$ $E° = +1.49 V$

Net: $2 HOCl(aq) \longrightarrow HClO_2(aq) + Cl^-(aq) + H^+(aq)$ $E°_{cell} = -0.14 V$
The reaction does not go to completion. In fact, it favors the reactant side.

20. Iodide ion is slowly oxidized to iodine, which is yellow-brown in aqueous solution, by oxygen in the air.
Oxidation: $2 I^-(aq) \longrightarrow I_2(aq) + 2 e^-$ $-E° = -0.535 V$
Reduction: $O_2(g) + 4 H^+(aq) + 4 e^- \longrightarrow 2 H_2O$ $E° = +1.229 V$

Net: $4 I^-(aq) + O_2(g) + 4 H^+(aq) \longrightarrow I_2(aq) + 2 H_2O$ $E°_{cell} = +0.694 V$

21. We first list values of these properties for the other halogens.

		F	Cl	Br	I
(a)	Covalent radius:	64	99.5	114	133 pm
(b)	Ionic radius:	136	181	195	216 pm
(c)	First ionization energy:	1681	1251	1140	1008 kJ/mol
(d)	electron affinity:	–322	–349	–325	–295 kJ/mol
(e)	electronegativity:	4.0	3.2	3.0	2.7
(f)	standard reduction potential:	+2.886 V	+1.358 V	+1.065 V	+0.535 V

We can do a reasonable job of predicting the properties of astatine by simply looking at the difference between Br and I and assuming that the same difference exists between I and At.

(a) Covalent radius: $\approx$152 pm   **(b)** Ionic radius: $\approx$237 pm
(c) 1st ionization energy: $\approx$880 kJ/mol   **(d)** Eln. affinity: $\approx$–265 kJ/mol
(e) electronegativity: $\approx$2.5   **(e)** Std. redn pot'l: $\approx$0.20 V to 0.30 V

Depending on the technique that you use, you may arrive at different answers. For example, –259 kJ/mol is a reasonable estimation of the electron affinity by the following reasoning. The difference between the value for Cl and Br is 24 kJ/mol, the difference between the values for Br and I is 30 kJ/mol, so the difference between I and At is about 36 kJ/mol.

22. (a) $\overline{F}$—$\widehat{Br}$—$\overline{F}$ **(b)** $\overline{F}$——$\overset{\overline{F}}{\underset{\overline{F}}{I}}$——$\overline{F}$ **(c)** $[\overline{Cl}$—$\widehat{I}$—$\overline{Cl}]^-$ **(d)** $\left[\overset{\overline{F}}{\underset{\overline{Cl}}{Cl}{-}I{-}\overline{Cl}}\right]$

(a) BrF_3 has two lone pairs and three atoms attached to the central Br atom. It has a T-shape.
(b) IF_5 has one lone pair and five atoms attached to the central I atom. It has a square pyramidal shape.
(c) ICl_2^- has two atoms and three lone pairs attached to the central I atom. It is linear.
(d) Cl_3IF^- has four atoms and two lone pairs attached to the central I atom. It is square planar.

23. $MnF_6^{2-} + 2\,SbF_5 \longrightarrow MnF_4 + 2\,SbF_6^-$ $2\,MnF_4 \longrightarrow 2\,MnF_3 + F_2(g)$

24. (a)   mass $F_2 = 1\,km^3 \times \left(\dfrac{1000\ m}{1\ km} \times \dfrac{100\ cm}{1\ m}\right)^3 \times \dfrac{1.03\ g}{1\ cm^3} \times \dfrac{1\ lb}{454\ g} \times \dfrac{1\ ton}{2000\ lb} \times \dfrac{1\ g\ F^-}{1\ ton}$

$= 1 \times 10^9\ g\ F_2 = 1 \times 10^6\ kg\ F_2$

(b) Bromine is extracted by displacing it from solution with $Cl_2(g)$. Since there is no chemical oxidizing agent that is stronger than F_2, this method of displacement would not work for F_2. Even if there were a chemical oxidizing agent stronger than F_2, it would displace O_2 before it displaced F_2. Obtaining F_2 would require electrolysis of its molten salts, obtained by evaporating the seawater.

25.   mass $F_2 = 1.00 \times 10^3\ kg\ rock \times \dfrac{1000\ g}{1\ kg} \times \dfrac{1\ mol\ 3Ca_3(PO_4)_2{\cdot}CaF_2}{1009\ g} \times \dfrac{2\ mol\ F}{1\ mol\ 3Ca_3(PO_4)_2{\cdot}CaF_2}$

$\times \dfrac{19.00\ g\ F}{1\ mol\ F} \times \dfrac{1\ kg}{1000\ g} = 37.66\ kg\ fluorine$

Oxygen

26.

Reaction:	$H_2O_2(aq) + H_2O$	$\rightleftharpoons$	$HO_2^-(aq)$	$+$	$H_3O^+(aq)$
Initial:	1.0 M				
Changes:	$-x$ M		$+x$ M		$+x$ M
Equil:	$(1.0 - x)$M		x M		x M

$K_a = \dfrac{[H_3O^+][HO_2^-]}{[H_2O_2]} = 10^{-11.75} = 1.8 \times 10^{-12} = \dfrac{x \cdot x}{1.0 - x} \approx \dfrac{x^2}{1.0}$

$x = \sqrt{1.8 \times 10^{-12}} = 1.3 \times 10^{-6}\ M = [H_3O^+]$ $pH = -\log(1.3 \times 10^{-6}) = 5.89$

27. (a) $3\ |\overline{O}{=}\overline{O}| \longrightarrow 2\ O{\cdots}O{\cdots}O$

Bonds broken: $3\ O{=}O = 3 \times 498\ kJ/mol = 1494\ kJ/mol$ Bonds formed: $4\ O{\cdots}O$
$\Delta H° = +285\ kJ/mol = $ bonds broken – bonds formed $= 1494\ kJ/mol - 4\ O{\cdots}O$
$4\ O{\cdots}O = 1494\ kJ/mol - 285\ kJ/mol = 1209\ kJ/mol$ $O{\cdots}O = 1209\ kJ/mol \div 4 = 302\ kJ/mol$

(b) $3\,|\bar{\text{O}}{=}\bar{\text{O}}| \longrightarrow 2\,|\bar{\text{O}}{-}\bar{\text{O}}{=}\bar{\text{O}}$ Average bond energy $= \dfrac{\text{O}{=}\text{O} + \text{O}{-}\text{O}}{2} = \dfrac{498 + 142}{2} = 320 \text{ kJ/mol}$

Since bond energies actually are average values taken from many sources, the result of 320 kJ/mol is in reasonably good agreement with the specific value of 302 kJ/mol obtained in part (a).

28. (a) H_2S, while polar, forms only weak hydrogen bonds. H_2O forms much stronger hydrogen bonds, which are much more difficult to disrupt, leading to a higher boiling point.

(b) All electrons are paired in O_3, producing a diamagnetic molecule. $\bar{\text{O}}{-}\bar{\text{O}}{=}\bar{\text{O}}$

(c) $\bar{\text{O}}{=}\bar{\text{O}}$ $|\bar{\text{O}}{-}\bar{\text{O}}{=}\bar{\text{O}}$ $\text{H}{-}\bar{\text{O}}{-}\bar{\text{O}}{-}\text{H}$

The O—O bond in O_2 is a double bond, which should be short (121 pm). That in O_3 is a "one-and-a-half" bond, of intermediate length (128 pm). And that of H_2O_2 is a longer (148 pm) single bond.

Sulfur

29. (a) ZnS   zinc sulfide   **(b)** $KHSO_3$ potassium hydrogen sulfite
(c) $K_2S_2O_3$ potassium thiosulfate   **(c)** SF_4 sulfur tetrafluoride

30. (a) $FeS(s) + 2\ HCl(aq) \longrightarrow FeCl_2(aq) + H_2S(g)$ MnS(s), ZnS(s), etc. also are possible.
(b) $CaSO_3(s) + 2\ HCl(aq) \longrightarrow CaCl_2(aq) + H_2O + SO_2(g)$
(c) Oxidation: $SO_2(aq) + 2\ H_2O \longrightarrow SO_4^{2-}(aq) + 4\ H^+(aq) + 2\ e^-$
Reduction: $\underline{MnO_2(s) + 4\ H^+(aq) + 2\ e^- \longrightarrow Mn^{2+}(aq) + 2\ H_2O}$

Net: $SO_2(aq) + MnO_2(s) \longrightarrow Mn^{2+}(aq) + SO_4^{2-}(aq)$
(d) Oxidation: $S_2O_3^{2-}(aq) + H_2O \longrightarrow 2\ SO_2(g) + 2\ H^+(aq) + 4\ e^-$
Reduction: $\underline{S_2O_3^{2-}(aq) + 6\ H^+(aq) + 4\ e^- \longrightarrow 2\ S(s) + 3\ H_2O}$

Net: $S_2O_3^{2-}(aq) + 2\ H^+(aq) \longrightarrow S(s) + SO_2(g) + H_2O$

31. (a) $S(s) + O_2(g) \longrightarrow SO_2(g)$ $2\ Na(s) + 2\ H_2O \longrightarrow 2\ NaOH(aq) + H_2(g)$
$2\ NaOH(aq) + SO_2(g) \longrightarrow Na_2SO_3(aq) + H_2O$
(b) $S(s) + O_2(g) \longrightarrow SO_2(g)$
$SO_2(g) + 2\ H_2O + Cl_2(g) \longrightarrow SO_4^{2-}(aq) + 2\ Cl^-(aq) + 4\ H^+(aq)$ $(E^\circ_{cell} = +1.19 \text{ V})$
$2\ Na(s) + 2\ H_2O \longrightarrow 2\ NaOH(aq) + H_2(g)$
$2\ NaOH(aq) + 2\ H^+(aq) + SO_4^{2-}(aq) \longrightarrow Na_2SO_4(aq) + 2\ H_2O$
(c) $S(s) + O_2(g) \longrightarrow SO_2(g)$ $2\ Na(s) + 2\ H_2O \longrightarrow 2\ NaOH(aq) + H_2(g)$
$2\ NaOH(aq) + SO_2(g) \longrightarrow Na_2SO_3(aq) + H_2O$
$Na_2SO_3(aq) + S(s) \longrightarrow NaS_2O_3(aq)$ (Boil the reactants in an alkline solution.)

32. The decomposition of thiosulfate ion is promoted in an acidic solution. Thus, simply adding a strong acid, such as HCl(aq), to the solid will result in no reaction for Na_2SO_4. But HCl(aq) added to $Na_2S_2O_3$ results in the production of $SO_2(g)$ and the formation of a S(s) precipitate.
$$S_2O_3^{2-}(aq) + 2\ H^+(aq) \longrightarrow S(s) + SO_2(g) + H_2O$$

33. Neither $Na^+(aq)$ nor $HSO_4^-(aq)$ will hydrolyze, being respectively the cation of a strong base and the anion of a strong acid. But $HSO_4^-(aq)$ will ionize further, $K_2 = 1.1 \times 10^{-2}$ for $HSO_4^-(aq)$. We set up the situation, and solve the quadratic equation to obtain $[H_3O^+]$.

$$[HSO_4^-] = \frac{12.5 \text{ g NaHSO}_4}{250.0 \text{ mL soln}} \times \frac{1000 \text{ mL}}{1 \text{ L soln}} \times \frac{1 \text{ mol NaHSO}_4}{120.1 \text{ g NaHSO}_4} \times \frac{1 \text{ mol HSO}_4^-}{1 \text{ mol NaHSO}_4} = 0.416 \text{ M}$$

Reaction:	$HSO_4^-(aq) + H_2O$	$\rightleftharpoons$	$SO_4^{2-}(aq)$ +	$H_3O^+(aq)$
Initial:	0.416 M			
Changes:	$-x$ M		$+x$ M	$+x$ M
Equil:	$(0.416 - x)$M		x M	x M

$$K_2 = \frac{[H_3O^+][SO_4^{2-}]}{[HSO_4^-]} = 0.011 = \frac{x^2}{0.416 - x} \qquad x^2 = 0.0046 - 0.011\,x \qquad 0 = x^2 + 0.011\,x - 0.0046$$

$$x = \frac{-b \pm \sqrt{b^2 - 4ac}}{2a} = \frac{-0.011 \pm \sqrt{1.2 \times 10^{-4} + 1.8 \times 10^{-2}}}{2} = 0.063 = [H_3O^+]$$

$$pH = -\log(0.063) = 1.20$$

34. We combine the two half-equations and obtain their resultant standard potential by combining their ΔG°'s.

$$2\,\{SO_4^{2-}(aq) + 4\,H^+(aq) + 2\,e^- \longrightarrow 2\,H_2O + SO_2(g)\} \qquad \Delta G^{\circ} = -4\mathcal{F}(+0.17\ V)$$

$$2\,SO_2(g) + 2\,H^+(aq) + 4\,e^- \longrightarrow S_2O_3^{2-}(aq) + H_2O \qquad \Delta G^{\circ} = -4\mathcal{F}(-0.40\ V)$$

$$2\,SO_4^{2-}(aq) + 10\,H^+(aq) + 8\,e^- \longrightarrow S_2O_3^{2-}(aq) + 5\,H_2O \qquad \Delta G^{\circ} = -8\mathcal{F}E^{\circ}$$

$$E^{\circ} = \frac{4(0.17\ V) + 4(-0.40\ V)}{8} = -0.12\ V$$

35. Oxidation: $\{SO_3^{2-}(aq) + H_2O \longrightarrow SO_4^{2-}(aq) + 2\,H^+(aq) + 2\,e^-\} \qquad \times 5$

Reduction: $\{MnO_4^-(aq) + 8\,H^+(aq) + 5\,e^- \longrightarrow Mn^{2+}(aq) + 4\,H_2O\} \qquad \times 2$

Net: $\quad 5\,SO_3^{2-}(aq) + 2\,MnO_4^-(aq) + 6\,H^+(aq) \longrightarrow 5\,SO_4^{2-}(aq) + 2\,Mn^{2+}(aq) + 3\,H_2O$

$$\text{mass Na}_2SO_3 = 26.50\ \text{mL} \times \frac{1\ L}{1000\ \text{mL}} \times \frac{0.0510\ \text{mol MnO}_4^-}{1\ L\ \text{soln}} \times \frac{5\ \text{mol SO}_3^{2-}}{2\ \text{mol MnO}_4^-} \times \frac{1\ \text{mol Na}_2SO_3}{1\ \text{mol SO}_3^{2-}}$$

$$\times \frac{126.0\ \text{g Na}_2SO_3}{1\ \text{mol Na}_2SO_3} = 0.426\ \text{g Na}_2SO_3$$

36. Sulfites are easily oxidized to sulfates by, for example, O_2 in the atmosphere. On the other hand, there are no oxidizing agents naturally available in reasonable concentrations that can oxidize SO_4^{2-} to a higher oxidation state, such as in $S_2O_8^{2-}$. Similarly, the atmosphere of earth is an oxidizing one, there are not reducing agents present to reduce SO_4^{2-} to a species with a lower oxidation state, except in localized areas.

37. $\text{volume H}_2SO_4 = 2.2 \times 10^6\ \text{tons coal} \times \dfrac{2000\ \text{lb}}{1\ \text{ton}} \times \dfrac{454\ \text{g}}{1\ \text{lb}} \times \dfrac{3.5\ \text{g S}}{100.0\ \text{g coal}} \times \dfrac{1\ \text{mol S}}{32.1\ \text{g S}} \times \dfrac{1\ \text{mol SO}_2}{1\ \text{mol S}}$

$$\times \frac{90.\ \text{mol SO}_2\ \text{converted}}{100.\ \text{mol SO}_2\ \text{emitted}} \times \frac{1\ \text{mol H}_2SO_4}{1\ \text{mol SO}_2} \times \frac{98.1\ \text{g H}_2SO_4}{1\ \text{mol H}_2SO_4} \times \frac{100.\ \text{g acid}}{98\ \text{g H}_2SO_4}$$

$$\times \frac{1\ \text{cm}^3\ \text{acid}}{1.84\ \text{g acid}} \times \frac{1\ L\ \text{acid}}{1000\ \text{cm}^3\ \text{acid}} = 1.1 \times 10^8\ L\ \text{acid}$$

Nitrogen

38. (a) $2\,NO_2(g) \rightleftharpoons N_2O_4(g)$ **(b)** $NH_4NO_3(s) \xrightarrow{\ \Delta\ } N_2O(g) + 2\,H_2O(g)$

(c) $2\,NH_3(g) + H_2SO_4(aq) \longrightarrow (NH_4)_2SO_4(aq)$

(d) $3\,Ag(s) + 4\,H^+(aq) + 4\,NO_3^-(aq) \longrightarrow 3\,AgNO_3(aq) + NO(g) + 2\,H_2O$

(e) $(CH_3)_2NNH_2(l) + 4\,O_2(g) \longrightarrow 2\,CO_2(g) + 4\,H_2O(l) + N_2(g)$

39. (a) In N_2H_4 there are $2 \times 5 + 4 \times 1 = 14$ valence electrons = 7 pairs.

(b) In HN_3 there are $1 + 3 \times 5 = 16$ valence electrons = 8 pairs.

(a) $H\!-\!\bar{N}\!-\!\bar{N}\!-\!H$ (b) $H\!-\!\bar{N}\!=\!N\!=\!\bar{N} \leftrightarrow H\!-\!\underline{N}\!\equiv\!N|$
      $\;\;\;|\;\;\;|$
      $\;\;H\;\;H$

40. (a) The phosphate ion hydrolyzes extensively. $PO_4^{3-}(aq) + H_2O \longrightarrow HPO_4^{2-}(aq) + OH^-(aq)$

(b) The dihydrogen phosphate ion both ionizes and hydrolyzes, with the ionization being the more effective.

Ionization: $H_2PO_4^-(aq) + H_2O \longrightarrow HPO_4^{2-}(aq) + H_3O^+(aq) \qquad K_a = 6.3 \times 10^{-8}$

Hydrolysis: $H_2PO_4^-(aq) + H_2O \longrightarrow H_3PO_4(aq) + OH^-(aq) \qquad K_b = 1.4 \times 10^{-12}$

Here, K_b equals K_w/K_1, where K_1 is the first ionization constant of phosphoric acid.

41. (a) HPO_4^{2-} hydrogen phosphate ion   **(c)** $H_6P_4O_{13}$ tetrapolyphosphoric acid

(b) $Ca_2P_2O_7$ calcium pyrophosphate or calcium diphosphate

42. Each simple cubic unit cell has one Po atom at each of its eight corners, but each corner is shared among eight unit cells. Thus, there is a total of one Po atom per unit cell. The edge of that unit cell is 335 pm. From this information we obtain the density of polonium.

$$\text{density} = \frac{1 \text{ Po atom} \times \dfrac{1 \text{ mol Po}}{6.022 \times 10^{23} \text{ Po atoms}} \times \dfrac{209 \text{ g Po}}{1 \text{ mol Po}}}{\left(335 \text{ pm} \times \dfrac{1 \text{ m}}{10^{12} \text{ pm}} \times \dfrac{100 \text{ cm}}{1 \text{ m}}\right)^3} = 9.23 \text{ g/cm}^3$$

43. hyponitrous acid, a weak diprotic acid   nitramide, contains the amide group

$$\text{H}\!-\!\overline{\text{O}}\!-\!\overline{\text{N}}\!=\!\overline{\text{N}}\!-\!\overline{\text{O}}\!-\!\text{H} \qquad\qquad \text{H}\!-\!\overline{\text{N}}\!-\!\overline{\text{O}}\!-\!\overline{\text{N}}\!=\!\overline{\text{O}} \quad\text{or}\quad \text{H}\!-\!\overline{\text{N}}\!-\!\text{N}\!=\!\overline{\text{O}}$$

Both Lewis structures for nitramide are plausible based on the information supplied. To choose between them would require further information, such as whether the molecule contains a nitrogen-nitrogen bond. Experimental evidence indicates the structure to be the one with the nitrogen-nitrogen bond.

Carbon and Silicon

44. In the sense that diamonds react imperceptibly slow at room temperature (either with oxygen to form carbon dioxide, or in its transformation to the more stable graphite), it is true that "diamonds last forever." However, at elevated temperatures, diamond will burn to form $CO_2(g)$ and thus the statement is false. Also, the transformation C(diamond) $\longrightarrow$ C(graphite) might occur more rapidly under other conditions. And, of course, eventually the transition occurs.

45. **(1)** $2\ CH_4(s) + S_8(g) \longrightarrow 2\ CS_2(g) + 4\ H_2S(g)$

   (2) $CS_2(g) + 3\ Cl_2(g) \longrightarrow CCl_4(l) + S_2Cl_2(l)$

   (3) $4\ CS_2(g) + 8\ S_2Cl_2(g) \longrightarrow 4\ CCl_4(l) + 3\ S_8$

46. A silane is a silicon-hydrogen compound, with the general formula Si_nH_{2n+2}. A silanol is a compound in which one of the hydrogens of silane is replaced by an —OH group. Then, the general formula becomes $Si_nH_{2n+1}(OH)$. In both of these classes of compounds the number of silicon atoms, n, ranges from 1 to 6. Silicones are produced when silanols condense into chains, with the elimination of a water molecule between every two silanol molecules.

HO—Si_nH_{2n}—OH + HO—Si_nH_{2n}—OH $\longrightarrow$ HO—Si_nH_{2n}—O—Si_nH_{2n}—OH + H_2O

47. **(a)** $(CH_3)_3SiCl + H_2O \longrightarrow (CH_3)_3Si$—OH + HCl

   2 $(CH_3)_3Si$—OH $\longrightarrow (CH_3)_3Si$—O—$Si(CH_3)_3 + H_2O$

   (b) A silicone polymer does not form from $(CH_3)_3Si$—Cl. Only a dimer is produced.

   (c) The product that results from the treatment of H_3CSiCl_3 is two long Si—O—Si chains, with CH_3 groups on the outside, and joined by Si—O—Si bridges. Part of one of these chains and the bridges are shown below.

$$
\begin{array}{cccccc}
| & | & | & | & | & | \\
\text{O} & \text{O} & \text{O} & \text{O} & \text{O} & \text{O} \\
| & | & | & | & | & | \\
-\text{O}-\text{Si}-\text{O}-\text{Si}-\text{O}-\text{Si}-\text{O}-\text{Si}-\text{O}-\text{Si}-\text{O}-\text{Si}-\text{O}- \\
| & | & | & | & | & | \\
\text{CH}_3 & \text{CH}_3 & \text{CH}_3 & \text{CH}_3 & \text{CH}_3 & \text{CH}_3
\end{array}
$$

48. Both the alkali metal carbonates and the alkali metal silicates are soluble in water. Hence, they also are soluble in acids. However, the carbonates will produce gaseous carbon dioxide—CO_2—on reaction with acid ($CO_3^{2-} + 2\ H^+ \longrightarrow$ "H_2CO_3" $\longrightarrow H_2O + CO_2$), while the silicates will produce silica—SiO_2— in an analogous reaction ($SiO_4^{4-} + 4\ H^+ \longrightarrow$ "H_4SiO_4" $\longrightarrow 2\ H_2O + SiO_2$). The silica is produced in many forms depending on reaction conditions: a colloidal dispersion, a gelatinous precipitate, or a semisolid gel. Silicates of cations other than those of alkali metals are insoluble in water, as are the analogous carbonates.

However, the carbonates will dissolve in acids (witness the reaction of acid rain on limesone carvings and marble statues, both forms of $CaCO_3$), while the silicates will not. (Silicate rocks are not significantly affected by acid rain.)

Boron

49. (a) B_4H_{10} contains a total of $4 \times 3 + 10 \times 1 = 22$ valence electrons or 11 pairs. Ten of these pairs could be allocated to form 10 B—H bonds, leaving but one pair to bond the four B atoms together, clearly an electron deficient situation.

(b) In our analysis in the first part, we noted that the four B atoms had but one electron pair to bond them together. To bond these four atoms into a chain requires three electron pairs. Since each electron pair in a bridging bond replaces two "normal" bonds, there must be at least two bridging bonds in the B_4H_{10} molecules. By analogy with B_2H_6, we might write the structure below left. But this structure uses only a total of 20 electrons. (The bridge bonds are shown as dots, normal bonds—electron pairs—as dashes.) In the structure at right below, we have retained some of the form of B_2H_6, and produced a compound with the formula B_4H_{10} and 11 electron pairs. (The experimentally determined structure of B_4H_{10} consists of a four–membered ring of alternating B and H atoms, held together by bridging bonds. Two of the B atoms have two H atoms bonds to each of them by normal covalent bonds. The other two B atoms have one H atom covalently bonded to each. One final B—B bond joins these last two B atoms, across the diameter of the ring.)

(c) C_4H_{10} contains a total of $4 \times 4 + 10 \times 1 = 26$ valence electrons or 13 pairs. A plausible Lewis structure

follows.

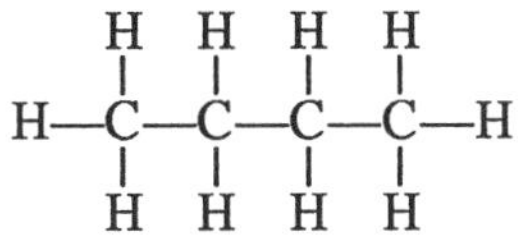

```
          H   H   H   H
          |   |   |   |
      H—C—C—C—C—H
          |   |   |   |
          H   H   H   H
```

50. (a)

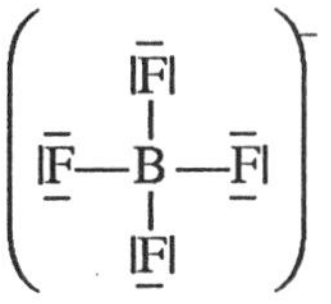

```
        ⎛      |F̄|      ⎞ ⁻
        ⎜       |       ⎟
        ⎜  |F̄—B—F̄|    ⎟
        ⎜       |       ⎟
        ⎝      |F̄|      ⎠
```

(b)

```
      |F̄|  H   H   H
       |   |   |   |
  |F̄—B←N—C—C—H
       |   |   |   |
      |F̄|  H   H   H
```

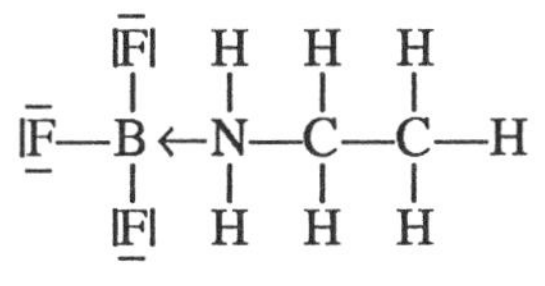

24 THE TRANSITION ELEMENTS

REVIEW QUESTIONS

1. **(a)** A domain is a region within a metal within which the magnetic moments of the metal atoms align. In some ways, it behaves as a small magnet.
   (b) Floatation is a process of separating metal ore from the extra rock, called *gangue*. It involves agitating the crushed rock and ore with water and a detergent. The ore particles stick to the bubbles and are skimmed off.
   (c) Leaching refers to treating ore-bearing material with a chemical solution to dissolve the desired metallic elements. The solution generally is trickled through the material, and the resulting metal-bearing solution collected at the bottom.
   (d) An amalgam is an alloy of another metal in mercury.

2. **(a)** The lanthanide contraction refers to the steady decrease in the size of atoms and (particularly) +3 ions as one proceeds across the lanthanide series from La through Lu.
   (b) Zone refining uses a ring oven to melt a portion of a cylindrical metal bar. Impurities are more soluble in the molten metal and tend to remain in the molten portion. The molten region then is swept down the bar to the end, carrying the impurities with it. The process can be repeated several times, concentrating impurities at one end of the bar, which is cut off and returned to be refined.
   (c) The basic oxygen process is a relatively recently developed technique of steel making in which molten pig iron is treated with oxygen gas (at about 10 atm pressure) and powdered limestone to reduce the carbon content and remove the undesireable impurities.
   (d) By the inertness of gold, we mean the nonreactivity of the metal with most chemical reagents. Gold will react with aqua regia, three parts hydrochloric acid and one part nitric acid, and with $F_2(g)$, but with little else.

3. **(a)** Paramagnetism indicates that atoms, molecules, or ions have one or more unpaired electrons. In Ferromagnetism, the unpaired electrons on several adjacent atoms or ions align with each other, producing a stronger magnetic field.
   (b) Roasting refers to heating a concentrated ore, usually to convert hydroxides, carbonates, or sulfides to oxides. Refining is the process of purifying the crude metal that results from the reduction of the ore.
   (c) In hydrometallurgy, reactions in aqueous solution are used in concentration, reduction, and/or refining. In pyrometallurgy, the process of reduction and/or refining are carried out at high temperatures.
   (d) Chromate ion is CrO_4^{2-}; dichromate ion is $Cr_2O_7^{2-}$.

4. **(a)** $Sc(OH)_3$ scandium hydroxide   **(b)** Cu_2O copper(I) oxide
   (c) $TiCl_4$ titanium(IV) chloride *or* titanium tetrachloride   **(d)** V_2O_5 vanadium(V) oxide
   (e) K_2CrO_4 potassium chromate   **(f)** K_2MnO_4 potassium manganate

5. **(a)** CrO_3 chromium(VI) oxide   **(b)** $FeSiO_3$ iron(II) silicate
   (c) $BaCr_2O_7$ barium dichromate   **(d)** $CuCN$ copper(I) cyanide
   (e) $CoCl_2 \cdot 6H_2O$ cobalt(II) chloride hexahydrate

6. **(a)** Pig iron is the iron obtained from the blast furnace. Once poured into molds it is called cast iron. "It contains about 95% Fe, 3% to 4% C, and various other impurities."
   (b) Ferromanganese is an alloy of iron and manganese, produced by reducing the mixed oxides of the two metals with carbon.

(c) Chromite iron is the principal ore of chromium, $Fe(CrO_2)_2$

(d) Brass is a mixture of copper and zinc, with small quantities of Sn, Pb, and Fe.

(e) Aqua regia is a mixture of three parts HCl(aq) and one part HNO_3(aq). It both is an oxidizing acid and can form complex ions with cations in solution, making it the only acid that can dissolve difficultly soluble substances, such as Au and HgS.

(f) Blister copper is the impure copper that results from the reduction process. The blisters are frozen bubbles of the SO_2(g) that forms during that process.

(g) Stainless steel contains iron, and a significant proportion (about 10% each) of chromium and nickel. It is not ferromagnetic.

7. (a) $TiCl_4(g) + 4\ Na(l) \xrightarrow{\Delta} Ti(s) + 4\ NaCl(l)$

(b) $Cr_2O_3(s) + 2\ Al(s) \xrightarrow{\Delta} 2\ Cr(l) + Al_2O_3(s)$

(c) $Ag(s) + HCl(aq) \longrightarrow$ no reaction

(d) $K_2Cr_2O_7(aq) + 2\ KOH(aq) \longrightarrow 2\ K_2CrO_4(aq) + H_2O$

(e) $MnO_2(s) + 2\ C(s) \xrightarrow{\Delta} Mn(l) + 2\ CO(g)$

8. (a) Oxidation: $\{Fe_2S_3(s) + 6\ OH^-(aq) \longrightarrow 2\ Fe(OH)_3(s) + 3\ S(s) + 6\ e^-\}\quad \times 2$

Reduction: $\{O_2(g) + 2\ H_2O + 4\ e^- \longrightarrow 4\ OH^-(aq)\}\qquad\qquad \times 3$

Net: $2\ Fe_2S_3(s) + 3\ O_2(g) + 6\ H_2O \longrightarrow 4\ Fe(OH)_3(s) + 6\ S(s)$

(b) Oxidation: $\{Mn^{2+}(aq) + 4\ H_2O \longrightarrow MnO_4^-(aq) + 8\ H^+(aq) + 5\ e^-\} \times 2$

Reduction: $\{S_2O_8^{2-}(aq) + 2\ e^- \longrightarrow 2\ SO_4^{2-}(aq)\}\qquad\qquad \times 5$

Net: $2\ Mn^{2+}(aq) + 8\ H_2O + 5\ S_2O_8^{2-}(aq) \longrightarrow 2\ MnO_4^-(aq) + 16\ H^+(aq) + 10\ SO_4^{2-}(aq)$

(c) Oxidation: $\{Ag(s) + 2\ CN^-(aq) \longrightarrow [Ag(CN)_2]^-(aq) + e^-\}\qquad\qquad \times 4$

Reduction: $O_2(g) + 2\ H_2O + 4\ e^- \longrightarrow 4\ OH^-(aq)$

Net: $4\ Ag(s) + 8\ CN^-(aq) + O_2(g) + 2\ H_2O \longrightarrow 4\ [Ag(CN)_2]^-(aq) + 4\ OH^-(aq)$

9. (a) $3\ Cu(s) + 8\ H^+(aq) + 2\ NO_3^-(aq) \longrightarrow 3\ Cu^{2+}(aq) + 2\ NO(g) + 4\ H_2O$
Virtually any first period transition metal can be substituted for Cu.

(b) $Cr_2O_3(s) + 2\ OH^-(aq)\ 3\ H_2O \longrightarrow 2\ Cr(OH)_4^-(aq)$
The oxide must be amphoteric. Thus, Sc_2O_3, TiO_2, ZrO_2, and ZnO could be substituted for Cr_2O_3.

(c) $2\ La(s) + 6\ HCl(aq) \longrightarrow 2\ LaCl_3(aq) + 3\ H_2(g)$
Any lanthanide or actinide element can be substituted for lanthanum.

10. (a) Ti $[Ar]\ _{3d}\boxed{\uparrow\,|\,\uparrow\,|\,\,|\,\,|\,\,}\ _{4s}\boxed{\uparrow\downarrow}$ (b) V^{3+} $[Ar]\ _{3d}\boxed{\uparrow\,|\,\uparrow\,|\,\,|\,\,|\,\,}\ _{4s}\boxed{\ }$

(c) Cr^{2+} $[Ar]\ _{3d}\boxed{\uparrow\,|\,\uparrow\,|\,\uparrow\,|\,\uparrow\,|\,\,}\ _{4s}\boxed{\ }$ (d) Mn^{4+} $[Ar]\ _{3d}\boxed{\uparrow\,|\,\uparrow\,|\,\uparrow\,|\,\,|\,\,}\ _{4s}\boxed{\ }$

(e) Mn^{2+} $[Ar]\ _{3d}\boxed{\uparrow\,|\,\uparrow\,|\,\uparrow\,|\,\uparrow\,|\,\uparrow}\ _{4s}\boxed{\ }$ (f) Fe^{3+} $[Ar]\ _{3d}\boxed{\uparrow\,|\,\uparrow\,|\,\uparrow\,|\,\uparrow\,|\,\uparrow}\ _{4s}\boxed{\ }$

11. We first give the orbital diagram for each of the species, and then count the number of unpaired electrons.

(a) Fe $[Ar]\ _{3d}\boxed{\uparrow\downarrow\,|\,\uparrow\,|\,\uparrow\,|\,\uparrow\,|\,\uparrow}\ _{4s}\boxed{\uparrow\downarrow}$ 4 unpaired electrons

(b) Sc^{3+} $[Ar]\ _{3d}\boxed{\,\,|\,\,|\,\,|\,\,|\,\,}\ _{4s}\boxed{\ }$ 0 unpaired electrons

(c) Ti^{2+} $[Ar]\ _{3d}\boxed{\uparrow\,|\,\uparrow\,|\,\,|\,\,|\,\,}\ _{4s}\boxed{\ }$ 2 unpaired electrons

(d) Mn^{4+} $[Ar]\ _{3d}\boxed{\uparrow\,|\,\uparrow\,|\,\uparrow\,|\,\,|\,\,}\ _{4s}\boxed{\ }$ 3 unpaired electrons

(e) Cr $[Ar]\ _{3d}\boxed{\uparrow\,|\,\uparrow\,|\,\uparrow\,|\,\uparrow\,|\,\uparrow}\ _{4s}\boxed{\uparrow}$ 6 unpaired electrons

(f) Cu^{2+} $[Ar]\ _{3d}\boxed{\uparrow\downarrow\,|\,\uparrow\downarrow\,|\,\uparrow\downarrow\,|\,\uparrow\downarrow\,|\,\uparrow}\ _{4s}\boxed{\ }$ 1 unpaired electron

Finally, arrange in order of decreasing number of unpaired electrons: $Cr > Fe > Mn^{4+} > Ti^{2+} > Cu^{2+} > Sc^{3+}$

12. (1) Transition metals tend to have <u>higher melting points</u> than representative metals. (2) Since they are metals, transition elements have fairly <u>low ionization energies</u>. (3) The ions of transition metals often <u>are colored</u> in aqueous solution. (4) Because they are metals and thus readily form cations, they have <u>negative standard reduction potentials</u>. Their compounds often have unpaired electrons because of the diversity of d-electron configurations, and thus (5) they often <u>are paramagnetic</u>.

13. Sc^{3+} has the electron configuration of the preceding noble gas (Ar). This is obviously an electron configuration in which all electrons are paired, a diamagnetic species. All of the other species have at least one unpaired electron, as orbital diagrams show.

Cr^{2+} [Ar] $_{3d}$ ↑ ↑ ↑ ↑ $_{4s}$ ☐ Fe^{3+} [Ar] $_{3d}$ ↑ ↑ ↑ ↑ ↑ $_{4s}$ ☐

Sc^{3+} [Ar] $_{3d}$ ☐ ☐ ☐ ☐ ☐ $_{4s}$ ☐ Cu^{2+} [Ar] $_{3d}$ ↑↓ ↑↓ ↑↓ ↑↓ ↑ $_{4s}$ ☐

14. Ti cannot display a +6 oxidation state, because in order for Ti to display a +6 oxidation state, two electrons would have to be removed from the noble gas electron configuration of Ar. This is exceedingly unlikely.

15. The ion that is the best oxidizing agent in aqueous solution is the ion that can most readily be reduced. And we can assess that property either by standard reduction potentials—$E°[Ag^+(aq)|Ag(s)] = +0.800$ V, $E°[Cu^{2+}(aq)|Cu(s)] = +0.337$ V, $E°[Zn^{2+}(aq)|Zn(s)] = -0.763$ V, $E°[Na^+(aq)|Na(s)] = -2.714$ V—with the most positive value indicating the best oxidizing agent, or from our own experience with the reverse reaction. We know that silver is the least likely of all the metals to be oxidized, and thus its ion is the most readily reduced.

16. +3 is probably the most stable oxidation state of iron because in this oxidation state iron has a d^5 electron configuration. As we have seen before, an electron configuration containing a half-filled subshell is unexpectedly stable energetically, compared to other configurations with partially filled subshells. In the cases of cobalt and nickel, the ions with d^5 configurations are Co^{3+} and Ni^{4+}. The slight advantage of a d^5 configuration is not sufficient to stabilize ions with such high charges, although Co^{3+} is stable under certain conditions.

EXERCISES

Properties of the Transition Elements

17. A given representative metal displays one oxidation state, usually equal to its family number in the periodic table. Exceptions are elements such as Tl (+1 and +3), Pb (+2 and +4), and Sn (+2 and +4) in which the lower oxidation state represents a pair of s electrons not being ionized (a so-called "inert pair").

    Representative metals do not form a wide variety of complex ions, with Al^{3+}, Sn^{2+}, Sn^{4+}, and Pb^{2+} being major exceptions. On the other hand, most transition metal ions form an extensive variety of complex ions; it is, in fact, the unusual transition metal cation that does not form complex ions.

    Most compounds of representative metals are colorless; exceptions occur when the anion is colored. On the other hand, many of the compounds of transition metal cations are colored.

    Virtually every representative metal cation has no unpaired electrons and hence is diamagnetic. On the other hand, many transition metals cations have one or more unpaired electrons and therefore are paramagnetic.

18. When an electron is added to a representative element to create the element of next highest atomic number, this electron is added to the outer shell of the atom, far from the nucleus. Thus, it has a major influence on the size of the atom. However, when an electron is added to a transition metal atom to create the atom of next highest atomic number, it is added to the electronic shell inside the outermost. The electron thus has been added to a position close to the nucleus to which it is attracted quite strongly and thus it has small effect on the size of the atom.

19. Manganese seems to exhibit the greatest number of oxidation states among the first-row transition elements, every oxidation state from +2 ($MnSO_4$) to +7 ($KMnO_4$). In achieving these oxidation states, Mn loses first the two $4s$ electrons, and then successively the five $3d$ electrons. Since oxidation states higher than +7 seemingly are not sustainable among the first-row transition elements, the range of states observed for Mn is the most extensive.

20. As we proceed from Sc to Cr the valence electron configuration has an increasing number of unpaired electrons, which are capable of forming bonds to adjacent atoms. As we continue beyond Cr, however, these electrons become paired, and the resulting atoms are less able to form bonds with their neighbors.

21. **(a)** As we proceed across the transition series, electrons are being added to a shell next to the valence shell. Thus, the electronic character of the outside of each atom is changing somewhat, leading to changes in their chemical properties. (Recall that this valence shell consists of only two $4s$ electrons.) As we proceed across the lanthanides, on the other hand, the electrons are being added to a shell two removed from the outermost shell; there is less of an effect on the electronic character of the valence shell. (Recall also that there are 10 electrons in subshells outside the one being filled: $5s^2$, $5p^6$, and $6s^2$.)

(b) The greater ease of forming lanthanide cations compared to forming transition metal cations, is due to the larger size of lanthanide atoms. The valence electrons of these larger atoms are much further from the nucleus, much less strongly attracted to the positive charge of the nucleus, and thus are removed much more readily.

22. Paramagnetism—the basis of ferromagnetism—means simply that the atoms or ions in a compound possess one or more unpaired electrons. Since each electron has an associated spin, these unpaired electrons act similarly to small magnets. In ferromagnetism, these small magnets are close enough to each other to line up together and reinforce each other in magnetic domains and thus create a larger magnetic effect than the rather disorganized arrangement of paramagnetic atoms. Magnetic domains can only be produced by species having ionic radii corresponding to those of iron, cobalt, and nickel.

23. The reason why the radii of Pd (138 pm) and Pt (139 pm) are so similar, and so different from the radius of Ni (125 pm) is because the lanthanide series intrudes between Pd and Pt. Because of the lanthanide contraction, elements in the second transition row are almost identical in size to their cogeners (family members) in the third transition row.

24. (a) $Sc(OH)_3(s) + 3\ H^+(aq) \longrightarrow Sc^{3+}(aq) + 3\ H_2O$

(b) $2\ Sc_2O_3(l,\ in\ Na_3ScF_6) + 3\ C(s) \xrightarrow{\ electrolysis\ } 4\ Sc(l) + 3\ CO_2(g)$

[By analogy to the equation for the electrolytic production of Al]

25. We write some of the following reactions as total equations rather than as net ionic equations, so that the reagents used are indicated.

(a) $FeS(s) + 2\ HCl(aq) \longrightarrow FeCl_2(aq) + H_2S(g)$

$4\ Fe^{2+}(aq) + O_2(g) + 4\ H^+(aq) \longrightarrow 4\ Fe^{3+}(aq) + 2\ H_2O$

$Fe^{3+}(aq) + 3\ Na^+(aq) + 3\ OH^-(aq) \longrightarrow Fe(OH)_3(s) + 3\ Na^+(aq)$

(b) $CaCO_3(s) + 2\ HCl(aq) \longrightarrow CaCl_2(aq) + H_2O + CO_2(g)$

$2\ CaCl_2(aq) + K_2Cr_2O_7(aq) + 2\ NaOH(aq) \longrightarrow 2\ CaCrO_4(s) + 2\ KCl(aq) + 2\ NaCl(aq) + H_2O$

(c) $(NH_4)_2Cr_2O_7(s) \xrightarrow{\ \Delta\ } N_2(g) + 4\ H_2O(g) + Cr_2O_3(s)$

$Cr_2O_3(s) + 6\ HCl(aq) \longrightarrow CrCl_3(aq) + 3\ H_2O$

26. (a) $TiO_2(s) + 2\ KOH(l) \longrightarrow K_2TiO_3(l) + H_2O(g)$

(b) $Cr(s) + 2\ HCl(aq) \longrightarrow Cr^{2+}(aq) + 2\ Cl^-(aq) + H_2(g)$

(c) $4\ Cr^{2+}(aq) + O_2(g) + 4\ H^+(aq) \longrightarrow 4\ Cr^{3+}(aq) + 2\ H_2O$

(d) $3\ Fe^{2+}(aq) + MnO_4^-(aq) + 4\ H^+(aq) \longrightarrow 3\ Fe^{3+}(aq) + MnO_2(s) + 2\ H_2O$

(e) $Ag(s) + 2\ HNO_3(aq) \longrightarrow AgNO_3(aq) + NO_2(g) + H_2O$

Oxidation-reduction

27. (a) Reduction: $VO^{2+}(aq) + 2\ H^+(aq) + e^- \longrightarrow V^{3+}(aq) + H_2O$

(b) Oxidation: $Cr^{2+}(aq) \longrightarrow Cr^{3+}(aq) + e^-$

(c) Oxidation: $Fe(OH)_3(s) + 5\ OH^-(aq) \longrightarrow FeO_4^{2-}(aq) + 4\ H_2O + 3\ e^-$

(d) Reduction: $[Ag(CN)_2]^-(aq) + e^- \longrightarrow Ag(s) + 2\ CN^-(aq)$

28. (a) First we need the reduction potential for the couple $VO_2^+(aq)|V^{2+}(aq)$

$$VO_2^+(aq) + 2\ H^+(aq) + e^- \longrightarrow VO^{2+}(aq) + H_2O \qquad \Delta G^\circ = -1\ \mathcal{F}\ (+1.000\ V)$$

$$VO^{2+}(aq) + 2\ H^+(aq) + e^- \longrightarrow V^{3+}(aq) + H_2O \qquad \Delta G^\circ = -1\ \mathcal{F}\ (+0.337\ V)$$

$$V^{3+}(aq) + e^- \longrightarrow V^{2+}(aq) \qquad \Delta G^\circ = -1\ \mathcal{F}\ (-0.255\ V)$$

$$\overline{VO_2^+(aq) + 4\ H^+(aq) + 3\ e^- \longrightarrow V^{2+}(aq) + 2\ H_2O \qquad \Delta G^\circ = -3\ \mathcal{F} E^\circ}$$

$$E° = \frac{1.000 \text{ V} + 0.337 \text{ V} - 0.255 \text{ V}}{3} = +0.361 \text{ V} \quad \text{We analyze the oxidation-reduction reaction.}$$

Oxidation:	$\{2 \text{ Br}^-(aq) \longrightarrow \text{Br}_2(l) + 2 \text{ e}^-\}$	$\times 3$	$-E° = -1.065 \text{ V}$
Reduction:	$\{\text{VO}_2{}^+(aq) + 4 \text{ H}^+(aq) + 3 \text{ e}^- \longrightarrow \text{V}^{2+}(aq) + 2 \text{ H}_2\text{O}\}$	$\times 2$	$E° = +0.361 \text{ V}$

Net: $\quad 6 \text{ Br}^-(aq) + 2 \text{ VO}_2{}^+(aq) + 8 \text{ H}^+(aq) \longrightarrow 3 \text{ Br}_2(l) + 2 \text{ V}^{2+}(aq) + 4 \text{ H}_2\text{O} \quad E°_{cell} = -0.704 \text{ V}$

This reaction does not occur to a significant extent as written.

(b) Oxidation: $\text{Fe}^{2+}(aq) \longrightarrow \text{Fe}^{3+}(aq) + \text{e}^- \qquad\qquad -E° = -0.771 \text{ V}$

Reduction: $\text{VO}_2{}^+(aq) + 2 \text{ H}^+(aq) + \text{e}^- \longrightarrow \text{VO}^{2+}(aq) + \text{H}_2\text{O} \qquad E° = +1.000 \text{ V}$

Net: $\quad \text{Fe}^{2+}(aq) + \text{VO}_2{}^+(aq) + 2 \text{ H}^+(aq) \longrightarrow \text{Fe}^{3+}(aq) + \text{VO}^{2+}(aq) + \text{H}_2\text{O} \quad E°_{cell} = +0.229 \text{ V}$

This reaction does occur to a significant extent under standard conditions.

(c) Oxidation: $\text{H}_2\text{O}_2(aq) \longrightarrow 2 \text{ H}^+(aq) + 2 \text{ e}^- + \text{O}_2(g) \qquad\qquad -E° = -0.695 \text{ V}$

Reduction: $\text{MnO}_2(s) + 4 \text{ H}^+(aq) + 2 \text{ e}^- \longrightarrow \text{Mn}^{2+}(aq) + 2 \text{ H}_2\text{O} \qquad E° = +1.23 \text{ V}$

Net: $\quad \text{H}_2\text{O}_2(aq) + \text{MnO}_2(s) + 2 \text{ H}^+(aq) \longrightarrow \text{O}_2(g) + \text{Mn}^{2+}(aq) + 2 \text{ H}_2\text{O} \quad E°_{cell} = +0.54 \text{ V}$

This reaction does occur to a significant extent under standard conditions.

29. When a species acts as a reducing agent, it is oxidized. From Table 21-1 we obtain the potentials for each of the following couples.

$-E°[\text{Zn}^{2+}(aq)/\text{Zn}(s)] = +0.763 \text{ V} \qquad -E°[\text{Sn}^{4+}(aq)/\text{Sn}^{2+}(aq)] = -0.154 \text{ V} \qquad -E°[\text{I}_2(s)/\text{I}^-(aq)] = -0.535 \text{ V}$

Each of these potentials is combined with the reduction potential cited in each part. If the resulting value of $E°_{cell}$ is positive, then the reducing agent will be effective in accomplishing the desired reduction, which we indicate with "yes"; if not, we write "no".

(a) $E°[\text{Cr}_2\text{O}_7{}^{2-}(aq)/\text{Cr}^{3+}(aq)] = +1.33 \text{ V}$

$E°_{cell} = E°[\text{Cr}_2\text{O}_7{}^{2-}(aq)/\text{Cr}^{3+}(aq)] - E°[\text{Zn}^{2+}(aq)/\text{Zn}(s)] = +1.33 \text{ V} + 0.763 \text{ V} = +2.09 \text{ V}$ yes

$E°_{cell} = E°[\text{Cr}_2\text{O}_7{}^{2-}(aq)/\text{Cr}^{3+}(aq)] - E°[\text{Sn}^{4+}(aq)/\text{Sn}^{2+}(aq)] = +1.33 \text{ V} - 0.154 \text{ V} = +1.18 \text{ V}$ yes

$E°_{cell} = E°[\text{Cr}_2\text{O}_7{}^{2-}(aq)/\text{Cr}^{3+}(aq)] - E°[\text{I}_2(s)/\text{I}^-(aq)] = +1.33 \text{ V} - 0.535 \text{ V} = +0.80 \text{ V}$ yes

(b) $E°[\text{Cr}^{3+}(aq)/\text{Cr}^{2+}(aq)] = -0.407 \text{ V}$

$E°_{cell} = E°[\text{Cr}^{3+}(aq)/\text{Cr}^{2+}(aq)] - E°[\text{Zn}^{2+}(aq)/\text{Zn}(s)] = -0.424 \text{ V} + 0.763 \text{ V} = +0.339 \text{ V}$ yes

$E°_{cell} = E°[\text{Cr}^{3+}(aq)/\text{Cr}^{2+}(aq)] - E°[\text{Sn}^{4+}(aq)/\text{Sn}^{2+}(aq)] = -0.424 \text{ V} - 0.154 \text{ V} = -0.578 \text{ V}$ no

$E°_{cell} = E°[\text{Cr}^{3+}(aq)/\text{Cr}^{2+}(aq)] - E°[\text{I}_2(s)/\text{I}^-(aq)] = -0.424 \text{ V} - 0.535 \text{ V} = -0.959 \text{ V}$ no

(c) $E°[\text{SO}_4{}^{2-}(aq)/\text{SO}_2(g)] = +0.17 \text{ V}$

$E°_{cell} = E°[\text{SO}_4{}^{2-}(aq)/\text{SO}_2(g)] - E°[\text{Zn}^{2+}(aq)/\text{Zn}(s)] = +0.17 \text{ V} + 0.763 \text{ V} = +0.93 \text{ V}$ yes

$E°_{cell} = E°[\text{SO}_4{}^{2-}(aq)/\text{SO}_2(g)] - E°[\text{Sn}^{4+}(aq)/\text{Sn}^{2+}(aq)] = +0.17 \text{ V} - 0.154 \text{ V} = +0.02 \text{ V}$ yes

$E°_{cell} = E°[\text{SO}_4{}^{2-}(aq)/\text{SO}_2(g)] - E°[\text{I}_2(s)/\text{I}^-(aq)] = +0.17 \text{ V} - 0.535 \text{ V} = -0.36 \text{ V}$ no

30. We simply combine the $\text{MnO}_4{}^-(aq)/\text{MnO}_2(s)$ couple with the $\text{MnO}_2(s)/\text{Mn}^{2+}(aq)$ couple.

$\text{MnO}_4{}^-(aq) + 4 \text{ H}^+(aq) + 3 \text{ e}^- \longrightarrow \text{MnO}_2(s) + 2 \text{ H}_2\text{O} \qquad \Delta G° = -3 \, \mathcal{F} \, (+1.70 \text{ V})$

$\text{MnO}_2(s) + 4 \text{ H}^+(aq) + 2 \text{ e}^- \longrightarrow \text{Mn}^{2+}(aq) + 2 \text{ H}_2\text{O} \qquad \Delta G° = -2 \, \mathcal{F} \, (+1.23 \text{ V})$

$\text{MnO}_4{}^-(aq) + 8 \text{ H}^+(aq) + 5 \text{ e}^- \longrightarrow \text{Mn}^{2+}(aq) + 4 \text{ H}_2\text{O} \qquad \Delta G° = -5 \, \mathcal{F} \, E°$

$$E° = \frac{(3 \times +1.70 \text{ V}) + (2 \times +1.23 \text{ V})}{5} = +1.51 \text{ V}$$

This value of $E°$ is precisely the same as the value given in Table 21-1.

31. We simply combine the two half-reactions that are indicated and compute the cell voltage.

(a) Oxidation: $\text{Mn}^{3+}(aq) + 2 \text{ H}_2\text{O} \longrightarrow \text{MnO}_2(s) + 4 \text{ H}^+(aq) + \text{e}^- \qquad -E° = -0.95 \text{ V}$

Reduction: $\text{Mn}^{3+}(aq) + \text{e}^- \longrightarrow \text{Mn}^{2+}(aq) \qquad\qquad\qquad\qquad\quad E° = +1.49 \text{ V}$

Net: $\quad 2 \text{ Mn}^{3+}(aq) + 2 \text{ H}_2\text{O} \longrightarrow \text{MnO}_2(s) + \text{Mn}^{2+}(aq) + 4 \text{ H}^+(aq) \quad E°_{cell} = +0.54 \text{ V}$

A positive cell voltage indicates that the reaction is spontaneous and the $\text{Mn}^{3+}(aq)$ disproportionates.

(b) Oxidation: $\{\text{MnO}_4{}^{2-}(aq) \longrightarrow \text{MnO}_4{}^-(aq) + \text{e}^-\} \qquad \times 2 \qquad -E° = -0.56 \text{ V}$

Reduction: $\text{MnO}_4{}^{2-}(aq) + 4 \text{ H}^+(aq) + 2 \text{ e}^- \longrightarrow \text{MnO}_2(s) + 2 \text{ H}_2\text{O} \qquad E° = +2.26 \text{ V}$

Net: $\quad 3 \text{ MnO}_4{}^{2-}(aq) + 4 \text{ H}^+(aq) \longrightarrow 2 \text{ MnO}_4{}^-(aq) + \text{MnO}_2(s) + 2 \text{ H}_2\text{O} \quad E°_{cell} = +1.70 \text{ V}$

Again, a positive cell voltage indicates that the reaction is spontaneous and that $MnO_4^{2-}(aq)$ disproportionates in acidic solution.

32. The reduction half-equation are significantly different in alkaline solution, as is its half-cell potential.

Oxidation: $MnO_4^{2-}(aq) \longrightarrow MnO_4^-(aq) + e^-$ $\qquad -E° = -0.56$ V

Reduction: $MnO_4^{2-}(aq) + H_2O + e^- \longrightarrow MnO_3^-(aq) + 2\ OH^-(aq)$ $\qquad E° = +0.3$ V

Net: $\qquad 2\ MnO_4^{2-}(aq) + H_2O \longrightarrow MnO_4^-(aq) + MnO_3^-(aq) + 2\ OH^-(aq)$ $\qquad E°_{cell} = -0.3$ V

The negative value of the cell voltage indicates that $MnO_4^{2-}(aq)$ is stable in alkaline solution, possibly because of the availability of the Mn(V) species in alkaline solution.

33. We would expect $Fe^{2+}(aq)$ to be oxidized to the +3 cation by $O_2(g)$ in acidic solution. One reason for this expectation is the fact that Fe^{3+} has a d^5 electron configuration, and we have previously seen that a half-filled subshell is unusually stable. We combine half-cell potentials.

Oxidation: $\{Fe^{2+}(aq) \longrightarrow Fe^{3+}(aq) + e^-\} \qquad \times 4$ $\qquad -E° = -0.771$ V

Reduction: $O_2(g) + 4\ H^+(aq) + 4\ e^- \longrightarrow 2\ H_2O$ $\qquad E° = +1.229$ V

Net: $\qquad 4\ Fe^{2+}(aq) + O_2(g) + 4\ H^+(aq) \longrightarrow 4\ Fe^{3+}(aq) + 2\ H_2O$ $\qquad E°_{cell} = +0.458$ V

34. The couple that we seek must have a half-cell potential of such a size that a positive sum results when this half-cell potential (1) is combined with $E°[VO^{2+}(aq)|V^{3+}(aq)] = +0.337$ V, but (2) a negative sum must be produced when this half-cell potential is combined with $E°[V^{3+}(aq)|V^{2+}(aq)] = -0.255$ V, or (3) when it is combined with $E°[V^{2+}(aq)|V(s)] = -1.18$ V. Now recall that we are speaking of an *oxidation* reaction for this half-cell. Thus, our three relationships are, in terms of the *reduction* potential, $E°$,

(1) $-E° + 0.337$ V > 0 $\qquad$ (2) $-E° - 0.255 < 0$ $\qquad$ (3) $-E° - 1.18$ V < 0

Multiply each relationship by -1, which changes the direction of the inequality, and all signs.

(1) $+E° - 0.337$ V < 0 $\qquad$ (2) $+E° + 0.255 > 0$ $\qquad$ (3) $+E° + 1.18$ V > 0

Solve each inequality: (1) $E° < +0.337$ V $\qquad$ (2) $E° > -0.255$ V $\qquad$ (3) $E° > -1.18$ V

Thus, possible reducing couples from Table 21-1 have $+0.255$ V $> -E° > -0.337$ V and include:

$-E°[Sn(s)|Sn^{2+}(aq)] = +0.137$ V $\qquad -E°[Pb(s)|Pb^{2+}(aq)] = +0.125$ V $\quad -E°[H_2(g)|H^+(aq)] = +0.000$ V

$-E°[H_2S(g)|S(s)] = -0.14$ V $\qquad -E°[Sn^{2+}(aq)|Sn^{4+}(aq)] = -0.154$ V

$-E°[SO_2(g)|SO_4^{2-}(aq)] = -0.17$ V $\qquad -E°[Cu(s)|Cu^{2+}(aq)] = -0.337$ V

35. We first determine the mass of the excess oxalic acid.

$$\text{mass } H_2C_2O_4 = 0.03006\ L \times \frac{0.1000\ \text{mol } MnO_4^-}{1\ L\ \text{soln}} \times \frac{5\ \text{mol } H_2C_2O_4}{2\ \text{mol } MnO_4^-} \times \frac{126.07\ g\ H_2C_2O_4 \cdot 2H_2O}{1\ \text{mol } H_2C_2O_4 \cdot 2H_2O}$$

$$= 0.9474\ g\ H_2C_2O_4$$

Thus, the mass of oxalic acid that reacted with MnO_2 is 1.651 g $- 0.9474$ g $= 0.704$ g $H_2C_2O_4 \cdot 2H_2O$

Now we can determine the mass of MnO_2 in the ore.

$$\text{mass } MnO_2 = 0.704\ g\ H_2C_2O_4 \times \frac{1\ \text{mol } H_2C_2O_4 \cdot 2H_2O}{126.07\ g\ H_2C_2O_4 \cdot 2H_2O} \times \frac{1\ \text{mol } MnO_2}{1\ \text{mol } H_2C_2O_4} \times \frac{86.94\ g\ MnO_2}{1\ \text{mol } MnO_2}$$

$$= 0.485\ g\ MnO_2$$

$$\%MnO_2 = \frac{0.485\ g\ MnO_2}{0.589\ g\ \text{ore}} \times 100\% = 82.3\%$$

Chromium and Chromium Compounds

36. Orange dichromate ion is in equilibrium with yellow chromate ion in aqueous solution.

$Cr_2O_7^{2-}(aq) + H_2O \rightleftharpoons 2\ CrO_4^{2-}(aq) + 2\ H^+(aq)$

The chromate ion in solution then reacts with lead(II) ion to form a precipitate of yellow lead(II) dichromate.

$Pb^{2+}(aq) + CrO_4^{2-}(aq) \rightleftharpoons PbCrO_4(s)$

$PbCrO_4(s)$ will form until $[H^+]$ from the first equilibrium increases to a significant value and both equilibria are simultaneously satisfied.

37. The initial dissolving reaction forms orange dichromate ion.

$2\ BaCrO_4(s) + 2\ HCl(aq) \rightleftharpoons 2\ Ba^{2+}(aq) + Cr_2O_7^{2-}(aq) + 2\ Cl^-(aq) + H_2O$

Dichromate ion is a good oxidizing agent, sufficiently strong to oxidize Cl⁻(aq) to Cl_2(g) if the concentrations of reactants are high, the solution is acidic, and the product Cl_2(g) is allowed to escape.

$6\ Cl^-(aq) + Cr_2O_7^{2-}(aq) + 14\ H^+(aq) \rightleftharpoons 3\ Cl_2(g) + 2\ Cr^{3+}(aq) + 7\ H_2O$

Chromium(III) can hydrolyze in solution to produce green $[Cr(OH)_4]^-$(aq), but the solution needs to be alkaline for this to occur. A more likely source of the green color is a complex ion such as $[Cr(H_2O)_4Cl_2]^+$.

38. Oxidation: $\{Zn(s) \longrightarrow Zn^{2+}(aq) + 2\ e^-\}$ $\qquad \times 3$

Reduction: $Cr_2O_7^{2-}$(aq, orange) $+ 14\ H^+(aq) + 6\ e^- \longrightarrow 2\ Cr^{3+}$(aq, green) $+ 7\ H_2O$

Net: $\qquad 3\ Zn(s) + Cr_2O_7^{2-}(aq) + 14\ H^+(aq) \longrightarrow 3\ Zn^{2+}(aq) + 2\ Cr^{3+}(aq) + 7\ H_2O$

Oxidation: $Zn(s) \longrightarrow Zn^{2+}(aq) + 2\ e^-$

Reduction: $\{Cr^{3+}$(aq, green) $+ e^- \longrightarrow Cr^{2+}$(aq, blue)$\}$ $\qquad \times 2$

Net: $\qquad Zn(s) + 2\ Cr^{3+}(aq) \longrightarrow Zn^{2+}(aq) + 2\ Cr^{2+}(aq)$

The green color is due to a chloro complex of Cr^{3+}(aq), such as $[Cr(H_2O)_4Cl_2]^+$(aq).

Oxidation: $\{Cr^{2+}$(aq, blue) $\longrightarrow Cr^{3+}$(aq, green) $+ e^-\}$ $\qquad \times 4$

Reduction: $O_2(g) + 4\ H^+(aq) + 4\ e^- \longrightarrow 2\ H_2O$

Net: $\qquad 4\ Cr^{2+}(aq) + O_2(g) + 4\ H^+(aq) \longrightarrow 4\ Cr^{3+}(aq) + 2\ H_2O$

39. We know the following about chromate and dichromate ions.
   a. Chromate ion predominates in alkaline solutions; dichromate ion in acidic solutions.
   b. Many chromate compounds are insoluble, exceptions being the chromates of ammonium ion and the alkali metal (group 1A) cations.
   c. Most dichromate compounds are soluble in water.
   d. Dichromate has a large positive reduction potential, +1.33 V to Cr^{3+}(aq).
   e. Chromate has a slightly negative reduction potential, –0.11 V to $Cr(OH)_3$(s).

All of these facts contribute to reasonableness in the behavior cited. Recall that oxoanions are more effective oxidizing agents in acidic than in alkaline solution. Thus, we would expect that the species that predominates in acidic solution to be the better oxidizing agent. Also note that most precipitation occurs effectively in alkaline solution. In fact, adding an acid to a compound is often an effective way of dissolving a water-insoluble compound. Thus, we expect to see the form that predominates in alkaline solution be the most effective precipitating agent. Notice also that CrO_4^{2-} is smaller than is $Cr_2O_7^{2-}$, giving it a higher lattice energy in its compounds and making those compounds harder to dissolve.

40. CO_2(g), as the oxide of a nonmetal, is an acid anhydride. Its function is to make the solution acidic. A resonable equation for the reaction that occurs follows.

$2\ CrO_4^{2-}(aq) + 2\ H^+(aq) \rightleftharpoons Cr_2O_7^{2-}(aq) + H_2O$
$2\ H_2O + 2\ CO_2(aq) \rightleftharpoons 2\ H^+(aq) + 2\ HCO_3^-(aq)$
$\overline{2\ CrO_4^{2-}(aq) + 2\ CO_2(aq) + H_2O \rightleftharpoons Cr_2O_7^{2-}(aq) + 2\ HCO_3^-(aq)}$

41. Simple substitution into expression (24.18) yields $[Cr_2O_7^{2-}]$ in each case. In fact, the expression is readily solved for the desired concentration: $[Cr_2O_7^{2-}] = 3.2 \times 10^{14}\ [H^+]^2\ [CrO_4^{2-}]^2$ In each case, we use the value of pH to determine $[H^+] = 10^{-pH}$.

(a) $[Cr_2O_7^{2-}] = 3.2 \times 10^{14}\ (10^{-6.62})^2\ (0.20)^2 = 7.4 \times 10^{-3}$ M

(b) $[Cr_2O_7^{2-}] = 3.2 \times 10^{14}\ (10^{-8.85})^2\ (0.20)^2 = 2.6 \times 10^{-7}$ M

42. We use expression (24.18).

$\dfrac{[Cr_2O_7^{2-}]}{[CrO_4^{2-}]^2} = 3.2 \times 10^{14}\ [H^+]^2 = 3.2 \times 10^{14}\ (10^{57.55})^2 = 3.2 \times 10^{14}\ (2.8 \times 10^{-6})^2 = 2.5 \times 10^3$

$[CrO_4^{2-}]_{initial} = \dfrac{1.505\ g\ Na_2CrO_4}{0.345\ L\ soln} \times \dfrac{1\ mol\ Na_2CrO_4}{161.97\ g\ Na_2CrO_4} \times \dfrac{1\ mol\ CrO_4^{2-}}{1\ mol\ Na_2CrO_4} = 0.0269$ M

Reaction: $\qquad 2\ CrO_4^{2-}(aq) + 2\ H^+(aq) \rightleftharpoons Cr_2O_7^{2-}(aq) + H_2O$
Initial: $\qquad\quad$ 0.0269 M
Changes: $\qquad$ –2x M $\qquad\qquad\qquad\qquad\quad$ +x M
Equil: $\qquad\quad$ (0.0269 – 2x)M $\qquad\qquad\qquad\quad\ x$ M

$\dfrac{[Cr_2O_7^{2-}]}{[CrO_4^{2-}]^2} = 2.5 \times 10^3 = \dfrac{x}{(0.0269 - 2x)^2}$ $\qquad$ $x = 2.5 \times 10^3 \,(0.000724 - 0.108x + 4x^2)$

$x = 1.81 - 270\,x + 1.0 \times 10^4\,x^2$ $\qquad\qquad$ $1.0 \times 10^4\,x^2 - 27_1\,x + 1.80 = 0$

$x = \dfrac{-b \pm \sqrt{b^2 - 4ac}}{2a} = \dfrac{27_1 \pm \sqrt{7.3 \times 10^4 - 7.2 \times 10^4}}{2 \times 10^4} = 0.015\ \text{M},\ 0.012\ \text{M}$

We choose the second root because the first root gives a negative $[CrO_4^{2-}]$.

$[Cr_2O_7^{2-}] = x = 0.012\ \text{M}$ $\qquad\qquad$ $[CrO_4^{2-}] = 0.0269 - 2x = 0.0029\ \text{M}$

There is a significant rounding error in this problem.

43. First we compute the amount of Cr(s) deposited.

$\text{amount Cr} = \left(0.0010\ \text{mm} \times \dfrac{1\ \text{cm}}{10\ \text{mm}} \times 22.7\ \text{cm}^2\right) \times \dfrac{7.14\ \text{g}}{1\ \text{cm}^3} \times \dfrac{1\ \text{mol Cr}}{52.0\ \text{g Cr}} = 3.1 \times 10^{-4}\ \text{mol Cr}$

Recall that the deposition of each mole of chromium requires 6 moles of electrons. We now compute the time required to deposit the 3.1×10^{-4} mol Cr.

$\text{time} = 3.1 \times 10^{-4}\ \text{mol Cr} \times \dfrac{6\ \text{mol e}^-}{1\ \text{mol Cr}} \times \dfrac{96485\ \text{C}}{1\ \text{mol e}^-} \times \dfrac{1\ \text{s}}{4.2\ \text{C}} = 43\ \text{s}$

The Coinage Metals

44. (a) $Cu^{2+}(aq) + H_2(g) \longrightarrow Cu(s) + 2\,H^+(aq)$ $\qquad$ **(b)** $Au^+(aq) + Fe^{2+}(aq) \longrightarrow Au(s) + Fe^{3+}(aq)$

$\quad\ $ **(c)** $2\,Cu^{2+}(aq) + SO_2(g) + 2\,H_2O \longrightarrow 2\,Cu^+(aq) + SO_4^{2-}(aq) + 4\,H^+(aq)$

45. In either case, the acid acts as an oxidizing agent and dissolves silver; the gold is unaffected.

Oxidation: $\{Ag(s) \longrightarrow Ag^+(aq) + e^-\}$ $\qquad\qquad\qquad\qquad$ $\times 3$

Reduction: $NO_3^-(aq) + 4\,H^+(aq) + 3\,e^- \longrightarrow NO(g) + 2\,H_2O$

Net: $\qquad$ $3\,Ag(s) + NO_3^-(aq) + 4\,H^+(aq) \longrightarrow 3\,Ag^+(aq) + NO(g) + 2\,H_2O$

Oxidation: $\{Ag(s) \longrightarrow Ag^+(aq) + e^-\}$ $\qquad\qquad\qquad\qquad$ $\times 2$

Reduction: $SO_4^{2-}(aq) + 4\,H^+(aq) + 2\,e^- \longrightarrow SO_2(g) + 2\,H_2O$

Net: $\qquad$ $2\,Ag(s) + SO_4^{2-}(aq) + 4\,H^+(aq) \longrightarrow Ag^+(aq) + SO_2(g) + 2\,H_2O$

46. There are two reasons why Au is soluble in aqua regia while Ag is not. First, Ag^+—the oxidation product—forms a very insoluble chloride, AgCl, which probably adheres to the surface of the metal and prevents further reaction. $AuCl_3$ is not noted as being an insoluble chloride. Second, gold(III) forms a very stable complex ion with chloride ion, $[AuCl_4]^-$. This complex ion is much more stable than the corresponding dichloroargentate ion, $[AgCl_2]^-$.

47. We know that $E^\circ_{cell} = +0.37$ V for the reaction $\ 2\,Cu^+(aq) \longrightarrow Cu(s) + Cu^{2+}(aq)$. We wish to evaluate the equilibrium constant for this reaction. $\ \Delta G^\circ = -RT \ln K_{eq} = -n\mathscr{F} E^\circ_{cell}$

$\ln K_{eq} = \dfrac{n\mathscr{F} E^\circ_{cell}}{RT} = \dfrac{1\ \text{mol e}^- \times \dfrac{96485\ \text{C}}{1\ \text{mol e}^-} \times +0.37\ \text{V}}{\dfrac{8.3145\ \text{J}}{\text{mol K}} \times 298.15\ \text{K}} = 14._4$ $\qquad$ $K_{eq} = e^{14.4} = 1._8 \times 10^6$

Group 2B Metals

48. We calculate the wavelength of light absorbed in order to promote an electron across each band gap.

First a few relationships. $\quad E_{mole} = \mathscr{N}_A E_{photon}$ $\qquad E_{photon} = h\nu$ $\qquad c = \nu\lambda$ or $\nu = c/\lambda$

Then, some algebra. $\quad E_{mole} = \mathscr{N}_A E_{photon} = \mathscr{N}_A h\nu = \mathscr{N}_A hc/\lambda$ $\quad$ or $\quad \lambda = \mathscr{N}_A hc/E_{mole}$

For ZnO, $\lambda = \dfrac{6.022 \times 10^{23}\ \text{mol}^{-1} \times 6.626 \times 10^{-34}\ \text{J s} \times 2.998 \times 10^8\ \text{m s}^{-1}}{290 \times 10^3\ \text{J mol}^{-1}} \times \dfrac{10^9\ \text{nm}}{1\ \text{m}}$

$\qquad = \dfrac{1.196 \times 10^8\ \text{J mol}^{-1}\ \text{nm}}{290 \times 10^3\ \text{J mol}^{-1}} = 413\ \text{nm}$ $\qquad$ violet light

For CdS, $\lambda = \dfrac{1.196 \times 10^8 \text{ J mol}^{-1} \text{ nm}}{250 \times 10^3 \text{ J mol}^{-1}} = 479$ nm   blue light

The blue light absorbed by CdS is subtracted from the white light incident on the surface of the solid. The remaining reflected light is colored, yellow in this case. When the violet light is subtracted from the white light incident on the ZnO surface, the reflected light appears white.

25 COMPLEX IONS

AND COORDINATION COMPOUNDS

REVIEW QUESTIONS

1. **(a)** The coordination number of a complex refers to the number of points of attachment for ligands on the central atom. If all ligands are monodentate, the coordination number equals the number of ligands attached to the central atom.

(b) Δ is the crystal field splitting energy: the energy difference between the lower- and higher-energy d orbitals in a crystal field splitting diagram.

(c) An aqua complex is one with water molecule(s) as the ligand(s).

(d) An enantiomer is one of two optical isomers: molecules of identical formula and bonding which are nonsuperimposable mirror images of each other.

2. **(a)** The spectrochemical series is a listing of the common ligands, ordered from strongest to weakest Lewis base, that is, from those creating the strongest to those creating the weakest ligand field.

(b) Crystal field theory explains the magnetism and colors of complex ions in terms of the difference in energy of d electrons created by the electrostatic field imposed by point charges at the locations of the ligands.

(c) Optical isomerism refers to two compounds that differ in physical and chemical properties only in the direction in which each rotates the plane of polarized light.

(d) Structural isomerism refers to differences in the ligands that are bonded to the central atom, and in the ligand atoms that are directly attached to the central atom.

3. **(a)** The coordination number of a central atom is the number of sites where ligands are bonded to that central atom. It is thus also the number of unidentate ligands bonded to a central atom. It also is known as the secondary valence. The oxidation number, or primary valence, is equal to the charge of the isolated central metal ion.

(b) A unidentate ligand is one that has only one point of attachment to the central metal atom. A multidentate ligand bonds to the central metal atom at two or more sites.

(c) In a *cis* isomer, two ligands have a small angular distance between their bonds to the central atom. In a *trans* isomer, two ligands are on opposite sides of the central atom.

(d) dextrorotatory and levorotatory isomers differ in the direction in which they rotate the plane of polarized light. A dextrorotatory isomer rotates this plane to the right (clockwise), while an levorotatory isomer rotates it to the left (counterclockwise).

(e) In a high-spin complex, the d electrons of the central atom are unpaired as much as possible; spread out in all five of the d orbitals. In a low-spin complex, the d electrons are grouped into two or three orbitals, thus pairing their spins and making the total net spin as small as possible.

4. **(a)** $[Co(NH_3)_2Cl_4]^{2-}$ diamminetetrachlorocobaltate(II)

(b) $[Mn(CN)_6]^{3-}$ hexacyanomanganate(III)

(c) $[Cr(en)_3][Ni(CN)_5]$ tris(ethylenediamine)chromium(III) pentacyanonickelate(II)

5. **(a)** $[Ni(NH_3)_6]^{2+}$ The coordination number of Ni is 6; there are six unidentate NH_3 ligands attached to Ni. Since the NH_3 ligand does not have a charge, the oxidation state of nickel is +2, the same as that of the complex ion; hexaamminenickel(II) ion

(b) $[AlF_6]^{3-}$ The coordination number of Al is six; F^- is monodentate. Each F^- has a -1 charge; thus the oxidation state of Al is +3; hexafluoroaluminate(III) ion

    (c) $[Cu(CN)_4]^{2-}$ The coordination number of Cu is 4; CN^- is monodendate. CN^- has a -1 charge; thus the oxidations state of Cu is $+2$; tetracyanocuprate(II) ion

    (d) $[Cr(NH_3)_3Br_3]$ The coordination number of Cr is 6; both NH_3 and Br^- are monodentate. NH_3 has no charge and Br^- has a -1 charge. Thus the oxidation state of chromium is $+3$; triamminetribromochromium(III)

    (e) $[Fe(ox)_3]^{3-}$ The coordination number of Fe is 6; oxalate is bidentate. $C_2O_4{}^{2-}$ (ox) has a -2 charge; thus the oxidation state of iron is $+3$; tris(oxalato)ferrate(III)

    (f) $[Ag(S_2O_3)_2]^{3-}$ The coordination number of Ag is 2; $S_2O_3{}^{2-}$ is monodentate. $S_2O_3{}^{2-}$ has a -2 charge; thus the oxidation state of silver is $+1$; dithiosulfatoargentate(I)

6. **(a)** $[Ag(NH_3)_2]^+$ diamminesilver(I) ion
    (b) $[Fe(H_2O)_5OH]^{2+}$ pentaaquahydroxoiron(III) ion
    (c) $[ZnCl_4]^{2-}$ tetrachlorozincate(II) ion
    (d) $[Pt(en)_2]^{2+}$ bis(ethylenediamine)platinum(II) ion
    (e) $[Co(NH_3)_4(NO_2)Cl]^+$ tetraamminechloronitrocobalt(III) ion

7. **(a)** $[Co(NH_3)_5Br]SO_4$ pentaamminebromocobalt(III) sulfate
    (b) $[Co(NH_3)_5SO_4]Br$ pentaamminesulfatocobalt(III) bromide
    (c) $[Cr(NH_3)_6][Co(CN)_6]$ hexaamminechromium(III) hexacyanocobaltate(III)
    (d) $Na_3[Co(NO_2)_6]$ sodium hexanitrocobaltate(III)
    (e) $[Co(en)_3]Cl_3$ tris(ethylenediamine)cobalt(III) chloride

8. **(a)** $[Ag(CN)_2]^-$ dicyanoargentate(I) ion
    (b) $[Ni(NH_3)_2Cl_4]^{2-}$ diamminetetrachloronickelate(II) ion
    (c) $[Cu(en)_3]SO_4$ tris(ethylenediammine)copper(II) sulfate
    (d) $Na[Al(H_2O)_2(OH)_4]$ sodium diaquatetrahydroxoaluminate(III)

9. The Lewis structures are grouped together at the end.
    (a) H_2O has $2 \times 1 + 6 = 8$ valence electrons, or 8 pairs.
    (b) OH^- has $1 + 6 + 1$(for charge) $= 8$ valence electrons, or 4 pairs.
    (c) ONO^- has $2 \times 6 + 5 + 1 = 18$ valence electrons, or 9 pairs.
           The structure has a -1 formal charge on the single-bonded oxygen.
    (d) SCN^- has $6 + 4 + 5 + 1 = 16$ valence electrons, or 8 pairs.
           This structure, appropriately, gives a -1 formal charge to N.

    (a) $H\!-\!\overline{\underline{O}}\!-\!H$ **(b)** $[H\!-\!\overline{\underline{O}}\mskip1mu|]^-$ **(c)** $[\underline{\overline{O}}\!-\!\overline{N}\!=\!\underline{\overline{O}}]^-$ **(d)** $[\underline{\overline{S}}\!=\!C\!=\!\underline{\overline{N}}]^-$

10. **(a)** iron(III) chloride hexahydrate $FeCl_3 \cdot 6H_2O$ $[Fe(H_2O)_6]Cl_3$
    (b) cobalt(II) hexachloroplatinate(IV) hexahydrate $Co[PtCl_6] \cdot 6H_2O$ $[Co(H_2O)_6][PtCl_6]$

11. *trans*-$[Cr(NH_3)_4ClOH]^+$ 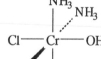   *trans*–tetraamminechlorohydroxochromium(III) ion

12. **(a)** $[PtCl_4]^{2-}$ tetrachloroplatinate(IV)
    (b) $[Fe(en)Cl_4]^-$ tetrachloro(ethylenediamine)ferrate(III)
    (c) *cis*-$[Fe(en)(ox)Cl_2]^-$ *cis*-dichloro(ethylenediamine)(oxalato)ferrate(III)

13. We assume that all of these complexes are octahedral in shape.

  (a) $[Co(NH_3)_5H_2O]^{3+}$ has only one isomer. Other supposed isomers result by simply rotating this isomer.

  (b) $[Co(NH_3)_4(H_2O)_2]^{3+}$ has two isomers, a *cis*-isomer (drawn on the left below) and a *trans*-isomer (drawn on the right). The two H_2O ligands are 90° from each other (*cis*-) or 180° from each other (*trans*-).

  (c) $[Co(NH_3)_3(H_2O)_3]^{3+}$ has two isomers, a *fac*-isomer (in which the three NH_3 ligands are 90° from each other, drawn at left below; also denoted a *cis*- isomer) and a *mer*-isomer (in which two of the NH_3 ligands are 180° from each other, on opposite sides of the central Co atom, drawn at right below; also denoted a *trans*-isomer).

  (d) $[Co(NH_3)_2(H_2O)_4]^{3+}$ has two isomers, a *cis*-isomer (drawn at left below) and a *trans*-isomer drawn on the right. In this case it is the two NH_3 groups that are 90° from each other (*cis*-) or 180° from each other (*trans*-).

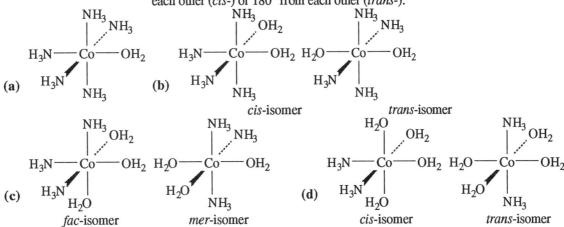

14. **(a)** $[Zn(NH_3)_4][CuCl_4]$ can display coordination isomerism. Another isomer is $[Cu(NH_3)_4][ZnCl_4]$.
  (b) $[Fe(CN)_5SCN]^{4-}$ displays linkage isomerism. The other isomer is $[Fe(CN)_5NCS]^{4-}$
  (c) $[Ni(NH_3)_5Cl]^+$ does not display isomerism.
  (d) $[Pt(py)Cl_3]^-$ does not display isomerism.
  (e) $[Cr(NH_3)_3(OH)_3]^-$ displays geometric isomerism. There is a *fac*-isomer (at left below) and a *mer*-isomer (on the right). These could also be labeled *cis*- and *trans*-, respectively.

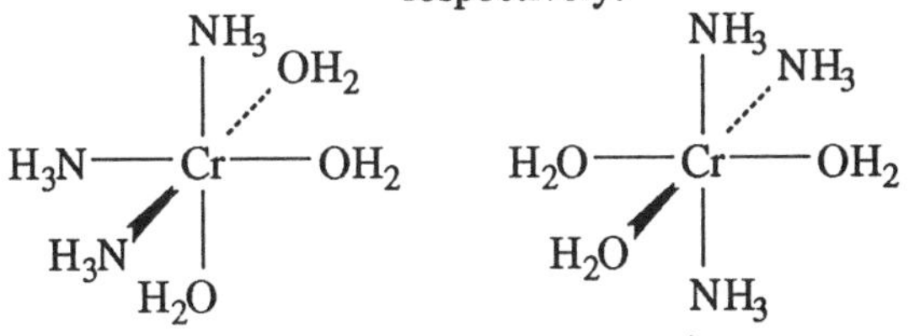

  (fac-isomer and mer-isomer structures)

  fac-isomer   *mer*-isomer

15. The yellow color indicates that the complex absorbs blue light, while the blue color indicates that the complex absorbs red and yellow light. Blue light has more energy per photon than does red and yellow light. The complex with the larger value of Δ will absorb the higher energy light. The ethylenediamine ligand produces a larger crystal field splitting energy than does the aqua ligand, according to the spectrochemical series. Thus, the yellow complex is $[Co(en)_3]^{3+}$ and the blue complex is $[Co(H_2O)_6]^{3+}$

EXERCISES

Nomenclature

16. **(a)** $[Co(NH_3)_4(H_2O)(OH)]^{2+}$ tetraammineaquahydroxocobalt(III) ion
  (b) $[Co(NH_3)_3(—NO_2)_3]$ triamminetrinitrocobalt(III)
  (c) $[Pt(NH_3)_4][PtCl_6]$ tetraammineplatinum(II) hexachloroplatinate(IV)

   (d) $[Fe(ox)_2(H_2O)_2]^-$ diaquabis(oxalato)ferrate(III) ion
   (e) $[Fe(py)(CN)_5]^{3-}$ pentacyanopyridineferrate(II) ion
   (f) $Ag_2[HgI_4]$ silver(I) tetraiodomercurate(II)

17. (a) $K_4[Fe(CN)_6]$ potassium hexacyanoferrate(II)
    (b) $[Cu(en)_2]^{2+}$ bis(ethylenediamine)copper(II) ion
    (c) $[Al(H_2O)_4(OH)_2]Cl$ tetraaquadihydroxoaluminum(III) chloride
    (d) $[Cr(en)_2NH_3Cl]SO_4$ amminechlorobis(ethylenediammine)chromium(III) sulfate
    (e) $[Fe(en)_3]_3[Fe(CN)_6]_2$ tris(ethylenediamine)iron(II) hexacyanoferrate(III)

18. (a)   cupric tetraammine ion   tetraamminecopper(II) ion $[Cu(NH_3)_4]^{2+}$
    (b)   dichlorotetraamminecobaltic chloride   tetraamminedichlorocobalt(III) chloride $[Co(NH_3)_4Cl_2]Cl$
    (c)   platinic(IV) hexachloride ion   hexachloroplatinate(IV) ion $[PtCl_6]^{2-}$
    (d)   disodium copper tetrachloride   sodium tetrachlorocuprate(II) $Na_2[CuCl_4]$

Bonding and Structure in Complex Ions

19. We assume that $[PtCl_4]^{2-}$ is square planar by analogy with $[Ni(CN)_4]^{2-}$, investigated in Example 25-5. The other two complex ions are octahedral.

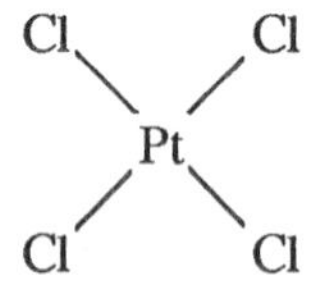

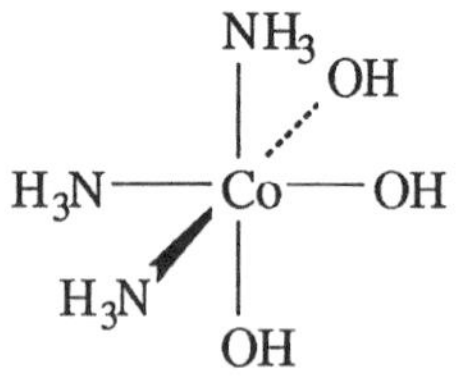

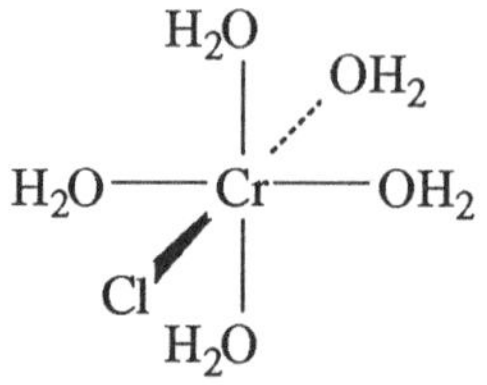

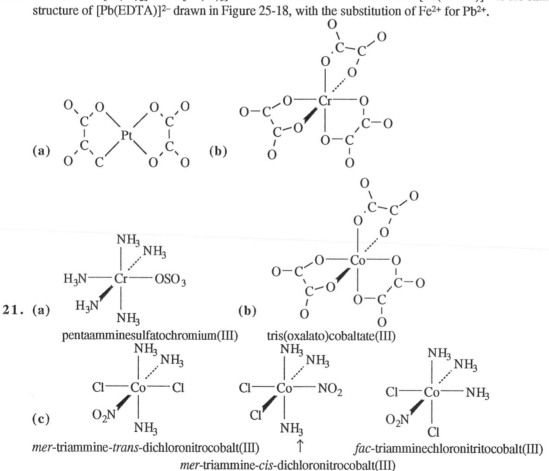

20. Structures of $[Pt(ox)_2]^{2-}$ and $[Cr(ox)_3]^{3-}$ are drawn below. The structure of $[Fe(EDTA)]^{2-}$ is the same as the structure of $[Pb(EDTA)]^{2-}$ drawn in Figure 25-18, with the substitution of Fe^{2+} for Pb^{2+}.

21. (a) pentaamminesulfatochromium(III)   **(b)** tris(oxalato)cobaltate(III)

(c) *mer*-triammine-*trans*-dichloronitrocobalt(III)   ↑   *fac*-triamminechloronitritocobalt(III)

                      mer-triammine-*cis*-dichloronitrocobalt(III)

Isomerism

22. (a) *cis-trans* isomerism cannot occur with tetrahedral structures because all of the ligands are separated by the same angular distance from each other. One ligand cannot be on the other side of the central atom from another.

(b) Square planar structures can show *cis-trans* isomerism. Examples are drawn below, with the *cis*-isomer drawn on the left, and the *trans*-isomer drawn on the right.

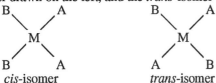

cis-isomer   trans-isomer

(c) Linear structures do not display *cis-trans* isomerism; there is only one way to bond the two ligands to the central atom.

23. All of the isomers are drawn together after the answer to this question.

(a) $[Cr(NH_3)_5OH]^{2+}$ has one isomer.

(b) $[Cr(NH_3)_3(H_2O)(OH)_2]^+$ has three isomers.

(c) $[Cr(en)_2Cl_2]^+$ has two geometric isomers, *cis-* and *trans-*.

(d) $[Cr(en)Cl_4]^-$ has only one isomer since the en ligand cannot bond *trans* to the central atom.

(e) $[Cr(en)_3]^{3+}$ has only one geometric isomer; it has two optical isomers.

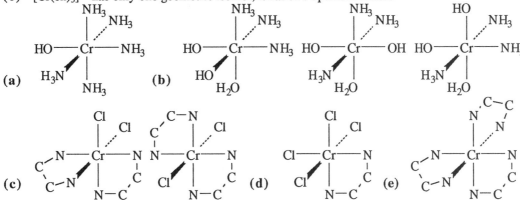

24. There are a total of four coordination isomers. They are listed below. We assume that the oxidation state of each central metal ion is +3.

$[Co(en)_3][Cr(ox)_3]$ tris(ethylenediamine)cobalt(III) tris(oxalato)chromate(III)

$[Co(en)_2(ox)][Cr(en)(ox)_2]$ bis(ethylenediamine)(oxalato)cobalt(III)
                                 (ethylenediamine)bis(oxalato)chromate(III)

$[Cr(en)_2(ox)][Co(en)(ox)_2]$ bis(ethylenediamine)(oxalato)chromium(III)
                                 (ethylenediamine)bis(oxalato)cobaltate(III)

$[Cr(en)_3][Co(ox)_3]$ tris(ethylenediamine)chromium(III) tris(oxalato)cobaltate(III)

25. (a) The three geometrical isomers of $[Co(NH_3)_4Cl_2]^+$ are sketched below. For clarity, we have shown only the chloro ligands.

(b) Attempts to produce optical isomers of $[Co(en)_3]^{3+}$ are shown below. The ethylenediamine ligand is shown as an arc in each structure. The only successful attempt occurs when the ligand connects the diagonal corners of a face, which may be too long a distance to span.

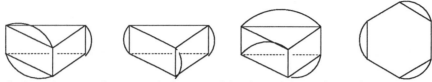

A planar hexagonal structure is drawn at right above; it has only one isomer.

26. The *cis*-dichlorobis(ethylenediamine)cobalt(III) ion is optically active. The two optical isomers are drawn below. But the *trans*-isomer is not optically active. We can tell that in advance by noting that the ion and its mirror image are superimposable.

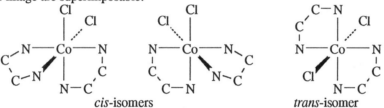

cis-isomers   *trans*-isomer

27. (a) There are three different square planar isomers, with D, C, and B, respectively, *trans* to the A ligand. They are drawn below.

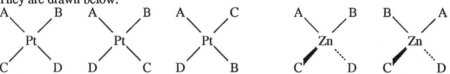

(b) Tetrahedral $[ZnABCD]^{2+}$ does display optical isomerism. The two optical isomers are drawn above.

28. (a) and (b) are mirror images of each other; but they are superimposable. They are not isomers. In fact, (a) and (b) are identical. (d) is a geometric isomer of (a). (c) is not an isomer, but a distinctly different compound.

Crystal Field Theory

29. In crystal field theory, the five *d* orbitals of a central transition metal ion are split into two (or more) groups of different energies. The energy spacing between these groups corresponds to the energy of a photon of visible light. Thus, the complex ion will absorb light with energy corresponding to this spacing. If white light is incident on the complex ion, the light remaining after absorption will be missing some of its components. That is light of certain wavelengths (corresponding to the energies absorbed) will no longer be present in the formerly white light. The resulting light is colored. For example, if blue light is absorbed from white light, the remaining light will be yellow in color.

30. Both of the central atoms have the same oxidation state, +3. We give the electron configuration of the central atom to the left, then the completed crystal field diagram in the center, and finally the number of unpaired electrons.

   (a) The chloro ligand is a weak field ligand in the spectrochemical series.

       Mo^{3+} [Kr] $4d^3$ ☐☐   weak field

               [Kr] $_{4d}$ ⊡⊡⊡☐☐   ⊡ ⊡ ⊡   3 unpaired electrons; paramagnetic

   (b) The ethylenediamine ligand is a strong field ligand in the spectrochemical series.

       Co^{3+} [Ar] $3d^6$ ☐☐   strong field

               [Ar] $_{3d}$ ⧳⊡⊡⊡☐   ⧳ ⧳ ⧳   no unpaired electrons; diamagnetic

31. We begin with the orbital diagram for Cr^{2+} [Ar] $_{3d}$ ⊡⊡⊡⊡☐

The strong field and weak field diagrams for octahedral complexes are given below, along with the number of unpaired electrons in each case.

strong field   ☐☐   weak field   ⊡☐

            ⧳ ⊡ ⊡   2 unpaired electrons   ⊡ ⊡ ⊡   4 unpaired electrons

32. Again, we begin with the orbital diagram for Cr^{3+} [Ar] $3d$ [↑|↑|↑| |]
The strong field and weak field diagrams for octahedral complexes are given below, along with the number of unpaired electrons in each case.

strong field   [|]   weak field   [|]
[↑|↑|↑]   3 unpaired electrons   [↑|↑|↑]   3 unpaired electrons

Clearly, the number of unpaired electrons is the same in each case.

33. The difference in color is due to the difference in the value of Δ, the ligand field splitting energy. When the value of Δ is high, short wavelength light, which has a blue color, is absorbed, and the substance or its solution appears yellow. On the other hand, when the value of Δ is low, light of long wavelength, which has a red or yellow color, is absorbed, and the substance or its solution appears blue. The cyano ligand is a strong field ligand, producing a large value of Δ, and thus yellow complexes. On the other hand, the aqua and chloro ligands are weak field ligands, producing a small value of Δ, and blue complexes.

34. **(a)** In $[CoCl_4]^{2-}$ the oxidation state of cobalt is +2. Chloro is a weak field ligand. The electron configuration of Co^{2+} is   [Ar] $3d^7$ or   [Ar] $3d$ [↑↓|↑↓|↑|↑|↑]   The tetrahedral ligand field diagram is given below.
weak field   [↑|↑|↑]
[↑↓|↑↓]   3 unpaired electrons

(b) In $[Cu(py)_4]^{2+}$ the oxidation state of copper is +2. Pyridine is a strong field ligand. The electron configuration of Cu^{2+} is   [Ar] $3d^9$ or   [Ar] $3d$ [↑↓|↑↓|↑↓|↑↓|↑]   There is no possible way in which an odd number of electrons can be paired up, without at least one electron being unpaired. $[Cu(py)_4]^{2+}$ is paramagnetic.

(c) In $[Mn(CN)_6]^{3-}$ the oxidation state of manganese is +3. Cyano is a strong field ligand. The electron configuration of Mn^{3+} is   [Ar] $3d^4$ or   [Ar] $3d$ [↑|↑|↑|↑|]   The ligand field diagram is given below, to the left.
In $[FeCl_4]^-$ the oxidation state of iron is +3. Chloro is a weak field ligand. The electron configuration of Fe^{3+} is   [Ar] $3d^5$ or   [Ar] $3d$ [↑|↑|↑|↑|↑]   The ligand field diagram is given below to the right.
strong field   [|]   weak field   [↑|↑|↑]
[↑↓|↑|↑]   2 unpaired electrons   [↑|↑]   5 unpaired electrons

There are more unpaired electrons in $[FeCl_4]^-$ than in $[Mn(CN)_6]^{3-}$.

35. The electron configuration of Ni^{2+} is   [Ar] $3d^8$ or   [Ar] $3d$ [↑↓|↑↓|↑↓|↑|↑]   Ammine is a strong field ligand. The ligand field diagrams are drawn below, octahedral to the left, tetrahedral in the center and square planar to the right.

octahedral   [↑|↑]   tetrahedral   [↑↓|↑|↑]   square planar   [ ]
[↑↓|↑↓|↑↓]   [↑↓|↑↓]   [↑↓]
                                                                          [↑↓]
                                                                          [↑↓|↑↓]

Since each ion has the same number of unpaired electrons (that is, 2 unpaired electrons), we cannot use magnetic properties to determine whether the ammine complex of nickel(II) is octahedral or tetrahedral. But we can determine if the complex is square planar, since the square planar complex is diamagnetic with zero unpaired electrons.

36. The difference is due to the fact that $[Fe(CN_6)]^{4-}$ is a stong field complex ion, while $[Fe(H_2O)_6]^{2+}$ is a weak field complex ion. The electron configurations for an iron atom and an iron(II) ion, and the ligand field diagrams for the two complex ions are given below.

iron atom   [Ar] $3d$ [↑↓|↑|↑|↑|↑] $4s$ [↑↓]
iron(II) ion   [Ar] $3d$ [↑↓|↑|↑|↑|↑] $4s$ [ ]

$[Fe(CN)_6]^{4-}$ [|] $[Fe(H_2O)_6]^{2+}$ [↑|↑]
                          [↑↓|↑↓|↑↓]   [↑↓|↑|↑]

Complex Ion Equilibria

37. **(a)** $Zn(OH)_2(s) + 4\,NH_3(aq) \rightleftharpoons [Zn(NH_3)_4]^{2+}(aq) + 2\,OH^-(aq)$
   (b) $Cu^{2+}(aq) + 2\,OH^-(aq) \rightleftharpoons Cu(OH)_2(s)$

The blue color is most likely due to some unreacted $[Cu(H_2O)_4]^{2+}$(aq, pale blue)

$Cu(OH)_2(s) + 4 NH_3(aq) \rightleftharpoons [Cu(NH_3)_4]^{2+}$(aq, dark blue) $+2 OH^-$(aq)

$[Cu(NH_3)_4]^{2+}$(aq) $+ 4 H_3O^+$(aq) $\rightleftharpoons [Cu(H_2O)_4]^{2+}$(aq) $+ 4 NH_4^+$(aq)

(c) $CuCl_2(s) + 2 Cl^-$(aq) $\rightleftharpoons [CuCl_4]^{2-}$(aq, yellow)

$2 [CuCl_4]^{2-}$(aq) $+ 4 H_2O \rightleftharpoons [CuCl_4]^{2-}$(aq, yellow) $+ [Cu(H_2O)_4]^{2+}$(aq, pale blue) $+ 4 Cl^-$(aq)

or $\rightleftharpoons 2 [Cu(H_2O)_2Cl_2]$(aq, green) $+ 4 Cl^-$(aq)

Either $[CuCl_4]^{2-}$(aq) $+ 4 H_2O \rightleftharpoons [Cu(H_2O)_4]^{2+}$(aq) $+ 4 Cl^-$(aq)

Or $[Cu(H_2O)_2Cl_2]$(aq) $+ 2 H_2O \rightleftharpoons [Cu(H_2O)_4]^{2+}$(aq) $+ 2 Cl^-$(aq)

38. $[Co(en)_3]^{3+}$ should have the largest overall K_f value. We expect a complex ion with multidentate ligands to have a larger value for its formation constant than complexes that contain only unidentate ligands. This is an expression of the chelate effect. Once one end of a multidentate ligand becomes attached to the central metal, the attachment of the remaining electron pairs is relatively easy because they already are close to the central metal (and do not have to migrate in from a distant point in the solution).

39. First: $[Fe(H_2O)_6]^{3+}$(aq) $+ en$(aq) $\rightleftharpoons [Fe(H_2O)_4(en)]^{3+}$(aq) $+ 2 H_2O$ $K_1 = 10^{4.34}$

Second: $[Fe(H_2O)_4(en)]^{3+}$(aq) $+ en$(aq) $\rightleftharpoons [Fe(H_2O)_2(en)_2]^{3+}$(aq) $+ 2 H_2O$ $K_2 = 10^{3.31}$

Third: $[Fe(H_2O)_2(en)_2]^{3+}$(aq) $+ en$(aq) $\rightleftharpoons [Fe(en)_3]^{3+}$(aq) $+ 2 H_2O$ $K_3 = 10^{2.05}$

Net: $[Fe(H_2O)_6]^{3+}$(aq) $+ 3 en$(aq) $\rightleftharpoons [Fe(en)_3]^{3+}$(aq) $+ 6 H_2O$ $K_f = K_1 \times K_2 \times K_3$

$\log K_f = 4.34 + 3.31 + 2.05 = 9.70$ $K_f = 10^{9.70} = 5.0 \times 10^9 = \beta_3$

40. (a) Since the overall formation constant is the product of the individual stepwise formation constants, the logarithm of the overall formation constant is the sum of the logarithms of the stepwise formation constants.

$\log K_f = \log K_1 + \log K_2 + \log K_3 + \log K_4 = 2.80 + 1.60 + 0.49 + 0.73 = 5.62$

$K_f = 10^{5.62} = 4.2 \times 10^5$

(b) If a 99% conversion to the chloro complex is achieved, $[[CuCl_4]^{2-}] = 0.99 \times 0.10$ M $= 0.099$ M, and $[[Cu(H_2O)_4]^{2+}] = 0.01 \times 0.10$ M $= 0.0010$ M. We substitute these values into the formation constant expression and solve for $[Cl^-]$

$$K_f = \frac{[[CuCl_4]^{2-}]}{[[Cu(H_2O)_4]^{2+}] [Cl^-]^4} = 4.2 \times 10^5 = \frac{0.099 \text{ M}}{0.0010 \text{ M} [Cl^-]^4}$$

$$[Cl^-] = \sqrt[4]{\frac{0.099}{0.0010 \times 4.2 \times 10^5}} = 0.12 \text{ M}$$

This is the final concentration of free chloride ion in the solution. If we wish to account for all the chloride ion that has been added, we must include the chloride ion present in the complex ion.

total $[Cl^-] = 0.12$ M free $Cl^- + (4 \times 0.099$ M) bound $Cl^- = 0.52$ M

41. The blue solution is that of $[Cr(H_2O_6)]^{2+}$. This is quickly oxidized to Cr^{3+} by $O_2(g)$ from the atmosphere. The green color is due to $[Cr(H_2O)_4Cl_2]^{3+}$.

$4 [Cr(H_2O)_6]^{2+}$(aq, blue) $+ 4 H^+$(aq) $+ 8 Cl^-$(aq) $+ O_2 \longrightarrow 4 [Cr(H_2O)_4Cl_2]^+$(aq, green) $+ 10 H_2O$

Over a period of time, we might expect volatile $HCl(g)$ to escape, leading to complex ions with more H_2O and less Cl^-. H^+(aq) $+ Cl^-$(aq) $\rightleftharpoons HCl(g)$

$[Cr(H_2O)_4Cl_2]^+$(aq, green) $+ H_2O \rightleftharpoons [Cr(H_2O)_5Cl]^{2+}$(aq, blue-green) $+ Cl^-$(aq)

$[Cr(H_2O)_5Cl]^{2+}$(aq, blue-green) $+ H_2O \rightleftharpoons [Cr(H_2O)_6]^{3+}$(aq, violet) $+ Cl^-$(aq)

Actually, to ensure the these final two reactions occur, and more rapidly, it would be helpful to dilute the solution with water after chromium metal has dissolved.

42. (a) $[Cu(H_2O)_2(NH_3)_2]^{2+}$

$\beta_2 = K_1 \times K_2 = 1.9 \times 10^4 \times 3.9 \times 10^3 = 7.4 \times 10^7$

(b) $[Zn(NH_3)_4]^{2+}$

$\beta_4 = K_1 \times K_2 \times K_3 \times K_4 = 3.9 \times 10^2 \times 2.1 \times 10^2 \times 1.0 \times 10^2 \times 50. = 4.1 \times 10^8$

(c) $[Ni(H_2O)_2(NH_3)_4]^{2+}$

$\beta_4 = K_1 \times K_2 \times K_3 \times K_4 = 6.3 \times 10^2 \times 1.7 \times 10^2 \times 54 \times 15 = 8.7 \times 10^7$

43. (a) Aluminum(III) forms a stable (and soluble) hydroxo complex but not a stable ammine complex.

$$[Al(H_2O)_3(OH)_3](s) + OH^-(aq) \rightleftharpoons [Al(H_2O)_2(OH)_4]^-(aq) + H_2O$$

(b) Although zinc(II) forms a soluble stable ammine complex ion, its formation constant is not sufficiently large to dissolve highly insoluble ZnS. However, it is sufficiently large to dissolve the moderately insoluble $ZnCO_3$. Said another way, ZnS does not produce sufficient $[Zn^{2+}]$ to permit the complex ion to form. $\qquad ZnCO_3(s) + 4\ NH_3(aq) \rightleftharpoons [Zn(NH_3)_4]^{2+}(aq) + CO_3^{2-}(aq)$

(c) Because of the large value of the formation constant for the complex ion, $[Co(NH_3)_6]^{3+}(aq)$, the concentration of free $Co^{3+}(aq)$ is too small to enable it to oxidize water to $O_2(g)$. Since there is not a complex ion present—except, of course, $[Co(H_2O)_6]^{3+}(aq)$—when $CoCl_3$ is dissolved in water, the $[Co^{3+}]$ is sufficiently high for the oxidation-reduction reaction to be spontaneous.

$$4\ Co^{3+}(aq) + 2\ H_2O \longrightarrow 4\ Co^{2+}(aq) + 4\ H^+(aq) + O_2(g)$$

Acid-Base Properties

44. $[Al(H_2O)_6]^{3+}(aq)$ is capable of ionizing: $\quad [Al(H_2O)_6]^{3+}(aq) + H_2O \rightleftharpoons [Al(H_2O)_5OH]^{2+}(aq) + H_3O^+(aq)$
The value of its ionization constant ($pK_a = 5.01$, from a handbook) approximates that of acetic acid.

45. (a) $[Cr(H_2O)_5OH]^{2+}(aq) + OH^-(aq) \longrightarrow [Cr(H_2O)_4(OH)_2]^+(aq) + H_2O$

(b) $[Cr(H_2O)_5OH]^{2+}(aq) + H_3O^+(aq) \longrightarrow [Cr(H_2O)_6]^{3+}(aq) + H_2O$

46. (a)

Reaction:	$[Fe(H_2O)_6]^{3+}(aq) + H_2O \rightleftharpoons$	$[Fe(H_2O)_5OH]^{2+}(aq)$	$+$	$H_3O^+(aq)$
Initial:	0.100 M			
Changes:	$-x$ M	$+x$ M		$+x$ M
Equil:	$(0.100 - x)$M	x M		x M

$$K_a = \frac{[[Fe(H_2O)_5OH]^{2+}][H_3O^+]}{[[Fe(H_2O)_6]^{3+}]} = 9 \times 10^{-4} = \frac{x \cdot x}{0.100 - x} \approx \frac{x^2}{0.100}$$

$$x = \sqrt{0.100 \times 9 \times 10^{-4}} = 9 \times 10^{-3}\ M = [H_3O^+] \quad pH = -\log(9 \times 10^{-3}) = 2.0$$

(b)

Reaction:	$[Fe(H_2O)_6]^{3+}(aq) + H_2O \rightleftharpoons$	$[Fe(H_2O)_5OH]^{2+}(aq)$	$+$	$H_3O^+(aq)$
Initial:	0.100 M			0.100
Changes:	$-x$ M	$+x$ M		$+x$ M
Equil:	$(0.100 - x)$M	x M		$(0.100 + x)$M

$$K_a = \frac{[[Fe(H_2O)_5OH]^{2+}][H_3O^+]}{[[Fe(H_2O)_6]^{3+}]} = 9 \times 10^{-4} = \frac{x(0.100 + x)}{0.100 - x} \approx \frac{0.100x}{0.100}$$

$$x = 9 \times 10^{-4}\ M = [[Fe(H_2O)_5OH]^{2+}]$$

(c) We simply substitute $[[Fe(H_2O)_5OH]^{2+}] = 1 \times 10^{-6}$ into the K_a expression, with $[[Fe(H_2O)_6]^{3+}] = 0.100$ M, and determine $[H_3O^+]$.

$$K_a = \frac{[[Fe(H_2O)_5OH]^{2+}][H_3O^+]}{[[Fe(H_2O)_6]^{3+}]} = 9 \times 10^{-4} = \frac{1 \times 10^{-6}\ [H_3O^+]}{0.100}$$

$$[H_3O^+] = \frac{0.100 \times 9 \times 10^{-4}}{1 \times 10^{-6}} = 90\ M$$

This is an impossibly high $[H_3O^+]$. It is impossible to maintain $[[Fe(H_2O)_5OH]^{2+}] = 1 \times 10^{-6}$ M.

47. For $[Ca(EDTA)]^{2-}$, $K_f = 4 \times 10^{10}$ and for $[Mg(EDTA)]^{2-}$, $K_f = 4 \times 10^8$. In Table 19-1, the least soluble calcium compound is $CaCO_3$, $K_{sp} = 2.8 \times 10^{-9}$, and the least soluble magnesium compound is $Mg_3(PO_4)_2$, $K_{sp} = 1 \times 10^{-25}$. Let us determine the equilibrium constants for adding carbonate ion to a solution of $[Ca(EDTA)]^{2-}(aq)$ and for adding phosphate ion to $[Mg(EDTA)]^{2-}(aq)$.

Instability:	$[Ca(EDTA)]^{2-}(aq) \rightleftharpoons Ca^{2+}(aq) + EDTA^{4-}(aq)$	$K_i = 1/4 \times 10^{10}$
Precipitation:	$Ca^{2+}(aq) + CO_3^{2-}(aq) \rightleftharpoons CaCO_3(s)$	$K_{ppt} = 1/2.8 \times 10^{-9}$
Net:	$[Ca(EDTA)]^{2-} + CO_3^{2-}(aq) \rightleftharpoons CaCO_3(s) + EDTA^{4-}(aq)$	$K = K_i \times K_{ppt}$

$$K = \frac{1}{4 \times 10^{10} \times 2.8 \times 10^{-9}} = 9 \times 10^{-3}$$

The small value of the equilibrium constant indicates that this reaction does not proceed very far toward products.

Instability:	$\{[Mg(EDTA)]^{2-}(aq) \rightleftharpoons Mg^{2+}(aq) + EDTA^{4-}(aq)\} \times 3$	$K_i^3 = 1/(4 \times 10^8)^3$
Precipitation:	$3\ Mg^{2+}(aq) + 2\ PO_4^{3-}(aq) \rightleftharpoons Mg_3(PO_4)_2(s)$	$K_{ppt} = 1/1 \times 10^{-25}$

Net: $\qquad 3\ [Mg(EDTA)]^{2-} + 2\ PO_4^{3-}(aq) \rightleftharpoons Mg_3(PO_4)_2(s) + 3\ EDTA^{4-}(aq) \qquad K = K_i^3 \times K_{ppt}$

$$K = \dfrac{1}{(4 \times 10^8)^3 \times 1 \times 10^{-25}} = 6$$

This is not such a small value, but again we do not expect the formation of much product, particularly if the $[EDTA^{4-}]$ is kept high. We can approximate what the concentration of precipitating anion must be in each case, assuming that the concentration of complex ion is 0.10 M and that of $[EDTA^{4-}]$ also is 0.10 M.

Reaction: $\quad [Ca(EDTA)]^{2-} + CO_3^{2-}(aq) \rightleftharpoons CaCO_3(s) + EDTA^{4-}(aq)$

$$K = \dfrac{[EDTA^{4-}]}{[[Ca(EDTA)]^{2-}][CO_3^{2-}]} = 9 \times 10^{-3} = \dfrac{0.10\ M}{0.10\ M\ [CO_3^{2-}]} \qquad [CO_3^{2-}] = 1 \times 10^2\ M$$

This is an impossibly high $[CO_3^{2-}]$.

Reaction: $\quad 3\ [Mg(EDTA)]^{2-} + 2\ PO_4^{3-}(aq) \rightleftharpoons Mg_3(PO_4)_2(s) + 3\ EDTA^{4-}(aq)$

$$K = \dfrac{[EDTA^{4-}]^3}{[[Mg(EDTA)]^{2-}]^3[PO_4^{3-}]^2} = 0.1_6 = \dfrac{(0.10\ M)^3}{(0.10\ M)^3\ [PO_4^{3-}]^2} \qquad [PO_4^{3-}] = 2._5\ M$$

Although this $[PO_4^{3-}]$ is not impossibly high, it is unlikely that it will occur without the deliberate addition of phosphate ion to the water.

Alternatively, we can substitute $[[M(EDTA)]^{2-}] = 0.10\ M$ and $[EDTA^{4-}] = 0.10\ M$ into the formation constant expression: $\quad K_f = \dfrac{[[M(EDTA)]^{2-}]}{[M^{2+}][EDTA^{4-}]}$ This substitution gives $[M^{2+}] = \dfrac{1}{K_f}$ and thus $[Ca^{2+}]$ $= 2.5 \times 10^{-11}\ M$ and $[Mg^{2+}] = 2.5 \times 10^{-9}\ M$, concentrations which do not lead to the formation of precipitates unless the concentrations of anions are substantial. Specifically, the required anion concentrations are $[CO_3^{2-}] = 1.1 \times 10^2\ M$ for $CaCO_3$, and $[PO_4^{3-}] = 2._5\ M$, just as computed above.

48. Let us first determine the concentration of uncomplexed $Pb^{2+}(aq)$. Because of the large value of the equilibrium constant, we assume that most of the lead(II) is present as $[Pb(EDTA)]^{2-}(aq)$. We attain equilibrium from that point. Remember that $[EDTA^{4-}]$ remains constant at 0.20 M

Reaction:	$Pb^{2+}(aq)$	$+$	$EDTA^{4-}(aq)$	$\rightleftharpoons$	$[Pb(EDTA)]^{2-}(aq)$
Initial:	0 M		0.20 M		0.010 M
Changes:	$+x$ M				$-x$ M
Equil:	x M		0.20 M		$(0.010 - x)$M

$$K_f = \dfrac{[[Pb(EDTA)]^{2-}]}{[Pb^{2+}][EDTA^{4-}]} = 2 \times 10^{18} = \dfrac{0.010 - x}{0.20\ x} \approx \dfrac{0.010}{0.20\ x} \qquad x = \dfrac{0.010}{0.20 \times 2 \times 10^{18}} = 3 \times 10^{-20}\ M$$

From Chapter 19, for PbS $K_{spa} = 3 \times 10^{-7}$. We substitute in the values that we have. Recall $[H_2S] = 0.10$ M.

$$K_{spa} = 3 \times 10^{-7} = \dfrac{[Pb^{2+}][H_2S]}{[H_3O^+]^2} = \dfrac{(3 \times 10^{-20})(0.10)}{[H_3O^+]^2} \qquad [H_3O^+] = \sqrt{\dfrac{(3 \times 10^{-20})(0.10)}{3 \times 10^{-7}}} = 1 \times 10^{-7}$$

This situation is a bit unusual. We would expect that the solution would have to be somewhat alkaline in order to maintain $EDTA^{4-}$ in its anion form. If that were not the case, however, then we would expect precipitation to occur.

Applications

49. (a) Solubility: $\quad AgBr(s) \rightleftharpoons Ag^+(aq) + Br^-(aq) \qquad\qquad K_{sp} = 5.0 \times 10^{-13}$

Formation: $\quad Ag^+(aq) + 2\ S_2O_3^{2-}(aq) \rightleftharpoons [Ag(S_2O_3)_2]^{3-}(aq) \qquad K_f = 1.7 \times 10^{13}$

Net: $\quad AgBr(s) + 2\ S_2O_3^{2-}(aq) \rightleftharpoons [Ag(S_2O_3)_2]^{3-}(aq) \qquad K = K_{sp} \times K_f$

$K = 5.0 \times 10^{-13} \times 1.7 \times 10^{13} = 8.5$

With a reasonably high $[S_2O_3^{2-}]$, this reaction will go essentially to completion.

(b) $NH_3(aq)$ cannot be used in the fixing of photographic film because of the relatively small value of K_f for $[Ag(NH_3)_2]^+(aq)$, $K_f = 1.6 \times 10^7$. This would produce a value of $K = 8.0 \times 10^{-6}$ in the expression above, too small to indicate a reaction that goes to completion.

50. Oxidation: $HO_2^-(aq) + OH^-(aq) \longrightarrow H_2O + O_2(g) + 2\ e^-$ $-E° = +0.076\ V$

Reduction: $\{[Co(NH_3)_6]^{3+}(aq) + e^- \longrightarrow [Co(NH_3)_6]^{2+}(aq)\} \times 2$ $E° = +0.10\ V$

Net: $HO_2^-(aq) + OH^-(aq) + 2\ [Co(NH_3)_6]^{3+}(aq) \longrightarrow 2\ [Co(NH_3)_6]^{2+}(aq) + H_2O + O_2(g)$

$E°_{cell} = +0.076\ V + 0.10\ V = +0.18\ V$

The positive value of the standard cell potential indicates that this is a spontaneous reaction.

51. First we need to write the reduction half-reaction in each case.

(a) $[Cu(H_2O)_4]^{2+}(aq) + 2\ e^- \longrightarrow Cu(s) + 4\ H_2O$

$$\text{no. mol } e^- = 0.347\ h \times \frac{3600\ s}{1\ h} \times \frac{2.13\ C}{1\ s} \times \frac{1\ mol\ e}{96485\ C} = 0.0276\ mol\ e^-$$

$$\text{mass Cu} = 0.0276\ mol\ e^- \times \frac{1\ mol\ Cu}{2\ mol\ e^-} \times \frac{63.55\ g\ Cu}{1\ mol\ e^-} = 0.876\ g\ Cu$$

(b) $[Cu(CN)_4]^{3-}(aq) + e^- \longrightarrow Cu(s) + 4\ CN^-(aq)$

$$\text{mass Cu} = 0.0276\ mol\ e^- \times \frac{1\ mol\ Cu}{1\ mol\ e^-} \times \frac{63.55\ g\ Cu}{1\ mol\ e^-} = 1.75\ g\ Cu$$

The reason for the difference is due to the number of moles of electrons requires to reduce a copper(II) complex (2 mol e-/mol complex ion) compared to the number required to reduce a copper(I) complex (1 mol e-/mol complex ion).

26 NUCLEAR CHEMISTRY

REVIEW QUESTIONS

1. (a) α refers to an alpha particle, a helium-4 nucleus: $^4_2\text{He}^{2+}$
    (b) β⁻ refers to a beta particle, that is, an electron.
    (c) β⁺ refers to a positron, identical in all respects to an electron but with opposite (positive) charge.
    (d) γ refers to gamma radiation, electromagnetic radiation with an energy measured in MeV per photon. In contrast, visible light has energies measured in eV per photon.
    (e) $t_{1/2}$ indicates half-life, the time needed for half of a sample to decay radioactively.

2. (a) A radioactive decay series is the series of successive products produced starting from one nuclide, which decays to a second nuclide, which in turn decays to yet a third, and so forth, until a stable nuclide is produced.
    (b) A charged-particle accelerator is a device which produces particles of very high speeds. It imparts energy and thus speed to these particles by using the influence of either an electric field or a magnetic field on the charge of the particle.
    (c) The neutron-to-proton ratio is the quotient of the number of neutrons divided by the number of protons. It is a general indication of the stability of a nuclide (when combined with atomic number). Certainly, for light elements (up to about $Z = 20$), a neutron-to-proton ration of 1-to-1 is that of a stable nuclide.
    (d) The mass-energy relationship is that first promulgated by Einstein: $E = mc^2$.
    (e) Background radiation is that radiation consistently present on earth from natural sources; cosmic rays, radioactive elements in the soil and the air, etc.

3. (a) Both electrons and positrons have the same mass, about 0.00055 u. However, the electron is negatively charged, while the positron has the same charge but a positive one.
    (b) The half-life, $t_{1/2}$, of a radioactive nuclide is the time necessary for half of one sample to decay. The decay constant is the rate constant for that first-order decay process: $\lambda = 0.693/t_{1/2}$.
    (c) The mass defect is the difference between the sum of the protons, neutrons, and electrons that constitute a nuclide and the nuclidic mass. The binding energy is that defect expressed as an energy.
    (d) Nuclear fission refers to the process in which the nucleus of an atom is split into smaller fragments, usually with the release of energy. Nuclear fusion refers to the process in which two lighter nuclei are combined into a nucleus of higher atomic number.
    (e) Primary ionization refers to those ions created by collision of the radiation with the atoms of the sample. However, when these atoms ionize they give off electrons, which often are energetic enough to produce further ionization, called secondary ionization, in the sample.

4. (a) γ rays penetrate through matter the greatest distance, largely because they are uncharged and thus do not interact with matter extensively.
    (b) α particles have the greatest ionizing power, principally because they have the largest charge and mass.
    (c) β particles are deflected the most in a magetic field, because of their small mass and relatively large charge, that is, because they have the largest charge-to-mass ratio.

5. (a) $^{32}_{16}\text{S} \longrightarrow ^{32}_{17}\text{Cl} + ^{\,0}_{-1}\beta$ (b) $^{14}_{8}\text{O} \longrightarrow ^{14}_{7}\text{N} + ^{\,0}_{+1}\beta$

(c) $^{235}_{92}U \longrightarrow {}^{231}_{90}Th + {}^4_2He$ **(d)** $^{214}_{83}Bi \longrightarrow {}^{214}_{84}Po + {}^0_{-1}\beta$

6. **(a)** $^{23}_{11}Na + {}^2_1H \longrightarrow {}^{24}_{11}Na + {}^1_1H$ **(b)** $^{59}_{27}Co + {}^1_0n \longrightarrow {}^{56}_{25}Mn + {}^4_2He$

     (c) $^{238}_{92}U + {}^2_1H \longrightarrow {}^{240}_{94}Pu + {}^0_{-1}\beta$ **(d)** $^{246}_{96}Cm + {}^{13}_6C \longrightarrow {}^{254}_{102}No + 5\,{}^1_0n$

     (e) $^{238}_{92}U + {}^{14}_7N \longrightarrow {}^{246}_{99}Es + 6\,{}^1_0n$

7. **(a)** $^{230}Th \longrightarrow {}^{226}Ra + {}^4He$ **(b)** $^{54}Co \longrightarrow {}^{54}Fe + \beta^+$

     (c) $^{232}Th + {}^4He \longrightarrow {}^{232}U + 4\,{}^1n$ **(d)** $^2H + {}^2H \longrightarrow {}^3He + {}^1n$

     (e) $^{241}_{95}Am + {}^4_2He \longrightarrow {}^{243}_{97}Bk + 2\,{}^1_0n$

8. **(a)** Since the decay constant is inversely related to the half-life, the nuclide with the smallest half-life also has the largest value of its decay constant. This is the nuclide $^{214}_{84}Po$ with a half-life of 1.64×10^{-4} s.

     (b) The nuclide that displays a 75% reduction in its radioactivity has passed through two half-lives in a period of two days. Thus, this is the nuclide with a half-life of approximately one day. This is the nuclide $^{28}_{12}Mg$, with a half-life of 21 h.

     (c) If more than 99% of the radioactivity is lost, less than 1% remains. Thus $(\frac{1}{2})^n < 0.010$. Now, when $n = 7$, $(\frac{1}{2})^n = 0.0078$. Thus, seven half-lives have elapsed in one month, and each half-life approximates 4.3 days. The longest lived nuclide that fits this description is $^{222}_{86}Rn$, which has a half-life of 3.823 days. Of course, all other nuclides with shorter half-lives also meet this criterion, specifically the following nuclides. $^{13}_8O$ (8.7×10^{-3} s), $^{28}_{12}Mg$ (21 h), $^{80}_{35}Br$ (17.6 min), and $^{214}_{84}Po$ (1.64×10^{-4} s).

9. Since $64 = 2^6$, six half-lives have elapsed in 12.0 h, and each half-life equals 2.00 h. The half-life of isotope B thus is 3.0 h. Now, since $32 = 2^5$, five half-lives must elapse before the decay rate of isotope B falls to $\frac{1}{32}$ of its original value. Thus, the time elapsed for this amount of decay is as follows.

time elapsed $= 5$ half-lives $\times \dfrac{3.0\ h}{1\ \text{half-life}} = 15\ h$

10. We use expression (26.13), substituting the disintegration rate for the number of atoms, since we recognize that in this first-order reaction the rate is directly proportional to the amount of reactant, that is, the number of atoms. (All radioactive decay processes are first-order reactions.) We also use equation (26.14), rearranged to $\lambda = \dfrac{0.693}{t_{1/2}}$. Thus $\ln \dfrac{N_t}{N_0} = -\dfrac{0.693\ t}{t_{1/2}}$

     (a) $\ln \dfrac{115\ \text{dis/min}}{1.00 \times 10^3\ \text{dis/min}} = -\dfrac{0.693\ t}{87.9\ d} = -2.163$ $t = \dfrac{2.163 \times 87.9\ d}{0.693} = 274\ d$

     (b) $\ln \dfrac{86\ \text{dis/min}}{1.00 \times 10^3\ \text{dis/min}} = -\dfrac{0.693\ t}{87.9\ d} = -2.45$ $t = \dfrac{2.45 \times 87.9\ d}{0.693} = 311\ d$

     (c) $\ln \dfrac{43\ \text{dis/min}}{1.00 \times 10^3\ \text{dis/min}} = -\dfrac{0.693\ t}{87.9\ d} = -3.15$ $t = \dfrac{3.15 \times 87.9\ d}{0.693} = 400.\ d$

11. The principal equation that we shall employ is $E = mc^2$, along with conversion factors.

     (a) $E = 1.05 \times 10^{-23}\ g \times \dfrac{1\ kg}{1000\ g} \times (3.00 \times 10^8\ m/s)^2 = 9.45 \times 10^{-10}\ kg\ m^2\ s^{-2} = 9.45 \times 10^{-10}\ J$

     (b) $E = 4.0015\ u \times \dfrac{931.5016\ MeV}{1\ u} = 3727.4\ MeV$

     (c) no. neutrons $= 1.50 \times 10^6\ MeV \times \dfrac{1\ u}{931.5016\ MeV} \times \dfrac{1\ \text{neutron}}{1.0087\ u} = 1.60 \times 10^3$ neutrons.

12. Mass of individual particles $= \left(10\ e^- \times \dfrac{0.00055\ u}{1\ e^-}\right) \times \left(10\ p \times \dfrac{1.0073\ u}{1\ p}\right) \times \left(10\ n \times \dfrac{1.0087\ u}{1\ n}\right)$

                           $= 0.0055\ u + 10.073\ u + 10.087\ u = 20.166\ u$

$\dfrac{\text{Binding energy}}{\text{nucleon}} = \dfrac{20.166\ u - 19.99244\ u}{20\ \text{nucleons}} \times \dfrac{931.5016\ MeV}{1\ u} = 8.10\ MeV$

13. **(c)** ^{80}Br does not occur naturally; it has an odd number of protons (35), an odd number of neutrons (45).

     (d) ^{132}Cs also does not occur naturally, for the same reason.

14. (a) A radioisotope with a long half-life is giving off few disintegrations per second, and thus is not exceedingly hazardous unless a large quantity of it is present. Radioisotopes with short half-lives, however, give off large quantities of radiation in a short time, but they also "burn themselves out" quickly and thus are hazardous for only a short time. One the other hand, radioisotopes with an intermediate half-life are both giving off reasonably large quantities of radiation and also are in existence for relatively long times.

(b) The radioisotopes that are hazardous from a distance are those that give off γ or high energy β radiation, radiation with long or moderate penetrating power. On the other hand, α particles are not hazardous at long distances because of their short penetrating power, but they are highly ionizing, and thus do significant damage when they are encountered at short distances.

(c) Potassium-40 is a radioisotope that decays by β emission to argon-40. This accounts for a large majority of the argon in the atmosphere. Helium-4 is also produced by radioactive decay, but it is so light that it escapes from the atmosphere at a much faster rate than does argon-40.

(d) Francium is produced by radioactive decay, and thus we should not expect it to occur where we find the other alkali metal compounds. It also is quite short-lived and thus there is not much present on the earth at any one time.

(e) Fusion involves joining positively charged nuclei together. Such a process requires quite energetic nuclei. This, in turn, means that the nuclei must be at very high temperatures.

EXERCISES

Radioactive Processes

15. (a) $^{234}_{94}\text{Pu} \longrightarrow ^{230}_{92}\text{U} + ^{4}_{2}\text{He}$ **(b)** $^{248}_{97}\text{Bk} \longrightarrow ^{248}_{98}\text{Cf} + ^{0}_{-1}\text{e}$

(c) $^{196}_{82}\text{Pb} + ^{0}_{-1}\text{e} \longrightarrow ^{196}_{81}\text{Tl}$ $^{196}_{81}\text{Tl} + ^{0}_{-1}\text{e} \longrightarrow ^{196}_{80}\text{Hg}$

(d) $^{214}_{82}\text{Pb} \longrightarrow ^{214}_{83}\text{Bi} + ^{0}_{-1}\text{e}$ $^{214}_{83}\text{Bi} \longrightarrow ^{214}_{84}\text{Po} + ^{0}_{-1}\text{e}$

(e) $^{226}_{88}\text{Ra} \longrightarrow ^{222}_{86}\text{Rn} + ^{4}_{2}\text{He}$ $^{222}_{86}\text{Rn} \longrightarrow ^{218}_{84}\text{Po} + ^{4}_{2}\text{He}$ $^{218}_{84}\text{Po} \longrightarrow ^{214}_{82}\text{Pb} + ^{4}_{2}\text{He}$

(f) $^{69}_{33}\text{As} \longrightarrow ^{69}_{32}\text{Ge} + ^{0}_{+1}\text{e}$

16. In isotopes of high atomic number, stable nuclides are characterized by a neutron-to-proton ratio that is greater than 1 and that increases with increasing atomic number. Naturally occurring isotopes of high atomic number decrease their atomic number by losing an alpha particle, which has a neutron-to-proton ratio of 1. This leaves the neutron-to-proton ratio of the daughter higher than that of the parent, when it should be slightly lower. In order to redress this, the number of neutrons needs to be decreased, the number of protons increases. Beta emission accomplishes this.

In contrast, artificially produced isotopes have no definite neutron-to-proton ratio. Thus, sometimes, the number of neutrons needs to be decreased, which is accomplished by beta emission, while at other times the number of protons needs to be decreased, which is accomplished by positron emission.

Radioactive Decay Series

17. We first write conventional nuclear reactions for each step in the decay series.

$^{232}_{90}\text{Th} \longrightarrow ^{228}_{88}\text{Ra} + ^{4}_{2}\text{He}$ $^{228}_{88}\text{Ra} \longrightarrow ^{228}_{89}\text{Ac} + ^{0}_{-1}\text{e}$ $^{228}_{89}\text{Ac} \longrightarrow ^{228}_{90}\text{Th} + ^{0}_{-1}\text{e}$

$^{228}_{90}\text{Th} \longrightarrow ^{224}_{88}\text{Ra} + ^{4}_{2}\text{He}$ $^{224}_{88}\text{Ra} \longrightarrow ^{220}_{86}\text{Rn} + ^{4}_{2}\text{He}$ $^{220}_{86}\text{Rn} \longrightarrow ^{216}_{84}\text{Po} + ^{4}_{2}\text{He}$

Now for a branch in the series

these two $^{216}_{84}\text{Po} \longrightarrow ^{212}_{82}\text{Pb} + ^{4}_{2}\text{He}$ $^{212}_{82}\text{Pb} \longrightarrow ^{212}_{83}\text{Bi} + ^{0}_{-1}\text{e}$

or these $^{216}_{84}\text{Po} \longrightarrow ^{216}_{85}\text{At} + ^{0}_{-1}\text{e}$ $^{216}_{85}\text{At} \longrightarrow ^{212}_{83}\text{Bi} + ^{4}_{2}\text{He}$

And now a second branch

these two $^{212}_{83}\text{Bi} \longrightarrow ^{208}_{81}\text{Tl} + ^{4}_{2}\text{He}$ $^{208}_{81}\text{Tl} \longrightarrow ^{208}_{82}\text{Pb} + ^{0}_{-1}\text{e}$

or these $^{212}_{83}\text{Bi} \longrightarrow ^{212}_{84}\text{Po} + ^{0}_{-1}\text{e}$ $^{216}_{84}\text{Po} \longrightarrow ^{208}_{82}\text{Pb} + ^{4}_{2}\text{He}$

Both branches end with the isotope $^{208}_{82}\text{Pb}$. The graph, similar to Figure 26-2, is drawn below.

18. In Figure 26-2, only the following mass numbers are represented: 206, 210, 214, 218, 222, 226, 230, 234, and 238. We see that these mass numbers are separated from each other by 4 units. The first of them, 206, equals $(4 \times 51) + 2$, that is $4n + 2$, where $n = 51$.

19. The series to which each nuclide belongs is determined by dividing its mass number by 4 and obtaining the remainder.
  (a) The mass number of $^{214}_{83}\text{Bi}$ is 214, and the remainder of its division by 4 is 2. This nuclide is a member of the $4n + 2$ series.
  (b) The mass number of $^{216}_{84}\text{Po}$ is 216, and the remainder of its division by 4 is 0. This nuclide is a member of the $4n$ series.
  (c) The mass number of $^{215}_{85}\text{At}$ is 215, and the remainder of its division by 4 is 3. This nuclide is a member of the $4n + 3$ series.
  (d) The mass number of $^{235}_{92}\text{U}$ is 235, and the remainder of its division by 4 is 3. This nuclide is a member of the $4n + 3$ series.

Nuclear Reactions

20. (a) $^{7}_{3}\text{Li} + ^{1}_{1}\text{H} \longrightarrow ^{8}_{4}\text{Be} + \gamma$ (b) $^{33}_{16}\text{S} + ^{1}_{0}\text{n} \longrightarrow ^{33}_{15}\text{P} + ^{1}_{1}\text{H}$
  (c) $^{239}_{94}\text{Pu} + ^{4}_{2}\text{He} \longrightarrow ^{242}_{96}\text{Cm} + ^{1}_{0}\text{n}$ (d) $^{238}_{92}\text{U} + ^{4}_{2}\text{He} \longrightarrow ^{239}_{94}\text{Pu} + 3\,^{1}_{0}\text{n}$

21. The series begins with uranium-238, and ends with lead-206.
$^{238}_{92}\text{U} \longrightarrow ^{234}_{90}\text{Th} + ^{4}_{2}\text{He}$ $^{234}_{90}\text{Th} \longrightarrow ^{234}_{91}\text{Pa} + ^{0}_{-1}\text{e}$ $^{234}_{91}\text{Pa} \longrightarrow ^{234}_{92}\text{U} + ^{0}_{-1}\text{e}$
$^{234}_{92}\text{U} \longrightarrow ^{230}_{90}\text{Th} + ^{4}_{2}\text{He}$ $^{230}_{90}\text{Th} \longrightarrow ^{226}_{88}\text{Ra} + ^{4}_{2}\text{He}$ $^{226}_{88}\text{Ra} \longrightarrow ^{222}_{86}\text{Rn} + ^{4}_{2}\text{He}$
$^{222}_{86}\text{Rn} \longrightarrow ^{218}_{84}\text{Po} + ^{4}_{2}\text{He}$
Now the series branches
these two $^{218}_{84}\text{Po} \longrightarrow ^{214}_{82}\text{Pb} + ^{4}_{2}\text{He}$ $^{214}_{82}\text{Pb} \longrightarrow ^{214}_{83}\text{Bi} + ^{0}_{-1}\text{e}$
or these $^{218}_{84}\text{Po} \longrightarrow ^{218}_{85}\text{At} + ^{0}_{-1}\text{e}$ $^{218}_{85}\text{At} \longrightarrow ^{214}_{83}\text{Bi} + ^{4}_{2}\text{He}$
And it branches yet again
these two $^{214}_{83}\text{Bi} \longrightarrow ^{210}_{81}\text{Tl} + ^{4}_{2}\text{He}$ $^{210}_{81}\text{Tl} \longrightarrow ^{210}_{82}\text{Pb} + ^{0}_{-1}\text{e}$
or these $^{214}_{83}\text{Bi} \longrightarrow ^{214}_{84}\text{Po} + ^{0}_{-1}\text{e}$ $^{214}_{84}\text{Po} \longrightarrow ^{210}_{82}\text{Pb} + ^{4}_{2}\text{He}$
Either of these branches is followed by a beta decay. $^{210}_{82}\text{Pb} \longrightarrow ^{210}_{83}\text{Bi} + ^{0}_{-1}\text{e}$
And finally by one last branch.
these two $^{210}_{83}\text{Bi} \longrightarrow ^{206}_{81}\text{Tl} + ^{4}_{2}\text{He}$ $^{206}_{81}\text{Tl} \longrightarrow ^{206}_{82}\text{Pb} + ^{0}_{-1}\text{e}$
or these $^{210}_{83}\text{Bi} \longrightarrow ^{210}_{84}\text{Po} + ^{0}_{-1}\text{e}$ $^{210}_{84}\text{Po} \longrightarrow ^{206}_{82}\text{Pb} + ^{4}_{2}\text{He}$

Rate of Radioactive Decay

22. We use expression (26.14) to determine λ and then expression (26.12) to determine the number of atoms.

$$\lambda = \frac{0.693}{5.2 \text{ y}} \times \frac{1 \text{ y}}{365.25 \text{ d}} \times \frac{1 \text{ d}}{24 \text{ h}} \times \frac{1 \text{ h}}{60 \text{ min}} = 2.5 \times 10^{-7} \text{ min}^{-1}$$

$$N = \frac{\text{rate of decay}}{\lambda} = \frac{185 \text{ atoms/min}}{2.5 \times 10^{-7} \text{ min}^{-1}} = 7.4 \times 10^{8} \ {}^{60}_{27}\text{Co atoms}$$

23. Since this is a first order reaction (as are all radioactive decay processes) the rate of decay is directly proportional to the number of atoms. We therefore use expression (26.13), with rates substituted for numbers of atoms. $\ln \dfrac{R_t}{R_o} = -\lambda t = \ln \dfrac{101 \text{ dis/min}}{185 \text{ dis/min}} = -2.5 \times 10^{-7} \text{ min}^{-1} \ t = -0.605$

$$t = \frac{0.605}{2.5 \times 10^{-7} \text{ min}^{-1}} = 2.4 \times 10^{6} \text{ min} \times \frac{1 \text{ h}}{60 \text{ min}} \times \frac{1 \text{ d}}{24 \text{ h}} \times \frac{1 \text{ y}}{365.25 \text{ d}} = 4.6 \text{ y}$$

24. Begin by determining the decay constant from the half-life. $\lambda = \dfrac{0.693}{t_{1/2}} = \dfrac{0.693}{14.2 \text{ d}} = 0.0488 \text{ d}^{-1}$

$$\ln \frac{N_t}{N_o} = -\lambda t = \ln \frac{1.00 \text{ atoms}}{1.00 \times 10^{3} \text{ atoms}} = -0.0488 \text{ d}^{-1} \ t = -6.908 \qquad t = \frac{6.908}{0.0488 \text{ d}^{-1}} = 142 \text{ d}$$

25. Let us use the first and the last values to determine the decay constant.

$$\ln \frac{R_t}{R_o} = -\lambda t = \ln \frac{138 \text{ cpm}}{1000 \text{ cpm}} = -\lambda \ 250 \text{ h} = -1.981 \qquad \lambda = \frac{1.981}{250 \text{ h}} = 0.00792 \text{ h}^{-1}$$

$$t_{1/2} = \frac{0.693}{\lambda} = \frac{0.693}{0.00792 \text{ h}^{-1}} = 87.5 \text{ h}$$

A slightly different value of $t_{1/2}$ may result from other combinations of R_o and R_t.

26. First calculate the decay constant. $\lambda = \dfrac{0.693}{1.7 \times 10^{7} \text{ y}} \times \dfrac{1 \text{ y}}{365.25 \text{ d}} \times \dfrac{1 \text{ d}}{24 \text{ h}} \times \dfrac{1 \text{ h}}{3600 \text{ s}} = 1.3 \times 10^{-15} \text{ s}^{-1}$

$$N = 1.00 \text{ mg } {}^{129}\text{I} \times \frac{1 \text{ g}}{1000 \text{ mg}} \times \frac{1 \text{ mol } {}^{129}\text{I}}{129 \text{ g}} \times \frac{6.022 \times 10^{23} \text{ atoms}}{1 \text{ mol } {}^{129}\text{I}} = 4.67 \times 10^{18} \ {}^{129}\text{I atoms}$$

decay rate $= \lambda N = 1.3 \times 10^{-15} \text{ s}^{-1} \times 4.67 \times 10^{18} \text{ atoms} = 6.1 \times 10^{3} \text{ dis/s}$

27. First we determine the decay constant. $\lambda = \dfrac{0.693}{1.39 \times 10^{10} \text{ y}} = 4.99 \times 10^{-11} \text{ y}^{-1}$

Then we can determine the ratio of the number of thorium atoms after (N_t) 4.5×10^{9} y to the number (N_0) initially. $\ln \dfrac{N_t}{N_0} = -kt = -(4.99 \times 10^{-11} \text{ y}^{-1})(4.5 \times 10^{9} \text{ y}) = -0.22 \qquad \dfrac{N_t}{N_0} = 0.80$

Thus, for every mole of ^{232}Th present initially, there is present after 4.5×10^{9} y, 0.80 mol ^{232}Th amd 0.20 mol ^{208}Pb. From this information, we can compute the mass ratio.

$$\frac{0.20 \text{ mol } {}^{208}\text{Pb}}{0.80 \text{ mol } {}^{232}\text{Th}} \times \frac{1 \text{ mol } {}^{232}\text{Th}}{232 \text{ g } {}^{232}\text{Th}} \times \frac{208 \text{ g } {}^{208}\text{Pb}}{1 \text{ mol } {}^{208}\text{Pb}} = \frac{0.22 \text{ g } {}^{208}\text{Pb}}{1 \text{ g } {}^{232}\text{Th}}$$

28. First we determine the decay constant. $\lambda = \dfrac{0.693}{1.4 \times 10^{10} \text{ y}} = 5.0 \times 10^{-11} \text{ y}^{-1}$

The rock currently contains 1.00 g ^{232}Th and 0.14 g ^{208}Pb. We can calculate the mass of ^{232}Th that must have been present to produce this 0.14 g ^{208}Pb, and from that find the original mass of ^{232}Th.

original mass ^{232}Th $= 1.00 \text{ g } {}^{232}\text{Th now} + \left(0.14 \text{ g } {}^{208}\text{Pb} \times \dfrac{232 \text{ g } {}^{232}\text{Th}}{208 \text{ g } {}^{208}\text{Pb}}\right) = (1.00 + 0.16) \text{ g} = 1.16 \text{ g}$

$$\ln \frac{N_t}{N_0} = -\lambda t = \ln \frac{1.00 \text{ g } {}^{232}\text{Th now}}{1.16 \text{ g originally}} = -0.148 = -5.0 \times 10^{-11} \text{ y}^{-1} \ t; \quad t = \frac{0.148}{5.0 \times 10^{-11} \text{ y}^{-1}} = 3.0 \times 10^{9} \text{ y}$$

Radiocarbon Dating

29. Again we use expression (26.13) and (26.14) to determine the time elapsed. The initial rate of decay is about 15 dis/min. First we compute the decay constant. $\lambda = \dfrac{0.693}{5730 \text{ y}} = 1.21 \times 10^{-4} \text{ y}^{-1}$

$$\ln \frac{12 \text{ dis/min}}{15 \text{ dis/min}} = -0.223 = -\lambda t \qquad t = \frac{0.223}{1.21 \times 10^{-4} \text{ y}^{-1}} = 1.84 \times 10^{3} \text{ y}$$

The object is less than 2000 years old, obviously not dating from the pyramid era, about 3000 B.C.

30. We use the value of λ from the previous exercise.

$$\ln \frac{R_t}{R_o} = -\lambda t = -(1.21 \times 10^{-4} \text{ y}^{-1})t = \ln \frac{0.03 \text{ dis min}^{-1} \text{ g}^{-1}}{15 \text{ dis min}^{-1} \text{ g}^{-1}} = -6.2$$

$$t = \frac{6.2}{1.21 \times 10^{-4} \text{ y}^{-1}} = 5.2 \times 10^4 \text{ y}$$

Energetics of Nuclear Reactions

31. The mass defect is the difference between the mass of the nuclide and the sum of the masses of its constituent particles. The binding energy is this mass defect expressed as an energy.

particle mass = $8 p + 8 n + 8 e = 8 (p + n + e) = 8 (1.0073 + 1.0087 + 0.0005486) \text{ u} = 16.1324 \text{ u}$

mass defect = $16.1324 \text{ u} - 15.99491 \text{ u} = 0.1375 \text{ u}$

$$\text{binding energy per nucleon} = \frac{0.1375 \text{ u} \times \dfrac{931.5 \text{ MeV}}{1 \text{ u}}}{16 \text{ nucleons}} = 8.005 \text{ MeV/nucleon}$$

32. mass defect = $(10.01294 \text{ u} + 4.00260 \text{ u}) - (13.00335 \text{ u} + 1.00783 \text{ u}) = 0.00436 \text{ u}$

$$\text{energy} = 0.00436 \text{ u} \times \frac{931.5016 \text{ MeV}}{1 \text{ u}} = 4.06 \text{ MeV}$$

33. mass defect = $(6.01513 \text{ u} + 1.008665 \text{ u}) - (4.00260 \text{ u} + 3.01604 \text{ u}) = 0.00516 \text{ u}$

$$\text{energy} = 0.00516 \text{ u} \times \frac{931.5016 \text{ MeV}}{1 \text{ u}} = 4.81 \text{ MeV}$$

34. $\text{energy} = 0.05 \text{ MeV} \times \dfrac{1 \times 10^6 \text{ eV}}{1 \text{ MeV}} \times \dfrac{1.602 \times 10^{-19} \text{ J}}{1 \text{ eV}} = 8 \times 10^{-15} \text{ J} = \dfrac{hc}{\lambda}$

$$\lambda = \frac{(6.63 \times 10^{-34} \text{ J·s}) (3.00 \times 10^8 \text{ m/s})}{8 \times 10^{-15} \text{ J}} = 2 \times 10^{-11} \text{ M} \times \frac{1 \text{ nm}}{1 \times 10^{-9} \text{ m}} = 0.02 \text{ nm}$$

Nuclear Stability

35. **(a)** We expect ^{20}Ne to be more stable than ^{22}Ne. A neutron-to-proton ratio of 1-to-1 is associated with stability for elements of low atomic number (with $Z \le 20$).

(b) We expect ^{18}O to be more stable than ^{17}O. An even number of protons and an even number of neutrons is associated with a stable istope.

(c) We expect ^{7}Li to be more stable than ^{6}Li. Both isotopes have an odd number of protons, but only ^{7}Li has an even number of neutrons.

36. β^- emission has the effect of "converting" a neutron to a proton. β^+ emission, on the other hand, has the effect of "converting" a proton to a neutron.

(a) The most stable isotope of phosphorus would be ^{31}P, with a neutron-to-proton ratio of close to 1-to-1 and an even number of neutrons. Thus, ^{29}P has "too few" neutrons, or too many protons. It should decay by β^+ emission. In contrast, ^{33}P has "too many" neutrons, or "too few" protons. Therefore, ^{33}P should decay by β^- emission.

(b) Based on the atomic weight of I (126.90447), we expect the isotopes of iodine to have mass numbers close to 127. This means that ^{120}I has "too few" neutrons and therefore should decay by β^+ emission, whereas ^{134}I has "too many" neutrons (or "too few" protons) and therefore should decay by β^- emission.

37. β^- emission has the effect of converting a neutron to a proton, while β^+ emission has the effect of converting a proton to a neutron.

(a) Based on the fact that elements of low atomic number have about the same number of protons as neutrons, $^{28}_{15}$P—with 15 protons and 13 neutrons—has too few neutrons. Therefore, it should decay by β^+ emission.

(b) Again based on the fact that elements of low atomic number have about the same number of protons as neutrons, $^{45}_{19}K$—with 19 protons and 26 neutrons—has too many neutrons. Therefore, it should decay by β^- emission.

(c) Based on the atomic weight of zinc (65.39) we expect most of its isotopes to have about 36 neutrons. There are 42 neutrons in $^{72}_{30}Zn$, more than we expect. Thus we expect this nuclide to decay by β^- emission.

38. In the cases where rounding off the atomic weight produces that mass number of the most stable isotope, there often is but one stable isotope. This often is the case when the atomic number of the element is an odd number. Such is the case with ^{39}K (Z = 19) and ^{87}Rb (Z = 37), but not ^{88}Sr (Z = 38). In the cases where this technique of rounding does not work, there are two or more stable isotopes of significant abundance. Note that the rounding off does not work in those cases where it predicts a nuclide with an odd number of neutraons and an odd number of protons (such as ^{64}Cu with 29 protons and 35 neutrons), whereas the rounding off technique works when the predicted nuclide has either an even number of protons, an even number of neutrons, or both.

39. A "doubly magic" nuclide is one in which the atomic number is a magic number (2, 8, 20, 28, 50, 82, 114) and the number of neutrons also is a magic number (2, 8, 20, 28, 50, 82, 126, 184, 196). The nuclides that fit this description are the following.

nuclide	^{4}He	^{16}O	^{40}Ca	^{56}Ni	^{208}Pb
no. of protons	2	8	20	28	82
no. of neutrons	2	8	20	28	126

40. Each α particle contains two protons and has a mass number of 4. Thus each α particle emission reduces the mass number by 4 and the atomic number by 2. The emission of 8 α particles would reduce the mass number by 32 and the atomic number by 16. Thus the overall reaction would be as follows.

$$^{238}_{92}U \longrightarrow 8\,^4_2He + ^{206}_{76}Os \quad \text{(76 protons and 130 neutrons)}$$

In Figure 26-7, a nuclide with 76 protons and 130 neutrons lies above, to the left of the belt of stability; it is radioactive.

Fission and Fusion

41. We notice in Figure 26-6 that the curve of binding energy against mass number increases more rapidly when small nuclei (with $A <\approx 60$) are fused into large ones than it does in the opposite direction, when large nuclei are fragmented into smaller ones.

42. We use the conversion factor between number of curies and mass of ^{131}I which was developed in the Summarizing Example.

$$\text{no. g I-131} = 170 \text{ curies} \times \frac{8.07 \times 10^{-6} \text{ g I-131}}{1 \text{ curie}} = 1.37 \times 10^{-3} \text{ g} = 1.37 \text{ mg}$$

43. We use $\Delta H^\circ_f[CO_2(g)] = -393.51$ kJ/mol as the heat of combustion of 1 mole of carbon. In the text, the energy produced by the fusion of 1.00 g ^{235}U is determined as 8.20×10^7 kJ.

$$\text{no. metric tons coal} = 1.00 \text{ kg } ^{235}U \times \frac{1000 \text{ g}}{1 \text{ kg}} \times \frac{8.20 \times 10^7 \text{ kJ}}{1.00 \text{ g } ^{235}U} \times \frac{1 \text{ mol C}}{393.5 \text{ kJ}} \times \frac{12.01 \text{ g C}}{1 \text{ mol C}}$$
$$\times \frac{1.00 \text{ g coal}}{0.85 \text{ g C}} \times \frac{1 \text{ kg}}{1000 \text{ g}} \times \frac{1 \text{ metric ton}}{1000 \text{ kg}} = 2.9 \times 10^3 \text{ metric tons (Mg)}$$

Effect of Radiation on Matter

44. The term "rem" is an acronym for "radiation equivalent–man," and takes into account the quantity of biological damage done by a given dosage of radiation. On the other hand, the rad is the dosage of radiation that will place 0.010 J of energy into each kg of irradiated matter. Thus, for living tissue, the rem give us a good idea of how much tissue damage a certain kind and quantity of radiation damage will do. But, for nonliving materials, the rad is often preferred, and indeed is often the only unit of utility.

45. Low-level radiation is very close in its dosage to background radiation and one problem is to separate out the effects of the two sources (low-level and background). The other problem is that low-level radiation does not produce severe damage in a short period of time. Thus the effects of low-level radiation will only accumulate over a long time period. Of course other effects, such as chemical and biological toxins, will also be effective over these time periods, and we have to try to separate these two types of effects. (There also is the genetic heritage of the organism to consider, of course.)

46. One reason why ^{90}Sr is hazardous is because strontium is in the same family of the periodic table as calcium, and hence often reacts in a similar fashion to calcium. One site where calcium is incorporated into the body is in bones, where it resides for a long time. Strontium is expected to behave in a similar fashion. Thus, it will be retained in the body for a long time. Bone is an especially dangerous place for a radioisotope to be present—even if it has low penetrating power, as do β rays—because blood cells are produced in bone marrow.

Application of Radioisotopes

47. If a small amount of tritium, as $T_2(g)$ or $^3H_2(g)$, were mixed with the $H_2(g)$ supply, it would behave chemically in much the same way as $H_2(g)$. But the small leak would be readily detected with a radiation detector.

48. In neutron activation analysis, the sample is bombarded with neutrons. Radioisotopes are produced by this process. These radioisotopes can be easily detected even in very small quantities, much smaller than the quantities that can be detected by conventional means of quantitative analysis. These radioisotopes are produced in quantities that are proportional to the quantity of each element originally present in the sample. And each radioisotope is characteristic of the element from which it was produced by neutron bombardment. Even microscopic samples can be analyzed by this technique. Finally, neutron activation analysis is a nondestructive technique, while the conventional techniques of precipitation or titration require that the sample, or at least part of it, be destroyed.

49. The recovered sample will be radioactive. When NaCl(s) and $NaNO_3$(s) are dissolved in solution, the ions (Na^+, Cl^-, and NO_3^-) are free to move throughout the solution. A given anion does not remain associated with a particular cation. Thus, all the anions and cations are shuffled and some of the radioactive ^{24}Na will end up in the crystallized $NaNO_3$.

50. We would expect the tritium label to appear in both the $NH_3(g)$ and the H_2O. When NH_4^+(aq) is formed, one of the four chemically and spatially equivalent H atoms is a tritium atom, In the subsequent reaction with NaOH to form $NH_3(g)$ and H_2O there are three chances in four that a tritium atom will remain attached to N in NH_3 and one chance in four that a tritium ion will unite with a hydroxide ion to form H_2O.

27 ORGANIC CHEMISTRY

1. **(a)** *t* is the symbol for tertiary; it refers to a carbon atom to which three other carbon atoms are singly bonded.

    (b) R— is the symbol for an alkyl group, a carbon-hydrogen chain with no multiple bonds.

    (c) A hexagon with an inscribed circle is the symbol for a benzene ring, the molecule with formula C_6H_6.

    (d) A carbonyl group is a carbon atom with an oxygen double bonded to it, $\diagdown C=O$.

    (e) A primary amine is an —NH_2 group with an attached hydrocarbon chain.

2. **(a)** In a substitution reaction, one atom or group of atoms replaces another atom or group of atoms.

    (b) The octane rating of gasoline is a measure of its quality. Gasoline's octane number is equal to the percent of isooctane in an isooctane-*n*-heptane mixture that has the same combustion characteristics as the gasoline.

    (c) Stereoisomerism refers to isomers that have similar bonding, but for which the substituent groups are oriented differently. It includes geometric isomerism and optical isomerism.

    (d) An ortho, para director is a group attached to an aromatic ring that causes substitution reactions to occur preferentially at positions ortho- and para- to it.

    (e) Condensation polymerization occurs when the monomers join to the polymer chain by the reaction of two functional groups and the elimination of a small molecule such as water.

3. **(a)** An alkane is a hydrocarbon in which there are no multiple bonds; in an alkene there is at least one double bond.

    (b) An aliphatic compound is an alkane, a compound of carbon and hydrogen in which all bonds are single bonds and in which the carbon skeleton consists of straight or branched chains. In an aromatic compound there is at least one benzene ring.

    (c) An alcohol is an aliphatic compound with an attached —OH group, a phenol has the —OH group attached to a benzene ring.

    (d) An ether has a C—O—C linkage. An ester has a $C—O—\overset{\overset{\displaystyle O}{\|}}{C}—$ linkage.

    (e) An amine contains N attached to a C atom; ammonia is NH_3.

4. **(a)** Skeletal isomers differ in how the atoms are joined together, the carbon chains that are present.

    (b) Positional isomers differ in the location of a functional group along the carbon chain of the molecule.

    (c) *cis-*, *trans-* isomers differ in the arrangement of atoms around a double bond.

    (d) ortho-, meta-, and para- isomers differ in the location of substituent groups on an aromatic ring.

5. Just in terms of straight and branched chains, C_4H_8 has more isomers than C_4H_{10}, as shown by the structures below, in which the C_4H_8 structures are at right. In addition, only C_4H_8 can form rings: one four-membered ring isomer and one three-membered ring isomer.

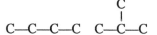

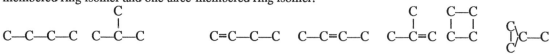

6. Cyclobutane has the formula C_4H_8; there is one carbon for every two hydrogens. $CH_3CH{=}CHCH_3$ has the formula C_4H_8, and $CH_3C{\equiv}CCH_3$ has the formula C_4H_6. Only in $CH_3CH{=}CHCH_3$ is there the same carbon-to-hydrogen ratio as in cyclobutane.

7. (a) CH₃ĊHCH₃ (with Br above)   alkyl halide (bromide)

(b) CH₃CH₂COH (with O above)   carboxylic acid

(c) C₆H₅CH₂CH (with O above)   aldehyde

(d) (CH₃)₂CHCH₂—O—CH₃   ether

(e) CH₃CCH₂CH₃ (with O above)   ketone

(f) CH₃CH(NH₂)CH₂CH₃   amine

(g) HO—C₆H₅—OH   alcohol phenyl group

(h) CH₃COCH₃ (with O above)   ester

8. (a) 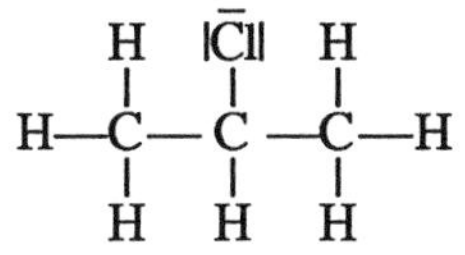 H—C—C—C—H (with H, Cl, H above and H, H, H below)

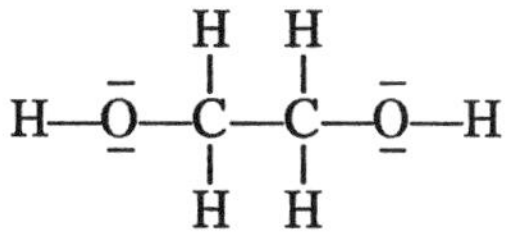

(b) H—O—C—C—O—H (with H, H above and H, H below)

(c) H—C—C=O (with H above and H, H below)

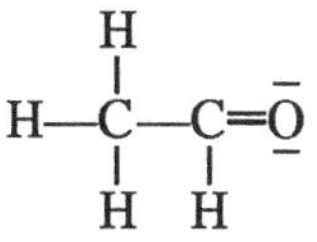

9. We draw only the carbon and chlorine atoms in each structure. Remember that there are four bonds to each carbon atom; the bonds not shown in these structures are C—H bonds.

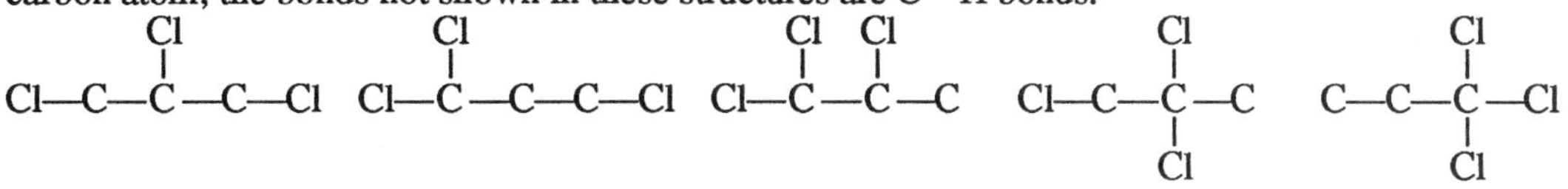

Cl—C—C—C—Cl (Cl above 2nd C)   Cl—C—C—C—Cl (Cl above 2nd C)   Cl—C—C—C (Cl, Cl above 1st, 2nd C)   Cl—C—C—C (Cl above 2nd C, Cl below 2nd C)   C—C—C—Cl (Cl above 2nd C, Cl below 2nd C)

10. (a) These are not isomers, since they have different formulas: C_4H_{10} and C_4H_8.

(b) These two compounds are skeletal isomers. (Only the carbon skeleton is shown below).

C—C—C—C—C—C—C (with C above 6th C)   C—C—C—C—C—C—C (with C above 6th C)

(c) These two compounds are identical.

(d) These two compounds are identical; simply rotated.

(e) These two compounds are identical; simply rotated.

(f) These are two ortho-para isomers, ortho-nitrophenol on the left, and para-nitrophenol on the right.

11. (a) The longest chain is eight carbons long, the two substituent groups are methyl groups, and they are both attached to the number 3 carbon atom. This is 3,3-dimethyloctane.

(b) The longest carbon chain is three carbons long, the two substituent groups are methyl groups, and they are both attached to the number 2 carbon atom. This is 2,2-dimethylpropane.

(c) The longest carbon chain is 7 carbon atoms long, there are two chloro groups attached to carbon atoms 2 and 3, and an ethyl group attached to carbon 5. This is 2,3-dichloro-5-ethylheptane.

12. (a) There are 2 bromo groups at the 1 and 4 positions on a benzene ring. This is 1,4-dibromobenzene or, more appropriately, *para*-dibromobenzene.

(b) There is a methyl group at position 1 on a benzene ring, and an amine group at position 2. This is *ortho*-aminotoluene or *ortho*-methylaniline.

(c) There is a —COOH group at position 1 on the benzene ring, and a —NO₂ group at position 3, or meta to the —COOH group. This is 3-nitrobenzoic acid or *m*-nitrobenzoic acid.

13. In the structural formulas drawn below, we omit the hydrogen atoms. Remember that there are four bonds to each C atom, The bonds that are not shown are C—H bonds.

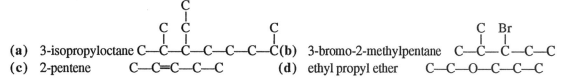

(a) 3-isopropyloctane C—C—C—C—C—C—C (with C, C above 2nd, 3rd C and C above that)   (b) 3-bromo-2-methylpentane C—C—C—C—C (with C, Br above)

(c) 2-pentene   C—C=C—C—C   (d) ethyl propyl ether   C—C—O—C—C—C

(e) *p*-bromophenol   Br—⟨O⟩—OH

14. (a) isopropyl alcohol
        $CH_3CH(OH)CH_3$

   (b) 1,1,1-trichlorodifluoroethane
        $CClF_2CH_3$

   (c) 2-methyl-1,3-butadiene
        $CH_2=C(CH_3)CH=CH_2$

15. (a)

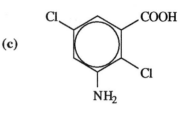

   (b) NO_2—⟨O⟩—OH

   (c)

16. Methyl ethyl ketone has the abbreviated structure C—C—C—C (with O double-bonded to the second carbon)   Thus, the alcohol from which it is produced must have an —OH group on carbon 2 and it must be a four-carbon-atom straight-chain alcohol. Of the alcohols listed, 2-butanol satisfies these criteria.

17. (a) $CH_3CH_2CH=CH_2$ $\xrightarrow{H_2O, H_2SO_4}$ $CH_3CH_2CH(OH)CH_3$

   (b) $CH_3CH_2CH_3 + Cl_2$ $\xrightarrow{h\nu}$ $CH_3CH_2CH_2Cl + CH_3CHClCH_3$ + other chlorinated propanes

   (c) ⟨O⟩—COOH $+ (CH_3)_2OH$ $\xrightarrow{\Delta}$ ⟨O⟩—$COOCH(CH_3)_2$

   (d) $CH_3CH(OH)CH_2CH_3$ $\xrightarrow{Cr_2O_7^{2-}, H^+}$ CH_3—C—CH_2CH_3 (with O double-bonded to the central C)

18. (a) C_8H_{18} would have a higher boiling point that C_6H_{12} because its London forces are stronger since C_8H_{18} is a larger molecule. There are no dipole-dipole forces or hydrogen bonds between these molecules.

   (b) Both C_3H_7OH and $C_7H_{15}OH$ are alcohols with the ability to hydrogen bond, but $C_7H_{15}OH$ has a much longer nonpolar hydrocarbon chain, interfering with its aqueous solubility. C_3H_7OH is more water soluble.

   (c) The functional groups determine the acidity of these two otherwise alike molecules. The aldehyde hydrogen is not notably acidic while that carboxylic hydrogen is. C_6H_5COOH is more acidic in aqueous solution.

EXERCISES

Organic Structures

19. In the following structural formulas, the hydrogen atoms are omitted for simplicity. Remember that there are four bonds to each carbon atom. The missing bonds are C—H bonds.

   (a) $CH_3CH_2CH_2CHBrCH_3$

        C—C—C—C—C (with Br on the fourth C)

   (b) $(CH_3)_2CHCH_2CH_2CH(CH_3)CH_2CH$

        C—C—C—C—C—C—C (with C branches on second and fifth C)

   (c) $(CH_3)_3CCH_2CH(CH_3)CH_2CH_2CH_3$

        C—C—C—C—C—C—C (with C, C branches on second C and C on fourth C)

   (d) $CH_3CH_2CH(CH_3)C(C_2H_5)=CH_2$

        C—C—C—C=C (with C branch and C—C branches)

20. (a) Each carbon atom is sp^3 hybridized. All of the C—H bonds in the structure (drawn below) are sigma bonds, between the $1s$ orbital of H and the sp^3 orbital of C. The C—C bond is between sp^3 orbitals on each C atom.

   (b) Each carbon atom is sp^2 hybridized. All of the C—H bonds in the structure (drawn below) are sigma bonds between the $1s$ orbital of H and the sp^2 orbital of C. The C—Cl bond is between the sp^2 orbital on C and the $3p$ orbital on Cl. The C=C double bond is composed of a sigma bond between the sp^2 orbitals on each C atom and a pi bond between the $2p_z$ orbitals on those C atoms.

(c) The left-most C atom (in the structure drawn below) is sp^3 hybridized, and the C—H bonds to that C atom are between the sp^3 orbitals on C and the $1s$ orbital on H. The other two C atoms are sp hybridized. The right-hand C—H bond is between the sp orbital on C and the $1s$ orbital on H. The C≡C triple bond is composed of one sigma bond formed by overlap of sp orbitals, one from each C atom, and two pi bonds, each formed by the overlap of two $2p$ orbitals, one from each C atom (that is a $2p_y$—$2p_y$ overlap and a $2p_z$—$2p_z$ overlap).

(d) The left- and right-most C atoms in the structure (drawn below) are sp^3 hybridized. All C—H bonds are sigma bonds formed by the overlap of an sp^3 orbital on C with a $1s$ orbital on H. The central C atom is sp^2 hybridized; both C—C bonds are sigma bonds, formed by the overplap between the sp^3 orbital on the terminal C atom and an sp^2 orbital on the central C atom. The C=O double bond is composed of a sigma bond between the sp^2 orbital on the central C atom and a $2p_y$ orbital on the O atom and a pi bond between the $2p_z$ orbital on the central C atom and the $2p_z$ orbital on the O atom.

(e) Both C atoms in the structure (drawn below) are sp^3 hybridized. All C—H bonds are sigma bonds, formed by the overlap between an sp^3 orbital on C with a $1s$ orbital on H. N also is sp^3 hybridized and the two N—H bonds are sigma bonds, formed by the overlap between an sp^3 orbital on N with a $1s$ orbital on H. The C—C bond is a sigma bond, formed by the overlap of an sp^3 orbital on each C. The C—N bond is formed by the overlap between and sp^3 orbital on C with an sp^3 orbital on N.

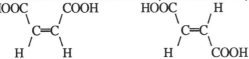

Isomers

21. Skeletal isomers differ from each other in the length of their carbon atom chains and in the length of the side chains. The carbon skeleton differs between these isomers. Positional isomers differ in the location or position where functional groups are attached to the carbon skeleton. Geometric isomers differ in whether two substituents are on the same side of the molecule or on opposite sides of the molecule from each other; usually they are on opposite sides or the same side of a double bond.

(a) The chloro group is attached to the terminal carbon in $CH_3CH_2CH_2Cl$ to the central carbon atom in $CH_3CHClCH_3$. These are positional isomers.

(b) $CH_3CH(CH_3)CH_2CH_3$, methylbutane, and $CH_3(CH_2)_3CH_3$, pentane, are skeletal isomers.

(c) The two chloro groups in CHCl=CHCl are on different carbon atoms. In $CH_2=CCl_2$ they are on the same atoms. These are positional isomers.

(d) Ortho-aminotoluene and meta-aminotoluene differ in the position of the —NH_2 group on the benzene ring. They are positional isomers.

(e) These two compounds differ in the location of the two —COOH groups; the left-hand one has the two groups on the same side of the double bond (the *cis* isomer) and the right-hand one has the two groups on opposite sides of the double bond (the *trans* isomer). These are geometric isomers.

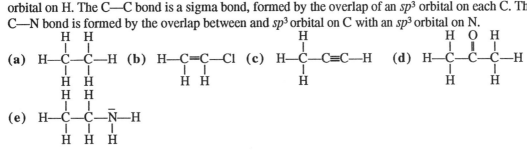

22. We show only the carbon atom skeleton in each case. Remember that there are four bonds to carbon. The bonds that are not indicated in these structures are C—H bonds.

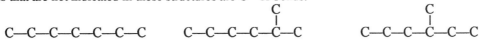

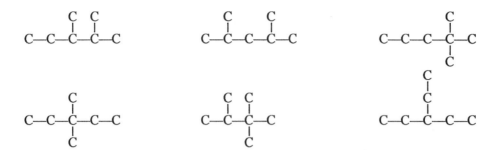

23. In the structures below we have omitted the hydrogen atoms. Remember that there are four bonds to each C atoms. The bonds that are missing are C—H bonds.

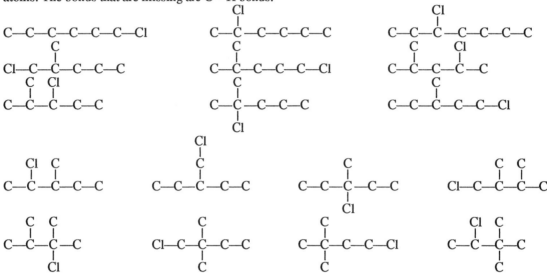

Functional Groups

24. (a) A carbonyl group is $-\overset{\overset{\displaystyle O}{\|}}{C}-$, while a carboxyl group is $-\overset{\overset{\displaystyle O}{\|}}{C}-OH$. The essential difference between them is the hydroxyl group, —OH.

(b) An amine group is —NH₂ (alkyl or aryl groups may be substituted for the hydrogens), while an **amide** group is $-\overset{\overset{\displaystyle O}{\|}}{C}-NH_2$. The essential difference is the carbonyl group, $-\overset{\overset{\displaystyle O}{\|}}{C}-$.

(c) An acid (that is, carboxylic acid) group is $-\overset{\overset{\displaystyle O}{\|}}{C}-OH$, while an acyl group is $R-\overset{\overset{\displaystyle O}{\|}}{C}-$. The essential difference is the presence of the alkyl group, —R, and the removal of the —OH group.

25. (a) aromatic nitro compound $C_6H_5NO_2$ nitrobenzene
    (b) aliphatic amine $C_2H_5NH_2$ ethylamine
    (c) chlorophenol $p\text{-ClC}_6H_4OH$ *para*-chlorophenol   (drawn below)
    (d) aliphatic diol $HOCH_2CH_2OH$ 1,2-ethanediol
    (e) unsaturated aliphatic alcohol $CH_2{=}CH(CH_2)_2OH$ 3-butene-1-ol
    (f) alicyclic ketone $C_6H_{11}O$ cyclohexanone
    (g) halogenated alkane $CH_3CHICH_2CH_3$ 2-iodobutane
    (h) aromatic dicarboxylic acid $o\text{-}C_6H_4(COOH)_2$ *ortho*-phthalic acid   (drawn below)

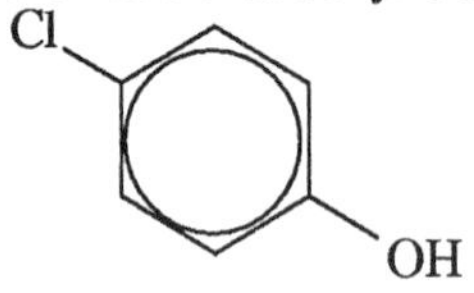

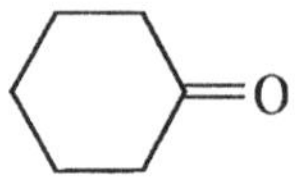

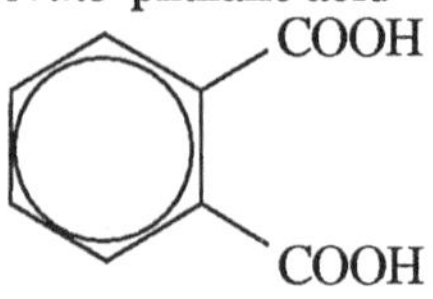

Nomenclature and Formulas

26. **(a)** The longest carbon chain has four carbon atoms and there is a methyl group attached to carbon 2. This is 2-methylbutane.

(b) The longest chain is three carbons long, there is a double bond between carbons 1 and 2, and a methyl group attached to carbon 2. This is 2-methylpropene.

(c) There is a three-carbon ring, to which a methyl group is attached. This is methylcyclopropane.

(d) The longest chain is 5 carbons long, there is a triple bond between carbons 2 and 3 and a methyl group attached to carbon 4. This is 4-methyl-2-pentyne.

(e) The longest chain is 6 carbons long. There are two methyl groups attached to carbons 3 and 4. This compound is 3,4-dimethylhexane.

(f) The longest carbon chain that contains the double bond is 5 carbons long. The double bond is between carbons 1 and 2. There is a propyl group attached to carbon 2, and two methyl groups attached to carbons 3 and 4. This is 3,4-dimethyl-2-propyl-1pentene.

27. Again we do not show the hydrogen atoms in the structures below. But we realize that there are four bonds to each C atom and the missing bonds are C—H bonds.

(a) isobutane

(b) cyclohexene

(c) 3-hexyne

(d) 2-butanol

(e) isopropyl methyl ether

(f) propionaldehyde

(g) *t*-butyl chloride

(h) diethylmethylamine

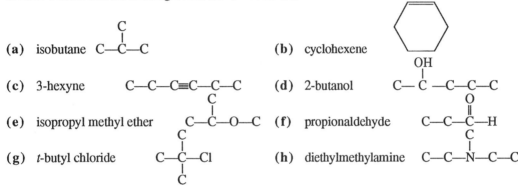

28. **(a)** The name pentene is insufficient. 1-pentene is $CH_2{=}CHCH_2CH_2CH_3$ and 2-pentene is $CH_3CH{=}CHCH_2CH_3$.

(b) The name butanone is sufficiently precise. There is only one four-carbon ketone.

(c) The name butyl alcohol is insufficiently precise. There are numerous butanols. 1-butanol is $HOCH_2CH_2CH_2CH_3$, 2-butanol is $CH_3CH(OH)CH_2CH_3$, isobutanol is $HOCH_2CH(CH_3)_2$, and *t*-butanol is $(CH_3)_3COH$.

(d) The name methylaniline is not sufficiently precise. It specifies $CH_3{—}C_6H_4{—}NH_2$, with no relative locations of the $—NH_2$ and $—CH_3$ substituents.

(e) The name methylcyclopentane is sufficiently precise. It does not matter where on a five-carbon ring with only C-C single bonds a methyl group is placed.

29. **(a)** 3-pentene specifies $CH_3CH_2CH{=}CHCH_3$. The proper name for this compound is 2-pentene. The number for the functional group should be as small as possible.

(b) Pentadiene is insufficiently precise. Possible pentadienes are 1,2 pentadiene, $CH_2{=}C{=}CHCH_2CH_3$, 1,3-pentadiene, $CH_2{=}CHCH{=}CHCH_3$, and 2,3 pentadiene, $CH_3CH{=}C{=}CHCH_3$.

(c) 1-propanone is incorrect. There cannot be a ketone on the first carbon of a chain. The compound specified is either propanaldehyde, CH_3CH_2CHO, or propanone (2-propanone), CH_3COCH_3 (acetone).

(d) Bromopropane is insufficiently precise. It could be either 1-bromopropane, $BrCH_2CH_2CH_3$, or 2-bromopropane, $CH_3CHBrCH_3$.

(e) Although 2,6-dichlorobenzene conveys enough information, the proper name for the compound is meta-dichlorobenzene, or 1,3-dichlorobenzene. Substituent numbers should be as small as possible.

(f) 2-methyl-3-pentyne is $(CH_3)_2CHC{\equiv}CCH_3$. The proper name for this compound is 4-methyl-2-pentyne. The number of the triple bond should be as small as possible.

(g) 2-methyl-4-butyloctane is $(CH_3)_2CHCH_2CH(C_4H_9)CH_2CH_2CH_2CH_3$. The proper name for this compound is 5-isobutylnonane. The carbon chain should be as long as possible.

(h) 4,4-dimethly-5-ethyl-1-hexyne is CH≡CCH₂C(CH₃)₂CH(C₂H₅)CH₃. The proper name is 4,4,5-trimethylheptyne. The carbon chain containing the functional group should be as long as possible.

30. (a) 2,4,6-trinitrotoluene

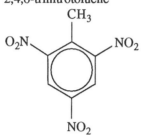

(b) methylsalicylate

(c) 2-hydroxy-1,2,3-propanetricarboxylic acid   HOOCCH₂C(OH)(COOH)CH₂COOH
(d) *o-t*-butylphenol

(e) 1-phenyl-2-aminopropane   C₆H₅CH₂CH(NH₂)CH₃
(f) 2-methylheptadecane   (CH₃)₂CH(CH₂)₁₄CH₃

Experimental Determination of Formulas

31. no. mmol C = 577 mg CO₂ × $\dfrac{1\ \text{mmol CO}_2}{44.01\ \text{mg CO}_2}$ × $\dfrac{1\ \text{mmol C}}{1\ \text{mmol CO}_2}$ = 13.1 mmol C

no. mmol H = 236 mg H₂O × $\dfrac{1\ \text{mmol H}_2\text{O}}{18.02\ \text{mg H}_2\text{O}}$ × $\dfrac{2\ \text{mmol H}}{1\ \text{mmol H}_2\text{O}}$ = 26.2 mmol H

The empirical formula of the compound is CH₂ since there are 2 mmol H for every mmol C.
The empirical molar mass is 14, which is one-fourth of the molar mass. Thus, the molecular formula must be (CH₂)₄ or C₄H₈.

Alkanes

32. We show only the carbon atom skeleton in each case. Remember that there are four bonds to carbon. The bonds that are not indicated in these structures are C—H bonds. Boiling points are given in parentheses after the name of each compound.

butane (–0.5°C)   C—C—C—C   hexane (68.7°C)   C—C—C—C—C—C

isobutane (–11.7°C)   C—C—C   3-methylpentane (63.3°C)

pentane (36.1)   C—C—C—C—C   isohexane (58.0°C)

isopentane (27.9°C)   C—C—C—C   2,3-dimethylbutane (58.0°C)

neopentane (9.5°C)   C—C—C   neohexane (49.7°C)

As we proceed down each list of alkanes, the structures become increasingly compact. Also, the boiling points decrease.

33. (a) An alkane with a molecular mass of 44 u must be a propane, since 3 carbons have a molecular mass of 36 u. There is only one propane, *n*-propane, and it has the condensed formula CH₃CH₂CH₃. Its two monochlorination products are CH₃CH₂CH₂Cl and CH₃CHClCH₃

(b) An alkane with a molecular mass of 58 u cannot be a pentane, since five carbons have a mass of 60 u, so it must be a butane. Both normal butane and isobutane have only two monobromination products. From *n*-butane, CH₃CH₂CH₂CH₃   CH₃CH₂CH₂CH₂Br and CH₃CH₂CHBrCH₃

From iso-butane, $CH_3CH(CH_3)_2$ $BrCH_2CH(CH_3)_2$ and $CH_3CBr(CH_3)_2$

34. Chlorination of CH_4 proceeds by a free radical chain reaction. Methyl radicals are produced in the course of this reaction and it is not unlikely that two methyl radicals will unite in a chain termination step to form an ethane molecule. $\cdot CH_3 + \cdot CH_3 \longrightarrow CH_3CH_3$ The resulting ethane molecule then can be attacked by a chlorine molecule to produce a molecule of monochloroethane. $CH_3CH_3 + Cl_2 \longrightarrow CH_3CH_2Cl + HCl$

Alkenes

35. In the case of ethene there are only two carbon atoms between which there can be a double bond. Thus, specifying the compound as 1-ethene is unnecessary. In the case of propene, there can be a double bond only between the central carbon atom and a terminal carbon atom. Thus here also specifying the compound as 1-propene is unnecessary. The case of butene is different, however, since 1-butene, $CH_2=CHCH_2CH_3$, is distinct from 2-butene, $CH_3CH=CHCH_3$.

36. In an alkene there is a $C=C$ double bond. On the other hand, in an cyclic alkane there are no double bonds, but rather a chain of carbon atoms joined at the ends into a ring.

37. We show only the carbon atom skeleton for each product. Remember that there are four bonds to carbon. The bonds that are not indicated in these structures are C—H bonds.
   (a) $CH_2=CHCH_3 + H_2 \xrightarrow{Pt, \ heat} C-C-C$
   (b) $CH_3CHOHCH_2CH_3 \xrightarrow{H_2SO_4, \ heat} CH_3CH=CHCH_3$

38. (a) $CH_3CCl=CH_2 + HCl \longrightarrow CH_3CCl_2CH_3$
   (b) $CH_3C\equiv CH + HCN \longrightarrow CH_3C(CN)=CH_2$
   (c) $CH_3CH=C(CH_3)_2 + HCl \longrightarrow CH_3CH_2CCl(CH_3)_2$
   (d)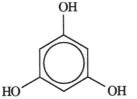

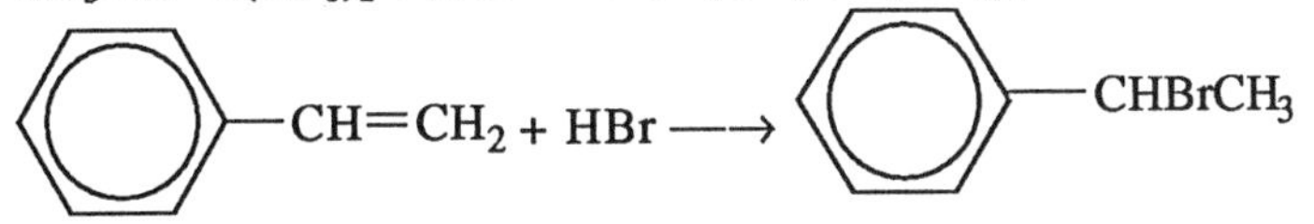

Aromatic Compounds

39. (a) meta-dinitrobenzene 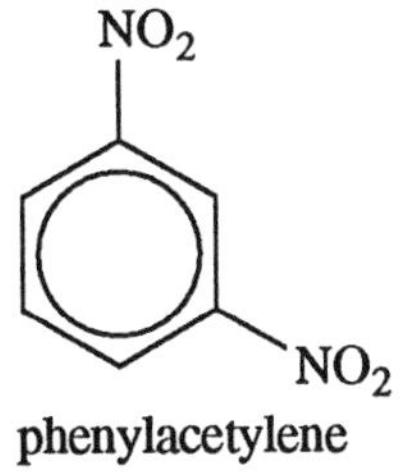
NO₂... (b) 3,5-dihydroxyphenol 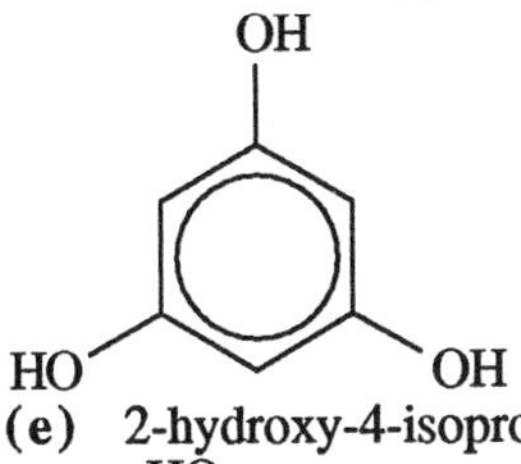 (c) *p*-phenylphenol

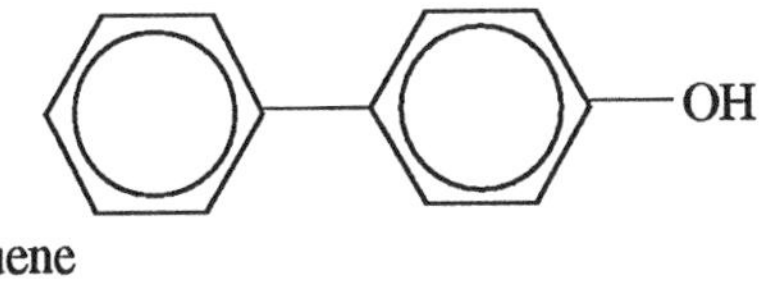

(d) phenylacetylene 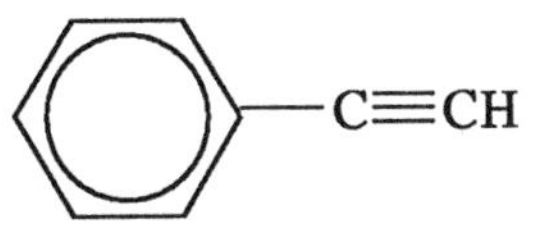   (e) 2-hydroxy-4-isopropyltoluene

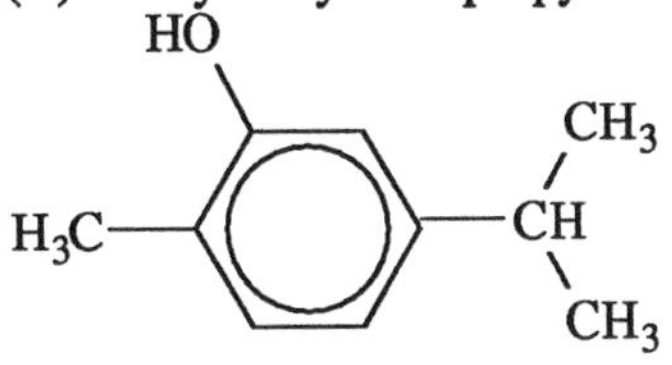

40. (a) Nitro groups are meta directors. Thus the only product formed is 5-bromo-1,3-dinitrobenzene, drawn below.
   (b) Amine is an ortho, para director. There are two products formed: *ortho*-bromoaniline and *para*-bromoaniline, both of which are drawn below.

(a)

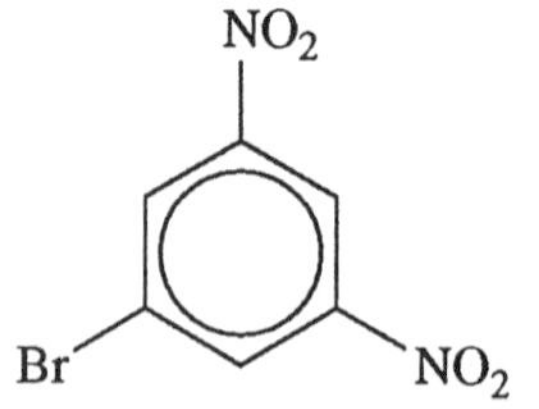

(b)

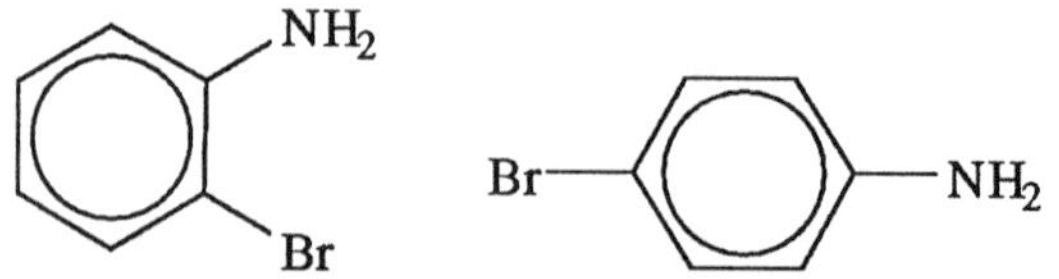

(c) Both an ether and a bromo group are ortho, para directors, but an ether is a stronger ortho, para director. The principal product we expect is 2,4-dibromoanisole, drawn below.

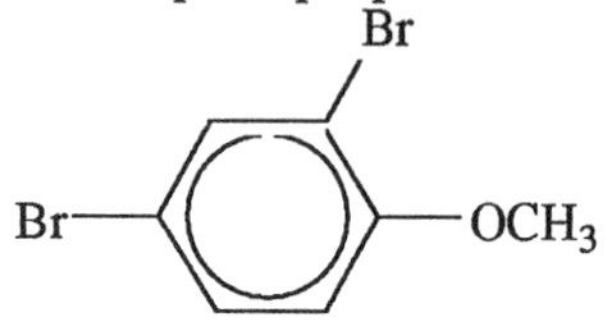

Organic Reactions

41. (a) In an aliphatic substitution reaction, an atom (usually H) of an alkane molecule is replaced by another atom or group of atoms. An example is equation (27.1).

(b) In an aromatic substitution reaction, an atom (usually H) of a phenyl group is replaced by another atom or group of atoms. Examples are in equations (27.7) and (27.8).

(c) In an addition reaction, the result is that a small molecule breaks into two parts and the fragments add, one to each of the carbon atoms of a double bond. An example is equation (27.9).

(d) In an elimination reaction , two molecules join together, with the simultaneous removal of at atom or a group of atoms from these two (larger) molecules. These removed pieces join together to form a small molecule. An example is equation (27.11).

42. (a) $CH_3CH_2CH_2CH_2CH_2OH \xrightarrow{Cr_2O_7^{2-},\ H^+} CH_3CH_2CH_2CH_2COOH$

(b) $CH_3CH_2CH_2COOH + HOCH_2CH_3 \xrightarrow{H^+} CH_3CH_2CH_2\overset{\overset{O}{\|}}{C}CH_2CH_3 + H_2O$

(c) $CH_3CH_2C(CH_3){=}CH_2 + H_2O \xrightarrow{H_2SO_4} CH_3CH_2\underset{\underset{OH}{|}}{\overset{\overset{CH_3}{|}}{C}}{-}CH_3$

43. (a) $HC{\equiv}CH \xrightarrow{Br_2} CHBr{=}CHBr \xrightarrow{Br_2} CHBr_2CHBr_2$

(b) $HC{\equiv}CH \xrightarrow{Pt,\ H_2} CH_2{=}CH_2 \xrightarrow{H_2O,\ H_2SO_4} CH_3CH_2OH \xrightarrow{Cr_2O_7^{2-},\ H^+} CH_3CHO$

Polymerization Reactions

44. An ester linkage is formed by the condensation of a carboxylic acid and an alcohol. Accompanying this condensation is the elimination of a water molecule between the ester and the carboxylic acid. Dacron is formed by the condensation of a *di*carboxylic acid with a *diol.* Thus, Dacron contains ester linkages. Because there are a large number of these ester linkages in Dacron—joining many subunits together— it is appropriate to call the polymer a polyester.

To determine the percent of oxygen in Dacron, we refer to the basic unit of the polymer, which is shown in Table 27-5. This unit has the formula $C_{10}H_8O_4$. $\%O = \dfrac{(4 \times 16.00)\ g\ O}{192.2\ g\ polymer} \times 100\% = 33.30\%O$

45. The polymerization of 1,6-hexanediamine with sebacyl chloride proceeds in the following manner. The italicized *H* and *Cl* atoms are removed in this condensation reaction.

$$H{-}\overset{\overset{H}{|}}{N}{-}(CH_2)_6{-}\overset{\overset{H}{|}}{N}{-}\mathit{H} + \mathit{Cl}{-}\overset{\overset{O}{\|}}{C}{-}(CH_2)_8{-}\overset{\overset{O}{\|}}{C}{-}\mathit{Cl} + \mathit{H}{-}\overset{\overset{H}{|}}{N}{-}(CH_2)_6{-}\overset{\overset{H}{|}}{N}{-}\mathit{H} + \mathit{Cl}{-}\overset{\overset{O}{\|}}{C}{-}(CH_2)_8{-}\overset{\overset{O}{\|}}{C}{-}Cl$$

$$\longrightarrow \left({-}\overset{\overset{O}{\|}}{C}{-}(CH_2)_8{-}\overset{\overset{O}{\|}}{C}{-}\overset{\overset{H}{|}}{N}{-}(CH_2)_6{-}\overset{\overset{H}{|}}{N}{-}\right)_x$$

46. In order to form long-chain molecules, every monomer must have at least two functional groups, one on each end of the molecule. Ethyl alcohol has only one functional group, a hydroxy group (—OH). It cannot participate in a polymerization reaction with dimethyl terephthalate.

47. In a simple molecular substance like benzene, all molecules are identical (C_6H_6). No matter how many of these molecules are present in one sample, any other sample with the same number of molecules has the same mass. The mass in grams of one mole of molecules—the molar mass—is a unique quantity. In a polymer the situation is quite different. The number of monomer units in a polymer chain is not a constant number but is widely variable. Thus individual polymer molecules differ very much in mass. Thus, a mass of a mole of their molecules also is quite variable. However, when we take a sample for analysis, we obtain many molecules of each chain length or molecular mass. The resulting determination of molar mass obtains the average mass of all of these different sized polymer molecules.

28 CHEMISTRY OF THE LIVING STATE

REVIEW QUESTIONS

1. **(a)** (+) is another way of designating a dextrorotatory compound.
   (b) L indicates that, in the Fisher projection of the compound, the —OH group on the penultimate carbon atom is to the left and the —H group is to the right.
   (c) A sugar is a carbohydrate that is either a monosaccharide or an oligosaccharide.
   (d) An α-amino acid is a carboxylic acid that has an amine group on the carbon next to the —COOH group, that is, on the α carbon atom. Glycine (H_2NCH_2COOH) is the simplest α-amino acid.
   (e) The pH at which the zwitterion form of an amino acid predominates in solution is known as the isoelectric point. The isoelectric point of glycine is $pI = 5.97$.

2. **(a)** Saponification is the hydrolysis of a glyceride in alkaline solution to produce glycerine and soaps: salts of fatty acids.
   (b) A chiral carbon atom is one that exhibits optical isomerism, usually one to which four different groups are attached.
   (c) A racemic mixture is one composed of an optically active compound and its enantiomer. Since these two compounds rotate polarized light by the same amount but in opposite directions, such a mixture does not rotate the plane of polarized light.
   (d) The denaturation of a protein is the process in which the protein is treated with somewhat harsh conditions and loses at least some of its secondary or tertiary structure, temporarily or permanently, leading to a loss of biological activity.

3. **(a)** A fat is a triglyceride in which the fatty acid chains are saturated hydrocarbon chains. Fats are solids at room temperature. An oil is a triglyceride in which the fatty acid chains are to some degree unsaturated. Oils are liquids at room temperature.
   (b) Enantiomers are optically active isomers of a compound that are mirror images of each other. Diastereomers are optically active isomers that are not mirror images.
   (c) The primary structure of a protein refers to the sequence of amino acids. The secondary structure describes the shape of that polypeptide chain.
   (d) ADP is adenosine diphosphate, ATP is adenosine triphosphate. They differ by a phosphate group.
   (e) DNA is deoxyribonucleic acid, RNA is ribonucleic acid. They differ by an O atom. DNA is found in the cell nucleus, RNA in the cytoplasm.

4. **(a)** $C_{17}H_{35}COOH$ is stearic acid. $C_{17}H_{33}COOH$ is oleic acid. $C_{11}H_{23}COOH$ is lauric acid. Thus, the given compound is glyceryl laurooleostearate.
   (b) $C_{17}H_{31}COOH$ is linoleic acid. Thus, the compound is glyceryl trilinoleate or trilinolein.
   (c) $C_{13}H_{27}COOH$ is myristic acid. Thus, the compound is sodium myristate.

5. **(a)** glyceryl lauromyristolinoleate

$$CH_2O-\overset{\overset{\displaystyle O}{\|}}{C}-(CH_2)_{10}CH_3$$
$$CHO-\overset{\overset{\displaystyle O}{\|}}{C}-(CH_2)_{12}CH_3$$
$$CH_2O-\overset{\overset{\displaystyle O}{\|}}{C}-(CH_2)_7-CH=CH-CH_2-CH=CH(CH_2)_4CH_3$$

(b) trilaurin

$$CH_2O-\overset{\overset{\displaystyle O}{\|}}{C}-(CH_2)_{10}CH_3$$
$$CHO-\overset{\overset{\displaystyle O}{\|}}{C}-(CH_2)_{10}CH_3$$
$$CH_2O-\overset{\overset{\displaystyle O}{\|}}{C}-(CH_2)_{10}CH_3$$

(c) potassium palmitate $CH_3(CH_2)_{14}—\overset{\overset{\displaystyle O}{\|}}{C}—O^-\,K^+$

(d) butyl linoleate $CH_3(CH_2)_3—O—\overset{\overset{\displaystyle O}{\|}}{C}—(CH_2)_7—CH{=}CH—CH_2—CH{=}CH—(CH_2)_4—CH_3$

6. A DL mixture contains equal amount of both enantiomers. This is also known as a racemic mixture. The optical activities of the two enantiomers cancel each other out, and the mixture rotates the plane of polarized light neither to the right nor to the left.

7. **(a)** D-(–)-arabinose is the optical isomer of L-(+)-arabinose. Its structure is drawn below.
(b) A diastereomer of L-(+)-arabinose is a molecule that is its optical isomer, but not its mirror image. There are several such diastereomers, one of which is drawn below.

(a)
H—C=O
HO—C—H
H—C—OH
H—C—OH
CH₂OH

(b)
H—C=O
HO—C—H
HO—C—H
H—C—OH
CH₂OH

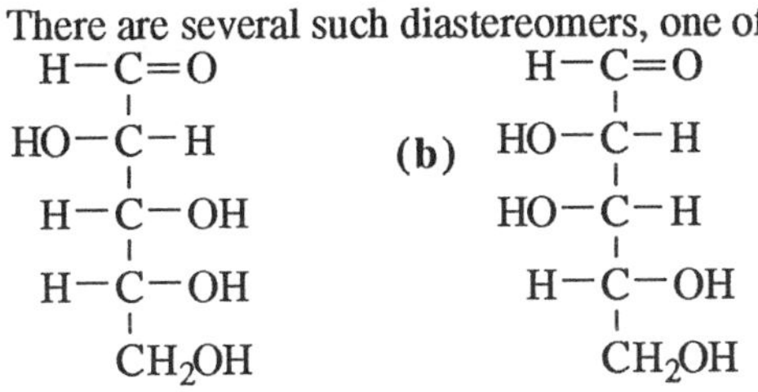

8. The pI of phenylalanine is 5.48. Thus, phenylalanine is in the form of a cation in 1.0 M HCl, in the form of an anion in 1.0 M NaOH, and in the form of a zwitterion at pH = 5.5. These three structures are drawn below.

(a) $\bigcirc—CH_2{-}\overset{\overset{\displaystyle NH_3^+\ Cl^-}{|}}{CH}\,COOH$
(b) $\bigcirc—CH_2{-}\overset{\overset{\displaystyle NH_2}{|}}{CH}\,COO^-\,Na^+$
(c) $\bigcirc—CH_2{-}\overset{\overset{\displaystyle NH_3^+}{|}}{CH}\,COO^-$

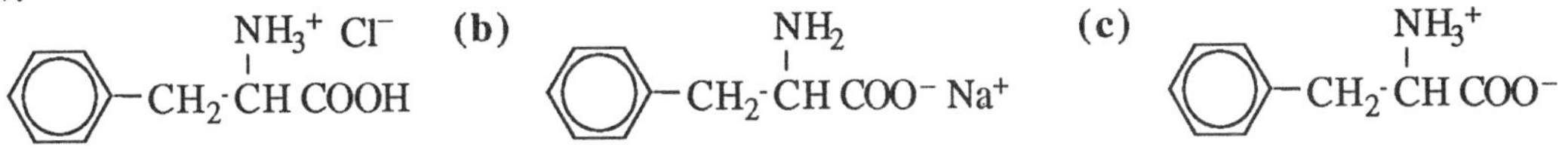

9. **(a)** glycylmethionine

$H_2C—\overset{\overset{\displaystyle O}{\|}}{C}{-}NH\underset{\underset{\displaystyle CH_2OH}{|}}{CH}{-}\overset{\overset{\displaystyle O}{\|}}{C}—OH$
$\underset{\displaystyle NH_2}{|}$

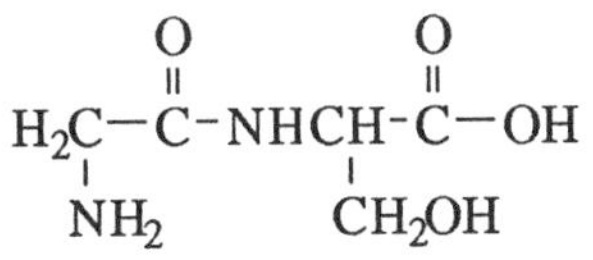

(b) isoleucylleucylserine

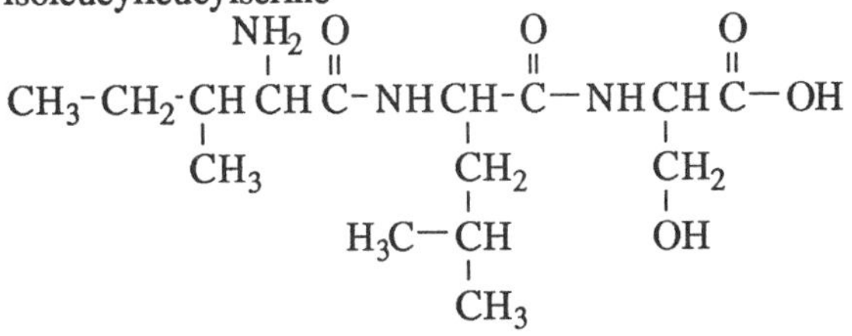

$CH_3{-}CH_2{-}\overset{\overset{\displaystyle NH_2}{|}}{CH}\underset{\underset{\displaystyle CH_3}{|}}{CH}\overset{\overset{\displaystyle O}{\|}}{C}{-}NH\underset{\underset{\underset{\underset{\displaystyle CH_3}{|}}{H_3C—CH}}{|}}{\underset{\displaystyle CH_2}{|}}{CH}{-}\overset{\overset{\displaystyle O}{\|}}{C}—NH\underset{\underset{\underset{\displaystyle OH}{|}}{CH_2}}{CH}\overset{\overset{\displaystyle O}{\|}}{C}—OH$

10. **(a)** $H_2N{-}CH_2{-}\overset{\overset{\displaystyle O}{\|}}{C}—NH\underset{\underset{\displaystyle CH_3}{|}}{CH}\overset{\overset{\displaystyle O}{\|}}{C}{-}NH\underset{\underset{\underset{\displaystyle OH}{|}}{CH_2}}{CH}{-}\overset{\overset{\displaystyle O}{\|}}{C}{-}NH\underset{\underset{\underset{\displaystyle CH_3}{|}}{CHOH}}{CH}\overset{\overset{\displaystyle O}{\|}}{C}—OH$
(b) glycylalanylserylthreonine

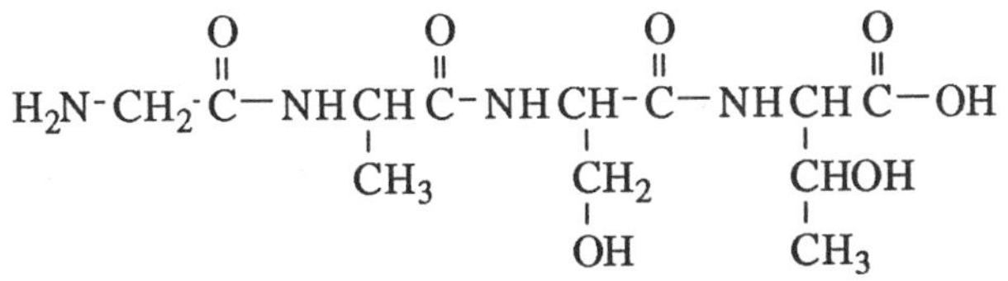

11. Heat can cause some of the very weak bonds that hold proteins together into their secondary and tertiary structure to break and re-form, altering that structure. pH can alter the ionic structure of the amino acids in a protein (protonating —NH₂ or deprotonating —COOH) and these different ionic forms also can have an effect on the secondary and tertiary structure of a protein. Finally, metal ions can bond to the sites responsible for secondary and tertiary structure or they can bond strongly to the active site of the enzyme. Alteration of the structure of the protein or blockage of the active site will affect the activity of an enzyme.

Enzymes are so specific in the reactions they catalyze because the active site is a three dimensional pocket into which the substrate (the reactant) fits precisely. The surface of platinum, on the other hand, is simply a two-dimensional surface requiring little of the shape of the reactant, other than its atoms be spaced appropriately so that they can bond to that surface, usually two at a time. (And this spacing requirement can often be accomodated in a number of ways.)

12. The bases are on the right side of the diagram. From the top down, they are: adenine (purine), uracil (pyrimidine), guanine (purine), and cytosine (pyrimidine). The pentose sugars have the five-membered rings as their carbon skeletons; they are ribose sugars, and hence this is a chain of RNA. The phosphate groups are the five-membered groups centered on P atoms and surrounded by O atoms.

EXERCISES

Structure and Composition of the Cell

13. The volume of a cylinder is given by $V = \pi r^2 h = \pi d^2 h/4$

$$V = [3.14159\ (1 \times 10^{-6}\ m)^2\ (2 \times 10^{-6}\ m) \div 4] \times \frac{1000\ L}{1\ m^3} = 1.6 \times 10^{-15}\ L$$

The volume of the solution in the cell is $V_{soln} = 0.80 \times 1.6 \times 10^{-15}\ L = 1.3 \times 10^{-15}\ L$

(a) $[H^+] = 10^{-6.4} = 4 \times 10^{-7}\ M$

$$\text{no. } H^+ \text{ ions} = 1.3 \times 10^{-15}\ L \times \frac{4 \times 10^{-7}\ mol\ H^+\ ions}{1\ L\ soln} \times \frac{6.022 \times 10^{23}\ ions}{1\ mol\ ions} = 3 \times 10^2\ H^+ \text{ ions}$$

(b) $$\text{no. } K^+ \text{ ions} = 1.3 \times 10^{-15}\ L \times \frac{1.5 \times 10^{-4}\ mol\ K^+\ ions}{1\ L\ soln} \times \frac{6.022 \times 10^{23}\ ions}{1\ mol\ ions} = 1.2 \times 10^5\ K^+ \text{ ions}$$

14. Mass of all lipid molecules $= 0.02 \times 2 \times 10^{-12}\ g = 4 \times 10^{-14}\ g$

$$\text{no. of lipid molecules} = 4 \times 10^{-14}\ g \times \frac{1\ lipid\ molecule}{700\ u} \times \frac{1\ u}{1.66 \times 10^{-24}\ g} = 3 \times 10^7\ \text{lipid molecules}$$

15. mass of protein in cytoplasm $= 0.15 \times 0.90 \times 2 \times 10^{-12}\ g = 2._7 \times 10^{-13}\ g$

$$\text{no. of protein molecules} = 2._7 \times 10^{-13}\ g \times \frac{1\ mol\ protein}{3 \times 10^4\ g} \times \frac{6.022 \times 10^{23}\ molecules}{1\ mol\ protein}$$

$$= 5 \times 10^6\ \text{protein molecules}$$

16. DNA length $= 4.5 \times 10^6$ mononucleotides $\times \dfrac{450\ pm}{1\ mononucleotide} \times \dfrac{10^{-12}\ m}{1\ pm} = 2 \times 10^{-3}\ m = 2\ mm$

2 mm $= 2 \times 10^3\ \mu m$, the length of the stretched out DNA is one thousand times 2 μm, the length of the cell. Thus, the DNA must be wrapped up, or coiled, within the cell.

Lipids

17. **(a)** Lipids are those compounds in living tissue that are soluble in nonpolar solvents, such as hydrocarbon liquids. Fats, oils, waxes, and triglycerides all are lipids.
   (b) A triglyceride is an ester of the trihydroxy alcohol glycerol, $CH_2OHCHOHCH_2OH$, and three long-chain carboxylic acids. Tristearin is an example of a triglyceride, as is glyceryl laurooleostearate.
   (c) A simple glyceride is a triglyceride (see the answer to part c) in which all three fatty acids are the same. Tristearin in an example of a simple glyceride.
   (d) In a mixed glyceride, the three fatty acids are not all the same. Glyceryl laurooleostearate is an example of a mixed glyceride.
   (e) A fatty acid is a carboxylic acid with a long hydrocarbon chain. Stearic acid, $CH_3(CH_2)_{16}COOH$, is an example of a fatty acid.
   (f) A soap is a compound of a fatty acid anion and a cation such as Na^+ or K^+. Sodium stearate is an example of a soap, $Na^+\ {}^-OOC(CH_2)_{16}CH_3$.

18. Polyunsaturated fatty acids are characterized by a large number of $C{=}C$ double bonds in their hydrocarbon chain. Stearic acid has no $C{=}C$ double bonds and therefore is not unsaturated, let alone polyunsaturated. But eleostearic acid has three $C{=}C$ double bonds and thus is polyunsaturated. Polyunsaturated fatty acids are recommended in dietary programs since saturated fats are linked to a high incidence of heart disease. Of the lipids listed in Table 28-2, safflower oil has the highest percentage of unsaturated fatty acids, predominately linoleic acid, a diunsaturated fatty acid.

19. Safflower oil contains a larger percentage of the di-unsaturated fatty acid, linoleic acid (75-80%) than does corn oil (34-62%). It also contains a smaller percentage of saturated fatty acids, particularly palmitic acid (6-7%) than does corn oil (8-12%). And the two oils contain about the same proportion of the mono-unsaturated fatty acid, oleic acid. Consequently, safflower oil should consume the greater amount of $H_2(g)$ in its complete hydrogenation to a solid fat.

20. trilaurin

$CH_2OOC(CH_2)_{10}CH_3$
|
$CHOOC(CH_2)_{10}CH_3$
|
$CH_2OOC(CH_2)_{10}CH_3$

saponification products of trilaurin: lauric acid and glycerol

$CH_2OHCHOHCH_2OH$

$Na^+ {}^-OOC(CH_2)_{10}CH_3$

21. First we write the equation for the saponification reaction.

$CH_2OOCC_{15}H_{31}$
|
$CHOOCC_{15}H_{31}$ $+ 3\ NaOH \longrightarrow CH_2OHCHOHCH_2OH + 3\ Na^+ {}^-OOCC_{15}H_{31}$
|
$CH_2OOCC_{15}H_{31}$

$$\text{mass of soap} = 125\ \text{g triglyceride} \times \frac{1\ \text{mol triglyceride}}{807.3\ \text{g triglyceride}} \times \frac{3\ \text{mol soap}}{1\ \text{mol triglyceride}} \times \frac{278.4\ \text{g soap}}{1\ \text{mol soap}}$$

$$= 129\ \text{g soap}$$

Carbohydrates

22. (a) A monosaccharide is the simplest carbohydrate. Glucose, galactose, fructose, and mannose are examples of monosaccharides.

(b) A disaccharide is two monosaccharides joined together. The most familiar disaccharide is sucrose, common table sugar, a disaccharide composed of glucose and fructose.

(c) An oligosaccharide is several monosaccharide units (from two to ten) joined together.

(d) A polysaccharide is many monosaccharide units joined together. Two common polysaccharides are starch and cellulose.

(e) Sugar is a common name given to monosaccharides and oligosaccharides.

(f) The term glycose is another term for carbohydrate.

(g) An aldose is a monosaccharide in which the C=O group is at the terminal carbon atom; glucose is an aldose.

(h) A ketose is a monosaccharide in which the C=O group is on one of the central atoms of the molecule.

(i) A pentose is a five-carbon monosaccharide; fructose is a pentose.

(j) A hexose is a six-carbon monosaccharide; glucose is a hexose.

23. A reducing sugar has a sufficient amount of the straight-chain form present at equilibrium that the sugar will reduce $Cu^{2+}(aq)$ to insoluble, red $Cu_2O(s)$. The free aldehyde group must be available to reduce the copper(II) ion.

24. (a) A dextrorotatory compound is one that rotates the plane of polarized light to the right: clockwise.

(b) A levorotatory compound is one that rotates the plane of polarized light to the left: counterclockwise.

(c) A racemic mixture is one composed of an optically active compound and its enantiomer. Since these two compounds rotate polarized light by the same amount but in opposite directions, such a mixture does not rotate the plane of polarized light.

(d) Two compounds that are optical isomers of each other—they have different locations of the substituent groups around their chiral carbons—but which are not mirror images of each other are diastereomers.

(e) (+) is another way of designating a dextrorotatory compound.

(f) (−) is another way of designating a levorotatory compound.

(g) D indicates that, in the Fisher projection of the compound, the —OH group on the penultimate carbon atom is to the right and the —H group is to the left.

25. The structure of L-glucose is as follows

Since D-glucose is dextrorotatory,

L-glucose is levorotatory

H—C=O
|
HO—C—H
|
H—C—OH
|
HO—C—H
|
HO—C—H
|
CH_2OH

26. Enantiomers are alike in all respects, including in the degreee which they rotate polarized light. They differ only in the direction in which this rotation occurs. Since α-glucose and β-glucose rotate the plane of polarized light by different degrees, they are diastereomers, not enantiomers.

27. We let x represent the fraction of α-D-glucose. Then $1 - x$ is the fraction of β-D-glucose.

$+52.7° = x (112°) + (1 - x)(18.7°) = 112°x + 18.7° - 18.7°x = 93°x + 18.7°$

$x = \dfrac{52.7° - 18.7°}{93°} = 0.37$ The solution is 37% α-D-glucose, and thus 63% β-D-glucose.

Amino acids, Polypeptides, and Proteins

28. **(a)** An α-amino acid is a carboxylic acid that has an amine group on the carbon next to the —COOH group, that is, on the α carbon atom. Glycine (H_2NCH_2COOH) is the simplest α-amino acid.

(b) A zwitterion is a form of an amino acid where the amine group is protonated (—NH_3^+) amd the carboxyl group is deprotonated (—COO^-). The zwitterion form of glycine is $^+H_3NCH_2COO^-$.

(c) The pH at which the zwitterion form of an amino acid predominates in solution is known as the isoelectric point. The isoelectric point of glycine is pI = 5.97.

(d) The peptide bond is the bond that forms between the carbonyl group of one amino acid and the amine group of another, with the elimination of a water molecule between them. The peptide bond between two glycine molecules is shown as a bold dash (—) in the structure below.

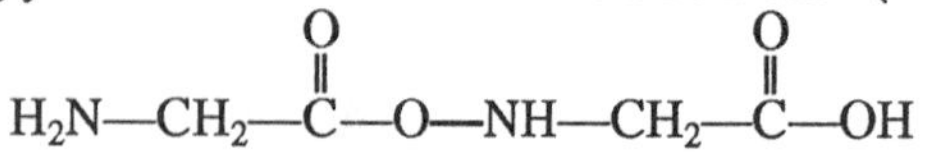

$$H_2N—CH_2—\overset{\overset{O}{\|}}{C}—O—NH—CH_2—\overset{\overset{O}{\|}}{C}—OH$$

(e) A polypeptide is a long chain of amino acids, joined together by peptide bonds.

(f) A protein is another name for a polypeptide, but there is a distinction that often is drawn. Proteins are longer chains than are polypeptides, and proteins are biologically active.

(g) The N-terminal amino acid in a polypeptide is the one at the end of a molecule where there is a free NH_2 group at the end of the polypeptide chain. The N-terminal amino acid is at the left end of the structure of diglycine above.

(h) An α helix is the natural secondary structure adopted by many proteins. It is rather like a spiral rising upward to the right (that is, clockwise as viewed from the bottom). This is a right-handed screw. The alpha helix is shown in Figure 28-9.

(i) Denaturation is that process in which at least some of the structure of a protein is disrupted, either thermally (with heat), mechanically, or by changing the pH or changing the ionic strength of the medium in which the protein is enveloped. Denaturation is accompanied by a decrease in the biologicaly activity. Denaturation can be temporary or permanent.

29. pH = 6.3 is the isoelectric point of proline. Thus proline will not migrate. But pH = 6.3 is more acidic than the isoelectric point of lysine (pI = 9.94). Thus, lysine is positively charged in this solution and will migrate toward the negatively charged cathode. And pH = 6.3 is less acidic than the isoelectric point of glutamic acid (pI = 3.22). Glutamic acid, therefore, is negatively charged in this solution and will migrate toward the positively charged anode.

30. **(a)** in strongly acidic solution   **(b)** at the isoelectric point **(c)** in strongly basic solution

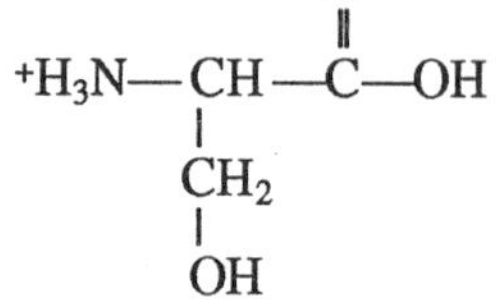

$$^+H_3N—\underset{\underset{\underset{OH}{|}}{\underset{CH_2}{|}}}{CH}—\overset{\overset{O}{\|}}{C}—OH$$

$^+H_3NCH(CH_2OH)COOH$

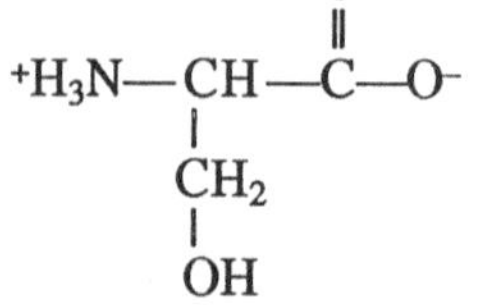

$$^+H_3N—\underset{\underset{\underset{OH}{|}}{\underset{CH_2}{|}}}{CH}—\overset{\overset{O}{\|}}{C}—O^-$$

$^+H_3NCH(CH_2OH)COO^-$

$$H_2N—\underset{\underset{\underset{OH}{|}}{\underset{CH_2}{|}}}{CH}—\overset{\overset{O}{\|}}{C}—O^-$$

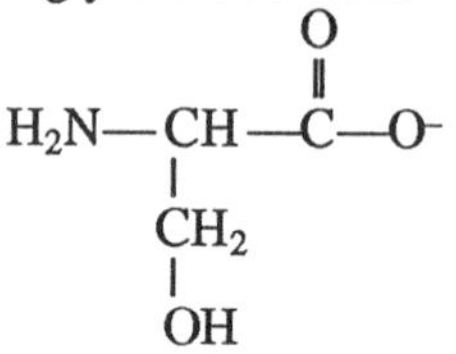

$H_2NCH(CH_2OH)COO^-$

31. (a) The structures of the tripeptides that contain one each of the amino acids alanine, serine, and lysine are drawn below.

Lys-Ser-Ala

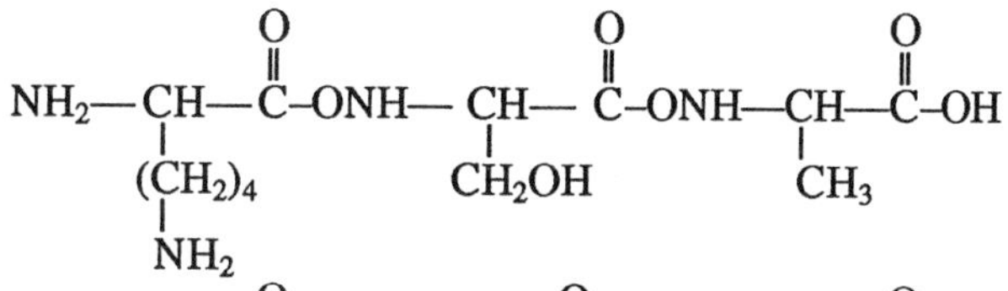

$$NH_2-CH-\overset{\overset{\displaystyle O}{\|}}{C}-ONH-CH-\overset{\overset{\displaystyle O}{\|}}{C}-ONH-CH-\overset{\overset{\displaystyle O}{\|}}{C}-OH$$

with side chains: $(CH_2)_4$—NH_2 ; CH_2OH ; CH_3

Lys-Ala-Ser

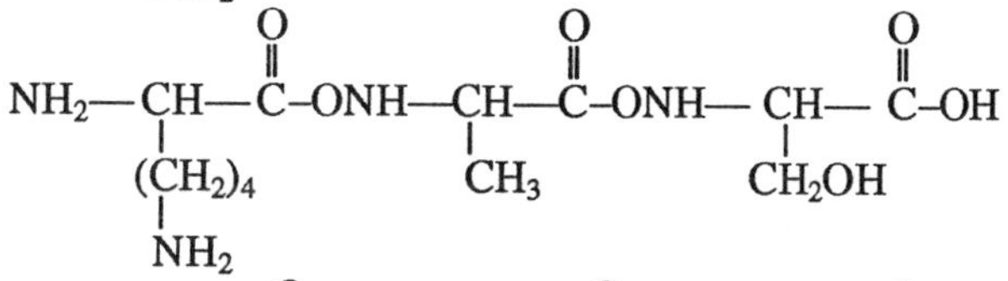

$$NH_2-CH-\overset{\overset{\displaystyle O}{\|}}{C}-ONH-CH-\overset{\overset{\displaystyle O}{\|}}{C}-ONH-CH-\overset{\overset{\displaystyle O}{\|}}{C}-OH$$

side chains: $(CH_2)_4$—NH_2 ; CH_3 ; CH_2OH

Ser-Lys-Ala

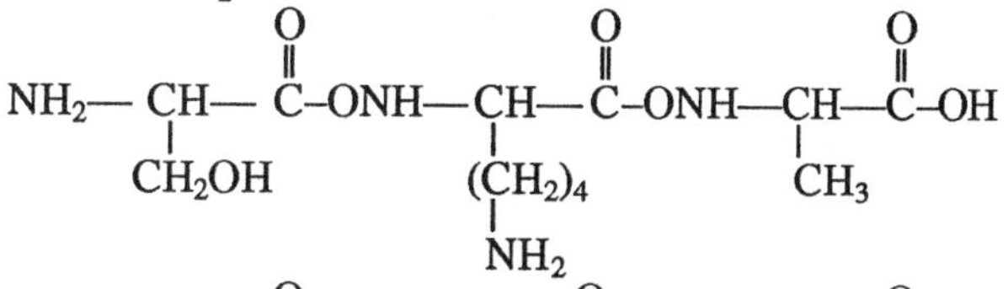

$$NH_2-CH-\overset{\overset{\displaystyle O}{\|}}{C}-ONH-CH-\overset{\overset{\displaystyle O}{\|}}{C}-ONH-CH-\overset{\overset{\displaystyle O}{\|}}{C}-OH$$

side chains: CH_2OH ; $(CH_2)_4$—NH_2 ; CH_3

Ser-Ala-Lys

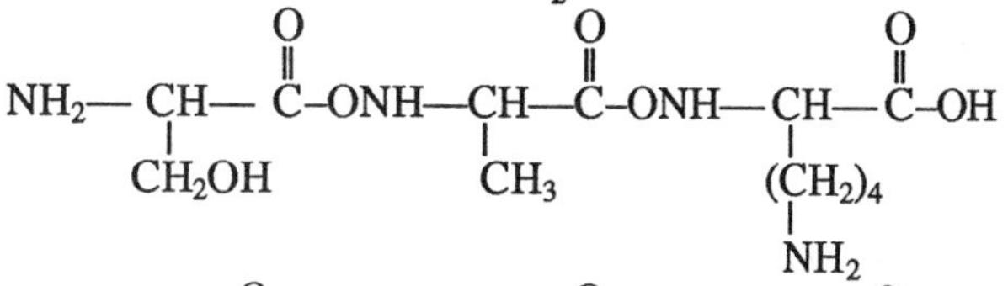

$$NH_2-CH-\overset{\overset{\displaystyle O}{\|}}{C}-ONH-CH-\overset{\overset{\displaystyle O}{\|}}{C}-ONH-CH-\overset{\overset{\displaystyle O}{\|}}{C}-OH$$

side chains: CH_2OH ; CH_3 ; $(CH_2)_4$—NH_2

Ala-Ser-Lys

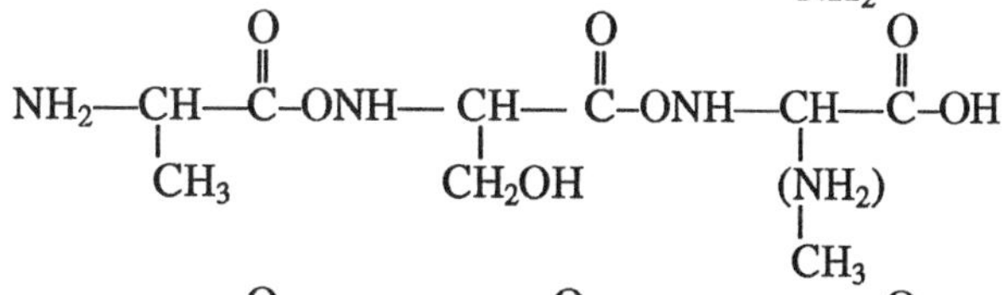

$$NH_2-CH-\overset{\overset{\displaystyle O}{\|}}{C}-ONH-CH-\overset{\overset{\displaystyle O}{\|}}{C}-ONH-CH-\overset{\overset{\displaystyle O}{\|}}{C}-OH$$

side chains: CH_3 ; CH_2OH ; (NH_2)—CH_3

Ala-Lys-Ser

$$NH_2-CH-\overset{\overset{\displaystyle O}{\|}}{C}-ONH-CH-\overset{\overset{\displaystyle O}{\|}}{C}-ONH-CH-\overset{\overset{\displaystyle O}{\|}}{C}-OH$$

side chains: CH_3 ; (NH_2)—CH_3 ; CH_2OH

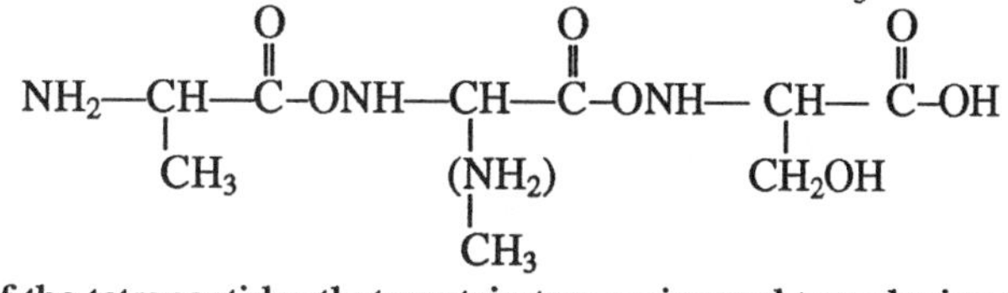

(b) The structures of the tetrapeptides that contain two serine and two alanine amino acids each are drawn below.

Ala-Ser-Ala-Ser

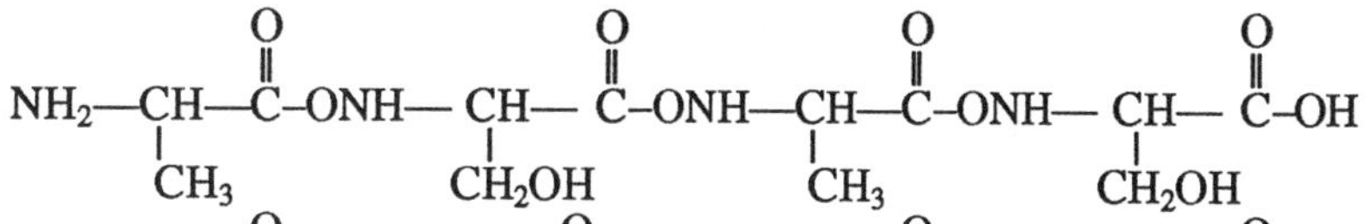

$$NH_2-CH-\overset{\overset{\displaystyle O}{\|}}{C}-ONH-CH-\overset{\overset{\displaystyle O}{\|}}{C}-ONH-CH-\overset{\overset{\displaystyle O}{\|}}{C}-ONH-CH-\overset{\overset{\displaystyle O}{\|}}{C}-OH$$

side chains: CH_3 ; CH_2OH ; CH_3 ; CH_2OH

Ala-Ala-Ser-Ser

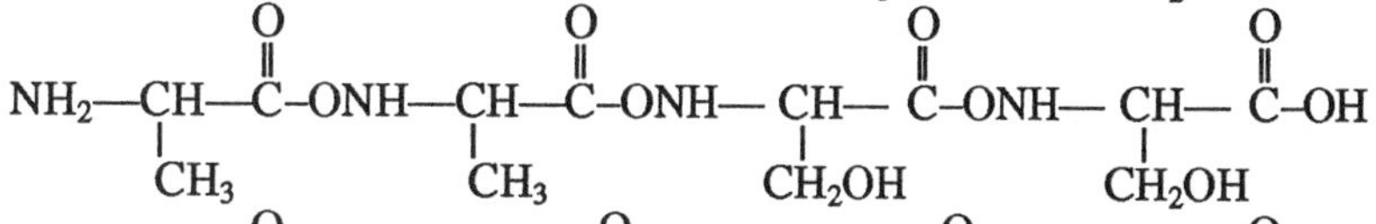

$$NH_2-CH-\overset{\overset{\displaystyle O}{\|}}{C}-ONH-CH-\overset{\overset{\displaystyle O}{\|}}{C}-ONH-CH-\overset{\overset{\displaystyle O}{\|}}{C}-ONH-CH-\overset{\overset{\displaystyle O}{\|}}{C}-OH$$

side chains: CH_3 ; CH_3 ; CH_2OH ; CH_2OH

Ala-Ser-Ser-Ala

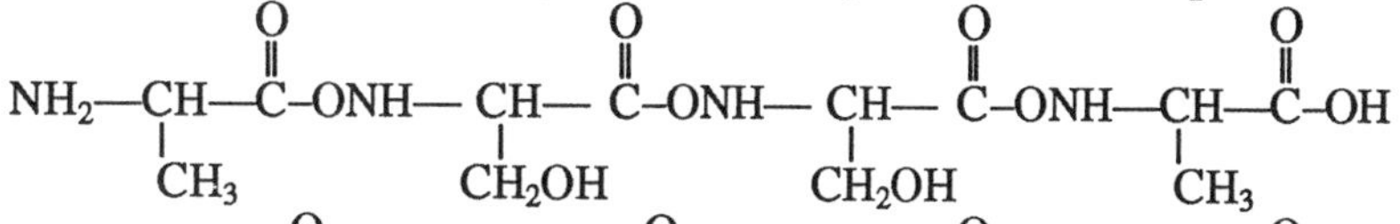

$$NH_2-CH-\overset{\overset{\displaystyle O}{\|}}{C}-ONH-CH-\overset{\overset{\displaystyle O}{\|}}{C}-ONH-CH-\overset{\overset{\displaystyle O}{\|}}{C}-ONH-CH-\overset{\overset{\displaystyle O}{\|}}{C}-OH$$

side chains: CH_3 ; CH_2OH ; CH_2OH ; CH_3

Ser-Ser-Ala-Ala

$$NH_2-CH-\overset{\overset{\displaystyle O}{\|}}{C}-ONH-CH-\overset{\overset{\displaystyle O}{\|}}{C}-ONH-CH-\overset{\overset{\displaystyle O}{\|}}{C}-ONH-CH-\overset{\overset{\displaystyle O}{\|}}{C}-OH$$

side chains: CH_2OH ; CH_2OH ; CH_3 ; CH_3

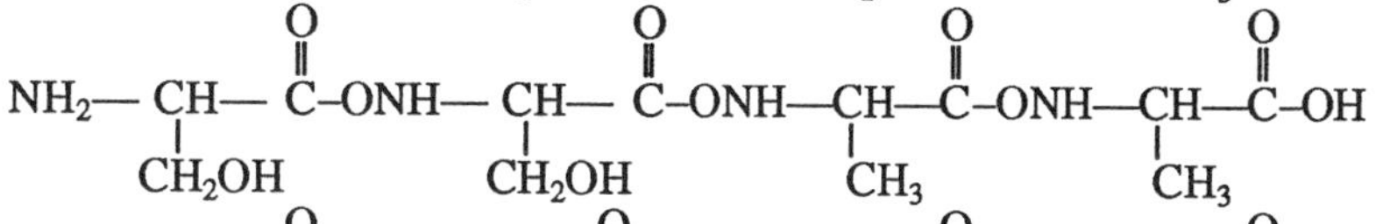

Ser-Ala-Ser-Ala

$$NH_2-CH-\overset{\overset{\displaystyle O}{\|}}{C}-ONH-CH-\overset{\overset{\displaystyle O}{\|}}{C}-ONH-CH-\overset{\overset{\displaystyle O}{\|}}{C}-ONH-CH-\overset{\overset{\displaystyle O}{\|}}{C}-OH$$

side chains: CH_2OH ; CH_3 ; CH_2OH ; CH_3

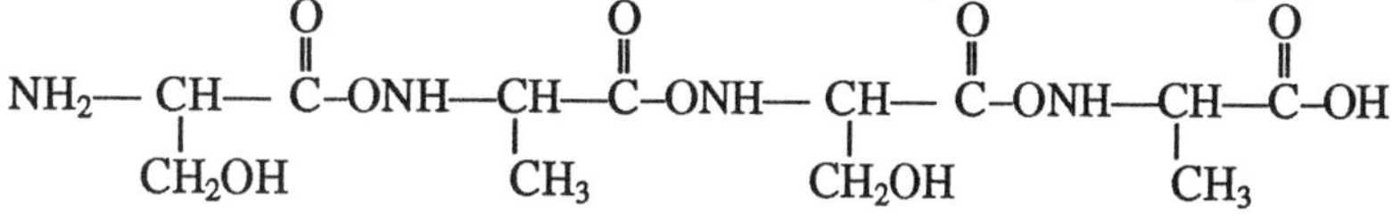

$$\text{Ser-Ala-Ala-Ser} \quad NH_2\!-\!CH\!-\!\overset{\overset{\displaystyle O}{\|}}{C}\!-\!ONH\!-\!CH\!-\!\overset{\overset{\displaystyle O}{\|}}{C}\!-\!ONH\!-\!CH\!-\!\overset{\overset{\displaystyle O}{\|}}{C}\!-\!ONH\!-\!CH\!-\!\overset{\overset{\displaystyle O}{\|}}{C}\!-\!OH$$
$$\underset{CH_2OH}{|} \qquad\qquad \underset{CH_3}{|} \qquad\qquad \underset{CH_3}{|} \qquad\qquad \underset{CH_2OH}{|}$$

32. (a) We put the fragments together as follows, starting from the Ala end, simply placing them down in a matching pattern. We do not assume that the fragments are given with the N-terminal end first.

Fragments: Ala   Ser
                   Ser   Gly   Val
                          Gly   Val   Thr
                                Val   Thr
                                Val   Thr   Leu

Result: Ala   Ser   Gly   Val   Thr   Leu

(b) alanyl-seryl-glycyl-valyl-threonyl-leucine, which is properly written without the hyphens, in the following manner.   alanylserylglycylvalylthreonylleucine

33. The *primary* structure of an amino acid is the sequence of amino acids in the chain of the polypeptide. The *secondary* structure describes how the protein chain is folded, coiled, or convoluted. Possible methods include α-helices and pleated sheets. This secondary structure is held together principally by hydrogen bonds. The *tertiary* structure of a protein refers to how different parts of the molecules, often quite distant from each other, are held together into crudely spherical shapes. Although hydrogen bonding is involved here as well, both disulfide linkages and salt linkages are responsible as well for tertiary structure. Finally, *quaternary* structure refers to how two or more protein molecules pack together into a larger protein complex. Not all proteins have a quaternary structure since many proteins have but one polypeptide chain.

34. $\mathfrak{M} = \dfrac{100.00 \text{ g hemoglobin}}{0.34 \text{ g Fe}} \times \dfrac{55.85 \text{ g Fe}}{1 \text{ mol Fe}} \times \dfrac{4 \text{ mol Fe}}{1 \text{ mol hemoglobin}} = 6.6 \times 10^4$ g/mol hemoglobin.

35. If we assume that there is only one active site per enzyme and that a silver ion is necessary to deactivate each active site, we obtain the molar mass of the protein as follows.

$$\text{molar mass} = \dfrac{1.0 \text{ mg}}{0.346 \text{ }\mu\text{mol Ag}^+} \times \dfrac{1 \text{ }\mu\text{mol}}{10^{-6} \text{ mol}} \times \dfrac{1 \text{ mol Ag}^+}{1 \text{ mol protein}} \times \dfrac{1 \text{ g}}{1000 \text{ mg}} = 2.9 \times 10^3 \text{ g/mol}$$

This is a minimum value because we have assumed that 1 Ag^+ ion is all that is necessary to denature each protein molecule. If more than one silver ion were required, the next to last factor in the calculation above would be larger than 1 mol Ag^+/1 mol protein, and the molar mass would increase correspondingly.

36. The sole difference between sickle cell hemoglobin and normal hemoglobin is due to the substitution of one amino acid for another (valine for glutamic acid) at only two positions in the entire protein. We can think of the fault as being due to the mistaken incorporation of one molecule for another during protein synthesis. This is the reason why the name "molecular disease" is apt in this case.

Metabolism

37. (a) An anabolic process is a metabolic process in which molecules are synthesized.

(b) A catabolic process is a metabolic process in which complex molecules are broken down into simpler ones, often with the production of energy.

(c) An endergonic process is one for which the standard free energy change is positive, that is, free energy is required for the process to occur, or is consumed during the process.

(d) ADP is an acronym for adenosine diphosphate, a phosphorylated nucleotide which can be further phosphorylated to store energy.

(e) ATP is an acronym for adenosine triphosphate, the phosphate ester of ADP, and in which energy is stored by the organism.

38. The energy stored in ATP amounts to 33.5 kJ/mol. Thus, stored in 15 mol of ATP is the following quantity of energy.

$$\text{energy} = 15 \text{ mol ATP} \times \dfrac{33.5 \text{ kJ}}{1 \text{ mol ATP}} = 502 \text{ kJ} \qquad \% \text{ efficiency} = \dfrac{502 \text{ kJ stored}}{837 \text{ kJ produced}} \times 100\% = 60.0\%$$

Nucleic Acids

39. The two major types of amino acids are DNA, deoxyribonucleic acid, and RNA, ribonucleic acid. Both of them contain phosphate groups. These phosphate groups alternate with sugars in forming the backbone of the molecule. These sugars are deoxyribose in the case of DNA and ribose in the case of RNA. Attached to each sugar is a purine or a pyrimidine base. The purine bases are adenine and guanine. One pyrimidine base is cytosine. In the case of RNA the other pyrimidine base is uracil, while in the case of DNA the other pyrimidine base is thymine.

40. The "thread of life" is an apt name for DNA, being both a literal and a figurative description. It is literal in that it is thread-like—long and narrow—in shape and is an essential molecule for life. It is figurative in that DNA is essential for the continuance of life and runs like a thread through all stages in the life of the organism, from origin through growth and reproduction to final death.

41. The complementary sequence to AGC is TCG. The hydrogen bonding is shown in the drawing below. One polynucleotide chain is completely shown, as is the hydrogen bonding to the bases in the other polynucleotide chain. Because of the distortions that result from depicting a three-dimensional structure in two dimensions, the hydrogen bonds themselves are distorted (they are all of approximately equal length) and the second sugar phosphate chain has been omitted.